DSP for MATLAB™ and LabVIEW™

SYNTHESIS LECTURES ON SIGNAL PROCESSING

Editor
José Moura, Carnegie Mellon University

Nonlinear Source Separation
Luis B. Almeida
2006

Spectral Analysis of Signals
Yanwei Wang, Jian Li, and Petre Stoica
2006

DSP for MATLAB™ and LabVIEW™

Forester W. Isen

www.morganclaypool.com

ISBN: 9781608450688 paperback

A Publication in the Morgan & Claypool Publishers series
SYNTHESIS LECTURES ON SIGNAL PROCESSING

Series Editor: José Moura, Carnegie Mellon University

Series ISSN
Synthesis Lectures on Signal Processing
Print 1932-1236

DSP for MATLAB™ and LabVIEW™

Forester W. Isen

SYNTHESIS LECTURES ON SIGNAL PROCESSING

MORGAN & CLAYPOOL PUBLISHERS

ABSTRACT

DSP for MATLAB™ and LabVIEW™ covers basic digital signal processing in a practical and accessible manner, and, as it includes all essential foundation mathematics as well, is suitable for both academic and professional use. As the title of the book implies, the scripts given in the text and supplied in a downloadable .zip file (available via the internet at **http://www.morganclaypool.com/page/isen**) are suitable for use with both MATLAB, a product of The Mathworks, Inc., and LabVIEW, a product of National Instruments, Inc. The text is well-illustrated with examples involving practical computation using m-code or MathScript (as m-code is usually referred to in LabVIEW-based literature), and LabVIEW VIs. There is also an ample supply of exercises, which consist of a mixture of paper-and-pencil exercises for simple computations, and script-writing projects having various levels of difficulty, from simple, requiring perhaps ten minutes to accomplish, to challenging, requiring several hours to accomplish.

Part I of the book consists of four chapters. The first chapter gives a brief overview of the field of digital signal processing. This is followed by a chapter detailing many useful signals and concepts, including convolution, recursion, difference equations, etc. The third chapter covers conversion from the continuous domain to the discrete domain and back (i.e., analog-to-digital and digital-to-analog conversion), aliasing, the Nyquist rate, normalized frequency, and conversion from one sample rate to another. The fourth chapter of Part I introduces the reader to many important principles of signal processing, including correlation, the correlation sequence, the Real DFT, correlation by convolution, matched filtering, estimation of frequency response, simple FIR lowpass, bandpass, and highpass filters, and simple (one or two pole) IIR filters. Part II of the book is devoted to discrete frequency transforms. After an overview of the Fourier and Laplace families of transforms, the text covers the DTFT (Discrete Time Fourier Transform), the z-transform (including various filter topologies such as Direct, Cascade, Parallel, and Lattice forms), Discrete Fourier Series (DFS), and the DFT (Discrete Fourier Transform). Part III of the book covers FIR and IIR design, including general principles of FIR design, characteristics of four types of linear-phase FIR, Windowed Ideal Lowpass filter design, Frequency Sampling design with optimized transition band coefficients, Equiripple FIR design of all standard filter types, and Classical IIR design. The Impulse Invariance and Bilinear Transform methods for converting an analog filter design to an equivalent digital filter design are also discussed in detail. Part IV of the book, LMS Adaptive Filtering, explains cost functions, performance surfaces, gradient estimation using coefficient perturbation, and the Least Mean Square (LMS) coefficient update algorithm. The issues of stability, convergence speed, and narrow-bandwidth signals are covered in a practical manner, with many illustrative scripts. Use of LMS adaptive filtering in various filtering applications and topologies is explored, including Active Noise Cancellation (ANC), system or plant modeling, periodic component elimination, Adaptive Line Enhancement (ALE), interference cancellation, echo cancellation, and equalization/deconvolution.

KEYWORDS

Higher-Level Terms:
MATLAB, LabVIEW, DSP (Digital Signal Processing), Sampling, LTI Systems, Analog-to-Digital, Digital-to-Analog, FIR, IIR, DFT, Time Domain, Frequency Domain, Aliasing, Binary Numbers, MathScript, Discrete Time Fourier Transform (DTFT), z-Transform, Discrete Fourier Transform (DFT), Fast Fourier Transform (FFT), Goertzel Algorithm, Discrete Fourier Series (DFS), Discrete Frequency Transform, FIR Design, Classical IIR Design, Windowed Ideal Lowpass, Frequency Sampling, Equiripple, Remez Exchange, LMS Adaptive Filter, Least Mean Square, Active Noise Cancellation (ANC), Deconvolution, Equalization, Inverse Filtering, Interference Cancellation, Echo Cancellation, Dereverberation, Adaptive Line Enhancer (ALE)

Lower-Level Terms:
Correlation, Convolution, Matched Filtering, Orthogonality, Interpolation, Decimation, Mu-Law, Stability, Causality, Difference Equations, Zero-Order Hold, Direct Form, Direct Form Transposed, Parallel Form, Cascade Form, Lattice Form, Decimation-in-time (DIT), Comb, Moving Average, Linear Phase, Passband Ripple, Stopband Attenuation, Highpass, Bandpass, Bandstop, Notch, Hilbert Transformer, Differentiator, Inverse-DFT, Cosine/Sine Summation Formulas, Alternation Theorem, Linear Phase Form, Cascaded Linear Phase Form, Frequency Sampling Form, Butterworth, Chebyshev Type-I, Chebyshev Type-II, Elliptic, Cauer, Single-H, Dual-H, Gradient, Cost Function, Performance Surface, Coefficient Perturbation

This book is dedicated to

The following memorable teachers:
Louise Costa
Maxine Ropshaw
Cdr. Charles Bradimore Brouillette
Rudd Crawford, Jr.
Sheldon Sarnevitz
J. J. McCusker
Dr. C. W. Rector
Dr. Samuel Saul Saslaw
Dr. R. D. Shelton

and to
Virginia L. (Durham) (Isen) Bowles
Renee J. (Udelsen) Isen

and to the memory of
Cdr. Forester W. Isen, Sr. (1916–1978)
James Daniel Mudd (1912–1997)
Glenn Warren McWhorter (1932–2008)
Douglas Hunter (1941–1963)
Diane Satterwhite (1949–1961)
John G. Elsberry (1949–1971)
Amelia Megan Au (1967–2007)

and to following humane organizations,
and all who work for them and support them:
Rikki's Refuge (`www.rikkisrefuge.org`*), Orange, Virginia*
Humane Society of Fairfax County (`www.hsfc.org`*), Fairfax, Virginia*
The Washington Animal Rescue League (WARL)
(`www.warl.org`*), Washington, DC*
Angels for Animals (`www.angelsforanimals.org`*), Youngstown, Ohio*
and all similar organizations and persons

and the feline companions who so carefully watched over
the progress of the book during its evolution over the past eight years:
Richard (1985–2001)
Blackwythe (1989–2001)
Tiger (1995?–)
Scooter (1997–)
Mystique (2000–)
Scampy (2002–)
Nudgy (2003–)
Percy (2004–)

Contents

Preface

Preface

0.1 INTRODUCTION

The present book, *DSP for MATLAB™ and LabVIEW™* is a work of twelve chapters that covers basic digital signal processing in a practical and accessible manner, but which nonetheless includes all essential foundation mathematics. As the title of the book implies, the scripts given in the text and supplied in a downloadable .zip file (available via the internet at **http://www.morganclaypool.com/page/isen**) are suitable for both MATLAB, a product of The Mathworks, Inc., and LabVIEW, a product of National Instruments, Inc.

The text is well-illustrated with examples involving practical computation using m-code or MathScript (as m-code is usually referred to in LabVIEW-based literature), and LabVIEW VIs. There is also an ample supply of exercises, which consist of a mixture of paper-and-pencil exercises for simple computations, and script-writing projects having various levels of difficulty, from simple, requiring perhaps ten minutes to accomplish, to challenging, requiring several hours to accomplish.

0.2 THE FOUR PARTS OF THE BOOK

Part I of the book consists of four chapters. The first chapter gives a brief overview of the field of digital signal processing and its relation to continuous-domain (or analog) processing. This is followed by a chapter detailing many useful signals and concepts, including the complex exponential, convolution, recursion, difference equations, etc. The third chapter covers conversion from the continuous domain to the discrete domain and back (i.e., analog-to-digital and digital-to-analog conversion), aliasing, the Nyquist rate, normalized frequency, conversion from one sample rate to another, waveform generation at various sample rates from stored wave data, and Mu-law compression. The final chapter of Part I introduces the reader to many important principles of signal processing, including correlation, the correlation sequence, orthogonality of sinusoids, the Real DFT, use of orthogonality in signal transmission, correlation by convolution, matched filtering, estimation of frequency response, the construction of simple FIR passband filters (lowpass, bandpass, highpass) by superposition of orthogonal correlators, and simple (one or two pole) IIR filters.

Part II of the book is devoted to discrete frequency transforms. After an overview of the Fourier and Laplace families of transforms (for both the continuous and discrete time domains), the text covers the DTFT (Discrete Time Fourier Transform) and its inverse, including the frequency response of a linear, time invariant (LTI) system, the z-transform and its inverse (including use of the z-transform for computing frequency response, and various filter topologies including Direct,

Cascade, Parallel, and Lattice forms), Discrete Fourier Series (DFS), sampling in the z-domain, the DFT (Discrete Fourier Transform) and its inverse (including the Decimation-in-Time FFT, linear, periodic, and circular convolution as related to the DFT, DFT leakage and windowing, and the issues of bin width, sample rate, and sampling duration).

Part III of the book covers FIR and IIR design, including general principles of FIR design (effect of filter length and windowing), characteristics of four types of linear-phase FIR (symmetry, frequency response, etc.), Windowed Ideal Lowpass filter design (including construction of highpass, bandpass, and bandstop filters from ideal lowpass filters), Frequency Sampling design with optimized transition band coefficients, Equiripple FIR design of all standard filter types (including Hilbert Transformers and Differentiators), and Classical IIR design (including Butterworth, Chebyshev Type I and II, and Elliptic filters of all four standard passband types). The Impulse Invariance and Bilinear Transform methods for converting an analog filter design to an equivalent digital filter are also discussed in detail.

Part IV of the book, LMS Adaptive Filtering, begins by explaining cost functions and performance surfaces, moves on to the use of gradient estimation using coefficient perturbation, and finally reaches the elegant and computationally efficient Least Mean Square (LMS) coefficient update algorithm. The issues of stability, convergence speed, and narrow-bandwidth signals are covered in a practical manner, with many illustrative scripts. Use of LMS adaptive filtering in various filtering applications and topologies is explored, including Active Noise Cancellation (ANC), system or plant modeling, periodic component elimination, Adaptive Line Enhancement (ALE), interference cancellation, echo cancellation, and equalization/deconvolution.

0.3 ORIGIN AND EVOLUTION OF THE BOOK

The present book originated with an idea to provide a short, simple course for intellectual property specialists and engineers that would provide more explanation and illustration of the subject matter than that found in conventional academic books. This idea was conceived in the mid-to-late 1990's when I was introduced to MATLAB by Dan Hunter, whose graduate school days occurred after the advent of both MATLAB and LabVIEW (mine did not). About the time I was seriously exploring the use of MATLAB to update my own knowledge of signal processing, Dr. Jeffrey Gluck began giving an in-house course at the agency on the topics of convolutional coding, trellis coding, etc., thus inspiring me to do likewise in the basics of DSP, a topic more in-tune to the needs of the unit I was supervising at the time. Two short courses were taught at the agency in 1999 and 2000 by myself and several others, including Dr. Hal Zintel, David Knepper, and Dr. Pinchus Laufer. In these courses we stressed audio and speech topics in addition to basic signal processing concepts. Some time after this, I decided to develop a complete course in book form, the previous courses having consisted of an ad hoc pastiche of topics presented in summary form on slides, augmented with visual presentations generated by custom-written scripts for MATLAB. An early draft of the book was kindly reviewed by Motorola Patent Attorney Sylvia Y. Chen, which encouraged me to contact Tom

Robbins at Prentice-Hall concerning possible publication. By 2005, Tom was involved in starting a publishing operation at National Instruments, Inc., and introduced me to LabVIEW with the idea of possibly crafting the book to be compatible with LabVIEW. After review of an existing draft of the manuscript by a panel of three in early 2006, it was suggested that all essential foundation mathematics be included if academic as well as professional appeal was wanted. Fortunately, I had long since retired from the agency and was able to devote the considerable amount of time needed for such a project.

The result is a book that should have appeal in both academic and professional settings, as it includes essential mathematical formulas and concepts as well as simple or "first principle" explanations that help give the reader a gentler, more intuitive entry into the mathematical treatment.

Many thanks go not only to all those mentioned above, but to Joel Claypool of Morgan & Claypool for his enthusiasm and support in publishing the book, to Dr. C. L. Tondo and his troops for their skill and expertise in setting the book in LaTeX and creating the PDF files, to Dr. Ale Quesada for her skill and expertise in editing the various illustration files for the book, to Mike Jones of Morgan & Claypool for various web-related and administrative services, and to all those behind the scenes, whose names I have never heard, for their various contributions that made possible the publication of this book.

Forester W. Isen
April 2009

Part I

Fundamentals of Discrete Signal Processing

CHAPTER 1

An Overview of DSP

1.1 SIGNALS, WAVES, AND DIGITAL PROCESSING

Two of the human senses, sight and hearing, work via the detection of waves. Useful information from both light and sound is gained by detection of certain characteristics of these waves, such as frequency and amplitude. Modern telecommunication depends on transducing sound or light into electrical quantities such as voltage, and then processing the voltage in many different ways to enable the information to be reliably stored or conveyed to a distant place and then regenerated to imitate (i.e., reconstruct) the original sound or light phenomenon.

For example, in the case of sound, a microphone detects rapid pressure variations in air and converts those variations to an output voltage which varies in a manner proportional to the variation of pressure on the microphone's diaphragm. The varying voltage can be used to cut a corresponding wave into a wax disc, to record corresponding wave-like variations in magnetism onto a ferromagnetic wire or tape, to vary the opacity of a linear track along the edge of a celluloid film (i.e., the sound-track of a motion picture film) or perhaps to modulate a carrier wave for radio transmission.

In recent decades, signal processing and storage systems have been developed that use discrete samples of a signal rather than the entire continuous time domain (or **analog**) signal. Several useful definitions are as follows:

- A **sample** is the amplitude of an analog signal at an instant in time.

- A system that processes a signal in sampled form (i.e., a sequence of samples) is known as a **Discrete Time Signal Processing System**.

- In a **Digital Signal Processing** system, the samples are converted to numerical values, and the values (numbers) stored (usually in binary form), transmitted, or otherwise processed.

The difference between conventional analog systems and digital systems is illustrated in Fig. 1.1. At (a), a conventional analog system is shown, in which the signal from a microphone is sent directly to an analog recording device, such as a tape recorder, recorded at a certain tape speed, and then played back at the same speed some time later to reproduce the original sound. At (b), samples of the microphone signal are obtained by an **Analog-to-Digital Converter (ADC)**, which converts instantaneous voltages of the microphone signal to corresponding numerical values, which are stored in a digital memory, and can later be sent to a **Digital-to-Analog Converter (DAC)** to reconstruct the original sound.

In addition to recording and reproducing analog signals, most other kinds of processing which might be performed on an analog signal can also be performed on a sampled version of the signal

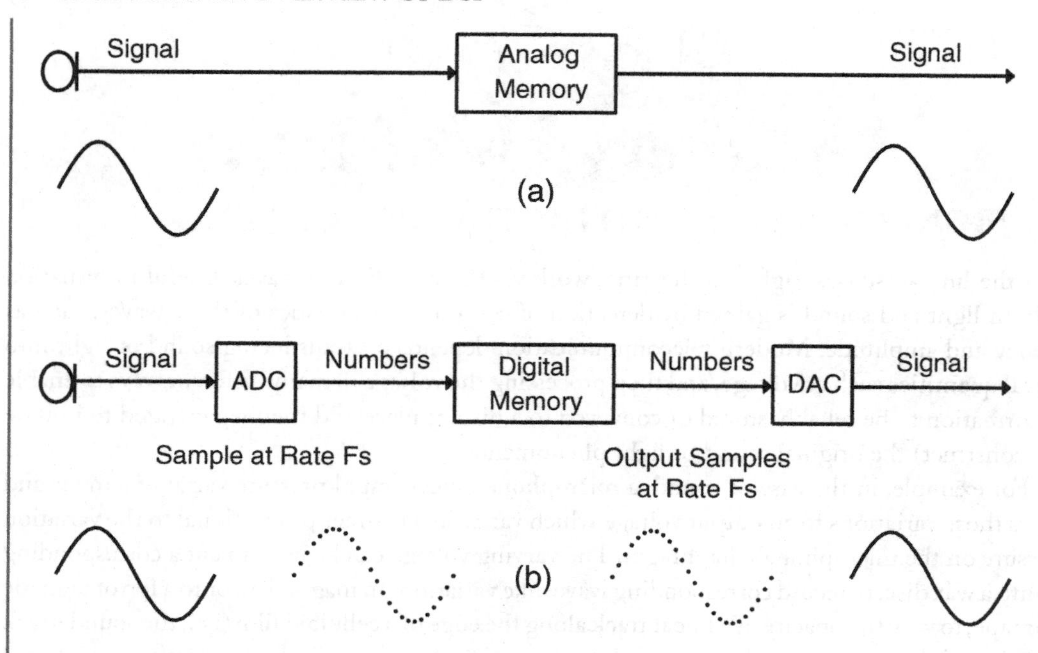

Figure 1.1: (a) Conventional analog recording and playback system; (b) A digital recording and playback system.

by using numerical algorithms. These can be categorized into two broad types of processing, time domain and frequency domain, which are discussed in more detail below.

1.2 ADVANTAGES OF DIGITAL PROCESSING

The reduction of continuous signals to sequences of numerical values (samples) that can be used to process and/or reconstruct the original signal, provides a number of benefits that cannot be achieved with continuous or analog signal processing. The following are some of the benefits of digital processing:

1. Analog hardware, such as amplifiers, filters, comparators, etc., is very susceptible to noise and deterioration through aging. Digital hardware works with only two signal levels rather than an infinite number, and hence has a high signal to noise ratio. As a result, there is little if any gradual deterioration of performance with age (although as with all things, digital hardware can suddenly and totally fail), and copies of signal files are generally perfect, absent component failure, media degeneration or damage, etc. This is not true with analog hardware and recording techniques, in which every copy introduces significant amounts of additional noise and distortion.

2. Analog hardware, for the most part, must be built for each processing function needed. With digital processing, each additional function needed can be implemented with a software module,

using the same piece of hardware, a digital computer. The computing power available to the average person has increased enormously in recent years, as evidenced by the incredible variety of inexpensive, high quality devices and techniques available. Hundreds of millions or billions of operations per second can be performed on a signal using digital hardware at reasonable expense; no reasonably-priced alternative exists using analog hardware and processing.

3. Analog signal storage is typically redundant, since wave-related signals (audio, video, etc.) are themselves typically redundant. For example, by taking into account this redundancy as well as the physiological limitations of human hearing, storage needs for audio signals can be reduced up to about 95%, using digitally-based compression techniques, such as MP3, AC3, AAC, etc.

4. Digital processing makes possible highly efficient security and error-correction coding. Using digital coding, it is possible, for example, for many signals to be transmitted at very low power and to share the same bandwidth. Modern cell phone techniques, such as CDMA (Code Division Multiple Access) rely heavily on advanced, digitally-based signal processing techniques to efficiently achieve both high quality and high security.

1.3 DSP NOMENCLATURE AND TOPICS

Figure 1.2 shows a broad overview of digital signal processing. Analog signals enter an ADC from the left, and samples exit the ADC from the right, and may be 1) processed strictly in the discrete time domain (in which samples represent the original signal at instants in time) or they may be 2) converted to a frequency domain representation (in which samples represent amplitudes of particular frequency components of the original signal) by a time-to-frequency transform, processed in the frequency domain, then converted back to the discrete time domain by a frequency-to-time transform. Discrete time domain samples are converted back to the continuous time domain by the DAC.

Note that a particular signal processing system might use only time domain processing, only frequency domain processing, or both time and frequency domain processing, so either or both of the signal processing paths shown in Fig. 1.2 may be taken in any given system.

1.3.1 TIME DOMAIN PROCESSING

Filtering, in general, whether it is done in the continuous domain or discrete domain, is one of the fundamental signal processing techniques; it can be used to separate signals by selecting or rejecting certain frequencies, enhance signals (such as with audio equalization, etc.), alter the phase characteristic, and so forth. Hence a major portion of the study of digital signal processing is devoted to digital filtering. Filtering in the continuous domain is performed using combinations of components such as inductors, capacitors, resistors, and in some cases active elements such as op amps, transistors, etc. Filtering in the discrete or digital domain is performed by mathematically manipulating or processing a sequence of samples of the signal using a discrete time processing system, which typically consists of registers or memory elements, delay elements, multipliers, and adders. Each of the preceding elements may be implemented as distinct pieces of hardware in an efficient arrangement designed to function for a particular purpose (often referred to as a **Pipeline**

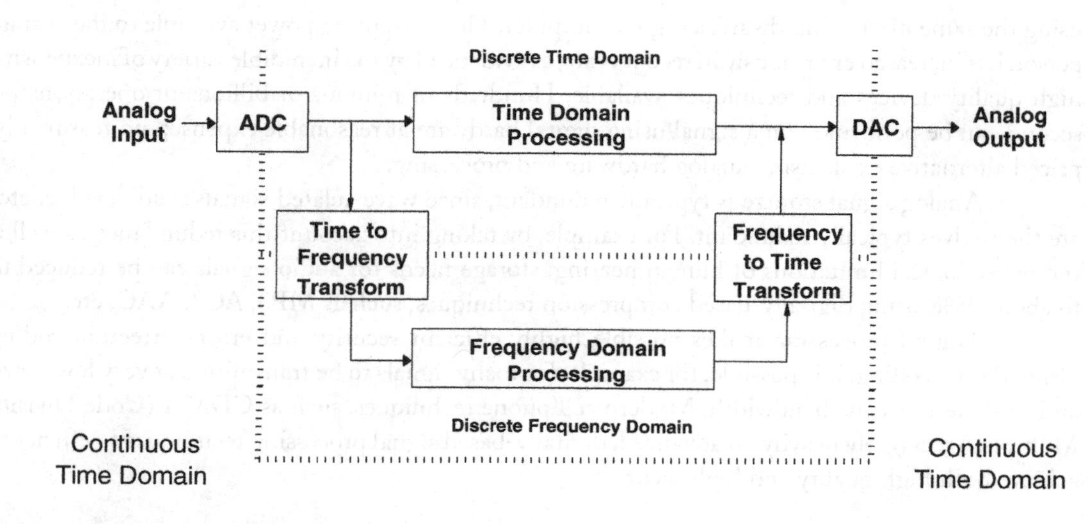

Figure 1.2: A broad, conceptual overview of digital signal processing.

Processor), or, the equivalent functions of all elements may be implemented on a general purpose computer by specifically designed software.

1.3.2 FREQUENCY TRANSFORMS

A time-to-frequency transform operates on a block of time domain samples and evaluates the frequency content thereof. A set of frequency coefficients is derived which can be used to quantify the amplitudes (and usually phases) of frequency components of the original signal, or the coefficients can be used to reconstruct the original time domain samples using an inverse transform (a frequency-to-time transform). The most well-known and widely-used of these transforms is the **Discrete Fourier Transform (DFT)**, usually implemented by the **FFT** (for **Fast Fourier Transform)**, the name of a class of algorithms that allow efficient computation of the DFT.

1.3.3 FREQUENCY DOMAIN PROCESSING

Most signal processing that can be done in the time domain can be also equivalently done in the frequency domain. Each domain has certain advantages for a given type of problem.

Time domain filtering, for example, can be performed using frequency transforms such as the DFT, and in certain cases efficiency can be greatly improved using this technique.

A second use is in digital filter design, in which the desired filter frequency response is specified in the frequency domain, i.e., as a set of DFT coefficients, for example.

Yet a third and very prevalent use is **Transform Coding**, in which signals are coded using a frequency transform (usually eliminating as much redundant information as possible) and then reconstructed from the transform coefficients. Transform Coding is a powerful tool for compression algorithms, such as those employed with MP3 (MPEG II, Level 3) for audio signals, JPEG, a common image compression format, etc. The use of such compression algorithms has revolutionized the audio and video fields, making storage of audio and video data very economical and deliverable via Internet.

CHAPTER 2

Discrete Signals and Concepts

2.1 OVERVIEW

If the study of digital signal processing is likened to a story, this chapter can be viewed as an introduction of the main characters in the story–they are the various types of signals (or their sampled versions, called sequences) and fundamental processes that we will see time and time again. Acquiring a good working knowledge of these is essential to understanding the rest of the story, just as is knowing the characters in a novel.

In the first part of the chapter, we introduce discrete sequence notation and many standard test signals including the unit impulse, the unit step, the exponential sequence (both real and complex), the chirp, etc., and we learn to add and multiply sequences that are offset in time. Any serious study of digital signal processing relies heavily on the representation of sinusoids by the complex exponential, and hence this is covered in detail in the chapter. In the latter part of the chapter, we introduce the concepts of linear, time-invariant (LTI) systems, convolution, stability and causality, the FIR, the IIR, and difference equations.

By the end of this chapter, the reader will be prepared for the next chapter in the story of DSP, namely, the process and requirements for obtaining sequences via sampling, formatting sample values in binary notation, converting sequences back into continuous domain signals, and changing the sample rate of a sequence.

2.2 SOFTWARE FOR USE WITH THIS BOOK

The software files needed for use with this book (consisting of m-code (.m) files, VI files (.vi), and related support files) are available for download from the following website:

http://www.morganclaypool.com/page/isen

The entire software package should be stored in a single folder on the user's computer, and the full file name of the folder must be placed on the MATLAB or LabVIEW search path in accordance with the instructions provided by the respective software vendor (in case you have encountered this notice before, which is repeated for convenience in each chapter of the book, the software download only needs to be done once, as files for the entire book are all contained in the one downloadable folder).

See Appendix A for more information.

2.3 DISCRETE SEQUENCE NOTATION

Digital Signal Processing must necessarily begin with a signal, and most signals, such as sound, images, etc., originate as continuous-valued (or analog) signals, and must be converted into a sequence of samples to be processed using digital techniques.

Figure 2.1 depicts a continuous-domain sine wave, with eight samples marked, sequentially obtained every 0.125 second. The signal values input to the ADC at sample times 0, 0.125, 0.25, 0.375, 0.5, 0.625, 0.75, etc., are 0, 0.707, 1, 0.707, 0, -0.707, -1, etc.

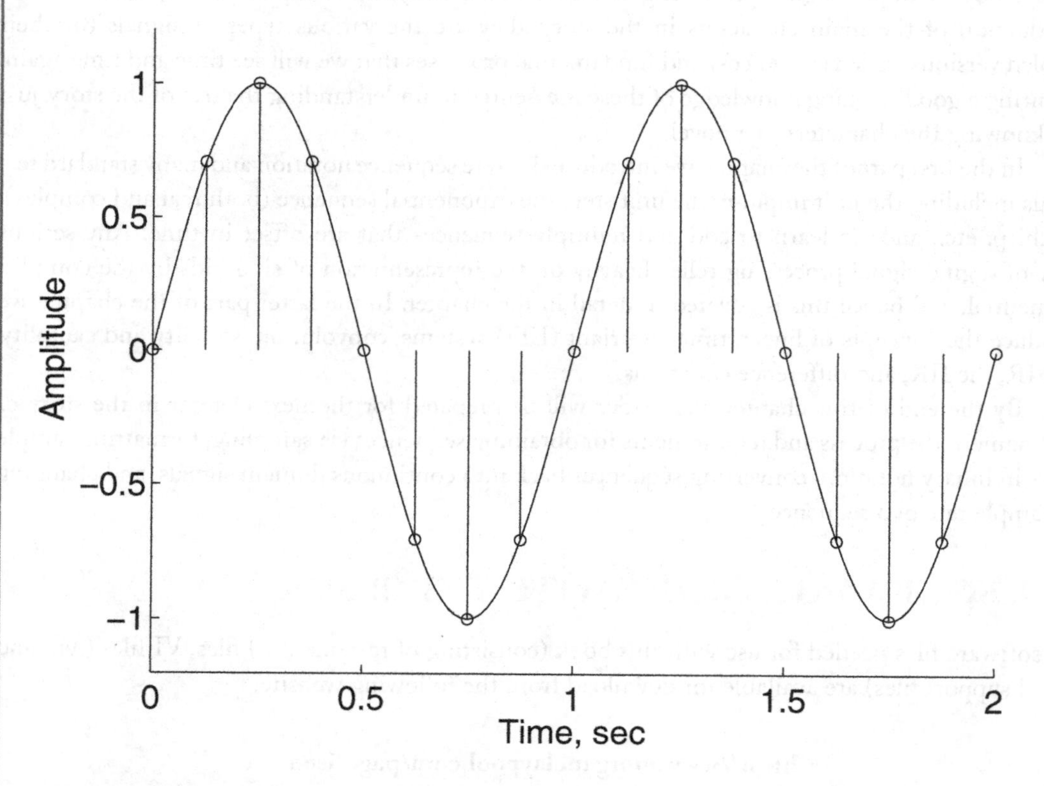

Figure 2.1: An analog or continuous-domain sine wave, with eight samples per second marked.

The samples within a given sample sequence are normally indexed by the numbers 0, 1, 2, etc., which represent multiples of the sample period T. For example, in Fig. 2.1, we note that the sample period is 0.125 second, and the actual sampling times are therefore 0 sec., 0.125 sec., 0.25 sec., etc. The continuous sine function shown has the value

$$f(t) = \sin(2\pi ft)$$

where t is time, f is frequency, and in this particular case, f = 1 Hz. Sampling occurs at times nT where n = 0, 1, 2,...and T = 0.125 second. The sample values of the sequence would therefore be sin(0), sin($2\pi(T)$), sin($2\pi(2T)$), sin($2\pi(3T)$), etc., and we would then say that $s[0] = 0$, $s[1] = 0.707$, $s[2] = 1.0$, $s[3] = 0.707$, etc. where $s[n]$ denotes the n-th sequence value, the amplitude of which is equal to the underlying continuous function at time nT (note that brackets surrounding a function argument mean that the argument can only assume discrete values, while parentheses surrounding an argument indicate that the argument's domain is continuous). We can also say that

$$s[n] = \sin[2\pi nT]$$

This sequence of values, the samples of the sine wave, can be used to completely reconstruct the original continuous domain sine wave using a DAC. There are, of course, a number of conditions to ensure that this is possible, and they are taken up in detail in the next chapter.

To compute and plot the sample values of a 2-Hz sine wave sampled every 0.05 second on the time interval 0 to 1.1 second, make the following MathScript call:

```
t = [0:0.05:1.1]; figure; stem(t,sin(2*pi*2*t))
```

where the t vector contains every sample time nT with T = 0.05. Alternatively, we might write

```
T = 0.05; n = 0:1:22; figure; stem(n*T,sin(2*pi*2*n*T))
```

both of which result in Fig. 2.2.

2.4 USEFUL SIGNALS, SEQUENCES, AND CONCEPTS

2.4.1 SINE AND COSINE

We saw above that a sine wave of frequency f periodically sampled at the time period T has the values

$$s[n] = \sin[2\pi f nT]$$

Once we have a sampled sine wave, we can mathematically express it without reference to the sampling period by defining the sequence length as N. We would then have, in general,

$$s[n] = \sin[2\pi nk/N]$$

where n is the sample index, which runs from 0 to $N-1$, and k is the number of cycles over the sequence length N. For the sample sequence marked in Fig. 2.1, we would have

$$s[n] = \sin[2\pi n2/16]$$

where we have noted that there are two full cycles of the sine over 16 samples (the 17th sample is the start of the third cycle). The correctness of this formula can be verified by noting that for

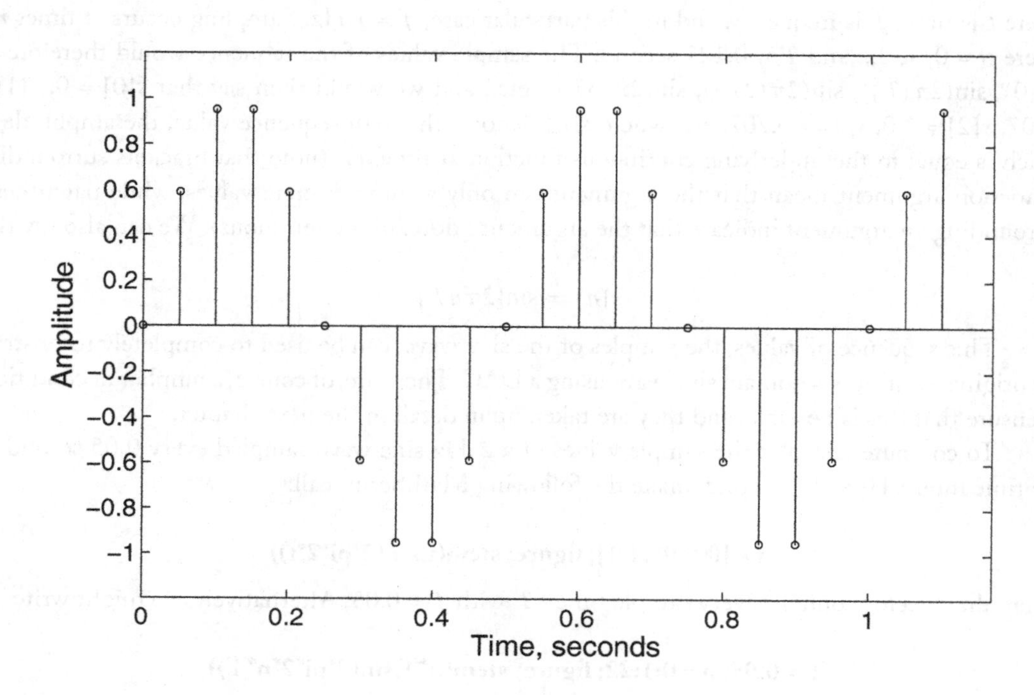

Figure 2.2: A plot of the samples of a sine wave having frequency 2 Hz, sampled every 0.05 second up to time 1.1 second.

the 17th sample, n = 16, and $s[16]$ = 0, as shown. Picking another sample, for n = 2, we get $s[2]$ = $\sin[2\pi(2)2/16] = \sin[\pi/2] = 1$, as shown.

A phase angle is sometimes needed, and is denoted θ by in the following expression:

$$s[n] = \sin[2\pi nk/N + \theta]$$

Note that if $\theta = \pi/2$, then

$$s[n] = \cos[2\pi nk/N]$$

We can illustrate this by generating and displaying a sine wave having three cycles over 18 samples, then the same sine wave, but with a phase angle of $\pi/2$ radians, and finally a cosine wave having three cycles over 18 samples and a zero phase angle. A suitable MathScript call, which results in Fig. 2.3, is

```
n = 0:1:17; y1 = sin(2*pi*n/18*3); subplot(311); stem(n,y1);
y2 = sin(2*pi*n/18*3 +pi/2); subplot(312); stem(n,y2);
y3 = cos(2*pi*n/18*3); subplot(313); stem(n,y3)
```

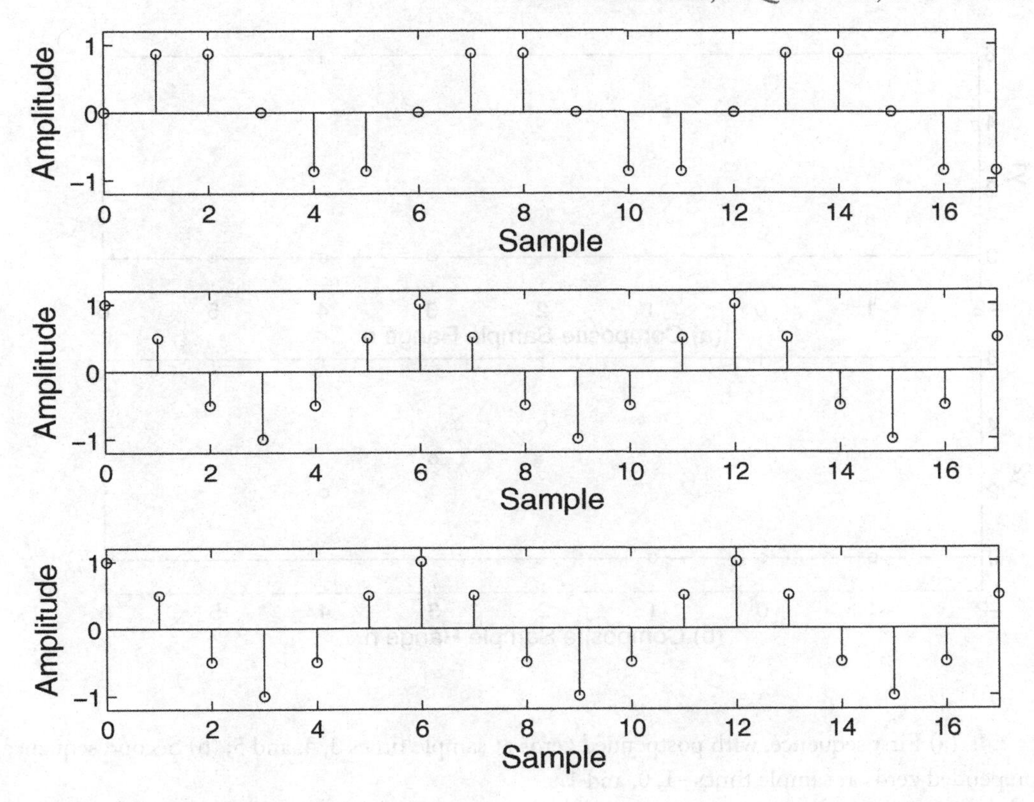

Figure 2.3: (a) Three cycles of a sine wave over 18 samples, with phase angle 0 radians; (b) Same as (a), with a phase angle of $\pi/2$ radians; (c) Three cycles of a cosine wave over 18 samples, with a phase angle of 0 radians.

2.4.2 SEQUENCE AND TIME POSITION VECTOR

Certain operations on two sequences, such as addition and multiplication, require that the sequences be of equal length, and that their proper positions in time be preserved.

Consider the sequence $x1$ = [1,2,3,4], which was sampled at sample time indices $n1$ = [-1,0,1,2], which we would like to add to sequence $x2$ = [4,3,2,1], which was sampled at time indices $n2$ = [2,3,4,5]. To make these two sequences equal in length, we'll prepend and postpend zeros as needed to result in two sequences of equal length that retain the proper time alignment. We see that the minimum time index is -1 and the maximum time index is 5. Since $x1$ starts at the minimum time index, we postpend zeros to it such that we would have $x1$ = [1,2,3,4,0,0,0], with corresponding time indices [-1,0,1,2,3,4,5]. Similarly, we prepend zeros so that $x2$ = [0,0,0,4,3,2,1], with the same total time or sample index range as the modified version of $x1$. Figure 2.4 depicts this process.

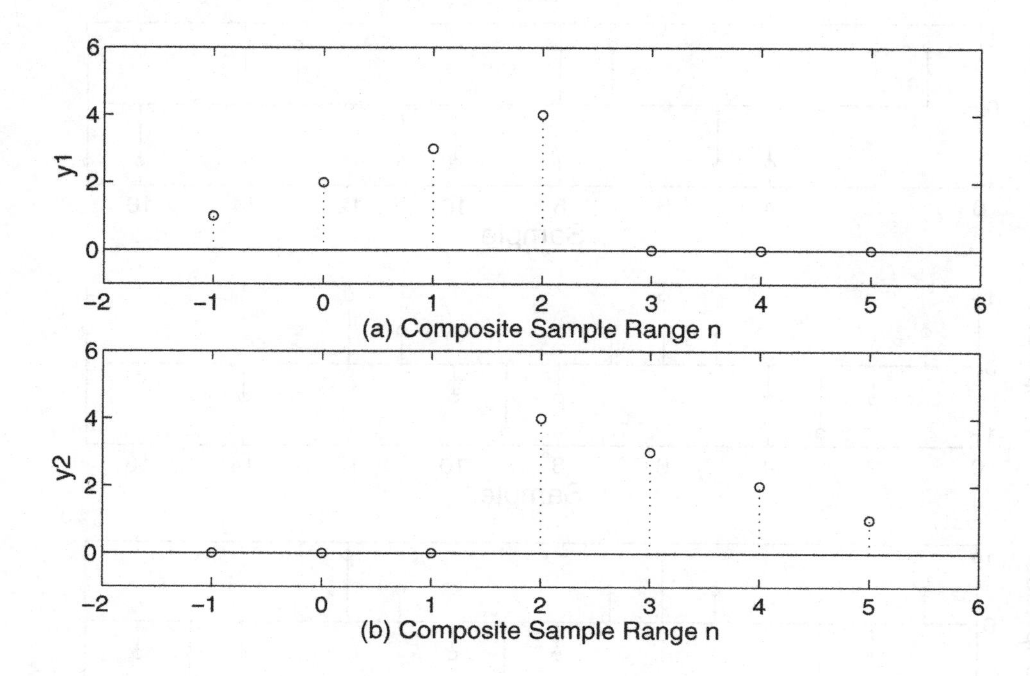

Figure 2.4: (a) First sequence, with postpended zeros at sample times 3, 4, and 5; (b) Second sequence, with prepended zeros at sample times -1, 0, and 1.

The sum is then

$$x1 + x2 = [1,2,3,4,0,0,0] + [0,0,0,4,3,2,1] = [1,2,3,8,3,2,1]$$

and has time indices [-1,0,1,2,3,4,5].

These two ideas, that sequences to be added or multiplied must be of equal length, but also properly time-aligned, lead us to write several MathScript functions that will automatically perform the needed adjustments and perform the arithmetic operation.

The following script will perform addition of offset sequences $y1$ and $y2$ that have respective time indices $n1$ and $n2$ using the method of prepending and postpending zeros.

```
function [y, nOut] = LVAddSeqs(y1,n1,y2,n2)
nOut = [min(min(n1),min(n2)):1:max(max(n1),max(n2))];
mnfv = min(nOut); mxfv = max(nOut);
y = [zeros(1,min(n1)-mnfv),y1,zeros(1,mxfv-max(n1))] + ...
[zeros(1,min(n2)-mnfv),y2,zeros(1,mxfv-max(n2))];
```

The function

$$[y, nOut] = LVMultSeqs(y1, n1, y2, n2)$$

works the same way, with the addition operator (+) in the final statement being replaced with the operator for multiplying two vectors on a sample-by-sample basis, a period following by an asterisk (.*).

We can illustrate use of the function $LVAddSeqs$ by using it to add the following sequences:

y1 = [3,-2,2], n1 = [-1,0,1], y2 = [1,0,-1], n2 = [0,1,2]

We make the call

[y, n] = LVAddSeqs([3,-2,2], [-1,0,1], [1,0,-1], [0,1,2])

which yields y = [3,-1,2,-1] and n = [-1,0,1,2].

We can illustrate use of the function $LVMultSeqs$ by using it to multiply the same sequences. We thus make the call

[y, n] = LVMultSeqs([3,-2,2], [-1,0,1], [1,0,-1], [0,1,2])

which yields y = [0,-2,0,0] and n = [-1,0,1,2].

2.4.3 THE UNIT IMPULSE (DELTA) FUNCTION

The **Unit Impulse** or **Delta Function is** defined as $\delta[n] = 1$ when $n = 0$ and 0 for all other values of n. The time of occurrence of the impulse can be shifted by a certain number of samples k using the notation $\delta[n-k]$ since the value of the function will only be 1 when n - k = 0.

The following function will plot a unit impulse at sample index n on the sample interval *Nlow* to *Nhigh*.

```
function LVPlotUnitImpSeq(n,Nlow,Nhigh)
xIndices = [Nlow:1:Nhigh];
xVals = zeros(1,length(xIndices));
xVals(find(xIndices-n==0))=1;
stem(xIndices,xVals)
```

An example MathScript call is

LVPlotUnitImpSeq(-2,-10,10)

A version of the script that returns the output sequence and its indices without plotting is

[xVals,xIndices] = LVUnitImpSeq(n,Nlow,Nhigh)

This version is useful for generating composite unit impulse sequences. For example, we can display, over the sample index interval -5 to 5, the output sequence

$$y[n] = 3\delta[n-2] - 2\delta[n+3]$$

by using the following m-code, which computes and displays the desired output sequence using the function *LVUnitImpSeq*, as shown in Fig. 2.5.

```
[y1,y1Ind] = LVUnitImpSeq(2,-5,5),
[y2,y2Ind] = LVUnitImpSeq(-3,-5,5),
y = 3*y1 - 2*y2, stem(y1Ind,y)
```

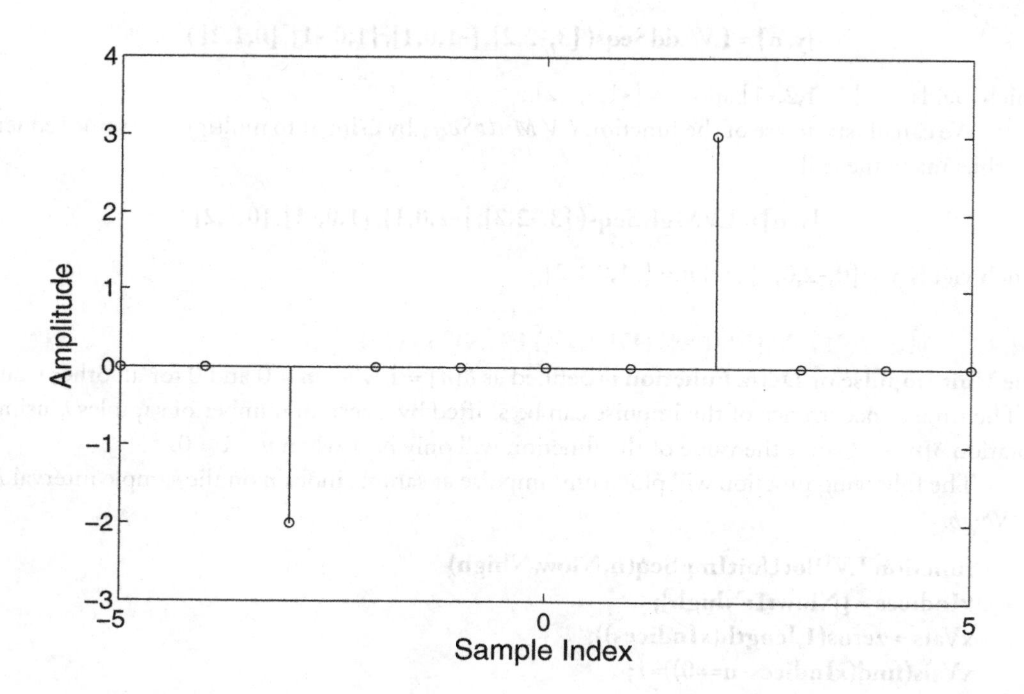

Figure 2.5: A graph of the function $y[n] = 3[n-2] - 2[n+3]$ for sample indices -5 to +5.

2.4.4 THE UNIT STEP FUNCTION

The **Unit Step Function is** defined as $u[n] = 1$ when $n \geq 0$ and 0 for all other values of n. The time of occurrence of the step (the value 1) can be shifted by a certain number of samples k using the notation $u[n-k]$ since the value of the function will only be 1 when n - $k \geq 0$.

The following function will plot a unit step at sample index n on the sample interval *Nlow* to *Nhigh*.

```
function LVPlotUnitStepSeq(n,Nlow,Nhigh)
xIndices = [Nlow:1:Nhigh];
yVals(1:1:length(xIndices)) = 0;
posZInd = find((xIndices-n)==0)
yVals(posZInd:1:length(xIndices)) = 1;
stem(xIndices,yVals)
```

An example MathScript call is

```
LVPlotUnitStepSeq(-2,-10,10)
```

A version of the script that returns the output sequence and its indices without plotting is

$$[yVals, xIndices] = LVUnitStepSeq(n, Nlow, Nhigh)$$

This version is useful for generating composite unit step sequences. For example, we can display, over the sample index interval [-10:10] the sequence $y[n]$, defined as follows,

$$y[n] = 3u[n-2] - 2u[n+3]$$

with the following m-code, which computes and plots $y[n]$, using the function *LVUnitStepSeq*:

```
[y1,y1Ind] = LVUnitStepSeq(2,-10,10);
[y2,y2Ind] = LVUnitStepSeq(-3,-10,10);
y = 3*y1 - 2*y2; stem(y1Ind,y)
```

As another example, we can express the four-sample sequence [1,1,1,1] having time vector [-1,0,1,2] as a sum of unit step sequences and verify the answer using MathScript. To start, we generate a unit step sequence starting at n = -1 and subtract from it a unit step sequence starting at n = 3:

$$y = u[n+1] - u[n-3]$$

To verify, we can modify the code from the previous example; the results are shown in Fig. 2.6.

```
[y1,y1Ind] = LVUnitStepSeq(-1,-10,10);
[y2,y2Ind] = LVUnitStepSeq(3,-10,10);
y = y1 - y2; stem(y1Ind,y)
```

2.4.5 REAL EXPONENTIAL SEQUENCE

A signal generated as

$$y[n] = a^n$$

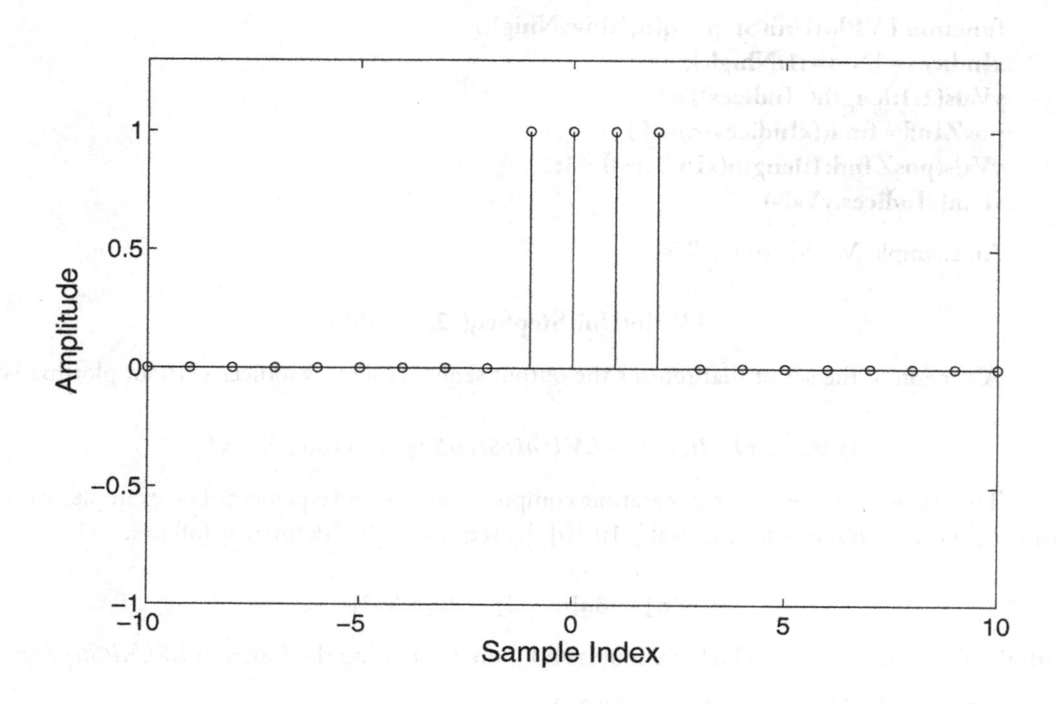

Figure 2.6: A plot over time indices -10 to +10 of the sequence defined as $y[n] = u[n+1] - u[n-3]$.

where a is a real number and n is real integer, produces a real sequence. In MathScript, to raise a number to a single power, use the "^" operator; to raise a number to a vector of powers, use the ".^" operator.

To illustrate this, we can generate and plot a real exponential sequence with $a = 2$ and $n =$ [0:1:6]. A suitable MathScript call is

y = 2.^([0:1:6]); figure; stem(y)

As a second example, we generate and plot the real exponential sequence with $a = 2$ and $n =$ [-6:1:0]. A suitable call is

n = [-6:1:0]; y = 2.^n; figure; stem(n,y)

the results of which are shown in Fig. 2.7.

2.4.6 PERIODIC SEQUENCES

A sequence that repeats itself exactly is called periodic. A periodic sequence can be generated from a given sequence S of length M by using the outer vector product of the sequence in column vector

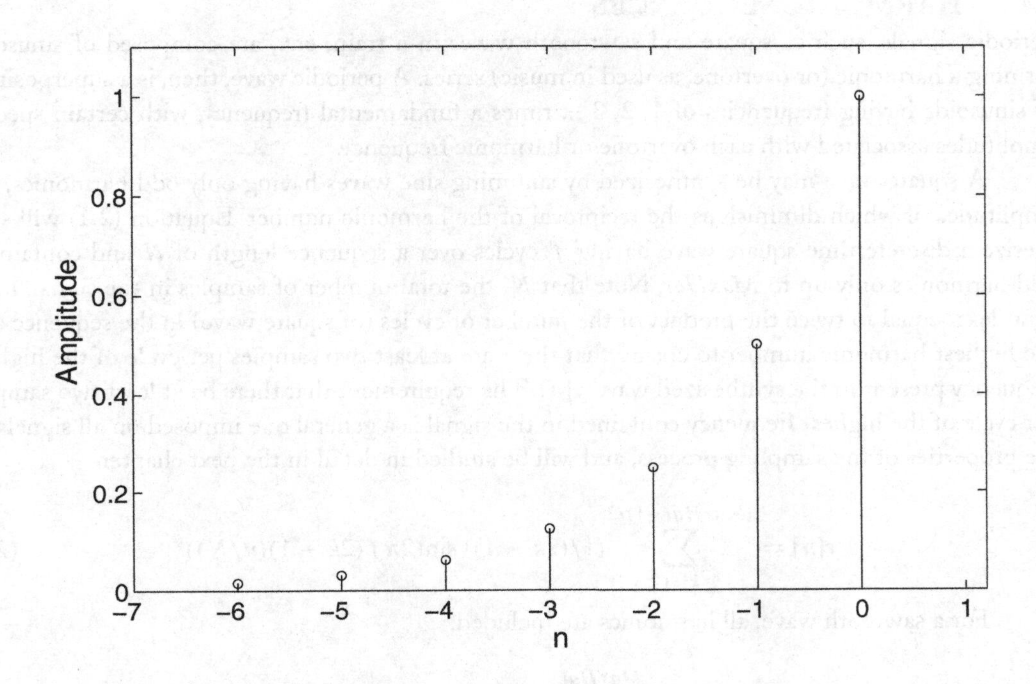

Figure 2.7: A real exponential sequence formed by raising the number 2 to the powers [-6:1:0].

form and a row vector of N ones. This generates an M-by-N matrix each column of which is the sequence S. The matrix can then be converted to a single column vector using MathScript's colon operator, and the resulting column vector is converted to a row vector by the transposition operator, the apostrophe.

The following function will generate n periods of the sequence y:

```
function nY = LVMakePeriodicSeq(y,N)
% LVMakePeriodicSeq([1 2 3 4],2)
y = y(:); nY = y*([ones(1,N)]); nY = nY(:)';
```

To illustrate use of the above, we will generate a sequence having three cycles of a cosine wave having a period of 11 samples. One period of the desired signal is

```
cos(2*pi*[0:1:10]/11)
```

and a suitable call that computes and plots the result is therefore

```
N = 3; y = [cos(2*pi*[0:1:10]/11)]';
nY = LVMakePeriodicSeq(y,N); stem(nY)
```

2.4.7 HARMONIC SEQUENCES

Periodic signals, such as square and sawtoooth waves in a train, etc., are composed of sinusoids forming a harmonic (or overtone, as used in music) series. A periodic wave, then, is a superposition of sinusoids having frequencies of 1, 2, 3 ... times a fundamental frequency, with certain specific amplitudes associated with each overtone or harmonic frequency.

A square wave may be synthesized by summing sine waves having only odd harmonics, the amplitudes of which diminish as the reciprocal of the harmonic number. Equation (2.1) will synthesize a discrete time square wave having f cycles over a sequence length of N and containing odd harmonics only up to *MaxHar*. Note that N, the total number of samples in the signal, must be at least equal to twice the product of the number of cycles (of square wave) in the sequence and the highest harmonic number to ensure that there are at least two samples per cycle of the highest frequency present in the synthesized wave $y[n]$. The requirement that there be at least two samples per cycle of the highest frequency contained in the signal is a general one imposed on all signals by the properties of the sampling process, and will be studied in detail in the next chapter.

$$y[n] = \sum_{k=1}^{(MaxHar+1)/2} (1/(2k-1))\sin(2\pi f(2k-1)(n/N)) \tag{2.1}$$

For a sawtooth wave, all harmonics are included:

$$y[n] = \sum_{k=1}^{MaxHar} (1/k)\sin(2\pi fk(n/N))$$

To illustrate the above formulas in terms of m-code, we will synthesize square and sawtooth waves having 10 cycles, up to the 99th harmonic.

We compute the necessary minimum sequence length as 2(10)(99) = 1980. The following MathScript call will synthesize a square wave up to the 99th harmonic; note that only odd harmonics are included, i.e., k = 1:2:99.

```
N=1980; n = 0:1:N; y = 0; for k = 1:2:99;
y = y + (1/k)*sin(2*pi*10*k*n/N); end; figure; plot(y)
```

For a sawtooth wave, the harmonic values are specified as 1:1:99, thus including both odd and even values:

```
N=1980; n = 0:1:N; y = 0; for k=1:1:99;
y = y + (1/k)*sin(2*pi*10*k*n/N); end; figure; plot(y)
```

The script

$$LVSynthPlotSquareSawtooth(WaveType, FundFreq, SR)$$

will synthesize square, sawtooth, and triangle waves one harmonic at a time (press any key for the next harmonic after making the initial call); specify *WaveType* as 1 for sawtooth, 2 for square, or 3 for triangle. The following call results in Fig. 2.8, which shows the synthesis of a 10-cycle square wave up to the first three harmonics (harmonics 1, 3, and 5).

LVSynthPlotSquareSawtooth(2,10,1000)

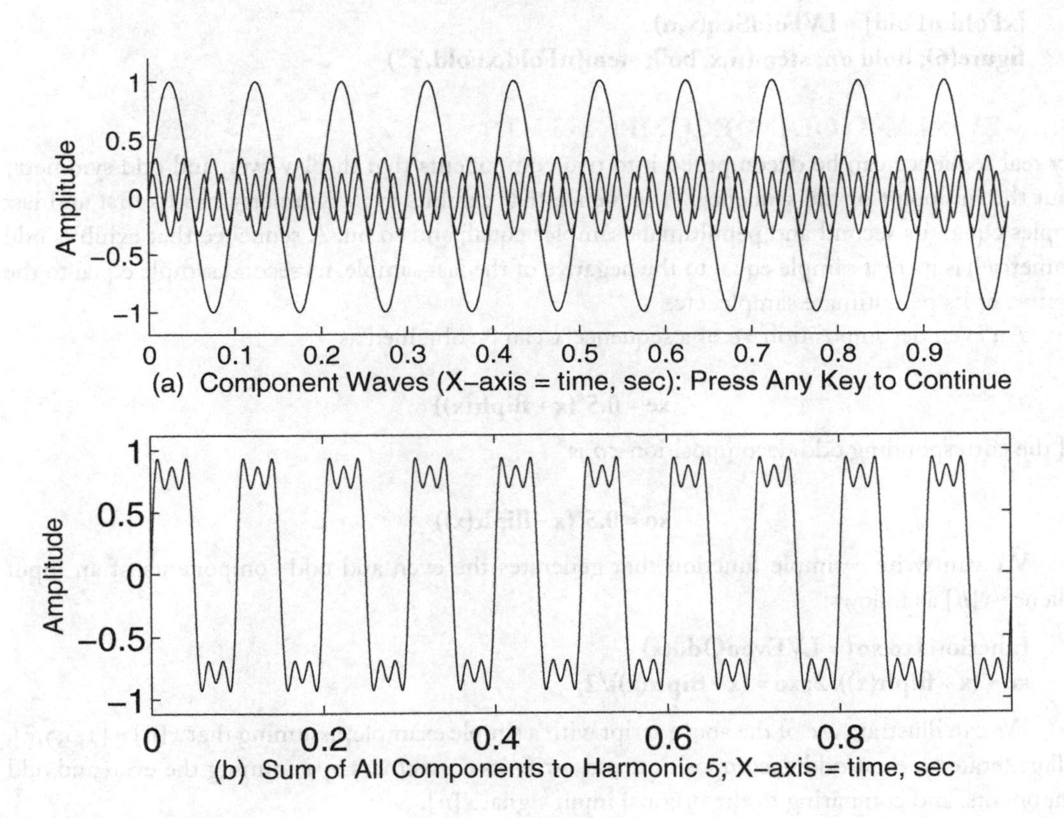

Figure 2.8: (a) The first three weighted harmonics of a square wave; (b) Superposition of the waves shown in (a).

2.4.8 FOLDED SEQUENCE

From time to time it is necessary to reverse a sequence in time, i.e., assuming that $x[n]$ = [1,2,3,4], the folded sequence would be $x[-n]$. The operation is essentially to flip the sequence from left to right around index zero. For example, the sequence [1,2,3,4] that has corresponding sample indices [3,4,5,6], when folded, results in the sequence [4,3,2,1] and corresponding indices [-6,-5,-4,-3].

To illustrate the above ideas, we can, for example, let $x[n]$ = [1,2,3,4] with corresponding sample indices n = [3,4,5,6], and compute $x[-n]$ using MathScript. We can write a simple script to accomplish the folding operation:

```
function [xFold,nFold] = LVFoldSeq(x,n)
xFold = fliplr(x), nFold = -fliplr(n)
```

For the problem at hand, we can then make the following call:

```
n = [3,4,5,6]; x=[1,2,3,4];
[xFold,nFold] = LVFoldSeq(x,n)
figure(6); hold on; stem(n,x,'bo'); stem(nFold,xFold,'r*')
```

2.4.9 EVEN AND ODD DECOMPOSITION

Any real sequence can be decomposed into two components that display even and odd symmetry about the midpoint of the sequence. A sequence that exhibits even symmetry has its first and last samples equal, its second and penultimate samples equal, and so on. A sequence that exhibits odd symmetry has its first sample equal to the negative of the last sample, its second sample equal to the negative of its penultimate sample, etc.

An even decomposition xe of a sequence x can be obtained as

xe = 0.5*(x + fliplr(x))

and the corresponding odd decomposition xo is

xo = 0.5*(x - fliplr(x))

We can write a simple function that generates the even and odd components of an input sequence $x[n]$ as follows:

```
function [xe,xo] = LVEvenOdd(x)
xe = (x + fliplr(x))/2; xo = (x - fliplr(x))/2;
```

We can illustrate use of the above script with a simple example; assuming that $x[n]$ = [1,3,5,7], we'll generate an even/odd decomposition and verify its correctness by summing the even and odd components, and comparing to the original input signal $x[n]$.

```
x = [1,3,5,7];
[xe,xo] = LVEvenOdd(x)
recon = xe + xo, diff = x - recon
```

From the above we get xe = [4,4,4,4] and xo = [-3,-1,1,3], the sum of which is [1,3,5,7], i.e., the original sequence x.

Another useful even/odd decomposition is defined such that

$$xe[n] = xe[-n]$$

and

$$xo[n] = -xo[-n]$$

In this case, the decompositions exhibit their symmetry about $n = 0$ rather than about the midpoint of the original sequence x. For example, if $x = [1,2,3,4]$ with corresponding sample indices $n = [3,4,5,6]$, a decomposition about $n = 0$ can be accomplished by padding x with zeros in such a manner to create a new sequence with time indices symmetrical about zero. In this case, the new sequence is

x = [zeros(1,9),1,2,3,4]

having sample indices [-6:1:6]. The new sequence x is then decomposed as above, i.e.,

```
x = [zeros(1,9),1,2,3,4];
[xe,xo] = LVEvenOdd(x)
recon = xe + xo, diff = x - recon
```

A script that performs an even-odd decomposition about zero and returns the even and odd parts and corresponding indices without plotting is

```
function [xe,xo,m] = LVEvenOddSymmZero(x,n)
m = -fliplr(n); m1=min([m,n]); m2 = max([m,n]); m = m1:m2;
nm = n(1)-m(1); n1 = 1:1:length(n); x1 = zeros(1,length(m));
x1(n1+nm) = x; xe = 0.5*(x1 + fliplr(x1));
xo = 0.5*(x1 - fliplr(x1));
```

The call

[xe,xo,m] = LVEvenOddSymmZero([1,2,3],[3,4,5])

for example, yields

xe = [1.5,1,0.5,0,0,0,0,0,0.5,1,1.5]

xo = [-1.5,-1,-0.5,0,0,0,0,0,0.5,1,1.5]

m = [-5:1:5]

The script (see exercises below)

$$LVxEvenOddAboutZero(x, n)$$

performs a symmetric-about-zero even-odd decomposition and plots the results. Figure 2.9, which was generated by making the script call

LVxEvenOddAboutZero([0.9.^([0:1:30])],[0:1:30])

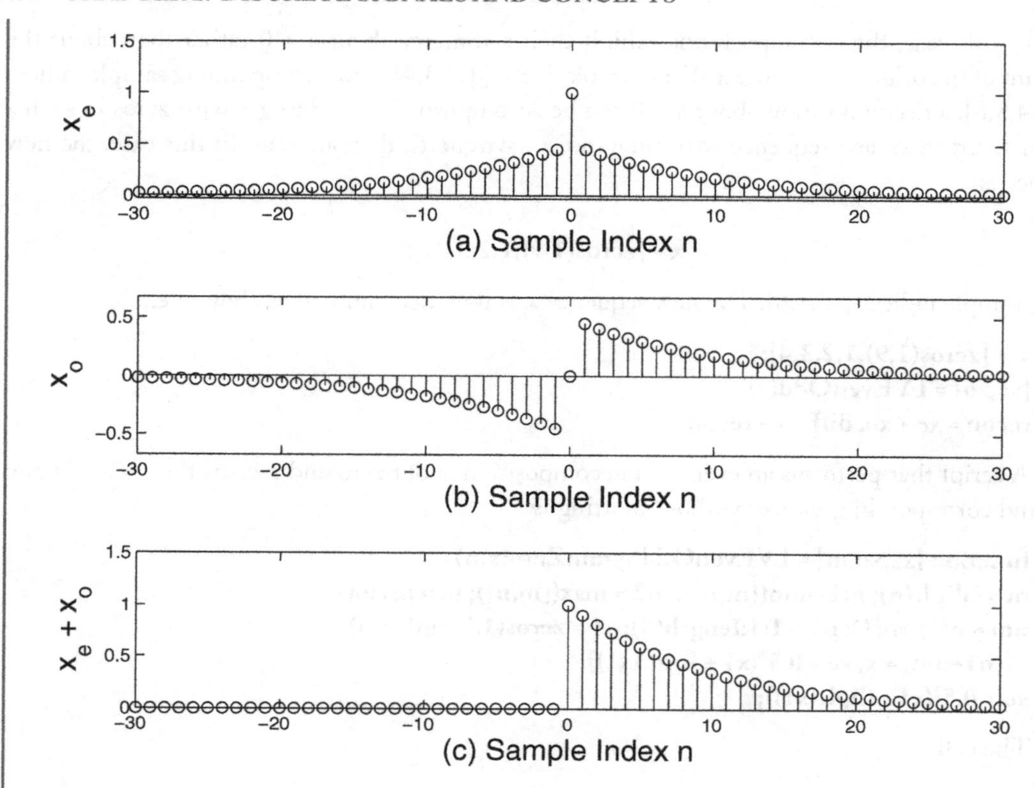

Figure 2.9: (a) Even component of a decaying exponential sequence; (b) Odd component of same; (c) Reconstruction of original exponential sequence, obtained by summing the even and odd components shown in (a) and (b).

shows the result of this process when applied to the sequence x having sample indices n = [0:1:30], where

x = 0.9.^([0:1:30])

2.4.10 GEOMETRIC SEQUENCE

The sum of a decreasing exponential sequence of numbers a^n, where $|a| < 1$, converges to the value $1/(1-a)$, i.e.,

$$\sum_{n=0}^{\infty} a^n \rightarrow \frac{1}{1-a} \tag{2.2}$$

A more general statement of this proposition is that

$$\sum_{n=N}^{\infty} a^n \rightarrow \frac{a^N}{1-a} \tag{2.3}$$

which allows computation of the sum starting from a value of n greater than 0.

Another way of thinking of this is that the sum of a geometric sequence a^n (we assume $|a| < 1$) is its first term divided by one minus the convergence ratio R, where

$$R = a^n / a^{n-1}$$

For example, we can determine the sum of the following sequence using Eq. (2.3), where $N = 0$:

$$1 + 1/2 + 1/4 + 1/8 + ...$$

Here we see that $a = 1/2$ since $a^0 = 1$, $a^1 = 1/2$, etc., so the sum is

$$\frac{1}{1-1/2} = \frac{1}{1/2} = 2$$

We can determine the sum of the following geometric sequence, for example,

$$1/3 + 1/9 + 1/27 + ...$$

using Eq. (2.3) as

$$\frac{1/3}{1-1/3} = 1/2$$

We can verify this result with the simple MathScript call

```
format long; n = 1:1:50; ans = sum((1/3).^n)
```

which yields ans = 0.500000000000000.

Note that it was only necessary to use the first 50 terms of the infinite sequence to obtain a value close (in this case equal within the limitations of accuracy imposed by the computer) to the theoretical value. As a approaches unity in value, more terms are needed to obtain a sum close to the theoretical value.

Sometimes the sum of a finite number of terms of such a sequence is needed. Supposing that the sum of the first N - 1 terms is needed; we can subtract the sum for terms N to ∞ from the sum for all terms, i.e.,

$$\sum_{n=0}^{N-1} a^n = \sum_{n=0}^{\infty} a^n - \sum_{n=N}^{\infty} a^n = \frac{1}{1-a} - \frac{a^N}{1-a} = \frac{1-a^N}{1-a}$$

2.4.11 RANDOM OR NOISE SEQUENCES

Noise is an ever-present background signal in communications systems. It is generated by many natural sources such as the Sun and Jupiter, lightning, many man-made sources, by active devices in electronic systems, etc. Noise assumes random values over time (rather than predictable values such as those of a sine wave, for example) which are described using statistics such as the probability density function, mean, standard deviation, etc.

It is often necessary to simulate noise in signals, and MathScript can be used to generate random sequence values using the functions

$$rand(m, n) \text{ or } randn(m, n)$$

where m and n are dimensions of the matrix of random numbers to be created.

The first function above generates a random signal having uniform distribution over the interval from 0 to 1; the second function above generates a signal having a Gaussian (or normal) distribution with a mean of 0 and standard deviation of 1 .

As an m-code example, we'll generate a signal containing noise of standard deviation 0.125 and a cosine of frequency 11 over 128 samples, and plot the result. The result from running the following m-code is shown in Fig. 2.10.

```
n = 0.125*randn(1,128); c = cos(2*pi*11*[0:127]/128);
figure(3); subplot(311); stem(c);
subplot(312); stem(n)
subplot(313); stem(n+c)
```

2.4.12 CHIRP

A sinusoid, such as a cosine wave, having a frequency that continuously increases with time, is expressed in the continuous domain as

$$y = \cos(\beta t^2)$$

Since the sampled version would have discrete sample times at nT, we would have

$$y[n] = \cos[\beta n^2 T^2]$$

Figure 2.11 shows a sampled chirp with β = 49, T = 1/256, and n = 0:1:255.

The chirp is a useful signal for testing the frequency response of a system such as a filter. A similar continuous domain technique is the use of a sweep generator to reveal the frequency response of analog circuits, such as the video intermediate frequency circuits in TV sets.

Other common uses for the chirped sinusoid are radar and ultrasonic imaging In both cases, a chirp is transmitted toward a target, with the expectation of receiving a reflection at a later time. Since the time of transmission of any frequency in the chirp is known, and the frequency and time received are known for any reflection, the difference in time between the transmission and reception

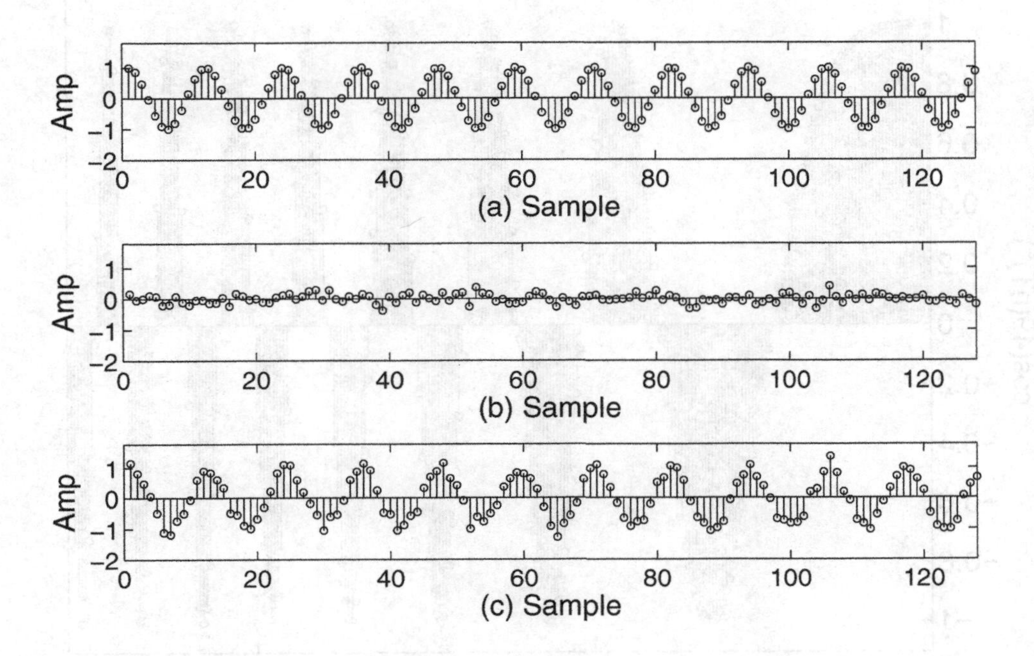

Figure 2.10: (a) A cosine sequence of amplitude 1.0; (b) Gaussian or white noise having standard deviation of 0.125; (c) The sum of the sequences at (a) and (b).

times is directly available. Since the velocity of the transmitted wave is known, the distance between the transmitter/receiver and the point of reflection on the target object can be readily determined.

MathScript's chirp function, in its simplest form, is

$$y = chirp(t, f0, t1, f1)$$

where t is a discrete time vector, $f0$ and $f1$ are the start and end frequencies, respectively, and $t1$ is the time at which frequency $f1$ occurs.

As an m-code example, we can generate a chirp that starts at frequency 0 and ends at frequency 50, over 1101 samples:

```
y = chirp([0:1:1100]/1100,0,1,50); figure; plot(y)
```

2.4.13 COMPLEX POWER SEQUENCE

While it is assumed that the reader's background encompasses complex numbers as part of a basic knowledge of continuous signals and systems, a brief summary of the common complex definitions

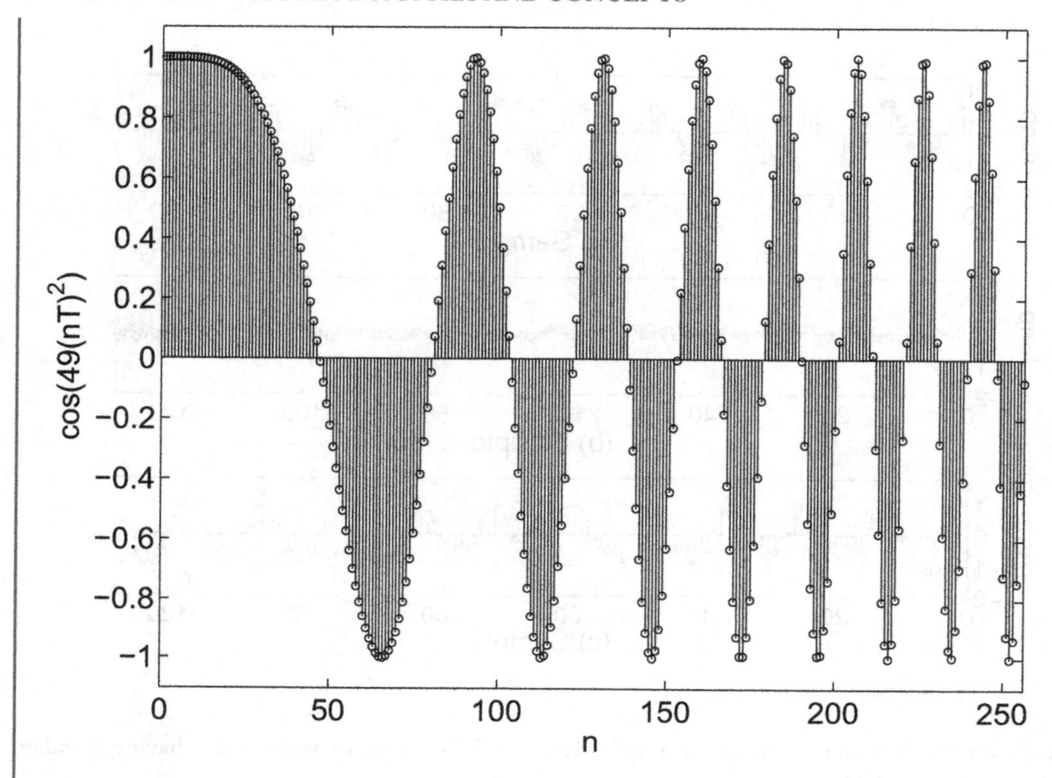

Figure 2.11: A stem plot of a sampled chirp.

and operations is found in the Appendices, which summary should provide a sufficient background for the following discussion, in which we present discrete signal sequence generation and representation using complex numbers.

An exponential of the form

$$y = e^{jx}$$

where e is the base of the natural logarithm system and j is the square root of negative one, generates a complex number lying on the unit circle in the complex plane, i.e., the number has a magnitude of 1.0. Such an exponential is equivalent to

$$e^{jx} = \cos(x) + j \sin(x)$$

As an m-code example, we can generate a complex number having a magnitude of 1.0 and lying at an angle of 45 degrees relative to the real axis of the complex plane with the following call:

y = exp(j*2*pi*(1/8))

A complex number that is repeatedly multiplied by itself generates a sequence of numbers (or samples) having real and imaginary parts which respectively define cosine and sine waves. Think of two complex numbers in polar form: the product has a magnitude equal to the product of the two magnitudes, and an angle equal to the sum of the angles. From this it can be seen that repeatedly multiplying a complex number by itself results in a sequence of complex numbers whose angles progress around the origin of the complex plane by equal increments, and the real and imaginary parts of which form, respectively, a sampled cosine sequence and a sampled sine sequence. If n represents a vector of powers, such as $0{:}1{:}N$, for example, then the complex power sequence is

$$(A\angle\theta)^n = (Ae^{j\theta})^n = A^n e^{jn\theta} = A^n(\cos n\theta + j\sin n\theta)$$

The script

$$LVxComplexPowerSeries(cn, maxPwr)$$

(see exercises below) generates a complex power sequence of the complex number *cn*, raised to the powers 0:1:*maxPwr*.

Figure 2.12, which was created using the script just mentioned with the call

LVxComplexPowerSeries(0.99*exp(j*pi/18),40)

shows the real and imaginary parts of an entire sequence of complex numbers created by raising the original complex number W (magnitude of 0.99 at an angle of 10 degrees ($\pi/18$ radians)) to the powers 0 to 40. Note that the real part, at (c), is a cosine, and the imaginary part, at (d), is a sine wave.

Let's compute powers 0:1:3 for the complex number [0 + j] and describe or characterize the resultant real and imaginary parts. The power sequence is $[j^0, j^1, j^2, j^3]$, which reduces to [1, j, -1, -j], with the real parts being [1,0,-1,0] and the imaginary parts being [0,1,0,-1]. These may be described as four-sample, single-cycle cosine and sine waves. Another way to write this would be

y = cos(2*pi*(0:1:3)/4) + j*sin(2*pi*(0:1:3)/4)

which returns the following:

y = [1,(0 +1i),(-1 + 0i),(-0 - 1i)]

Let's compute the complex power sequence W^n where n = 0:1:4 and $W = (\sqrt{2}/2)(1 + j)$. Note initially that $W = 1\angle 45$. Then $W^{0:1:4} = [1, 1\angle 45, 1\angle 90, 1\angle 135, 1\angle 180]$, which reduces to

[1, 0.707(1+j), j, 0.707(-1+j), -1]

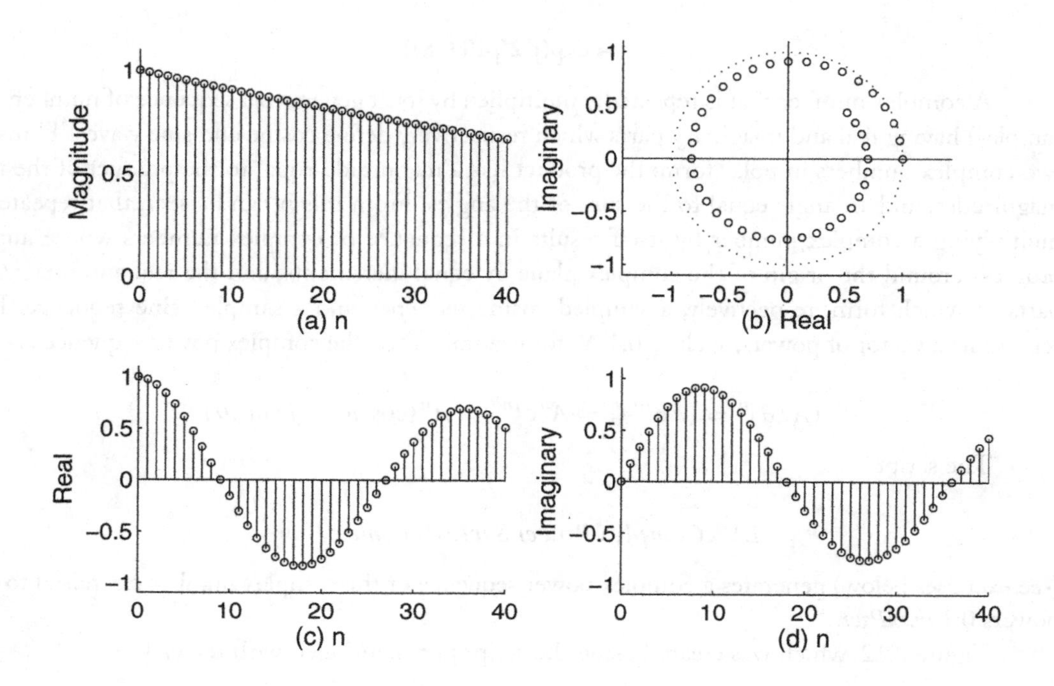

Figure 2.12: (a) Magnitude of $W = (0.99^{*}\exp(j\pi/18))^{n}$ where n = 0:1:40; (b) Plot of entire power sequence in complex plane; (c) Real part of entire power sequence of W for powers 0 to 40; (d) Imaginary part of entire power sequence of W for powers 0 to 40.

To compute the expression using m-code, make the call

```
n = 0:1:4; W = (sqrt(2)/2)*(1+j); y = W.^n
```

As a final example, we'll compute the complex sequence values for

$$e^{-j2\pi nk/N}$$

where $n = [0,1,2,3]$, $N = 4$, and $k = 2$. This reduces to

$$\cos(\pi(0:1:3)) + j\sin(\pi(0:1:3))$$

which yields zero for all the imaginary components and for the real components we get [1,-1,1,-1]. This can be verified by making the call

```
cos(pi*(0:1:3)) + j*sin(pi*(0:1:3))
```

2.4.14 SPECIFIC FREQUENCY GENERATION

- For a given sequence length N, by choosing

$$W = M \exp(j2\pi k/N)$$

the power sequence

$$W^n = [M \exp(j2\pi k/N)]^n = M^n(\cos(2\pi nk/N) + j \sin(2\pi nk/N)) \quad (2.4)$$

will define a complex sinusoid having k cycles over N samples after each power n of W from 0 to N-1 has been evaluated. Note that the complex exponential sequence grows or decays with each succeeding sample according to the value of the magnitude M. If $|M|$ = 1, the sequence has a constant, unity-amplitude; if $|M| < 1$, the sequence decays, and if $|M| > 1$, the sequence grows in amplitude with each succeeding sample.

```
function [seqCos,seqSin] = LVGenFreq(M,k,N)
% [seqCos,seqSin] = LVGenFreq(1,2,8)
n = [0:1:N-1]; arg = 2*pi*k/N;
mags = (M.^n); maxmags = max(mags);
W2n = mags.*exp(j*arg).^n;
seqCos = real(W2n); seqSin = imag(W2n);
figure(66); subplot(211); stem(seqCos);
subplot(212); stem(seqSin)
```

To illustrate use of the above script, we'll generate cosine and sine waves, having peak-to-peak amplitudes of 2 (i.e., amplitudes of unity), and having 7.5 cycles over 73 samples. A peak-to-peak amplitude of 2 means a variation in amplitude from -1 to +1, and hence an amplitude of 1.0, i.e., in the following call we set M = 1.

```
[seqCos,seqSin] = LVGenFreq(1,7.5,73);
```

The result is shown in Fig. 2.13.

To illustrate the generation of a growing complex exponential, we'll generate a cosine, sine pair having a frequency of 3 cycles over 240 samples, and which increases in amplitude by a factor of 1.5 per cycle. Since there are 240/3 samples per cycle, we take the 80th root of 1.5 as M. Code that makes the computation of M easier and more flexible would be

```
F=3;N=240;
[seqCos,seqSin] = LVGenFreq(1.5^(F/N),F,N)
```

The relationship given in Eq. (2.4) can be readily verified graphically and numerically using the following script:

```
function LVPowerSeriesEquiv(M,k,N)
% LVPowerSeriesEquiv(0.9,3,64)
```

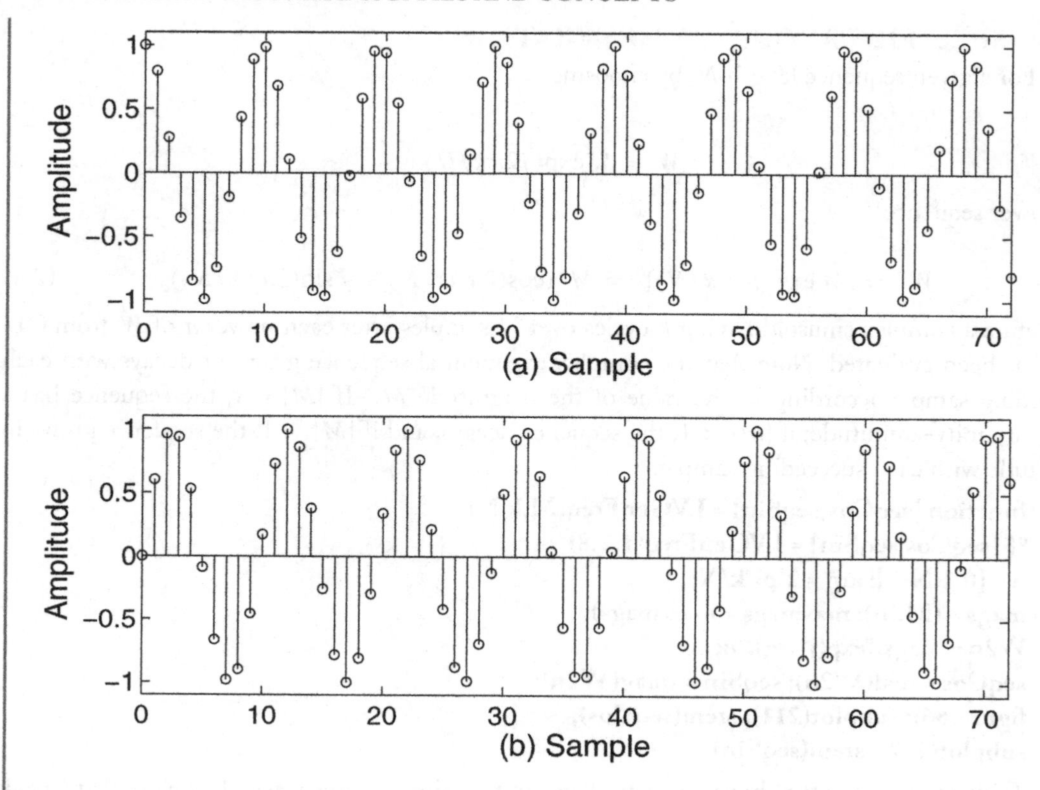

Figure 2.13: (a) The real part of a complex exponential series having 7.5 cycles per 73 samples; (b) The imaginary part of a complex exponential series having 7.5 cycles per 73 samples.

```
n = [0:1:N-1]; arg = 2*pi*k/N;
mags = (M.^n); maxmags = max(mags);
W2n = mags.*exp(j*arg).^n;
rightS = mags.*(cos(n*arg) + j*sin(n*arg));
figure(19); clf; hold on;
plot(real(W2n),imag(W2n),'bo');
plot(real(rightS),imag(rightS),'rx');
grid on; xlabel('Real'); ylabel('Imag')
axis([-maxmags,maxmags,-maxmags,maxmags])
```

Figure 2.15 shows the result from making the call

LVPowerSeriesEquiv(0.9,3,32)

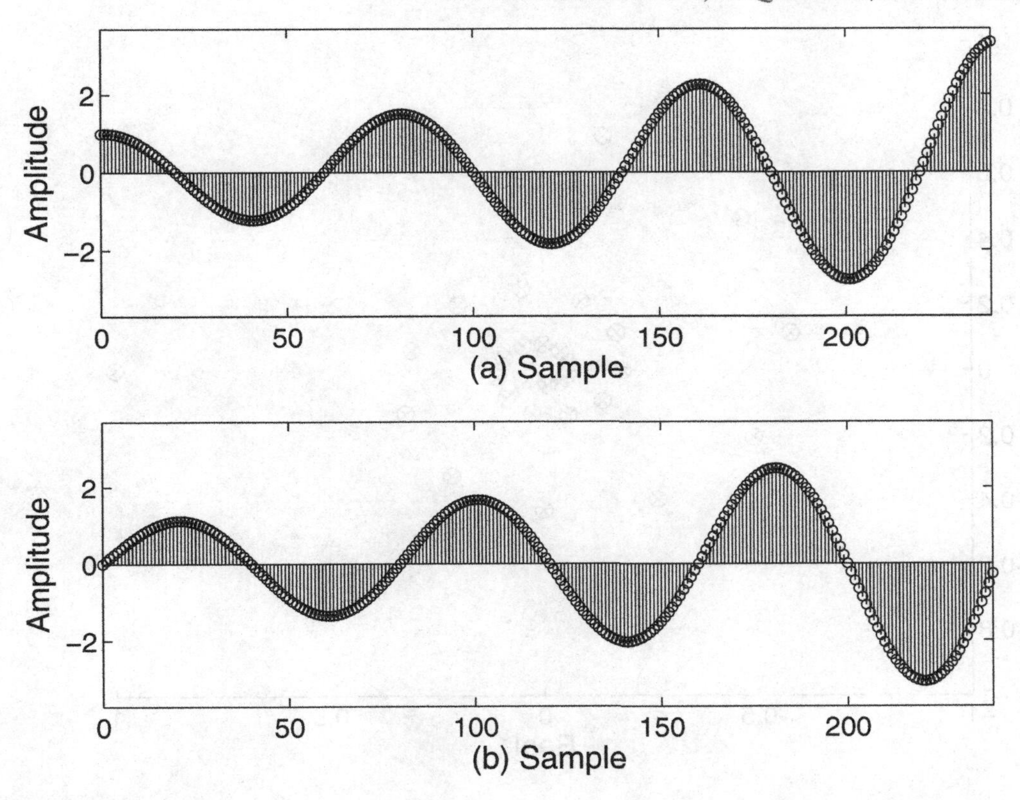

Figure 2.14: (a) The real part of a complex exponential series having one cycle per eight samples, growing in amplitude by a factor of 1.5 per cycle; (b) The imaginary part of a complex exponential series having one cycle per eight samples, growing in amplitude by a factor of 1.5 per cycle.

As a final illustration, we generate 2 cycles of a unity-amplitude cosine over 8 samples, using complex exponentials as follows

real(exp(j*2*pi*2*(0:1:7)/8))

and then compute the same complex exponential using the Euler identity

$$\cos(\theta) = (e^{j\theta} + e^{-j\theta})/2$$

for which a suitable call would be

(exp(j*2*pi*2*(0:1:7)/8) + exp(-j*2*pi*2*(0:1:7)/8))/2

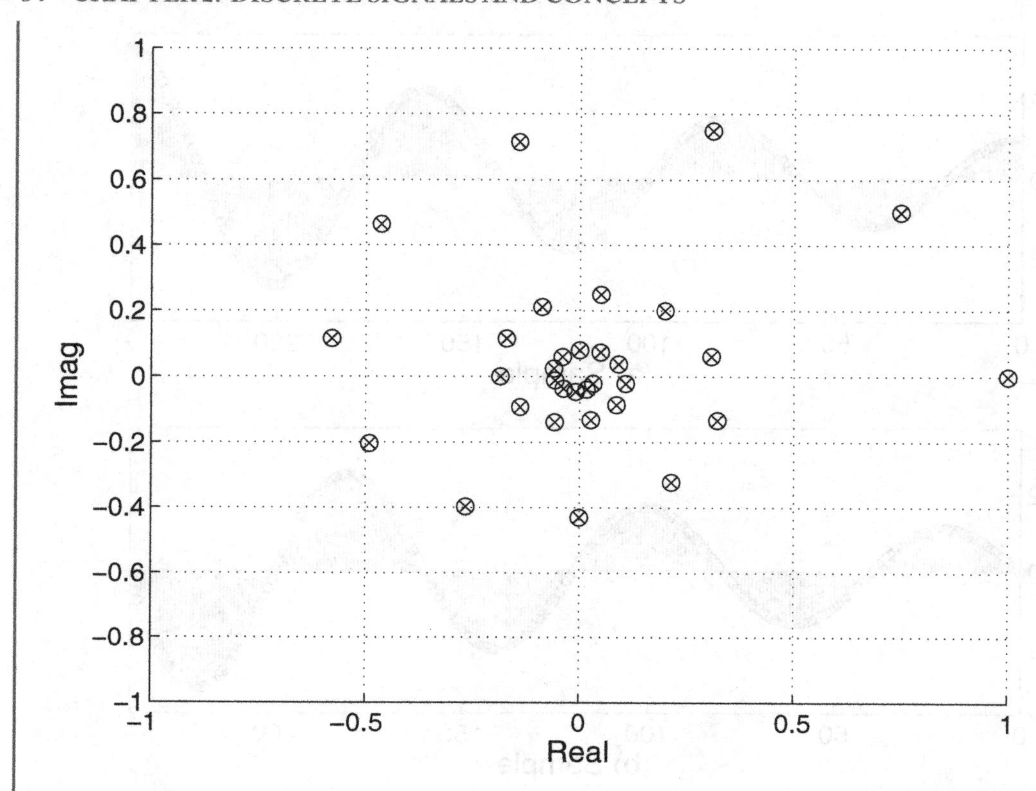

Figure 2.15: The complex power series $0.9^n \exp(j2\pi 3/N)^n$ where $N = 32$ and $n = 0{:}1{:}N\text{-}1$, plotted as circles, and the same series, computed as $0.9^n(\cos(2\pi n3/N) + j\sin(2\pi n3/N))$, plotted as x's, which lie inside the circles since the two methods of computation are equivalent.

2.4.15 ENERGY OF A SIGNAL

The energy of a sequence $x[n]$ is defined as

$$E = \sum_{n=-\infty}^{\infty} x[n]x^*[n] = \sum_{n=-\infty}^{\infty} |x[n]|^2$$

where $x^*[n]$ is the complex conjugate of $x[n]$. If E is finite, $x[n]$ is called an **Energy Sequence.**

2.4.16 POWER OF A SIGNAL

The power of a signal over a number of samples is defined as

$$P = \frac{1}{(2N+1)} \sum_{n=-N}^{N} |x[n]|^2$$

A signal having finite power is called a **Power Signal**.

2.5 DISCRETE TIME SYSTEMS

2.5.1 LTI SYSTEMS

A processing system that receives an input sample sequence $x[n]$ and produces an output sequence $y[n]$ in response is called a **Discrete Time System**. If we denote a discrete time system by the operator DTS, we can then state this in symbolic form:

$$y[n] = DTS\ [x[n]]$$

A number of common signal processes and/or equivalent structures, such as FIR and IIR filtering constitute discrete time systems; they also possess two important properties, namely, 1) Time or Shift Invariance, and 2) Linearity.

A discrete time system DTS is said to be **Shift Invariant**, **Time Invariant**, or **Stationary** if, assuming that the input sequence $x[n]$ produces the output sequence $y[n]$, a shifted version of the input sequence, $x[n-s]$ produces the output sequence $y[n-s]$, for any shift of time s. Stated symbolically, this would be

$$DTS\ [x[n-s]] = y[n-s]$$

A discrete time system DTS that generates the output sequences $y_1[n]$ and $y_2[n]$ in response, respectively, to the input sequences $x_1[n]$ and $x_2[n]$ is said to be **Linear** if

$$DTS\ [ax_1[n] + bx_2[n]] = ay_1[n] + by_2[n]$$

where a and b are constants. This is called the **Principle of Superposition.**

A system that is both shift or time invariant and linear will produce the same output sequence $y[n]$ in response to the sequence $x[n]$ regardless of any shift in time of n samples. Such systems are referred to as **Linear, Time Invariant (LTI)** systems.

Example 2.1. Demonstrate linearity and time invariance for the system below using MathScript.

$$y[n] = 2x[n]$$

We begin with code to compute $y[n] = 2x[n]$ where $x[n]$ can be scaled by the constant A. The code below generates a cosine of frequency F, scaled in amplitude by A as $x[n]$, computes $y[n]$, and then plots $x[n]$ and $y[n]$. You can change the scaling constant A and note the linear change in the output, i.e., if the input signal is scaled by A, so is the output signal (comparison of the results from running the two example calls given in the script above will demonstrate the scaling property).

```
function LVScCosine(A,N,F)
% LVScCosine(1,128,3)
% LVScCosine(2,128,3)
t = [0:1:N-1]/N; x = [A*cos(2*pi*F*t)]; y = 2*x;
subplot(2,1,1); stem(x); subplot(2,1,2); stem(y)
```

Shift invariance can be demonstrated by delaying the input signal and noting that the output is just a shifted version of the output corresponding to the undelayed input signal. For example, we can write a script similar to the above one that inserts a delay Del (a number of samples valued at zero) before the cosine sequence. The reader should run both of the calls given in the script below to verify the shift invariance property.

```
function LVScDelCosine(A,N,F,Del)
% LVScDelCosine(1,128,3,0)
% LVScDelCosine(1,128,3,30)
t = [0:1:N-1]/N; x = [zeros(1,Del),A*cos(2*pi*F*t)];
y = 2*x; figure(14); subplot(2,1,1);
stem(x); subplot(2,1,2); stem(y)
```

We now provide a script that will implement a simple LTI system based on scaling and delaying an input signal multiple times and adding the delayed, scaled versions together. The input argument $LTICoeff$ is a row vector of coefficients, the first one of which weights the input signal, $x[n]$, the second one of which weights the input signal delayed by one sample, i.e., $x[n-1]$, and so on.

```
function [yC,nC] = LV_LTIofX(LTICoeff,x)
% [yC,nC] = LV_LTIofX([1,-2,1],cos(2*pi*12*[0:1:63]/64) )
x1 = LTICoeff(1)*x;
nC = [0:1:length(x1)-1]; yC = x1;
if length(LTICoeff)< 2
return; end
for LTICoeffCtr = 2:1:length(LTICoeff)
xC = [LTICoeff(LTICoeffCtr)*x];
newnC = [0:1:length(xC)-1] + (LTICoeffCtr-1);
[yC, nC] = LVAddSeqs(yC,nC,xC,newnC); end
```

Example 2.2. If $x = \cos(2\pi[0:1:63]/64)$, compute $y[n]$ for the LTI system defined by

$$y[n] = LTI(x) = 2x[n] - x[n-1]$$

We make the call

[yC,nC] = LV_LTIofX([2,-1],cos(2*pi*[0:1:63]/64)); stem(nC,yC)

the results of which are shown in Fig. 2.16.

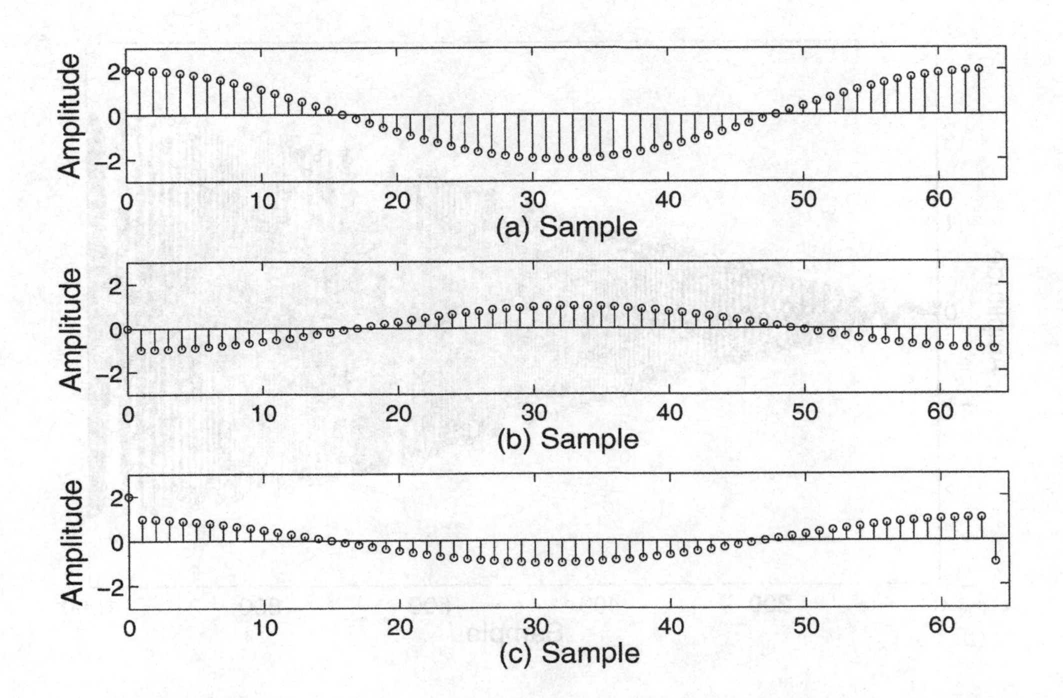

Figure 2.16: (a) The sequence $2\cos(2\pi t)$ where t = [0:1:63]/64; (b) The sequence $-\cos(2\pi t)$, delayed by one sample; (c) The sum or superposition of the sequences at (a) and (b).

Example 2.3. In this example, we will see how a simple LTI system can have a frequency-selective capability. Determine the response $y[n]$ of the LTI system defined as

$$y[n] = 0.1x[n] - x[n-1] + x[n-2] - 0.1x[n-3]$$

where the test signal $x[n]$ is a linear chirp, sampled at 1000 Hz, of one second duration, that changes frequency linearly from 0 to 500 Hz.

We can use the script *LV_LTIofX* with the following call

[yC,nC] = LV_LTIofX([0.1,-1,1,-0.1],...
chirp([0:1:999]/1000,0,1,500));stem(nC,yC)

the results of which are shown in Fig. 2.17. We see that the simple four-sample LTI system has been able to process a chirp in such a way as to progressively emphasize higher frequencies. Thus this simple LTI system functions as a highpass filter.

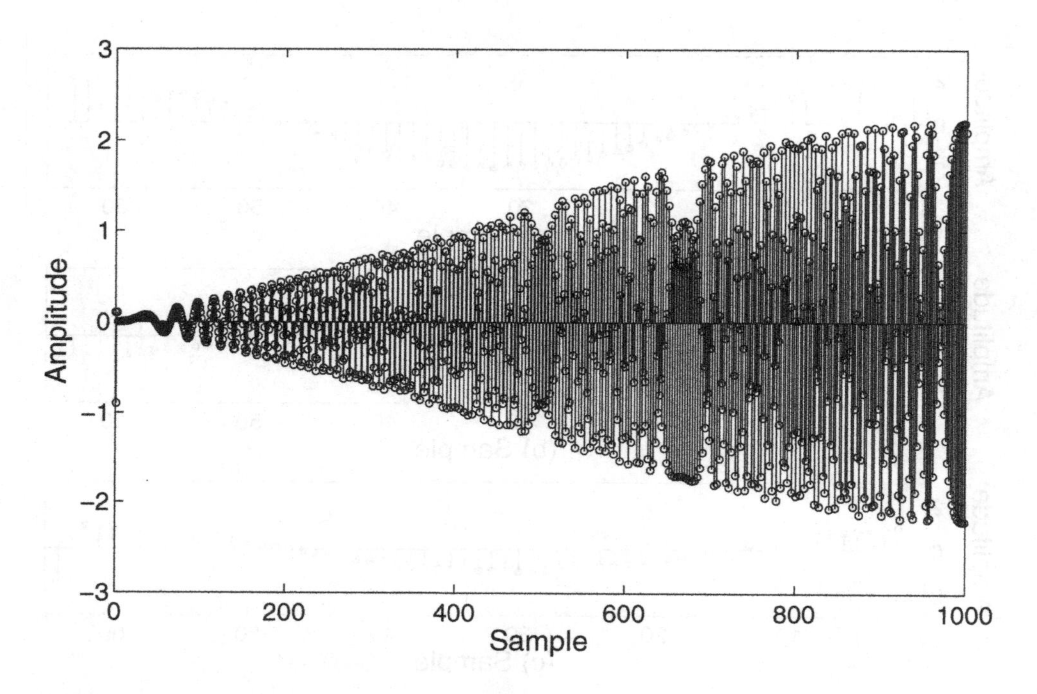

Figure 2.17: The result from convolving a linear chirp sampled at 1000 Hz of one second duration, having frequencies from 0 to 500 hz, with the LTI system defined by the coefficients [0.1,-1,1,-0.1].

The script (see exercises below)

$$LVxLinearab(a, b, f1, f2, N, Del, LTICoeff)$$

uses code similar to that above to compute $x_1[n]$, $y_1[n]$, $x_2[n]$, $y_2[n]$, $ax_1[n] + bx_2[n]$, $ay_1[n] + by_2[n]$, and $LTI\ [ax_1[n] + bx_2[n]]$, where LTI represents the system defined by *LTICoeff* just as for the script *LV_LTIofX*. Test signal $x_1[n]$ is a cosine of frequency $f1$, and test signal $x_2[n]$ is a sine of frequency $f2$, both over N samples. An arbitrary delay of Del samples can be inserted at the leading side of $x_1[n]$ and $x_2[n]$.

Figure 2.18 was generated by making the call

LVxLinearab(2,-3,13,5,128,0,[2,-1,1,2])

which demonstrates the superposition property for the LTI system defined as

$$y[n] = 2x[n] - x[n-1] + x[n-2] + 2x[n-3] \tag{2.5}$$

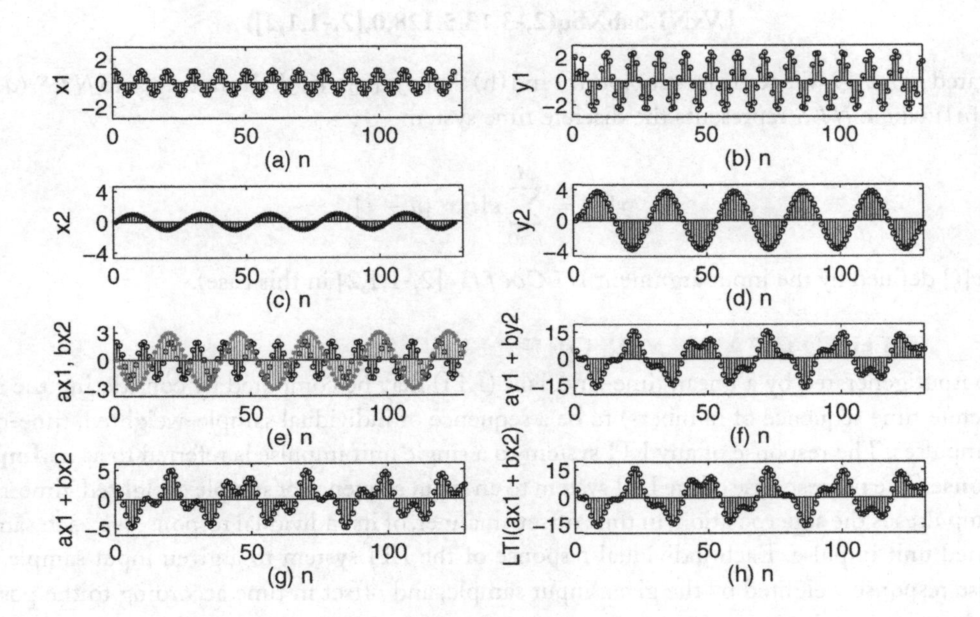

Figure 2.18: (a) $x_1[n]$; (b) $y_1[n]$; (c) $x_2[n]$; (d) $y_2[n]$; (e) $ax_1[n]$ (circles) and $bx_2[n]$ (stars); (f) $ay_1[n] + by_2[n]$; (g) $ax_1[n] + bx_2[n]$; (h) $LTI\ (ax_1[n] + bx_2[n])$.

Note that subplots (f) and (h) show, respectively, $ay_1[n] + by_2[n]$ and

$$LTI\ (ax_1[n] + bx_2[n])$$

where the LTI operator in this case represents the system Eq. (2.5). Subplots (f) and (h) show that

$$LTI\ (ax_1[n] + bx_2[n]) = ay_1[n] + by_2[n]$$

For contrast, let's consider the second order (i.e., nonlinear) system

$$y[n] = 2x^2[n] - x^2[n-1] + x^2[n-2] + 2x^2[n-3]$$

The script

$$LVxNLSabXSq(a, b, f1, f2, N, Del, NLCoeff)$$

(see exercises below) performs the superposition test on the (nonlinear) system

$$y[n] = c[0]x^2[n] + c[1]x^2[n-1] + c[2]x^2[n-2] + ...$$

where $c[n]$ are the elements of the input vector $NLCoeff$, and plots the results. The script call

LVxNLSabXSq(2,-3,13,5,128,0,[2,-1,1,2])

generated Fig. 2.19. We see from subplots (f) and (h) that $ay_1[n] + by_2[n]$ is not equal to $NLS\ (ax_1[n] + bx_2[n])$ where NLS represents the discrete time system

$$y[n] = \sum_{i=0}^{M} c[i]x^2[n-i]$$

with $c[i]$ defined by the input argument $NLCoeff$(=[2,-1,1,2] in this case).

2.5.2 METHOD OF ANALYSIS OF LTI SYSTEMS

The output generated by a linear, time-invariant (LTI) may be computed by considering the input (a discrete time sequence of numbers) to be a sequence of individual sample-weighted, time-offset unit impulses. The response of any LTI system to a single unit impulse is referred to as its **Impulse Response.** The net response of the LTI system to an input sequence of sample-weighted, time-offset unit impulses is the superposition, in time offset- manner, of its individual responses to each sample-weighted unit impulse. Each individual response of the LTI system to a given input sample is its impulse response weighted by the given input sample, and offset in time according to the position of the input sample in time.

Figure 2.20 depicts this process for the three-sample signal sequence [1,-0.5,0.75], and an LTI system having the impulse response 0.7^n, where n = 0:1:∞. We cannot, obviously, perform the superposition for all n (i.e., an infinite number of values), so we illustrate the process for a few values of n.

The process above, summing delayed, sample-weighted versions of the impulse response to obtain the net output, can be performed according to the following formula, where the two sequences involved are denoted $h[n]$ and $x[n]$:

$$y[k] = \sum_{n=-\infty}^{\infty} x[n]h[k-n] \tag{2.6}$$

Equation (2.6) is called the **convolution formula.**

Example 2.4. Let $x[n] = [1, -0.5, 0.75]$ and $h[n] = 0.7^n$ Compute the first 3 output values of y as shown in subplot (h) of Fig. 2.20, using Eqn. (2.6).

We note that $x[n] = 0$ for $n < 0$ and $n > 2$. We also note that $h[n] = 0$ for values of $n < 0$. We set the range of k as 0:1:2 to compute the first three output samples, and as a result, n = 0:1:2,

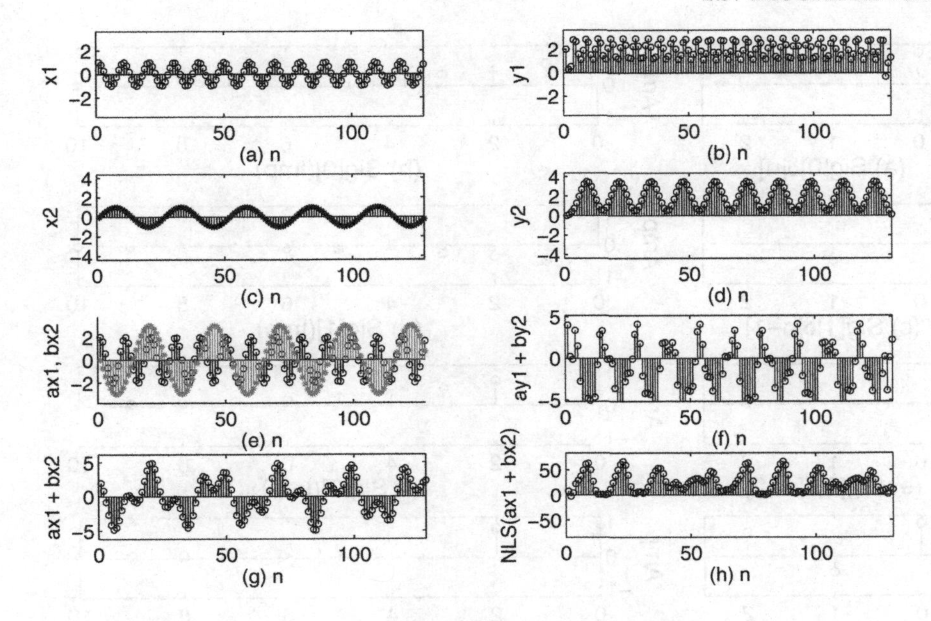

Figure 2.19: (a) $x_1[n]$; (b) $y_1[n]$; (c) $x_2[n]$; (d) $y_2[n]$; (e) $ax_1[n]$ (circles) and $bx_2[n]$ (stars); (f) $ay_1[n] + by_2[n]$; (g) $ax_1[n] + bx_2[n]$; (h) $NLS\ (ax_1[n] + bx_2[n])$.

which may be explained as follows: since the maximum value of k we will compute is 2, we need not exceed $n = 2$ since if n exceeds 2, $k - n$ is less than zero and as a result, $h[n] = 0$. To compute $y[k]$ for higher values of k, a correspondingly larger range for n is needed.

We get

$$y[0] = \sum_{n=0}^{2} x[n]h[0-n]$$

The sum above is

$$y[0] = x[0]h[0] + x[1]h[-1] + x[2]h[-2] = x[0]h[0] = 1$$

$$y[1] = x[0]h[1] + x[1]h[0] = (1)(0.7) + (-0.5)(1) = 0.2$$

$$y[2] = x[0]h[2] + x[1]h[1] + x[2]h[0] = 0.49 + (-0.5)(0.7) + 0.75 = 0.89$$

Example 2.5. For the sequences above, compute $y[k]$ for $k = 3$.

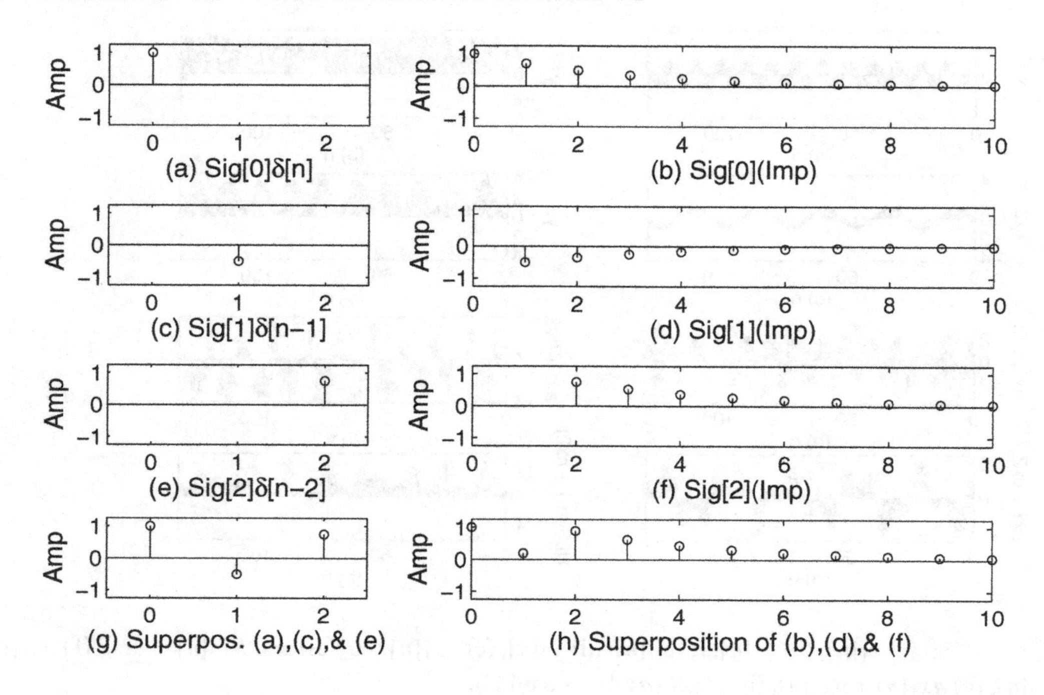

Figure 2.20: (a) First sample of signal, multiplied by $\delta(n)$; (b) Impulse response, weighted by first signal sample; (c) Second signal sample; (d) Impulse response, scaled by second signal sample, and delayed by one sample; (e) Third signal sample; (f) Impulse response scaled by third signal sample, delayed by two samples; (g) Input signal, the superposition of its components shown in (a), (c), and (e); (h) The convolution, i.e., the superposition of responses shown in (b), (d), and (f).

We set $k = 3$ and $n = 0{:}1{:}3$. Then

$$y[3] = x[0]h[3] + x[1]h[2] + x[2]h[1] + x[3]h[0]$$

which is

$$1(0.343) + (-0.5)(0.49) + 0.75(0.7) + 0(1) = 0.623$$

Example 2.6. Use m-code to compute the first 10 output values of the convolution of the two sequences $[1, -0.5, 0.75]$ and 0.7^n.

MathScript provides the function

$$conv(x, y)$$

which convolves the two sequences x and y. We make the call

conv([1 -0.5 0.75],[0.7.^(0:1:9)])

If the roles of the two sequences above had been reversed, that is to say, if we had defined the sequence [1,-0.5, 0.75] as the impulse response $h[n]$ in Eq. (2.6), and $x[n]$ = 0.7.^(0:1:9), the result would be as shown in Fig. 2.21. Note that the first three samples of the convolution sequence are the same as shown in Fig. 2.20.

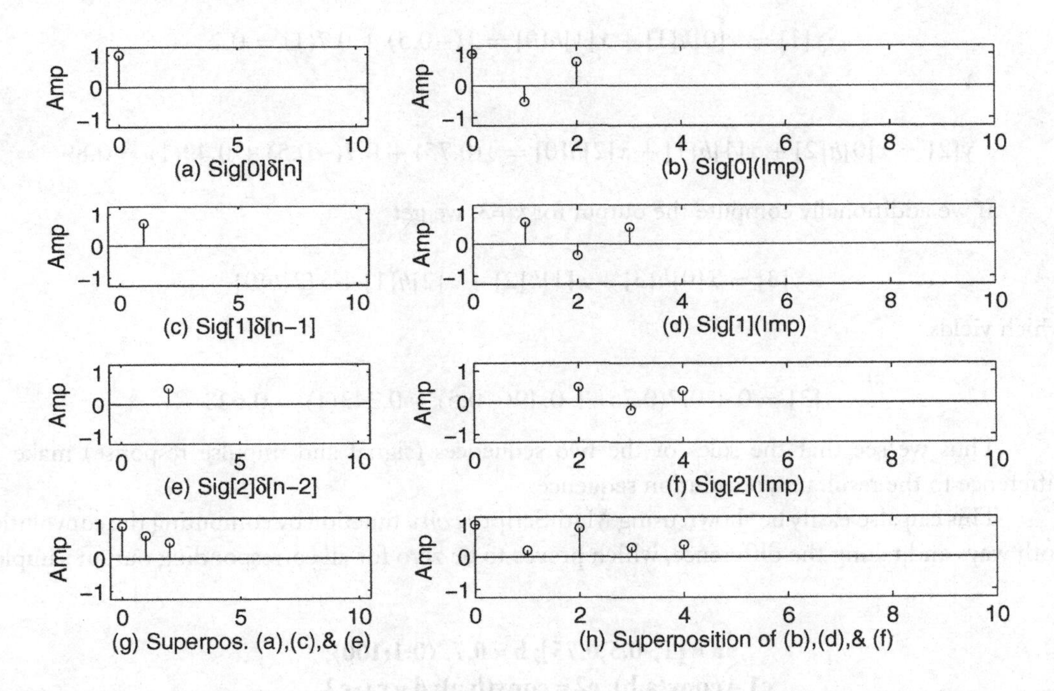

Figure 2.21: The convolution depicted in Fig. 2.20, with the roles of signal and impulse response reversed. (a) First signal sample, multiplied by $\delta\ (n)$; (b) Impulse response, scaled by first sample of signal; (c) Second sample of signal; (d) Impulse response, scaled by second signal sample, delayed by one sample; (e) Third sample of signal; (f) Impulse response scaled by third signal sample, delayed by two samples; (g) Input signal, the superposition of its components shown in (a), (c), and (e); (h) The convolution, i.e., the superposition of responses shown in (b), (d), and (f).

Example 2.7. In this example, we'll reverse the role of signal and impulse response and show that the convolution sequence is the same. Let $x[n]$ = 0.7.^(0 : 1 : 9) and $h[n] = [1, -0.5, 0.75]$. Compute the first three samples of the convolution sequence.

Note that $h[n] = 0$ for $n < 0$ and $n > 2$. We set the range of k as 0:1:2, and n = 0:1:2. We get

$$y[0] = \sum_{n=0}^{2} x[n]h[0-n]$$

The sum above is, for k = 0:1:2

$$y[0] = x[0]h[0] + x[1]h[-1] + x[2]h[-2] = x[0]h[0] = 1$$

$$y[1] = x[0]h[1] + x[1]h[0] = 1(-0.5) + 0.7(1) = 0.2$$

$$y[2] = x[0]h[2] + x[1]h[1] + x[2]h[0] = 1(0.75) + 0.7(-0.5) + 0.49(1) = 0.89$$

If we additionally compute the output for k =3, we get

$$y[3] = x[0]h[3] + x[1]h[2] + x[2]h[1] + x[3]h[0]$$

which yields

$$y[3] = 0 + 0.7(0.75) + 0.49(-0.5) + 0.343(1) = 0.623$$

Thus we see that the roles of the two sequences (signal and impulse response) make no difference to the resultant convolution sequence.

This can also easily be shown using MathScript's *conv* function by computing the convolution both ways and taking the difference, which proves to be zero for all corresponding output samples.

```
a = [1,-0.5,0.75]; b = 0.7.^(0:1:100);
c1 = conv(a,b); c2 = conv(b,a); d = c1-c2
```

2.5.3 GRAPHIC METHOD

An easy way to visualize and perform convolution is by time-reversing one of the two sequences and passing it through the other sequence from the left, one sample at a time. Each convolution output sample is computed by multiplying all overlapping samples and adding the products. Figures 2.22 and 2.23 show the two (equivalent) graphic orientations to compute the convolution sequence of two sequences. In Fig. 2.22, the first sequence is 0.8.^(0:1:9) and the second sequence is 0.5*[(-0.7).^[0:1:9]], and in Fig. 2.23, the roles are reversed. Inspection of subplots (c)-(f) shows that each of the four convolution values that can be computed from the illustrated overlapping sequences must be identical for the two figures since exactly the same samples from each sequence are overlapping, i.e., either sequence may be time reversed and moved through the other from left to right with the same computational result.

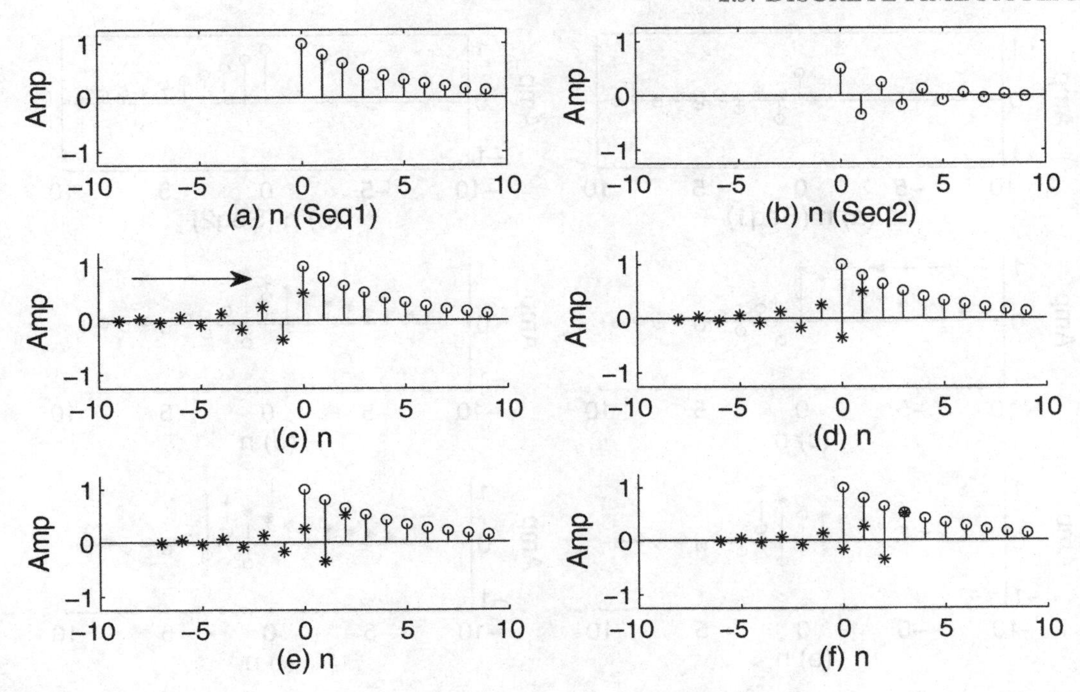

Figure 2.22: (a) First Sequence; (b) Second Sequence; (c) Second sequence time reversed (TR) and oriented to compute the first value of the convolution sequence (arrow shows direction the second sequence will slide, sample-by-sample, to perform convolution); (d) TR second sequence oriented to compute the second value of the convolution sequence; (e) TR second sequence oriented to compute the third value of the convolution sequence; (f) TR second sequence oriented to compute the fourth value of the convolution sequence.

Example 2.8. Compute the first four values of the convolution of the sequences $[1, -1, 1, -1, 1]$ and $[1, 0.5, 0.25, 0.125]$ using the graphic visualization method.

Figure 2.24 illustrates the process. Flipping the second sequence from right to left, we get for the first convolution sequence value (1)(1) =1, the second value is (1)(-1) + (0.5)(1) = -0.5, the third value is (1)(1) + (0.5)(-1) + (0.25)(1) = 0.75, and the fourth value is (1)(-1) + (0.5)(1) + (0.25)(-1) + (0.125)(1) = -0.625.

We can check this by making the call

y = conv([1,0.5,0.25,0.125],[1,-1,1,-1,1])

which yields (with redundant zeros eliminated for brevity)

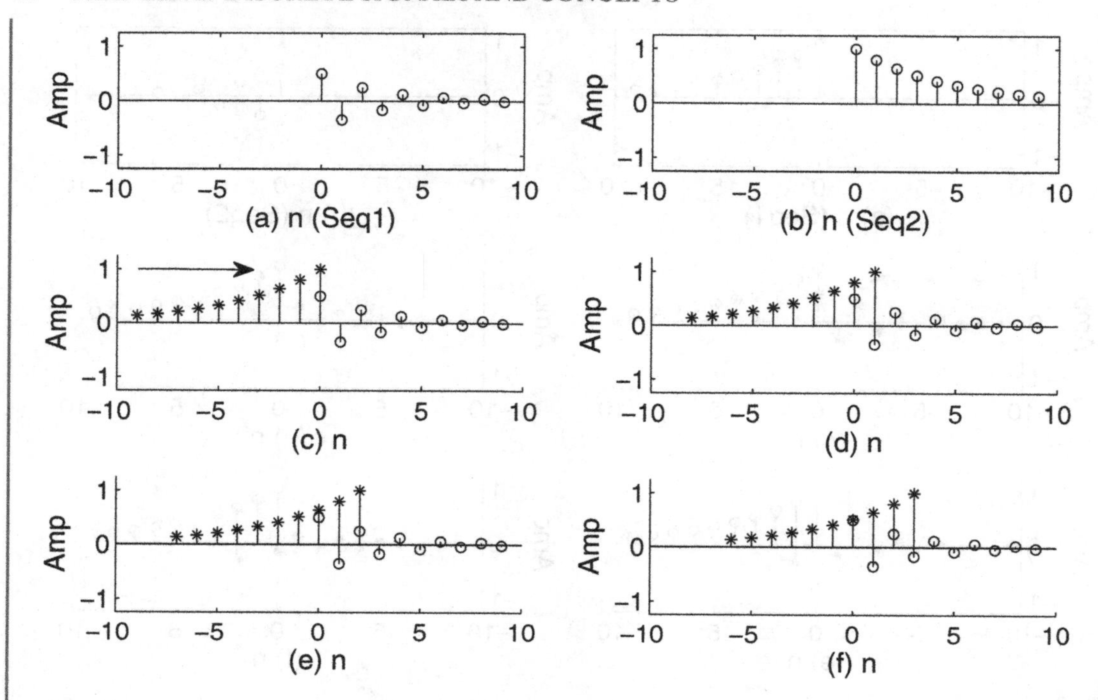

Figure 2.23: (a) First Sequence (second sequence in previous figure); (b) Second Sequence (first sequence in previous figure); (c) Second sequence time reversed (TR) and oriented to compute the first value of the convolution sequence (arrow shows direction the second sequence will slide, sample-by-sample, to perform convolution); (d) TR second sequence oriented to compute the second value of the convolution sequence; (e) TR second sequence oriented to compute the third value of the convolution sequence; (f) TR second sequence oriented to compute the fourth value of the convolution sequence.

$$y = 1, -0.5, 0.75, -0.625, 0.625, 0.375, 0.125, 0.125$$

2.5.4 A FEW PROPERTIES OF CONVOLUTION

Let's use the symbol $\circledast$ to represent convolution. Then we can compactly represent the convolution $y[n]$ of two sequences $h[n]$ and $x[n]$ as

$$y[n] = h[n] \circledast x[n]$$

Convolution is linear, so

$$y[n] = h[n] \circledast ax[n] = a(h[n] \circledast x[n])$$

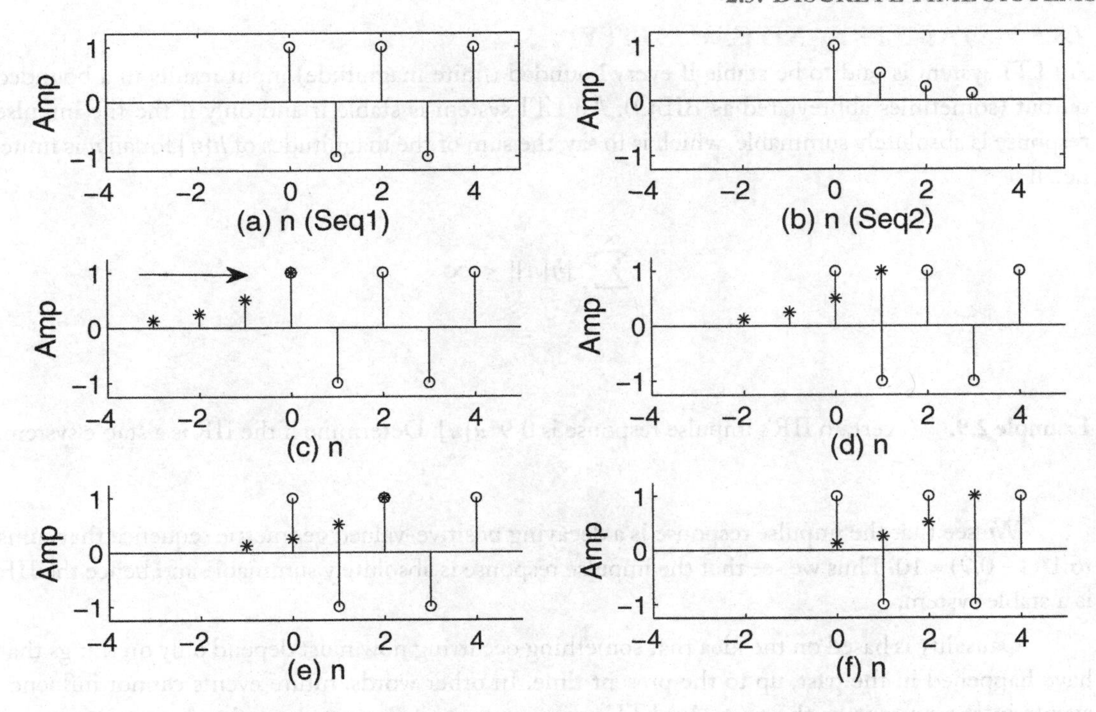

Figure 2.24: (a) First Sequence; (b) Second Sequence; (c) Second sequence time reversed (TR) and oriented to compute the first value of the convolution sequence (arrow shows direction the second sequence will slide, sample-by-sample, to perform convolution); (d) TR second sequence oriented to compute the second value of the convolution sequence; (e) TR second sequence oriented to compute the third value of the convolution sequence; (f) TR second sequence oriented to compute the fourth value of the convolution sequence.

where a is a constant.

The distributive property applies to convolution: for two sequences $x_1[n]$ and $x_2[n]$

$$h[n] \circledast (x_1[n] + x_2[n]) = h[n] \circledast x_1[n] + h[n] \circledast x_2[n]$$

The commutative property also applies, so

$$y[n] = h[n] \circledast x[n] = x[n] \circledast h[n]$$

which we showed by example above.

2.5.5 STABILITY AND CAUSALITY

An LTI system is said to be stable if every bounded (finite magnitude) input results in a bounded output (sometimes abbreviated as BIBO). An LTI system is stable if and only if the the impulse response is absolutely summable, which is to say, the sum of the magnitudes of $h[n]$ for all n is finite, i.e., if

$$\sum_{n=-\infty}^{\infty} |h[n]| < \infty$$

Example 2.9. A certain IIR's impulse response is $0.9^n u[n]$. Determine if the IIR is a stable system.

We see that the impulse response is a decaying positive-valued geometric sequence that sums to 1/(1 - 0.9) = 10. Thus we see that the impulse response is absolutely summable and hence the IIR is a stable system.

Causality is based on the idea that something occurring now must depend only on things that have happened in the past, up to the present time. In other words, future events cannot influence events in the present or the past. An LTI system output $y[n]$ must depend only on previous or current values of input and output. More particularly, a system is causal if

$$h[n] = 0 \text{ if } n < 0$$

Example 2.10. Determine if the following impulse response is causal:

$$h[n] = 0.5^{n+1} u[n+1]$$

Since the nonzero portion of the sequence begins at n = -1, $h[n]$ is not causal.

2.5.6 LTI SYSTEM AS A FILTER

An LTI system that has been designed to achieve a particular purpose or perform a given function, such as frequency selection or attenuation, is called a **Filter**. There are two basic types of digital filter, the FIR filter and the IIR filter, each having certain advantages and disadvantages that determine suitability for a given use.

The FIR
The Finite Impulse Response (FIR), or Transversal Filter, comprises structures or algorithms that produce output samples that are computed using only the current and previous input samples. The response of such a system to a unit impulse sequence is finite, and hence such a system is called a Finite Impulse Response filter. Figure 2.25 shows a typical example, having an arbitrary number of M delay stages, each denoted by the letter D. The cascaded delay stages act like a bucket brigade, transporting each sample, as it enters from the left, one delay stage at a time to the right. As shown, a sample sequence $s[n]$ is passing through the filter. The output of each delay stage is scaled by a multiplier according to the coefficients b_i, and all products summed yield the output. Any number of delay stages may be used, as few as one stage being possible. The larger the number of delay stages and multipliers, the greater can be the frequency selectivity.

The output of an FIR is, in general, computed using convolution. The manner of determining what an FIR's coefficients b_i should be to achieve a certain signal processing purpose is the subject of FIR design, which is covered extensively in Part III of the book (Chapters 8–10).

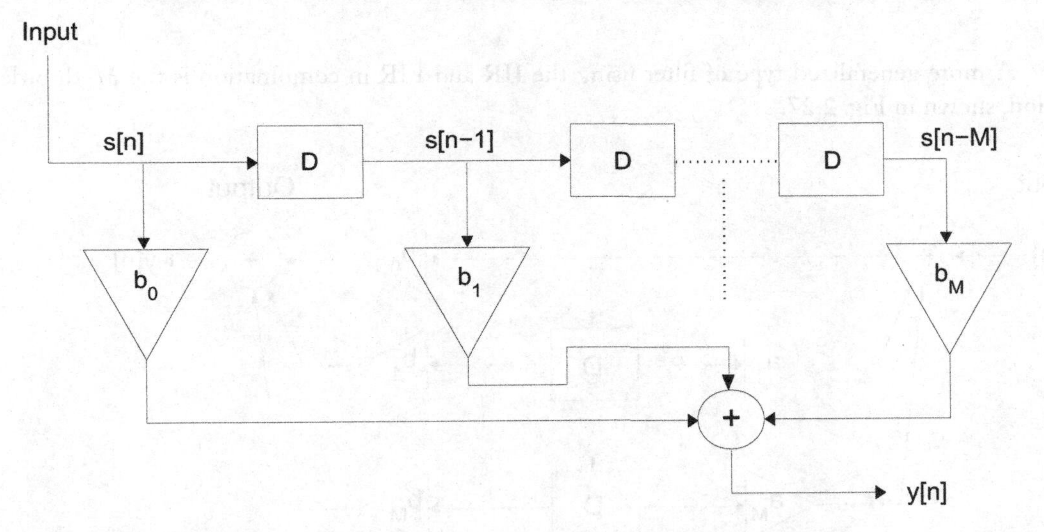

Figure 2.25: A generalized finite impulse response (FIR) filter structure. Note that the filter has M delay stages (each marked with the letter **D**) and $M + 1$ coefficient multipliers.

The IIR
The second type of basic digital filter is the **Infinite Impulse Response (IIR)** filter, which produces output samples based on the current and possibly previous input samples and previous values of its own output. This feedback process is usually referred to as a **Recursive Process**, or **Recursion**, and produces, in general, a unit impulse response which is infinite in extent. When the impulse response decays away to zero over time, the filter is stable. Figure 2.26 shows the simplest possible such filter, having one stage of delay and feedback.

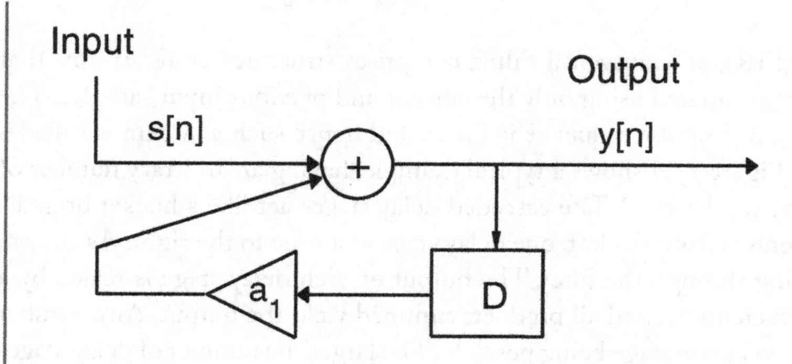

Figure 2.26: A simple recursive filter structure having a summing junction and a single feedback stage comprised of a one-sample delay element and a scaler.

A more generalized type of filter using the IIR and FIR in combination is the M-th order section, shown in Fig. 2.27.

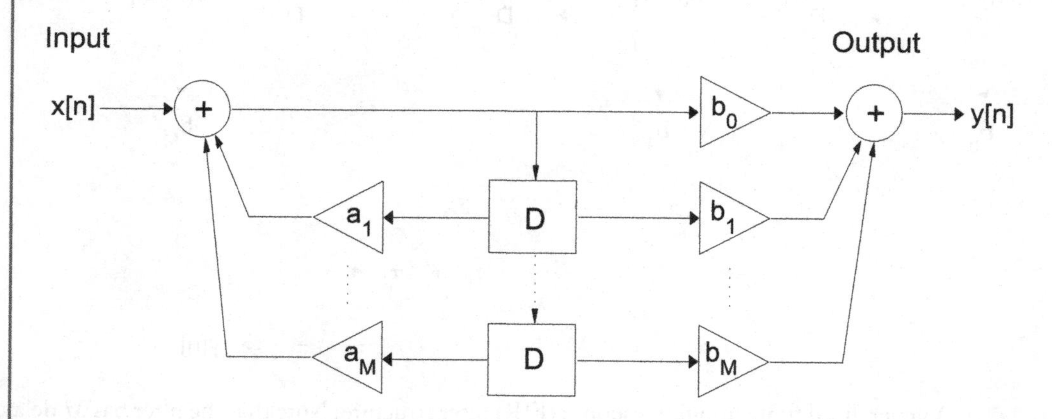

Figure 2.27: A generalized M-th order digital filter utilizing both recursive and nonrecursive computation.

Design of IIRs is covered in Chapter 10.

2.5.7 DIFFERENCE EQUATIONS

LTI systems can also be represented by a constant-coefficient equation that permits sequential computation of the system's output. If $x[n]$ represents an input sequence and $y[n]$ represents the output sequence of the system, an FIR can be represented by the difference equation

$$y[n] = \sum_{m=0}^{M} b_m x[n-m]$$

Example 2.11. Compute the response to the sequence $x[n] = ones(1,4)$ of the FIR represented by the difference equation given below (assume that $x[n] = 0$ for $n < 0$).

$$y[n] = x[n] - x[n-1]$$

The sequence of computation is from $n = 0$ forward in time:

y[0] = x[0] - x[-1] = 1
y[1] = x[1] - x[0] = 1 - 1 = 0
y[2] = x[2] - x[1] = 1 - 1 = 0
y[3] = x[3] - x[2] = 1 - 1 = 0
y[4] = x[4] - x[3] = 0 -1 = -1

To do the above using MathScript, the following code is one possibility. The reader should study the code and be able to explain the purpose of or need for 1) extending x to a length of five by adding one zero-valued sample, 2) the statement y(1) = x(1), and 3) running n effectively from 1 to 5 rather than 0 to 4.

```
x = [ones(1,4) 0]; y(1) = x(1); for n = 2:1:5; y(n) = x(n) - x(n-1); end; ans = y
```

A basic difference equation for an IIR can be written as

$$y[n] = x[n] - \sum_{p=1}^{N} a_p y[n-p]$$

In this equation, the output $y[n]$ depends on the current value of the input $x[n]$ and previous values of the output $y[n]$, such as $y[n-1]$, etc.

Example 2.12. Compute the first four values of the impulse response of the IIR whose difference equation is

$$y[n] = x[n] + 0.9y[n-1]$$

We use $x[n]$ = [1,0,0,0] and get

y[0] = x[0] + 0.9y[-1] = 1 + 0 = 1
y[1] = x[1] + 0.9y[0] = 0 + 0.9 = 0.9
y[2] = x[2] + 0.9y[1] = 0 + (0.9)(0.9) = 0.81
y[3] = x[3] + 0.9y[2] = 0 + (0.9)(0.81) = 0.729

A more generic form of the difference equation which includes both previous inputs and previous outputs is

$$y[n] = \sum_{m=0}^{M} b_m x[n-m] - \sum_{p=1}^{N} a_p y[n-p] \tag{2.7}$$

MathScript provides the function

$$filter(b, a, x)$$

which accepts an input vector x and evaluates the difference equation formed from coefficients b and a.

Example 2.13. A certain LTI system is defined by b = [1] and a = [1, −1.27, 0.81]. Compute and plot the first 30 samples of the response of the system to a unit impulse, i.e., the system's impulse response.

We make the call

```
y = filter([1],[1,-1.27, 0.81],[1,zeros(1,29)]); figure; stem(y)
```

which results in Fig. 2.28.

Example 2.14. Write the difference equation corresponding to the coefficients in the previous example.

From Eqn. (2.7) we get

$$y[n] = x[n] - (-1.27y[n-1] + 0.81y[n-2]) \tag{2.8}$$

which yields

$$y[n] = x[n] + 1.27y[n-1] - 0.81y[n-2]) \tag{2.9}$$

Example 2.15. A certain filter is defined by b = [0.0466, 0.1863, 0.2795, 0.1863, 0.0466] and a = [1, −0.7821, 0.68, −0.1827, 0.0301]. Compute the filter's response to a linear chirp having frequencies varying linearly from 0 to 500 Hz.

The given coefficients were chosen to yield a smooth lowpass effect. The details of how to design such a filter are found in Chapter 10, Classical IIR Design. Much preparation in the intervening chapters will be needed to prepare the student for the study of IIR design. The point of this exercise is to show how a small number of properly chosen coefficients can give a very desirable filtering result. We run the following m-code, which results in Fig. 2.29.

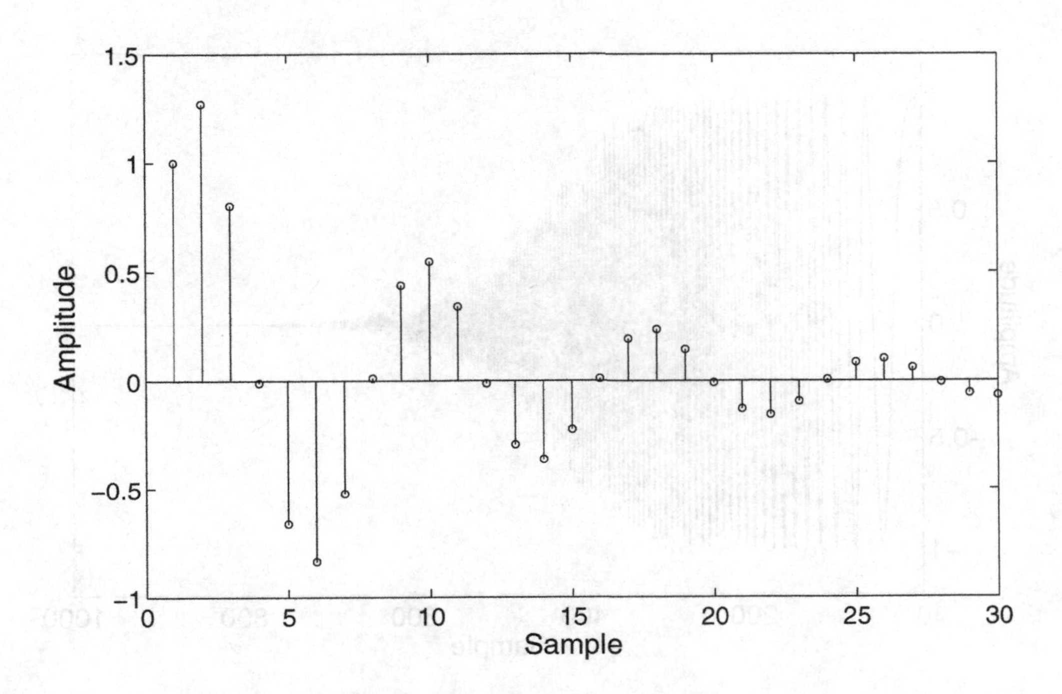

Figure 2.28: The first 30 samples of the impulse response of an LTI system defined by the coefficients $b = [1]$ and $a = [1, -1.27, 0.81]$.

```
b = [ 0.0466, 0.1863, 0.2795, 0.1863, 0.0466];
a = [1, -0.7821, 0.68, -0.1827, 0.0301];
y = filter([b],[a],[chirp([0:1:999]/1000,0,1,500)]); plot(y)
xlabel('Sample'); ylabel('Amplitude')
```

2.6 REFERENCES

[1] James H. McClellan, Ronald W. Schaefer, and Mark A. Yoder, *Signal Processing First*, Pearson Prentice Hall, Upper Saddle River, New Jersey, 2003.

[2] John G. Proakis and Dimitris G. Manolakis, *Digital Signal Processing, Principles, Algorithms, and Applications, Third Edition*, Prentice Hall, Upper Saddle River, New Jersey, 1996.

[3] Vinay K. Ingle and John G. Proakis, *Digital Signal Processing Using MATLAB V.4*, PWS Publishing Company, Boston, 1997.

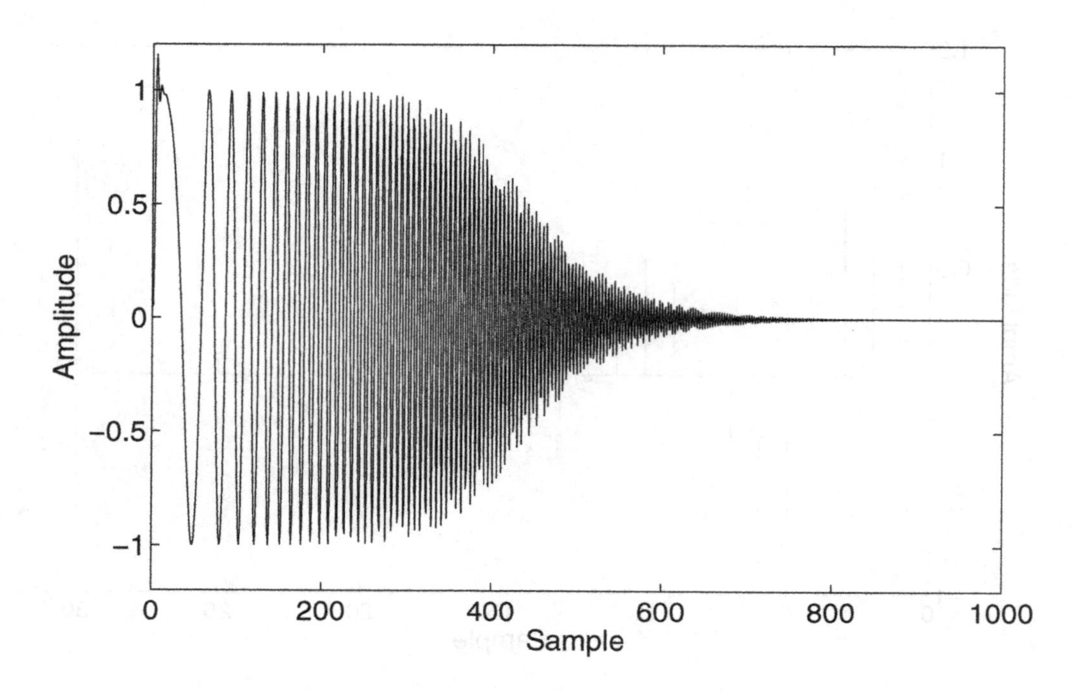

Figure 2.29: The chirp response of a filter having both b and a coefficients, chosen to yield a smooth lowpass effect.

[4] Richard G. Lyons, *Understanding Digital Signal Processing, Second Edition*, Prentice Hall, Upper Saddle River, New Jersey 2004.

2.7 EXERCISES

1. Compute and display the following sequences: 1) a sine wave having frequency = 4.5 Hz and a phase angle of 60 degrees, sampled at a rate of 2400 Hz for 2.5 seconds, beginning at time t = 0.0 second. Display the result two different ways, one way using time for the horizontal axis, and the other using sample index for the horizontal axis; 2) Repeat the above, but assume that sampling begins at t = -1.25 second rather than at t = 0.0 second.

2. Perform addition on the following pair of sequences using paper and pencil and the method of prepending and postpending zeros, then verify your answer using the script *LVAddSeqs*: 1) x1 = [1,-2,6,4], n1 = [-9,-8,-7,-6], x2 = [6,2,-3, -1], n2 = [0,1,2,3].

3. Perform addition on the following pair of sequences using the method of prepending and post-pending zeros. Use m-code, and proceed by creating a figure with three subplots. In the first subplot, plot the first sequence with proper time axes; in the second subplot, plot the second sequence with

the proper number of prepended zeros so that the two sequences are properly time-aligned, then plot the sum in the third subplot. Complete by verifying your answer using *AddSeqs*.

$$\sin(2\pi(5)(-1 : 0.01 : 1))$$

$$\cos(2\pi(2.75)(0 : 0.01 : 1))$$

4. Express the infinite sequence $y = 0.9^n$ (where n = 0:1:∞) as a sum of weighted unit impulse sequences.

5. Express the eight sample rectangular signal, [1,1,1,1,1,1,1,1] as a sum of unit step functions.

6. Express the sequence [0,1,2,-2,-2,-2 ...] as a sum of unit step functions, where the steady value -2 continues to $n = \infty$.

7. Compute the first 100 values of the sequence

$$y = 0.9^n u[n] - 0.9^{n-2} u[n-2]$$

8. Compute and display eleven periods of the sequence

$$y = u[n] - u[n-8] + \delta[n-4]$$

9. For

$$y[n] = 0.8^{n-3} u[n-3]$$

compute $y[-n]$ for n = -20:-1:0.

10. Let the sequence $y[n]$ = [1,2,3,4,5,6] with indices [2:1:7]. Graph the following sequences over the range n =-10:1:10.

(a) $y[n-1]$
(b) $y[n+2]$
(c) $y[1-n]$
(d) $y[-3-n]$

11. Decompose

$$y = \cos(2\pi(0 : 1 : 16)/16)$$

into even and odd components having the same length as y.

12. Decompose

$$y = \cos(2\pi(0 : 1 : 16)/16)$$

into even and odd components that are symmetrical about $n = 0$ (note that for y itself, the first sample is at $n = 0$).

13. Write a script that can generate Fig. 2.9 when called with the following statement, where the first argument is the sequence $y[n]$ and the second argument is the corresponding vector of sample indices.

LVxEvenOddAboutZero([0.9.^([0:1:30])],[0:1:30])

Your script should be general enough to process a call such as

LVxEvenOddAboutZero([0.8.^([4:1:30])],[-4:1:30])

14. Compute the sum of the geometric sequence 0.95^n for $n = 10$ to ∞.

15. Compute and display a signal consisting of a chirp added to noise having normal (Gaussian) distribution. The sequence should be 1024 samples long, the chirp should start at time 0.0 second and frequency 0 Hz and end at time 2.0 second at frequency 10 Hz. The noise should have a standard deviation of 1.0. Perform again for noise with standard deviations of 0.125, 0.25, 0.5, 1.0, and 2.0.

16. Write the m-code for the script

$$LVxLinearab(a, b, f1, f2, N, Del, LTICoeff)$$

as described in the text, producing the plots shown and described in Fig. 2.18, and conforming to the following function definition:

```
function LVxLinearab(a,b,f1,f2,N,Del,LTICoeff)
% Demonstrates the principle of superposition, i.e.,
% LTI(ax1 + bx2) = aLTI(x1) + bLTI(x2) where LTI is a
% linear time invariant operator defined by LTICoeff, the
% coefficients of c[n] which weight an input sequence and delayed
% versions thereof, i.e., y = LTI(x) = c[0]x[n] + c[1]x[n-1] + ..
% a and b are constants, and f1 and f2 are frequencies of cosine waves
% that are used as x1 and x2
% N is the length of the test sequences x1 and x2, and Del is a number of
% samples of delay to impose on x1 and x2 to demonstrate shift invariance.
% Test calls:
% LVxLinearab(2,5,3,5,128,0,[2])
% LVxLinearab(2,-3,13,5,128,0,[2,-1,1,2])
```

17. Write the m-code for the script

$$LVxNLSabXSq(a, b, f1, f2, N, Del, NLCoeff)$$

as described in the text and which produces the plots shown and described in Fig. 2.19.

```
function LVxNLSabXSq(a,b,f1,f2,N,Del,NLCoeff)
% Demonstrates that the principle of superposition is not true for
% a nonlinear system defined by NLSCoeff, the
% coefficients of which weight an input sequence and delayed
% versions thereof raised to the second power, i.e.,
% NLS(x) = c[0]x^2[n] + c[1]x^2[n-1] + c[2]x^2[n-2] + ...
% where c{n} are the members of the vector NLSCoeff.
% a and b are constants, and f1 and f2 are frequencies of cosine
% and sine waves, respectively, that are used as x1 and x2 in
% the superposition test i.e., does NLS(ax1 + bx2) =
% aNLS(x1)+ bNLS(x2)?
% N is the length of the test sequences x1 and x2, and Del is
% a number of samples of delay to impose on x1 and x2
% test for shift invariance.
% Test call:
% LVxNLSabXSq(2,-3,13,5,128,0,[2,-1,1,2])
```

The writing of the script can be modularized by first writing a script that will take the input coefficients $NLCoeff$ and generate the system output for a given input or test sequence $x[n]$:

```
function [yC,nC] = LVxNLSofXSq(NLCoeff,x)
% Delays, weights, and sums the square of the
% input sequence x ( = x[n] with n = 0:1:length(x)-1)
% according to yC = c(0)*x(n).^2 + c(1)*x(n-1).^ 2 + ...
% with NLCoeff (= [c[0],c[1],c[2],...]) and
% nC are the sample indices of yC.
% Test call:
% [yC,nC] = LVxNLSofXSq([1,-2,1],cos(2*pi*12*[0:1:63]/64) )
```

18. The real part of a power sequence generated from a certain complex number z having magnitude 1.0, results in 8 cycles of a cosine wave over a total of 32 samples. What is the value of z?

19. What complex number of magnitude 1.0 will generate two cycles of a complex sinusoid when raised to the power sequence 0:1:11?

20. How many cycles of a complex sinusoid are generated when the complex number

$$\cos(\pi/180) + j\sin(\pi/180)$$

is raised to the power sequence $n = 0{:}1{:}539$?

21. Write a script conforming to the following call syntax

$$LVxComplexPowerSeries(cn, maxPwr)$$

and which creates the plots shown in Fig. 2.15, where cn is a complex number which is raised to the powers 0:1:$maxPwr$.

```
function LVxComplexPowerSeries(cn,maxPwr)
% Raises the complex number cn to the powers
% 0:1:maxPwr and plots the magnitude, the real
% part, the imaginary part, and real v. imaginary parts.
% Test calls:
% LVxComplexPowerSeries(0.69*(1 + j),50)
% LVxComplexPowerSeries(0.99*exp(j*pi/18),40)
```

22. Determine if the following difference equations represent stable LTI systems or not:

$$y[n] = x[n] + y[n-1]$$

$$y[n] = x[n] + 1.05y[n-1]$$

$$y[n] = x[n] + 0.95y[n-1]$$

$$y[n] = x[n] + 1.2y[n-2]$$

$$y[n] = x[n] - 1.2y[n-2]$$

$$y[n] = x[n] - 1.8y[n-1] - 0.8y[n-2]$$

$$y[n] = x[n] - 1.8y[n-1] + 0.8y[n-2]$$

$$y[n] = x[n] + 1.27y[n-1] - 0.81y[n-2]$$

$$y[n] = x[n] - 1.27y[n-1] + 0.81y[n-2]$$

23. Compute and plot the response of the following systems ((a) through (e) below) to each of the following three signals:

$$x[n] = u[n] - u[n-32]$$

and

x(n) = [1, zeros(1,100)]

and

x(n) = chirp([0:1/1000:1],0,1,500)

Be sure to review how to convert between b and a coefficients suitable for a call to the function $filter$ and the coefficients in a difference equation. Note particularly Eqs. (2.8) and (2.9). For situations where the system is an FIR, you can also use the script LV_LTIofX.

a) The system defined by the difference equation

$$y[n] = x[n] + x[n-2]$$

b) The system defined by the difference equation

$$y[n] = x[n] - 0.95y[n-2]$$

c) The system defined by the coefficients a = [1] and

$$b = [0.1667, 0.5, 0.5, 0.1667]$$

d) The system defined by the coefficients b = [1] and

$$a = [1, 0, 0.3333]$$

e) The system defined by the coefficients

$$b = [0.1667, 0.5, 0.5, 0.1667]$$

$$a = [1, 0, 0.3333]$$

24. Use paper and pencil and the graphical method to compute the first five values of the convolution sequence of the following sequence pairs, then check your answers by using the MathScript function $conv$.

(a) **[(-1).^(0:1:7)], [0.5*ones(1,10)]**
(b) **[0.1,0.7,1,0.7,0.1], [(-1).^(0:1:9)]**
(c) **[1,1], [(-0.9*j).^(0:1:7)]**
(d) **[(exp(j*pi)).^(0:1:9)], [ones(1,3)]**

25. Verify that the commutative property of convolution holds true for the argument pairs given below, using the script LV_LTIofX, and then, for each argument pair, repeat the exercise using the MathScript function $conv$. Plot results for comparison.

(a) chirp([0:1/99:1],0,1,50) and [1,1];
(b) chirp([0:1/99:1],0,1,50) and [1,0,1];
(c) [1,0,-1] and chirp([0:1/999:1],0,1,500)

(d) [1,0,-1] and chirp([0:1/999:1],-500,1,500)

26. Does a linear system of the form

$$y = kx + c$$

where k and c are constants, and x is an independent variable obey the law of superposition, i.e., is it true that

$$y(ax_1 + bx_2) = ay(x_1) + by(x_2)$$

where a and b are constants? Prove your answer.

27. Sound travelling from a point of origin in a room to a listening point typically takes many transmission paths, including a direct path, and a number of reflected paths. We wish to determine how the sound quality at the listening point is affected by the multiple transmission paths, i.e., assuming that the source emits a linear chirp of unity amplitude covering all the frequencies of interest to us, what will the frequency response at the listening point be? Assume that the transmission paths from the source point to the listening point can be modeled as an LTI system, and that we can sample a microphone output at the listening point and plot the received signal over time. Since the chirp frequency increases linearly over time, a plot of received amplitude versus time is equivalent to one of received amplitude versus frequency.

We can simulate this experiment by knowing the various path lengths and the speed of sound, which allows us to compute the sound transit times for the direct and reflected paths. Assume that the hypothetical microphone output used at the listening point is sampled at 10 kHz, and the speed of sound in air is 1080 ft/sec., and that we can model the transfer function with the direct and two reflected paths. To perform the simulation, follow these steps:

1) Compute all path lengths in terms of time, and since frequency response is not dependent on bulk delay, only on the time (or phase) differences between the interfering signals, the transit time of the direct (i.e., shortest) path can be subtracted from all transit times.

2) Specify each of the reflected (i.e., delayed) paths as a number of equivalent samples of delay, based on the sample rate at the microphone. Then we can specify the system as a set of delays and appropriate amplitudes, and use the function LV_LTIofX to compute the output.

3) A typical call to LV_LTIofX should be of the form

$$[yC, nC] = LV\_LTIofX(LTICoeff, x)$$

where x is a linear chirp over a duration of one second, having 10,000 samples and chirping from 0 to 5 kHz, and the input argument $LTICoeff$ is of the form

[1,zeros(1,n1-1),RG1,zeros(1,(n2-n1)-1),RG2]

where $n1$ is the number of samples of delay between the time of arrival (TOA) of the direct wave and the TOA of the first reflected wave, $(n2 - n1)$ is the additional delay from the TOA of the first reflected wave to the TOA of the second reflected wave, $RG1$ is the gain of the first reflected path relative to that of the direct path, which is defined as 1.0, and $RG2$ is the gain of the second reflected path relative to that of the direct path.

The following test parameters consist of three path lengths in feet and the relative amplitudes of each wave upon arrival at the listening point. For each set of test parameters, determine and make the appropriate call to LV_LTIofX, and plot the output yC versus frequency.

(a) Path Lens = [2.16, 2.268, 2.376]; RelAmps = [1,0.98,0.96];
(b) Path Lens = [8.64, 8.748, 8.856]; RelAmps = [1,0.98,0.96];
(c) Path Lens = [10.8, 11.664, 14.148]; RelAmps = [1,0.96,0.88];
(d) Path Lens = [75.6, 86.076, 98.388]; RelAmps = [1,0.93,0.81];
(e) Path Lens = [54, 54.108, 54.216]; RelAmps = [1,1,1];
(f) Path Lens = [54, 54.108, 54.216]; RelAmps = [1,0,1]; (1st refl path damped)

Now repeat the exercise, but instead of using a simple linear chirp, use the following complex chirp:

t = [0:1/9999:1]; fMx = 5000;
x = chirp(t,0,1,fMx) + j*chirp(t,0,1,fMx,'linear',90);

Also, instead of using the script LV_LTIofX, use the function $filter$, with input argument $a = 1$ and b equal to the input argument $LTICoeff$ as would have been used with LV_LTIofX. Since the result is complex, plot the absolute value of the result of the filtering operation.

Figure 2.30 shows the result from this alternate procedure, using the parameters specified at (e) above. Figure 2.31 (resulting from the call at (f) above) shows what happens when the first reflected path is severely attenuated, as perhaps by an acoustic absorber inserted into the path to alter the frequency response at the listening point.

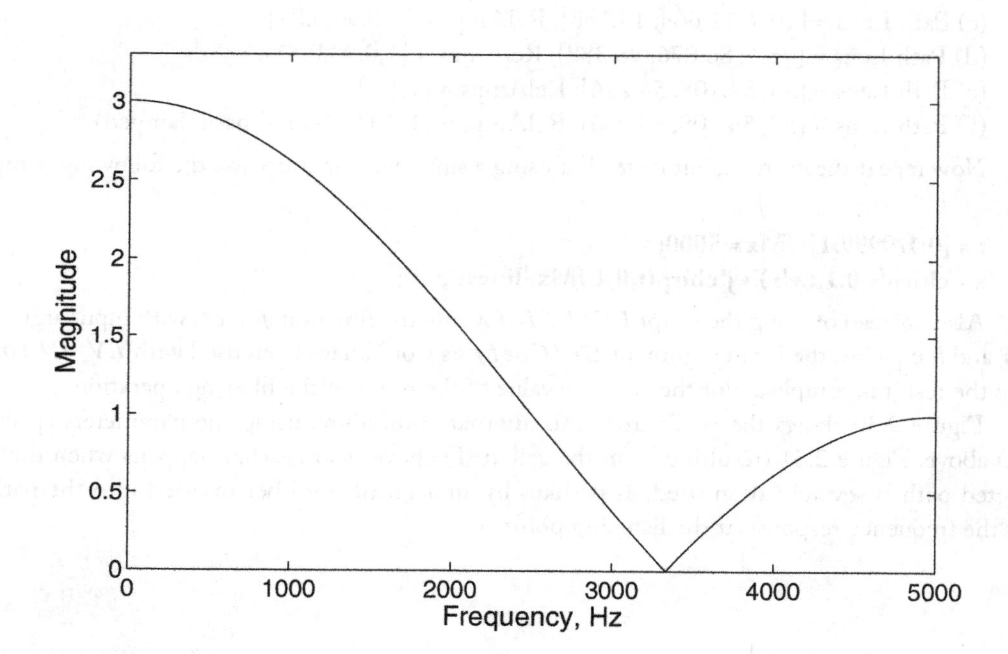

Figure 2.30: An estimate of the frequency response between source and listening points in a room, modeled with a direct and two reflected paths. The first reflected path is only one sample longer in duration than the direct path. The second reflected path is only one sample of delay longer than the first reflected path. Both reflected waves arrive at the same amplitude as that of the direct wave.

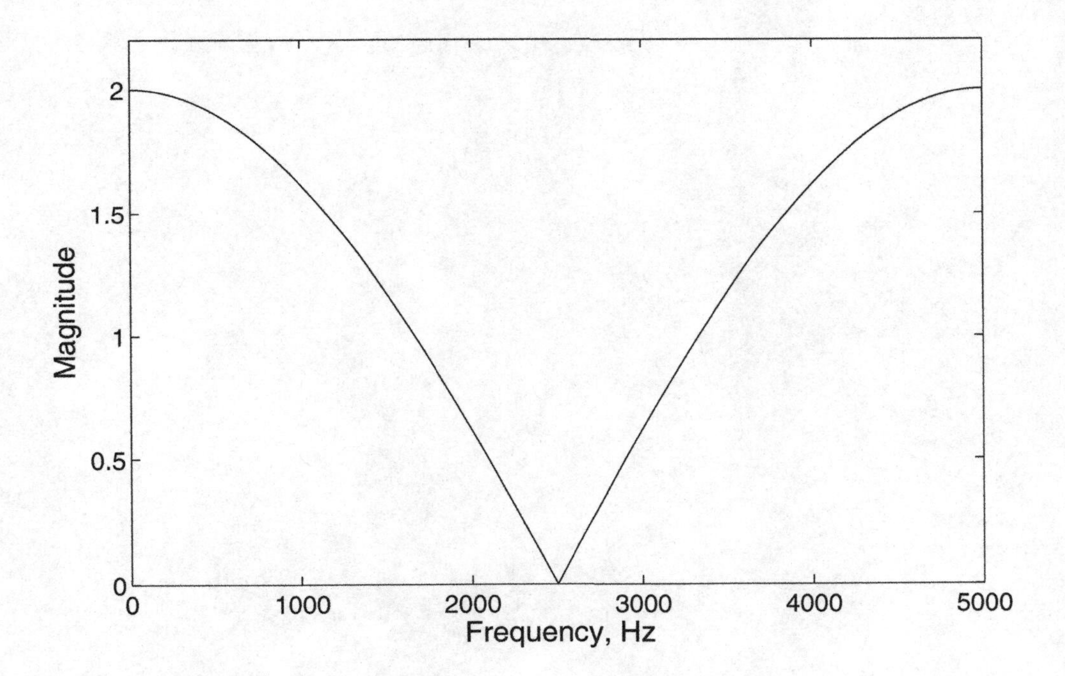

Figure 2.31: An estimate of the frequency response between source and listening points in a room, modeled with a direct and two reflected paths. The first reflected path is one sample longer in duration than the direct path, but has been hypothetically severely attenuated (to amplitude zero) with an acoustic absorber to observe the effect on frequency response. The second reflected path is only one sample of delay longer than the first reflected path. The second reflected wave arrives at the same amplitude as that of the direct wave.

CHAPTER 3
Sampling and Binary Representation

3.1 OVERVIEW

Having equipped ourselves with the knowledge of a number of common signal types and the basic concepts of LTI systems in the previous chapter, we continue our study of DSP with the conversion of an analog signal from the continuous domain to the discrete or sampled domain, and back again. Real-world analog signals (such as audio and video signals) first enter the digital realm via the process of sampling. We begin our discussion with two very important requirements of sampling, namely, the minimum acceptable or **Nyquist** sample rate and the need for bandlimiting a signal prior to sampling. With these all-important principles established, we discuss normalized frequency, which is the basis for evaluating and describing frequency content and response in the digital domain. Nyquist rate and normalized frequency are among the most fundamental of concepts associated with digital signal processing, and little further discussion of the topic can be meaningfully had until the reader understands them well.

In this chapter, in addition to the Nyquist rate and normalized frequency, we also discuss the basics of binary counting and formats and analog-to-digital and digital-to-analog conversion. These are important topics since they refer to implementation, which is always with limited precision representation of numbers. The theory of digital signal processing, such as digital filtering, the DFT, etc., is usually taught as though all numbers are of perfect accuracy or infinite precision. When digital signal processing algorithms are implemented on a computer, for example, all numbers are stored and all computations are made with only finite precision, which can adversely affect results. Much of the DSP literature concerns these issues, and thus familiarity with the principles and nomenclature of conversion and quantization is essential.

3.2 SOFTWARE FOR USE WITH THIS BOOK

The software files needed for use with this book (consisting of m-code (.m) files, VI files (.vi), and related support files) are available for download from the following website:

http://www.morganclaypool.com/page/isen

The entire software package should be stored in a single folder on the user's computer, and the full file name of the folder must be placed on the MATLAB or LabVIEW search path in accordance with the instructions provided by the respective software vendor (in case you have encountered this notice before, which is repeated for convenience in each chapter of the book, the software download only needs to be done once, as files for the entire book are all contained in the one downloadable folder).

See Appendix A for more information.

3.3 ALIASING

In 1928, Nyquist, working at the Bell Telephone Laboratories, discovered that in order to adequately reconstruct a sinusoid, it was only necessary to obtain two samples of each cycle of the sinusoid. So if we have a continuous-valued voltage representing a single frequency sinusoid, we need to obtain amplitude samples of the signal twice per cycle. If we sample regularly at equal intervals, we can describe the sampling operation as operating at a certain frequency, and obviously this frequency, F_S, will have to be at least twice the frequency of the sinusoid we are sampling.

- If a sinusoid is sampled fewer than two times per cycle, a phenomenon called **Aliasing** will occur, and the sampled signal cannot be properly reconstructed. When aliasing occurs, a signal's original, pre-sampling frequency generally appears in the sampler output as a different apparent, or aliased, frequency.
- In signals containing many frequencies, the sampling rate must be at least twice the highest frequency in the signal–this ensures that each frequency in the signal will be sampled at least twice per cycle.

The preceding statement leads to the question, "how do you know what the highest frequency is in the signal you are quantizing?" The general answer is that the only way to know is to completely control the situation by filtering the analog signal before you sample it; you would use an analog (continuous domain) lowpass filter with a cutoff frequency at half the sampling frequency. This ensures that in fact the sampling rate is more than twice the highest frequency in the signal.

Such an analog filter is called an **Anti-Aliasing Filter**, or sometimes, simply an **Aliasing Filter**. The usual arrangement is shown in Fig. 3.1; the time domain signal, which might have unlimited or unknown bandwidth, passes through an anti-aliasing filter in which all frequencies above one-half the sampling rate are removed. From there, the actual sampling operation is performed: a switch is momentarily closed and the instantaneous amplitude of the signal is stored or held on a capacitor while the unity-gain-buffered output of the Sample-and-Hold is fed to an Analog-to-Digital Converter (ADC), an example of which we'll cover in detail after completing our discussion of aliasing.

Figure 3.2 shows what happens when an 8 Hz sine wave is sampled at a rate of 9 Hz rather than the minimum acceptable value of 16 Hz–the output sequence looks just like the output sequence that would have been generated from sampling a 1 Hz-inverted-phase-sine wave at a rate of 9 Hz.

- An aliased frequency in a digital sequence has forever lost its original identity. There are, assuming an unlimited bandwidth frequency being input to a sampler without an anti-aliasing

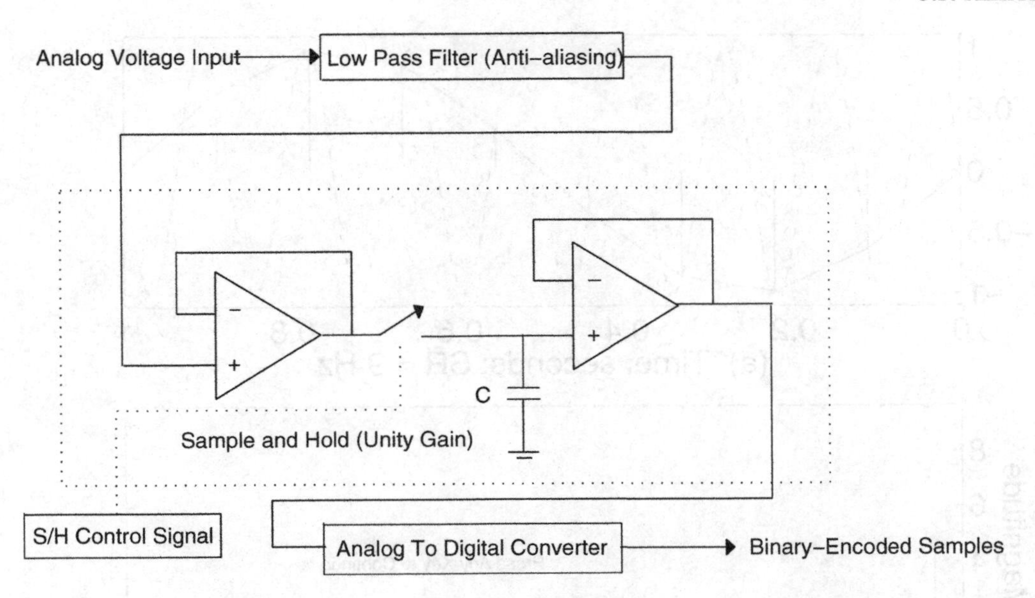

Figure 3.1: A typical sampling arrangement, showing an anti-aliasing filter followed by a sample-and-hold, which captures and holds an analog value to be digitized by the ADC.

filter, literally an infinite number of frequencies which could be aliased into even a single frequency lying below the Nyquist rate (half the sampling rate). In this situation, once a frequency becomes aliased, there is no way to reverse the aliasing process and determine which of the infinite number of possible source frequencies was in fact the original frequency.

- Samples are like snapshots or individual frames in a motion picture of action which is continuous in nature. If we take 30 snapshots or frames per second of human actors, we are certain that we have a good idea of everything that takes place in the scene simply because human beings cannot move fast enough to do anything of significance in the time interval between any two adjacent snapshots or frames (note that the human eye itself takes snapshots or samples of visually received information and sends them one after another to the brain). Imagine then if we were to lower the frame rate from one frame (or snapshot) every 30th of a second to, say, one frame every five seconds. At this low rate, it is clear that the actors could engage in a huge range of activities and complete them between frames. From looking at the sequence of frames, each one five seconds apart, we could not say what had taken place between frames.

Likewise, when sampling a waveform, too low a sampling rate for the frequencies present in the waveform causes loss of vital information as to what went on between samples.

Example 3.1. Illustrate the effect of sampling an 8 Hz sinusoid at sample rates varying from 240 Hz downward to 8 Hz.

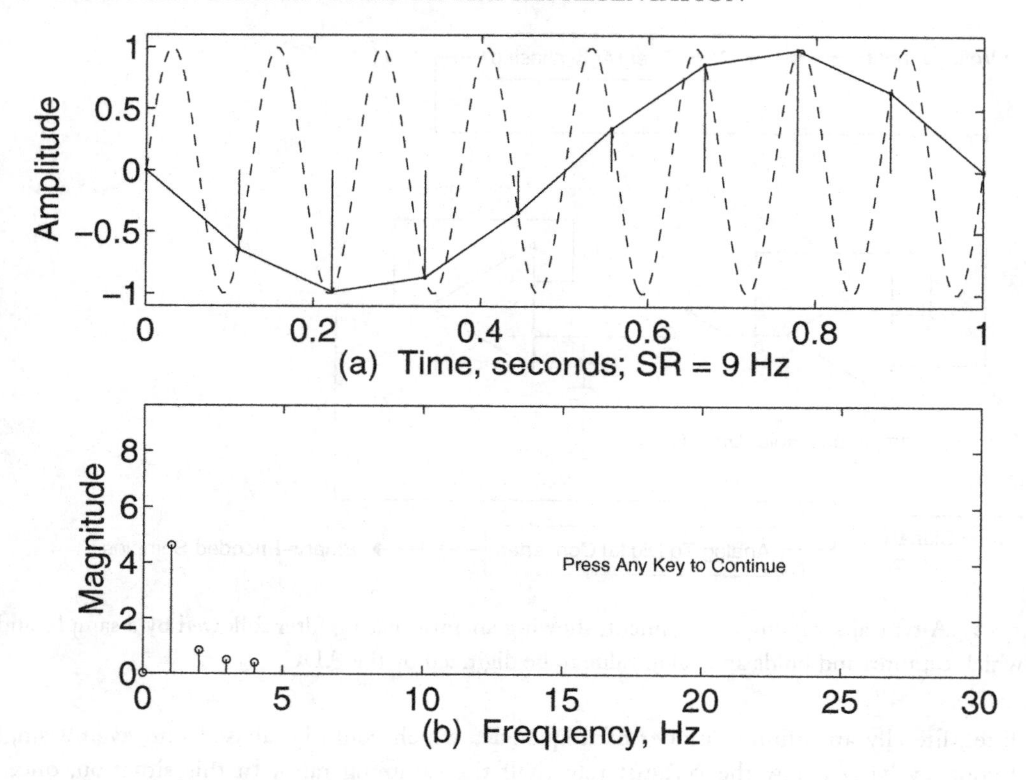

Figure 3.2: (a) Eight cycles of a sine wave, sampled at only 9 Hz; in order to avoid aliasing, at least 16 samples should have been taken. Due to the phenomenon of aliasing, the nine samples give the impression that the original signal was actually a one cycle sine wave. (b) A frequency content plot, obtained using the Discrete Fourier Transform, confirms what the eye sees in (a).

The script

LVAliasingMovieSine

performs a demonstration of the effect of various sampling rates applied to eight cycles of a sinusoid, partly automated and partly requiring user manual input. A similar demonstration is given by the VI

$$DemoAliasingMovieSineVI$$

which requires on-screen manual variation of the sample rate F_S of an 8 Hz sine wave between F_S = 120 Hz and F_S = 8 Hz.

Figure 3.3, plot (a), shows eight cycles of a sine wave in which the sample rate is 240, or about 30 samples per cycle.

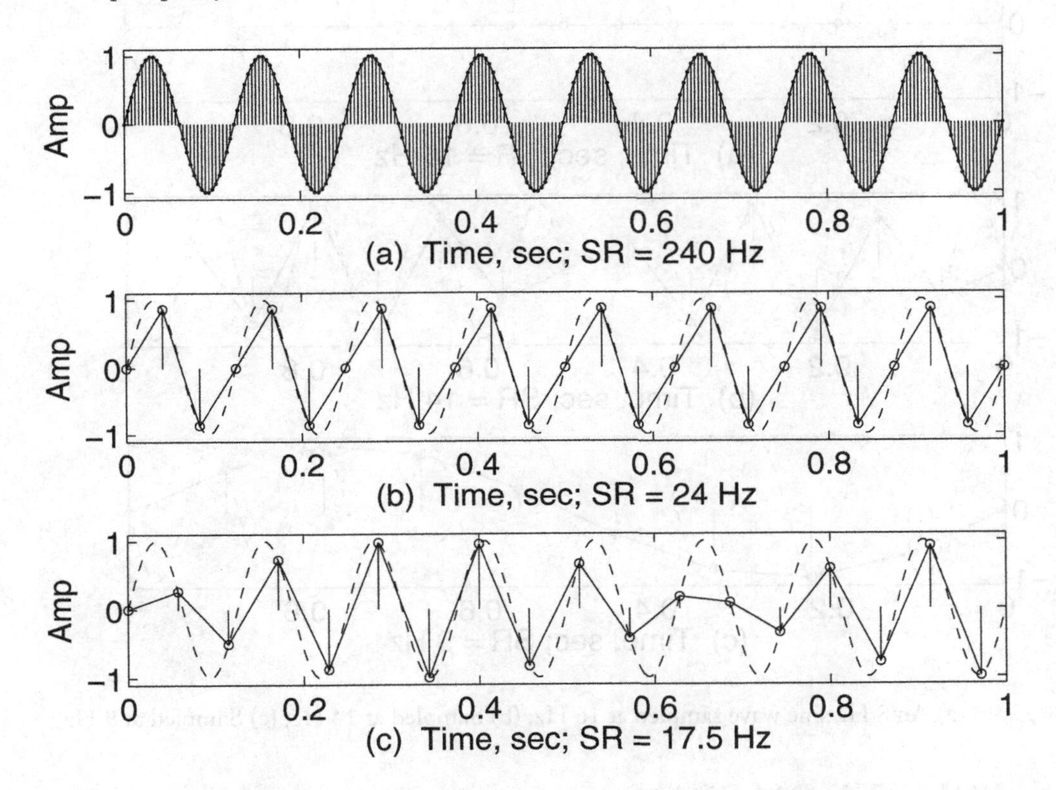

Figure 3.3: (a) Eight cycles of a sine wave taken over a 1 second period, with a stem plot of 240 samples thereof superimposed; (b) The same eight-cycle sine, sampled at a rate of 24 Hz; (c) Sampled at 17.5 Hz.

If instead of a 240 Hz sampling rate, we use 24 Hz instead, we can still see what appears to be eight cycles of a waveform, since there are three samples per cycle (at 0, 120, and 240 degrees).

At a sample rate of 17.5 Hz, it is still possible to see eight cycles. When the sample rate falls to exactly 16 Hz (Fig. 3.4, plot (a), exactly two samples per cycle), we have the unfortunate situation that the sampling operation was synchronized with the wave at phase zero, yielding only samples of zero amplitude, and resulting in the spectrum having no apparent content.

When the sample rate falls to 14 Hz, a careful counting of the number of apparent cycles yields an apparent 6 Hz wave, not 8 Hz, as shown in Fig. 3.4, plot (b). At a sampling rate of 9 Hz, we see the samples outlining a perfect inverted-phase 1 Hz wave!

You may have noticed some apparent relationship between input frequency, sampling rate, and apparent output frequency. For example, with an input frequency of 8 Hz and a sampling rate of 9 Hz, we saw an apparent output frequency of 1 Hz, with the sine wave's phase inverted.

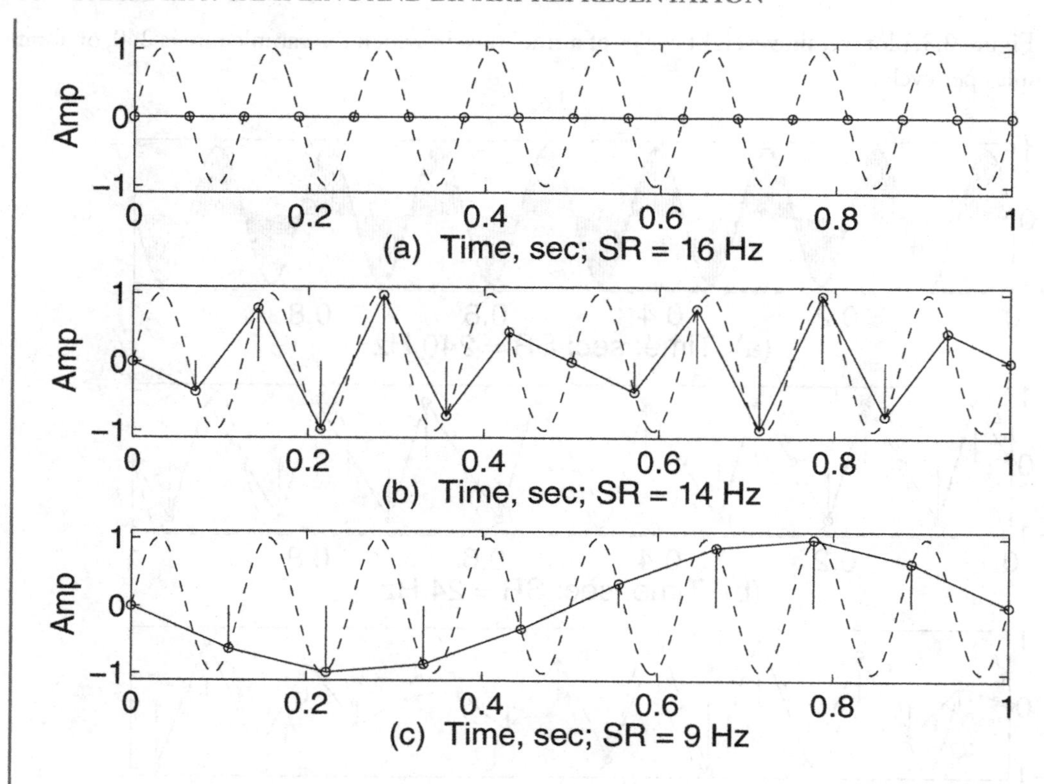

Figure 3.4: (a) An 8 Hz sine wave sampled at 16 Hz; (b) Sampled at 14 Hz; (c) Sampled at 9 Hz.

3.4 FOLDING DIAGRAM

- For any given sampling rate, a **Folding Diagram** that shows the periodic nature of the sampling function can be constructed. Such a diagram allows easy determination of the apparent output frequency of the sampler for a given input frequency. The sampling function, as illustrated by a folding diagram, "folds" input frequencies around odd multiples of half the sampling rate.

Example 3.2. Construct a Folding Diagram for a 9 Hz sampling rate and determine the output frequency and phase if an 8 Hz signal is input to the sampler.

Figure 3.5, plot (a), shows a standard Folding Diagram using a sampling rate of 9 Hz. Use is self-explanatory, except for the phase of the output signal compared to the input signal. When the slope of the folding diagram is positive, the output frequency is in-phase with the input, and when the slope of the folding diagram is negative, the output frequency's phase is inverted when compared to the input signal's phase.

Thus we see in Fig. 3.5, plot (a), a sampling rate of 9 Hz, an input frequency of 8 Hz, and an apparent output frequency of 1 Hz, which is phase reversed since the input frequency lies between one-half the sampling rate and the sampling rate, i.e., the folding diagram has a negative slope for an input frequency of 8 Hz.

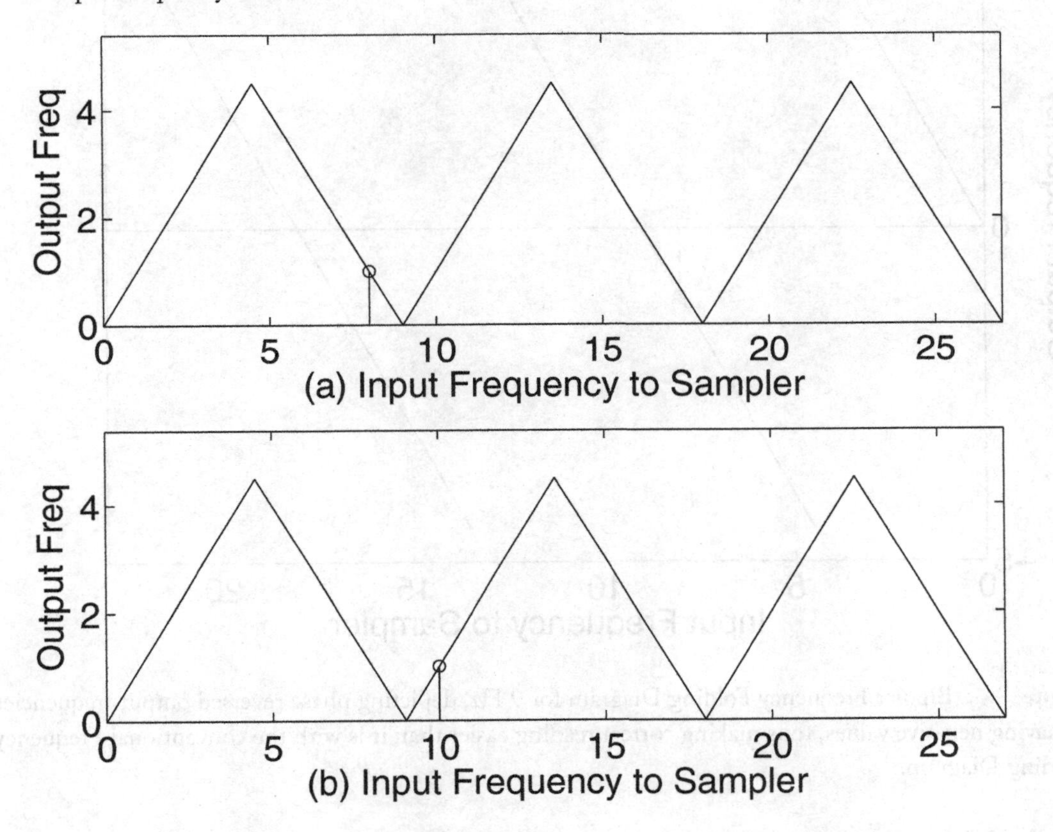

Figure 3.5: (a) The Frequency Folding Diagram for a sampling rate of 9 Hz, with an 8 Hz input signal to the sampler; (b) Folding Diagram for a 9 Hz sampling rate, with a 10 Hz input signal.

You can see in Fig. 3.5, plot (b), an input frequency *greater* than the sampling rate by, say, 1 Hz, still shows up in the output with an apparent frequency of 1 Hz, according the Folding Diagram for a 9 Hz sampling rate. In this case, however, its phase will be the same as that of the input signal at 10 Hz.

An easier way to see the phase reversal is to use the Folding Diagram shown in Fig. 3.6, in which the "downside" areas of input frequency that result in phase reversal are graphed such that the output frequency has a negative sign, indicating phase reversal.

In this diagram, if the input frequency is a multiple of half the sampling frequency, the output frequency sign is indeterminate. The actual output frequency will be equal to the input frequency,

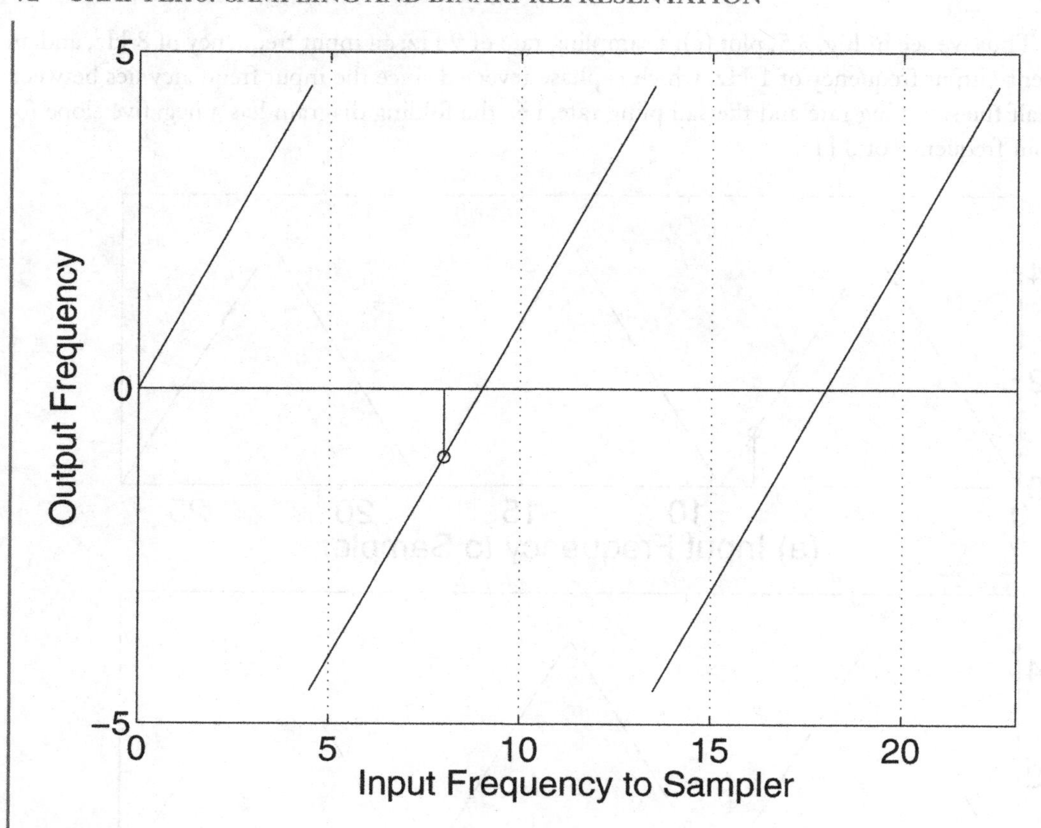

Figure 3.6: Bipolar Frequency Folding Diagram for 9 Hz, depicting phase reversed output frequencies as having negative values, thus making correct reading easier than it is with the conventional Frequency Folding Diagram.

with a variable amplitude, except for the case in which the input frequency bears that special phase relationship (as seen in Fig. 3.4), plot (a), in which case all samples have a uniform value of zero, conveying no frequency information.

To see the effect of aliasing, you can run the script

$$LVAliasing(SR, Freq)$$

and try various frequencies *Freq* with a given sampling rate *SR* to verify the correctness of the Folding Diagram.

The script evaluates the expression

```
sin(2*pi*(0:1/SR:1-1/SR)*Freq)
```

for user-selected values of SR (Sample Rate) and $Freq$ (sinusoid frequency).

You can start out with the call

LVAliasing(100,2)

which results in Fig. 3.7, plot (a). A sample rate of 100 Hz with a frequency of 102 Hz is shown in plot (b). Note that the two plots are identical. The result would also have been identical if the frequency had been 202 Hz, or 502 Hz, etc. Rerun *LVAliasing* with the sampling rate as 100 Hz, and the frequency as 98 Hz. The result is shown in Fig. 3.7, plot (c)—an apparent 2 Hz sine wave, but phase inverted, as would have been predicted by a folding diagram based on a 100 Hz sample rate.

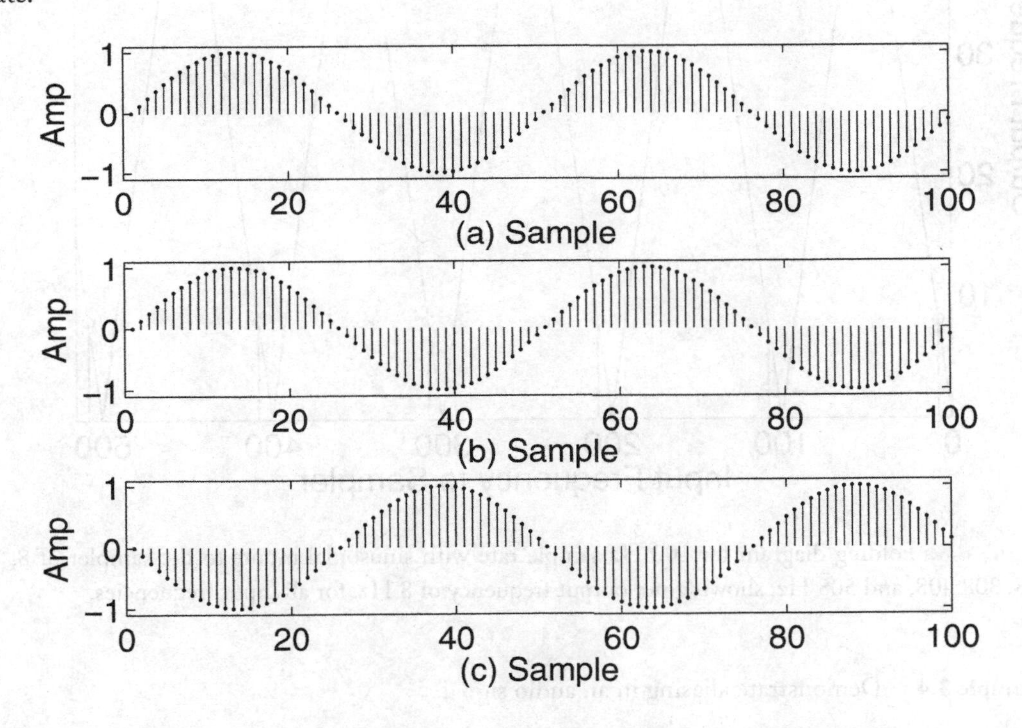

Figure 3.7: (a) A sine wave with a frequency of 2 Hz sampled at 100 Hz; (b) A sine wave with a frequency of 102 Hz sampled at 100 Hz; (c) A sine wave with a frequency of 98 Hz sampled at 100 Hz.

Example 3.3. Construct a folding diagram for a sampling rate of 100 Hz, and plot input frequencies at 8 Hz, 108 Hz, 208 Hz, 308 Hz, 408 Hz, and 508 Hz.

Figure 3.8 shows the result; note that all input frequencies map to the same apparent output frequency, 8 Hz (marked with a horizontal dotted line). Note further that only input frequencies lying between 0 and 50 Hz map to themselves as output frequencies. All input frequencies higher than 50 Hz appear in the output as aliased frequencies.

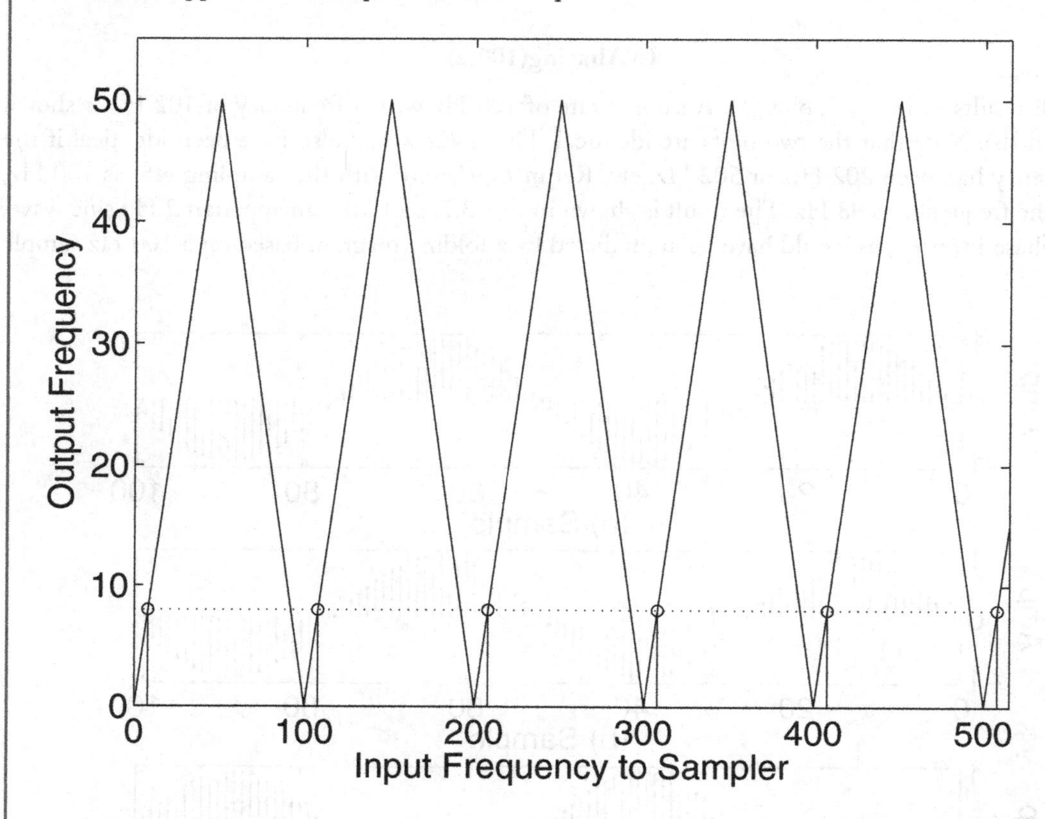

Figure 3.8: Folding diagram for 100 Hz sample rate with sinusoidal inputs to the sampler of 8, 108, 208, 308 408, and 508 Hz, showing net output frequency of 8 Hz. for all input frequencies.

Example 3.4. Demonstrate aliasing in an audio signal.

The script

ML_AliasingChirpAudio(SR,StartFreq,EndFreq)

has been provided to both illustrate and make audible aliasing in an audio signal. A variable sample rate SR, and the lowest frequency $StartFreq$ and highest frequency $EndFreq$ of a linear chirp are specified as the input arguments.

Figure 3.9, plot (a) shows the result of the call

ML_AliasingChirpAudio(3000,0,1500)

which specifies a lower chirp limit *StartFreq* of 0 Hz and an upper chirp limit *EndFreq* of 1500 Hz, which is right at the Nyquist rate of 3000/2 = 1500 Hz. You can see a smooth frequency increase in the spectrogram in plot (b). Frequency 1.0 represents half the sampling frequency (3 kHz), or 1500 Hz in this case. If you have a sound card on your computer, the chirp should sound automatically when the call above is made.

- To use the script *ML_AliasingChirpAudio* with LabVIEW, restrict the value of SR to one of the following values: 8000, 11025, 22050, 44100.

In Fig. 3.9, plot (c), the lower chirp limit is 0 Hz, and the upper chirp limit is 3000 Hz, with the same 3000 Hz sampling rate. You can see a smooth frequency increase in the spectrogram (plot (d)) up to the midpoint in time, and then the apparent frequency smoothly decreases to the starting frequency. Listening to the chirp frequency go up and then suddenly reverse should fix in your mind what happens in aliasing.

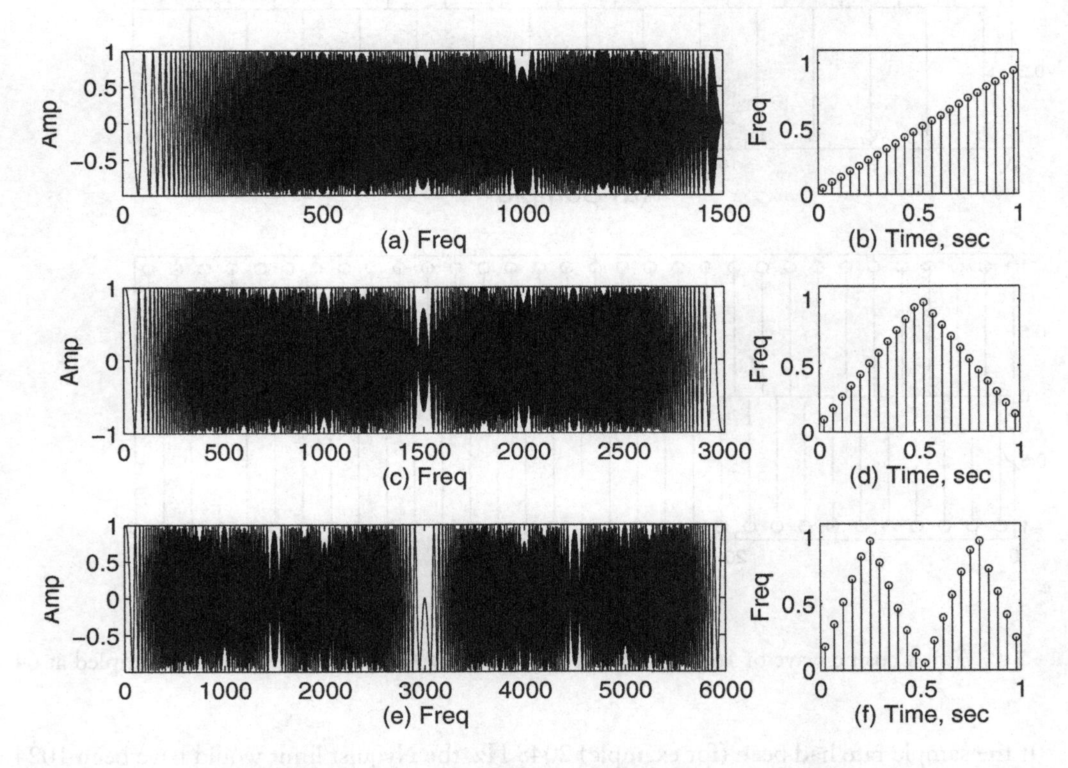

Figure 3.9: (a) Chirp from 0 Hz to 1500 Hz, sampled at 3000 Hz; (b) Spectrogram (frequency versus time) of (a); (c) Chirp from 0 Hz to 3000 Hz, sampled at 3000 Hz; (d) Spectrogram of (c); (e) Chirp from 0 Hz to 6000 Hz, sampled at 3000 Hz; (f) Spectrogram of (e).

Running the experiment one more time, let's use 6000 Hz as the upper chirp frequency, with the same sampling frequency of 3000 Hz. The result is shown in plot (e), with the spectrogram in plot (f).

3.5 NORMALIZED FREQUENCY

Consider a 16-Hz cosine wave sampled at 32 Hz for one second, with proper anti-aliasing. The resulting sequence would look like Fig. 3.10, plot (a). On the other hand, a 32-Hz cosine sampled at 64 Hz results in the same apparent signal, an alternation between +1 and -1, as shown in Fig. 3.10, plot (b).

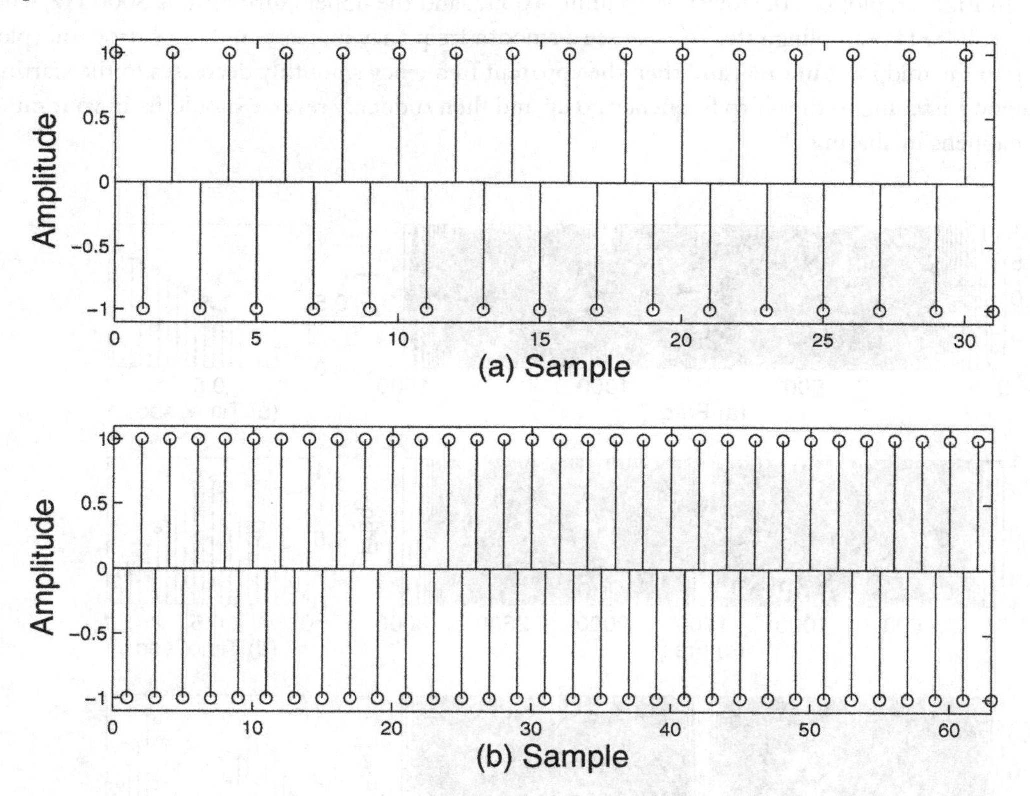

Figure 3.10: (a) A cosine wave of 16 Hz sampled at 32 Hz; (b) A cosine wave of 32 Hz sampled at 64 Hz.

If the sample rate had been (for example) 2048 Hz, the Nyquist limit would have been 1024 Hz, and a 1024 Hz cosine in the original analog signal would show up as an alternation with a two-sample period. If instead, the cosine's frequency had been one-half the Nyquist limit (in this case 512 Hz), there would have been four samples per cycle, and so on.

- **Two sinusoids, sampled at different rates, that bear the same frequency relative to their respective Nyquist rates, produce sample sequences having the same frequency content.**
- **Two sinusoids, sampled at different rates, that bear the same frequency and phase relative to their respective Nyquist rates, produce essentially the same sample sequences.**

Example 3.5. Verify the above statements using MathScript.

The m-code

```
N=32; FrN2=0.5; t=0:1/N:1-1/N;
figure; stem(cos(2*pi*t*FrN2*(N/2)))
```

allows you to verify this by holding *FrN2* (a fraction of the Nyquist rate *N/2*, such as 0.5, etc.) constant, while changing the sample rate *N*. In the above call, the phase angle (0) remains the same as *N* changes, and the resulting sequences will be identical except for total number of samples. Note that *FrN2* = 1 generates the Nyquist rate.

That the two sinusoids must have the same phase to produce apparently identical sequences may be observed by modifying the above call by inserting a phase angle in the cosine argument, and changing it from one sampling operation to the next. For example, run this m-code

```
pa = 1.5; N = 32; fN2 = 0.25; t = 0:1/N:1-1/N;
figure; stem(cos(2*pi*t*fN2*N/2+pa))
```

and note the resulting sample sequence. Then follow with the call below and note the result.

```
pa = 1; N = 64; fN2 = 0.25; t = 0:1/N:1-1/N;
figure; stem(cos(pi*t*fN2*N+pa))
```

Had the value of *pa* (phase angle) been the same in both experiments above, so would have been the resulting sequences. The sequences produced have the same apparent frequency content but different sample values (and hence different appearances when graphed) due to the phase difference.

We define **Normalized Frequency** as the original signal frequency divided by the Nyquist rate.

Definition 3.6.

$$F_{norm} = F_{orig} / F_{Nyquist}$$

Once an analog signal has been sampled, the resulting number sequence is divorced from real time, and the only way to reconstruct the original signal properly is to send the samples to a DAC at the original sampling rate. Thus it is critical to know what the original sampling rate was for purposes of reconstruction as a real time signal.

When dealing with just the sequence itself, it is natural to speak of the frequency components relative to the Nyquist limit.

- It is important to understand the concept of normalized frequency since the behavior of sequences in digital filters is based not on original signal frequency (for that is not ascertainable from any information contained in the sequence itself), but on normalized frequency.

Not only is it standard in digital signal processing to express frequencies as a fraction of the Nyquist limit, it is also typical to express normalized frequencies in radians. For example, letting k = 0 in the expression

$$W^n = [\exp(j2\pi k/N)]^n$$

yields the complex exponential with zero frequency, which is composed of a cosine of constant amplitude 1.0 and a sine of constant amplitude 0.0. If N is even, letting $k = N/2$ yields the net radian argument of π (180 degrees) and the complex exponential series

$$W^n = [\exp(j\pi)]^n = (-1)^n$$

which is a cosine wave at the Nyquist limit frequency.

Thus we see that radian argument 0 generates a complex exponential with frequency 0, the radian argument π generates the Nyquist limit frequency, and radian arguments between 0 and π generate proportional frequencies therebetween. For example, the radian argument $\pi/2$ yields the complex exponential having a frequency one-half that of the Nyquist limit, that is to say, one cycle every four samples. The radian argument $\pi/4$ yields a frequency one-quarter of the Nyquist limit or one cycle every eight samples, and so forth.

Hence the normalized frequency 1.0 (the Nyquist limit) may be taken as a short form of the radian argument 1.0 times π, the normalized frequency of 0.5 represents the radian argument $\pi/2$, and so forth.

- Radian arguments between 0 and π radians correspond to normalized frequencies between 0 and 1, i.e., between DC and the Nyquist limit frequency (half the sampling rate).

Example 3.7. A sequence of length eight samples has within it one cycle of a sine wave. What is the normalized frequency of the sine wave for the given sequence length?

A single-cycle length of two samples represents the Nyquist limit frequency for any sequence. For a length-eight sequence, a single cycle sinusoid would therefore be at one-quarter of the Nyquist frequency. Since Nyquist is represented by the radian argument π, the correct radian frequency is $\pi/4$ or 0.25π, and the normalized frequency is 0.25.

Example 3.8. A sequence of length nine samples has within it 1.77 cycles of a cosine wave. What is the normalized frequency of the cosine wave?

The Nyquist frequency is one-half the sequence length or 4.5; the normalized frequency is therefore 1.77/4.5 = 0.3933. We can construct and display the actual wave with the call

figure; stem(cos(2*pi*1.77*(0:1:8)/9))

Example 3.9. Generate, in a sequence length of 9, a sine wave of normalized frequency 0.5 (or radian frequency of 0.5π).

A plot of the signal can be obtained by making the call

figure; stem(sin(2*pi*(0:1:8)/9*(0.5*4.5)))

The frequency argument for the call is constructed as the product of the normalized frequency 0.5 and the Nyquist limit frequency, which is 9/2 = 4.5 for this example.

Example 3.10. Generate the complex exponential having a normalized frequency of 0.333π in a length-eleven sequence, and plot the imaginary part.

The following is a suitable call:

figure; stem(imag(exp(j*2*pi*(0:1:10)/11*(0.333*5.5))))

- The VI

DemoComplexPowerSeriesVI

allows you to use the mouse to move the cursor in the complex plane and to view the corresponding complex power series. As you move the cursor, the complex number corresponding to the cursor position is used as the kernel of a complex power series, the real and imaginary parts of which are displayed in real time. Figure 3.11 shows an example of the VI in use.

Example 3.11. Use the VI *DemoComplexPowerSeriesVI* to generate and display the real and imaginary parts of a complex power sequence generated from a complex number having a magnitude of 0.9 and a normalized frequency of 0.25.

A normalized frequency of 0.25 implies a frequency of one-quarter of the Nyquist rate. The angle is thus $\pi/4$ radians, yielding the complex number 0.707 + j0.707. After scaling by the magnitude of 0.9, we get the complex number 0.636 + j0.636, or

0.9*exp(j*pi*0.25)

To see the power sequence generated from this number, use the VI

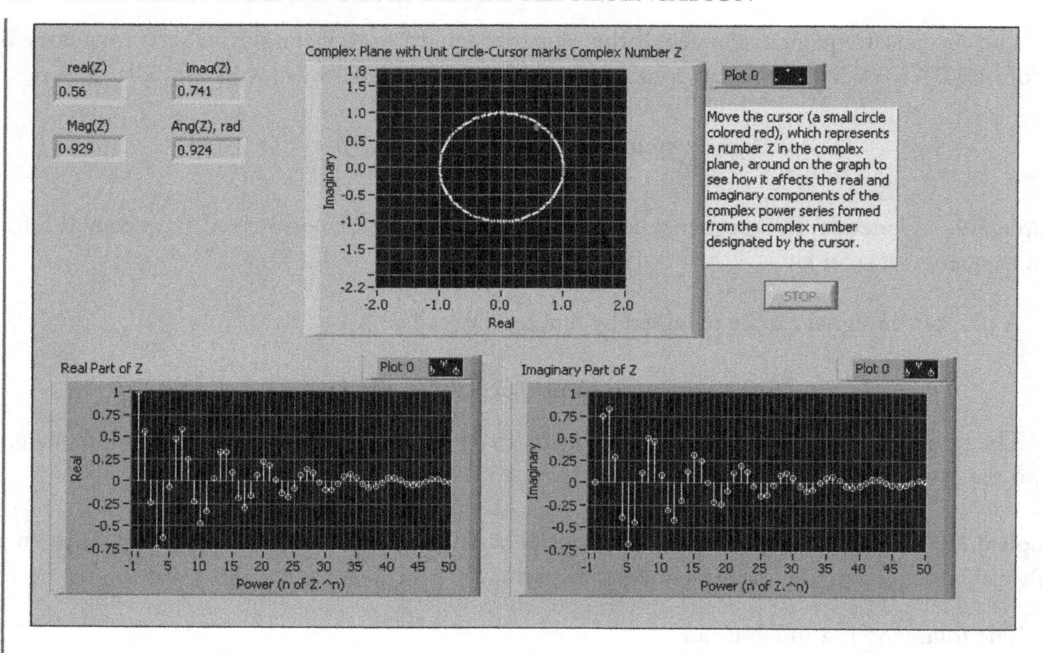

Figure 3.11: A VI that demonstrates the relation between a complex number and the power sequence arising from it. The complex number designated by the cursor position in the complex plane is used as the kernel of a power sequence (the first 50 powers are computed), and the real and imaginary parts are displayed, updated in real time as the user drags the cursor in the complex plane.

$$DemoComplexPowerSeriesVI$$

and adjust the cursor to obtain a position yielding magnitude 0.9 and normalized frequency 0.25, or alternatively, make the MathScript call

LVxComplexPowerSeries(0.636*(1+j), 50)

which displays the same information for the complex number 0.636*(1+j) , computed and displayed for the first 50 powers (creation of this script was part of the exercises of the previous chapter).

Figure 3.12 shows the display when the cursor is set at complex coordinates (0.636 + j*0.636). The resultant power sequence is a decaying complex sinusoid exhibiting one cycle over eight samples, i.e., one-quarter the frequency of the Nyquist limit.

Suitable m-code calls to plot the real and imaginary parts of a complex power sequence are, for the example at hand,

figure; stem(real((0.9*exp(j*pi*0.25)).^(0:1:29)))

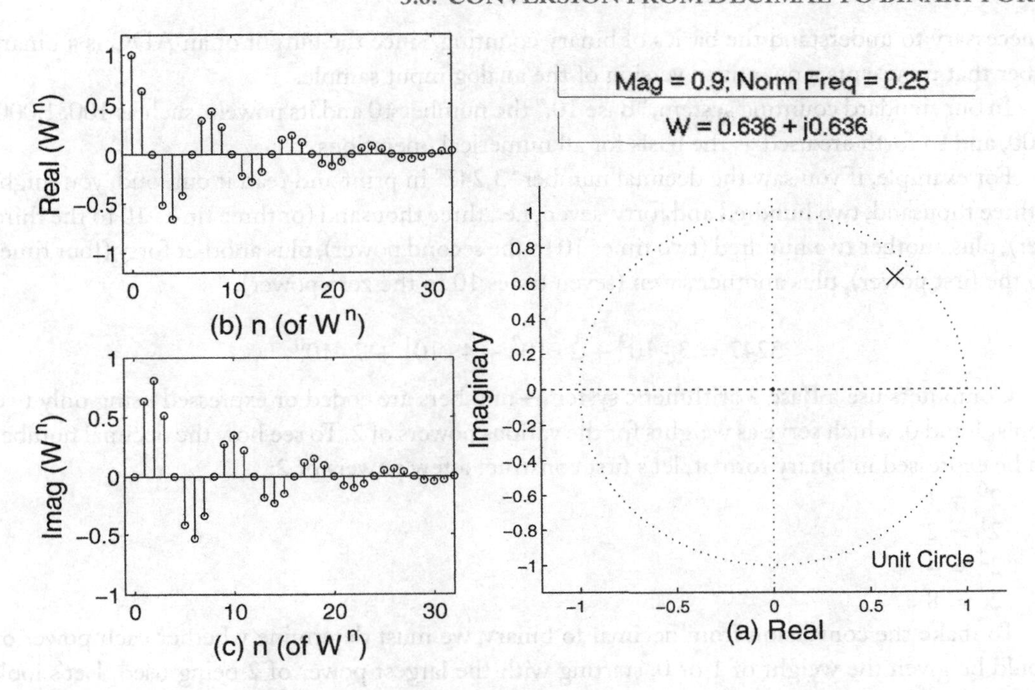

Figure 3.12: (a) Complex plane, showing the complex number having magnitude 0.9 and angle 45 degrees, i.e., 0.636 + j0.636; (b) Real part of complex power sequence; (c) Imaginary part of complex power sequence.

and

```
figure; stem(imag((0.9*exp(j*pi*0.25)).^(0:1:29)))
```

You can experiment by changing any of the magnitude, normalized frequency, or length of the exponent vector.

- A script that allows you to dynamically see the impulse response and frequency response generated by a pole or zero, or pair of either, as you move the cursor in the complex plane is

$$ML\_DragPoleZero$$

3.6 CONVERSION FROM DECIMAL TO BINARY FORMAT

Digital systems represent all signals as sequences of numbers, and those numbers are expressed in binary format using only two symbols, 1 and 0. Thus prior to discussing analog-to-digital conversion, it is necessary to understand the basics of binary counting, since the output of an ADC is a binary number that represents a quantized version of the analog input sample.

In our standard counting system, "Base 10," the number 10 and its powers, such as 100, 1,000, 10,000, and so forth are used as the basis for all numerical operations.

For example, if you saw the decimal number "3,247" in print and read it out loud, you might say "three thousand, two hundred and forty-seven, i.e., three thousand (or three times 10 to the third power), plus another two hundred (two times 10 to the second power), plus another forty (four times 10 to the first power), plus another seven (seven times 10 to the zero power).

$$3247 = 3 \cdot 10^3 + 2 \cdot 10^2 + 4 \cdot 10^1 + 7 \cdot 10^0$$

Computers use a Base 2 arithmetic system—numbers are coded or expressed using only two symbols, 1 and 0, which serve as weights for the various powers of 2. To see how the decimal number 9 can be expressed in binary format, let's first construct a few powers of 2:

$2^0 = 1$
$2^1 = 2$
$2^2 = 4$
$2^3 = 8$

To make the conversion from decimal to binary, we must determine whether each power of 2 should be given the weight of 1 or 0, starting with the largest power of 2 being used. Let's look ahead to the answer, which is [1 0 0 1] and note that

$$9 = 1 \cdot 2^3 + 0 \cdot 2^2 + 0 \cdot 2^1 + 1 \cdot 2^0 \tag{3.1}$$

In Eq. (3.1), we call the weighting values, 1 or 0, for each power of 2, **Binary Digits**, or more commonly **Bits**. In any binary number, the rightmost bit (the weight for 2^0) is called the **Least Significant Bit**, or **LSB**, while the leftmost bit is called the **Most Significant Bit**, or **MSB**.

To arrive at (3.1) algorithmically, we'll use a simple method called **Successive Approximation,** which is commonly used to convert analog or continuous domain sample values to binarily-quantized sample values, in which we consider whether or not each bit, starting with the MSB, should ultimately be given a weight of 1 or 0. An example should make the method clear.

Example 3.12. Convert decimal number 9 to binary notation using the Method of Successive Approximation.

We start by setting the MSB, 2^3 (= 8), to 1. Since $1 \cdot 2^3$ is less than 9, we retain the value 1 for the 2^3 bit. Proceeding to 2^2 (= 4), if we add 4 to the 8 we have ($8 = 1 \cdot 2^3$), we get 12 ($= 1 \cdot 2^3 + 1 \cdot 2^2$). But 12 exceeds the number to be converted, 9, so we reset the weight of 2^2 from 1 to 0. Proceeding to 2^1 (= 2), we see that the 8 we have ($1 \cdot 2^3 + 0 \cdot 2^2$) plus another 2 would result in a sum total of 10, so we reset the 2^2 weight to 0. Proceeding to 2^0 (= 1), we see that adding 1 to the current sum total of 8 gives the required 9, so we keep the 2^0 weight at 1. Thus we have 9 = [1 0 0 1] in binary notation.

3.7 QUANTIZATION ERROR

Note that so far we have been examining the conversion of integer values between decimal and binary formats. In general, the amplitudes of a sequence of samples will not always be integer multiples of the LSB, and thus there will be, in general, a difference between the quantized value and the original value. This difference is called the **Quantization Error** and may be made arbitrarily small by using a larger number of quantization bits. We will examine this in detail below after the presentation of a script to perform analog-to-digital conversion on a sequence and compute the quantization error.

For purposes of illustrating the general principles of ADC and DAC, we will continue using this integer format, i.e., quantizing signals as integral multiples of the LSB. There are, of course, binary formats that contain fractional parts as well as an integer part. See reference [1] for further details.

3.8 BINARY-TO-DECIMAL VIA ALGORITHM

Note that part of the process of successive approximation involves evaluating a test binary word's equivalent decimal value, i.e., a digital-to-analog step, which is the reverse of the overall procedure being conducted. Prior to presenting a program to convert an entire sequence of decimal numbers (or equivalently, from the realm of real-world signals, a sequence of analog or continuous-valued samples) to binary format, we'll describe several simple scripts to convert a binary representation back to decimal.

Example 3.13. Write a simple program that will convert a binary number to decimal form.

First, we receive an arbitrary length binary number bn, and then note that $bn(1)$ is the weight for $2^{LenBN-1}$, $bn(2)$ is the weight for $2^{LenBN-2}$, and so forth, where $LenBN$ is the length of bn. We then initialize the converted decimal number dn at 0. The loop then multiplies each bit by its corresponding power of two, and accumulates the sum.

```
function [dn] = LVBinary2Decimal(bn)
% [dn] = LVBinary2Decimal([1,0,1,0,1,0,1,0])
Lenbn = length(bn); dn = 0; for ctr = Lenbn:-1:1;
dn = dn + 2^(ctr-1)*bn(Lenbn-ctr+1); end;
```

A simplified (vectorized) equivalent of the above code is

```
function [dn] = LVBinary2DecimalVec(bn)
% [dn] = LVBinary2DecimalVec([1,0,1,0,1,0,1,0])
dn = sum(2.^(length(bn)-1:-1:0).*bn);
```

Yet another variation uses the inner or dot product of the power of two vector and the transpose (into a column vector) of the binary number:

```
function [dn] = LVBin2DecDotProd(bn)
% [dn] = LVBin2DecDotProd([1,0,1,0,1,0,1,0])
```

```
dn = 2.^(length(bn)-1:-1:0)*bn';
```

3.9 DECIMAL-TO-BINARY VIA ALGORITHM

Now we are ready, in the following example, to present a script for algorithmic conversion of a decimal number (zero or positive) to binary format.

Example 3.14. Write a script that will quantize an input decimal-valued sample to a given number of bits, and print out descriptions of each step taken during the successive approximation process.

```
function [BinOut, Err] = LVSuccessAppSingle(DecNum,Bits)
% [BinMat, Err] = LVSuccessAppSingle(5.75,4)
if length(DecNum)>1; error('DecNum must be scalar')
end
BinOut = zeros(1,Bits); BinWord = zeros(1,Bits);
for ctr = 1:1:Bits
Status = (['Setting Bit ',num2str(ctr),' to 1'])
BinWord(1,ctr) = 1
DecEquivCurrBinWord = LVBinary2DecimalVec(BinWord)
Status = (['Subtracting decimal equiv of BinWord from sample being quantized'])
diff = DecNum - DecEquivCurrBinWord
if diff < 0;
Status = (['Resetting Bit ',num2str(ctr),' to 0'])
BinWord(1,ctr) = 0;
end
end
Status = (['Final Binary Word:'])
BinOut(1,:) = BinWord;
Err = DecNum - LVBinary2DecimalVec(BinWord);
```

The following printout occurs after making the call

```
[BinMat] = LVSuccessAppSingle(1.26,2)
```

```
Status = Setting Bit 1 to 1
BinWord = 1 0
DecEquivCurrBinWord = 2
Status = Subtracting decimal equiv of BinWord from sample being quantized
diff = -0.7400
Status = Resetting Bit 1 to 0
Status = Setting Bit 2 to 1
```

```
BinWord = 0 1
DecEquivCurrBinWord = 1
Status = Subtracting decimal equiv of BinWord from sample being quantized
diff = 0.2600
Status = Final Binary Word:
BinMat = 0 1
Err = 0.2600
```

3.10 OFFSET TO INPUT TO REDUCE ERROR

The method of successive approximation as described above always chooses bit weights which add up to a quantized value which is less than or equal to the actual value of the input signal. As a result, the approximation of the quantized signal to the actual signal is biased an average of one-half LSB toward zero, and the maximum quantization error is 1.0 times the LSB.

Example 3.15. Demonstrate that a bias to the input signal of 0.5 LSB away from zero results in a maximum quantization error of one-half LSB rather than 1.0 LSB.

We present a script that can convert an entire sequence of decimal numbers to binary equivalents, quantized to a specified number of bits, and where a bias of one-half LSB to the input signal can be optionally added. This script is not the most efficient one possible; in the exercises below, the reader is presented with the task of writing a script that will conduct the successive approximation technique simultaneously on an entire input sequence, thus eliminating the outer loop, which can make the computation very slow for long sequences.

```
function [BinMat,Err] = LVSuccessApp(DecNum,MaxBits,LSBBias)
% [BinMat,Err] = LVSuccessApp(7.5*[sin(0.5*pi*[0:1/18:1])]+7.5,4,1);
BinMat = zeros(length(DecNum),MaxBits);
DecOutMat = zeros(1,length(DecNum));
if LSBBias==0; xDecNum = DecNum;
else; xDecNum = DecNum + 0.5; end
for DecNumCtr = 1:1:length(DecNum)
BinTstWord = zeros(1,MaxBits);
for ctr = 1:1:MaxBits; BinTstWord(1,ctr) = 1;
DecEquivCurrBinTstWord = LVBinary2DecimalVec(BinTstWord);
diff = xDecNum(DecNumCtr) - DecEquivCurrBinTstWord;
if diff < 0; BinTstWord(1,ctr) = 0;
elseif diff==0; break; else; end
end
BinMat(DecNumCtr,:) = BinTstWord;
DecOutMat(1,DecNumCtr) = LVBinary2DecimalVec(BinTstWord);
```

```
Err = DecNum-DecOutMat;
end
figure(78); clf; subplot(211); ldn = length(DecNum);
xvec = 0:1:ldn-1; hold on; plot(xvec,DecNum,'b:');
plot(xvec,DecNum,'bo'); stairs(xvec,DecOutMat,'r');
xlabel('(a) Sample'); ylabel('Amplitude')
subplot(212); stairs(xvec,Err,'k');
xlabel('(b) Sample'); ylabel('Error')
```

The call

```
[BinMat,Err] = LVSuccessApp(7.5*[sin(0.5*pi*[0:1/18:1])]+7.5,4,0);
```

results in Fig. 3.13, in which the quantized values are plotted in stairstep fashion rather than as discrete points. Note that the error is positive, and has a maximum amplitude of about 1.0, i.e., one LSB.

The following call specifies that prior to conversion, 0.5LSB (i.e., 0.5 in this case since the LSB is 1.0) be added to the signal:

```
[BinMat,Err] = LVSuccessApp(7.5*[sin(0.5*pi*[0:1/18:1])]+7.5,4,1);
```

The result is shown in Fig. 3.14, in which it can be seen that the error is both positive and negative, but does not exceed 0.5, i.e., one-half LSB.

3.11 CLIPPING

The call

```
[BinMat] = LVSuccessApp(10*[sin(2*pi*[0:1/18:1])]+10,4,1);
```

results in Fig. 3.15. As only four bits were specified for the quantization, we can only represent the numbers 0-15, whereas the input waveform ranges in value from 0 to 20. The algorithm outputs its maximum value of 15 (binary [1,1,1,1]) for signal values of 14.5 and higher; the output is said to be clipped since graphically it appears that its upper portions (samples 2 through (7) have simply been cut or clipped off.

The situation above can be remedied by increasing the number of bits from 4 to 5. We make the call

```
[BinMat] = LVSuccessApp(10*[sin(2*pi*[0:1/18:1])]+10,5,1);
```

which results in Fig. 3.16. For five bits, the possible output values range from 0 to 31, easily encompassing the signal at hand.

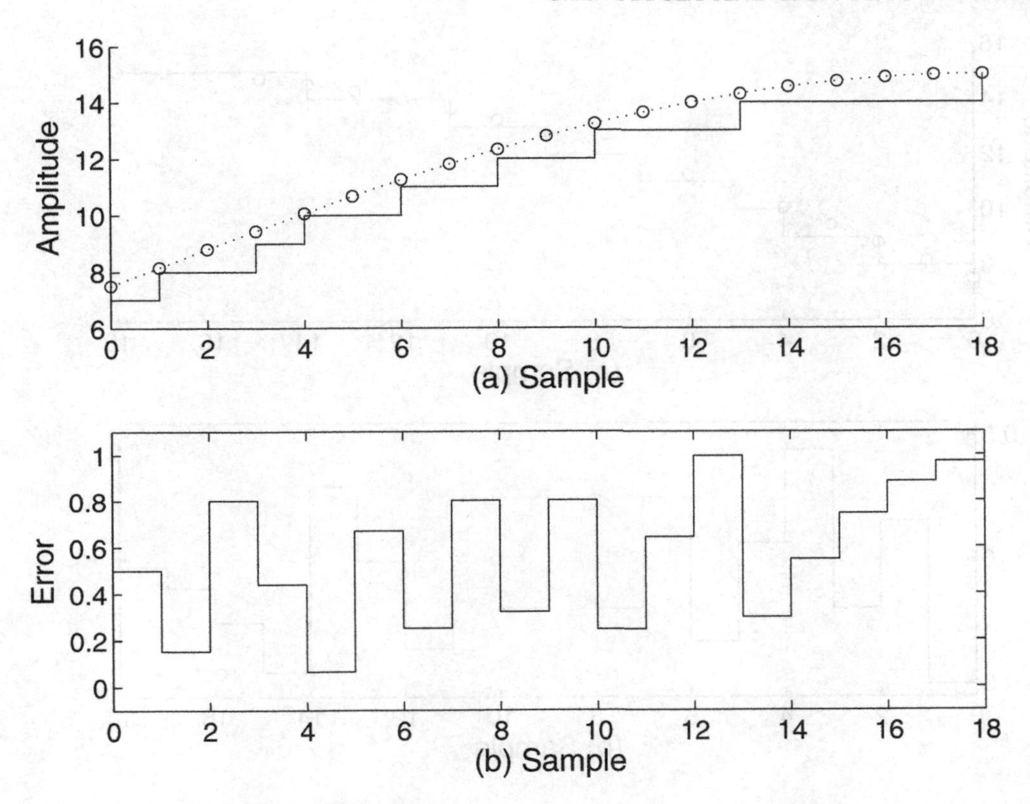

Figure 3.13: (a) Analog signal (dashed), analog sample values (circles), and quantized samples (stairstep); (b) Quantization error, showing a maximum magnitude equal to the LSB.

3.12 OFFSET AND SIGN-PLUS-MAGNITUDE

The method just discussed assumes all numbers to be converted are zero or positive. This is a common method of digitization, the **Offset Method (**sometimes called the **Unipolar** method**)**, in which a typical analog signal having both positive and negative voltage values is given a DC offset so that it is entirely nonnegative. Another format is the **Sign-Plus-Magnitude Method**, in which the most significant bit represents the sign of the number, and the remaining bits represent the magnitude.

Example 3.16. Assume that a binary number is in the sign-plus-magnitude format, with a sign bit value of 0 meaning positive and 1 meaning negative. Write a simple script to convert such a number to decimal equivalent.

Initially, we set aside the first bit of the binary number, which is the sign bit, and convert the remaining bits to decimal as before. If the sign bit has the value 1, the answer is multiplied by -1, otherwise, it is left positive.

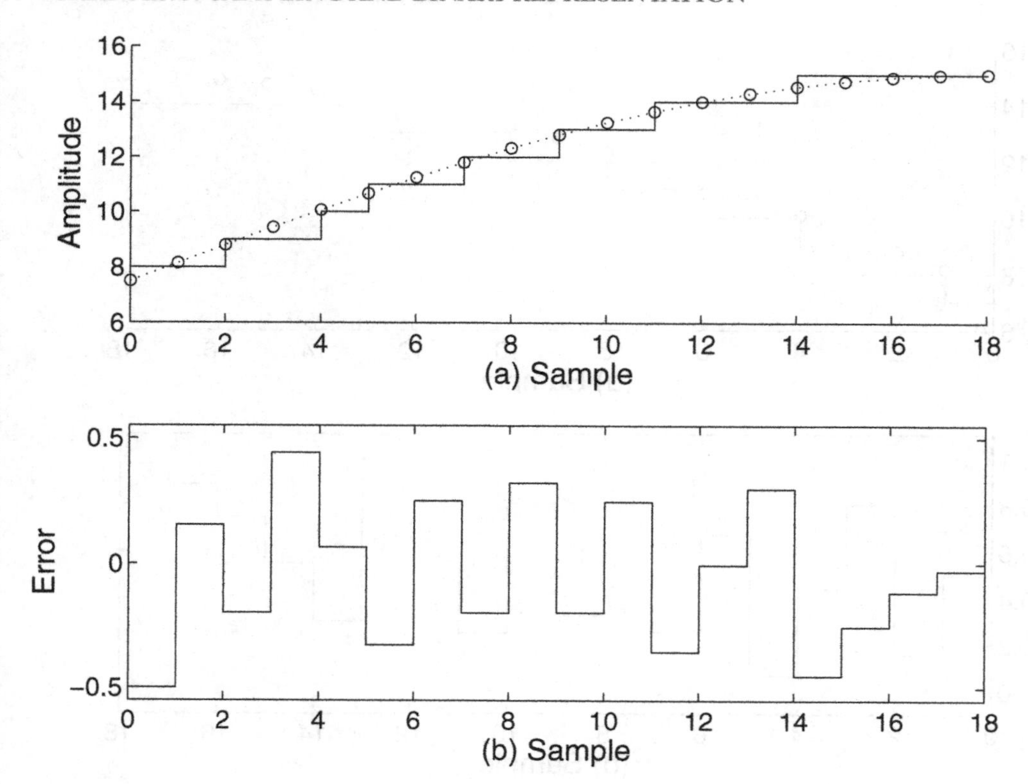

Figure 3.14: (a) Analog signal (dashed), analog sample values (circles), and quantized samples (stairstep); (b) Quantization error, showing a maximum magnitude error of 0.5 LSB.

```
function [dn] = LVSignPlusMag2Dec(bn)
% [dn] = LVSignPlusMag2Dec([1,0,1,0,1])
pbn = fliplr(bn(2:length(bn)));
dn = (-1)^(bn(1))*sum(2.^(0:1:length(pbn)-1).*pbn)
```

The script

$$BinOut = LVxBinaryCodeMethods(BitsQ, SR, Bias, Freq, Amp, CM, PlotType)$$

(see exercises below) affords the opportunity to experiment with the analog-to-digital conversion of a test sine wave, using either of the two formats mentioned above. It allows you to select the number of quantization bits $BitsQ$, the sample rate SR, amount of bias to the input signal $Bias$, test sine wave frequency $Freq$, test sine wave amplitude Amp, coding method CM, which includes

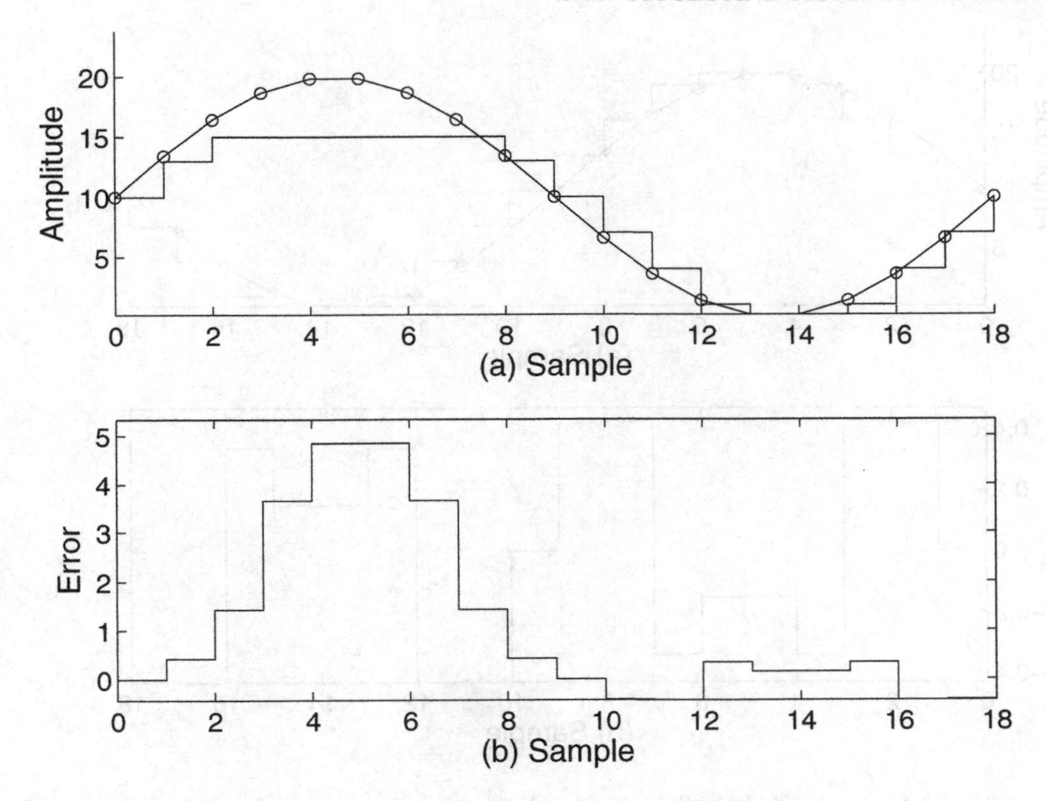

Figure 3.15: (a) Analog signal (dashed), analog sample values (circles), and quantized samples (stairstep). Due to clipping, the sample values for sample times 2-6 are limited to quantized level 15; (b) Error signal, difference between original signal and the quantized signal at sample times.

Sign-Plus-Magnitude and *Offset*, and *PlotType*, which displays the quantized output in volts or as multiples of the LSB.

Example 3.17. Demonstrate the Sign-Plus-Magnitude conversion method (or format) using the script *LVxBinaryCodeMethods.*

Figure 3.17, plots (a) and (b), show a portion of the result from using parameters of *SR* = 1000 Hz, *Freq* = 10 Hz, *Amp* = 170, *BitsQ* = 2 Bits, *Bias* = None, and *CM* = *Sign plus Mag,* and *PlotType* as volts. The call is

BinOut = LVxBinaryCodeMethods(2,1000,0,10,170,1,1)

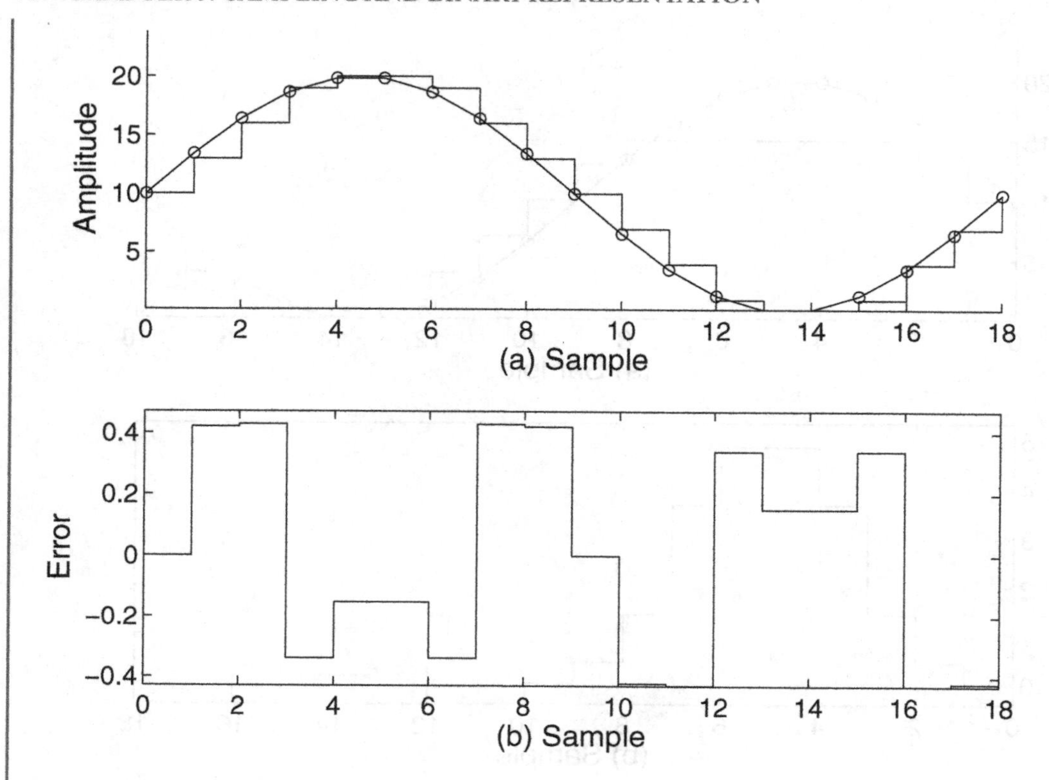

Figure 3.16: (a) Analog signal (dashed), analog sample values (circles), and quantized samples (stairstep); (b) Quantization error, showing relief from the clipping of the previous example, achieved by increasing the number of quantization bits, allowing for quantization up to decimal equivalent 31 ($2^5 - 1$).

Plot (a) of the figure shows the analog voltage and the discrete, quantized samples; plot (b) shows the noise voltage, which is the difference between the original analog signal and the quantized version.

The use of only two bits, no input bias, and sign-plus-magnitude coding results in a quantized signal which is almost completely noise (albeit in this case the noise component is highly periodic in nature). As can be seen in plot (a), only two samples are quantized at or near their true (original analog) voltages; all other quantized voltages are zero, resulting in a noise signal of very high amplitude relative to the original signal.

The situation can be considerably improved by biasing the input signal by one-half LSB away from zero. Plots (c) and (d) show the result.

Example 3.18. Demonstrate the Offset conversion method using the script *LVxBinaryCodeMethods*.

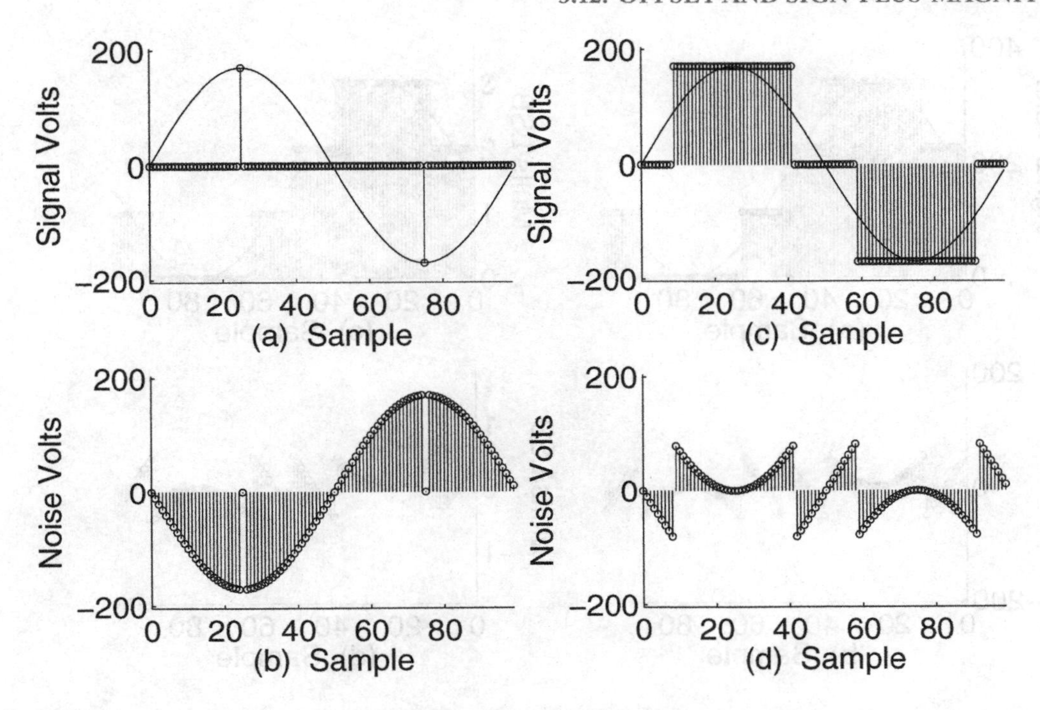

Figure 3.17: (a) A portion of an analog or continuous 10 Hz sine wave (solid); quantized using 2 bits (one sign and one magnitude), LSB = 170 volts, stem plot; (b) Quantization Noise; (c) Analog 10 Hz sine wave, solid; quantized using 2 bits (one sign and one magnitude), input biased one-half LSB away from zero for both positive and negative values, LSB = 170 volts; (d) Quantization Noise.

The offset method gives 2^n quantization levels, as opposed to the sign-plus-magnitude coding method, which yields only $2^n - 1$ distinct levels. Using *SR* = 1000 Hz, *Freq* = 10 Hz, *Amp* = 170, *BitsQ* = 2 Bits, *Bias* = One-Half LSB, and *CM* = Offset, and *PlotType* as volts, we get Fig. 3.18, plots (a) and (b). The call is

BinOut = LVxBinaryCodeMethods(2,1000,1,10,170,2,1)

In the previous figure, the quantized output was depicted in volts; the actual output of an ADC is a binary number giving the sample's amplitude as a multiple of the value of the LSB. In Fig. 3.18, plots (c) and (d), the quantized output and the quantization noise are both plotted as multiples of the value of the LSB, which is 113.333 volts, computed by dividing the peak amplitude (340 volts) of the input signal by $(2^2 - 1)$, which is the number of increments above zero needed to reach the peak amplitude from the minimum amplitude, which is set at zero in accordance with the Offset Coding Method.

In Fig. 3.18, plots (c) and (d), there are four possible binary outputs, and they are

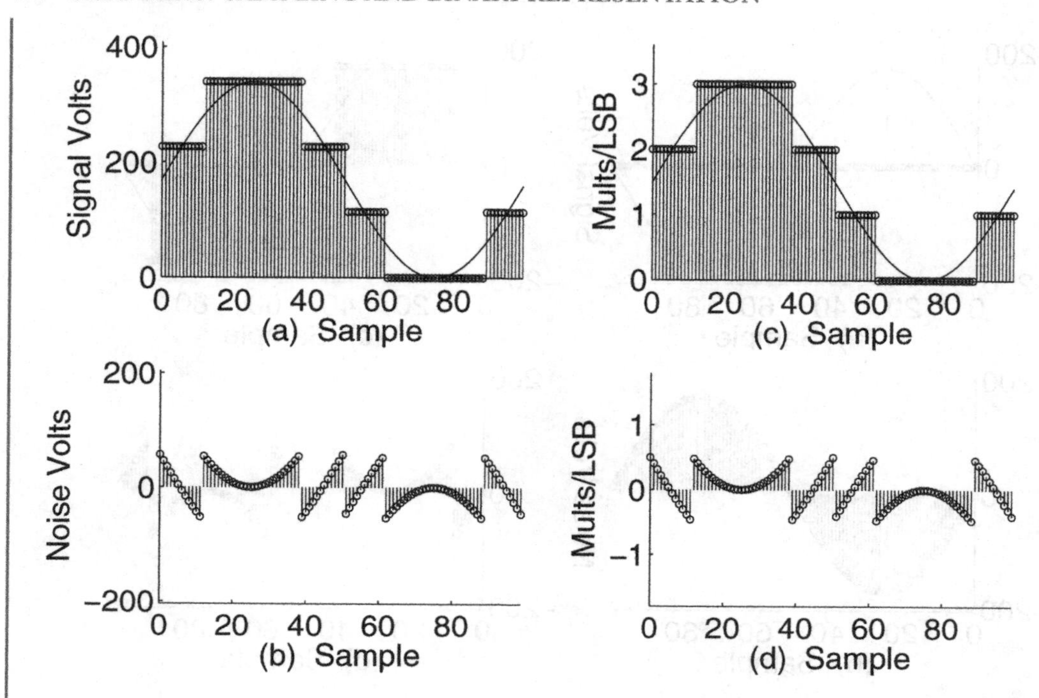

Figure 3.18: (a) A portion of an analog 10 Hz sine wave (solid), quantized using 2 bits, Offset Method, input biased one-half LSB, LSB = 113.33 volts (stem plot); (b) Quantization Noise; (c) Analog 10 Hz sine wave (solid), quantized using 2 bits, Offset Method, input biased one-half LSB, in Multiples of the LSB (stem plot); (d) Quantization Noise.

Binary 0 = [0 0] (0 times 113.33 Volts)
Binary 1 = [0 1] (1 times 113.33 Volts)
Binary 2 = [1 0] (2 times 113.33 Volts)
Binary 3 = [1 1] (3 times 113.33 Volts)

Note: For the following two examples, assume proper anti-aliasing filtering is used prior to any sampling and that no bias is applied to the input signal.

Example 3.19. An analog signal $x = \cos(2\pi(100)t)$ with t in seconds is sampled for one second at 200 Hz, with the first sample being taken at time 0, when the signal value is 1.0 volts. (a). What is the value of the analog signal at the first four sampling times? (b). For the same signal, suppose the sampling rate had been 400 Hz. In that case, what would have been the values of the analog signal at the first four sampling times? (c). Assume that the sampling is done by an ADC that accepts input voltages between 0 and 1 volt. What signal adjustments need to be made prior to applying the analog signal to the ADC's input?

(a) The call

x = cos(2*pi*100*(0:0.005:0.015))

gives the answer as [1,-1,1,-1];

(b) The call

x = cos(2*pi*100*(0:0.0025:0.0075))

gives the answer as [1,0,-1,0];

(c) To convert the analog input signal into one which matches the ADC's input voltage range, either multiply the signal by 0.5 and then add 0.5 volt to the result, or add 1.0 volt to the signal and multiply the result by 0.5.

Example 3.20. Assume the ADC in the preceding example, part (c), quantizes all voltages as positive offsets from 0 volts, and that a 1 volt input to the ADC yields a binary output of [1111]. Assume that the test signal is 100 Hz but that the sampling rate has been increased to 10, 000 Hz, giving rise to quantized sample values covering the entire 0 to 1 volt range. (a). If the binary output were [0001], what quantized input voltage to the ADC would be represented? (b). What range of analog signal voltages input to the ADC could have resulted in the binary output given in (a)? (c). What was the original signal voltage equivalent to the voltage determined in (a) above (remember that we had to shift and scale the original analog voltage prior to applying it to the ADC)?

(a) Since binary (offset) output [1111] (15 in decimal) is equivalent to a 1-volt analog input, it follows that binary output [0001] is equivalent to 1/15 volt;

(b) Since there is no bias to the input of 0.5 LSB, values of 1/15 volt up to just less than 2/15 volt will quantized to [0001];

(c) Since we added 1 volt and multiplied by 0.5 to get the analog input voltage, we take 1/15 volt, multiply by 2, and then subtract 1.0 volt to get -13/15 volt.

The following examples illustrate determination of the equivalent LSB voltage for a given ADC operation, and choosing the number of quantization bits to achieve a certain LSB voltage.

Example 3.21. Suppose that we have an ADC which will accept analog input voltages ranging between −170 and +170 volts, and that is capable of quantizing the input to 3 bits, one of which will be a sign bit, leaving two bits for magnitude. Determine the value of the LSB in volts.

The maximum is 170 volts, and we will have $2^3 - 1 = 7$ distinct levels (remember that we lose a level when using one bit for the sign) which can be specified by the 3 bit output. Note that we have three distinct magnitudes above zero available for positive values, and the same for negative values, and one level for zero, giving us $3 + 3 + 1 = 7$ distinct levels of quantization.

Doing the arithmetic, we can see that one bit is equivalent to (170 volts/3 levels) = 56.67 volts/level.

Figure 3.19, plots (a) and (b), show the result. The 3-bit quantized values differ by as much as one-half the LSB (56.67 divided by two, or 28.33 volts) from the correct value. In some applications, this might be of sufficient accuracy, but for "high end" applications, such as audio and video, 16 or more bits of quantization are necessary.

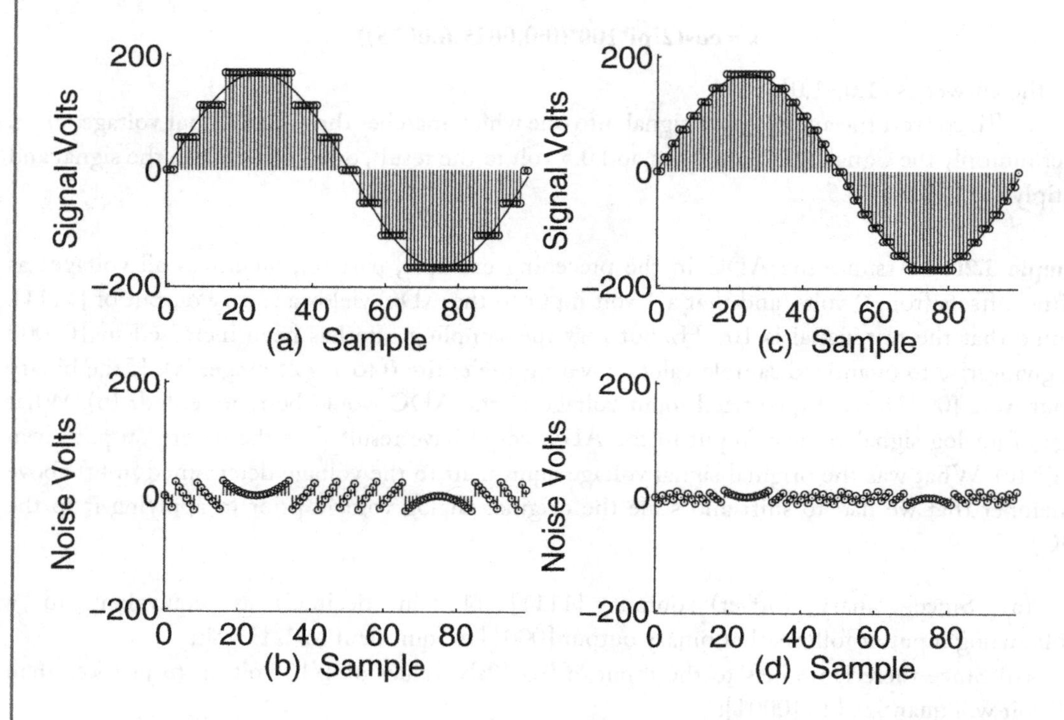

Figure 3.19: (a) A portion of an analog 10 Hz sine wave (solid), quantized using 3 bits (1 Sign, 2 Magnitude), input biased one-half LSB, LSB = 113.33 volts (stem plot); (b) Quantization Noise; (c) Analog 10 Hz sine wave (solid), quantized using 4 bits (1 Sign, 3 Magnitude), input biased one-half LSB (stem plot); (d) Quantization Noise.

Example 3.22. Suppose that it was required that the samples be accurate to the nearest 1.25 volts. Determine the total number of bits needed to achieve this.

We would need (170 volts/1.25 volts/level) = 136 levels per side, or 273 levels total. Investigating powers of two, we see that $2^8 = 256$ is inadequate, but $2^9 = 512$ will work—so, in addition to the sign bit, we need to employ nine bits of quantization, which will give the LSB as (170 volts/255 levels) = 0.667 volts/level.

Example 3.23. Suppose we wanted to quantize, using 3 bits and the Offset method, a 340 volt peak-to-peak sine wave (this would correspond to a sine wave varying between -170 and 170 volts which has been offset toward the positive by 170 volts such that there are no longer any negative voltages). What would be the value of the LSB? State the possible binary outputs and the corresponding input voltages they represent.

With Offset Coding, we get 2^3 distinct levels, one of which is the lowest level, zero. There are therefore $2^3 - 1 = 7$ levels to reach the highest level, which would yield the LSB as $340 \div 7 =$ 48.57 volts.

The possible binary output codes would be 000, 001, 010, 011, 100, 101, 110, 111, which would represent quantized voltages of 0, 48.57, 97.14, 145.71, 194.28, 242.85, 291.42, and 340.00 volts, respectively.

For simple input signals, the error signal is well-correlated with the input signal, as can be seen in Fig. 3.19, plots (c) and (d); when the input signal is complicated, the error signal will be less correlated with the input, i.e., appear more random. Noise that correlates well with the input tends to be tonal and more audible; there exist a number of methods to "whiten" the spectrum of the sampled signal by decorrelating the sampling noise. One such method, **Dithering**, is to add a small amount of random noise to the signal, which helps decorrelate sampling artifacts. Extensive discussions of dithering are found in [1] and [2].

Quantization using 4 bits still results in objectionably large amounts of quantization noise, as shown in Fig. 3.19, plots (c) and (d).

Note that no value of the signal to be quantized is ever further than one-half the LSB amount from one of the possible quantized output levels. This can be seen clearly in Fig. 3.18, plots (c) and (d). Recall that the LSB amount would be (170 volts/7 levels) = 24.2857 volts/level. Note that in Fig. 3.19, plot (d), the maximum noise amplitude is about half that, or a little over 12 volts.

As a good approximation, it can be said that each time the number of quantizing bits goes up by one, the maximum magnitude of quantization noise is halved, and the corresponding noise power is quartered, since power is proportional to the square of voltage or amplitude.

References [1] and [2] give extensive and accessible discussions of quantization noise.

3.13 DAC WITH VARIABLE LSB

The eventual goal of much signal processing, such as that of audio or video signals, is to convert digital samples back into an analog signal. This is the process of signal reconstruction, which is performed by sending the digital samples (binary numbers) to a DAC.

In addition, as we have seen, the heart of a successive approximation converter is a digital-to-analog converter.

Figure 3.20 depicts a typical arrangement for converting a sequence of digitized samples back into an analog signal. The latch holds a digital word (typically 4-24 bits) on the input of the DAC until the next digital word is issued from the digital sample source. Holding a value constant until the next one is retrieved is referred to as a **Zero Order Hold,** which is discussed later in this chapter.

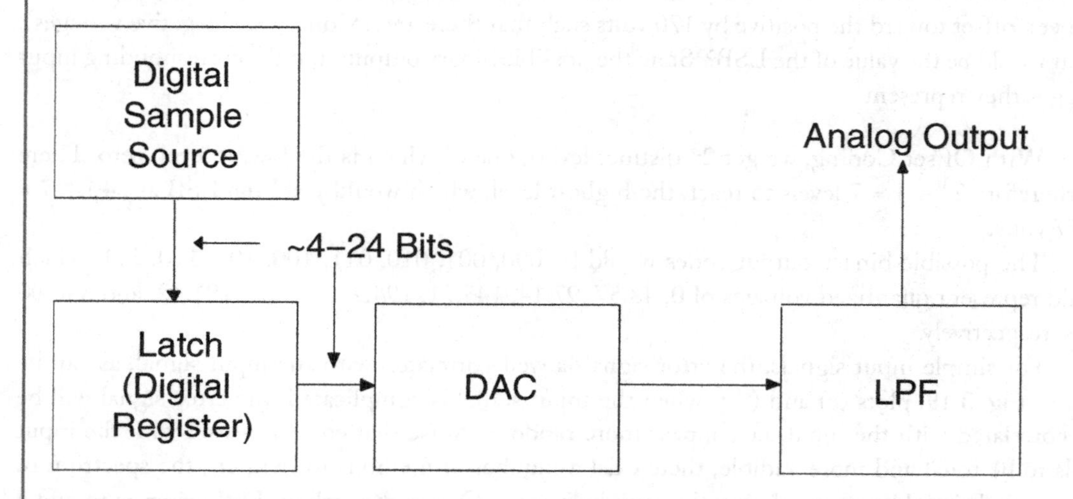

Figure 3.20: A representative arrangement for converting digitized samples to an analog signal. The digital path to the DAC is usually a number of bits as depicted, presented in parallel to the DAC, which outputs an analog signal on a single line.

A simple DAC is shown Fig. 3.21. A reference voltage feeds binarily-weighted resistors that feed the inverting input of an operational amplifier, which sums the currents and outputs a voltage proportional to the current according to the formula

$$V_{out} = V \cdot (Bit(3)(-R/R) + Bit(2)(-R/2R) + Bit(1)(-R/4R) + Bit(0)(-R/8R))$$

which reduces to

$$V_{out} = -V \cdot (Bit(3) + 0.5Bit(2) + 0.25Bit(1) + 0.125Bit(0))$$

where $Bit(0)$ means the value (1 or 0) of the 2^0 bit, and so forth. This scheme can be expanded to any number of needed bits.

Here we see that for the type of DAC shown in Fig. 3.21,

$$LSB_{volts} = -V/(2^{N-1}) \tag{3.2}$$

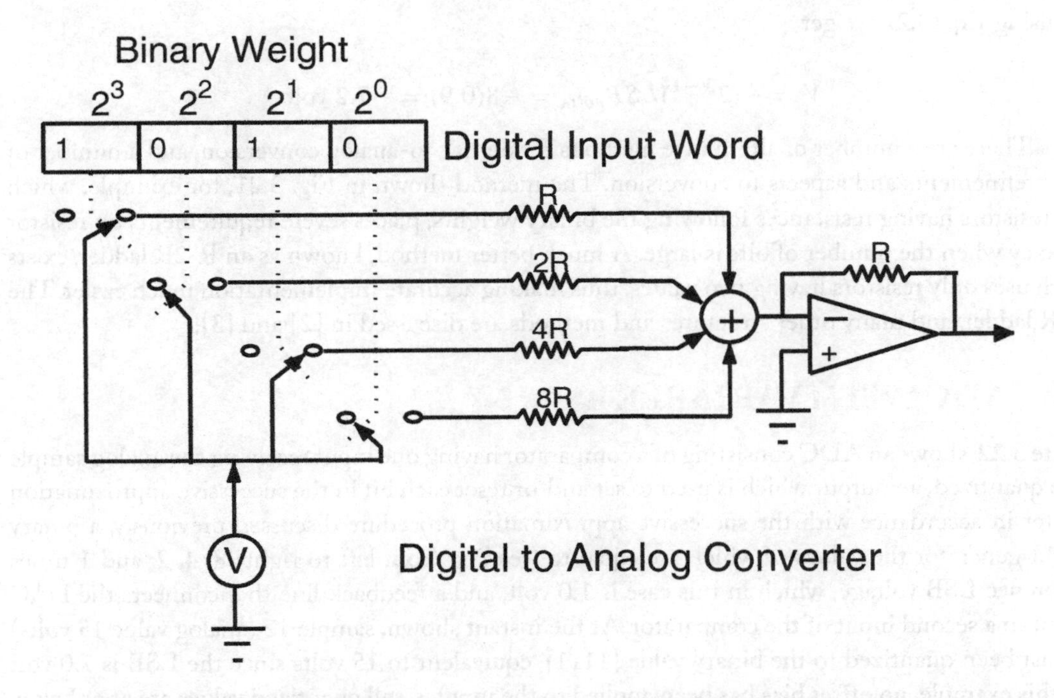

Figure 3.21: A simple 4-bit Digital-to-Analog converter. Note that the contributions from each resistor enter a summing junction, the output of which is connected to the negative input of the operational amplifier; the summing junction is shown for conceptual clarity only. In an actual device, the summation would be accomplished simply by connecting the leads from all resistances directly to the negative input of the operational amplifier.

where N is the number of quantization bits; hence the LSB value may be arbitrarily set by changing the reference voltage V. We can also note that when all bits are equal to 1, the maximum value of V_{out} is attained, and

$$LSB_{volts} = -(V_{out})\text{max}/(2^N - 1) \tag{3.3}$$

Example 3.24. A sequence that is in four-bit offset binary format is to be converted to the analog domain using a DAC of the type shown in Fig. 3.21, and it is desired that its analog output voltage range be 0 to +13.5 volts. Determine the necessary value of reference voltage V.

We note that for four bits, the highest level is 15 times the LSB value, so we get

$$LSB_{volts} = 13.5/15 = 0.9 \text{ volt}$$

and using Eq. (3.2), we get

$$V = -(2^{N-1})LSB_{volts} = -8(0.9) = -7.2 \text{ volt}$$

There are a number of alternative methods for digital-to-analog conversion, and a number of other refinements and aspects to conversion. The method shown in Fig. 3.21, for example, which uses resistors having resistances following the binary weights, places severe requirements on resistor accuracy when the number of bits is large. A much better method, known as an R-2R ladder, exists which uses only resistors having two values, thus making accurate implementation much easier. The R-2R ladder, and many other structures and methods are discussed in [2] and [3].

3.14 ADC WITH VARIABLE LSB

Figure 3.22 shows an ADC consisting of a comparator having one input receiving the analog sample to be quantized, an output which is used to set and/or reset each bit in the successive approximation register in accordance with the successive approximation procedure discussed previously, a binary weight generator the values of which are equal to (reading from left to right) 8, 4, 2, and 1 times a reference LSB voltage, which in this case is 1.0 volt, and a feedback line that connects the DAC output to a second input of the comparator. At the instant shown, sample 12 (analog value 15 volts) has just been quantized to the binary value [1111], equivalent to 15 volts since the LSB is 1.0 volt (for this example, no offset bias has been applied to the input, so all quantized values are at or below the sample values).

In Fig. 3.23, the LSB voltage has been changed to 0.75 volt. Since four bits are being used, the maximum voltage output of the DAC section is 15 times 0.75 = 11.25 volt. All input signal values at or above 11.25 volt will therefore be quantized at the highest binary output of the ADC, which is [1111]. Note that the Binary Weight Generator now bears weights proportional to the LSB value of 0.75 volt, i.e., 6, 3, 1.5, and 0.75 volt.

Figures 3.22 and 3.23 were created using the script

ML_SuccessiveApproximation

which, when called, opens up the GUI as shown in the figures just mentioned. The GUI allows selection of a number of parameters including automatic (fast or slow) or manual computation.

3.15 ZERO-ORDER HOLD CONVERSION

Figure 3.24 shows what the output of a DAC with a zero-order hold would look like prior to filtering with a low pass filter. In the Zero-Order Hold system, each DAC conversion value is held on the output of the DAC until the next sample arrives at the DAC and changes the output value.

The use of a zero-order hold in digital-to-analog conversion is perhaps the simplest or most practical method of conversion. The stairstep waveforms it generates contain the original sinusoids,

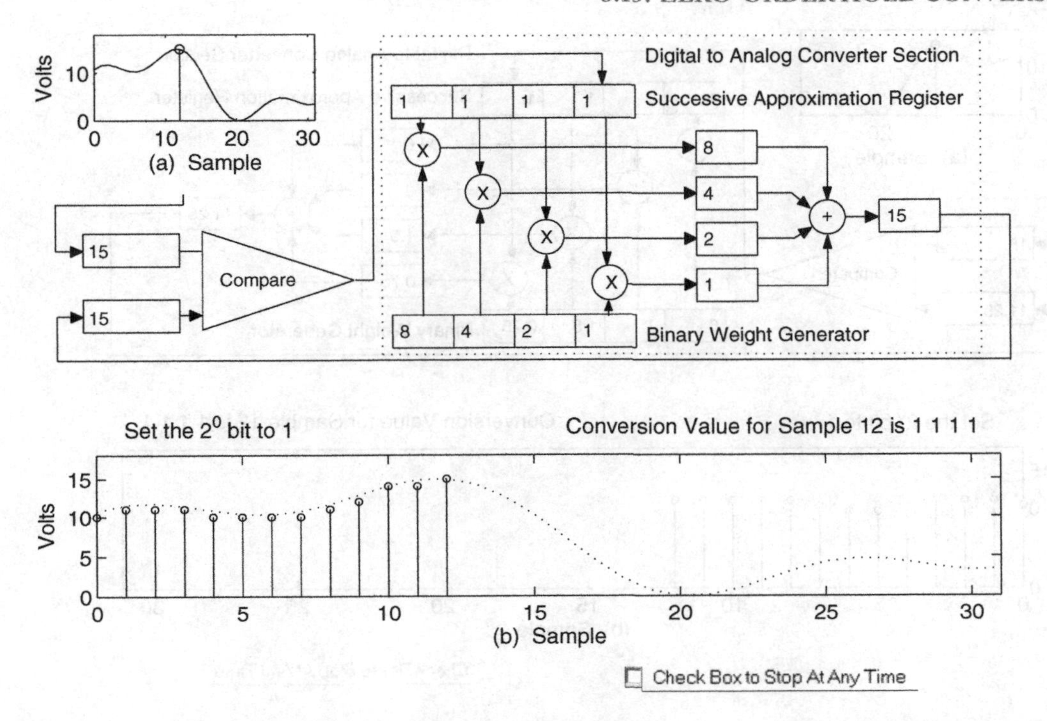

Figure 3.22: A schematic arrangement of an ADC having LSB = 1.0 volt, using the Offset method with no bias to the input signal; (a) Analog signal (solid), sample being quantized (circle); (b) Quantized samples.

plus overtones thereof (for a basic discussion of conversion methods other than by use of a zero-order hold, see reference [1]).

There are actually two entirely separately arising spectral impurities (frequency components that were not present in the original analog signal) in the output of a DAC having a zero-order hold: quantization noise, and stairstep noise. Even if the number of bits of quantization is very large, leading to effectively no quantization error, there will still be a very large stairstep component in the DAC output which was not in the original analog signal. Referring to a sine wave, for example, the amplitude of the stairstep component is large when the number of samples per cycle of the sine wave is low, and decreases as the number of samples per cycle increases.

Fortunately, the frequencies comprising the stairstep component all lie above the Nyquist limit and can be filtered out using a lowpass reconstruction filter designed especially for zero-order hold reconstruction (LPF in Fig. 3.20). Quantization noise, on the other hand, has components below the Nyquist limit, and thus cannot be eliminated with lowpass filtering. Zero-order hold conversion acts like a gently-sloped lowpass filter, so a special reconstruction filter must be used that cuts off at the Nyquist limit but emphasizes higher frequencies in the passband. Details may be found in [2].

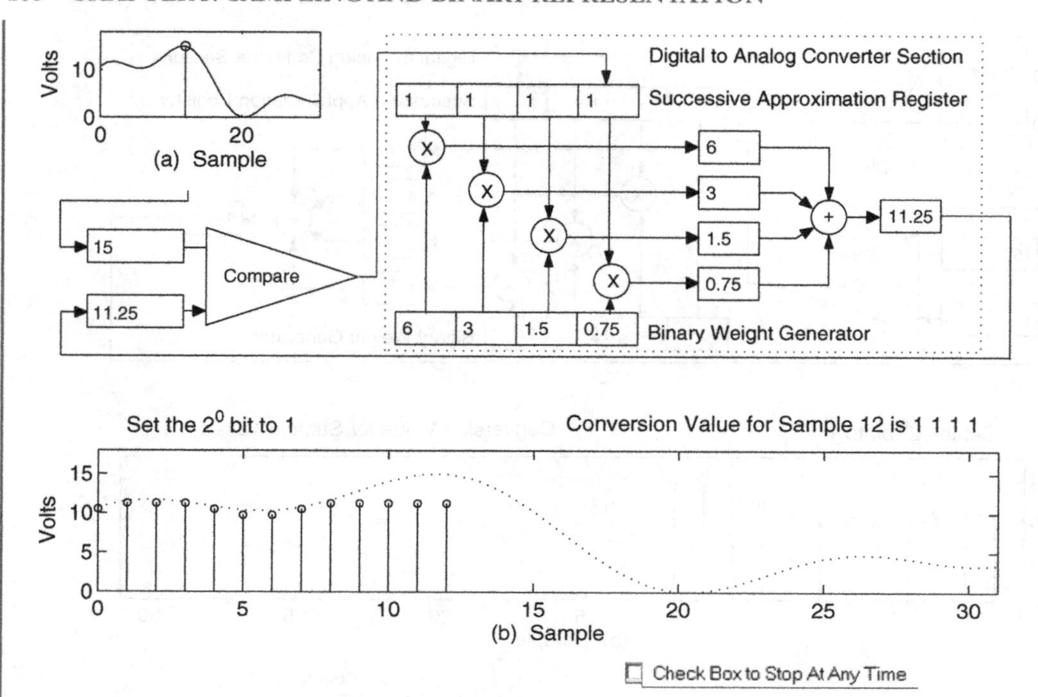

Figure 3.23: A schematic arrangement of an ADC having LSB = 0.75 volt, using the Offset method with no bias to the input signal; (a) Analog signal (solid), sample being quantized (circle); (b) Quantized samples.

3.16 CHANGING SAMPLE RATE

There are many different standard sample rates in use, and from time to time it is necessary to convert one sequence sampled at a first rate to an equivalent sequence sampled at a different rate. It might, for example, be necessary to add two sequences together, one of which was sampled at 11.025 kHz, and the other sampled at 22.05 kHz. To do this, one of the sequences must have its sample rate converted to match that of the other prior to addition.

In this section we look at two important processes, interpolation and decimation by whole number ratios, and the combination of the two techniques, which allows change of sample rate by a whole number ratio, such as 4:3, for example. We include a basic discussion of interpolation using the *sinc* function, and compare it to linear interpolation.

3.16.1 INTERPOLATION

From time to time, it is necessary to either increase the sampling rate of a given sequence, or decrease the frequencies contained in the sequence by a common factor. Let's consider a concrete example.

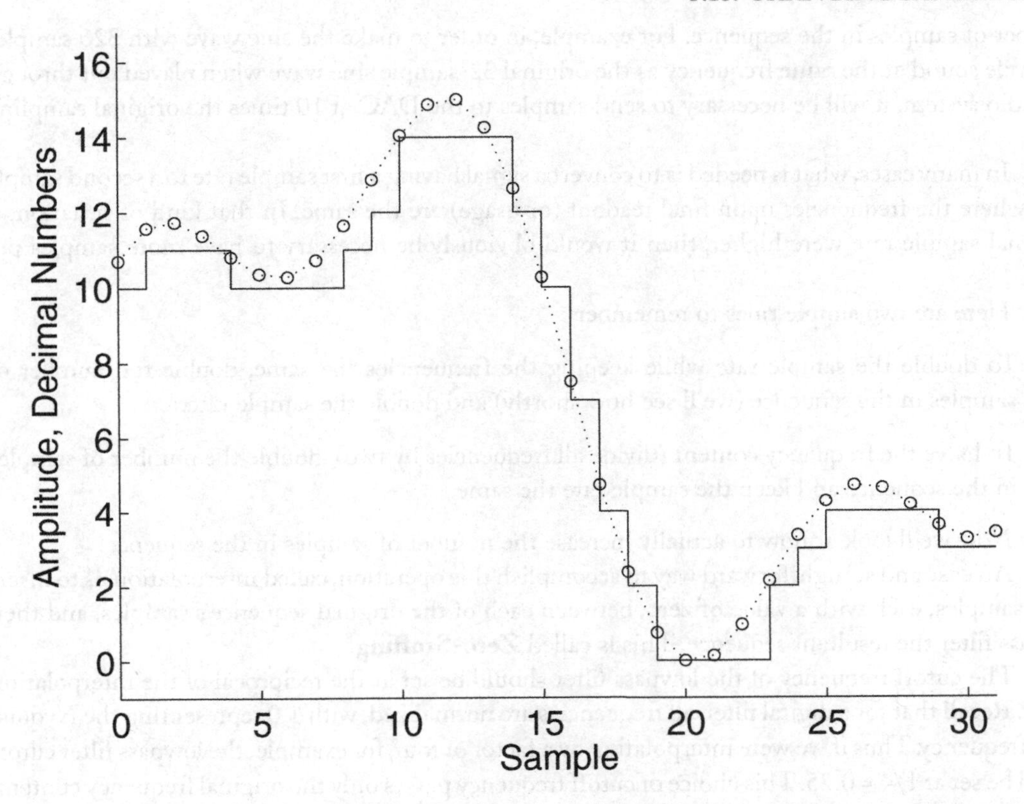

Figure 3.24: Original Analog Signal (Dashed); Ideal (i.e., Quantized to Infinite Precision) Samples (Circles); Quantized Samples at DAC output (Solid).

Supposing one second's worth of samples were taken at a sampling rate of 65,536 Hz of a signal composed of three sine waves having respective frequencies of 3000 Hz, 4000 Hz, and 5000 Hz (any other collection or mixture of frequencies would have been possible provided, of course, that the highest frequency was less than half of 65,536).

Now suppose we would like to divide all of these frequencies by a common factor, such as 10. This would reduce the three frequencies to 300, 400, and 500 Hz.

Thinking intuitively, one cycle of a sine wave occurring over 32 samples (for example), can have its apparent frequency reduced by a factor of 10 (for example) if something were done to make the same one cycle occur over 320 samples instead of 32, assuming that the sample rate remains the same. Thus it would now take 10 times as long to read out that one sine wave cycle, hence its frequency is now one-tenth of the original frequency.

If, on the other hand, we only want to increase the number of samples per cycle while keeping the same frequency, we must increase the sample rate by the same factor with which we increased the number of samples in the sequence. For example, in order to make the sine wave with 320 samples per cycle sound at the same frequency as the original 32-sample sine wave when played out through an audio system, it will be necessary to send samples to the DAC at 10 times the original sampling rate.

In many cases, what is needed is to convert a signal having a first sample rate to a second sample rate, where the frequencies upon final readout (or usage) are the same. In that kind of situation, if the final sample rate were higher, then it would obviously be necessary to have more samples per cycle.

Here are two simple rules to remember:

- To double the sample rate while keeping the frequencies the same, double the number of samples in the sequence (we'll see how shortly) and double the sample rate.
- To halve the frequency content (divide all frequencies by two), double the number of samples in the sequence and keep the sample rate the same.

Now we'll look at how to actually increase the number of samples in the sequence.

An easy and straightforward way to accomplish this operation, called interpolation, is to insert extra samples, each with a value of zero, between each of the original sequence's samples, and then lowpass filter the resultant sequence. This is called **Zero-Stuffing**.

The cutoff frequency of the lowpass filter should be set at the reciprocal of the interpolation factor. Recall that for a digital filter, all frequencies are normalized, with 1.0 representing the Nyquist limit frequency. Thus if we were interpolating by a factor of four, for example, the lowpass filter cutoff would be set at 1/4 = 0.25. This choice of cutoff frequency passes only the original frequency content, as effectively decreased by the interpolation factor.

The script

$$LV\_InterpGeneric(Freq, InterpFac)$$

demonstrates the principles involved.

Example 3.25. Demonstrate interpolation by a factor of 2 on a 4-cycle sine wave over 32 samples.

We call the script as follows:

LV_InterpGeneric(4,2)

This will show four cycles of a sine wave, as shown at (a) in Fig. 3.25, and will double the number of samples per cycle using the zero-stuffing technique. Plot (b) of Fig. 3.25 shows the original sequence with zero-valued samples inserted between each pair of original samples, and Plot (c) of Fig. 3.25 shows the result after lowpass filtering.

Figure 3.26 shows what happens from the frequency-domain point of view: at (a), where we see that the spectrum shows only a single frequency (4 Hz), corresponding to the four cycle sine

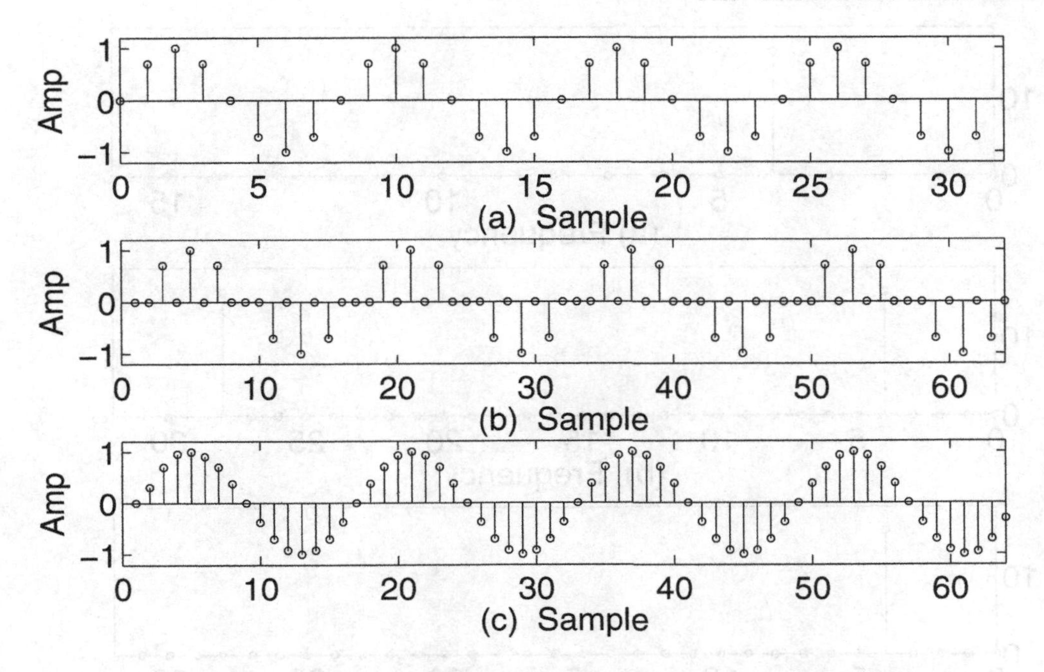

Figure 3.25: (a) Original sequence; (b) Sequence at (a) with one zero-valued sample inserted between each original sample; (c) Sequence at (b) after being lowpass-filtered.

wave shown in plot (a) of Fig. 3.25. After zero-stuffing, another frequency component appears at 28 Hz (plot (b) of Fig. 3.26). Thinking logically, we want only the 4 Hz component in the final sequence, so lowpass filtering to get rid of the 28 Hz should restore the signal to something that looks like a pure sine wave, which in fact it does. Plot (c) of Fig. 3.26 shows the resultant spectrum, which shows a few small amplitude components at frequencies other than four. These components correspond to the distortion which may be seen on the leading edge of the signal shown in plot (c) of Fig. 3.25. This distortion is merely the transient, nonvalid output of the lowpass filter used to eliminate the high frequency (28 Hz in this example) component.

Example 3.26. Demonstrate interpolation of a multi-frequency audio signal. Play the interpolated sequence at the original sample rate to show the frequency decrease, and play it at the new sample rate necessary to maintain the original output frequencies.

The script

$$LV\_InterpAudio(InterpFac)$$

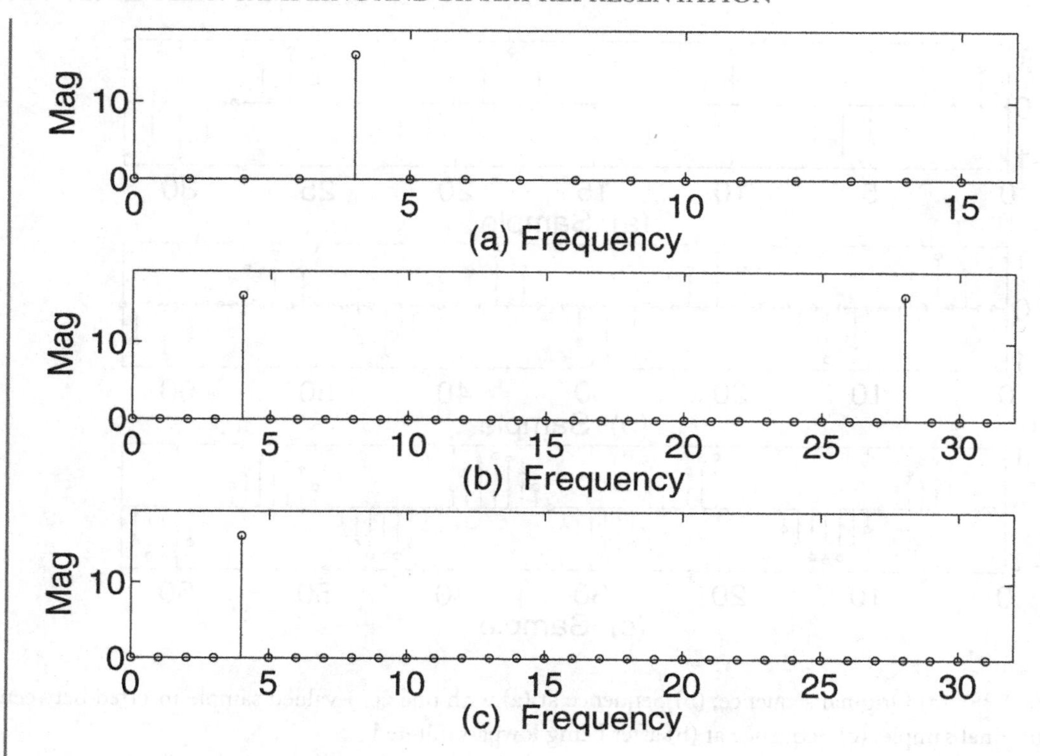

Figure 3.26: (a) Spectra of original sequence as shown in the previous figure; (b) Spectrum of the zero-stuffed sequence as shown in (b) in the previous figure; (c) Spectrum of the lowpass-filtered sequence as shown in (c) of the previous figure.

allows you to interpolate an audio signal which is a mixture of frequencies 400, 800, and 1200 Hz, sampled at 11025 Hz, and play the results. The value of input argument *InterpFac* is limited to 2 and 4.

A typical call might be

LV_InterpAudio(2)

which will stuff one sample valued at zero between each sample of the original test sequence, and then lowpass filter using a normalized cutoff frequency of 1/2.

Figure 3.27, in plot (a), shows the first 225 samples of the original sequence, about 8 cycles of the repetitive, composite signal, of which the lowest frequency is 400 Hz. Plot (b) of Fig. 3.27 shows the first 225 samples of the interpolated sequence, showing about four cycles of the frequency-reduced composite signal (now with the lowest frequency being 200 Hz).

Plot (a) of Fig. 3.28 shows the frequencies in the original sequence as 400, 800, and 1200 Hz; in plot (b), we see that the three frequencies have indeed been reduced by a factor of 2 to 200, 400,

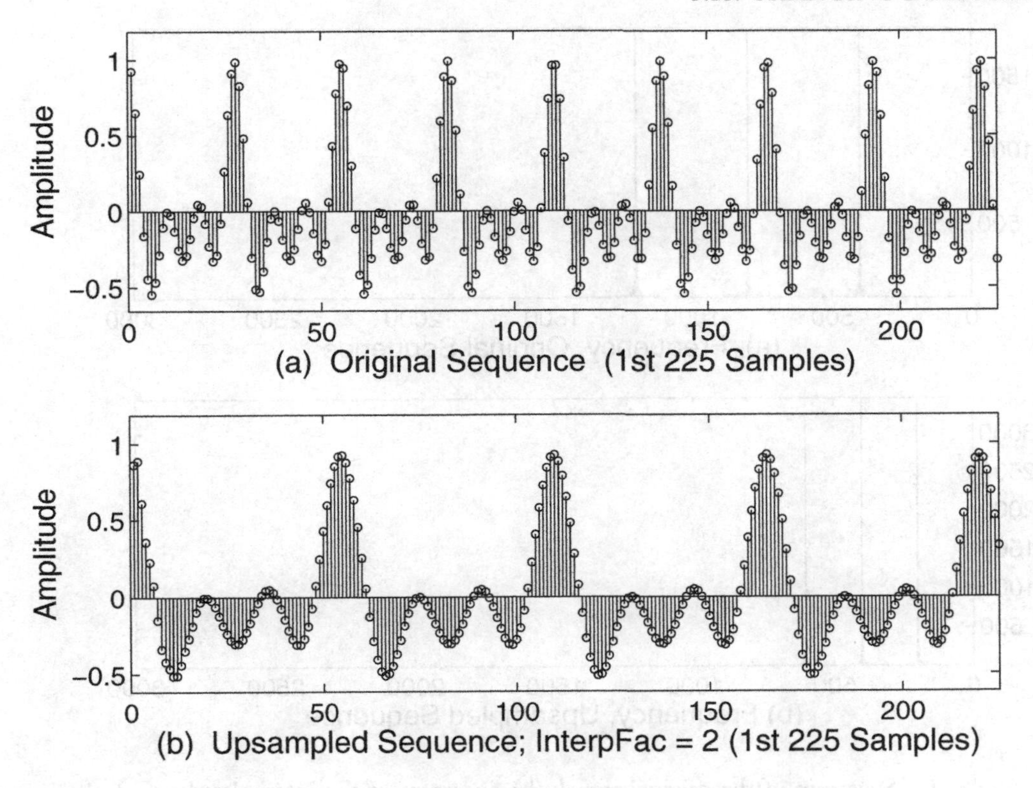

Figure 3.27: (a) A portion of the original sequence; (b) The same number of samples of the interpolated sequence.

and 600 Hz. In general, the frequencies in the spectrum of the upsampled sequence are decreased by the interpolation factor, assuming that the sampling rate remains the same. If the sampling rate were also increased by the same factor as the interpolation factor, then the frequencies would not change; there would merely be a larger number of samples per cycle.

For MATLAB users only, the script

$$ML\_InterpAudioPOC(InterpFac)$$

is very similar to $LV\_InterpAudio$, except that $InterpFac$ may assume the positive integral values 2, 3, 4, etc. Several of the windows that are opened by $ML\_InterpAudioPOC$ have push buttons that allow playing the input sequence at the original sample rate (11025 Hz) and the output sequence at the original or new sample rates.

- To maintain the same apparent output frequency after interpolation, compute the necessary post-interpolation sample rate SR_{New} as follows:

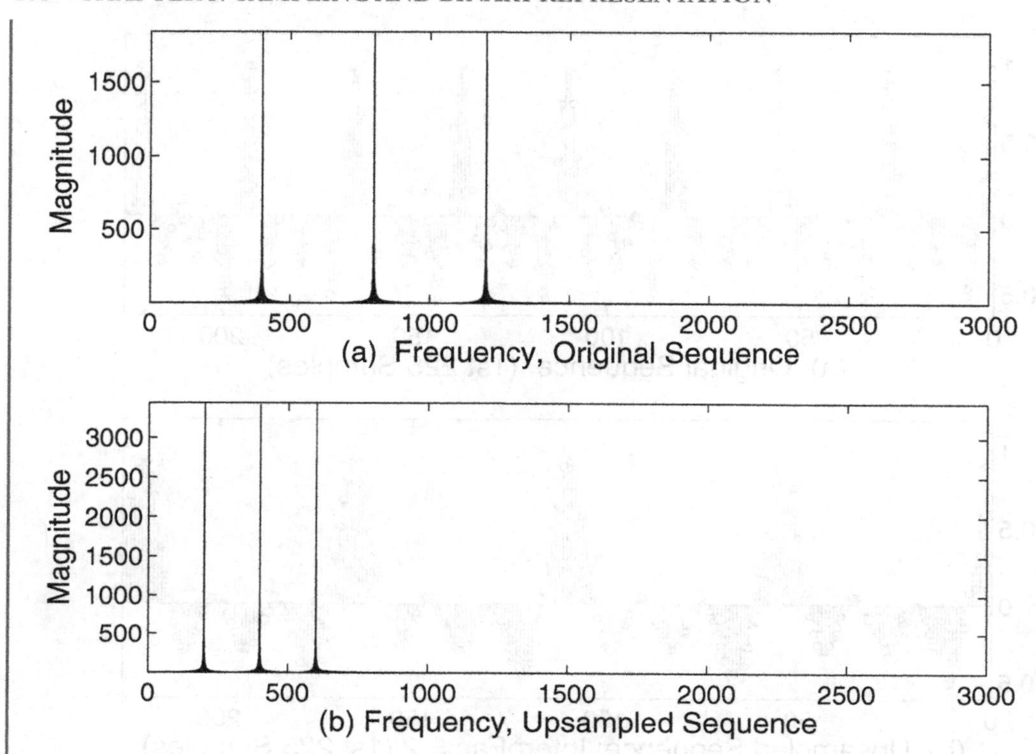

Figure 3.28: (a) Spectrum of the original signal; (b) Spectrum of the interpolated signal, showing a frequency reduction of a factor of two.

$$SR_{New} = SR_{Orig} \cdot I$$

where I is the interpolation factor and the original, pre-interpolation sample rate is SR_{Orig}.

- If the sample rate is not to be changed, then the post-interpolation apparent output frequency F_{New} is computed as follows:

$$F_{New} = \frac{F_{Orig}}{I}$$

where the original frequency is F_{Orig}.

3.16.2 DECIMATION

For the problem of upsampling or interpolation, we inserted extra samples between existing samples, and then lowpass filtered to limit the frequency content to that of the original signal.

For the problem of downsampling or decimation, a new sequence is created consisting of every n-th sample of the original sequence. This procedure decreases the sampling rate of the original signal, and hence lowpass filtering of the original signal must be performed prior to decimating.

Example 3.27. We have 10,000 samples, obtained at a sample rate of 10 kHz. Suppose the highest frequency contained in the signal is 4 kHz, and we wish to downsample by a factor of four. Describe the necessary filtering prior to decimation.

The new sampling rate is 2.5 khz. Thus the frequency content of the original signal must be filtered until the highest remaining frequency does not exceed one-half of 2.5 kHz, or 1.25 kHz. Thus the original signal must be severely lowpass filtered to avoid aliasing.

Example 3.28. Consider a different example, in which the original sequence is 50,000 samples, sampled at 50 kHz, with the highest frequency contained therein being 4 kHz. We wish to decimate by a factor of four. Describe the necessary filtering prior to decimation.

The new sample rate is 12.5 kHz, which is more than twice the highest frequency in the original sequence, so in this case we do not actually have to do any lowpass filtering prior to decimation, although lowpass filtering with the proper cutoff frequency should be conducted as a matter of course in any algorithm designed to cover a variety of situations, some of which might involve the new Nyquist rate falling below the highest signal frequency.

The following script allows you to decimate a test signal having multifrequency content.

$$LV\_DecimateAudio(DecimateFac)$$

The argument *DecimateFac* is the factor by which to decimate, and it may have values of 2 or 4. This is based on the use of a test signal comprising frequencies of 200, 400, and 600, and an original sampling rate of 44,100. The function

$$sound(y, Fs)$$

(where y is a vector of audio samples to be played, and Fs is the sample rate) is used to play the resultant sounds, and, in LabVIEW, it permits only a few sample rates, namely: 44,100 Hz, 22,050 Hz, 11,025 Hz, and 8,000 Hz. The script creates the original signal at a sample rate of 44,100 Hz, and allows you to decimate by factors of 2 or 4, thus reducing the sample rates to 22,050 Hz and 11,025 Hz, respectively. The script automatically sounds the original signal at 44,100 Hz, the decimated signal at 44,100, and then the decimated signal at the new sample rate, which is either 22,050 Hz or 11,025 Hz depending on whether *DecimateFac* was chosen as 2 or 4.

Figure 3.29 shows a result from the script call

LV_DecimateAudio(2)

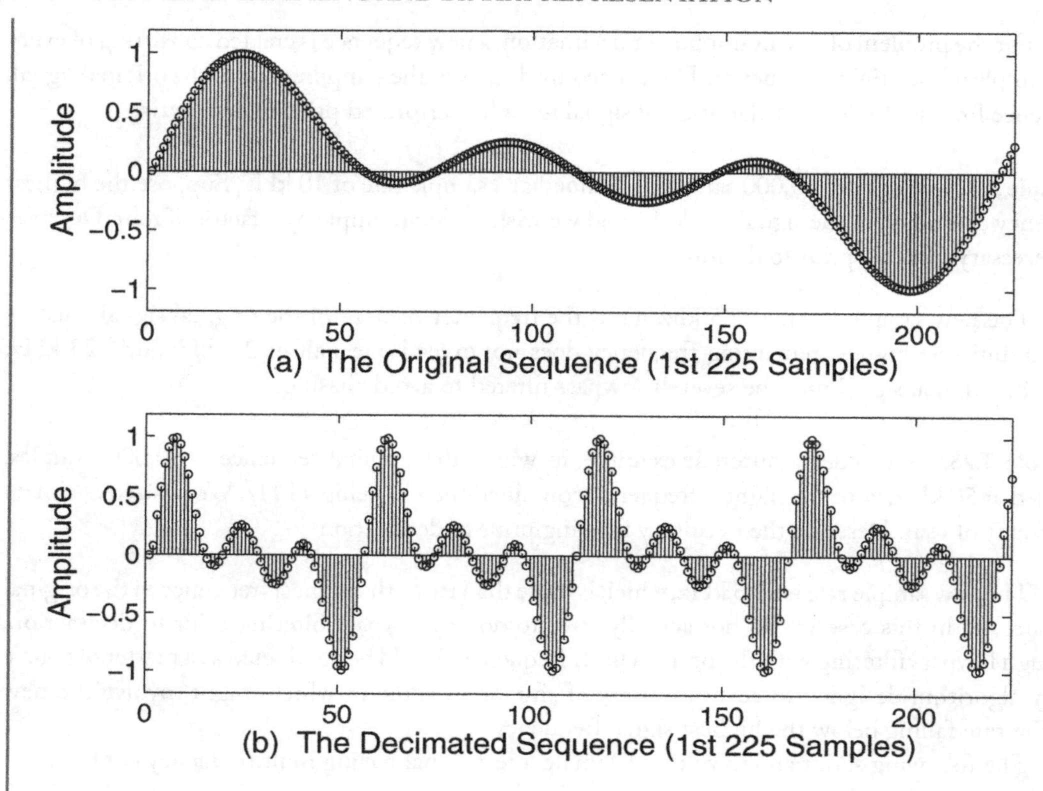

Figure 3.29: (a) Original sequence; (b) Result from taking every 2nd sample of sequence at (a), after suitably lowpass filtering.

which decimates the test audio signal by a factor of 2.

Decimation of a complex waveform can be used to decrease the number of samples per cycle, or increase the output frequency, depending on whether the sample rate decreases proportionately to the decimation factor, or whether it remains at the original rate. In the former case, the apparent output frequency is the same, but with fewer samples per cycle; in the latter case, the apparent output frequency increases by the decimation factor.

For MATLAB users, the script

$$ML\_DecimateAudioPOC(DecimateFac)$$

is similar to $LV\_DecimateAudio$, except that *DecimateFac* may take on the positive integral values 2, 3, 4, etc. Several of the windows opened by the scripts provide push buttons that may be used to play on command the original and decimated sequences at the original sample rate.

- To maintain the same apparent output frequency after decimation, compute the necessary post-decimation sample rate SR_{New} as follows:

$$SR_{New} = \frac{SR_{Orig}}{D}$$

where D is the decimation factor and the original, pre-decimation sample rate is SR_{Orig}.

- If the sample rate is not to be changed, then the post-decimation apparent output frequency F_{New} is computed as follows:

$$F_{New} = F_{Orig} \cdot D$$

3.16.3 COMBINING INTERPOLATION WITH DECIMATION

Based on what we've learned about interpolation and decimation, it should be obvious that we can alter sample rate by (theoretically) any whole number ratio. For example, if the original sample rate is 44,100 Hz, and we happened to want to change it to 33,075 Hz, this could be done by interplating by a factor of 3, and then decimating by a factor of 4. The steps must be done in this order: interpolation first, lowpass filtering, then decimation. In this way, only one lowpass filtering operation is necessary, after interpolation but before decimation. The lowpass filter cutoff is chosen as the more restrictive of the cutoff frequencies of the two lowpass filters that would have been required by the interpolation and decimation operations had they been performed by themselves.

The script

$$LV\_ChangeSampRateByRatio(OrigSR, InterpFac, DecimateFac)$$

allows for experimentation in changing sample rate using this method. If the value of *OrigSR* is one of the standard sampling rates, namely, 8 kHz, 11.025 kHz, 22.05 kHz, or 44.1 kHz, the original signal will be played as an audio signal. Likewise, if the final sample rate, computed according to the input arguments, is within 1 % of one of the standard sampling rates, the final or resampled signal will also be played out as an audio signal using the function *sound* (see the script for details). A typical call would be

LV_ChangeSampRateByRatio(8000,22,8)

which interpolates by a factor of 22, lowpass filters (with a normalized cutoff frequency equal to the reciprocal of the greater of the interpolation factor and the decimation factor, which in this case is 22), and then decimates by a factor of eight. Figure 3.30 shows leading portions of the input and output waveforms. The input waveform is a test signal consisting of three sinusoids have frequencies of 400, 800, and 1200 over a sequence length of 8000.

For MATLAB users, the script

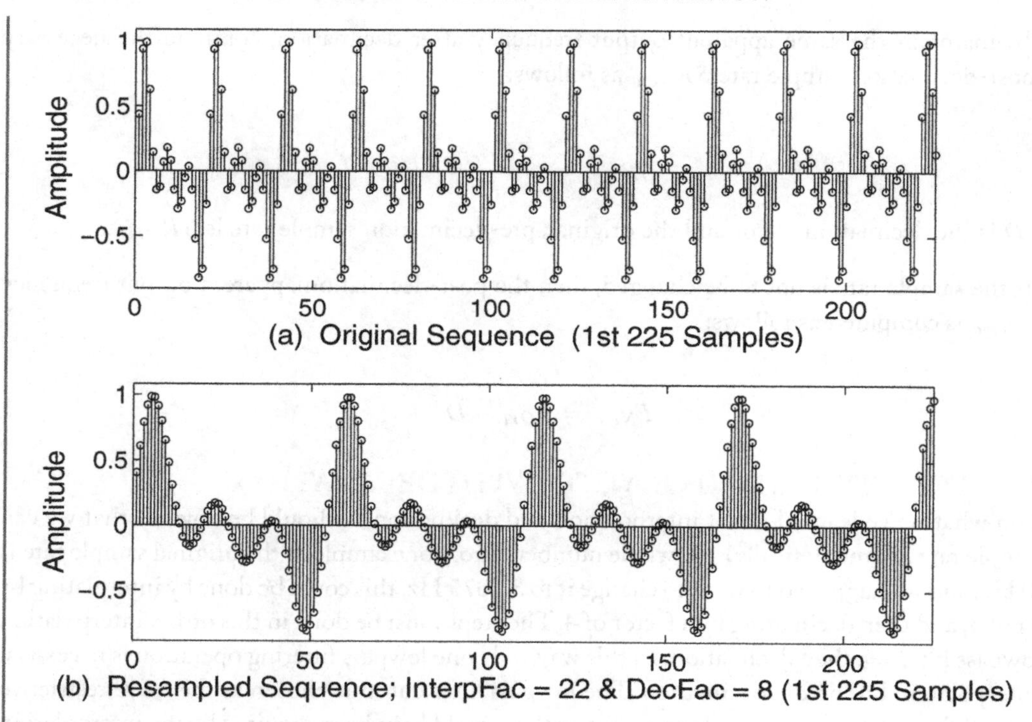

Figure 3.30: (a) Original sequence; (b) Result from interpolating by a factor of 22 and decimating by a factor of eight.

$$ML\_ChangeSampRateByRatioPOC(InterpFac, DecimateFac)$$

is similar to $LV\_ChangeSampRateByRatio$, except that the original sample rate is set at 44,100 Hz, and the final sample rate, which is (InterpFac/DecimateFac), may be other than a standard sampling rate (as mentioned above).

MathScript has sophisticated functions for interpolation, decimation, and the combination of the two, called resampling. The functions are *interp*, *decimate*, and *resample*. These functions take great pains to give the most accurate result possible, especially at the first and last samples of an input sequence. In using any of these functions, for example, you'll find that the output sequence is temporally aligned as best as possible with the input; that is to say, the waveform phase of the output sequence matches the waveform phase of the input sequence as closely as possible. Also, the lowpass filter length is made variable according to the input parameters.

3.16.4 BANDLIMITED INTERPOLATION USING THE SINC FUNCTION

According to the Shannon sampling theorem, a bandlimited continuous domain signal can be exactly reconstructed from its samples provided the sampling rate is at least twice the highest frequency contained in the bandlimited signal. Then

$$x(t) = \sum_{n=-\infty}^{\infty} x(nT_S)\frac{\sin((\pi/T_S)(t-nT_S))}{(\pi/T_S)(t-nT_S)}$$

Here the sinc functions are continuous functions ranging over all time, and exact reconstruction depends on all the samples, not just a few near a given sample. In practical terms, however, it is possible to reconstruct a densely-sampled version of the underlying bandlimited continuous domain signal using discrete, densely-sampled windowed sinc functions that are weighted by a finite number of signal samples and added in an offset manner. Such a reconstruction provides many interpolated samples between the signal samples, and thus is useful in problems involving change of sample rate.

The script (see exercises below)

LVxInterpolationViaSinc(N,SampDecRate,valTestSig)

creates a densely sampled waveform (a simulation of a continuous-domain signal) of length *N*, then decimates it by the factor *SampDecRate* to create a sample sequence from which the original densely sampled waveform is to be regenerated using sinc interpolation. Two methods are demonstrated.

One method is to create a sinc function of unit amplitude having a period of M samples and an overall length of perhaps 5^*M or more samples. The sinc function is then windowed. A sample sequence to be used to reconstruct a densely sampled version of its underlying bandlimited continuous domain signal has one-half the sinc period, less one, zeros inserted between each sample, and the zero-padded sequence is then linearly convolved with the sinc function. This is essentially the zero-stuffing method using a sinc filter as the lowpass filter. This method is no more efficient than linear convolution. It computes many values between samples, when, in most applications, only one or a few interpolated values are needed between any two samples.

Another method, which allows you to obtain a single interpolated value between two samples of a sequence is matrix convolution. A matrix is formed by placing a number of copies of a densely-sampled windowed sinc function as columns in a matrix, each sinc function offset from the previous one by $M/2$ samples (it is necessary to let M be even). A particular row of the matrix is selected (according to exactly where between two existing samples another sample value is needed) and then multiplied by the sample sequence (as a column vector) to give an interpolated value.

Figure 3.31, which results from the call

LVxInterpolationViaSinc(1000,100,1)

illustrates the matrix convolution method. For this example, which uses 10 contiguous samples from a sequence to perform a bandlimited, densely-sampled reconstruction of the underlying continuous domain signal, there are 10 matrix columns containing sinc functions, offset from each other.

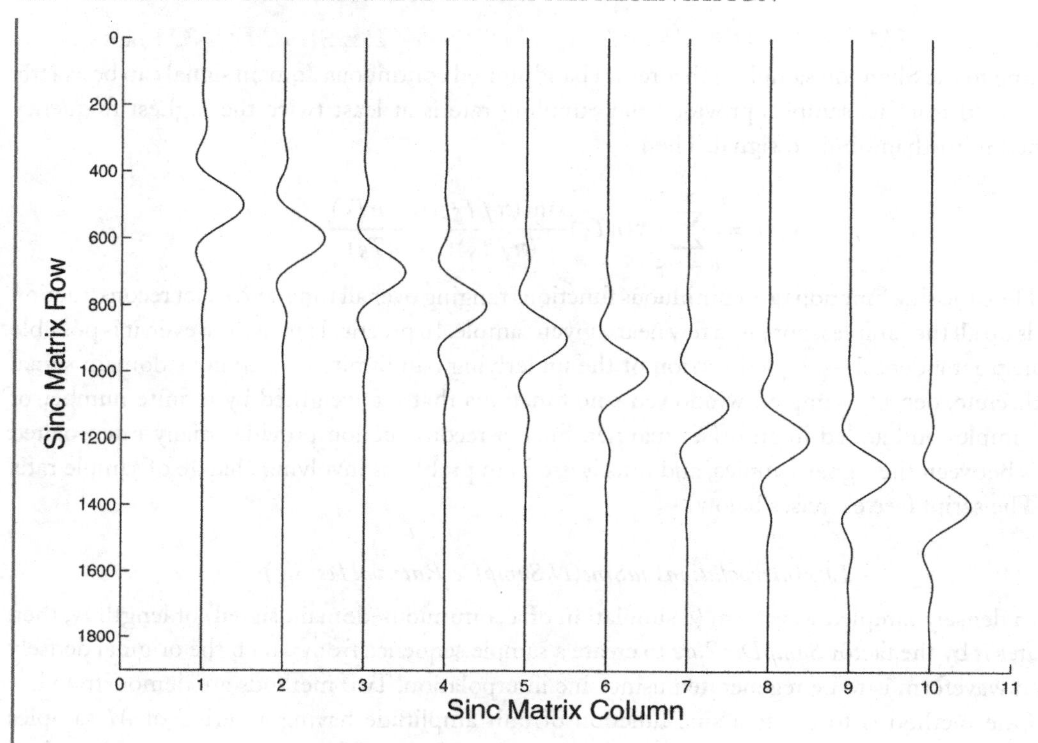

Figure 3.31: Sinc functions, offset from one another by 100 samples, depicted in 10 columns of a matrix suitable for bandlimited interpolation between samples of a sequence. Note that for display purposes only, all sinc amplitudes have been scaled by 0.5 to avoid overlap with adjacent plots. They are not so scaled in the actual matrix in the script.

Figure 3.32 shows the matrix columns after being weighted by the appropriate samples, and at column zero, the sum of all the weighted columns is shown, which forms the reconstruction.

Figure 3.33 shows more detail of the reconstruction: the simulated continuous-domain signal (solid line), 10 samples extracted therefrom by decimating the original signal by a factor of 100 (stem plot), and the reconstruction (dashed line) of the original signal according to the matrix convolution method described above.

The results of the linear convolution method are shown in Fig. 3.34. Note again that the best approximation to the original signal is found in the middle of the reconstruction.

Reference [4] discusses ideal sinc interpolation in Section 4.2, and covers sampling rate conversion in Chapter 10.

Example 3.29. Devise a simple (not necessarily efficient) method to convert a sequence sampled at 44,100 Hz to an equivalent one sampled at 48,000 Hz.

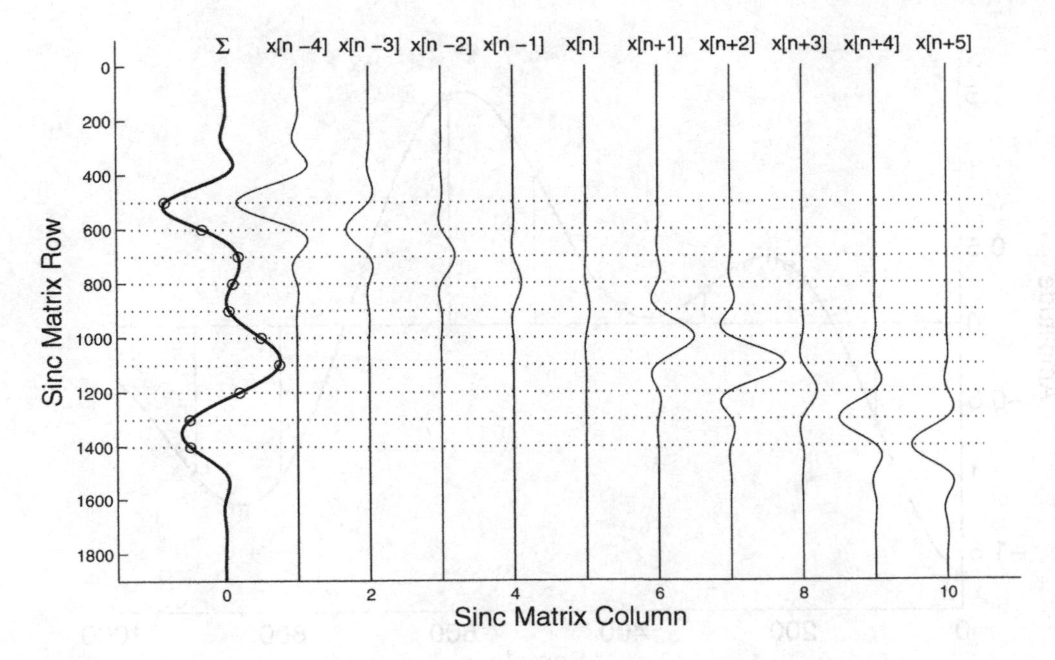

Figure 3.32: Sinc matrix columns, each weighted by the sample shown at the top of the column, and the sum of all columns (i.e., the interpolated reconstruction), at the left with the symbol Σ at the column head. The sample values corresponding to samples x[n-4] to x[n+5] are plotted as circles on the reconstruction, with x[n-4] being topmost. The dashed horizontal lines intersect the 10 matrix columns at the values that are summed to equal the corresponding interpolated value marked with a circle; the particular rows summed correspond to the sample values x[n-4] up to x[n+5]. Any row may be summed (as a single operation) to obtain an interpolated value between the original samples. Note that the smallest reconstruction errors occur in the middle of the reconstruction. Note also that all signal amplitudes have been scaled by 0.5 to avoid overlap with adjacent plots, but are not so scaled in the computation.

The sample rate ratio can be approximated as 160:147. Assuming that a sequence *Sig* sampled at 44.1 kHz has indices 1,2,3, ..., the indices of samples, had the sample rate been 48 kHz instead of 44.1 kHz, would be

1:147/160:length(Sig) = 1, 1.9188, 2.8375, 3.7562...

This generates a total number of new samples of 160/147*length(Sig), which, when read out at the new sample rate of 48 kHz, will produce the same apparent frequencies as the original sequence read out at 44.1 kHz. To obtain the sample values, sinc interpolation as discussed above can be used. The more densely sampled the sinc function, and the larger the number of original sequence samples included in the approximation, the more accurate will be the interpolated values.

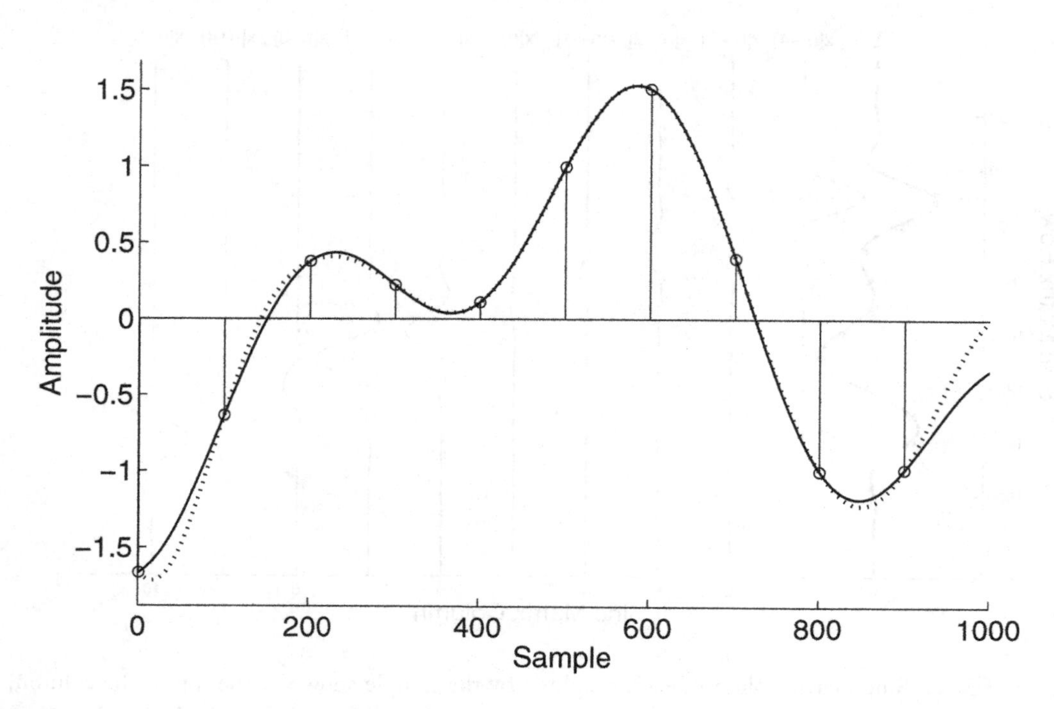

Figure 3.33: Continuous-domain signal (solid), simulated with 1000 samples; 10 samples of the (simulated) continuous-domain signal (stem plot); a reconstruction (dashed line) of the continuous-domain signal from the 10 samples. Note that the reconstruction is only a good approximation from about samples 400-700 of the original (simulated) continuous-domain signal.

The script (see exercises below)

$$LVxInterp8Kto11025(testSigType, tsFreq)$$

performs both linear and sinc interpolation of an audio signal to change its sample rate from 8 kHz to 11.025 kHz. The audio signal may be a cosine having a specified frequency, or the file *'drwatsonSR8K.wav'*.

Figures 3.35 and 3.36, respectively, show several samples from the interpolation of cosines of 250 Hz and 2450 Hz, respectively, obtained by making the calls

LVxInterp8Kto11025(0,250)
LVxInterp8Kto11025(0,2450)

Note that at the much lower frequency (i.e., low relative to the Nyquist rate), the sinc- and linear-interpolated samples do not differ much, whereas at the higher frequency, there are large discrepancies.

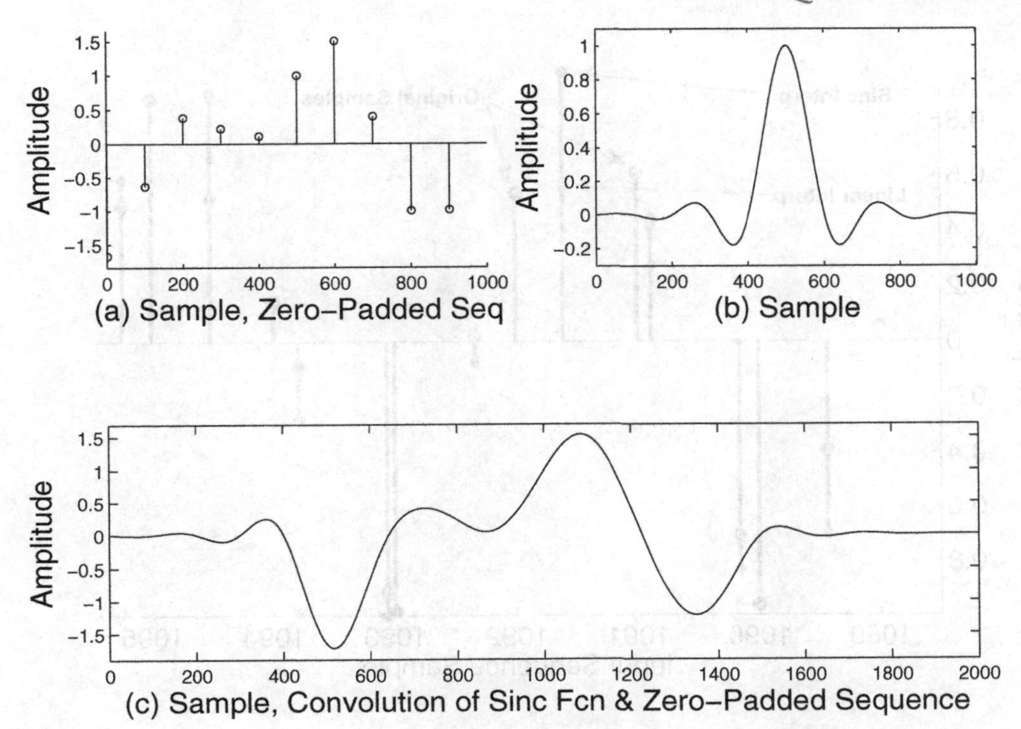

Figure 3.34: (a) The sample sequence, padded with 99 zeros between each sample; (b) The *sinc* function; (c) Convolution of waveforms at (a) and (b).

3.16.5 EFFICIENT METHODS FOR CHANGING SAMPLE RATE

More advanced (i.e., efficient) methods of interpolation exist which make interpolation and decimation by ratios of large numbers (such as 160:147) feasible for real time processing. References [7]-[9] discuss efficient sample rate conversion methods.

3.17 FREQUENCY GENERATION

Sinusoids of arbitrary frequency for use in music or the like can be generated by storing one or more cycles of a sinusoid in a memory and cyclically reading out samples. The effective frequency can be varied using decimation by, in general, nonintegral decimation factors. Linear interpolation between samples is often adequate, so the more costly sinc interpolation referred to above need not be used.

3.17.1 VARIABLE SR

A first method of readout is to sequentially and cyclically read out every sample in the ROM at a frequency (sample rate) which is a multiple of the number of samples in the ROM. This method

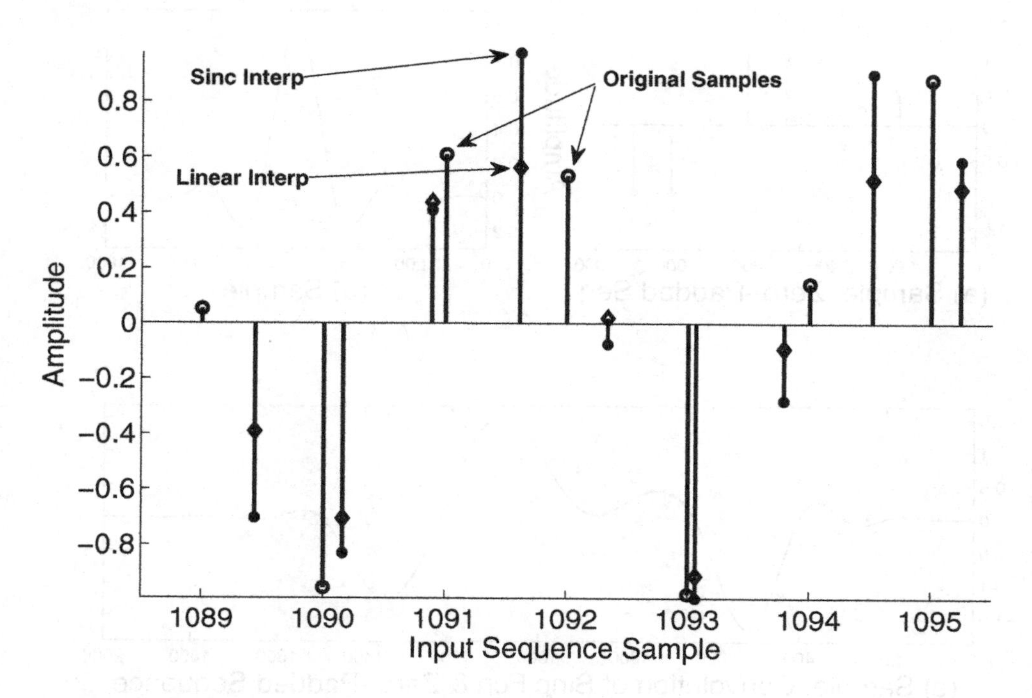

Figure 3.35: A plot of several samples from an audio sequence sampled at 8 kHz, with interpolated values at the new sample rate of 11025 Hz. Interpolated values computed using linear interpolation are plotted as stems with diamond heads, and those computed using sinc interpolation are plotted as stems with stars. The original samples, plotted at integral-valued index values, are plotted as stems headed by circles. Note that in some cases the sinc-interpolated value differs remarkably from the linear-interpolated value. The test signal consisted of a 2450 Hz cosine wave. The interpolated samples are shown prior to post-interpolation lowpass filtering.

is used principally in electronic musical instruments, where hardware or software is available to generate all the different sample rates needed. In this method, each sample of the ROM is read out in its turn, one after another. For example, if the ROM contains one cycle of a sine over 32 samples, and we want a 1 kHz output, we must read the ROM at 32 kHz. This is very straightforward. If we happened to have 2 cycles of a sinusoid in the ROM, over 32 samples, we would only need to have a sample readout rate of 16 kHz.

3.17.2 CONSTANT SR, INTEGRAL DECIMATION

Another method of readout from a ROM containing a sinusoid is to maintain a fixed sample rate SR, and to read out every n-th sample.

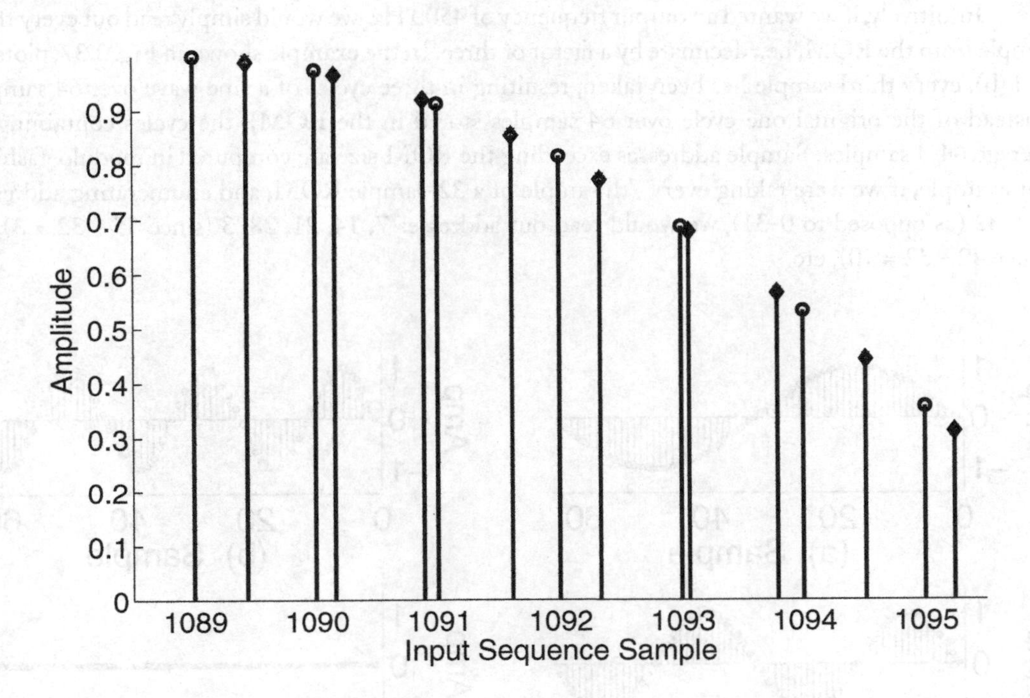

Figure 3.36: Original samples (at 8 kHz sample rate) and interpolated values (at 11025 Hz) of a 250 Hz cosine, same marker system as for the previous figure. Note that at this lower frequency there are many more samples per cycle of the signal, and the difference between linear and sinc interpolation is much less. The interpolated samples are shown prior to post-interpolation lowpass filtering.

A formula expressing the net output frequency of a sinusoid produced by reading each address sequentially from a ROM containing a sinusoid would be

$$F_{Out} = \frac{SR}{(N\ /\ F_{ROM})} = \frac{SR \cdot F_{ROM}}{N}$$

where SR is the rate at which samples are read from the ROM, N is the total number of samples in the ROM and F_{ROM} is the number of cycles of the sinusoid contained in the N samples of the ROM. Most usually, $F_{ROM} = 1$, but that need not be the case.

Consider the case of a 32-sample ROM containing one cycle of a sine wave. The ROM is sequentially read out, one address at a time, at a rate of 48 kHz, yielding a net output frequency of

$$F_{Out} = \frac{48000 \cdot 1}{32} = 1500 \text{ Hz}$$

Intuitively, if we wanted an output frequency of 4500 Hz, we would simply read out every third sample from the ROM, i.e., decimate by a factor of three. In the example shown in Fig. 3.37, plots (a) and (b), every third sample has been taken, resulting in three cycles of a sine wave over 64 samples (instead of the original one cycle over 64 samples stored in the ROM), the cycles containing an average 64/3 samples. Sample addresses exceeding the ROM size are computed in modulo-fashion. For example, if we were taking every 7th sample of a 32-sample ROM, and enumerating addresses as 1-32 (as opposed to 0-31), we would read out addresses 7, 14, 21, 28, 3 (since 35 - 32 = 3), 10 (since 42 - 32 = 10), etc.

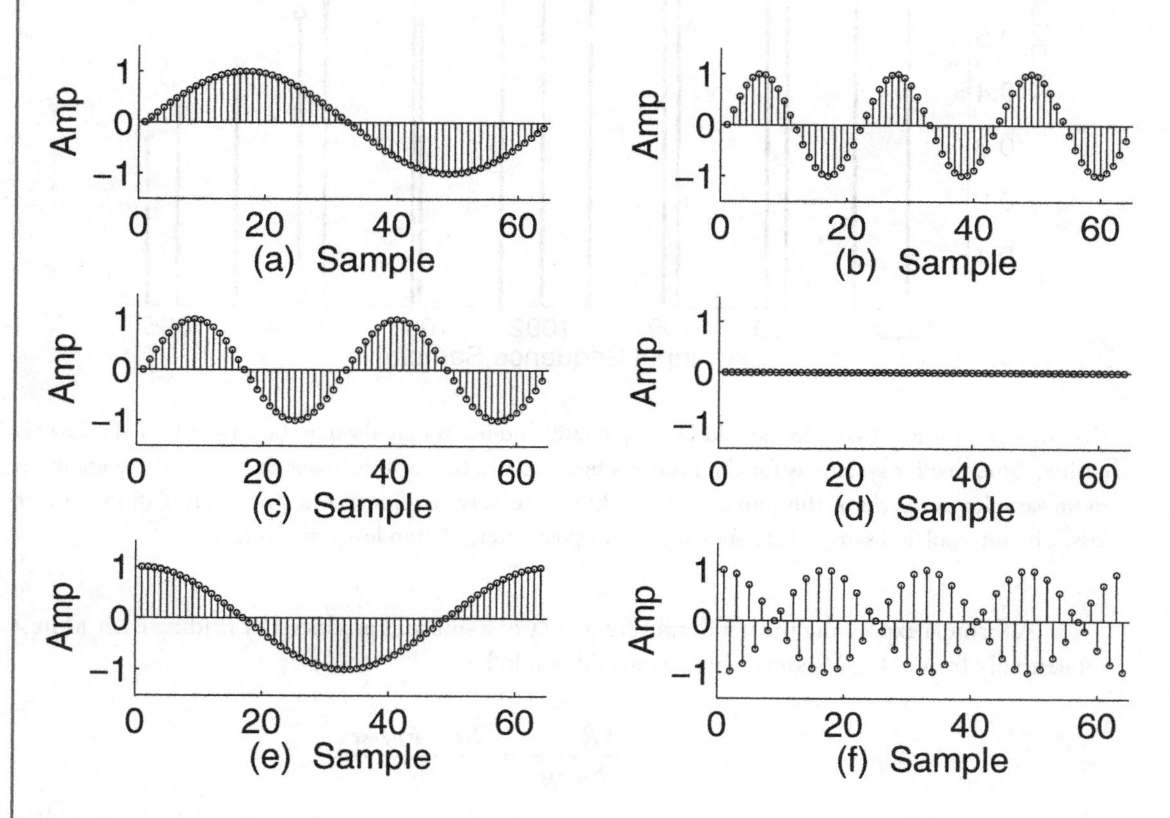

Figure 3.37: (a) A 64-sample sine wave; (b) The resultant waveform from reading out every third sample, i.e., sample indices 1, 4, 7, etc; (c) Original 64-sample sequence; (d) Decimated by a factor of 32 (only phases 0 and 180 degrees of the sine are read); (e) Original 64-sample sequence; (f) Result from taking every 34th sample.

From the above, we can therefore state a formula for the net output frequency when applying a decimation factor D.

$$F_{Out} = \frac{SR \cdot D \cdot F_{ROM}}{N} \tag{3.4}$$

The script

LVSineROReadout(N,F_ROM,D)

applies Eq. (3.4) to illustrate the principle of reading a sinusoid from a ROM using decimation. The call

LVSineROMReadout(32,1,3)

for example, results in Fig. 3.38.

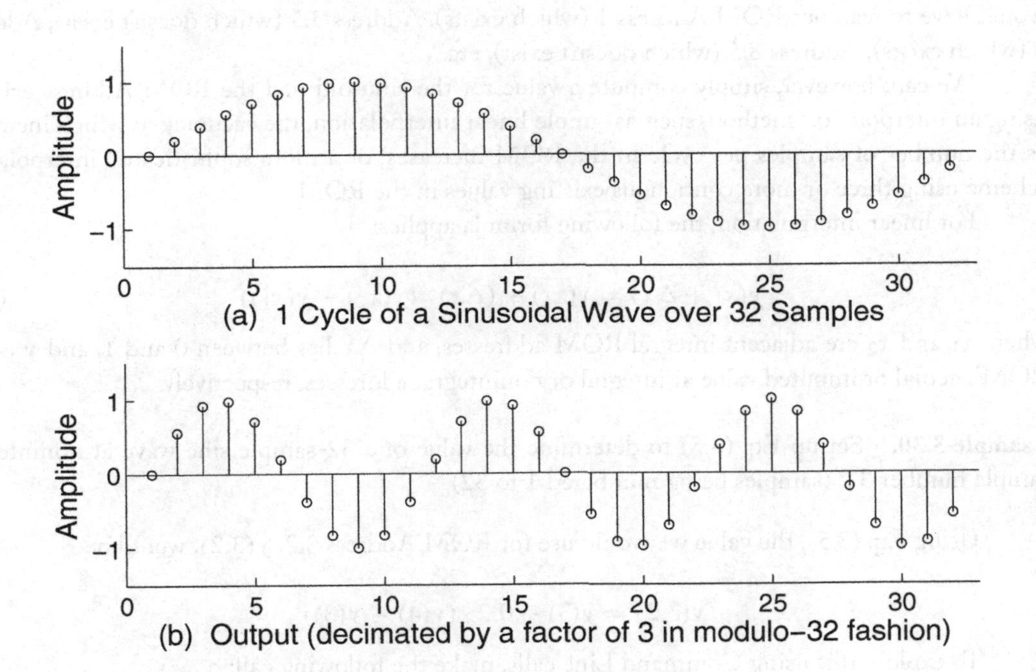

Figure 3.38: (a) One cycle sine wave over 32 samples; (b) Waveform at (a), decimated by a factor of 3.

As we saw in an earlier chapter, when sampling a sinusoid at exactly its Nyquist limit (2 samples per cycle), there can be trouble when the phase is not known or controlled. You'll be well-reminded of the dangers of sampling exactly at the Nyquist limit by looking at Fig. 3.37, plots (c) and (d), which depict the result from taking every 32nd sample from two cycles of a sine wave.

Now let's decimate such that the final sample rate is less than twice the highest frequency in the ROM. If we were to decimate by a factor of 34, the number of samples per cycle would be reduced to 64/34 = 1.882 which of course implies that aliasing will occur. The result from decimating a one cycle, 64-sample cosine by a factor of 34 is shown in Fig. 3.37, plots (e) and (f).

3.17.3 CONSTANT SR, NON-INTEGRAL DECIMATION

By removing the restriction that D be an integer in Eq. (3.4), it is possible to achieve any output frequency within the allowed range of output frequencies.

For example, let D equal a noninteger such as 2.5. Then we would have

$$F_{Out} = \frac{32000 \cdot 2.5 \cdot 1}{32} = 2500 \text{ Hz}$$

The problem with nonintegral decimation is that there are no ROM addresses other than the integers from 1 to N, the length of the ROM. For a decimation factor of 2.5, for example, we would have to read out ROM Address 1 (which exists), Address 3.5 (which doesn't exist), Address 6 (which exists), Address 8.5 (which doesn't exist), etc.

We can, however, simply compute a value for the sinusoid had the ROM Address existed, using an interpolation method, such as simple linear interpolation, the accuracy of which increases as the number of samples per cycle in the ROM increases, or a more sophisticated interpolation scheme using three or more contiguous existing values in the ROM.

For linear interpolation, the following formula applies:

$$y(x_1 + \Delta x) = y(x_1) + (\Delta x) \cdot (y(x_2) - y(x_1)) \tag{3.5}$$

where x_1 and x_2 are adjacent integral ROM addresses, and Δx lies between 0 and 1, and y is the ROM's actual or imputed value at integral or nonintegral addresses, respectively.

Example 3.30. Set up Eq. (3.5) to determine the value of a 32-sample sine wave at nonintegral sample number 3.2 (samples being numbered 1 to 32).

Using Eq. (3.5), the value we would use for ROM Address 3.2, $y(3.2)$, would be

$$y(3.2) = y(3) + 0.2 \cdot (y(4) - y(3))$$

To explore this using Command Line calls, make the following call:

y = sin(2*pi*(0:1:31) / 32); y3pt2 = y(3) + (0.2)*(y(4) - y(3))

The above call yields *y3pt2* = 0.4173. To see what the sine value would be using MathScript, make this call:

alty3pt2 = sin(2*pi*2.2 / 32)

which yields *alty3pt2* = 0.4187 (note that the third sample of y is for $t = 2/32$, since the time vector starts at 0/32, not 1/32).

For a closer approximation using linear interpolation for a sine, start with a longer sine. For example, computing sample 3.2 for a length 512 sine make the following calls:

y = sin(2*pi*(0:1:511) / 512); y3pt2 = y(3) + (0.2)*(y(4)-y(3))

and

alty3pt2 = sin(2*pi*2.2 / 511)

which yield *y3pt2* = 0.02699 and *alty3pt2* = 0.02704, which differ by about 5.3 parts in one hundred thousand. The error varies according to where in the cycle of the sinusoid you are interpolating.

The script (see exercises below)

$$LVxSineROMNonIntDecInterp(N, F\_ROM, D)$$

affords experimentation with nonintegral decimation of a sinusoid. For this script, D may have a nonintegral value. A typical call, which results in Fig. 3.39, would be

LVxSineROMNonIntDecInterp(32,1,2.5)

In plot (a) of Fig. 3.39, the known (ROM) sample values are plotted as stems with small circles, and the values computed by linear interpolation for a decimation factor of 2.5 (sample index numbers 1, 3.5, 6, 8.5, etc.) are plotted as stems with diamond heads. In the lower plot, the computed values marked with diamonds in the upper plot are extracted and plotted to show the decimated output sequence (the computation shown in the figure was stopped before the computed value of the address exceeded the ROM length in order to keep the displayed mixture of existing ROM samples and interpolated samples from becoming confusing).

The decimation factor may also be less than 1.0, which leads to a frequency reduction rather than a frequency increase. Figure 3.40 shows the first 32 samples of the decimated output sequence when the decimation factor is set at 0.5.

3.18 COMPRESSION

An interesting and useful way to reduce the apparent quantization noise when using a small number of bits (such as eight or fewer) is to use A-Law or μ-Law compression. These two schemes, which are very similar, compress an audio signal in analog form prior to digitization.

In an analog system, where μ-Law compression began, small signal amplitudes are boosted according to a logarithmic formula. The signal is then transmitted through a noisy channel. Since the (originally) lower level audio signals, during transmission, have a relatively high level, their signal to noise ratio emerging from the channel at the receiving end is much better than without compression.

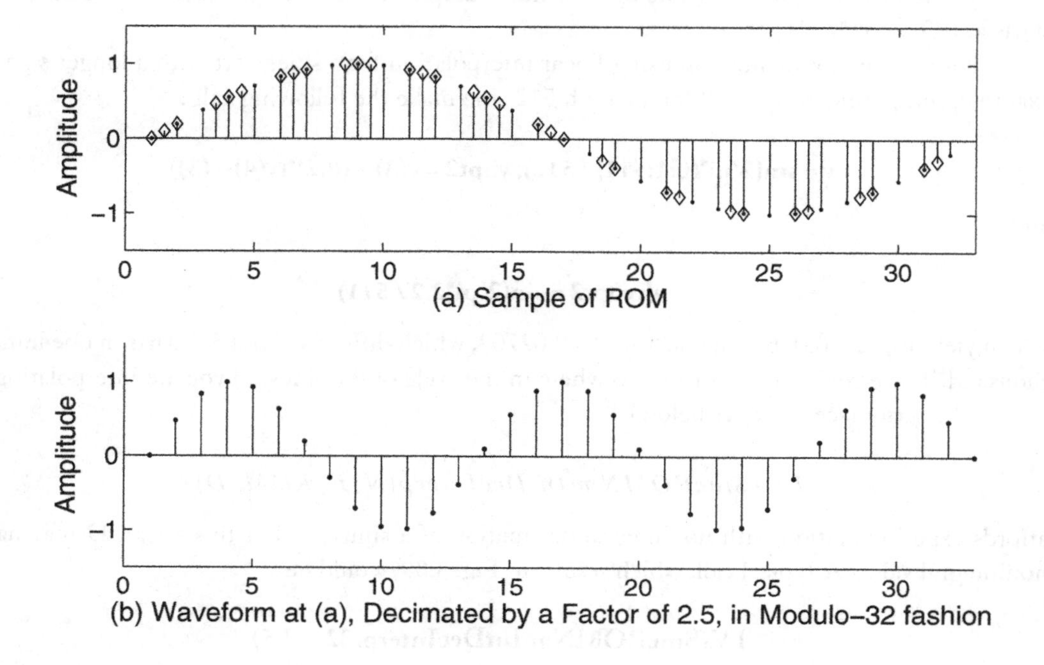

Figure 3.39: (a) One cycle of a 32-sample sinusoid, interpolated values (every 2.5 th sample) marked with diamonds; (b) Net decimated output sequence, formed by extracting the interpolated samples (marked with diamonds) from the upper plot.

The expansion process adjusts the gain of the signal inversely to the compression characteristic, but the improved signal-to-noise ratio remains after expansion.

Using a sampled signal sequence, the signal samples have their amplitudes adjusted (prior to being quantized) according to the following formula:

$$F(s[n]) = S_{MAX}(\log(1+\mu\frac{s[n]}{S_{MAX}})/\log(1+\mu))sgn(s[n]) \tag{3.6}$$

where S_{MAX} is the largest magnitude in the input signal sequence $s[n]$, sgn is the sign function, $F(s[n])$ is the compressed output sequence, and μ is a parameter which is typically chosen as 255.

In compression, smaller signal values are boosted significantly so that there are far fewer small values hovering near zero amplitude. This is because with a small number of bits, small amplitudes suffer considerably more than larger amplitudes. Pyschoacoustically speaking, larger amplitude sounds mask noise better than smaller amplitude sounds. A large step-size (and the attendant quantization noise) at low signal levels is much more audible than the same step-size employed at high amplitude levels. Additionally, small amplitudes which would be adequately encoded using a large

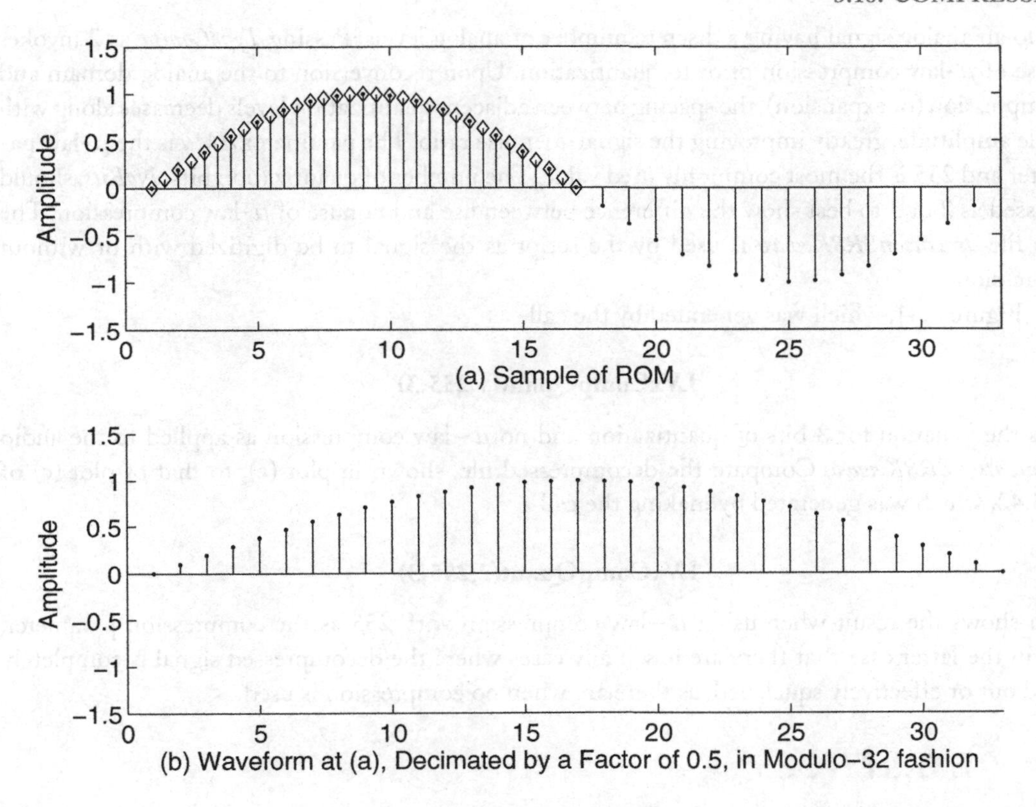

Figure 3.40: (a) One cycle of a 32-sample sinusoid, interpolated values (every half sample) marked with diamonds; (b) Net decimated output sequence, formed by extracting the interpolated samples (marked with diamonds) from the upper plot.

number of bits may actually be encoded as zero when using a small number of bits, leading to a squelch effect which can be disconcerting.

In μ-Law compression, the interesting result is that smaller signal amplitudes wind up being quantized (after reconstruction) with smaller amplitude differences between adjacent quantized levels than do larger signal amplitudes. As a result, when the signal amplitude is low, accompanying quantization noise is also less, and thus is better masked.

The script (see exercises below)

LVxCompQuant(TypeCompr,AMu,NoBits)

affords experimentation with μ-law compression as applied or not applied to an audio signal quantized to a specified number of bits. Passing *TypeCompr* as 1 results in no μ-law compression being applied, the audio signal is simply quantized to the specified number of bits and then converted back to an analog signal having a discrete number of analog levels. Passing *TypeCompr* as 2 invokes the use of μ-law compression prior to quantization. Upon reconversion to the analog domain and decompression (or expansion), the spacing between adjacent quantization levels decreases along with sample amplitude, greatly improving the signal-to-noise ratio. The parameter *AMu* is the μ-law parameter and 255 is the most commonly used value. The number of quantization bits, *NoBits*, should be passed as 2 or 3 to best show the difference between use and nonuse of μ-law compression. The audio file *drwatsonSR8K.wav* is used by the script as the signal to be digitized with or without compression.

Figure 3.41, which was generated by the call

LVxCompQuant(1,255,3)

shows the situation for 3 bits of quantization and no $\mu-$law compression as applied to the audio file *drwatsonSR8K.wav*. Compare the decompressed file, shown in plot (c), to that of plot (c) of Fig. 3.42, which was generated by making the call

LVxCompQuant(2,255,3)

which shows the result when using $\mu-$law compression with 255 as the compression parameter. Note in the latter case that there are few if any cases where the decompressed signal is completely zeroed out or effectively squelched, as there are when no compression is used.

3.19 REFERENCES

[1] James H. McClellan, Ronald W. Schaefer, and Mark A. Yoder, *Signal Processing First*, Pearson Prentice Hall, Upper Saddle River, New Jersey, 2003.

[2] Alan V. Oppenheim and Ronald W. Schaefer, *Discrete-Time Signal Processing*, Prentice-Hall, Englewood Cliffs, New Jersey, 1989.

[3] John G. Proakis and Dimitris G. Manolakis, *Digital Signal Processing, Principles, Algorithms, and Applications, Third Edition*, Prentice Hall, Upper Saddle River, New Jersey, 1996.

[4] Ken C. Pohlmann, *Principles of Digital Audio*, Second Edition, SAMS, Carmel, Indiana.

[5] John Watkinson, *The Art of Digital Audio, First Edition*, (Revised Reprint, 1991), Focal Press, Jordan Hill, Oxford, United Kingdom.

[6] James A. Blackburn, *Modern Instrumentation for Scientists and Engineers*, Springer-Verlag, New York.

[7] Chris Dick and Fred Harris, "FPGA Interpolators Using Polynomial Filters," *The 8th International Conference on Signal Processing Applications and Technology, Toronto, Canada*, September 11, 1998.

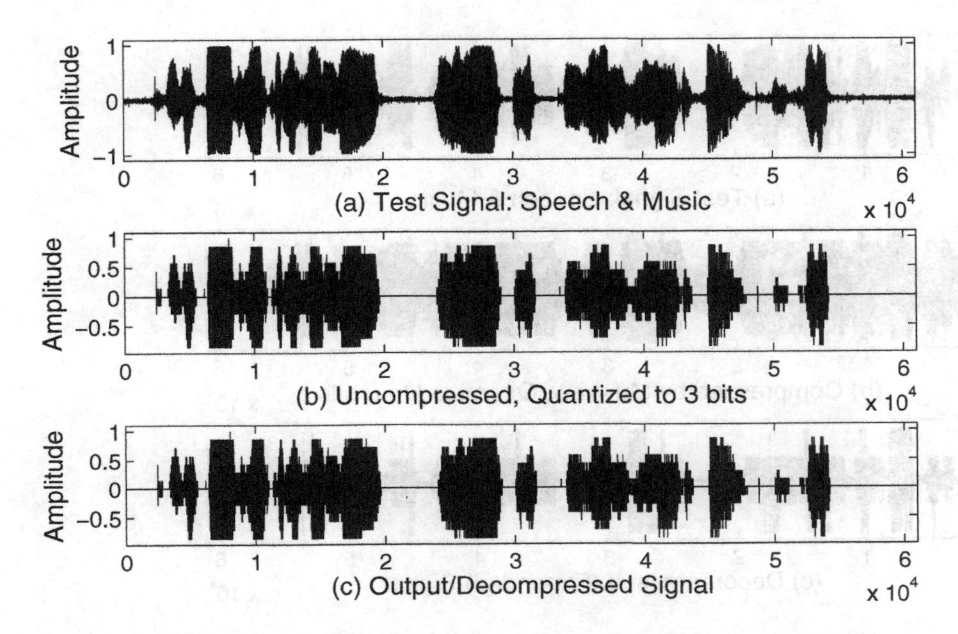

Figure 3.41: (a) Test signal (analog); (b) Test signal, without prior compression, quantized to 3 bits; (c) Analog signal reconstructed from quantized samples at (b), exhibiting a squelch effect, or total loss of reconstructed signal when the original signal level was very low.

[8] Valimaki and Laakso, "Principles of Fractional Delay Buffers," *IEEE International Conference on Acoustics, Speech, and Signal Processing (ICASSP'00), Istanbul, Turkey*, 5-9 June 2000.

[9] Mar et al, "A High Quality, Energy Optimized, Real-Time Sampling Rate Conversion Library for the StrongARM Microprocessor," *Mobile & Media Systems Laboratory, HP Laboratories Palo Alto*, HPL-2002-159, June 3, 2002.

3.20 EXERCISES

1. An ADC that operates at a rate of 10 kHz is not preceded by an anti-aliasing filter. Construct an appropriate folding diagram and give the apparent output frequency and phase (in-phase or phase-reversed) corresponding to each of the following input frequencies:
   (a) 4 kHz
   (b) 5 kHz
   (c) 9 kHz
   (d) 10 kHz
   (e) 16 kHz

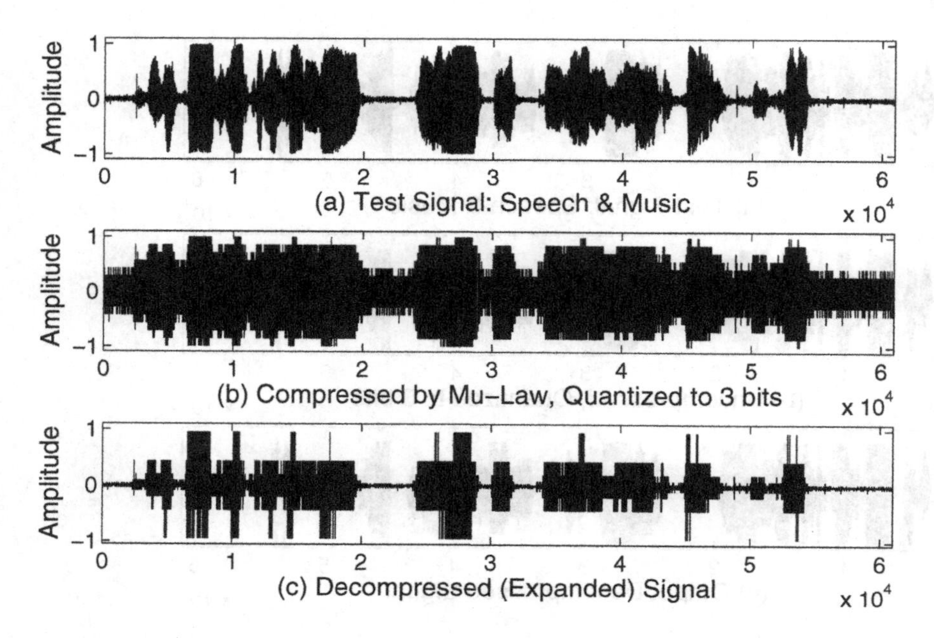

Figure 3.42: (a) Test signal (analog); (b) Quantized version of test signal, which was compressed using μ−law compression prior to being quantized; (c) Analog signal reconstructed from quantized samples at (b) by first converting from quantized form to analog, then decompressing the analog signal values. Note that the compression technique has prevented the squelch effect from occurring.

(f) 47 kHz
(g) 31 kHz

2. For each of the input frequencies given above, and using a sampling rate of 10 kHz, generate and plot the ADC's sequence of output values. To do this, set up a time vector with one second's worth of sampling times, and evaluate the values of a sine or cosine for the given input frequency, and plot the result. Note: it is helpful to plot only about the first 50 samples of the output rather than the entire output.

3. State what value of cutoff frequency an anti-aliasing filter should have for the ADC of the above exercises (i.e., an ADC operating at sampling rate of 10 kHz).

4. A certain system transmits any one of eight frequencies which are received at the input of an ADC operating at 10 kHz. An anti-aliasing filter cutting off above 10 kHz (not 5 kHz, the Nyquist rate) precedes the ADC. Using the folding diagram for 10 kHz, state the apparent frequency in the output corresponding to each of the following eight input frequencies: [1 kHz, 2kHz, 3kHz, 4 kHz, 6.5 kHz, 7.5 kHz, 8.5 kHz, and 9.5 kHz.

5. MathScript provides various functions that return b and a coefficients that can be used to implement filters of various types, such as lowpass, highpass, bandpass, and notch. The frequency arguments for such functions are given in normalized frequency. For example, the function

$$b = fir1(N, Wn)$$

where N is the filter order (one fewer than the filter or impulse response length that will be provided as b), and Wn is the normalized frequency specification. Lowpass and highpass filter impulse responses that have cutoffs at one-half the Nyquist rate, for example, and a length of 21 samples (for example), can be obtained by making the call

b = fir1(20,0.5)

for the lowpass filter, and

b = fir1(20,0.5,'high')

for the highpass filter.

Likewise, bandpass or notch filters of length 31, for example, and band limits of 0.4 and 0.6 (for example) can be obtained with the call

b = fir1(30,[0.4,0.6])

for the bandpass filter and

b = fir1(30,[0.4,0.6],'stop')

for the notch filter.

For the following sequence sample rates, desired filter length, and desired actual frequency cutoff limits, design the call for the $fir1$ function; verify the filter function by filtering a chirp of frequency range 0 to the respective Nyquist rate with the impulse response by using the MathScript function *filter*.

(a) SR = 25,000; filter length = 37; [7000] (lowpass)
(b) SR = 35,000; filter length = 47; [11000] (lowpass)
(c) SR = 44,100; filter length = 107; [5000,8000] (bandpass)
(d) SR = 48,000; filter length = 99; [65] (high)
(e) SR = 8,000; filter length = 47; [800,1200] (notch)

6. What normalized frequency will generate a complex exponential exhibiting one cycle for every

(a) 2 samples
(b) 4 samples
(c) 8 samples
(d) 17.5 samples

(e) 3.75 samples

For each of (a) through (e), devise a call to generate and plot the real and imaginary parts of a complex exponential having the respective stated period.

7. Using paper and pencil, convert the following decimal numbers to binary offset notation using 4 bits:

(a) 0; (b) 1; (c) 2; (d) 3; (e) 8; (f) 9; (g) 15

8. Using paper and pencil, convert the following binary numbers (in offset format) to decimal equivalent:

(a) [1001001]; (b) [10000001]; (c) [10101010]; (d) [11111111]

9. Write a script that implements the following function:

[OutputMat] = LVxDigitizePosNums(NumsToDig,NumBits)

in which *OutputMat* is a matrix of binary numbers, each row of which represents one number from the input vector *NumsToDig*, which consists of zero or positive decimal numbers to be converted to binary offset format, and NumBits is the minimum number of bits necessary to quantize all the decimal numbers in *NumsToDig* without clipping.

```
function [OutputMat] = LVxDigitizePosNums(xNumsToDig,nBits)
% Receives a vector of positive decimal numbers to convert to
% binary notation using nBits number of bits, and produces
% as Output a matrix each row of which is a binary
% representation of the corresponding input decimal number.
% Test call:
% LVxDigitizePosNums([0,1,34,23,2,17,254,255,127],8)
```

The following steps need to be completed:

(a) From the input arguments, generate a power-of-two vector of appropriate length;

(b) Initialize *OutputMat* as a matrix of zeros having a number of rows equal to the length of vector *NumsToDig*, and a number of columns equal to *NumBits*;

(c) To conduct the actual successive approximation algorithm, all the values in one column of *OutputMat* at a time are set to 1, and the following matrix equation is evaluated:

TestMat = OutputMat*xWtVec' - NumsToDig';

(d) The row indices must be found of entries in *TestMat* (which is a column vector) that are less than zero, and the corresponding values in *OutputMat* are then reset from 1 to zero.

(e) The operation proceeds in this manner from column to column until all columns have been experimentally set to 1, tested, and reset to zero as required. The final result is then the output of the function, *OutputMat*.

10. Write a script that can receive decimal numbers from 0.00 up to 10.00 and convert them to binary offset notation having 16 bits of precision, using the method of successive approximation. State the value of the LSB.

11. A certain audio signal has values ranging from -50 to +50 volts, and is to be quantized using an ADC that accepts an input voltage range of 0 to +10 volts. It is desired to quantize the audio signal to an accuracy of at least one part per thousand. Describe the signal adjustments necessary to make the audio signal suitable as an input signal, and determine the minimum number of bits of precision needed in the ADC. State the value of the resulting LSB in volts. What are the minimum and maximum binary output values from the ADC assuming offset notation?

12. Write a script that uses the provided wave file *drwatsonSR8K.wav* as a test signal, compresses it according to μ-Law compression as described above using the input parameter AMu as the compression parameter μ, quantizes it using a user-selectable number of bits $NoBits$, decompresses it, and allows you to play it through your computer's sound card.

LVxCompQuant(TypeCompr,AMu,NoBits)

For comparison purposes, you should be able to process and play the test signal with and without use of μ-Law compression according to the value of the input argument *TypeCompr*.

To perform the decompression, first generate the decompression formula by solving for $s[n]$ in (3.6); a helpful thing to do at the signal creation stage is to divide the signal by the maximum of its absolute value so that S_{MAX} =1, thus simplifying the algebra in the decompression formula.

Vectors may be played using the function

sound(AudioSig,NetSR)

where *AudioSig* is the vector, $NetSR$ is the sample rate at which it to be played. *NetSR* must be any of 8 kHz, 11.025 kHz, 22.05 kHz, or 44.1 kHz.

13. Write a script that conforms to the following call syntax

BinaryMatrix = LVxADCPosNegQuants(NumsToDig)

where *NumsToDig is* a decimal number array to be converted to a binary array, each binary number in the array having a single sign bit followed by N bits representing the magnitude, for a total of N + 1 bits per binary number. Positive decimal numbers are converted to binary representations and a sign bit of 0 is placed to the left of the magnitude's MSB. Negative numbers have their magnitudes converted to binary representations and a sign bit of 1 is used.

The input argument consists of a vector of real, positive and/or negative decimal numbers, and the function returns a matrix of binary numbers, each original decimal number occupying one row of the returned matrix. The function automatically determines how many bits will be required, which depends on the number with the largest magnitude in the vector of numbers to be digitized (an LSB value of 1.0 is assumed for simplification). For example, the call

LVxADCPosNegQuants([17,7,-100,77])

should return the following binary matrix:

$$\begin{bmatrix} 0 & 0 & 0 & 1 & 0 & 0 & 0 & 1 \\ 0 & 0 & 0 & 0 & 0 & 1 & 1 & 1 \\ 1 & 1 & 1 & 0 & 0 & 1 & 0 & 0 \\ 0 & 1 & 0 & 0 & 1 & 1 & 0 & 1 \end{bmatrix}$$

in which the first row is the binary representation of the first number in the vector of decimal numbers to be converted (17 in this case), and so forth. The leftmost bit of each row is the sign bit; positive numbers have "0" for a sign bit, and negative numbers have a "1."

14. Write a script that conforms to the following call syntax

$$BinOut = LVxBinaryCodeMethods(BitsQ, SR, Bias, Freq, Amp, CM, PlotType)$$

as described in the text and below:

```
function BinOut = LVxBinaryCodeMethods(BitsQ,SR,Bias,Freq,...
Amp,CodeMeth,PlotType)
% Quantizes a sine wave of amplitude Amp and frequency
% Freq using BitsQ number of quantization bits at a sample
% rate of SR.
% Bias: 0 for none, 1 for 1/2 LSB
% CodeMeth: 1 = Sign + Mag; 2 = Offset;
% PlotType: val 0 plot as multiples of LSB, or 1 to plot in volts
% Generates a figure with three subplots, the first is the
% (simulated) analog test signal, the second has an overlay
% of the simulated test signal and its quantized version, and the
% third is the quantization error
% Test calls:
% BinOut = LVxBinaryCodeMethods(4, 1000,0, 10,170,2,0)
% BinOut = LVxBinaryCodeMethods(2, 1000,0, 10,170,1,1)
% BinOut = LVxBinaryCodeMethods(3, 1000,1, 10,170,1,1)
```

15. Write the m-code for the script

$$LVxInterpolationViaSinc(N, SampDecRate, valTestSig)$$

which conforms to the syntax below, functions as described in the text, and which creates plots as shown in the text. Test it with the given sample calls:

```
function LVxInterpolationViaSinc(N,SampDecRate,valTestSig)
```

```
% N is the master sequence length, from which a densely-sampled
% test signal is generated. The master sequence is decimated by
% every SampDecRate samples to create the test sample
% sequence from which an interpolated version of the underlying
% bandlimited continuous domain signal will be generated.
% N must be an even integer, and SampDecRate must be an
% integer; valTestSig may be any integer from 1 to 5, such that
% 1 yields the waveform cos(2*pi*n*0.1) + 0.7*sin(2*pi*n*0.24);
% 2 yields DC
% 3 gives a single triangular waveform
% 4 gives two cycles of a triangular waveform
% 5 gives 0.5*cos(2*pi*n*0.125) + 1*sin(2*pi*n*0.22);
% where n = -N/2:1:N/2 if N is even
% Sample calls:
% LV_InterpolationViaSinc(1000,100,1)
% LV_InterpolationViaSinc(1000,50,1)
% LV_InterpolationViaSinc(1000,25,1)
% LV_InterpolationViaSinc(5000,100,1)
```

16. Write the m-code for the script

$$LVxSineROMNonIntDecInterp(N, F\_ROM, D)$$

as described and illustrated in the text. Test the script with the following calls:

(a) **LVxSineROMNonIntDecInterp(77,1,7.7)**
(b) **LVxSineROMNonIntDecInterp(23,2.5,0.6)**
(c) **LVxSineROMNonIntDecInterp(48,3,4)**

17. A chirped audio signal which is of one second duration and which sweeps from 0 Hz to 8000 Hz in one second is sampled at 2000 Hz. Compute and plot the resultant wave State at what times during the one second interval maxima and minima in observed (or audible or apparent) frequency occur.

18. Write the m-code for the following function, which uses sinc interpolation to convert an audio file sampled at 8 kHz to one sampled at 11.025 kHz.

```
function LVxInterp8Kto11025(testSigType,tsFreq)
% Converts an audio signal having a sample rate of 8 kHz to
% one sampled at 11.025 kHz using both sinc and linear
% interpolation. The audio files used can be either the audio
% file 'drwatsonSR8K.wav' or a cosine of user designated
% frequency. For sinc interpolation it uses 10 samples of the
% input signal and a 100 by 10 sinc interpolation matrix to
```

```
% generate sample values located at the fractional sample
% index values 1:320/441:length(OriginalAudioFile)
% To use 'drwatsonSR8K.wav', pass testSigType as 1 and
% tsFreq as [] or any number; to use a cosine, pass
% testSigType as 0 and pass tsFreq as the desired cosine
% frequency in Hz. The script automatically plays the original,
% sinc-interpolated, and linear interpolated signals using the
% Mathscript function sound., and plots, for a short sample
% interval, the original, sinc-interpolated, and linear interpolated
% signal values on one axis for comparison prior to
% post-interpolation lowpass filtering, which limits the
% passband of the interpolated signals to the original 4 kHz
% bandwidth.
% Test calls:
% LVxInterp8Kto11025(0,250)
% LVxInterp8Kto11025(0,2450)
% LVxInterp8Kto11025(1,[])
```

The sinc interpolation matrix may, for example, consist of rows 900-999 as seen in Fig. 3.31. As an example computation, if (say) a sample is needed at index 10.71, the ten columns of row 72 (the row index must start with one rather than zero) of the sinc matrix would be multiplied by original signal samples 6-15, respectively, with sample 10 (i.e., **floor(10.71)**) being multiplied by column 5 as shown in Fig. 3.31, and sample 11 (i.e., **ceil(10.71)**) being multiplied by column 6. The products are summed to obtain the interpolated value. The operation is actually the inner or dot product of the chosen sinc matrix row with a column vector selected from the signal to be interpolated.

It is only necessary to perform the interpolation for a limited number of samples to demonstrate efficacy; 1000-2000 samples are enough at a sample rate of 8kHz to hear that the original and interpolated versions have the same apparent frequency.

The interpolated samples should be lowpass filtered after interpolation to retain the original frequency limit, which is 4 kHz. The following code may be used:

```
[b,a] = cheby1(12,0.5,7.9/11.025);
SincInterpSig = filter(b,a,SincInterpSig);
LinInterpSig = filter(b,a,LinInterpSig);
```

CHAPTER 4

Transform and Filtering Principles

4.1 OVERVIEW

Having become acquainted in the last chapter with basic signal acquisition (ADC) and reconstruction techniques (DAC) and the all-important concepts of Nyquist rate and normalized frequency, we are now in a position to investigate the powerful principle of **Correlation**. Discrete frequency transforms, which can be used to determine the frequency content or response of a discrete signal, and time domain digital filters (i.e., the FIR and IIR), which can preferentially select or reject certain frequencies in a signal, function according to two principles of correlation–namely, respectively, the single-valued correlation of two equally-sized, overlappingly aligned waveforms, and the **Correlation Sequence**. Another principle, that of orthogonality, also plays an important role in frequency transforms, and this too will be explored to show why frequency transforms such as the DFT require that two correlations (or a single complex correlation) be performed for each frequency being tested.

We begin our discussion with an elementary concept of correlation, the correlation of two samples, and quickly move, step-by-step, to the correlation of a signal of unknown frequency content with two equally-sized orthogonal sinusoids (i.e., cosine-sine pairs) and, in short order, the real DFT. We then briefly illustrate the use of the property of orthogonality in signal transmission, a very interesting topic which illustrates the power of mathematics to allow intermixing and subsequent decoding of intelligence signals. From there we investigate the correlation sequence, performing correlation via convolution, and matched filtering. We then informally examine the frequency selective properties of the correlation sequence, and learn how to construct basic (although inefficient) filters. We learn the principle of sinusoidal fidelity and then determination of time delays between sequences using correlation, an application of correlation that is often used in echo canceller training and the like. For the final portion of the chapter, we investigate simple one- and two-pole IIRs with respect to stability and frequency response, and demonstrate how to generate IIRs having real-only filter coefficients by using poles in complex conjugate pairs.

By the end of this chapter, the reader will be prepared to undertake the study of discrete frequency transforms found in Part II of the book, which includes a general discussion of various transforms and detailed chapters on the Discrete Time Fourier Transform (DTFT), the Discrete Fourier Transform (DFT), and the z-Transform.

4.2 SOFTWARE FOR USE WITH THIS BOOK

The software files needed for use with this book (consisting of m-code (.m) files, VI files (.vi), and related support files) are available for download from the following website:

http://www.morganclaypool.com/page/isen

The entire software package should be stored in a single folder on the user's computer, and the full file name of the folder must be placed on the MATLAB or LabVIEW search path in accordance with the instructions provided by the respective software vendor (in case you have encountered this notice before, which is repeated for convenience in each chapter of the book, the software download only needs to be done once, as files for the entire book are all contained in the one downloadable folder).

See Appendix A for more information.

4.3 CORRELATION AT THE ZEROTH LAG (CZL)

A simple concept of numerical correlation is this: if two numbers are both positive, or both negative, they correlate well; if one is positive, and one is negative, they correlate strongly in a negative or opposite sense. A way to quantify this idea of correlation is simply to multiply the two numbers. If they are both positive, or both negative, the product is positive; if one number is positive, and one number is negative, the product is negative. Thus two numbers having the same sign have a positive correlation value, while two numbers having opposite signs have a negative correlation value. If one number is zero, the product is zero, and no correlation between the two numbers can be determined, that is, they are neither alike nor unalike; in this case they are termed "uncorrelated."

To correlate two waveforms (i.e., sequences) having the same length N, multiply corresponding samples and add up all the products. Imagine this as first laying one waveform on top of the other so that the first sample of one waveform lies atop the first sample of the other, and so forth. Then multiply each pair of corresponding samples and add the products.

Stated mathematically, the correlation of two sequences $A[n]$ and $B[n]$ having equal lengths N is

$$CZL = \sum_{n=0}^{N-1} A[n]B[n] \tag{4.1}$$

where CZL is the **Correlation at the Zeroth Lag** of sequences $A[n]$ and $B[n]$. The nomenclature *Zeroth Lag* is used in this book for clarity to denote this particular case, in which two sequences of equal length are correlated with no offset or delay (lag) relative to one another. In most books, the process represented by Eq. (4.1) is usually referred to simply as the "correlation" of sequences $A[n]$ and $B[n]$.

Example 4.1. Compute the CZL of the sequence $[ones(1, 16), zeros(1, 16)]$ with itself. Follow up by correlating the sequence with the negative of itself.

Figure 4.1, plots (a) and (b), show two instances of the subject 32-sample sequence. To compute the correlation value, figuratively lay the first waveform on top of the second. Multiply all overlapping samples, and add the products to get the answer, which is 16. Since the two waveforms of Fig. 4.1 are in fact the same, they have a high positive value of correlation, which is expected from the basic concept of correlation.

The sequence and its negative are shown in plots (c) and (d) of Fig. 4.1. The correlation value in this case is −16, indicating a strong anti-correlation.

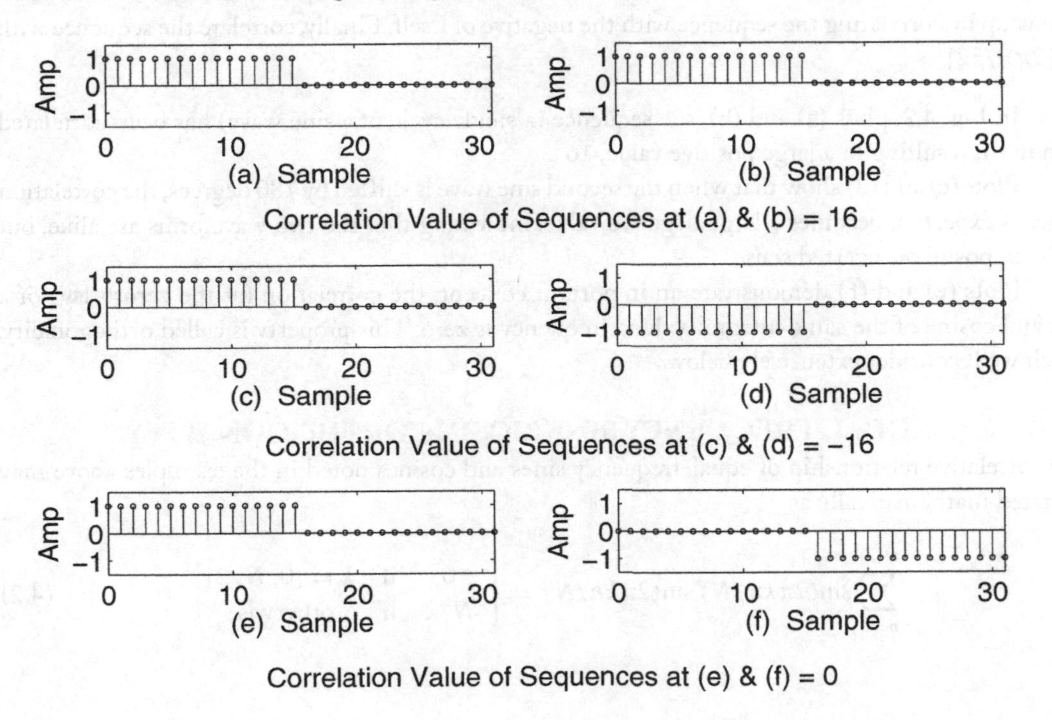

Figure 4.1: (a) First sequence; (b) Second sequence, identical to the 1st, leading to a large positive correlation; (c) First sequence; (d) Second sequence, the negative of the 1st, leading to a large negative correlation; (e) First sequence; (f) Second sequence, having no overlapping sample pairs with the first sequence that have nonzero products, leading to a correlation value of 0.

Example 4.2. Compute the CZL of the two sequences [ones(1, 16), zeros(1, 16)] and [zeros(1, 16), -ones(1, 16)].

The two sequences are shown in plots (e) and (f) of Fig. 4.1. The correlation value is zero since (when you figuratively lay one waveform atop the other) one value of each pair of overlapping samples is always zero, yielding a product of zero for all 32 sample pairs, summing to zero.

Example 4.3. Construct two sequences, two different ways, of length eight samples, using as values only 1 or -1, which yield a CZL of 0.

Two such possible sequences are: First pair: [-ones(1,8)] and [-ones(1,4), ones(1,4)]; a second pair might be: [-ones(1,8)] and [-1,-1,1,1,-1,-1,1,1]. The reader should be able to construct many more examples. For example, if, of the first pair, the first vector were [ones(1,8)] instead of [-ones(1,8)], the CZL would still be zero.

Example 4.4. Compute the correlation of the sequence $\sin[2\pi n/32]$ with itself ($n = 0:1:31$). Follow up by correlating the sequence with the negative of itself. Finally, correlate the sequence with $\cos[2\pi n/32]$.

In Fig. 4.2, plots (a) and (b), the sequence (a single cycle of a sine wave) has been correlated with itself; resulting in a large positive value, 16.

Plots (c) and (d) show that when the second sine wave is shifted by 180 degrees, the correlation value, as expected, becomes a large negative value, indicating that the two waveforms are alike, but in the opposite or inverted sense.

Plots (e) and (f) demonstrate an important concept: the correlation (at the zeroth lag) of a sine and cosine of the same integral-valued frequency is zero. This property is called orthogonality, which we'll consider extensively below.

4.3.1 CZL EQUAL-FREQUENCY SINE/COSINE ORTHOGONALITY

The correlative relationship of equal-frequency sines and cosines noted in the examples above may be stated mathematically as

$$\sum_{n=0}^{N-1} \sin(2\pi kn/N)\sin(2\pi kn/N) = \begin{cases} 0 & \text{if} \quad k = [0, N/2] \\ N/2 & \text{if} \quad \text{otherwise} \end{cases} \tag{4.2}$$

and

$$\sum_{n=0}^{N-1} \sin(2\pi kn/N)\cos(2\pi kn/N) = 0 \tag{4.3}$$

where k is an integer, N is the sequence length, and n = 0:1:$N-1$. The latter equation expresses the principle of orthogonality of equal integral-frequency sine and cosine waves, which is simply

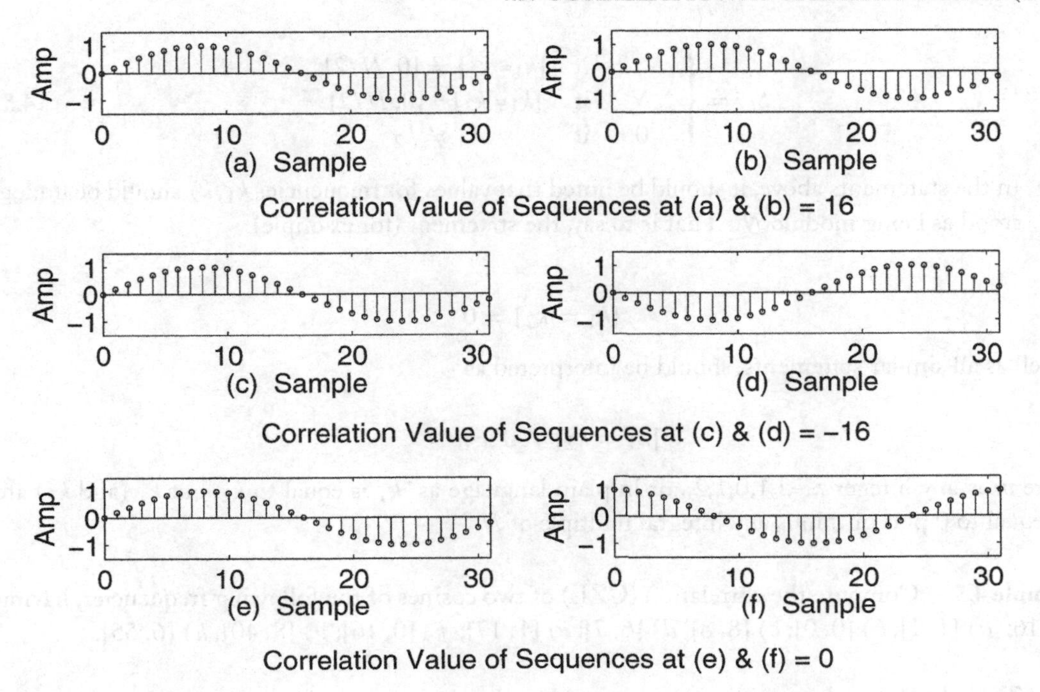

Figure 4.2: (a) First sequence, a sine wave; (b) Second sequence, identical to the first, yielding a large positive correlation; (c) First sequence; (d) Second sequence, the negative of the first, showing a large negative correlation; (e) First sequence; (f) Second sequence, a cosine wave, yielding a correlation value of 0.

that their correlation value over one period is zero. When cosines are used instead of sines, the relationship is

$$\sum_{n=0}^{N-1} \cos(2\pi kn/N)\cos(2\pi kn/N) = \begin{cases} N & \text{if } \; k = [0, N/2] \\ N/2 & \text{if } \quad \text{otherwise} \end{cases}$$

4.3.2 CZL OF SINUSOID PAIRS, ARBITRARY FREQUENCIES

Consider the following sum S_C where N is the number of samples in the sequence, n is the sample index, and k_1 and k_2 represent integer frequencies such as -3, 0,1, 2, etc.

$$S_C = \sum_{n=0}^{N-1} \cos[2\pi k_1 n/N]\cos[2\pi k_2 n/N] \tag{4.4}$$

yields

$$S_C = \begin{cases} N/2 & \text{if} \quad [k_1 = k_2] \neq [0, N/2] \\ N & \text{if} \quad [k_1 = k_2] = [0, N/2] \\ 0 & \text{if} \quad k_1 \neq k_2 \end{cases} \tag{4.5}$$

- In the statements above, it should be noted that values for frequencies k_1, k_2 should be understood as being modulo N. That is to say, the statement (for example)

$$[k_1 = k_2] \neq 0$$

as well as all similar statements, should be interpreted as

$$[k_1 = k_2] \neq 0 \pm mN$$

where m is any integer ...–2,1,0,1,2...or in plain language as "k_1 is equal to k_2, but k_1 (and k_2) are not equal to 0 plus or minus any integral multiple of N."

Example 4.5. Compute the correlation (CZL) of two cosines of the following frequencies, having N = 16: *a*) [1, 1]; *b*) [0, 0]; *c*) [8, 8]; *d*) [6, 7]; *e*) [1, 17]; *f*) [0, 16]; *g*) [8, 40]; *h*) [6, 55].

To make the computations easy, we provide a simple function:

```
function [CorC] = LVCorrCosinesZerothLag(k1,k2,N)
n = 0:1:N-1;
CorC = sum(cos(2*pi*n*k1/N).*cos(2*pi*n*k2/N));
```

and after making the appropriate calls, we get the following answers:

a) In this case, $k_1 = k_2 \neq [0, 8]$, yielding $N/2 = 8$ in this case
b) In this case, $k_1 = k_2 = 0$, yielding $N = 16$
c) In this case, $k_1 = k_2 = N/2$, yielding $N = 16$
d) In this case, $k_1 \neq k_2$ yielding 0
e) In this case, 17 (1+N) is equivalent to 1; the call

[CorC] = LVCorrCosinesZerothLag(1,17,16)

yields $CorC = 8$, just as for case (a) in which the frequencies were [1,1].

f) Note that 16 is (0 +N), so this is equivalent to case (b) and the call

[CorC] = LVCorrCosinesZerothLag(0,16,16)

yields $CorC = 16$, just as for case (b) in which the frequencies were [0,0].

g) Note that the second frequency, 40, is equal to (8+2N), so this is equivalent to case (c) in which the frequencies were [8,8]; the call

[CorC] = LVCorrCosinesZerothLag(8,40,16)

yields $CorC = 16$, just as for case (c) in which the frequencies were [8,8].

h) The frequency 55 is equivalent to (7+3N), so this is equivalent to case (d); the call

[CorC] = LVCorrCosinesZerothLag(6,55,16)

yields $CorC = 0$, just as for case (d) in which the frequencies were [6,7].

The CZL of two sine waves having arbitrary, integral-valued frequencies, defined as

$$S_S = \sum_{n=0}^{N-1} \sin(2\pi k_1 n/N)\sin(2\pi k_2 n/N)$$

yields

$$S_S = \begin{cases} N/2 & \text{if} \quad [k_1 = k_2] \neq [0, N/2] \\ 0 & \text{if} \quad [k_1 = k_2] = [0, N/2] \\ 0 & \text{if} \quad k_1 \neq k_2 \end{cases} \tag{4.6}$$

where k_1 and k_2 are integers and are modulo-N as described and illustrated above.

Figure 4.3 shows the values of S_C and S_S for the case of $k1 = 0$ with $k2$ varying from 0 to 5, while Fig. 4.4 shows S_C and S_S for the case of $k1 = 3$ with $k2$ varying from 0 to 5.

Example 4.6. Compute the CZL of two sine waves having frequencies of one and two cycles, respectively, over 32 samples.

Figure 4.5 shows one and two cycle sine waves at plots (a) and (b) respectively. Since the two sine waves differ in frequency by an integer, the CZL is zero, as displayed beneath the plots.

Plots (b) and (c) show another example in which the two sine waves have frequencies of three and five cycles over the sequence length, with the predicted CZL of 0.

Figure 4.5, plot (c), shows the CZL of a four cycle sine wave and one of 5.8 cycles. In this case, the CZL is not guaranteed to be zero, and in fact it is not equal to zero.

4.3.3 ORTHOGONALITY OF COMPLEX EXPONENTIALS

The sum of the product (i.e., the CZL) of two complex exponentials each having an integral number of cycles k_1, k_2, respectively, over the sequence length N, where k_1 and k_2 are modulo-N, is

$$\sum_{n=0}^{N-1} e^{j2\pi k_1 n/N} e^{-j2\pi k_2 n/N} = \begin{cases} 0 & \text{if} \quad k_1 \neq k_2 \\ N & \text{if} \quad \text{otherwise} \end{cases}$$

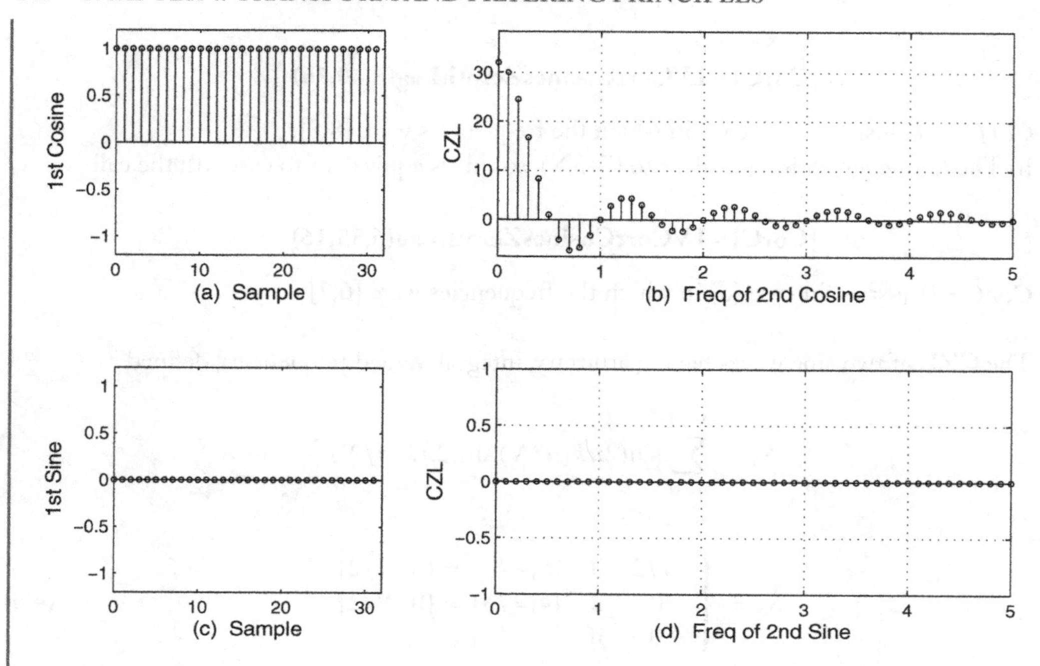

Figure 4.3: (a) First cosine, $k1 = 0$; (b) CZL between first and second cosines, the second cosine frequency $k2$ varying between 0 and 5 cycles per 32 samples; (c) First sine, $k1 = 0$; (d) CZL between first and second sines, the second sine frequency $k2$ varying between 0 and 5 cycles per 32 samples.

4.3.4 SUM OF SAMPLES OF SINGLE COMPLEX EXPONENTIAL

An interesting and useful observation is that if one of the complex exponentials has its frequency as zero, it is identically equal to 1.0, and the sum reduces to

$$\sum_{n=0}^{N-1} e^{\pm j2\pi k_1 n/N} = \begin{cases} N & \text{if} \quad k_1 = ... - N, 0, N, 2N, ... \\ 0 & \text{if} \quad \text{otherwise} \end{cases}$$

For the sum of the samples of a single complex exponential, use the following:

```
function s = LVSumCE(k,N)
% s = LVSumCE(2,32)
n = 0:1:N-1; s = sum(exp(j*2*pi*n*k/N))
```

The call

s = LVSumCE(2,32)

for example, yields the sum as s = -3.3307e-016, which differs from zero only by reason of computer roundoff error, whereas a call in which the frequency argument is equal to N, such as

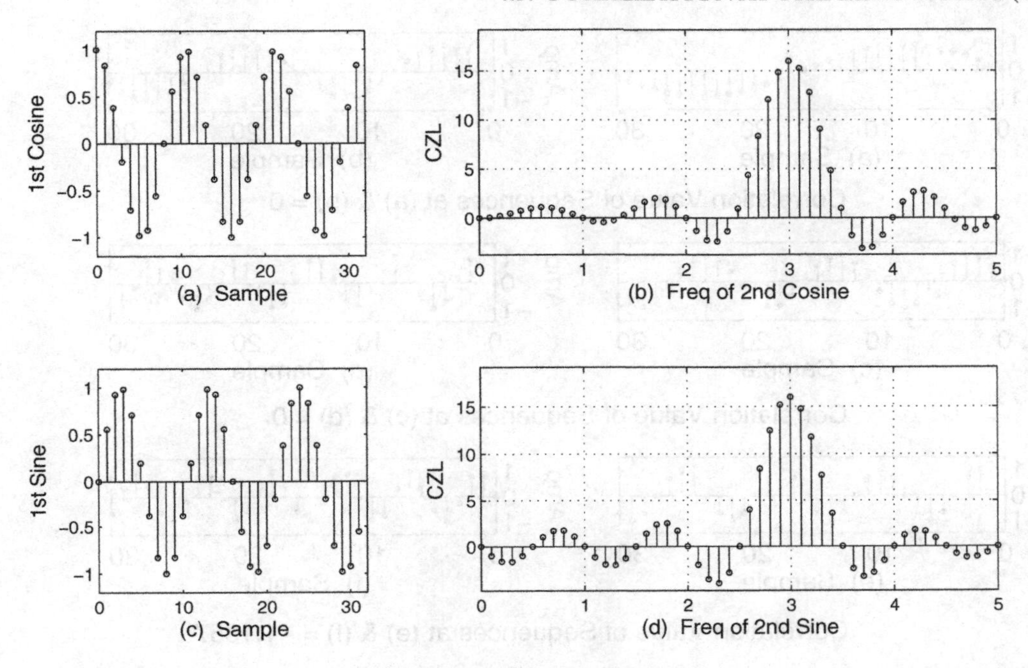

Figure 4.4: (a) First cosine, $k1 = 3$; (b) CZL between first and second cosines, the second cosine frequency $k2$ varying between 0 and 5 cycles per 32 samples; (c) First sine, $k1 = 3$; (d) CZL between first and second sines, the second sine frequency $k2$ varying between 0 and 5 cycles per 32 samples.

s = LVSumCE(32,32)

yields $s = 32$.

4.3.5 IDENTIFYING SPECIFIC SINUSOIDS IN A SIGNAL

We now consider the problem of identifying the presence or absence of sinusoids of specific frequency and phase in a signal. We begin the discussion with a specific problem set forth in the following example.

Example 4.7. A sequence $A[n]$ can assume the values of 0, $\sin[2\pi n/N + \theta]$, or $\sin[4\pi n/N + \theta]$ where n is the sample index, N is the sequence length, θ is a variable phase which may assume the values 0, $\pi/4$, $\pi/2$, $3\pi/4$, or π radians. Devise a way to determine the signal present in $A[n]$ using correlation.

Let's consider several cases to see what the difficulties are:

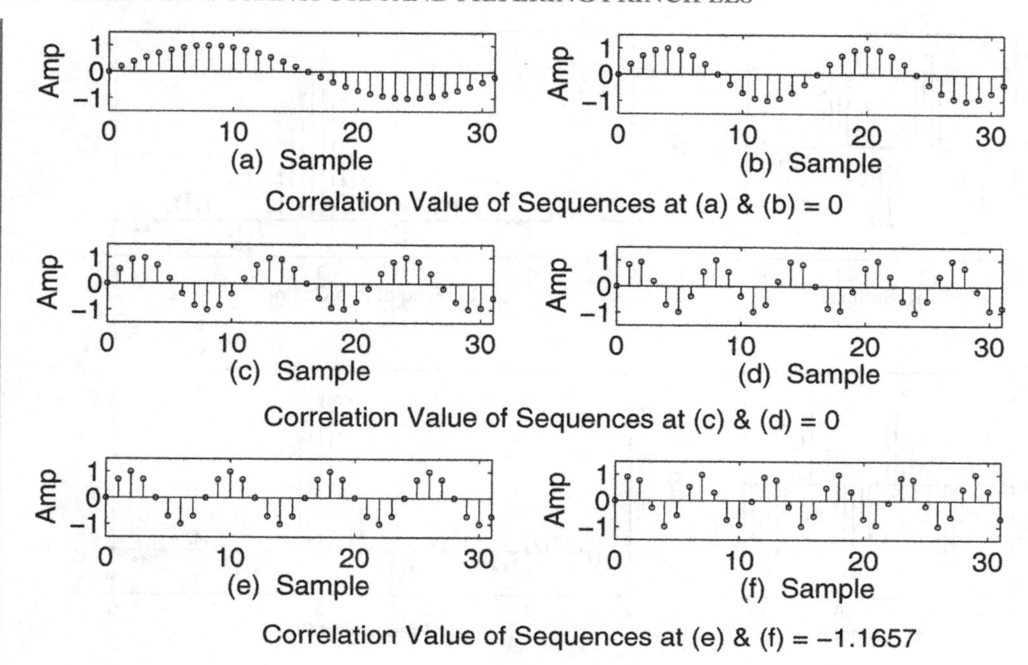

Figure 4.5: (a) One cycle of a sine wave; (b) Two cycles of a sine wave, with correlation at the zeroth lag (CZL) value of zero; (c) Three cycles of a sine wave; (d) Five cycles of a sine wave; (e) Four cycles of a sine wave; (f) Five point eight (5.8) cycles of a sine wave.

Suppose in one case that $A[n] = \sin[2\pi n/N + \theta]$ and $\theta = 0$. The correlation of $A[n]$ with a test sine wave $TS[n] = \sin[2\pi n/N]$ will yield a large positive correlation value. If, however, $\theta = \pi/2$ radians (90 degrees), then the same correlation would yield a value of zero due to orthogonality.

In another case, suppose $A[n] = 0$. The correlation of $A[n]$ with $TS[n]$ yields zero irrespective of the value of θ. Unfortunately, we don't know if that means the unknown signal is simply identically zero (at least at samples where the test sine wave is nonzero), or if the unknown signal happens to be a sinusoid that is 90 degrees out of phase with the test sine wave $TS[n]$.

In yet a third case, if $A[n] = \sin[4\pi n/N + \theta]$, any correlation with $TS[n]$ will also yield zero for any value of θ.

The solution to the problem, then, is to perform two correlations for each test frequency with $A[n]$, one with a test sine and one with a test cosine. In this manner, for each frequency, if the sinusoid of unknown phase is present, but 90 degrees out of phase with one test signal, it is perfectly in phase with the other test signal, thus avoiding the ambiguous correlation value of zero. $A[n]$, then, must be separately correlated with $\sin[2\pi n/N]$ and $\cos[2\pi n/N]$ to adequately detect the presence of $\sin[2\pi n/N + \theta]$, and $A[n]$ must also be separately correlated with $\sin[4\pi n/N]$ and $\cos[4\pi n/N]$ to detect the presence of $\sin[4\pi n/N + \theta]$. If all four correlations (CZLs) are equal to zero, then the signal $A[n]$ is presumed to be equal to zero, assuming that it could only assume the values given in the statement of the problem.

- For a given frequency, and any phase of the unknown signal, correlation values obtained by correlating the unknown with both test cosine and test sine waves having the same frequency as the unknown result that can be used to not only determine the phase of the unknown signal, but to actually completely reconstruct it. This technique is described and illustrated immediately below.

4.3.6 SINGLE FREQUENCY CORRELATION AND RECONSTRUCTION

For a real signal containing a single frequency, the original signal can be reconstructed from correlation values by a simple procedure. For example, if the unknown is

$$x[n] = \sin[2\pi kn/N + \theta]$$

then we compute

$$CZL_{COS} = \sum_{n=0}^{N-1} x[n]\cos[2\pi kn/N] \tag{4.7}$$

and

$$CZL_{SIN} = \sum_{n=0}^{N-1} x[n]\sin[2\pi kn/N] \tag{4.8}$$

where k is a real integer and N is the sequence length; then the original sinusoid can be reconstructed as

$$(1/N)(CZL_{COS}\cos[2\pi kn/N] + CZL_{SIN}\sin[2\pi kn/N]) \tag{4.9}$$

if k = 0 or $N/2$, or

$$(2/N)(CZL_{COS}\cos(2\pi kn/N) + CZL_{SIN}\sin(2\pi kn/N)) \tag{4.10}$$

otherwise.

The script (see exercises below)

$$LVxTestReconSineVariablePhase(k1, N, PhaseDeg)$$

allows you to enter a frequency $k1$ and length N to be used to construct three sinusoids, which are 1) a sine wave of arbitrary phase $PhaseDeg$ and frequency $k1$, and 2) a test correlator sine of frequency $k1$, and 3) a test correlator cosine of frequency $k1$.

The call

LVxTestReconSineVariablePhase(1,32,-145)

was used to generate Fig. 4.6, which shows the test correlators in plots (a) and (c), respectively, the sine of arbitrary phase in plot (b), and the perfectly reconstructed sine of arbitrary phase in plot (d).

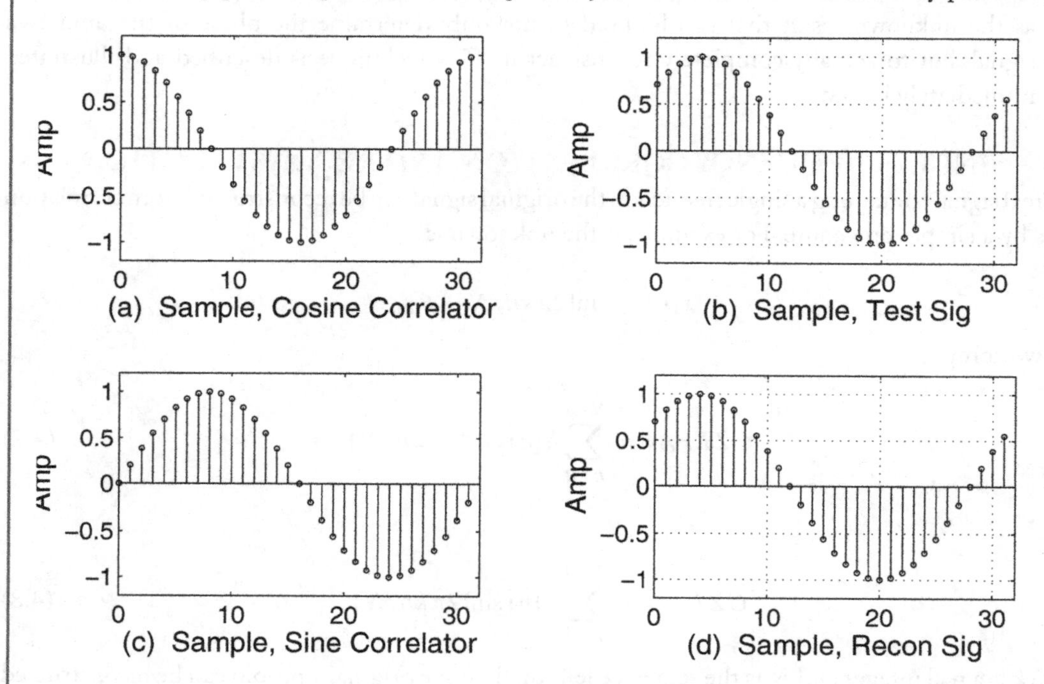

Figure 4.6: (a) and (c) Test Cosine and Sine correlators, respectively; (b) Test sinusoid of arbitrary phase; (d) Test sinusoid reconstructed from the test waveforms at (a) and (c) and the correlation coefficients.

Example 4.8. Note the fact that the reconstruction formula's weight is 1.0 if k is 0 or $N/2$, but 2.0 otherwise. This is in accordance with the correlation values shown in (4.5) and (4.6). Devise a Command Line call which can be used to illustrate the need for the two different coefficients used in reconstruction.

The following call

k = 0;C=sum(cos(2*pi*k*(0:1:7)/8).^2)

can be used, letting k vary as 0:1:7 in successive calls, and noting the value of C for each value of k. The function *cos* in the call should then be changed to *sin,* and the experiments performed again. You will observe that for the cosine function (*cos*), C is N for the case of $k = 0$ or 4 (i.e., $N/2$), but

C is equal to $N/2$ for other values of k. In the case of the sine (*sin*) function, C is zero when k is 0 or $N/2$ since the sine function is identically zero in those cases, and C is $N/2$ for other values of k. This information will be prove to be relevant in the very next section of this chapter as well as in Part II of the book, where we study the complex DFT.

4.3.7 MULTIPLE FREQUENCY CORRELATION AND RECONSTRUCTION

For a test signal containing a single frequency, a pair of CZLs using cosine and sine waves of the same frequency enables reconstruction of the original test signal using the correlation coefficients and the cosine and sine test waves as basis functions.

This can be extended to reconstruct a complete sequence of length N containing any frequency (integer- or noninteger- valued), between 0 and $N/2$. To achieve this, it is necessary, in general, to do cosine-sine CZLs for all nonaliased integral frequencies possible within the sequence length. For a sequence of length 8, for example, we would do CZLs using cosine-sine correlators of frequencies 0, 1, 2, 3, and 4.

The analysis formulas are

$$R[k] = CZL_{COS}[k] = \sum_{n=0}^{N-1} x[n]\cos[2\pi nk/N] \tag{4.11}$$

$$I[k] = CZL_{SIN}[k] = \mp\sum x[n]\sin[2\pi nk/N] \tag{4.12}$$

where N is the length of the correlating functions as well as the signal $x[n]$, n is the sample index, which runs from 0 to $N-1$,and k represents the frequency index, which assumes values of 0,1,.. $N/2$ for N even or 0,1,...(N-1)/2 if N is odd.

Equations (4.11) and (4.12) are usually called the Real DFT analysis formulas, and the variables $R[k]$ and $I[k]$ (or some recognizable variation thereof) are usually described as the real and imaginary parts of the (real) DFT.

The synthesis formula is

$$x[n] = (A_k/N)\sum_{k=0}^{K} R[k]\cos[2\pi nk/N] \mp I[k]\sin[2\pi nk/N] \tag{4.13}$$

where $n = 0, 1, \ldots N-1$ and $K = N/2$ for N even, and $(N-1)/2$ for N odd, and the constant A_k =1 if k = 0 or $N/2$, and A_k =2 otherwise.

The sign of $I[k]$ in the synthesis formula must match the sign employed in the analysis formula. It is standard in electrical engineering to use a negative sign for the sine-correlated (or imaginary) component in the analysis stage.

Note that the reconstructed wave is built one sample at a time by summing the contributions from each frequency for the one sample being computed. It is also possible, instead of computing one sample of output at a time, to compute and accumulate the contribution to the output of each harmonic for all N samples.

A script (see exercises below) that utilizes the formulas and principles discussed above to analyze and reconstruct (using both methods discussed immediately above) a sequence of length N is

$$LVxFreqTest(TestSeqType, N, UserTestSig, dispFreq)$$

where the argument *TestSeqType* represents a test sequence consisting of cosine and sine waves having particular frequencies, phases, and amplitudes (for a list of the possible parameter values, see the exercises below), N is the desired sequence length, *UserTestSeq* is a user-entered test sequence which will be utilized in the script when *TestSeqType* is passed as 7, and *dispFreq* is a particular test correlator frequency used to create two plots showing, respectively, the test signal and the test cosine of frequency *dispFreq*, and the test signal and the test sine of frequency *dispFreq*. Figure 4.7 shows, for the call

LVxFreqTest(5,32,[],1)

the plots of test signal versus the test correlator cosine having $dispFreq = 1$ at plot (a), and the test correlator sine having $dispFreq = 1$ at plot (b).

Figure 4.8 plots the cosine-correlated and sine-correlated coefficients after all correlations have been performed.

Finally, Fig. 4.9 shows the reconstruction process, in which the coefficients are used to reconstruct or synthesize the original signal. The script actually performs the reconstruction two ways and plots both results together on the same axis: in the first way, one sample at a time of $x[n]$ is computed using formula (4.13), and in the second method, all samples of $x[n]$ are computed at once for each basis cosine and sine, and all weighted basis cosines and sines are summed to get the result, which is identical to that obtained using the sample-by-sample method. The latter method, synthesis harmonic-by-harmonic, gives a more intuitive view of the reconstruction process, and thus the upper plot of Fig. 4.9 shows a 2-cycle cosine, and a 5-cycle cosine, each scaled by the amplitude of the corresponding correlation coefficient and the middle plot shows the same for the sine component, in this case a 1-cycle sine. The lower plot shows the original signal samples as circles, and the reconstructed samples are plotted as stars (since the reconstruction is essentially perfect using both methods, the stars are plotted at the centers of the corresponding circles).

As mentioned above, analysis formulas (4.11) and (4.12) and synthesis formula (4.13) form a version of the Discrete Fourier Transform (DFT) known as the **Real DFT** since only real arithmetic is used. The standard version of the DFT, which uses complex arithmetic, and which has far more utility than the Real DFT, is discussed extensively in Part II of the book. In Part II, the theoretical basis for both the Real and Complex DFTs will be taken up; our brief foray into the Real DFT has served to illustrate the basic underlying principle of standard frequency transforms such as the DFT, which is correlation between the signal and orthogonal correlator pairs of various frequencies.

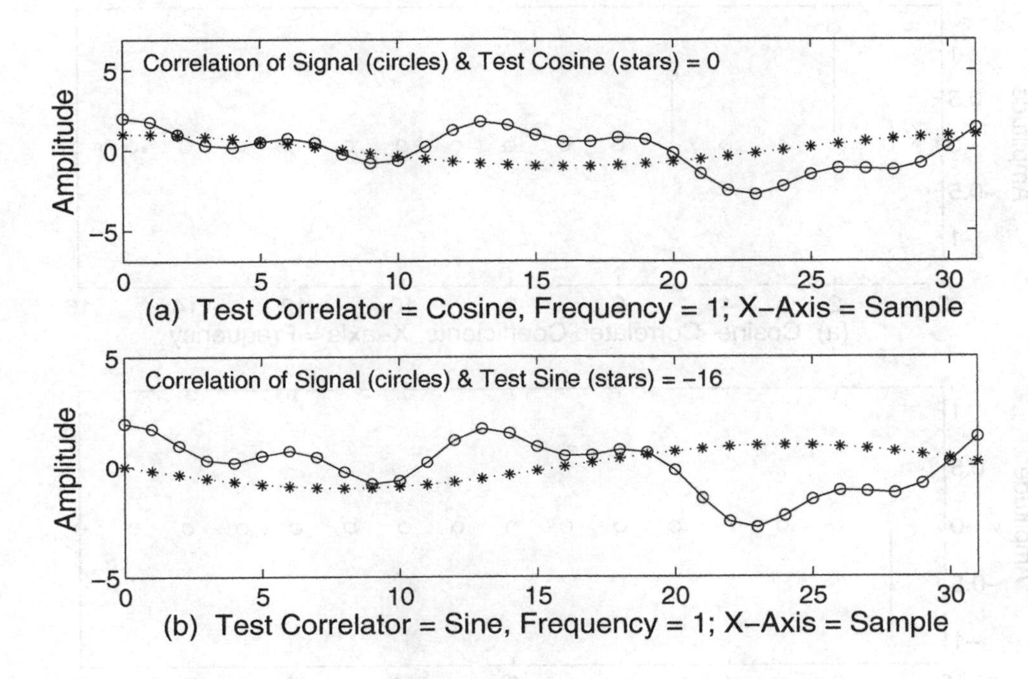

Figure 4.7: (a) Samples of test signal (circles, connected by solid line for visualization of waveform), test cosine correlator (stars, connected with dashed line for visualization of waveform); (b) Samples of test signal (circles, connected with solid line, as in (a)), test sine correlator (stars, connected with dashed line, as in (a)).

4.4 USING ORTHOGONALITY IN SIGNAL TRANSMISSION

The property of orthogonality of sinusoids, i.e., that

$$\sum_{n=0}^{N-1} \cos(2\pi nF/N)\sin(2\pi nF/N) = 0$$

can be put to remarkable use in encoding and transmitting signals. For example, suppose it is desired to transmit two real numbers A and B simultaneously within the same bandwidth. The numbers can be encoded as the amplitudes of *sine* and *cosine* functions having the same frequency F, where F cannot be equal to 0 or $N/2$:

$$S = A\cos(2\pi nF/N) - B\sin(2\pi nF/N)$$

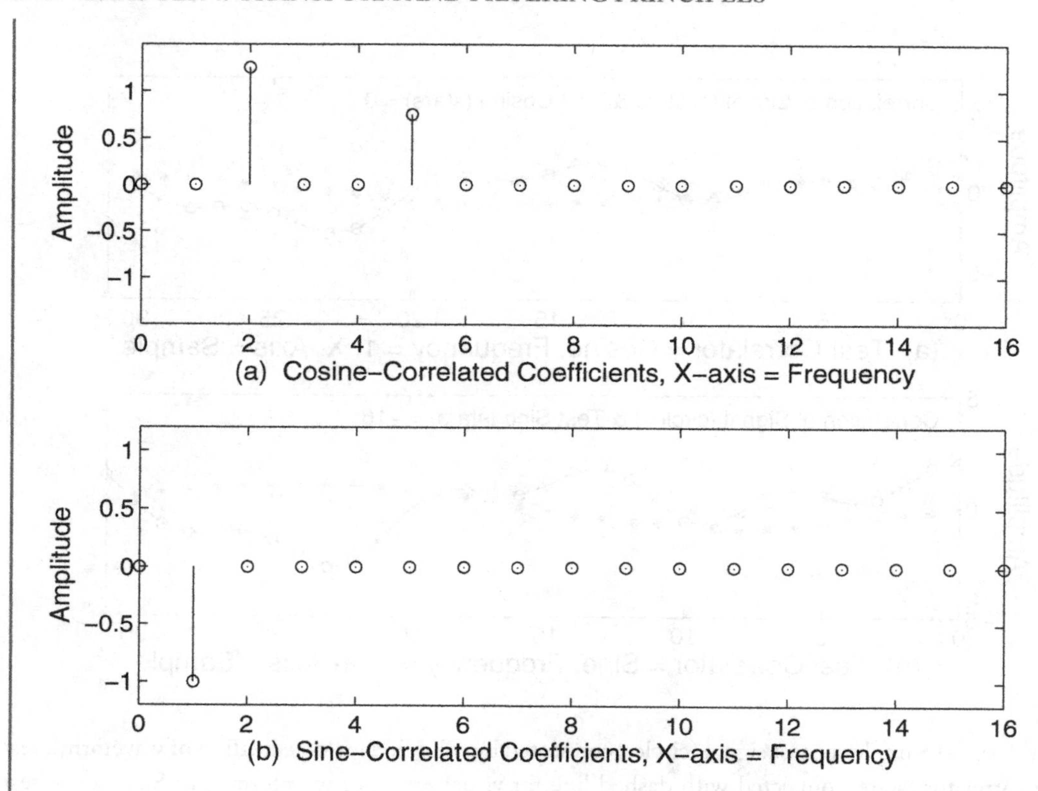

Figure 4.8: (a) Coefficients of cosine-correlated components (scaled according to the synthesis formula); (b) Coefficients of sine-correlated components (scaled according to the synthesis formula).

At the receiving end, because the two carrier waves $\cos(2\pi nF/N)$ and $\sin(2\pi nF/N)$ are orthogonal, it is possible to separately recover the amplitudes of each by multiplying S separately by each, and summing the products (i.e., computing the CZL).

$$S_{R1} = \sum_{n=0}^{N-1} S \cdot \cos(2\pi nF/N)$$

$$S_{R2} = \sum_{n=0}^{N-1} S \cdot \sin(2\pi nF/N)$$

Example 4.9. Let N = 20, F = 7, A = 2 and B = 5. Form the signal S and decode it in accordance with the equations above.

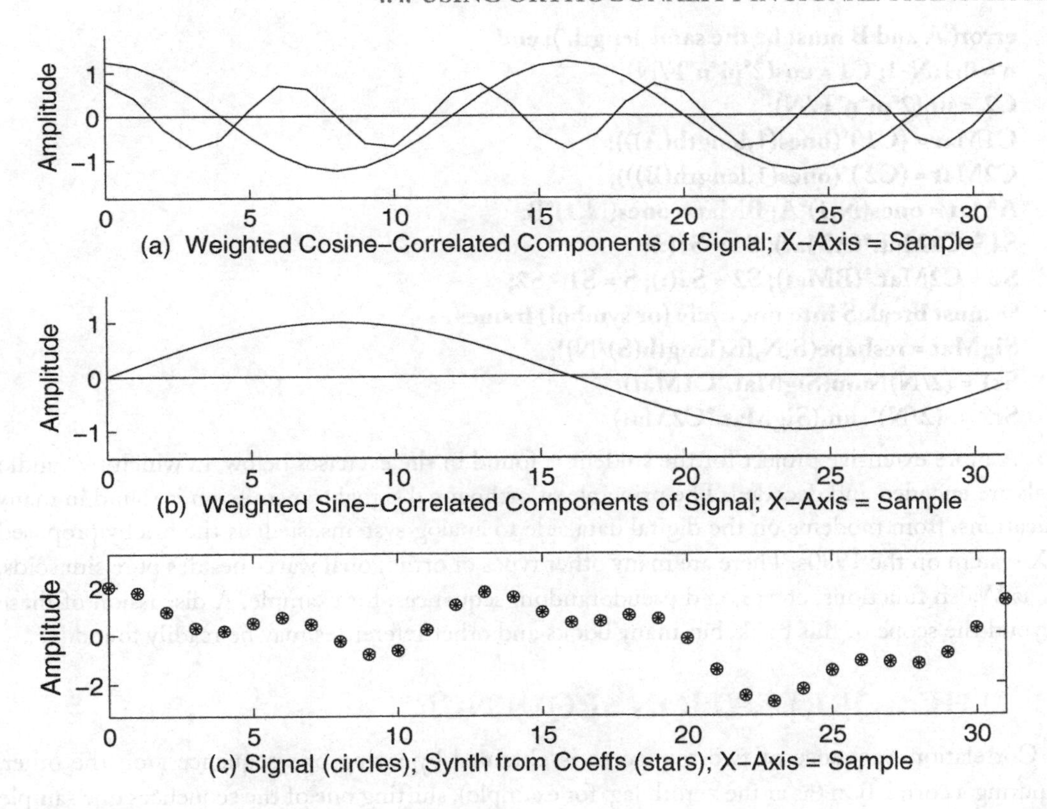

Figure 4.9: (a) Cosine-correlated components, each weighted with its respective coefficient (plotted as continuous functions for ease of visualization); (b) Sine-correlated component, weighted with its coefficient (plotted as continuous function for ease of visualization); (c) Sum of all components, plotted as stars, atop unknown signal's samples, plotted as circles.

A basic script to demonstrate this is

```
function LVOrthogSigXmissBasic(N,F,A,B)
% LVOrthogSigXmissBasic(20,7,2,5)
n = 0:1:N-1; C1 = cos(2*pi*n*F/N);
C2 = sin(2*pi*n*F/N); S = A*C1 - B*C2;
Sr1 = (2/N)*sum(S.*C1)
Sr2 = -(2/N)*sum(S.*C2)
```

A slightly longer script that allows A and B to be equal length row vectors of real numbers is

```
function [Sr1,Sr2] = LVOrthogSigXmiss(N,F,A,B)
% [Sr1,Sr2] = LVOrthogSigXmiss(20,7,[2,-1,3,0,7],[5,2,-6,3,1])
if ~(length(A)==length(B))
```

```
error('A and B must be the same length'); end
n = 0:1:N-1; C1 = cos(2*pi*n*F/N);
C2 = sin(2*pi*n*F/N);
C1Mat = (C1')*(ones(1,length(A)));
C2Mat = (C2')*(ones(1,length(B)));
AMat = ones(N,1)*A; BMat = ones(N,1)*B;
S1 = C1Mat.*(AMat); S1 = S1(:);
S2 = C2Mat.*(BMat); S2 = S2(:); S = S1 - S2;
% must break S into one cycle (or symbol) frames
SigMat = reshape(S,N,fix(length(S)/N));
Sr1 = (2/N)*sum(SigMat.*C1Mat)
Sr2 = -(2/N)*sum(SigMat.*C2Mat)
```

A more extensive project for the student is found in the exercises below, in which two audio signals are encoded and decoded. The principle of orthogonal signal transmission is found in many applications, from modems on the digital data side to analog systems, such as the briefly proposed FMX system on the 1980s. There are many other types of orthogonal waves besides pure sinusoids, such as Walsh functions, chirps, and pseudorandom sequences, for example. A discussion of these is beyond the scope of this book, but many books and other references may be readily found.

4.5 THE CORRELATION SEQUENCE

The Correlation Sequence of two sequences is obtained by laying one sequence atop the other, computing a correlation (as at the zeroth lag, for example), shifting one of the sequences one sample to the left while the other sequence remains in place, computing the correlation value again, shifting again, etc.

We can define the k-th value of the Correlation Sequence C as

$$C[k] = \sum_{n=0}^{N-1} A[n]B[n+k] \tag{4.14}$$

where k is the Lag Index. If sequences $A[n]$ and $B[n]$ are each eight samples in length, for example, C would be computed for values of k between -7 and +7. Note that in the formula above, $B[n+k]$ is defined as 0 when $n + k$ is less than 0 or greater than N - 1. Note also that $C[0]$ is the CZL.

Example 4.10. Use Eq. (4.14) to compute the correlation sequence of the following sequences: $[1, -1]$ and $[-2, 1, 3]$.

The valid index values for each sequence run from 0 to the length of the respective sequence, minus 1. We pick the shorter sequence length to set N = 2 and then determine the proper range of k as -1 to +2 (the range of k must include all values of k for which the sequences overlap by at least one sample). Then we get

$$C[-1] = A[0]B[-1] + A[1]B[0] = +2$$

$$C[0] = A[0]B[0] + A[1]B[1] = -3$$

$$C[1] = A[0]B[1] + A[1]B[2] = -2$$

$$C[2] = A[0]B[2] + A[1]B[3] = 3$$

Note that $B[-1]$ and $B[3]$ lie outside the valid index range (0 to N -1) and are defined as having the value 0.

It should be noted that the computations above may be viewed as the multiplication and summing of overlapping samples from the two sequences (or waveforms) when one waveform is held in place and the other slides over the first from right to left, one sample at a time.

Example 4.11. Show graphically how the correlation sequence of two rectangular sequences is computed according to Eq. (4.14).

Figure 4.10, plot (a) shows two sampled rectangles poised, just overlapping by one sample, to compute the correlation sequence. The rightmost one will "slide" over the leftmost one, one sample at a time, and the correlation value at each position is computed by multiplying all overlapping samples and adding all products. In this example, the amplitudes of the two sequences have been chosen so that all products are 1, making the arithmetic easy to do. Plot (b) shows the correlation sequence, plotted for Lag -7, the first Lag index at which the two rectangles overlap. The LabVIEW VIs

DemoCorrelationRectangles.vi

DemoCorrelationSines.vi

and

DemoCorrelationSineCosine.vi

implement correlation, respectively, of two rectangles, two sines, and sine and cosine, on a step-by-step basis, one sample at a time.

A script that implements the same demonstrations as given by the above-mentioned VIs is

ML_Correlation

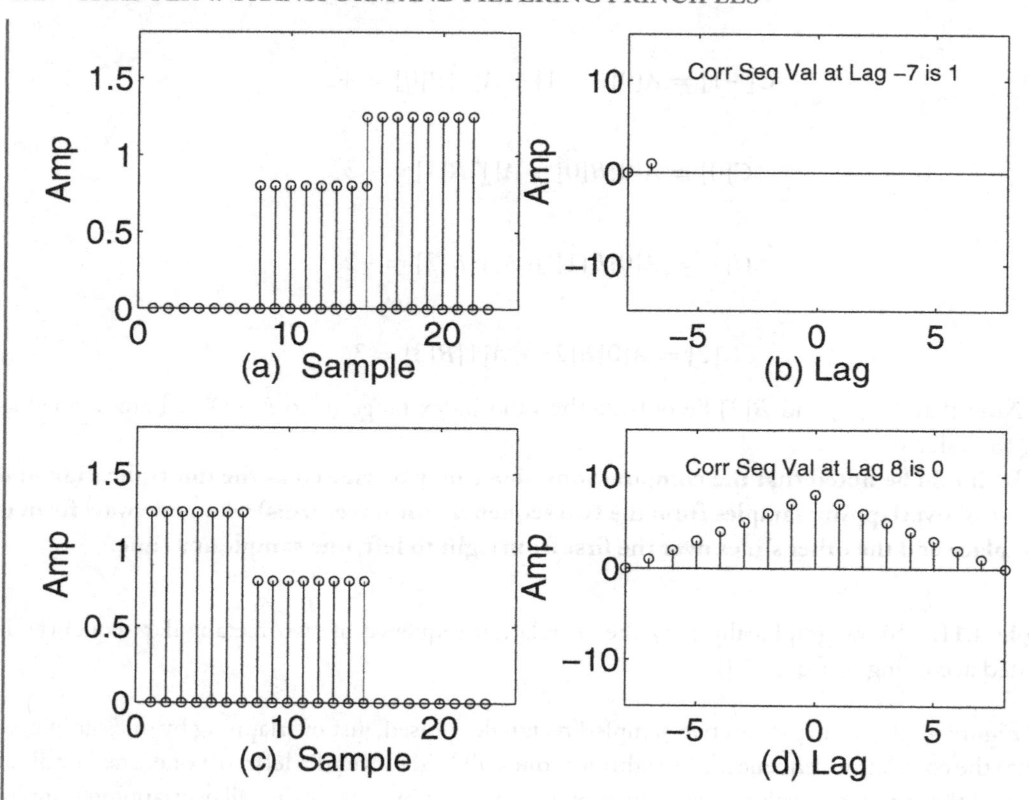

Figure 4.10: (a) Two rectangles just touching, ready to begin computing the correlation sequence; (b) First value of correlation sequence, plotted at Lag -7; (c) The two rectangles, after the one which was rightmost originally has been moved to the left one sample at a time to compute the correlation sequence; (d) Complete correlation sequence.

Figure 4.10, plot (c), shows the result after completely "sliding" the rightmost rectangle over the leftmost rectangle, and hence having computed the entire correlation sequence for the two rectangles.

Example 4.12. Using MathScript, compute and plot the correlation of the sequences $\sin(2\pi(0:1:7)/4)$ and $\cos(2\pi(0:1:7)/8)$.

MathScript provides the function *xcorr* to compute the correlation sequence; if two input arguments are provided, the **Cross-correlation** sequence is computed, which is just the correlation sequence as defined in Eq. (4.14). If only one input argument is provided, the correlation of the sequence with itself is computed, which is referred to as the **Auto-correlation** sequence, which is thus defined as

$$C[k] = \sum_{n=0}^{N-1} A[n]A[n+k] \tag{4.15}$$

We thus run the following m-code:

```
y = xcorr([sin(2*pi*(0:1:7)/8)],[cos(2*pi*(0:1:7)/8)])
figure; stem(y)
```

Example 4.13. Compute and plot the auto-correlation of the sequence $\sin(2\pi(0:1:7)/4)$.

We make the call

```
y = xcorr([sin(2*pi*2*(0:1:7)/8)]); figure; stem([-7:1:7],y)
```

which plots the correlation sequence versus lag number, with 0 representing the two sine sequences laying squarely atop one another without offset, i.e., the CZL. Figure 4.11 shows the results.

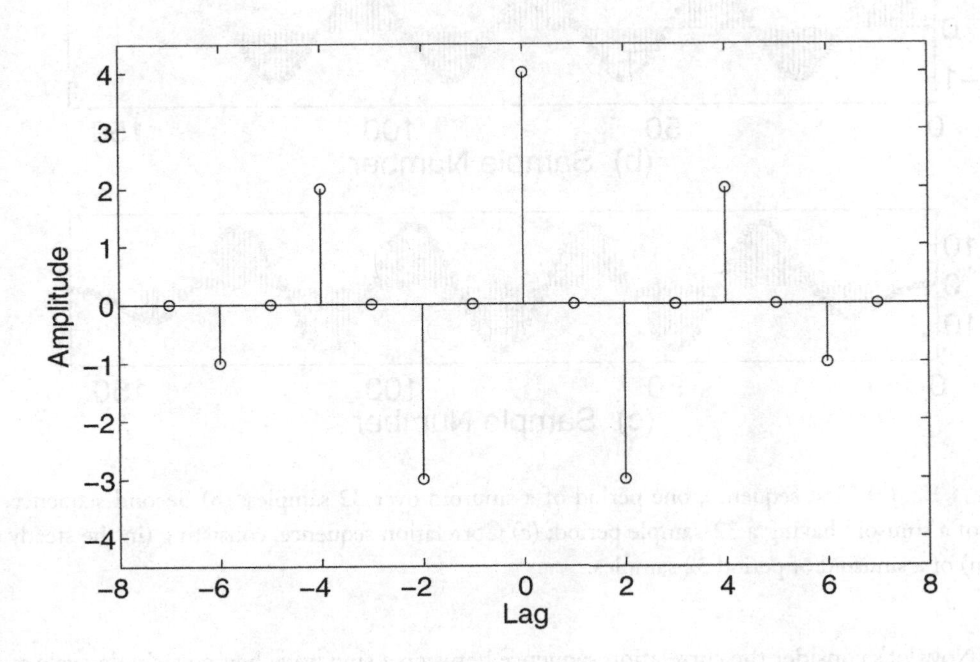

Figure 4.11: The autocorrelation sequence of the sequence $\sin(2\pi[0:1:7]/8)$ plotted against Lag number. Note that the largest value of positive correlation occurs at the zeroth lag when the waveform lies squarely atop itself. This results in every overlying sample pair having a positive product, which in turn results in a large positive sum or correlation value.

Let's compute the correlation sequence between a single cycle of a sine wave having a period of N samples and multiple cycles of a sine wave having a period of N samples. Figure 4.12 shows the correlation sequence generated by correlating a sine wave having a period of 32 samples with four cycles of a sine wave having the same period. Note that the correlation sequence comprises nonperiodic or transient "tail" portions at each end and a central portion that is sinusoidal and periodic over 32 samples.

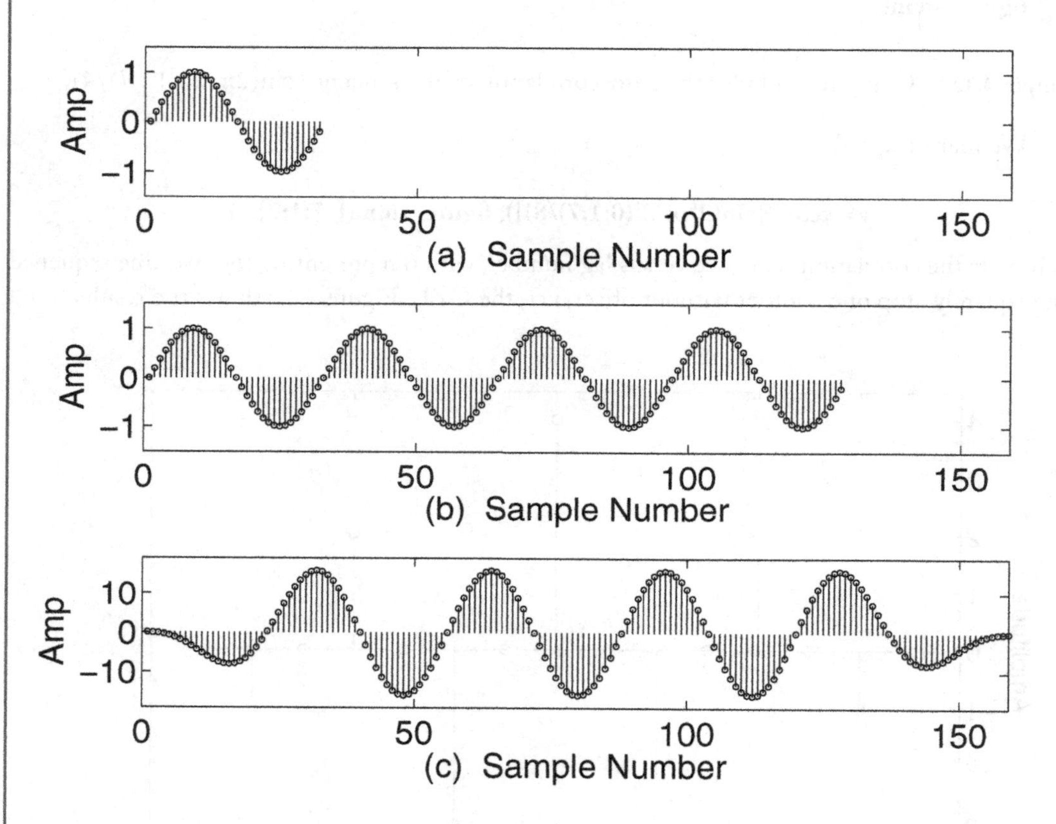

Figure 4.12: (a) First sequence, one period of a sinusoid over 32 samples; (b) Second sequence, four cycles of a sinusoid having a 32-sample period; (c) Correlation sequence, consisting (in the steady state portion) of a sinusoid of period 32 samples.

Now let's consider the correlation sequence between a sine wave having a single cycle over N samples and a sine wave having multiple cycles over N samples. Figure 4.13, shows the result from correlating a single cycle of a sine wave (of period $N = 32$) with a sequence of sine waves having two cycles per N samples. Once the correlation sequence has proceeded to a certain point (the 32nd sample), all samples of the first sequence are overlain with samples of the second sequence. From this time until the second waveform starts to "emerge" (i.e., leave at least one sample of the first sequence "uncovered" or "unmatched"), the correlation sequence value is zero due to orthogonality.

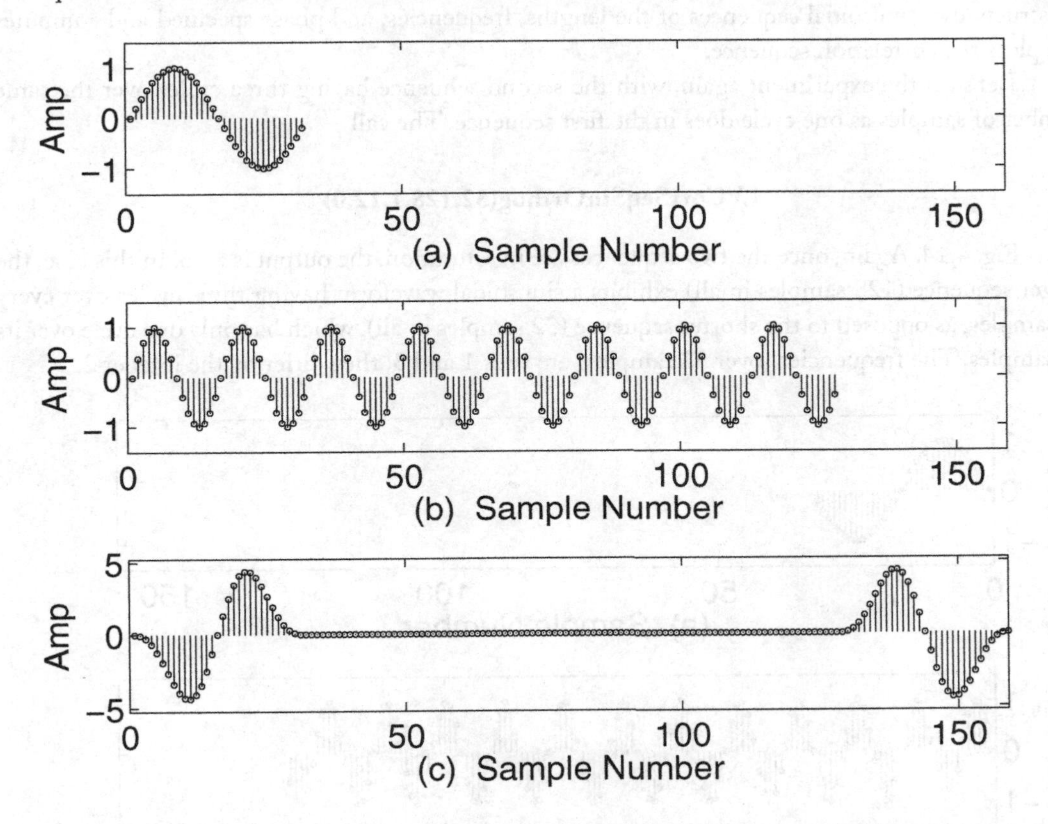

Figure 4.13: (a) First sinusoidal sequence, having one cycle per 32 samples; (b) Second sequence, having two cycles per 32 samples; (c) Correlation sequence of waveforms shown in (a) and (b).

For purposes of discussion, we will refer to the shorter sequence as the correlator, the longer sequence as the test or excitation sequence, and this fully "covered" state as "saturated." In the saturated state, the correlation sequence values reflect a "state-steady" response of the correlator to the test (or excitation) signal when the test signal is periodic.

Figure 4.13 was generated by the script

$$LVCorrSeqSinOrthog(LenSeq1, LenSeq2, Freq1, Freq2, phi)$$

and in specific, the call

LVCorrSeqSinOrthog(32,128,1,8,0)

The script receives as arguments two sequence lengths $LenSeq1$ and $LenSeq2$, two corresponding frequencies $Freq1$ and $Freq2$, and a phase angle for the second sequence. It then constructs two sinusoidal sequences of the lengths, frequencies, and phase specified and computes and plots the correlation sequence.

Let's do the experiment again, with the second sequence having three cycles over the same number of samples as one cycle does in the first sequence. The call

LVCorrSeqSinOrthog(32,128,1,12,0)

yields Fig. 4.14. Again, once the two sequences are in saturation, the output is zero. In this case, the longer sequence (128 samples in all) exhibits a sinusoidal waveform having three cycles over every 32 samples, as opposed to the shorter sequence (32 samples in all), which has only one cycle over its 32 samples. The frequencies (over 32 samples) are thus 1 and 3; they differ by the integer 2.

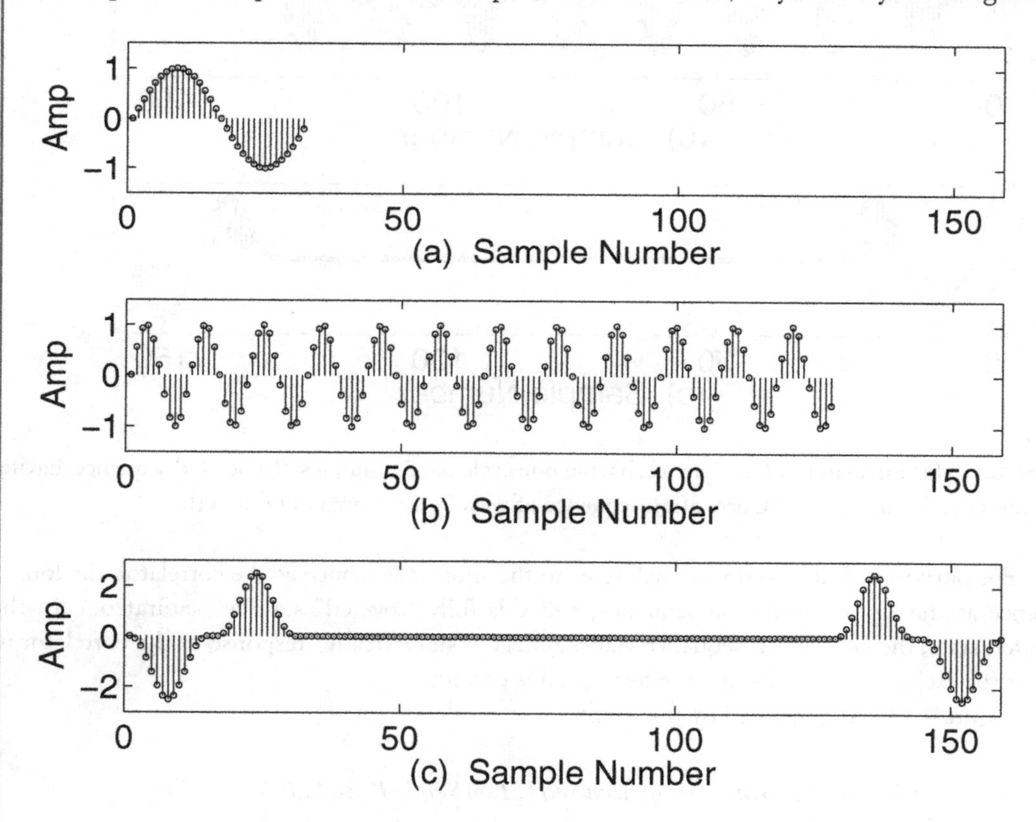

Figure 4.14: (a) First sinusoidal sequence, having one cycle per 32 samples; (b) Second sequence, having twelve cycles per 128 samples, or a net of 3 cycles per 32 samples; (c) Correlation sequence of waveforms shown in (a) and (b).

Example 4.14. Devise a sequence (i.e., correlator) that will yield steady state correlation sequence values of 0 when correlated with the sequence

[1,0,-1,0,1,0,-1,0,1,0,-1,0]

Observe that the given sequence is in fact the half-band frequency, a sinusoid showing one full cycle every four samples. Sinusoids showing 0 or 2 cycles over four samples will be orthogonal and will yield steady-state correlation values of zero. Two possible sequences are therefore [1,1,1,1] and [1,-1,1,-1]. Two more possible sequences are [-ones(1,4)] and [-1,1,-1,1].

Example 4.15. Devise several correlators each of which will eliminate from the correlation sequence (in steady state) most of the high frequency information in the following test sequence.

[1,1,1,1,-1,-1,-1,-1,1,-1,1,-1,1,-1,1,-1,1,1,1,1,-1,-1,-1,-1]

The high frequency information is at the Nyquist limit, and consists of the pattern [1,-1,1,-1]. Correlators such as [1,1] or [-1,-1] (i.e., DC) will eliminate the Nyquist limit frequency. Longer versions also work ([ones(1,4)]) or its negative.

4.6 CORRELATION VIA CONVOLUTION

Previously we've noted that the output of an LTI system can be computed by use of the convolution formula

$$y[k] = \sum_{n=-\infty}^{\infty} x[n]h[k-n]$$

Graphically, this may be likened to time-reversing one of the sequences and passing the left-most sequence (now having negative time indices since it has been time-reversed) to the right through the other sequence, sample-by-sample, and computing the sum of products of overlapping samples for each shift.

Correlation, as we have seen above, may be graphically likened to sliding one sequence to the left through the other, summing the products of overlapping samples at each shift to obtain the corresponding correlation value. Correlation can thus be computed as a convolution, by first time-reversing one of the sequences to be correlated prior to computing the convolution.

Example 4.16. Compute the correlation sequence of [5, 4, 3, 2, 1] and [1, 2, 3, 4, 5] using MathScript's function *xcorr* and again using its function *conv*, and compare the result.

The correlation sequence, obtained by the call

ycorr = xcorr([5,4,3,2,1],[1,2,3,4,5])

is [25, 40, 46, 44, 35, 20, 10, 4, 1].

The result from convolution, obtained by the call

yconv = conv([5,4,3,2,1], fliplr([1,2,3,4,5]))

yields the identical result.

Figure 4.15 illustrates the difference between convolution and correlation for the general case of a nonsymmetric impulse response, while Fig. 4.16 shows the same for the condition of a symmetric impulse response. The latter situation is quite common, since FIR impulse responses, with some exceptions, are usually designed to be symmetric.

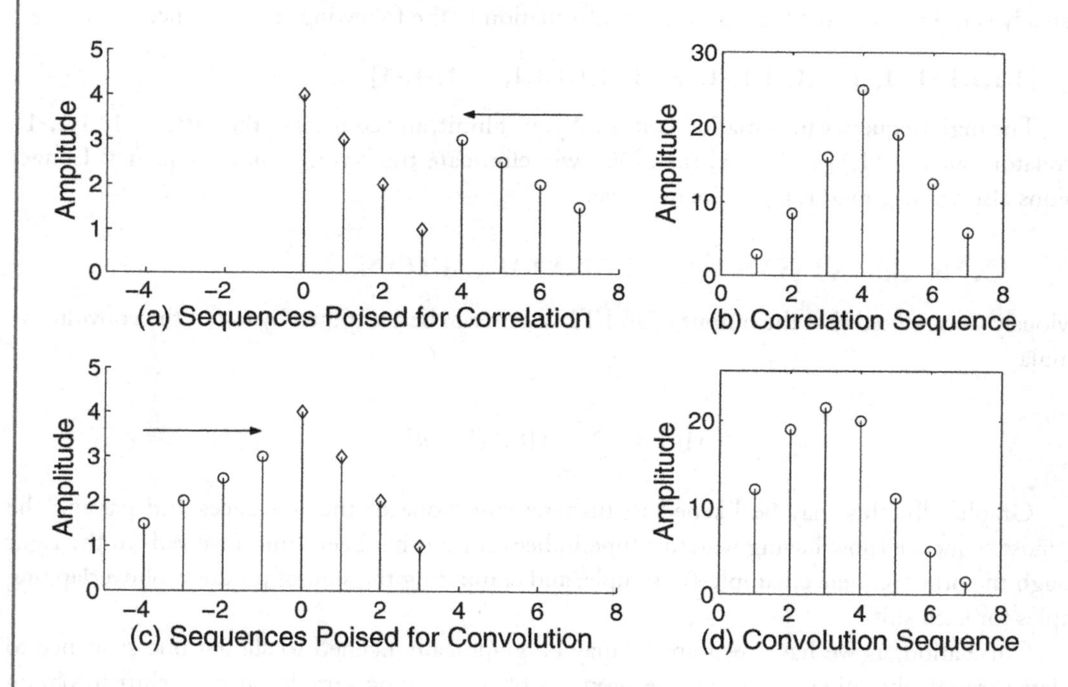

Figure 4.15: (a) Signal sequence, samples 4-7, poised to be correlated with Impulse Response, samples 0-3; (b) Correlation sequence of sequences in (a); (c) Signal, (samples -4 to -1), properly flipped to be convolved with Impulse Response, samples 0-3; (d) Convolution sequence of sequences in (c). The arrows over the signal sequences show the direction samples are shifted for the computation.

Figure 4.17 demonstrates that even though an impulse response may be asymmetric, the convolution and correlation sequences have the same magnitude of chirp response, i.e., the same magnitude of response to different frequencies. Note that there are some differences such as phase and the lead-in and lead-out transients, but the magnitude of the response to different frequencies is essentially the same.

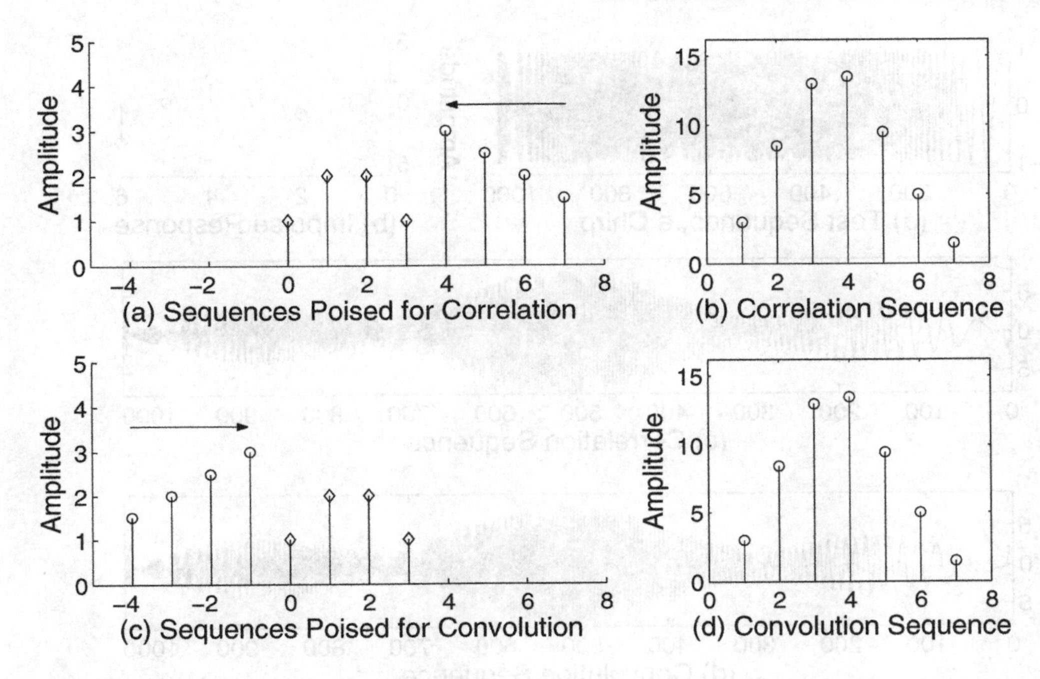

Figure 4.16: (a) Signal sequence, samples 4-7, poised to be correlated with Impulse Response, samples 0-3; (b) Correlation sequence of sequences in (a); (c) Signal, (samples -4 to -1), properly flipped to be convolved with Impulse Response, samples 0-3; (d) Convolution sequence of sequences in (c). Note the symmetrical Impulse Response and hence the identical correlation and convolution sequences.

Note that while the order of convolution makes no difference to the convolution sequence, i.e.,

$$y[k] = \sum_{n=-\infty}^{\infty} b[n]x[k-n] = \sum_{n=-\infty}^{\infty} x[n]b[k-n] \tag{4.16}$$

the order of correlation does make a difference. In general,

$$c_1[k] = \sum_{n=-\infty}^{\infty} b[n]x[n+k]$$

is a time-reversed version of

$$c_2[k] = \sum_{n=-\infty}^{\infty} x[n]b[n+k]$$

In what case would $c_1[k]$ equal $c_2[k]$?

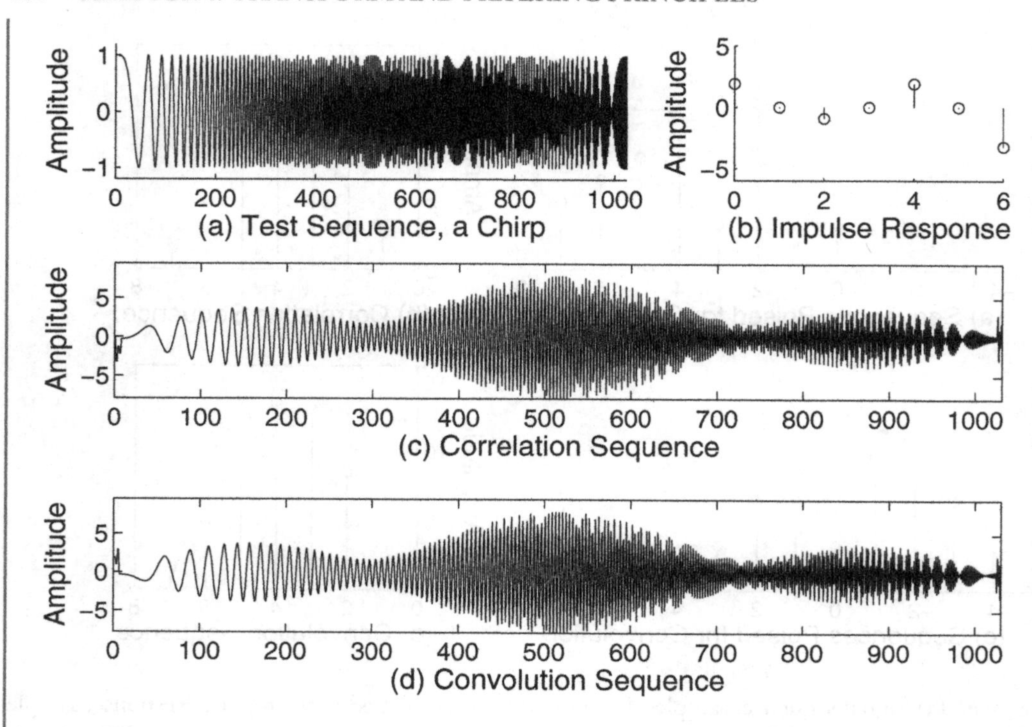

Figure 4.17: (a) Test Linear Chirp, 0 to 512 Hz in 1024 samples; (b) Asymmetric Impulse Response; (c) Correlation Sequence; (d) Convolution Sequence.

4.7 MATCHED FILTERING

Let's consider the problem of detecting an asymmetrically shaped time domain signal, such as a chirp. A good way to detect such a waveform in an incoming signal is to make sure that during convolution, it will correlate well with the impulse response being used. For this to happen, the impulse response will need to be a time-reversed version of the signal being sought. The script

$$LVxMatchedFilter(NoiseAmp, TstSeqLen, FlipImpResp)$$

(see exercises below) illustrates this point. A typical call which reverses the chirp for use as an impulse response in convolution is

LVxMatchedFilter(0.5,128,1)

where the parameter *FlipImpResp* is passed as 1 to reverse the chirp in time, or 0 to use it in non-time-reversed orientation as the filter impulse response. Figure 4.18, plot (a), shows the test sequence, a chirp being received with low frequencies occurring first, containing a large amount of noise, about to be convolved with an impulse response which is the chirp in non-time-reversed format. Plot (b) of the same figure shows the entire convolution sequence, in which no distinctive peak may be seen. Contrast this to the case in which the impulse response is the chirp in time-reversed format, as shown in Fig. 4.19.

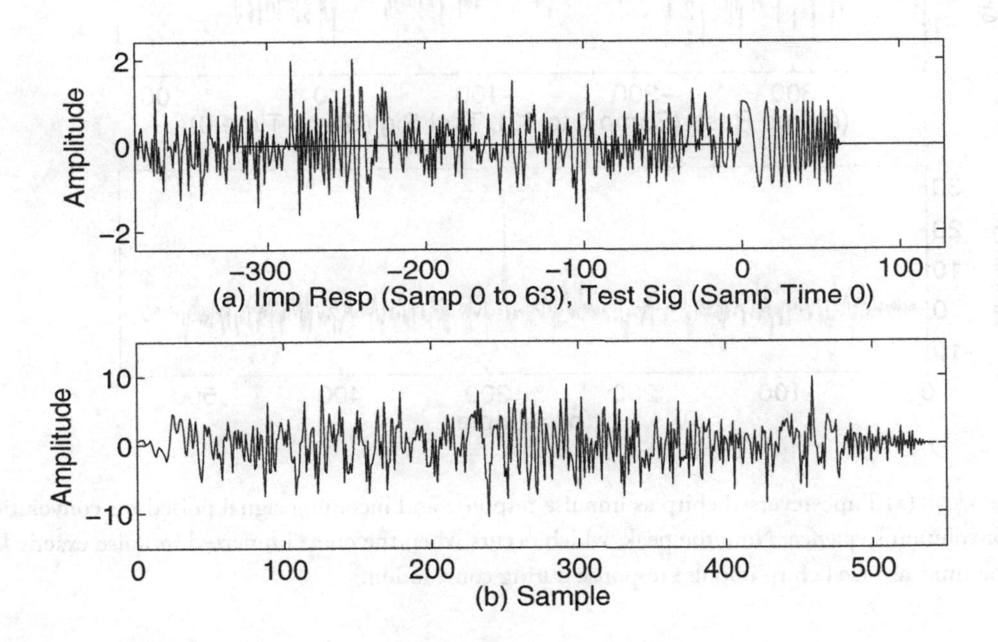

Figure 4.18: (a) Non-time-reversed chirp as impulse response (samples 0 to 63), and incoming signal (samples to left of sample 0); (b) Convolution sequence.

4.8 ESTIMATING FREQUENCY RESPONSE

We've seen above how to determine the frequency content or response of a signal at discrete integral frequencies using Eqs. (4.11) and (4.12). It is also possible to perform similar correlations at as many frequencies as desired between 0 and the Nyquist limit for the signal, leading to a more detailed estimate of the frequency response of a test signal when, for example, considered as a filter impulse response. The following code will estimate the frequency response at a desired number of evenly spaced frequency samples between 0 and the Nyquist limit for the test signal. The loop (four indented lines) may be replaced with the commented-out vectorized code line immediately following. The cosine- and sine-based correlations are done simultaneously with one operation by summing the two correlators after multiplying the sine correlator by $-j$. The result from making the call

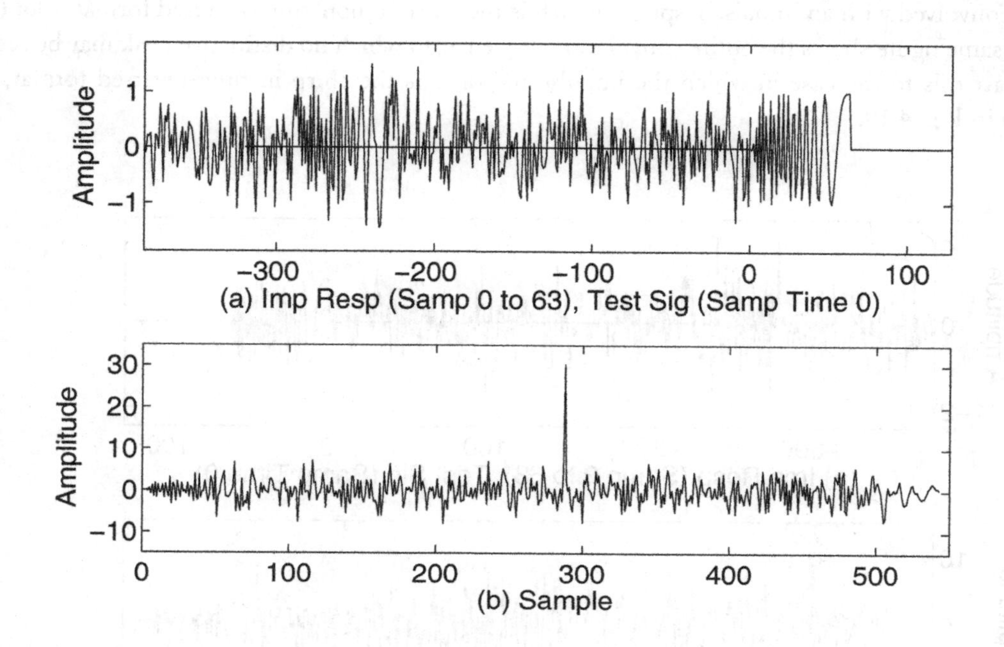

Figure 4.19: (a) Time-reversed chirp as impulse response and incoming signal poised for convolution; (b) Convolution sequence. Note the peak, which occurs when the chirp immersed in noise exactly lays atop the time-reversed chirp impulse response during convolution.

LVFreqResp([1.9,0,-0.9,0,1.9,0,-3.2],500)

is shown in Fig. 4.20 (the test signal is the same as that used in Fig. 4.17, in which the frequency response was estimated using a linear chirp).

```
function LVFreqResp(tstSig, NoFreqs)
% LVFreqResp([1.9, 0, -0.9, 0, 1.9, 0, -3.2], 500)
NyqLim = length(tstSig)/2; LTS = length(tstSig);
t = [0:1:(LTS-1)]/(LTS); FR = [];
frVec = 0:NyqLim/(NoFreqs-1):NyqLim;
for Freq = frVec    % FR via loop
testCorr = cos(2*pi*t*Freq) - j*sin(2*pi*t*Freq);
FR = [FR, sum(tstSig.*testCorr)];
end
% FR = exp(-j*(((2*pi*t)'*frVec)'))*(tstSig');
figure(9)
```

```
xvec = frVec/(frVec(length(frVec)));
plot(xvec,abs(FR));
xlabel('Normalized Frequency')
ylabel('Magnitude')
```

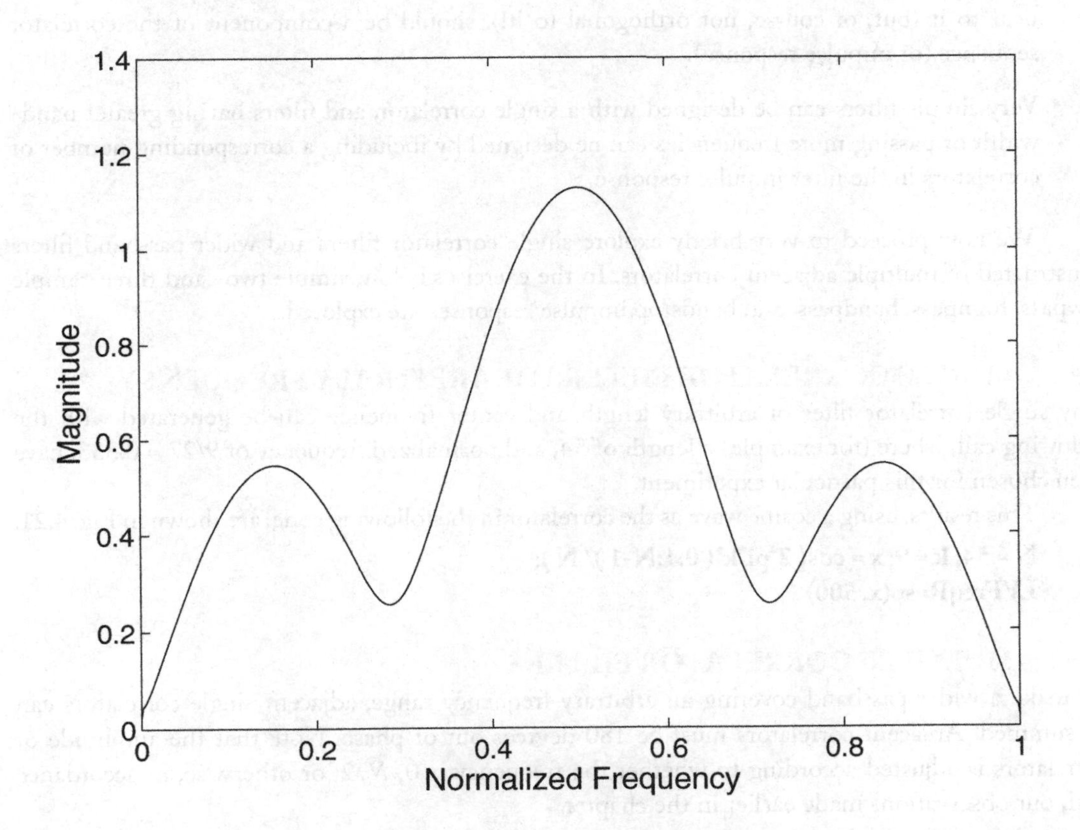

Figure 4.20: The magnitude of frequency response at 500 frequency samples of the sequence [1.9, 0, -0.9, 0, 1.9, 0, -3.2], estimated using cosine-sine correlator pairs at evenly spaced frequencies between 0 and 1.0, normalized frequency.

4.9 FREQUENCY SELECTIVITY

So far, we've seen that:

- A sinusoid of a given period will correlate very well with itself, yielding (in saturation or steady state) a sinusoidal correlation sequence having the same period.

- The correlation sequence of sinusoids that have different integral-valued frequencies over the same number of samples N is zero when in steady state.
- A broad principle is that if any particular frequency is desired to be passed from the test sequence to the output sequence with substantial amplitude, that frequency, or a frequency near to it (but, of course, not orthogonal to it!), should be a component of the correlator sequence (or impulse response).
- Very simple filters can be designed with a single correlator, and filters having greater bandwidth or passing more frequencies can be designed by including a corresponding number of correlators in the filter impulse response.

We now proceed to very briefly explore single correlator filters and wider passband filters constructed of multiple adjacent correlators. In the exercises below, simple two- and three-sample lowpass, highpass, bandpass, and bandstop impulse responses are explored.

4.9.1 SINGLE CORRELATOR FILTERS OF ARBITRARY FREQUENCY

Any single-correlator filter of arbitrary length and center frequency can be generated with the following call, where (for example) a length of 54, and normalized frequency of 9/27 = 0.333, have been chosen for this particular experiment.

This results, using a cosine wave as the correlator in the following code, are shown in Fig. 4.21.

```
N = 54; k = 9; x = cos( 2*pi*k*( 0:1:N-1 )/ N );
LVFreqResp(x, 500)
```

4.9.2 MULTIPLE CORRELATOR FILTERS

To make a wider passband covering an arbitrary frequency range, adjacent single correlators can be summed. Adjacent correlators must be 180 degrees out of phase. Note that the amplitude of correlators is adjusted according to whether the frequency is 0, $N/2$, or otherwise, in accordance with our observations made earlier in the chapter.

```
function Imp = LVBasicFiltMultCorr(N,LoK,HiK)
% Imp = LVBasicFiltMultCorr(31,0,7)
Imp = 0;
for k = LoK:1:HiK
if k==0|k==N/2
A = 1; else; A = 2; end
Imp = Imp + A*((-1)^k)*cos( 2*pi*k*( 0:1:N-1 )/ N );
end
LVFreqResp(Imp1, 500)
```

Three filters were generated using the above code with the following calls

```
Imp = LVBasicFiltMultCorr(30,0,4);
```

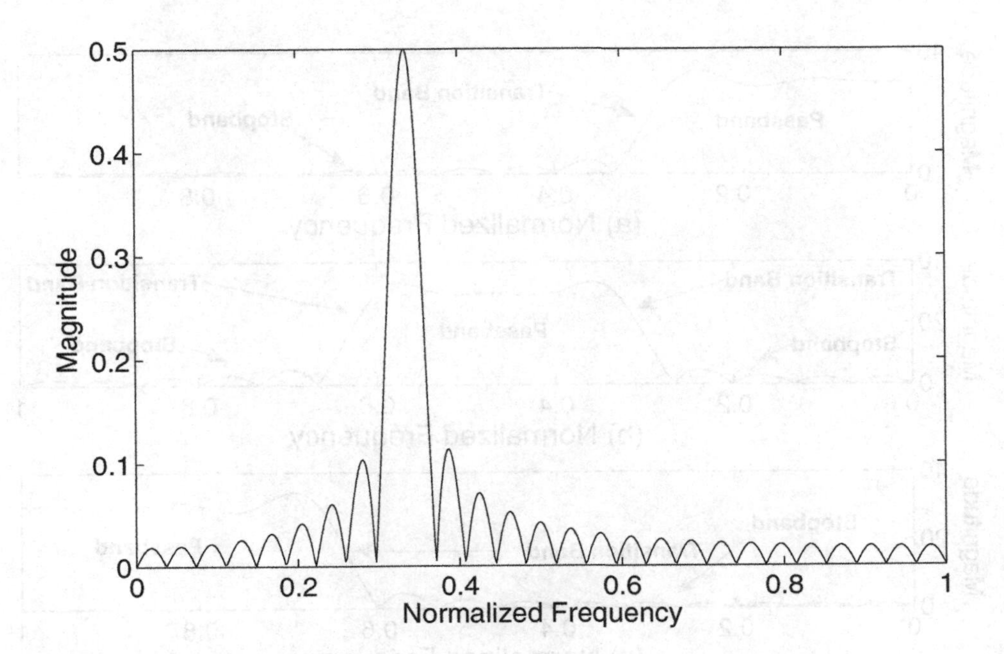

Figure 4.21: The frequency response of a particular single correlator filter.

```
Imp = LVBasicFiltMultCorr(30,5,9);
Imp = LVBasicFiltMultCorr(30,10,15);
```

the results of which are shown in Fig. 4.22.

4.9.3 DEFICIENCIES OF SIMPLE FILTERS

Although it is clear that the simple filters we have studied can provide basic filtering responses such as lowpass, etc., there was in actuality very little or no control over a number of parameters that are important. These parameters, which are discussed in detail in Part III of the book. include the amount of ripple (deviation from flatness) in the passband, the maximum response in the stopband(s), the steepness of roll off or transition from passband to stopband, and the phase response of the filter. The lack of adequate signal suppression in the stopbands, is clearly shown in Figs. 4.21 and 4.22, as is the large amount of passband ripple. These and the other deficiencies mentioned will be attacked using a variety of methods to effectively design excellent filters meeting user-given design specifications. The description of these methods requires a number of chapters, but with the brief look we have had at frequency-selective filtering using the fundamental idea of correlation, the reader should have little difficulty understanding the principles of operation and design of FIR filters when they are encountered in Part III of the book.

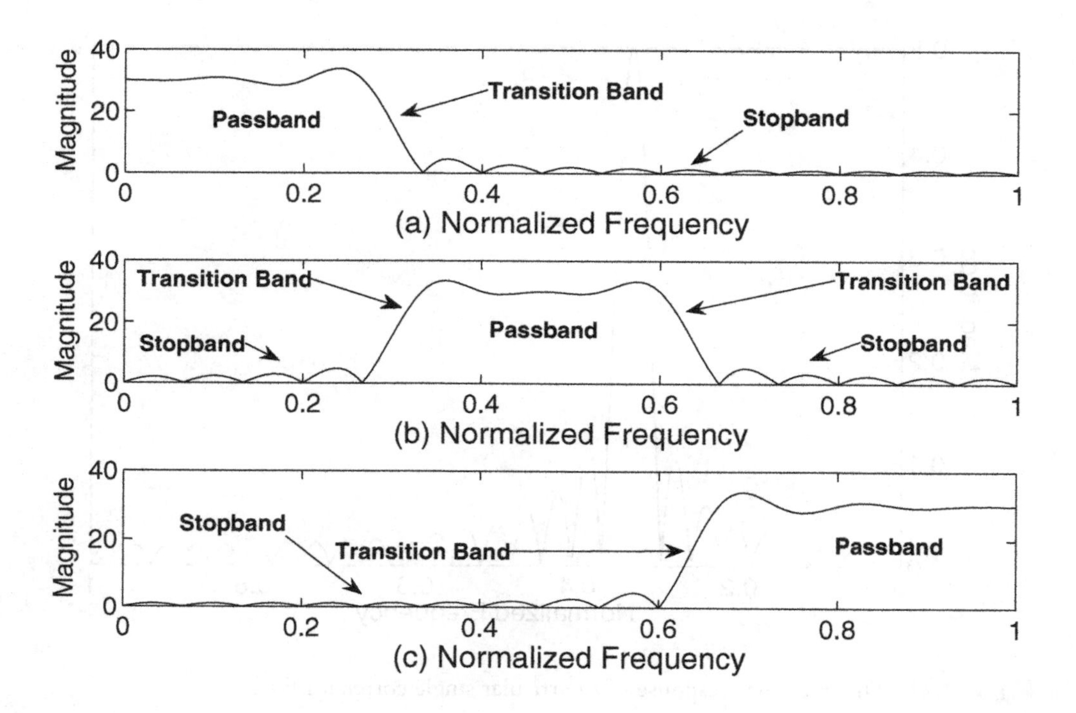

Figure 4.22: (a) Magnitude response of a simple lowpass filter made by summing adjacent cosine correlators; (b) Same, but bandpass; (c) Same, but highpass. For purposes of discussion, a "passband" can be defined as a range of frequencies over which the desired filter response is above a certain level, a "stopband" can be defined as a range of frequencies over which the desired filter response is below a certain level or near zero, and a "transition band" is a range of frequencies lying between a passband and a stopband.

4.10 SINUSOIDAL FIDELITY

We noted earlier that in saturation, the correlation sequence between a correlator and a periodic sinusoidal excitation signal contains a steady-state or periodic response in saturation. Now let's consider Fig. 4.23, which shows two sequences, a 32-sample sine wave correlator in plot (a), a 128-sample excitation sequence containing 11.7 cycles of a sine wave in plot (b), and the correlation sequence in plot (c).

Figure 4.23 clearly shows two transient portions, each of which has a length of one sample less than the length of the correlator. The remaining central portion of the correlation sequence constitutes the steady-state response. Note that the steady-state response contains only the frequency content of the test signal.

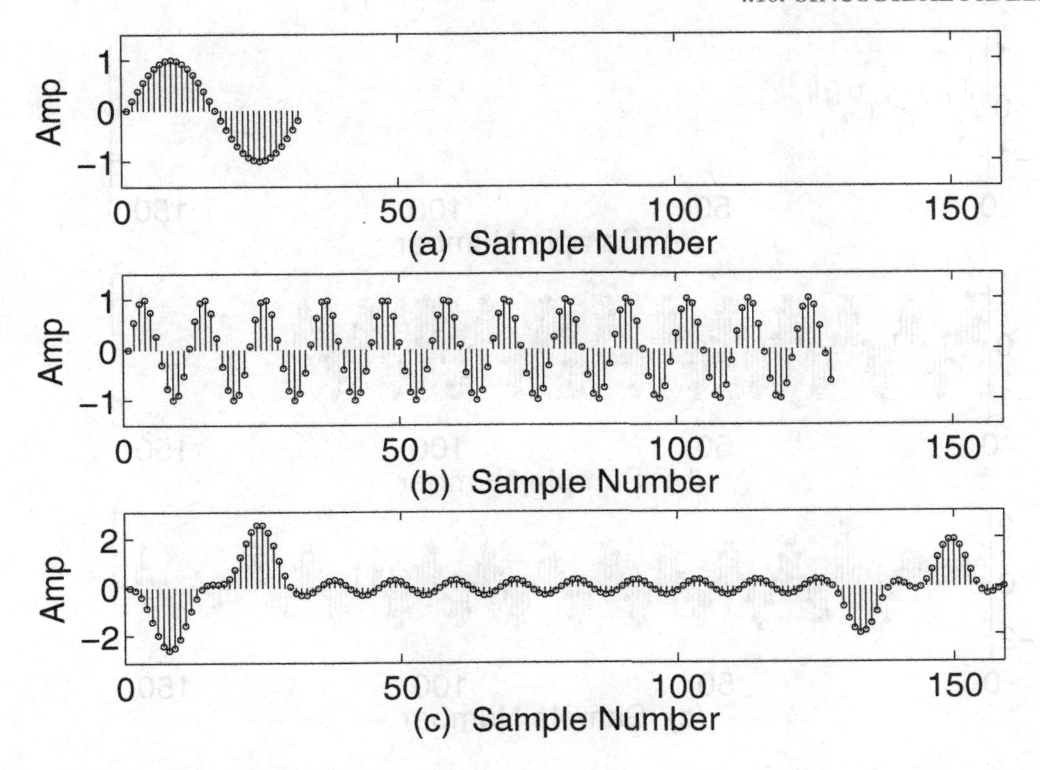

Figure 4.23: (a) Correlator; (b) Test signal; (c) Correlation sequence of sequences shown in (a) and (b).

The steady-state portion of the correlation sequence between a sinusoidal excitation signal and a correlator is a sinusoid with the same frequency (or period, in samples) of the sinusoid in the excitation sequence. In other words, what goes in is what comes out, frequency or period-wise, in steady-state. Don't forget that if the two sequences happen to have different integral numbers of cycles over the same period N, the steady-state output will be zero. The correlator need not be sinusoidal, it can be a periodic waveform or random noise, for example, and the same principle holds true, as shown in Fig. 4.24.

This sinusoid-goes-in-sinusoid-comes-out principle is usually referred to as the principle of **Sinusoidal Fidelity**, or **Sinusoidal Invariance.**

By way of contrast, Fig. 4.25 shows the situation when the correlator at plot (a) is a sinusoid and the excitation sequence (at plot (b)) is random noise; the correlation sequence (at plot (c)) is not periodic, since the excitation sequence is not periodic.

The principle of sinusoidal fidelity is only true for sinusoids, which are basic, or fundamental waves which do not break down further into constituent waves. For example, square and sawtooth waves as well as all other periodic waves, are made up of a superposition of harmonically related

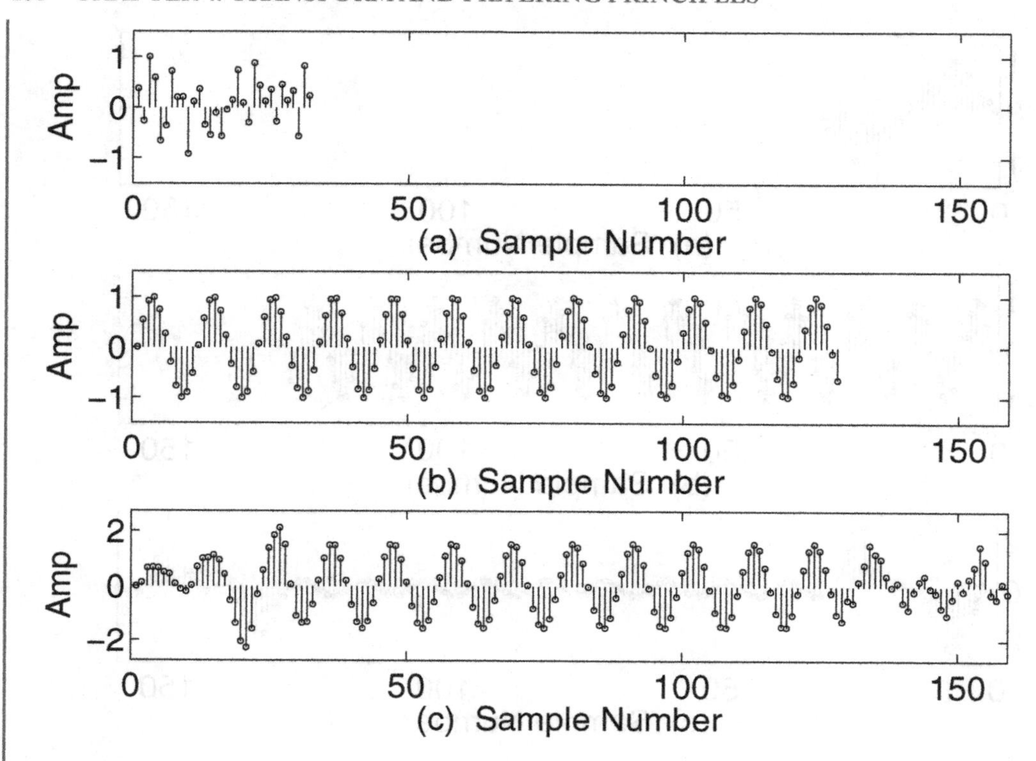

Figure 4.24: (a) Correlator, random noise; (b) Test signal; (c) Correlation sequence of sequences shown in (a) and (b).

sinusoids. A sawtooth, for example, requires that its constituent sinusoids have particular amplitudes and that they are all in-phase. In Fig. 4.26, plot (a) shows a correlator, 32 samples of random noise, plot (b) shows a sawtooth, periodic over 12 samples, and plot (c) shows the correlation sequence, in which the sinusoidal constituents of the sawtooth have had their amplitudes and phases (but not their frequencies) randomly shifted by the correlator, resulting in an output waveform having a random shape (peridocity, is, however, retained). For a pure (single-frequency) sinusoidal input, the output waveshape and periodicity are the same as the input, with only the amplitude and phase changing. However, a complex periodic waveform requires that the relative amplitudes and phases of its constituent sinusoids remain the same, otherwise the waveform loses its characteristic shape.

4.11 DETERMINATION OF TIME DELAY USING CORRELATION

Let's take a look at another use for the correlation sequence, estimating the time delay between two sequences.

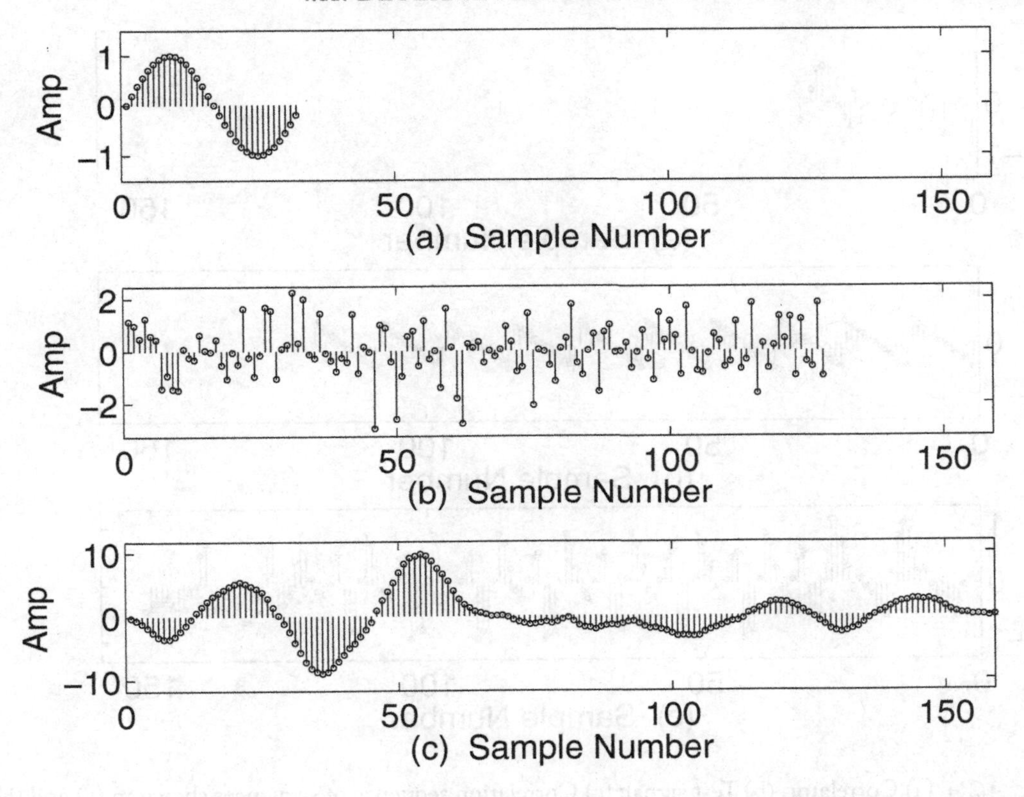

Figure 4.25: (a) Correlator, one cycle of a sine wave over 32 samples; (b) Excitation sequence, 128 samples of random noise; (c) Correlation sequence.

Figure 4.27 shows the basic setup: two microphones pick up the same sound, but they are at different distances from the sound source. Hence sound features found in the digital sequence from the second microphone will be delayed relative to those same features found in the first microphone's digital sequence. What is sought is the difference in arrival times of the same signal at the two microphones. This information can be used to estimate azimuth (angle) from a point between the microphones to the sound source. This kind of information can be used, for example, to automatically aim a microphone or video camera.

Figure 4.28 depicts two sequences as they were initially captured. The second sequence has several extra delayed versions of the original sound feature (a single cycle of a sine wave) to better illustrate the determination of time delay using the cross-correlation sequence. These delayed versions would be analogous to echoes or reflected waves. In a real situation, such echoes would likely be present in the first sequence as well, but they have been left out here so the correlation sequence

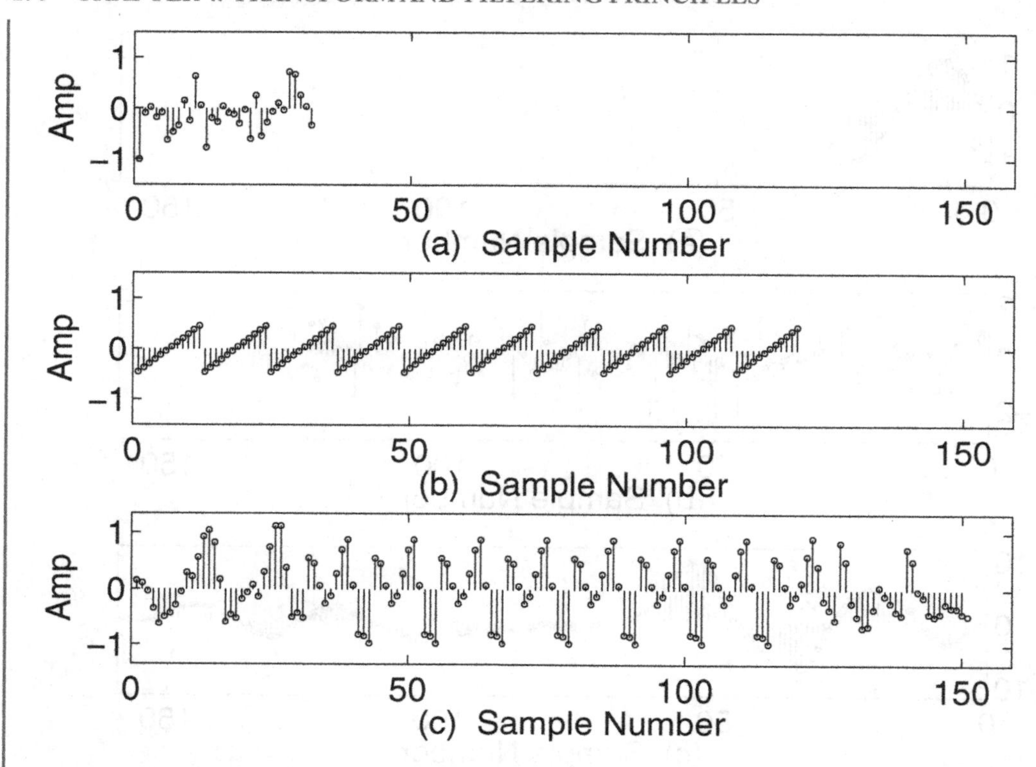

Figure 4.26: (a) Correlator; (b) Test signal; (c) Correlation sequence of sequences shown in (a) and (b).

will be simple to interpret. The use of directional microphones, very common in practice, would also make a significant difference in the content of sequences captured by the two microphones.

The script

LVxCorrDelayMeasure(0.1)

(see exercises below) steps through the computation of the cross-correlation sequence between the two hypothetical captured sound sequences as depicted in Fig. 4.28. The argument in parentheses in the function call specifies the amplitude of white noise added to the basic signal to better simulate a real situation.

Figure 4.29 shows the full cross-correlation sequence, which clearly shows the delays from the principle feature of the first sequence to the corresponding delayed versions found in the second sequence.

At Lag 35 (we started out with the two sequences overlapping, i.e., at Lag zero, and started shifting to the left, which increases the Lag index), the main sound feature has reached maximum correlation with the first sequence. At this point, if we know the sample rate, we can compute the time delay and also the path length difference between Paths 1 and 2 from the sound source to the

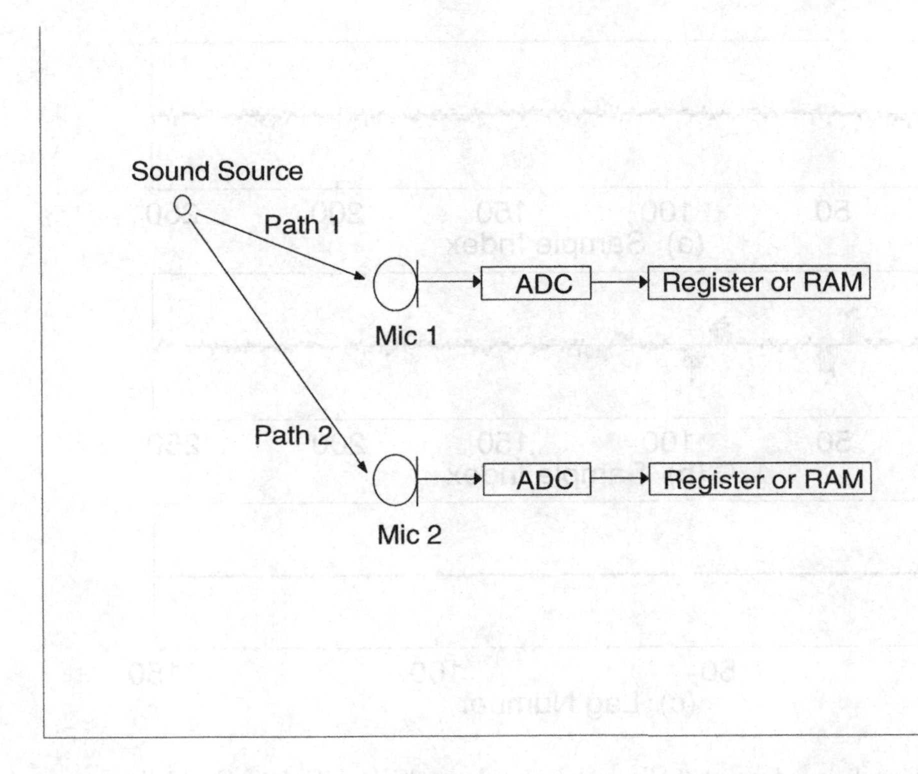

Figure 4.27: Two microphones receiving a sound in a room. The second microphone, further away from the sound source than the first microphone, receives the sounds later. The sounds are digitized and stored in the respective registers.

two microphones. For example, if the sample rate were 8 kHz, 35 samples of delay would yield about 0.0044 second of delay. Since sound travels at around 1080 ft/second in air, this implies that the path length difference in feet between paths 1 and 2 was

$$(0.0044 \text{ sec})(1080 \text{ ft/ sec}) = 4.725 \text{ ft}$$

4.12 THE SINGLE-POLE IIR

4.12.1 PHYSICAL ARRANGEMENT

Consider the arrangement shown in the left-hand portion of Fig. 4.30. An input signal (shown in plot (a)) enters a summing junction; the sum exits and becomes the output, but the output value also enters a delay, which has an input side and an output side. The value at the input side moves to the output side at every sample or clock time. The large triangle with a number in its interior is a gain block or multiplier, and the number in its interior is multiplied by the delayed output signal, and the

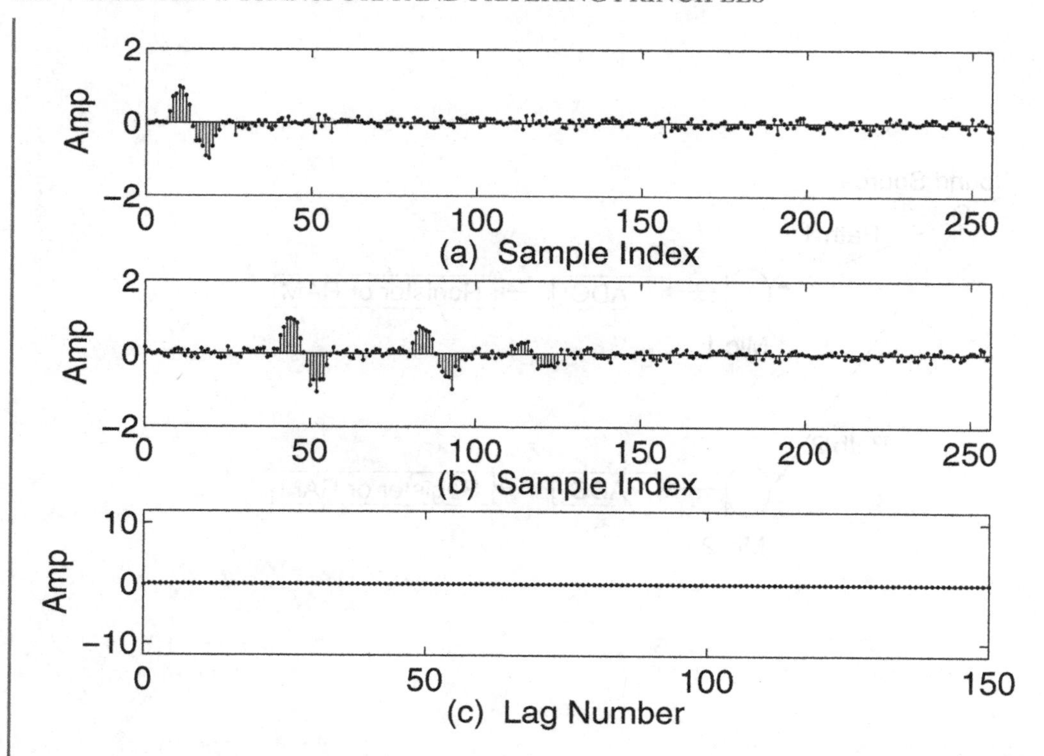

Figure 4.28: (a) First digitized sound; (b) Second digitized sound, not shifted; (c) Correlation sequence (initialized with zeros).

product is passed to the summing junction. The value of the gain is called the *pole*, and is in general, a complex number.

4.12.2 RECURSIVE COMPUTATION

In the single pole IIR, the current (or n^{th}) output of the filter is equal to the current (n^{th}) input, weighted by coefficient b, plus the previous (or $(n-1)^{th}$) output weighted by coefficient a (which, for this simple single pole case, is equal to the pole). This can be written as

$$y[n] = bx[n] + ay[n-1] \tag{4.17}$$

Example 4.17. Filter the sequence s = $[s_0]$ with a single pole IIR having b = 1 and a = p.

The filter impulse response is $h[n]$ =1, p, p^2, p^3, ...etc. and the output sequence is s_0, ps_0, p^2s_0, p^3s_0... , which is clearly the impulse response weighted by s_0, i.e., $s_0h[n]$.

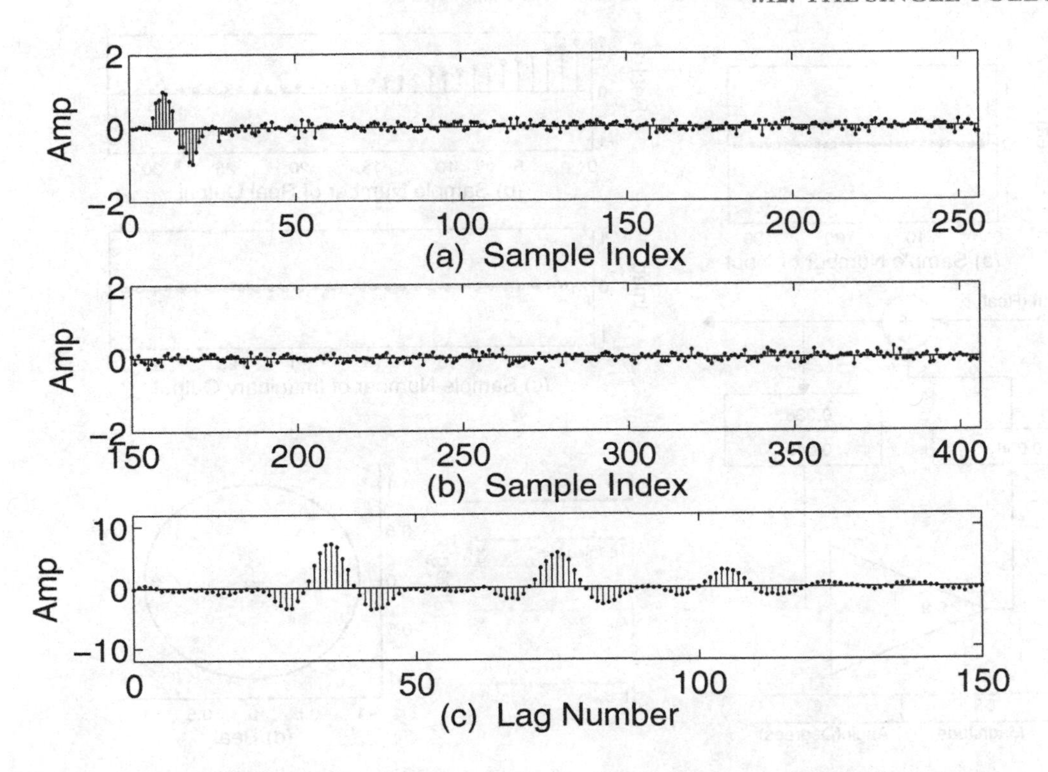

Figure 4.29: (a) First digitized sound; (b) Second digitized sound, shifted 150 samples to the left; (c) Correlation sequence (up to Lag 150).

If, for example, $s_0 = 2$, then the output sequence is $2, 2p, 2p^2, ... = 2h[n]$.

Example 4.18. Filter the sequence s = $[s_0\ s_1]$ with the same IIR as used immediately above and show that the output is the superposition of weighted, delayed versions of the filter impulse response.

The first few outputs are

$$s_0 h[0],\ s_1 h[0] + s_0 h[1],\ s_1 h[1] + s_0 h[2], \ldots \tag{4.18}$$

which can be seen as the sum of two sequences, the second delayed by one sample, namely,

$$s_0 h[0],\ s_0 h[1],\ s_0 h[2] \ldots = s_0 h[n]$$

and

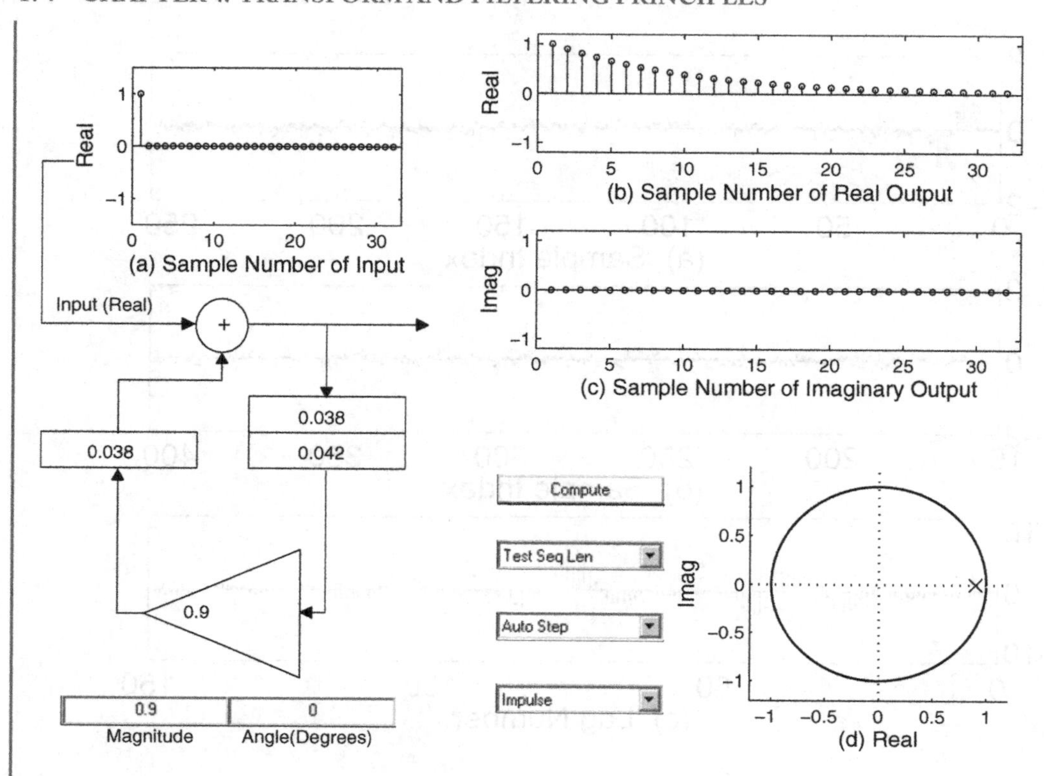

Figure 4.30: (a) Input sequence, a unit impulse; (b) Real part of output sequence, a decaying exponential; (c) Imaginary part of output sequence, identically zero; (d) Complex plane with unit circle and the pole plotted thereon.

$$0, s_1h[0], s_1h[1], s_1h[2]... = s_1h[n-1]$$

and thus it is apparent that the filter output is the superposition of sample-weighted, time-offset versions of the filter impulse response, i.e., the filter output is the convolution of the signal sequence and the filter's impulse response. Output sequence (4.18) can also be visualized as a convolution in which $h[n]$ is flipped from right to left and moved through the signal, i.e.,

$$y[k] = \sum_{n=0}^{1} x[n]h[k-n]$$

where the limits of n have been set appropriately for the signal length.

4.12.3 M-CODE IMPLEMENTATION

The following code shows how to implement the simple difference equation 4.17 in m-code, using values of x, b, p, and SR as shown:

```
SR = 24; b = 1; p = 0.8; y = zeros(1,SR);
x = [1,zeros(1,SR)]; y(1) = b*x(1);
for n = 2:1:SR
y(1,n) = b*x(1,n) + p*y(1,n - 1);
end; figure;
stem(y)
```

The *for* loop, of course, makes this an iterative or recursive computation. Hence the equation serves as a **recursive filter**, a term often used to describe IIR filters.

The function *filter* can also be used to compute the output of a single pole IIR using the following call syntax:

Output = filter(b,[1,-p],Input)

where $Input$ is a signal such as (for example) the unit impulse, unit step, a chirp, etc., and b and p are as used in Eq. (4.17).

4.12.4 IMPULSE RESPONSE, UNIT STEP RESPONSE, AND STABILITY

If you are at all familiar with feedback arrangements, you should suspect that if the feedback weight or gain (or in other words, the pole's magnitude) is too large, the filter will become unstable. To remain stable, the magnitude of the pole must be less than 1.0. In the following discussion, we use the impulse and unit step responses corresponding to poles of several magnitudes (< 1.0 and 1.0) to explore the issue of stability in the single-pole IIR.

Impulse Response

Figure 4.30 shows the result from using 0.9 as the value of the pole and an impulse as the input signal. The resultant output, the impulse response, ultimately decays away, and the filter's response to a bounded signal (one having only finite values) is stable. The n-th value of the impulse response $y[n]$ is

$$y[n] = p^n; n = 0 : 1 : \infty$$

- If $|p| < 1, |p|^n \to 0$ as $n \to \infty$
- If $|p| = 1, |p|^n = 1$ for all n
- If $|p| > 1, |p|^n \to \infty$ as $n \to \infty$

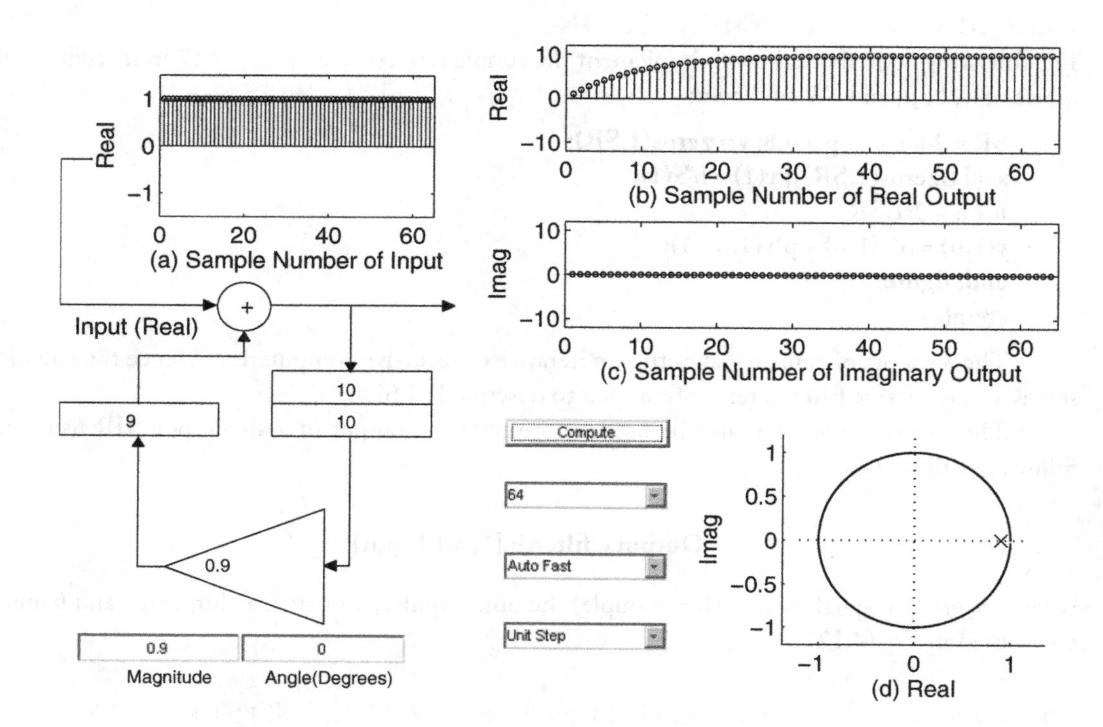

Figure 4.31: (a) Input sequence, a unit step sequence; (b) Real part of output sequence; (c) Imaginary part of output sequence; (d) The pole, plotted in the complex plane.

Unit Step Response

Figure 4.31 shows the unit step (i.e., DC or frequency 0) response of an IIR with a pole at 0.9. We can determine an expression for the steady-state unit step (DC) response of a single-pole filter by observing the form of the response to the unit step. If the value of the pole is p, then the sequence of output values is 1, $1 + p$, $1 + p + p^2$, etc., or at the N-th output

$$y[N] = \sum_{n=0}^{N} p^n \tag{4.19}$$

If $|p| < 1.0$, then $y[N]$ converges to

$$Y_{SS} = \frac{1}{1-p} \tag{4.20}$$

where Y_{SS} is the response when $N = \infty$,which we'll refer to as the steady state response. In reality, of course, N never reaches infinity, but the difference between the theoretical value of Y_{SS} and $y[N]$

can be made arbitrarily small by increasing N. In the case of Fig. 4.31, the steady state value is

$$Y_{SS} = \frac{1}{1 - 0.9} = 10$$

It can be seen that as p approaches 1.0, the steady state Unit Step response will approach infinity.

Example 4.19. Demonstrate with several example computations that Eq. (4.20) holds true not only for real poles, but also for complex poles having magnitude less than 1.0.

A simple method is to use the function *filter*; a suitable call for a single-pole IIR is

```
p = 0.9*j; y = filter([1], [1,-p], [ones(1,150)])
```

where p is the pole.

Another way is to write an expression which will convolve a unit step of significant length, say 200 samples, with a truncated version of the single-pole IIR's impulse response. A suitable call to create the impulse response and perform the convolution might be

```
p = 0.9*j; xp = 0:1:99; x = p.^xp; y = conv(x, ones(1,200))
```

For either of the two methods described above, examine the output sequence to find the steady-state value. If the magnitude of the pole is too close to 1.0, it may be necessary to use longer test sequences than those in the examples above. Once the steady-state value has been found, check it against the formula's prediction. In this case, (for p = 0.9*j), both methods produce, after a certain number of samples of output, the steady-state value of 0.5525 + j0.4973.

Stability

Figure 4.32 shows a single pole filter with borderline stability. In this case, the magnitude of the pole is exactly 1.0, and the impulse response does not decay away. This is analogous to an oscillator, in which a small initial disturbance creates a continuous output which does not decay away. The seriousness of the situation can be seen by using a unit step as the test signal. Figure 4.33 shows the result: a ramp which theoretically would simply continue to increase to infinity if the filter were allowed to run forever.

An accumulator is a digital register having a feedback arrangement that adds the output of the register to the current input, which is then stored in the register. The input signal is thus accumulated or integrated. The single pole IIR with pole value equal to 1.0 functions as an **Integrator**, which we see is not a stable system.

As another demonstration of the visual effect of instability, let's feed a chirp into a filter having a magnitude 1.0 pole (in this case at $\pi/4$ radians, or 45°). Figure 4.34 shows the result: the filter "rings," that is to say, once the chirp frequency gets near the pole's resonant frequency, the output

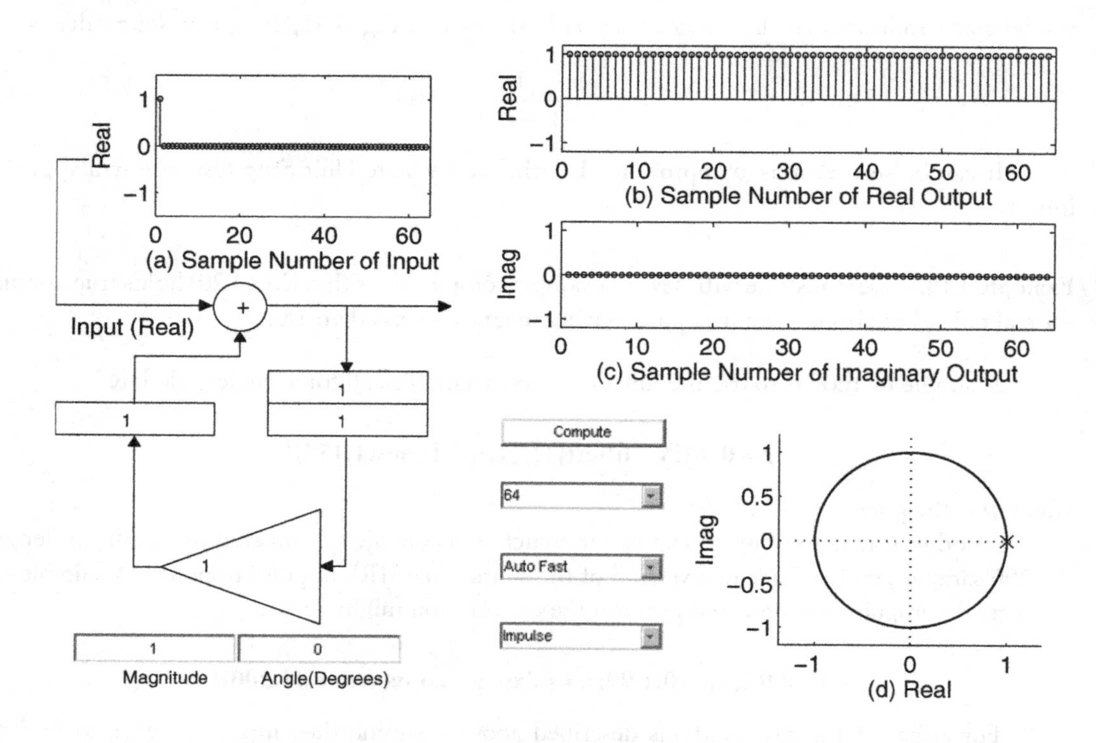

Figure 4.32: (a) Input sequence, a unit impulse sequence; (b) Real part of output sequence, a unit step; (c) Imaginary part of output sequence, identically zero; (d) The pole, plotted in the complex plane.

begins to oscillate and the filter seems to ignore the remainder of the input signal as the frequency passes beyond the pole's resonant frequency.

Reducing the pole's magnitude to 0.95, however, gives the filter a stable response, as shown in Fig. 4.35.

4.12.5 LEAKY INTEGRATOR

A single-pole IIR with a real pole at frequency zero having a magnitude less than 1.0 is often termed a **Leaky Integrator**, and finds frequent use as a signal averager. Often, when the leaky integrator is used to average a signal, the input signal is scaled before it enters the summing junction so that the steady state unit step response is 1.0. The equation of a Leaky Integrator as employed in signal averaging is

$$y[n] = \beta x[n] + (1 - \beta)y[n - 1] \tag{4.21}$$

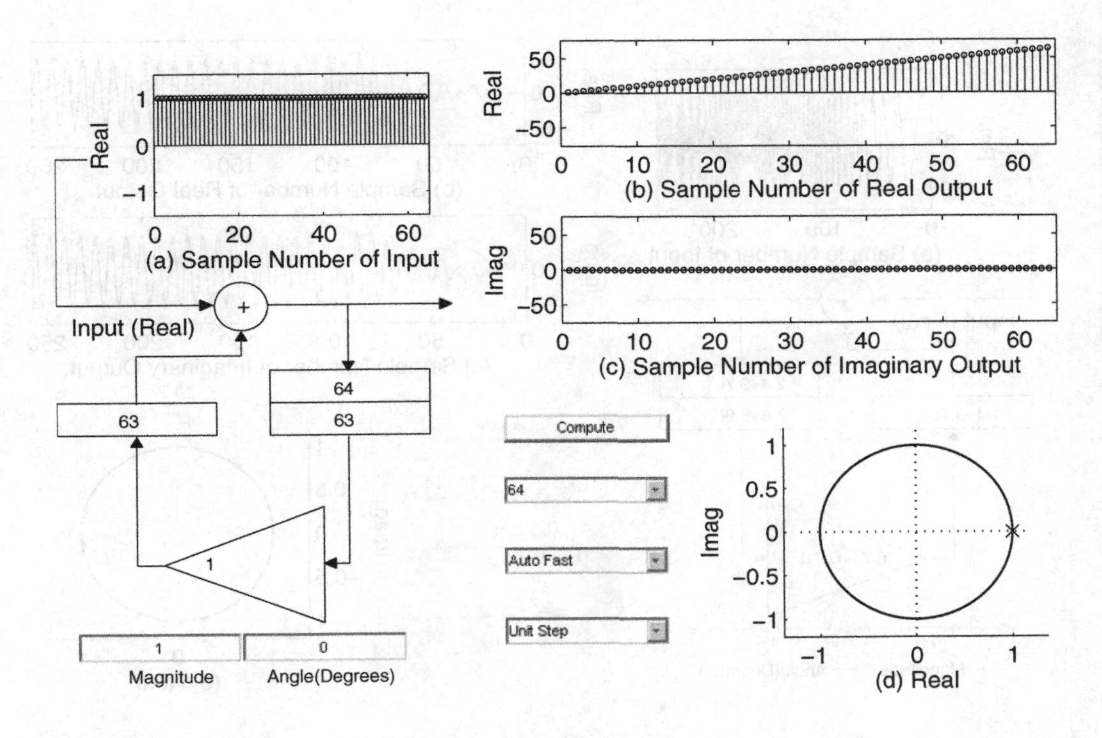

Figure 4.33: (a) Input sequence, a unit step sequence; (b) Real part of output sequence, an increasing ramp; (c) Imaginary part of output sequence, identically zero; (d) The pole, plotted in the complex plane.

where $0 < \beta < 1.0$. Thus in the example above, with the pole at 0.9, we have $(1-\beta) = 0.9$, which implies that $\beta = 0.1$, and the steady state unit step response is 1.0. When used as a signal averager, β is chosen according to the desired relative weights to be given to the current input sample and the past history of input sample values.

Example 4.20. Verify that Eq. (4.21) will result in a steady state value for y of 1.0 when $\beta = 0.1$ and $x[n]$ is a unit step.

We can use the *filter* function in a straightforward manner a first way

x = ones(1,100); y = filter(0.1,[1,-0.9],x); figure;stem(y)

or with an equivalent call, which prescales the unit step before filtering

x = 0.1*ones(1,100); y = filter(1,[1,-0.9],x); figure;stem(y)

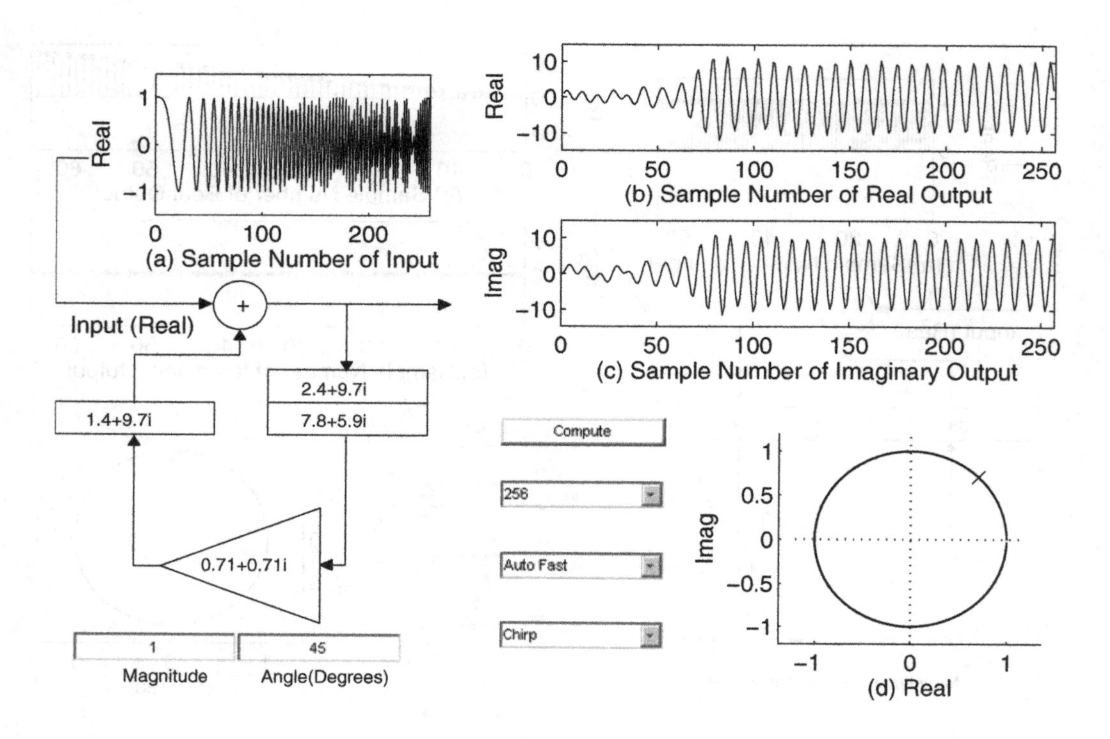

Figure 4.34: (a) Input sequence, a chirp; (b) Real part of output sequence, which is a continuous oscillation at the frequency of the pole; (c) Imaginary part of output sequence; (d) The pole, plotted in the complex plane.

Example 4.21. Consider the leaky integrator defined by the difference equation $y[n] = 0.1x[n] + 0.9y[n-1]$. For $y[100]$, determine what the relative weights are of $x[100]$, $x[99]$, and $x[98]$ as they appear in an expression for $y[100]$.

By writing out the equations for $y[98]$, $y[99]$, and $y[100]$ and substituting, we get the equation

$$y[100] = 0.1x[100] + 0.09x[99] + 0.081x[98] + 0.72y[97]$$

from which the relative weights can be seen.

4.12.6 FREQUENCY RESPONSE

Whenever the magnitude of the pole is less than 1, the impulse response has a geometrically decaying magnitude. Note that the pole need not be a real number, but a complex number works also, resulting in an impulse response which exhibits a decaying sinusoidal characteristic, with the frequency depending on the pole's angular location along the unit circle–a pole at 0 degrees yields

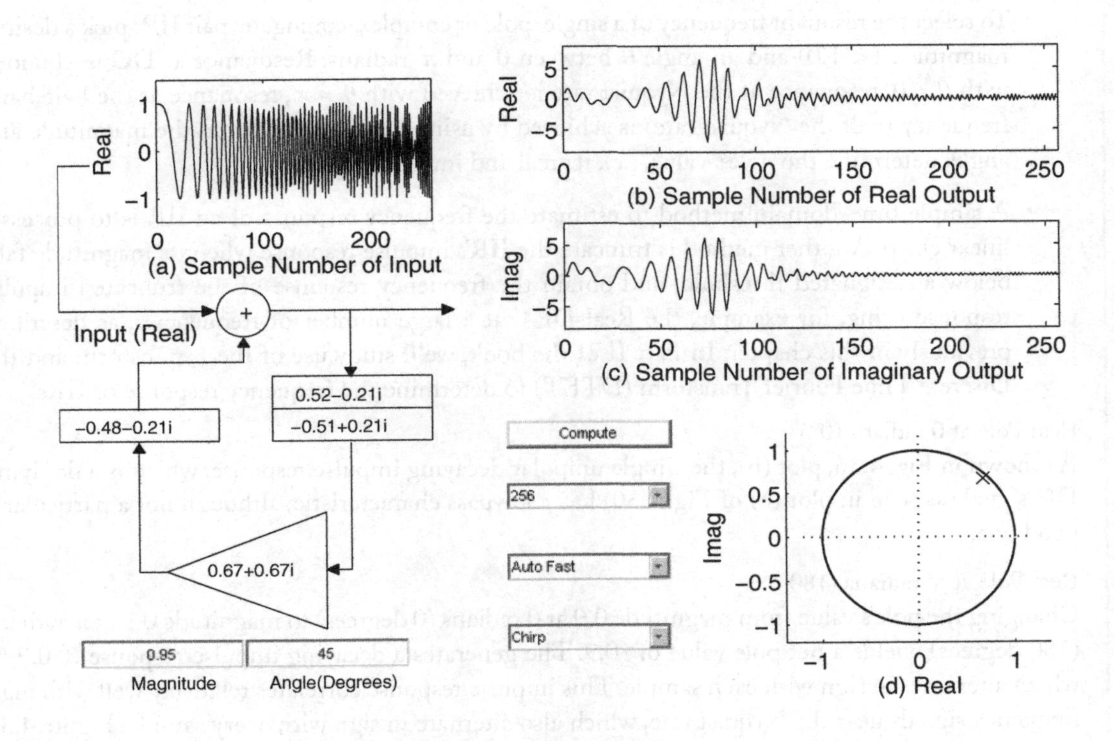

Figure 4.35: (a) Input sequence, a chirp; (b) Real part of output sequence, showing stable resonance at the frequency of the pole; (c) Imaginary part of output sequence; (d) The pole, plotted in the complex plane.

a decaying DC (or unipolar) impulse response, a pole at 180 degrees (π radians) yields the Nyquist frequency (one cycle per two samples), and poles in between yield proportional frequencies. A pole at 90 degrees, for example ($\pi/2$ radians), yields a decaying sinusoidal impulse response exhibiting one cycle every four samples.

- For most filtering projects, the input is real, and it is desired, to keep the hardware and/or software simple, that the output also be real. This implies that the coefficient or pole must also be real. However, for a single feedback delay and multiplier arrangement, the coefficient must be complex if resonant frequencies other than DC and the Nyquist limit frequency are to be had.

- The goal of real input, real coefficients, and a real output, at any frequency from DC to Nyquist can be attained by using one or more pairs of poles that are complex conjugates of each other. This cancels imaginary components in the output signal.

- To select the resonant frequency of a single-pole or complex-conjugate-pair IIR, pick a desired magnitude (< 1.0) and an angle θ between 0 and π radians. Resonance at DC is obtained with $\theta = 0$, resonance at the Nyquist rate is achieved with $\theta = \pi$, resonance at the half-band frequency (half the Nyquist rate) is achieved by using $\theta = \pi/2$, etc. From the magnitude and angle, determine the pole's value, i.e., its real and imaginary parts.

- A simple time-domain method to estimate the frequency response of an IIR is to process a linear chirp. Another method is truncate the IIR's impulse response when its magnitude falls below a designated threshold, and obtain the frequency response of the truncated impulse response using, for example, the Real DFT at a large number of frequencies, as described previously in this chapter. In Part II of the book, we'll study use of the z-transform and the Discrete Time Fourier Transform (DTFT) to determine the frequency response of IIRs.

Real Pole at 0 radians (0°)
As shown in Fig. 4.36, plot (b), the simple unipolar decaying impulse response, which is a decaying DC signal (as seen in plot (b) of Fig. 4.30) has a lowpass characteristic, although not a particularly good one.

Real Pole at π radians (180°)
Changing the pole's value from magnitude 0.9 at 0 radians (0 degrees) to magnitude 0.9 at π radians (180 degrees), yields a net pole value of -0.9. The generates a decaying impulse response ($(-0.9)^n$) which alternates in sign with each sample. This impulse response correlates relatively well with high frequency signals near the Nyquist rate, which also alternate in sign with every sample. Figure 4.37 shows the result; a real pole at -0.9 produces a crude, but recognizable, highpass filter.

Complex Pole
Poles whose angles lie between 0 and π radians generate impulse responses that correlate with various frequencies between DC and the Nyquist limit. Figure 4.38, plot (b), shows the chirp response resulting from selecting the pole at $\pi/2$ radians with a magnitude of 0.9.

- All of Figs. 4.30 through 4.38 were made using the script

$$ML\_SinglePole$$

which creates a GUI with a number of drop-down menus allowing selection of test signal type, manner of computation and display (auto-step, etc.) of output, and test sequence length. Two edit boxes allow entry of the value of the pole using polar coordinates, namely, magnitude and angle in degrees.

4.12.7 COMPLEX CONJUGATE POLES

As mentioned above, complex poles can be used in pairs to achieve a filter having real coefficients. Let's start out by determining the net impulse response of two cascaded single pole IIRs having complex conjugate poles. To do this, we convolve the impulse responses of each, which are

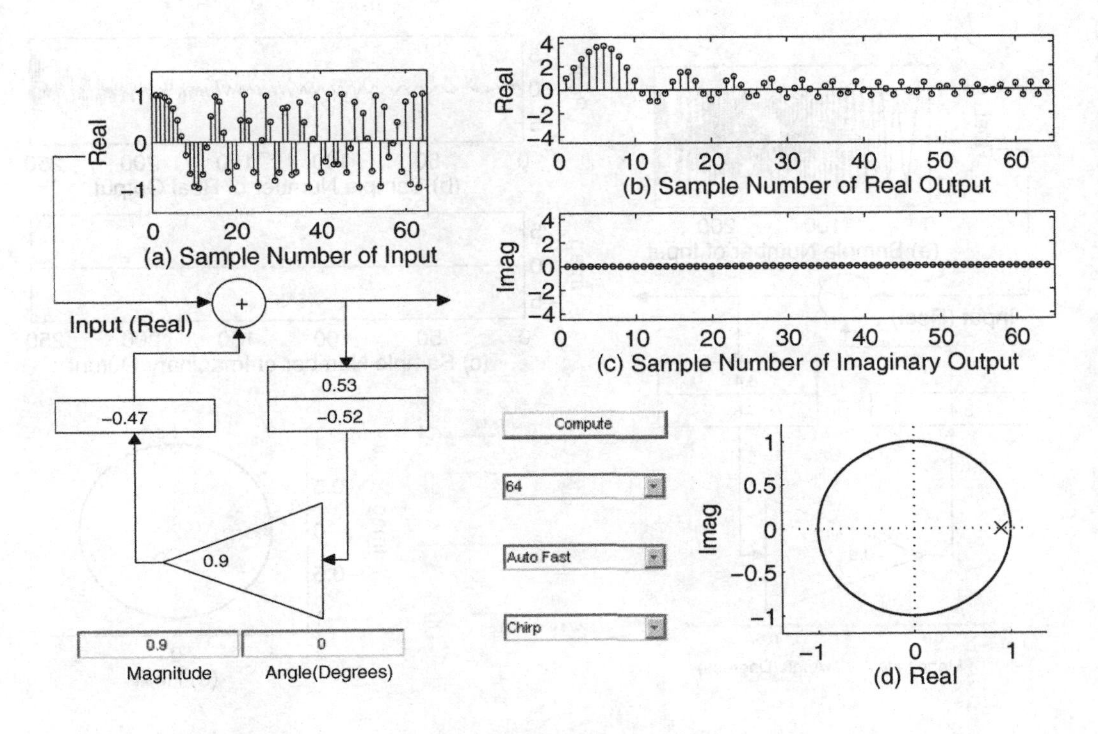

Figure 4.36: (a) Input sequence, a chirp; (b) Real part of output sequence, showing a lowpass effect; (c) Imaginary part of output sequence, identically zero; (d) The pole, plotted in the complex plane.

$$1, p, p^2, p^3, ...$$

and

$$1, p_c, p_c^2, p_c^3, ...$$

where p_c is the complex conjugate of p. The first few terms of the convolution are

$$1$$

$$p + p_c$$

$$p^2 + pp_c + p_c^2$$

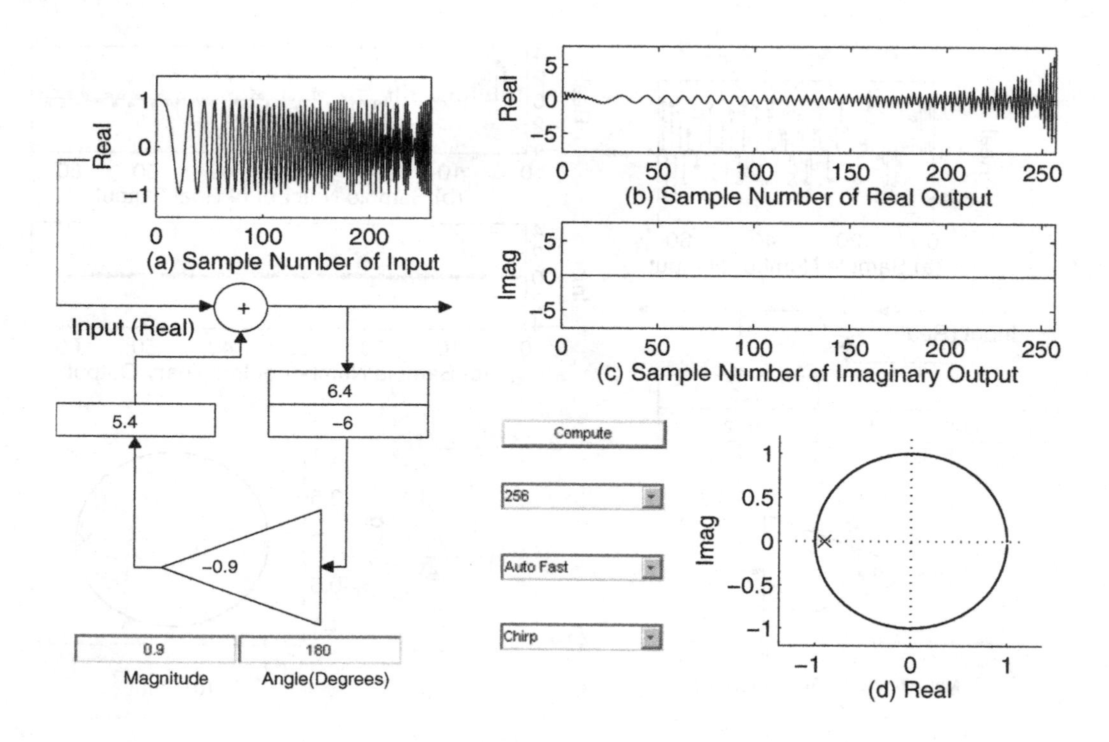

Figure 4.37: (a) Input sequence, a chirp; (b) Real part of output sequence, showing a highpass effect; (c) Imaginary part of output sequence, identically zero.

$$p^3 + p^2 p_c + p p_c^2 + p_c^3 \tag{4.22}$$

or in generic terms

$$c[n] = \sum_{m=0}^{n} p^{n-m} p_c^m \tag{4.23}$$

The symmetry of form of the terms of Eq. (4.23) results in a cancellation of imaginary components. For example, we see that

$$p^n + p_c^n = M^n \angle n\theta + M^n \angle(-n\theta)$$

where $p = M\angle\theta$, which reduces to

$$M^n(\exp(jn\theta) + \exp(-jn\theta)) = 2M^n \cos(n\theta)$$

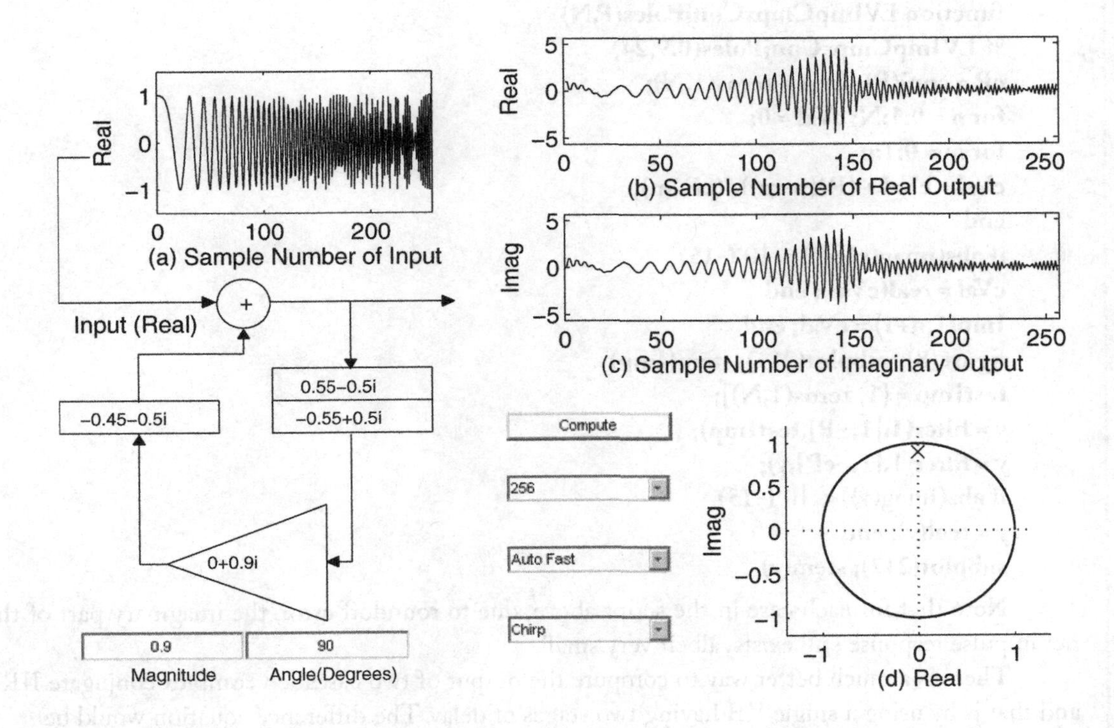

Figure 4.38: (a) Input sequence, a chirp; (b) Real part of output sequence, showing a bandpass effect centered at the halfband frequency; (c) Imaginary part of output sequence, showing a bandpass effect; (d) Unit circle and pole.

Similarly, pairs of terms such as

$$[p^2 p_c, pp_c^2]$$

sum to real numbers, and

$$pp_c = M\exp(j\theta)M\exp(-j\theta) = M^2\cos(0) = M^2$$

Example 4.22. Verify that the impulse response of two single pole IIRs having complex conjugate poles is real.

The following script does this in two ways. The first method uses Eq. (4.23), and the second method uses MathScript's *filter* function, computing the first filter's output, and then filtering that result with the second filter having the conjugate pole. The results are identical.

```
function LVImpCmpxConjPoles(P,N)
% LVImpCmpxConjPoles(0.9,24)
cP = conj(P); Imp = zeros(1,N);
for n = 0:1:N; cVal = 0;
for m = 0:1:n
cVal = cVal + (P.^(n-m)).*(cP.^m);
end
if abs(imag(cVal))< 10^(-15)
cVal = real(cVal); end
Imp(1,n+1) = cVal; end
figure(8); subplot(211); stem(Imp)
testImp = [1, zeros(1,N)];
y = filter(1,[1, -P],testImp);
y = filter(1,[1, -cP],y);
if abs(imag(y))< 10^(-15)
y = real(y); end
subplot(212); stem(y)
```

Note that for each case in the script above, due to roundoff error, the imaginary part of the net impulse response still exists, albeit very small.

There is a much better way to compute the output of two cascaded complex conjugate IIRs, and that is by using a single IIR having two stages of delay. The difference equation would be

$$y[n] = bx[n] + a_1 y[n-1] + a_2 y[n-2]$$

We can solve for b, a_1, and a_2 if we know the first few values of the impulse response as computed (for example) by the script above. For $p = 0.9$, for example, the first few values of the impulse response are

$$1, 1.8, 2.43, 2.916, ...$$

Assuming that $y[n]$ is identically zero for $n < 0$, we process a unit impulse ($x[n] = 1$ for $n = 0$, and 0 for all other n) to obtain the impulse response, the values of which we already know from the script above. Thus when $n = 0$, we get $x[0] = 1$, and $y[0] = 1$, requiring that $b = 1$. When $n = 1$, $x[1] = 0$, and $y[1] = 1.8$, which yields $a_1 = 1.8$. For $n = 2$, $x[2] = 0$, $y[2] = 2.43$, and we get $a_2 = -0.81$.

Thus the net difference equation is

$$y[n] = x[n] + 1.8y[n-1] - 0.81y[n-2] \tag{4.24}$$

Note that Eq. (4.24) requires no complex arithmetic at all, and the problem of roundoff error does not occur insofar as imaginary components are concerned. Eq. (4.24) only guarantees, in general, that the first several values of the impulse response will be generated since that is all that has been taken into account. In this case, since the impulse response in question was in fact generated using known poles, and there is no additive noise in the impulse response, the impulse response can be completely generated by solving the difference equation forward in time. In Part IV of the book, we'll investigate this type of process more completely with an algorithm known as Prony's Method, which will allow us to determine a set of coefficients that results in the closest fit to a sequence that might, for example, be quite long and contain noise.

We can use the $filter$ function with these coefficients to verify the result

```
x = [1,zeros(1,50)]; y = filter(1,[1,-1.8,0.81],x);
figure; stem(y)
```

A yet easier way to obtain the coefficients of the real filter that results from cascading two complex conjugate pole filters is to convolve the coefficient vectors of the two filters, i.e.,

$$Coeff_{CC} = conv([1, -p], [1, -p_c])$$

Example 4.23. Determine the real coefficients of the second order IIR that results from using the two complex conjugate poles

$$[0.65 + j0.65, 0.65 - j0.65]$$

We convolve the coefficient representations of each:

```
y = conv([1, -(0.672 +j*0.672)],[1, -(0.672 - j*0.672)])
```

which yields

```
y = [1, -1.344, 0.9034]
```

Since the poles have angles of $\pm\pi/4$ radians, we would expect the peak response of the IIR to lie at one-quarter of the Nyquist rate, or one-eighth of the sample rate. We can explore this with a script that receives one pole as its magnitude and angle in radians, computes the complex conjugate pole, computes the net real coefficients, and filters a linear chirp–a call that yields the answer for the specific problem at hand is

```
LVRealFiltfromCCPoles(0.95, pi/4)
```

which results in Fig. 4.39.

```
function LVRealFiltfromCCPoles(PoleMag,PoleAng)
% LVRealFiltfromCCPoles(0.95,pi/4)
Pole = PoleMag*exp(j*PoleAng); cPole = conj(Pole);
rcoeffs = conv([1, -(Pole)],[1, -(cPole)]); SR = 1024;
```

```
t = 0:1/(SR-1):1; x = chirp(t,0,1,SR/2);
y = filter([1],[rcoeffs],x); figure(8); plot(y);
xlabel('Sample'); ylabel('Amplitude')
```

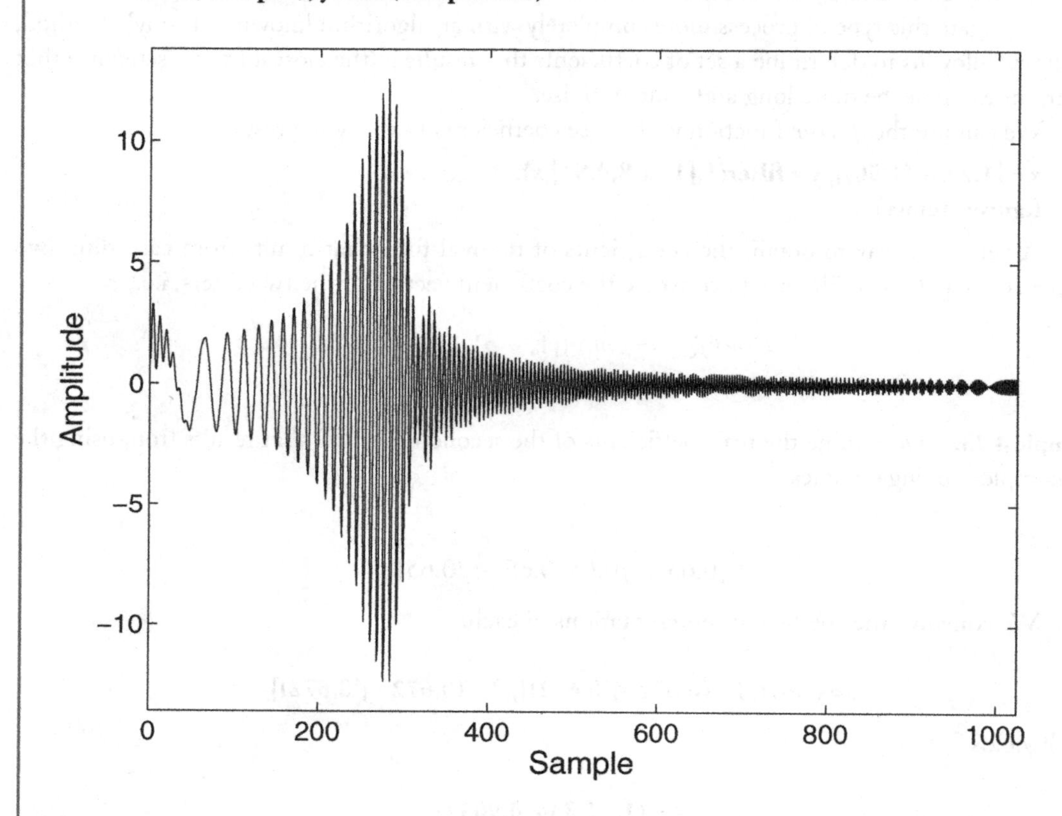

Figure 4.39: Convolution of a linear chirp with a second order all-real coefficient filter obtained by convolving the coefficient vectors of two complex conjugate single pole IIRs.

The results above will be again demonstrated and generalized when the topic of the z-transform is taken up in Part II of the book. We will also explore different filter topologies or implementations for a given net impulse response.

The LabVIEW VI

DemoDragPolesVI

allows you to select as a test signal a unit impulse, a unit step, or a chirp. The pole or complex conjugate pair of poles is specified by dragging a cursor in the z-plane. From these poles, the VI forms an IIR and filters the selected test signal. The real and imaginary outputs of the filter are plotted. The importance of using poles in complex conjugate pairs can readily be seen by alternately selecting "Single Pole" and "Complex Conjugate Pair" in the Mode Select box. The use of a complex conjugate pair of poles results in all-real filter coefficients and an all-real response to an all-real input signal.

A script that allows you to move the cursor in the complex plane and see the frequency and impulse responses arising from a single pole or a complex-conjugate pair of poles is

$$ML\_DragPoleZero$$

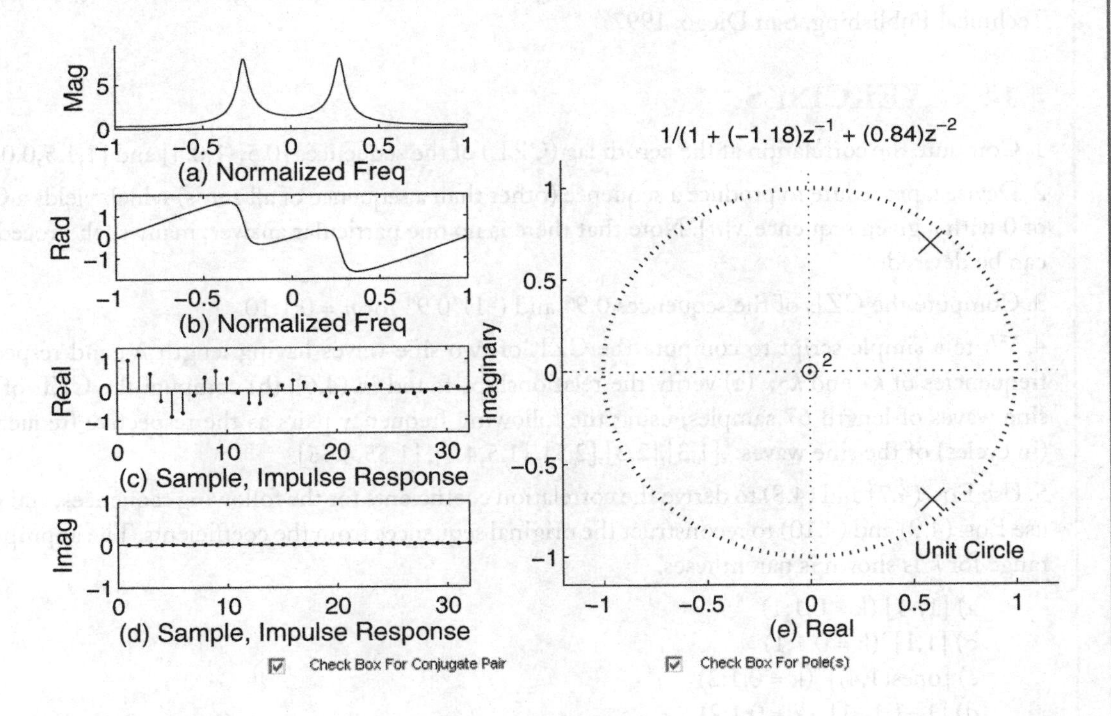

Figure 4.40: (a) Magnitude of frequency response of an LTI system constructed using the poles shown in plot (e); (b) Phase response of same; (c) Real part of impulse response of same; (d) Imaginary part of impulse response of same.

A snapshot of a typical display generated by $ML\_DragPoleZero$ is shown in Fig. 4.40. Note that a pair of complex-conjugate poles is specified and displayed (by the symbol 'X'), and the resulting impulse response, which is real-only, decays since the poles have magnitude less than 1.0.

4.13 REFERENCES

[1] John G. Proakis and Dimitris G. Manolakis, *Digital Signal Processing, Principles, Algorithms, and Applications, Third Edition*, Prentice Hall, Upper Saddle River, New Jersey, 1996.

[2] Alan V. Oppenheim and Ronald W. Schaefer, *Discrete-Time Signal Processing*, Prentice-Hall, Englewood Cliffs, New Jersey, 1989.

[3] James H. McClellan, Ronald W. Schaefer, and Mark A. Yoder, *Signal Processing First*, Pearson Prentice Hall, Upper Saddle River, New Jersey, 2003.

[4] Steven W. Smith, *The Scientist and Engineer's Guide to Digital Signal Processing*, California Technical Publishing, San Diego, 1997.

4.14 EXERCISES

1. Compute the correlation at the zeroth lag (CZL) of the sequences [0.5,-1,0,4] and [1,1.5,0,0.25].

2. Devise a procedure to produce a sequence (other than a sequence of all zeros) which yields a CZL of 0 with a given sequence $x[n]$. Note that there is no one particular answer; many such procedures can be devised.

3. Compute the CZL of the sequences 0.9^n and $(-1)^n 0.9^n$ for n = 0:1:10.

4. Write a simple script to compute the CZL of two sine waves having length N, and respective frequencies of k_1 and k_2; (a) verify the relationships found in (4.6); (b) compute the CZL of two sine waves of length 67 samples, using the following frequency pairs as the respective frequencies (in cycles) of the sine waves: .[1,3],[2,3],[2,5], [1.5,4.5], [1.55,4.55].

5. Use Eqs. (4.7) and (4.8) to derive the correlation coefficients for the following sequences, and then use Eqs. (4.9) and (4.10) to reconstruct the original sequences from the coefficients. The appropriate range for k is shown is parentheses.

   a) [1,-1] (k = 0:1:1)
   b) [1,1] (k = 0:1:1)
   c) [ones(1,4)] (k = 0:1:2)
   d) [1,-1,1,-1] (k = 0:1:2)

6. Determine the correct range for k and use the analysis Eqs. (4.7) and (4.8) to determine the correlation coefficients for the following sequences: [sin(2*pi*(3)*(0:1:7)/8)], [sin(2*pi*(3.1)*(0:1:7)/8)], and [rand(1,8)]. Reconstruct each sequence using the obtained coefficients and Eqs. (4.9) and (4.10).

7. Repeat the previous exercise using signal lengths of 9 instead of 8, and the appropriate range for k.

8. Using paper and pencil, manually compute the auto-correlation sequence of the signal

$$y = u[n] - u[n-8]$$

9. Use paper and pencil to compute the correlation sequences of the following:

   a) [2,0.5,1,-1.5] and [0.5,2,1,2/3].
   b) [5:-1:1] with itself.
   c) [5:-1:1] and [1:1:5]

10. Manually compute the cross-correlation sequence of the two sequences $\delta[n]$ and $\delta[n-5]$ over the interval n = 0:1:9. Repeat for the sequences $\delta[n]$ and $u[n-5]$.

11. Using paper and pencil, use the "sliding waveform method" to compute the convolution and correlation sequences of the following two sequences: [1:1:6] and [2,-1 4,1]. Use the appropriate MathScript functions to verify your results.

12. Verify, using paper and pencil, that the convolution of the sequences [1,2,3,4] and [-1,0.6,3] is the same whichever sequence is chosen to be "flipped" from right to left and moved through the other from the left. Compute the correlation sequence both ways, i.e., picking one sequence, then the other, to be the one that slides from right to left over the other, and verify that the two results are the retrograde of each other.

13. Write a script that will convolve a linear chirp having a lower frequency limit of 0 Hz and an upper frequency limit of 1000 Hz with the following impulse responses and plot the result.

   (a) [0.02,0.23,0.4981,0.23,0.02]
   (b) [0.0087,0,-0.2518,0.5138,-0.2518,0,0.0087]
   (c) [0.5,0,1,0,0.5]
   (d) [0.0284,0,-0.237,0,0.469,0,-0.237,0,0.0284]

   After performing the convolution, state what kind of filtering effect each impulse response has.

14. Repeat the previous exercise, using an upper frequency limit for the chirp of 5000 Hz, and compare the output plots to those of the previous exercise.

15. Repeat exercise 13 or 14 above using a complex chirp rather than a real chirp as the test signal, and plot the magnitude of the convolution as the estimate of frequency response. A complex linear chirp running from frequency 0 Hz to 2500 Hz, for example, may be generated by the following m-code:

```
SR = 5000; t = 0:1/(SR-1):1;
cmpxChrp = chirp(t,0,1,SR/2) + j*chirp(t,0,1,SR/2,'linear',90);
```

16. Plot the convolution of the first 100 samples of the following impulse responses with a linear chirp from 0 to 4000 Hz and characterize the results as to filter type (lowpass, highpass, etc.).

   (a) $h = 0.9^n u[n]5$
   (b) $h = (-0.1^n)0.9^n u[n]$
   (c) $h = (1 + (-0.1^n))0.9^n u[n]$
   (d) h = **0.5*exp(j*pi/2).^(0:8) + 0.5*exp(-j*pi/2).^(0:8)**

17. Write the m-code for the script

$$LVxMatchedFilter(NoiseAmp, TstSeqLen, FlipImpResp)$$

that generates a test signal of length $TstSeqLen$ consisting of a chirp of a given length (less than $TstSeqLen$) immersed in noise having amplitude $NoiseAmp$ and length $TstSeqLen$ which can be convolved with either the chirp of given length itself or a time-reversed version of it, according to the input argument *FlipImpResp*, which can assume the value 0 to use the non-time-reversed chirp as a filter impulse response or 1 to use the time-reversed chirp as the filter impulse response. Plot the impulse response being used and the test signal, poised to begin convolution, on a single plot, and the convolution sequence on a second plot. The function specification is as follows:

```
function LVxMatchedFilter(NoiseAmp,TstSeqLen,FlipImpResp)
% Forms a test impulse response, a chirp having a length equal
% to half of TstSeqLen, then builds a test sequence having
% length TstSeqLen and containing the chirp and noise of
% amplitude NoiseAmp. The test signal is then convolved with
% either the chirp or a time-reversed version of the chirp, and
% the results plotted to demonstrate the principle of matched
% filtering.
% FlipImpResp: Use 1 for Time-reversed, 0 for not time-reversed
% impulse response.
% Test calls:
% LVxMatchedFilter(0.5,128,0) % imp resp not time reversed
% LVxMatchedFilter(0.5,128,1) % imp resp time reversed
```

18. Write the m-code for the script

$$LVxTestReconSineVariablePhase(k1, N, PhaseDeg)$$

receives a frequency $k1$ and length N to be used to construct three sinusoids, which are 1) a sine wave of arbitrary phase $PhaseDeg$ and frequency $k1$, and 2) a test correlator sine of frequency $k1$ and 3) a test correlator cosine of frequency $k1$. Compute the correlation coefficients using Eqs. (4.7) and (4.8), and then reconstruct the sine wave of arbitrary phase $PhaseDeg$ and frequency $k1$ using Eqs. (4.9) and (4.10). Plot the test sine wave, the two correlators, and the reconstructed test sine wave. The function specification is as follows:

```
function LVxTestReconSineVariablePhase(k1,N,PhaseDegrees)
% Performs correlations at the zeroth lag between
% test cosine and sine waves of frequency k1, having N samples,
% and a sinusoid having a phase of PhaseDegrees, and then
% reconstructs the original sinusoid of phase equal to
% PhaseDegrees by using the CZL values and the test sine
% and cosine.
```

```
%Test calls:
% LVxTestReconSineVariablePhase(1,32,45)
% LVxTestReconSineVariablePhase(0,32,90)
% LVxTestReconSineVariablePhase(16,32,90)
% LVxTestReconSineVariablePhase(1,32,45)
```

19. Write a script that implements the function of the script *LVxFreqTest*. The script should receive the arguments shown in the function definition below, compute the Real DFT coefficients one-by-one using Eqs. (4.11) and (4.12), and then synthesize the original test signal from the coefficients using Eq. (4.13). Three figures should be created, the first one having two subplots and being the analysis window, in which a given set (real and imaginary) of analysis basis functions (i.e., test correlators) having frequency *dispFreq* is plotted against the test signal, and the correlation value according to the Real DFT analysis formula is computed and displayed. The second figure is the synthesis window and should have three subplots, the first two being an accumulation of weighted basis harmonics, and the third being the final reconstruction. The third figure has two subplots and should show the real and imaginary coefficients. Test your script with all six test signals. Examples of each of the three windows are shown in the text.

```
function LVxFreqTest(TestSignalType,N,UserTestSig,dispFreq)
% TestSignalType may be passed as 1-7, as follows:
% 1 yields TestSig = sin(2πt) + 1.25cos(4πt + 2π(60/360)) +
% 0.75cos(12πt + 2π(330/360))
% 2 yields TestSig = sin(4πt + π/6)
% 3 yields TestSig = sin(5.42πt)
% 4 yields TestSig = 0.25 + cos(2.62πt + π/2) + sin(5.42πt + π/6)
% 5 yields TestSig = sin(2πt) + 1.25cos(4πt)+ 0.75cos(10πt)
% 6 yields TestSig = sin(4πt);
% 7 uses the test signal supplied as the third argument. Pass this
% argument as [] when TestSignalType is other than 7.
% N is the test signal length and must be at least 2. When
% TestSignalType is passed as 7, the value passed for N is
% overridden by the length of UserTestSig. In this case, N may
% be passed as the empty matrix [] or an arbitrary number if desired.
% dispFreq is a particular correlator frequency which is used to
% create two plots, one showing the test signal chosen by
% TestSignalType and the test correlator cosine of frequency
% dispFreq, and another showing the test signal chosen by
% TestSignalType and the test correlator sine of frequency dispFreq.
% Test Calls:
% LVxFreqTest(5,32,[])
% LVxFreqTest(4,19,[])
```

```
% LVxFreqTest(7,11,[cos(2*pi*(2.6)*(0:1:10)/11)])
```

20. The script $LVFreqResp$ was presented earlier in the chapter. It evaluated the frequency response of a test signal of length N by correlating the test signal with many test correlators of length N having frequencies evenly spaced between 0 and π radians. In this exercise we develop a method which pads the test signal with zeros to a length equal to a user-desired correlator length, and then performs the Real DFT using the zero-padded test signal

```
function [FR] = LVxFreqRespND(tstSig, LenCorr)
% FR = LVxFreqRespND([ones(1,32)], 128)
% Pads tstSig with zeros to a length equal to LenCorr,
% then computes the Real DFT of the padded tstSig
% over frequencies from 0 to the maximum
% permissible frequency, which is LenCorr/2 if LenCorr
% is even, or (LenCorr-1)/2 if NoFreqs is odd.
% Delivers the output FR as the sum of the real correlation
% coefficients plus j times the imaginary correlation coeffs.
```

For each of the following test signals $tstSig$ and corresponding values of $LenCorr$, plot the magnitude of FR and compare results to those obtained by performing the same computation using the function fft, using the following m-code:

```
figure(55)
subplot(211)
[FR] = LVxFreqRespND(tstSig, LenCorr);
stem([0:1:length(FR)-1],abs(FR))
subplot(212)
fr = fft(tstSig, LenCorr);
if rem(LenCorr,2)==0
plotlim = LenCorr/2 + 1;
else
plotlim = (LenCorr-1)/2 + 1;
end
stem([0:1:plotlim-1],abs(fr(1,1:plotlim)));
```

Test Signals:

(a) tstSig = ones(1,9); LenCorr = 9
(b) tstSig = ones(1,9); LenCorr = 10
(c) tstSig = ones(1,9); LenCorr = 100
(d) tstSig = ones(1,32); LenCorr = 32
(e) tstSig = ones(1,32); LenCorr = 37
(f) tstSig = ones(1,32); LenCorr = 300

21. Write the m-code for the script

$$LVxCorrDelayMeasure(NoiseAmp)$$

Your implementation should have three plots, the first plot being a test signal consisting of a variable amount of noise mixed with a single cycle of a sine wave, which occurs early in the sequence, the second plot consisting of a second test signal consisting of a variable amount of noise and several time-offset cycles of a sine wave, and the third consisting of the cross correlation sequence of the two test signals. The sequence in the second plot should move to the left one sample at a time, and the third plot should be created one sample at a time as the sum of the products of all overlapping samples (i.e., having the same sample index) of the two test signals, as shown on the plots. Figures 4.28 and 4.29 show examples, at the beginning and end of the computation of the correlation sequence, respectively. The function specification is as follows:

```
function LVxCorrDelayMeasure(k)
% Demonstrates the principle of identifying the time
% delay between two signals which are correlated but
% offset in time. A certain amount of white noise
% (amplitude set by the value of k) is mixed into the signal.
% Typical Test call:
% LVxCorrDelayMeasure(0.1)
```

22. For an input signal consisting of a unit step, using pencil and paper and the difference equation

$$y[n] = x[n] + ay[n-1]$$

compute the first five output values of a single-pole IIR having the following pole values:

(a) $a = 0.99$
(b) $a = 1.0$
(c) $a = 1.01$

23. Repeat the previous exercise, but, using m-code, compute the first 100 values of output and plot the results on a single figure having three subplots.

24. For an input signal consisting of a unit impulse, using pencil and paper, compute the first five output values of a single-pole IIR having the following pole values:

(a) $a = 0.98$
(b) $a = 1.0$
(c) $a = 1.02$

25. Repeat the previous exercise, but, using m-code, compute the first 100 values of output and plot the results on a single figure having three subplots.

26. Consider a cascade of two single-pole filters each of which has a pole at 0.95. Compute the impulse response of the cascaded combination of IIRs the following two ways:

(a) Compute the first 100 samples of the impulse response of the first IIR, then, using the result as the input to the second filter, compute the net output, which is the net impulse response.

(b) Compute the first 100 samples of the net output of the composite filter made by 1) convolving the coefficient vectors for the two individual single pole filters to obtain a second-order coefficient vector, then 2) use the function *filter* to process a unit impulse of length 100 to obtain the impulse response; compare to the result obtained in (a).

27. Let two IIRs each be a properly-scaled leaky integrator with $\beta = 0.1$. Now construct a filter as the cascade connection of the two leaky integrators. Filter a cosine of frequency 2500 Hz and unity amplitude, sampled at 10,000 Hz with the cascade of two leaky integrators. Determine the steady state amplitude of the output.

28. Determine the correlation (CZL) for the following two sequences, for the values of N indicated, where $n = 0{:}1{:}N - 1$:

$$S_1 = \cos(2\pi n3/N) + \sin(2\pi n4/N)$$

$$S_2 = \cos(2\pi n4/N) + \sin(2\pi n3/N)$$

(a) $N = 0.5$;
(b) $N = 1$;
(c) $N = 2$;
(d) $N = 3$;
(e) $N = 6$;
(f) $N = 8$;

29. Write a script that encodes the audio files *drwatsonSR8K.wav* and *whoknowsSR8K.wav* on orthogonal sinusoidal carriers of the same frequency, creates a single signal by taking the difference between the two modulated carriers to create a transmission signal, and then decodes the transmission signal to produce the first audio signal, the second audio signal, or a combination of the two, according to the value of a decoding phase parameter supplied in the function call, according to the specification below. The script should play the decoded audio signal through the computer's sound card. The file *whoknowsSR8K.wav* is much longer than the file *drwatsonSR8K.wav*, so it should be truncated after reading to the same length as *drwatsonSR8K.wav*.

The script should create plots as shown and described in Fig. 4.41:

```
function LVxOrthogAudio(DecodePhi)
% Creates cosine and sine carriers of equal frequency,
% modulates each by a corresponding audio file,
% drwatsonSR8K.wav or whoknowsSR8K.wav, takes
% the difference to create a transmission signal, then
% decodes the transmission signal to produce one or the
% other of the encoded audio signals, according to the
% value of the variable DecodePhi in the function call.
% DecodePhi is the phase angle of the decoding carrier,
```

```
% and should be between 0 and pi/2; 0 will decode the sine
% carrier's audio, pi/2 will decode the cosine carrier's audio,
% and numbers in between 0 and pi/2 will cause a proportional
% mixture of the two audio audio signals to be decoded.
% Test calls:
% LVxOrthogAudio(0)
% LVxOrthogAudio(pi/2)
% LVxOrthogAudio(pi/4)
```

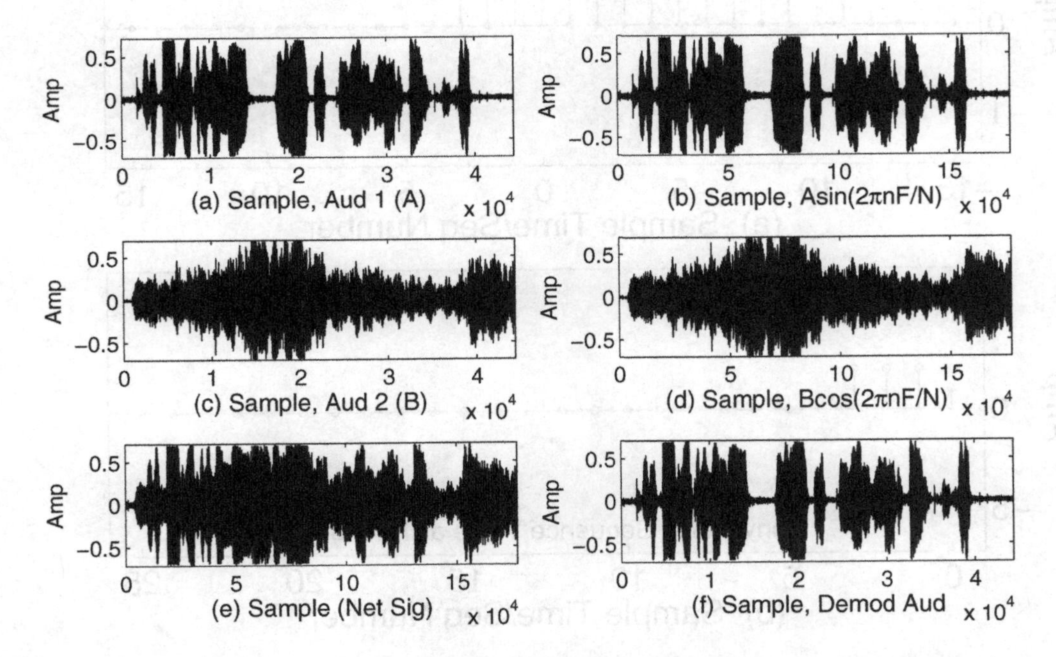

Figure 4.41: (a) First audio signal, *drwatsonSR8K.wav*; (b) Cosine carrier modulated with first audio signal; (c) Second audio signal, *whoknowsSR8K.wav*; (d) Sine carrier modulated with second audio signal; (e) Difference between cosine and sine modulated carriers, forming net transmission signal; (f) Demodulated signal using DemodPhi = 0, yielding the first audio signal.

30. Derive the 2-point impulse responses [1,1] (lowpass) and [1,-1] (highpass) using 2-point cosines. What are the frequencies of the correlators present in each impulse response? Use the script *LVFreqResp(tstSig, NoFreqs)* to evaluate the frequency response of each impulse response at 500 points.

31. Derive the 3-point impulse responses [1,0,-1] (bandpass) and [1,0,1] (bandstop) using cosines of length 4. What are the frequencies of the correlators present in each impulse response? Use the

script *LVFreqResp*(*tstSig, NoFreqs*) to evaluate the frequency response of each impulse response at 500 points.

32. Develop the script *LVxConvolution2PtLPF*(*Freq*) as described below, and which creates Fig. 4.42, which shows, in subplot (a), the position of the test signal as it moves sample-by-sample from left to right over the two point impulse response, computing the convolution sequence one sample at a

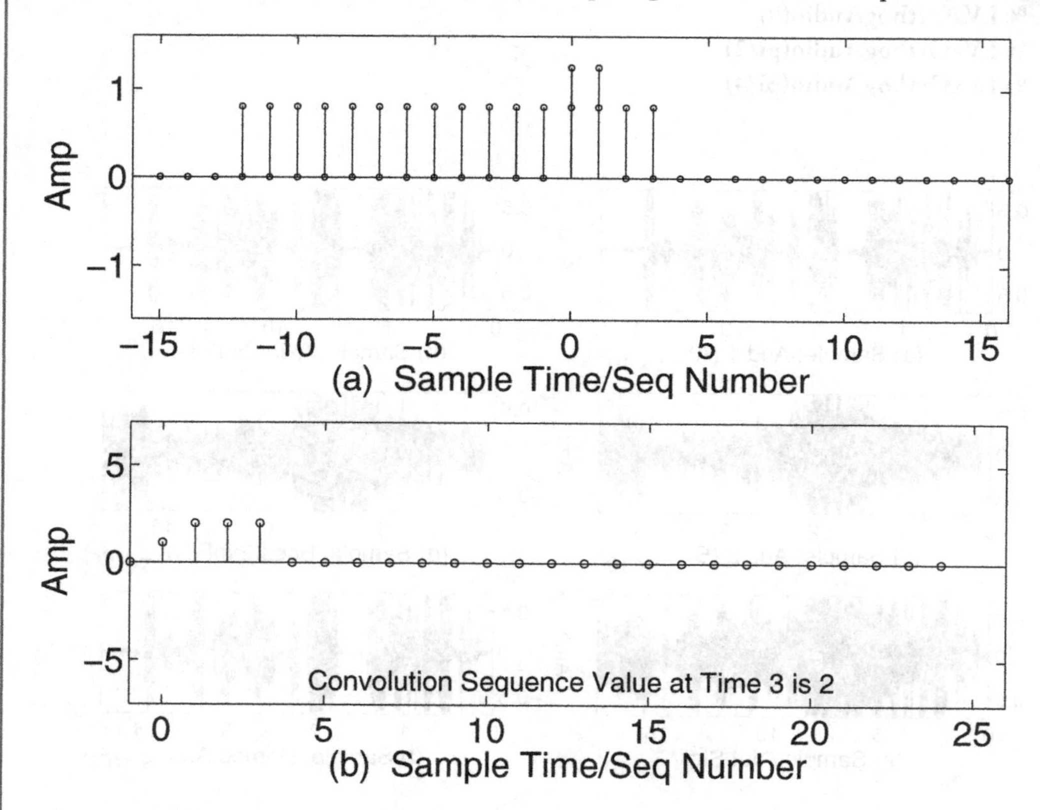

Figure 4.42: (a) Two-sample lowpass impulse response (amplitude 1.25, at sample indices 0 and 1) and DC test signal (amplitude 0.8), advanced from the left to sample index 3; (b) The convolution sequence up to sample index 3.

time, and displaying the convolution sequence in subplot (b). Observe and explain the results from the two sample calls given below.

```
LVxConvolution2PtLPF(Freq)
% Freq is the frequency of the test sinusoid of length 16 which
% will be convolved with the 2-point impulse response [1, 1].
% Values of Freq up to 8 will be nonaliased.
% Test calls:
```

```
% LVxConvolution2PtLPF(0)
% LVxConvolution2PtLPF(8)
```

33. Develop the script *LVxConvolution2PtHPF(Freq)* as described below, and which creates a figure similar to Fig. 4.42, (except that the impulse response is [1,-1] rather than [1,1]), which shows, in subplot (a), the position of the test signal as it moves sample-by-sample from left to right over the two point impulse response, computing the convolution sequence one sample at a time, and displaying the convolution sequence in subplot (b). Observe and explain the results from the two sample calls given below.

```
LVxConvolution2PtHPF(Freq)
% Freq is the frequency of the test sinusoid of length 16 which
% will be convolved with the 2-point impulse response [1,-1].
% Values of Freq up to 8 will be nonaliased.
% Test calls:
% LVxConvolution2PtHPF(1)
% LVxConvolution2PtHPF(8)
```

34. Estimate the frequency response at least 500 frequencies between 0 and π radians of the leaky integrator whose difference equation is

$$y[n] = \beta x[n] + (1 - \beta)y[n - 1]$$

for the following values of β:

(a) 0.01
(b) 0.05
(c) 0.1
(d) 0.5
(e) 0.9
(f) 0.99

35. Compute the convolution sequence of the two sequences [1,1,-1,-1] and [4,3,2,1,4,3,2,1,4,3,2,1,4,3,2,1], identify the transient and steady-state portions of the convolution sequence, and explain the frequency content of the steady-state portion.

Part II

Discrete Frequency Transforms

CHAPTER 5

The Discrete Time Fourier Transform

5.1 OVERVIEW

5.1.1 IN THIS PART OF THE BOOK

In this part of the book we take up discrete frequency transforms in detail, including an overview of many transforms, both continuous-domain and discrete-domain, followed in sequence by detailed discussions of a number of discrete transforms, knowledge of which is generally deemed essential in the signal processing field.

5.1.2 IN THIS CHAPTER

We are now prepared in this chapter to begin a detailed exploration of discrete frequency transforms. A number of such transforms exist, and we'll begin by summarizing all of the basic facts and comparing each to the others to better emphasize the important characteristics of each distinct transform. We include a brief mention of continuous signal domain transforms for background and perspective, but concentrate most of our effort on discrete signal transforms. All of the transforms we'll investigate, both the continuous and the discrete domain types, work on the same general concept–summing (or integrating in the case of continuous time signals) the product of the signal and orthogonal-pair correlating waveforms of different frequencies.

The transforms covered in detail in this book section are the **Discrete Time Fourier Transform (DTFT)**, which is covered in this chapter, the ***z*-Transform (*z*-T)**, covered in the following chapter, and the **Discrete Fourier Transform (DFT)**, covered in the third and final chapter of this part of the book.

By the end of this chapter, the reader will have learned how to evaluate the frequency response of an LTI system using the DTFT. This sets the stage for the following chapter, which discusses the more generalized z-transform, which is in widespread use in industry and academia as the standard method to describe the transfer function of an LTI system. This will be followed in the succeeding chapter by a detailed look at the workhorse of practical frequency domain work, the DFT and a fast implementation, the decimation-in-time FFT, as well as time domain convolution using the frequency domain, the Goertzel algorithm, computing the DTFT using the DFT, etc.

5.2 SOFTWARE FOR USE WITH THIS BOOK

The software files needed for use with this book (consisting of m-code (.m) files, VI files (.vi), and related support files) are available for download from the following website:

http://www.morganclaypool.com/page/isen

The entire software package should be stored in a single folder on the user's computer, and the full file name of the folder must be placed on the MATLAB or LabVIEW search path in accordance with the instructions provided by the respective software vendor (in case you have encountered this notice before, which is repeated for convenience in each chapter of the book, the software download only needs to be done once, as files for the entire book are all contained in the one downloadable folder).

See Appendix A for more information.

5.3 INTRODUCTION TO TRANSFORM FAMILIES

The chief differences among the transforms mentioned below involve whether they 1) operate on continuous or discrete time signals, 2) provide continuous or discrete frequency output, and 3) use constant unity-amplitude correlators (in the case of the Fourier family of transforms), or dynamic (decaying, steady-state, or growing correlators) in the case of the Laplace and z- transforms.

The following table summarizes the main characteristics of a number of well-known transforms with respect to the following categories: Input Signal Domain (continuous **C** or discrete **D** signal), Output (Frequency) Domain (produces continuous **C** or discrete **D** frequency output), and Correlator Magnitude (constant, unity magnitude for Fourier-based transforms, or variable magnitudes (decaying, growing, or constant unity) for Laplace Transform and z-Transform, accordingly as $e^{-\sigma t}$ or $|z|^n$, respectively.

Transform	Input	Output	Correlator Mag.
Laplace Transform	C	C	$e^{-\sigma t}$
Fourier Transform	C	C	1
Fourier Series	C	D	1
Discrete Time Fourier Transform	D	C	1
Discrete Fourier Series	D	D	1
Discrete Fourier Transform	D	D	1
z-transform	D	C	$\|z\|^n$

For purposes of discrete signal processing, what is needed is a numerically computable representation (transform) of the input sequence; that is to say, a representation which is itself a finite but complete representation that can be used to reenter the time domain, i.e., reconstruct the original signal. For transforms that are not computable in this sense, samples of the transform can be computed. Of all the transforms discussed in the following section of the chapter, only the **Discrete Fourier Series (DFS)** and Discrete Fourier Transform (DFT) are computable transforms in the sense mentioned above.

The use of dynamic correlators results in a transform that is an algebraic expression that implicitly or explicitly contains information on the system poles and zeros. The system response to signals other than steady-state, unity amplitude signals can readily be determined, although such transforms can also be evaluated to produce the same result provided by the Fourier Transform (in the case of the Laplace Transform) or the DTFT, DFS, and DFT (in the case of the z-Transform). Thus the Laplace Transform and z-Transform are more generalized transforms having great utility for system representation and computation of response to many types of signals from the continuous and discrete time domains, respectively.

While this book is concerned chiefly with discrete signal processing, we give here a brief discussion of certain continuous time transforms (Laplace, Fourier, Fourier Series) to serve as background or points of reference for the discrete transforms that will be discussed in more detail below and in chapters to follow.

5.3.1 FOURIER FAMILY (CONSTANT UNITY-MAGNITUDE CORRELATORS)

Fourier Transform

$$F(\omega) = \int_{-\infty}^{\infty} x(t)e^{-j\omega t}dt$$

The Fourier transform operates on continuous time, aperiodic signals and evaluates the frequency response in the continuous frequency domain. The correlators are complex exponentials having constant unity amplitude. Both t (time) and ω (frequency) run from negative infinity to positive infinity. The Fourier Transform is a reversible transform; the inverse transform is

$$x(t) = \int_{-\infty}^{\infty} F(\omega)e^{j\omega t}d\omega$$

Fourier Series

Many signals of interest are periodic, that is, they are composed of a harmonic series of cosines and sines. For a periodic, continuous time signal of infinite extent in time, a set of coefficients can be obtained based on a single period (between times t_o and $t_o + T$) of the signal $x(t)$:

$$c_k = \frac{1}{T}\int_{t_o}^{t_o+T} x(t)e^{-jk\omega_0 t}dt$$

where T is the reciprocal of the fundamental frequency F_0 and $k = 0, \pm 1, \pm 2, ...$

For real $x(t)$, c_k and c_{-k} are complex conjugates. If we say

$$c_k = |c_k|\, e^{j\theta_k}$$

then the original sequence can be reconstructed according to the formula

$$x(t) = c_0 + 2\sum |c_k| \cos(2\pi k F_0 t + \theta_k)$$

An equivalent expression is

$$x(t) = a_0 + \sum_{k=1}^{\infty} a_k \cos(2\pi k F_0 t) - b_k \sin(2\pi k F_0 t)$$

where $a_0 = c_0$, $a_k = 2\,|c_k| \cos \theta_k$, and $b_k = 2\,|c_k| \sin \theta_k$.

Discrete Time Fourier Transform (DTFT)

The DTFT is defined for the discrete time input signal $x[n]$ as

$$DTFT(x[n]) = X(e^{j\omega}) = \sum_{n=-\infty}^{\infty} x[n]e^{-j\omega n} \tag{5.1}$$

where ω (radian frequency) is a continuous function and runs from $-\pi$ to π, and $x[n]$ is absolutely summable, i.e.,

$$\sum_{n=-\infty}^{\infty} |x[n]| < \infty$$

The inverse DTFT (IDTFT) is defined as

$$x[n] = \frac{1}{2\pi} \int_{-\pi}^{\pi} X(e^{j\omega})e^{j\omega n} d\omega \tag{5.2}$$

DTFT theory will be discussed in detail below, while computation of the DTFT using the DFT will be discussed in the chapter on the DFT.

Discrete Fourier Series (DFS)

A periodic sequence $x[n]$ ($-\infty < n < \infty$) may be decomposed into component sequences that comprise a harmonic series of complex exponentials. Since a sampled sequence is bandlimited, it follows that the frequency of the highest harmonic is limited to the Nyquist rate. The normalized frequencies of the harmonics are $2\pi k/N$, with k = 0:1:N-1. Since the transform involves a finite number of frequencies to be evaluated, the DFS is a computable transform.

The DFS coefficients $\widetilde{X}[k]$ are

$$DFS(x[n]) = \widetilde{X}[k] = \sum_{n=0}^{N-1} \widetilde{x}[n]e^{-j2\pi kn/N} \tag{5.3}$$

where k is an integer and $\widetilde{x}[n]$ is one period of the periodic sequence $x[n]$. Typical ranges for k are: 0:1:N-1, or -N/2+1:1:N/2 for even length sequences, or -(N-1)/2:1:(N-1)/2 for odd length sequences.

Once having the DFS coefficients, the original sequence $x[n]$ can be reconstructed using the following formula:

$$x[n] = \frac{1}{N} \sum_{k=0}^{N-1} \widetilde{X}[k]e^{j2\pi kn/N} \tag{5.4}$$

The DFS, a computable transform, forms an important theoretical basis for the Discrete Fourier Transform and will be discussed in more detail in the chapter on the DFT.

Discrete Fourier Transform (DFT)

$$DFT[k] = \frac{1}{N}\sum_{n=0}^{N-1} x[n]e^{-j2\pi kn/N}$$

The DFT (a computable transform) operates on discrete, or sampled signals, and evaluates frequency response at a number (equal to about half the sequence length) of unique frequencies. The correlators are complex exponentials having constant, unity amplitude, and the frequencies k range from 0 to $N-1$ or, for even-length sequences, $-N/2+1$ to $N/2$ and $-N/2$ to $-N/2$ for odd-length DFTs.

The DFT (including its efficient implementation, the FFT) will be discussed extensively in the chapter that follows the chapter on the z-transform.

5.3.2 LAPLACE FAMILY (TIME-VARYING-MAGNITUDE CORRELATORS)

Laplace Transform (LT)

The LT is defined as

$$\pounds(s) = \int_{-\infty}^{\infty} x(t)e^{-st}dt = \int_{-\infty}^{\infty} x(t)e^{-\sigma t}e^{-j\omega t}dt$$

The LT is the standard frequency transform for use with continuous time domain signals and systems. The parameter s represents the complex number $\sigma + j\omega$, with σ, a real number, being a damping coefficient, and $j\omega$, an imaginary number, representing frequency. Both σ and ω run from negative infinity to positive infinity. The correlators generated by e^{-st} are complex exponentials having amplitudes that decay, grow, or retain unity-amplitude over time, depending on the value of σ. By varying both σ and ω, the poles and zeros of the signal or system can be identified. Results are graphed in the s-Domain (the complex plane), using rectangular coordinates with σ along the horizontal axis, and $j\omega$ along the vertical axis. The magnitude of the transform can be plotted along a third dimension in a 3-D plot if desired, but, more commonly, a 2-D plot is employed showing only the locations of poles and zeros.

The LT is a reversible transform, and can be used to solve differential equations, such as those representing circuits having inductance and capacitance, in the frequency domain. The time domain solution is then obtained by using the Inverse LT. The LT is used extensively for circuit analysis and representation in the continuous domain. We'll see later in the book that certain well known or classical IIRs (Butterworth, Chebyshev, etc) have been extensively developed in the continuous domain using LTs, and that the Laplace filter parameters can be converted to the digital domain to create an equivalent digital IIR.

Note that the FT results when $\sigma = 0$ in the Laplace transform. That is to say, when the damping coefficient is zero, the Laplace correlators are constant, unity-amplitude complex exponentials just as those of the FT. Information plotted along the Imaginary axis in the s- or Laplace domain is equivalent, then, to the FT.

z-Transform (**z-T**)

The z-transform (z-T) is a discrete time form of the Laplace Transform. For those readers not familiar with the LT, study of the z-T can prove helpful since many Laplace properties and transforms of common signals are analogous to those associated with the z-T. The z-T is defined as

$$X(z) = \sum_{n=-\infty}^{\infty} x[n]z^{-n}$$

The z-T converts a number sequence into an algebraic expression in z, and, in the reverse or inverse z-T, converts an algebraic expression in z into a sequence of numbers. The z-T is essentially a discrete-time version of the Laplace transform. The correlators are discrete-time complex exponentials with amplitudes that grow, shrink, or stay the same according to the value of z (a complex number) at which the transform is evaluated. Values of z having magnitudes < 1 result in a correlation of the signal with a decaying (discrete) complex exponential, evaluation with z having a magnitude equal to 1 results in a Fourier-like response, and evaluation with z having a magnitude greater than 1 results in a correlation of the signal with a growing discrete complex exponential. Results can be plotted in the z-Domain, which is the complex plane using polar coordinates of r and θ, where θ corresponds to normalized radian frequency and r to a damping factor. The unit circle in the z-domain corresponds to the imaginary axis in the s-Domain; the left hand s-plane marks a region of stable pole values in the s-plane that corresponds to the area inside the unit circle in the z-plane. As in the s-plane, the z-T magnitude may be plotted using a 3-D plot, or, using a 2-D plot, only the location of the poles and zeros may be plotted.

Among the discrete-signal transforms, the z-T is more general than the DTFT, since in the z-T, the test exponentials may also have decay factors (negative or positive). All of the members of the Fourier transform family use exponentials of constant unity amplitude to perform their correlations with the signal of unknown frequency content, and thus cannot give pole and zero locations as can the Laplace and z- transforms.

The z-transform will be discussed extensively in the next chapter.

Additional Transforms

There are also several other transforms that are derived from the DFT; namely, the Discrete Cosine Transform (DCT), and the Discrete Sine Transform (DST). These transforms use multiples of half-cycles of either the cosine or sine, respectively, as the correlators, rather than multiples of full cycles as in the DFT. The bin values are real only. A form of the DCT called the MDCT (Modified DCT) is used in certain audio compression algorithms such as MP3.

Reference [1] gives a thorough and very accessible development of the DFT, and also briefly discusses the Laplace transform, the z-transform, the DCT, and the DST.

5.4 THE DTFT

The DTFT provides a continuous frequency spectrum for a sampled signal, as opposed to a discrete frequency spectrum (in which only a finite number of frequency correlators are used, as is true of the DFT and DFS).

In Eq. (5.1), $x[n]$ is a sampled signal which may be of either finite or infinite extent, and ω, a continuous function of frequency which may assume values from $-\pi$ to π.

Example 5.1. Derive an algebraic expression for the DTFT of the function $0.9^n u[n]$, and then evaluate it numerically at frequencies between 0 and π radians, at intervals of 0.01 radian.

We must evaluate the expression

$$DTFT(x[n]) = \sum_{n=-\infty}^{\infty} x[n]e^{-j\omega n}$$

We note that $x[n] = 0$ for $n < 0$, that $x[n]$ itself is absolutely summable since it is a decreasing geometric series. The n-th term of the summation is

$$0.9^n e^{-j\omega n}$$

and we note that each successive term is arrived at by multiplying the previous term by the number $0.9e^{-j\omega}$ and thus the net sequence forms a geometric series the sum of which is

$$\frac{1}{(1-0.9e^{-j\omega})} \tag{5.5}$$

We can evaluate this expression at a finite number of values of ω and plot the result. The following code directly evaluates expression (5.5) at frequencies between 0 and π radians, as specified by the vector w:

```
w = 0:0.01:pi; DTFT = 1./(1-0.9*exp(-j*(w)));
figure(8); plot(w/(pi),abs(DTFT));
xlabel('Normalized Frequency'); ylabel('Magnitude')
```

Example 5.2. Evaluate and plot the magnitude of the DTFT of the following sequence: [1, 0, 1].

$$F(\omega) = \sum_{n=0}^{2} x[n]e^{-j\omega n} = \sum [1, 0, 1][e^{-j\omega 0}, e^{-j\omega 1}, e^{-j\omega 2}] = 1 + e^{-j\omega 2} \tag{5.6}$$

From our earlier work, we recognize the impulse response [1,0,1] as that of a simple bandstop filter. We can show that this is so by evaluating Eq. (5.6) at a large (but necessarily finite) number of values of ω with the following code, the results of which are shown in Fig. 5.1.

```
w = 0:0.01:pi; DTFT = 1+exp(-j*2*w);
figure(8); plot(w/(pi),abs(DTFT));
xlabel('Normalized Frequency'); ylabel('Magnitude')
```

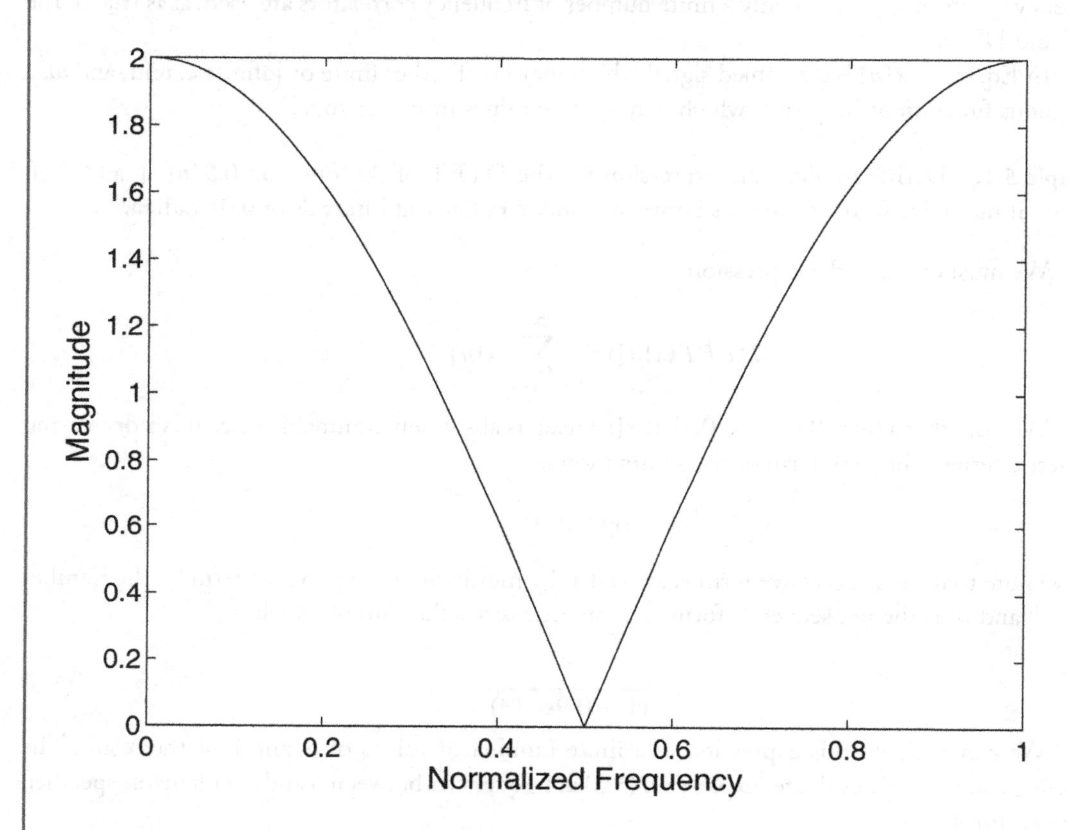

Figure 5.1: Magnitude of the DTFT of the simple notch filter [1, 0, 1].

Example 5.3. Write a script that will evaluate and plot the magnitude and phase of the DTFT for any sequence; test it for the sequences [1, 0, 1], [1, 0,-1], and [1, 0, 0, 1].

Such a script should allow one to specify how many samples M of the DTFT to compute over the interval 0 to $R\pi$, with $R = 1$ being suitable for real $x[n]$ and $R = 2$ being suitable for complex $x[n]$. Values of R greater than 2 allow demonstration of periodicity of the DTFT. The code creates an n-by-k matrix $dMat$ of complex correlators, where each column is a complex correlator of length n and frequency k. The DTFT is obtained by multiplying the signal vector x on the right by $dMat$. Each element in the resulting row vector of frequency responses is obtained as the inner or dot product of the signal vector x with a column of $dMat$.

```
function LV_DTFT_Basic(x,M,R)
% LV_DTFT_Basic([1,0,1],300,1)
N = length(x); W = exp(-j*R*pi/M); k = 0:1:M-1;
n = 0:1:N-1; dMat = W.^(n'*k); d = x*dMat; figure(9)
subplot(2,2,1); plot(R*[0:1:M-1]/M,abs(d));
grid; xlabel('Norm Freq'); ylabel('Mag')
subplot(2,2,2); plot(R*[0:1:M-1]/M,angle(d))
grid; xlabel('Norm Freq'); ylabel('Radians')
subplot(2,2,3); plot(R*[0:1:M-1]/M,real(d));
grid; xlabel('Norm Freq'); ylabel('Real')
subplot(2,2,4); plot(R*[0:1:M-1]/M,imag(d))
grid; xlabel('Norm Freq'); ylabel('Imag')
```

The result from making the call

LV_DTFT_Basic([1,0,1],300,1)

is shown in Fig. 5.2.

A more versatile version of the above code is the script (see exercises below)

$$LVxDTFT(x, n, M, R, FreqOpt, FigNo)$$

which, from sequence x having time indices n, computes M frequency samples over the interval $R\pi$, which can be computed symmetrically or asymmetrically with respect to frequency zero (*FreqOpt* = 1 for symmetrical, 2 for asymmetrical). The radian frequencies of evaluation would be, for the asymmetrical option

$$R\pi([0 : 1 : M - 1])/M$$

and for the symmetrical option

$$R\pi([-(M - 1)/2 : 1 : (M - 1)/2])/M \quad (M \text{ odd})$$

$$R\pi([-M/2 + 1 : 1 : M/2])/M \quad (M \text{ even})$$

The desired figure number is supplied as *FigNo*, an option allowing you to create different figures for comparison with each other. A typical call is

LVxDTFT([1,0,1],[0:1:2],300,2,1,88)

which results in Fig. 5.3.

A second script (for a complete description of input arguments, see exercises below)

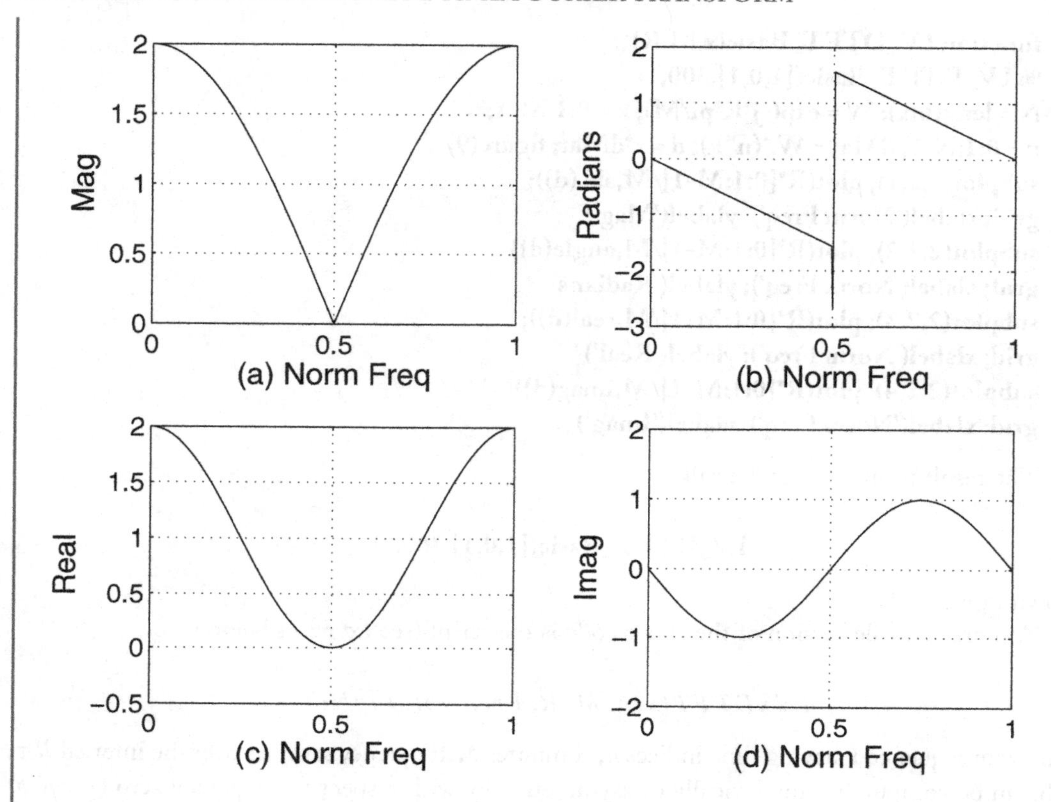

Figure 5.2: (a) Magnitude of DTFT of the sequence [1 0 1]; (b) Phase response of DTFT; (c) Real component of DTFT ; (d) Imaginary component of DTFT.

$$LVxDTFT\_MS(x, SampOffset, FreqOffsetExp, M, R, TimeOpt, FreqOpt)$$

allows you to enter one sequence, and the second sequence is created as a modification of the first, delayed by *SampOffset* samples and offset in frequency by the complex exponential *FreqOffsetExp*. Input arguments M and R are as described for the script *LVxDTFT*; $FreqOpt$ determines whether the DTFT is computed symmetrically or asymmetrically about frequency zero, as described above for the script *LVxDTFT*. For the asymmetrical time option (determined by the input argument *TimeOpt*), the sequence time indices of the first sequence are given as

$$n = 0 : 1 : N - 1;$$

where N is the length of x. For the symmetrical time index option, the time indices are given as

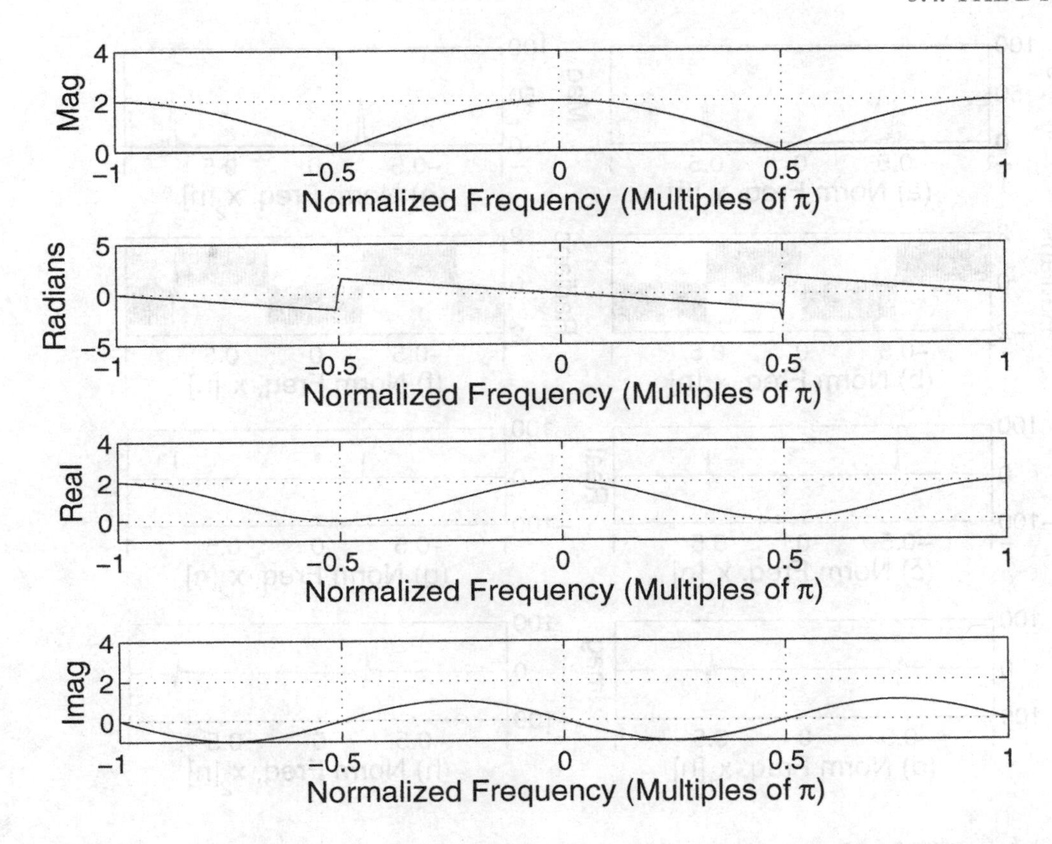

Figure 5.3: (a) Magnitude of DTFT of the sequence [1,0,1]; (b) Phase of DTFT of same; (c) Real part of DTFT of same; (d) Imaginary part of DTFT of same.

$$n = -(N-1)/2 : 1 : (N-1)/2 \quad (n \text{ odd})$$

$$n = -N/2 + 1 : 1 : N/2 \quad (n \text{ even})$$

This script is useful for demonstrating the effect on the DTFT of time and frequency shifts to a first test sequence. A typical call, which results in Fig. 5.4, is

```
nN = (0:1:100)/100;
LVxDTFT_MS([cos(2*pi*25*nN)],0,...
exp(j*2*pi*12.5*nN),200,2,2,1)
```

We will make use of these scripts shortly while studying the various properties of the DTFT.

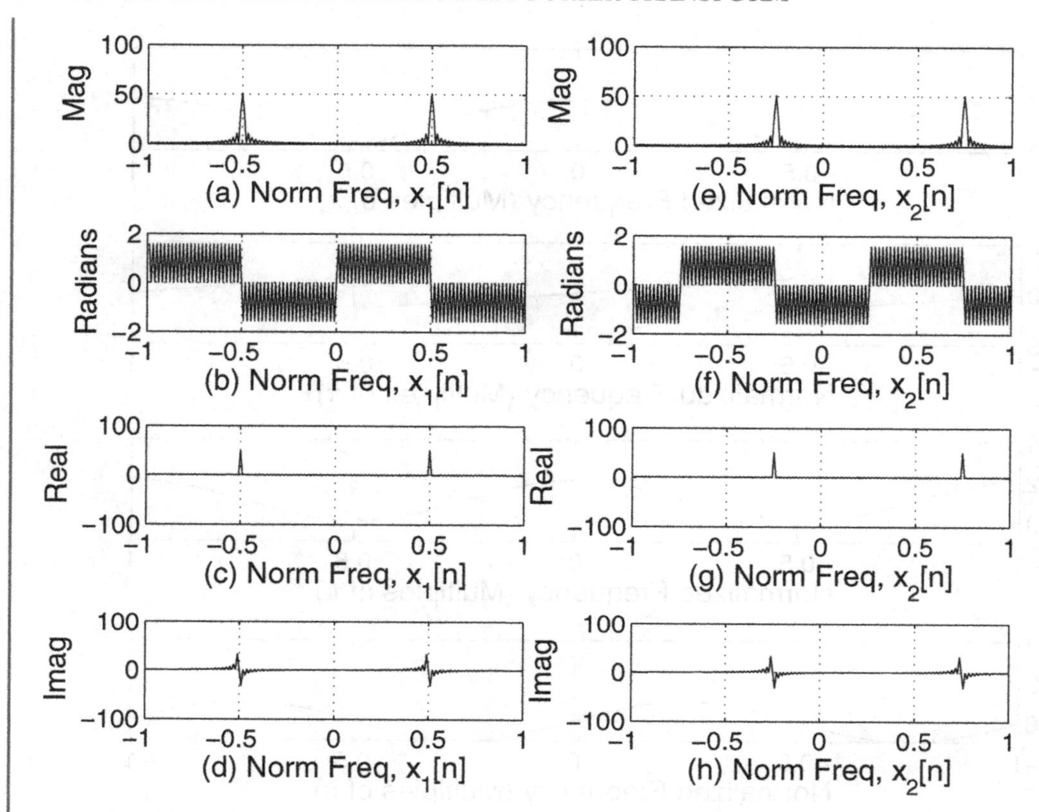

Figure 5.4: DTFT of first sequence, with its magnitude and phase and real and imaginary parts being shown, respectively, in plots (a)-(d); DTFT of second sequence, which is the first sequence offset in frequency by $\pi/4$ radian, with its magnitude, phase, and real and imaginary parts being shown, respectively, in plots (e)-(h). All frequencies are normalized, i.e., in units of π radians.

5.5 INVERSE DTFT

The Inverse DTFT, i.e., the original time domain sequence $x[n]$ from which a given DTFT was produced, can be reconstructed by evaluating the following integral:

$$x[n] = \frac{1}{2\pi} \int_{-\pi}^{\pi} X(e^{j\omega}) e^{j\omega n} d\omega \tag{5.7}$$

We will illustrate this with several examples, one analytic and the other numerical.

Example 5.4. Using Eq. (5.7), compute $x[0]$, $x[1]$, and $x[2]$ from the DTFT obtained in Eq. (5.6).

We get

$$x[0] = \frac{1}{2\pi}\int_{-\pi}^{\pi}(1+e^{-j\omega 2})e^{j\omega 0}d\omega$$

and

$$x[1] = \frac{1}{2\pi}\int_{-\pi}^{\pi}(1+e^{-j\omega 2})e^{j\omega 1}d\omega$$

and

$$x[2] = \frac{1}{2\pi}\int_{-\pi}^{\pi}(1+e^{-j\omega 2})e^{j\omega 2}d\omega$$

The formula for $x[0]$ reduces to

$$\frac{1}{2\pi}\int_{-\pi}^{\pi}(1+e^{-j\omega 2})d\omega = \frac{1}{2\pi}(\int_{-\pi}^{\pi}d\omega + \int_{-\pi}^{\pi}e^{-j\omega 2}d\omega)$$

which is

$$\frac{1}{2\pi}(\omega\mid_{-\pi}^{\pi} + \int_{-\pi}^{\pi}e^{-j\omega 2}d\omega) = 1+0 = 1$$

where we note that

$$\int_{-\pi}^{\pi}e^{\pm j\omega n}d\omega = \begin{cases} 2\pi & \text{if} \quad n=0 \\ 0 & \text{if} \quad n=\pm 1, \pm 2... \end{cases}$$

For $x[1]$ we get

$$x[1] = \frac{1}{2\pi}\int_{-\pi}^{\pi}(1+e^{-j\omega 2})e^{j\omega 1}d\omega = \frac{1}{2\pi}\int_{-\pi}^{\pi}(e^{j\omega 1}+e^{-j\omega 1})d\omega = 0$$

The formula for $x[2]$ is the same as that for $x[0]$ with the exception of the sign of the complex exponential, which does not affect the outcome. The reconstructed sequence is therefore [1,0,1], as expected.

Example 5.5. Using numerical integration, approximate the IDTFT that was determined analytically in the previous example.

Let's reformulate the code to obtain the DTFT from -pi to +pi, and to use a much finer sample spacing (this will improve the approximation to the true, continuous spectrum DTFT), and then perform the IDTFT one sample at a time in accordance with Eq. (5.2):

```
N=10^3; dw = 2*pi/N; w = -pi:dw:pi*(1-2/N);
DTFT = 1+exp(-j*2*w);
```

```
x0 = (1/(2*pi))*sum(DTFT.*exp(j*w*0)*dw)
x1 = (1/(2*pi))*sum(DTFT.*exp(j*w*1)*dw)
x2 = (1/(2*pi))*sum(DTFT.*exp(j*w*2)*dw)
```

Running the preceding code yields the following answer, which rounds to the original sequence, [1,0,1]:

```
x0 = 1.0000 - 0.0000i
x1 = -4.9127e-017
x2 = 1.0000 + 0.0000i
```

5.6 A FEW PROPERTIES OF THE DTFT

5.6.1 LINEARITY

The DTFT of a linear combination of two sequences $x_1[n]$ and $x_2[n]$ is equal to the sum of the individual responses, i.e.,

$$DTFT(ax_1[n] + bx_2[n]) = aDTFT(x_1[n]) + bDTFT(x_2[n])$$

5.6.2 CONJUGATE SYMMETRY FOR REAL $x[n]$

For real $x[n]$, the real part of the DTFT shows even symmetry ($X(e^{j\omega}) = X(e^{-j\omega})$), and the imaginary part shows odd symmetry ($X(e^{j\omega}) = -X(e^{-j\omega})$).

5.6.3 PERIODICITY

The DTFT of a sequence $x[n]$ repeats itself every 2π:

$$X(e^{jw}) = X(e^{j(w+2\pi n)})$$

where $n = 0, \pm 1, \pm 2...$

To illustrate this principle, consider the following: for a given sequence length, the Nyquist limit is half the sequence length, and this represents a frequency shift of π radians. To shift 2π radians therefore is to shift by a frequency equal to the sequence length. The following m-code, the results of which are illustrated in Fig. 5.5, verifies this property.

```
SR = 100; nN = (0:1:SR)/SR;
LVxDTFT_MS([cos(2*pi*25*nN)],0,exp(j*2*pi*SR*nN),200,2,2,1)
```

You can gain insight by noting that the code

```
nN = (0:1:100)/100; y = exp(j*2*pi*100*nN)
```

yields $y = ones(1, 101)$, which clearly transforms the original sequence into itself, i.e., the new sequence is the same as the old, and hence the DTFT is the same. In other words, the DTFT of a sequence repeats itself for every frequency shift of 2π radians of the original sequence.

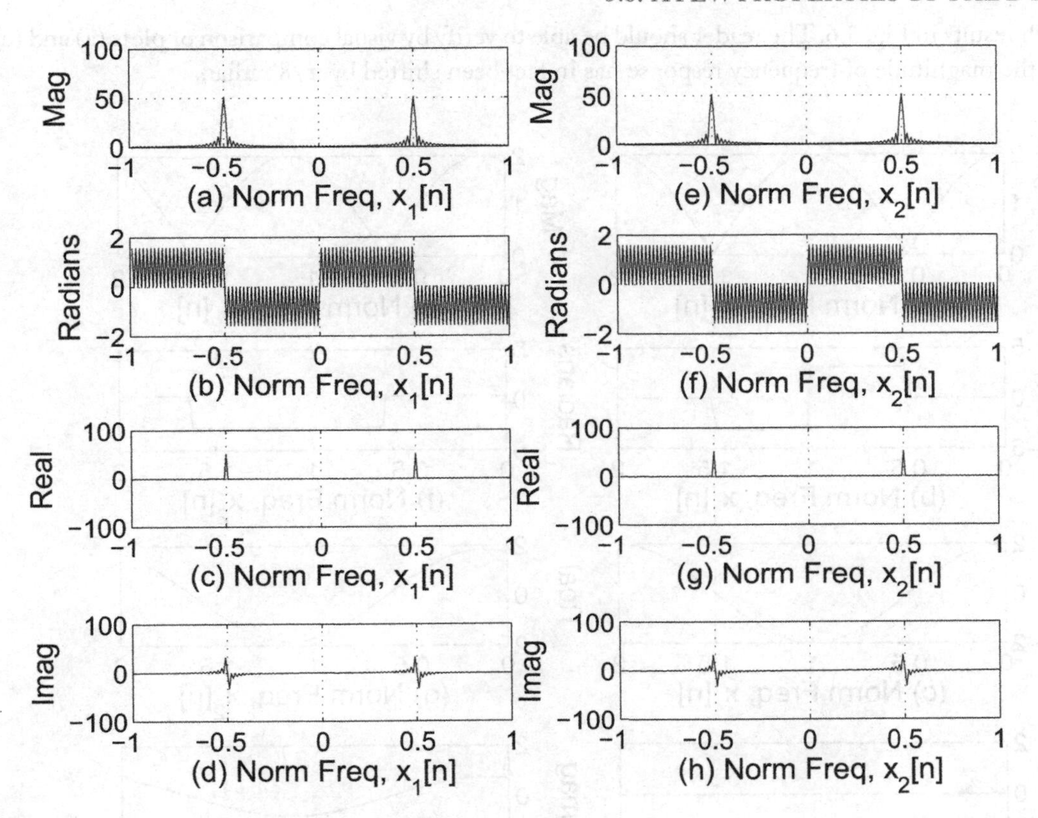

Figure 5.5: DTFT of first sequence, with its magnitude and phase and real and imaginary parts being shown, respectively, in plots (a)-(d); DTFT of second sequence, which is the first sequence offset in frequency by 2π radians, with its magnitude, phase, and real and imaginary parts being shown, respectively, in plots (e)-(h). All frequencies are normalized, i.e., in units of π radians.

5.6.4 SHIFT OF FREQUENCY

If the signal $x[n]$ is multiplied by a complex exponential of frequency F_0, the result is that the DTFT of $x[n]$ is shifted.

$$DTFT(x[n]e^{j\omega_0 n}) = X(e^{j(\omega-\omega_0)})$$

To demonstrate this property, we can use the script *LVxDTFT_MS*. We pick the short sequence [1,0,1] as $x[n]$, and specify no sample offset, but a frequency offset of $2\pi/16$ radians. We thus make the call

LVxDTFT_MS([1,0,1],0,exp(j*2*pi*1*(0:1:2)/16),100,2,1,2)

which results in Fig. 5.6. The reader should be able to verify by visual comparison of plots (a) and (e) that the magnitude of frequency response has in fact been shifted by $\pi/8$ radian.

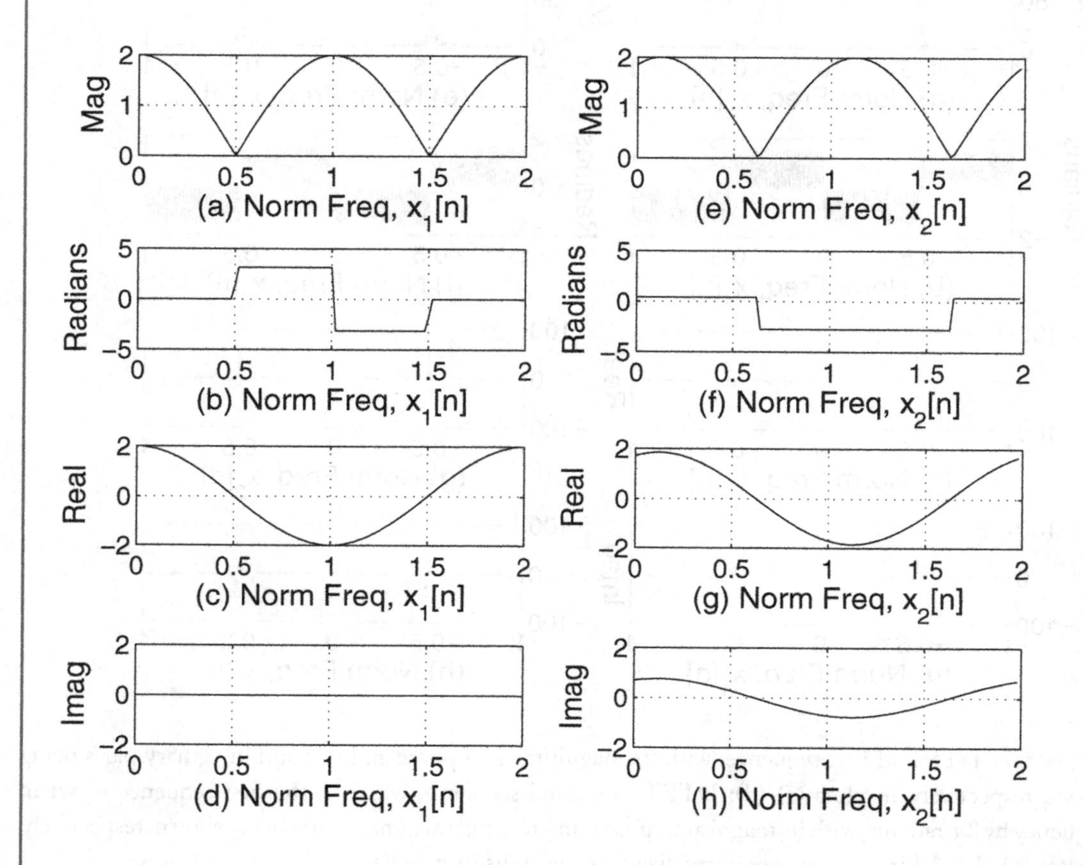

Figure 5.6: DTFT of first sequence, with its magnitude and phase and real and imaginary parts being shown, respectively, in plots (a)-(d); DTFT of second sequence, which is the first sequence offset in frequency by $\pi/8$ radian, with its magnitude, phase, and real and imaginary parts being shown, respectively, in plots (e)-(h). All frequencies are normalized, i.e., in units of π radians.

5.6.5 CONVOLUTION

The DTFT of the time domain convolution of two sequences is equal to the product of the DTFTs of the two sequences.

$$DTFT(x_1[n] * x_1[n]) = X_1(e^{j\omega})X_2(e^{j\omega})$$

Example 5.6. Consider the two sequences [1, 0, 1] and [1, 0, 0, −1]. Obtain the time domain convolution by taking the inverse DTFT of the product of the DTFTs of each sequence, and confirm the result using time domain convolution.

The DTFT of the first sequence is

$$X_1(e^{j\omega}) = 1 + e^{-j\omega 2}$$

and for the second sequence

$$X_2(e^{j\omega}) = 1 - e^{-j\omega 3}$$

The product is

$$1 + e^{-j\omega 2} - e^{-j\omega 3} - e^{-j\omega 5}$$

or stated more completely

$$1e^{-j\omega 0} + 0e^{-j\omega 1} + e^{-j\omega 2} - e^{-j\omega 3} + 0e^{-j\omega 4} - e^{-j\omega 5}$$

which can be recognized by inspection as the DTFT of the time domain sequence

$$[1, 0, 1, -1, 0, -1]$$

To confirm, we make the following call:

```
y = conv([1,0,1],[1,0,0,-1])
```

which produces the same result.

5.6.6 EVEN AND ODD COMPONENTS

If a real sequence $x[n]$ is decomposed into its even and odd components, and the DTFT taken of each, it will be found that the DTFT of the even part is all real and equal to the real part of $DTFT(x[n])$, while the DTFT of the odd part will be found to be imaginary only and equal to the imaginary part of $DTFT(x[n])$. Using $X(e^{j\omega})$, $XE(e^{j\omega})$, $XO(e^{j\omega})$ as the DTFT's of $x[n]$, $xe[n]$, and $xo[n]$, we would have

$$\text{Re}(X(e^{j\omega})) = \text{Re}(XE(e^{j\omega})) = XE(e^{j\omega})$$

and

$$\text{Im}(X(e^{j\omega})) = \text{Re}(XO(e^{j\omega})) = XO(e^{j\omega})$$

Example 5.7. Write a script that will demonstrate the above even-odd properties.

We can perform the even-odd decomposition using the script *LVEvenOddSymmZero* presented earlier, and then use the script *LVxDTFT* in three separate calls to open three separate windows to show $X(e^{j\omega})$, $XE(e^{j\omega})$, and $XO(e^{j\omega})$; the following script also reconstructs the original DTFT as the sum of its even and odd components, subtracts this from the DTFT of the original signal, and obtains the RMS error, which should prove to be essentially zero, within the limits of roundoff error.

```
x = [1:1:9]; [xe,xo,m] = LVEvenOddSymmZero(x,[0:1:8]);
d = LVxDTFT(x,[0:1:8],200,2,2,10);
de = LVxDTFT(xe,m,200,2,2,11);
do = LVxDTFT(xo,m,200,2,2,12);
RMS = sqrt((1/200)*sum((d - (de+do)).^2))
```

The value of RMS reported after running the above code was [1.0190e-016, -1.6063e-015i], which is essentially zero, within the limits of roundoff error.

5.6.7 MULTIPLICATION BY A RAMP

$$DTFT(nx[n]) = j\frac{dX(e^{j\omega})}{d\omega}$$

This property states that if the DTFT of $x[n]$ is $X(e^{j\omega})$, then the DTFT of $nx[n]$ is the derivative with respect to ω of $X(e^{j\omega})$ multiplied by j.

5.7 FREQUENCY RESPONSE OF AN LTI SYSTEM

5.7.1 FROM IMPULSE RESPONSE

An interesting and useful result occurs by convolving a complex exponential $e^{j\omega k}$ of radian frequency of $\omega = \omega_0$ with an LTI system having an impulse response represented by $h[n]$:

$$y[k] = \sum_{n=-\infty}^{\infty} h[n]e^{j\omega_0(k-n)} = (\sum h[n]e^{-j\omega_0 n})e^{j\omega_0 k} \tag{5.8}$$

The rightmost expression in Eq. (5.8) is the input signal $e^{j\omega_0 k}$scaled by the DTFT of $h[n]$ evaluated at ω_0. Since the DTFT evaluated at a single frequency is a complex number, it is sometimes convenient to represent it as a magnitude and phase angle. Thus for each frequency ω_0 we would have the result

$$y[k] = H(e^{j\omega_0})e^{j\omega_0 k} = (|H(e^{j\omega_0})|)(\angle H(e^{j\omega_0}))e^{j\omega_0 k} \tag{5.9}$$

An interpretation of this result is that a sinusoidal excitation of frequency ω_0 to an LTI system produces an output that is a sinusoid of the same frequency, scaled by the magnitude of the DTFT of $h[n]$ at ω_0, and phase shifted by the angle of the DTFT of $h[n]$ evaluated at ω_0. This is consistent with the principle of Sinusoidal Fidelity discussed earlier in the book.

Example 5.8. Consider the LTI system whose impulse response is $[1, 0, 1]$. Determine the magnitude of the response to the complex exponential.

```
exp(j*2*pi*(0:1:31)*5/32)
```

We note that the normalized frequency is 5/16 = 0.3125 and obtain the magnitude of H at the given frequency as 1.111 using the following code:

```
w = 5*pi/16; DTFT = 1+exp(-j*2*w);
magH = abs(DTFT)
```

To obtain the magnitude of response via time domain convolution, we make the following call:

```
tdMagResp = max(abs(conv([1,0,1],exp(j*2*pi*(0:1:31)*5/32))))
```

which produces the identical result, 1.111.

Equation (5.9) may be generalized and applied to real sinusoids. Thus the **steady state** response $y[n]$ of an LTI system to a cosine (or sine) of magnitude A, frequency ω_0 and phase angle ϕ, with the DTFT magnitude and angle being M and θ is

$$y[n] = MA\cos(\omega_0 n + \phi + \theta) \tag{5.10}$$

Example 5.9. Verify Eq. (5.10) using both cosine and sine waves having normalized frequencies of 0.53 for the LTI system whose impulse response is $[1, 0, 1]$.

We'll use a discrete cosine of the stated frequency with ϕ = 0 and M = 1, and then compute the output according to Eq. (5.10) and then by convolution with the impulse response. The plot will show that the results are identical except for the first two and last two samples (recall that Eq. (5.10) represents a steady state response). In the code below, you can substitute the sine function (sin) for the cosine function (cos) as well as change the frequency F and test signal length N.

```
F = 0.53; N = 128; w = F*pi; dtft = 1+exp(-j*2*w);
M = abs(dtft); theang = angle(dtft);
y1 = M*cos(2*pi*(0:1:N-1)*(N/2*F)/N + theang);
y2 = conv([1 0 1],cos(2*pi*(0:1:N-1)*(N/2*F)/N));
stem(y1,'bo'); hold on; stem(y2,'r*')
```

Since LTI systems obey the law of superposition, Eq. (5.10) can be generalized for an input signal comprising a sum or superposition of sinusoids:

$$y[n] = \sum_{k=1}^{K} M_k A_k \cos(\omega_k n + \phi_k + \theta_k) \tag{5.11}$$

5.7.2 FROM DIFFERENCE EQUATION

Thus far we have been using the impulse response of an LTI system to determine the system's frequency response; in many cases, a difference equation may be what is immediately available. While it is certainly possible to process a unit impulse using the difference equation and thus obtain the impulse response, it is possible to obtain the frequency response directly from the difference equation. Assuming that the difference equation is of the form

$$y[n] + \sum_{k=1}^{K} a_k y[n-k] = \sum_{m=0}^{M} b_m x[n-m] \tag{5.12}$$

and recalling from Eq. (5.9) that

$$y[n] = H(e^{j\omega})e^{j\omega n}$$

and substituting in Eq. (5.12), we get

$$H(e^{j\omega})e^{j\omega n} + \sum_{k=1}^{K} a_k H(e^{j\omega})e^{j\omega(n-k)} = \sum_{m=0}^{M} b_m e^{j\omega(n-m)}$$

which reduces to

$$H(e^{j\omega}) = \frac{\sum_{m=0}^{M} b_m e^{-j\omega m}}{1 + \sum_{k=1}^{K} a_k e^{-j\omega k}} \tag{5.13}$$

Example 5.10. Determine the frequency response of a certain LTI system that is defined by the following difference equation, then compute and plot the magnitude and phase response for $\omega = 0$ to 2π.

$$y[n] = x[n] + x[n-2] + 0.9y[n-1]$$

We first rewrite the equation to have all y terms on the left and all x terms on the right:

$$y[n] - 0.9y[n-1] = x[n] + x[n-2]$$

Then we get

$$H(e^{j\omega}) = \frac{1 + e^{-j\omega 2}}{1 - 0.9e^{-j\omega}}$$

which can be numerically evaluated using the following code:

```
N=10^3; dw=2*pi/N; w = 0:dw:2*pi-dw;
H = (1+exp(-j*2*w))./(1-0.9*exp(-j*w));
figure; subplot(2,1,1); plot(abs(H)); subplot(2,1,2); plot(angle(H))
```

5.8 REFERENCES

[1] William L. Briggs and Van Emden Henson, *The DFT, An Owner's Manual for the Discrete Fourier Transform*, SIAM, Philadelphia, 1995.

[2] Steven W. Smith, *The Scientist and Engineer's Guide to Digital Signal Processing*, California Technical Publishing, San Diego, 1997.

[3] Alan V. Oppenheim and Ronald W. Schaefer, *Discrete-Time Signal Processing*, Prentice-Hall, Englewood Cliffs, New Jersey, 1989.

[4] John G. Proakis and Dimitris G. Manolakis, *Digital Signal Processing, Principles, Algorithms, and Applications, Third Edition*, Prentice Hall, Upper Saddle River, New Jersey, 1996.

[5] Vinay K. Ingle and John G. Proakis, *Digital Signal Processing Using MATLAB V.4*, PWS Publishing Company, Boston, 1997.

5.9 EXERCISES

1. Determine analytically the DTFT of the following sequences:

(a) $0.95^n u[n]$ - $0.8^{n-1} u[n-1]$
(b) $\cos(2\pi(0{:}1{:}3)/4)$
(c) $(0.85)^{n-2} u[n-2]$
(d) $([0,0,(0.85)^{n-2} u[n-2]])u[n]$
(e) $(-0.9)^n u[n] + (0.7)^{n-3} u[n-3]$
(f) $n(u[n] - u[n-3])$

2. Compute the DTFTs of the following sequences and plot the magnitude and phase responses (use one of the scripts developed or presented earlier in this chapter, such as *LV_DTFT_Basic*). Be sure to increase the number of DTFT samples as the signal length increases. It's a good idea to use at least 10 times as many DTFT samples as the sequence length being evaluated. Consider plotting the magnitude on a logarithmic scale to see fine detail better. To avoid the problem of taking the logarithm of zero, add a small number such as 10^{-10} to the absolute value of the DTFT, then use the function $log10$ and multiply the result by 20.

(a) **cos(2*pi*k*(0:1:N-1)/N)** for $N = 8, 32, 128$ and $k = 0, 1$, and $N/2$.
(b) **[ones(1,10),zeros(1,20),ones(1,10),zeros(1,20)]**
(c) **b = fir1(21,0.5)**
(d) **[1,0,1]**
(e) **[1,0,1,0,1,0,1]**
(f) **[1,0,1,0,1,0,1,0,1,0,1,0,1]**
(g) **[1,0,1,0,1,0,1,0,1,0,1,0,1].*hamming(13)'**
(h) **[1,0,1,0,1,0,1,0,1,0,1,0,1].*blackman(13)'**
(i) **[1,0,1,0,1,0,1,0,1,0,1,0,1].*kaiser(13,5)'**
(j) **[1,0,-1,0,1,0,-1,0,1,0,-1,0,1].*hamming(13)'**
(k) **[1,0,-1,0,1,0,-1,0,1,0,-1,0,1].*blackman(13)'**
(l) **[1,0,-1,0,1,0,-1,0,1,0,-1,0,1].*kaiser(13,5)'**
(m) **[real(j.^(0:1:10))].*blackman(11)'**
(n) **[real(j.^(0:1:20))].*blackman(21)'**
(o) **[real(j.^(0:1:40))].*blackman(41)'**
(p) **[real(j.^(0:1:80))].*blackman(81)'**

3. Write a script that can receive b and a difference equation coefficients (according to Eq. (5.13)) in row vector form (normalized so $a_0 = 1$), compute, and display the following:

(a) The unit impulse response of the system defined by b and a.
(b) The unit step response of the system.
(c) The magnitude and phase of the DTFT.
(d) The response to a linear chirp of length 1024 samples and frequencies from 0 to 512 Hz (0 to π radians in normalized frequency).
(e) The magnitude of response to a complex linear chirp of length 1024 samples and frequencies from 0 to 512 Hz (0 to π radians in normalized frequency). Such a chirp can be generated by the following code:

```
N = 1024; t = 0:1/N:1; y = chirp(t,0,1,N/2) + ...
j*chirp(t,0,1,N/2,'linear',90)
```

(f) The response to a signal of length 1024 samples containing a cosine of frequency 128.
(g) The response to a signal of length 1024 samples containing a cosine of frequency 256.

Use the script to evaluate the LTI systems defined by the following difference equations or b and a coefficients. You should note that some of the systems are not stable. Compare the plot of DTFT magnitude to the two chirp responses for each of the difference equations below. Where can the steady-state magnitude of the responses specified in (b), (f), and (g) above be found on the DTFT magnitude plot? State whether each of the systems is stable or unstable.

(I) y[n] = x[n] + x[n-1] + 0.9y[n-1]
(II) y[n] = x[n] + 1.4y[n-1] - 0.81y[n-2]
(III) y[n] = x[n] + 1.4y[n-1] + 0.81y[n-2]
(IV) y[n] = x[n] - 2.45y[n-1] +2.37y[n-2] -0.945y[n-3]

(V) y[n] = x[n] +1.05y[n-1]
(VI) y[n] = 0.094x[n] + 0.3759x[n-1] + 0.5639x[n-2] + 0.3759x[n-3] + ...
0.094x[n-4] + 0.486y[n-2] + 0.0177y[n-4]
(VII) b = [0.6066,0,2.4264,0,3.6396,0,2.4264,0,0.6066];
a = [1,0,3.1004,0,3.7156,0,2.0314,0,0.4332]

4. For the following different functions, compute the DTFT, then plot the function on a first subplot and its DTFT magnitude on a second subplot. The $sinc$ function can be evaluated using the function $sinc(x)$. What kind of filter impulse response results from the functions below? What effect does the parameter c have, and what effect does the length of the vector n have?

(a) **n =-9:1:9; c = 0.8; y = sinc(c*n);**
(b) **n =-9:1:9; c = 0.4; y = sinc(c*n);**
(c) **n =-9:1:9; c = 0.2; y = sinc(c*n);**
(d) **n =-9:1:9; c = 0.1; y = sinc(c*n);**
(e) **n =-39:1:39; c = 0.8; y = sinc(c*n);**
(f) **n =-39:1:39; c = 0.4; y = sinc(c*n);**
(g) **n =-39:1:39; c = 0.2; y = sinc(c*n);**
(h) **n =-39:1:39; c = 0.1; y = sinc(c*n);**
(i) **n =-139:1:139; c = 0.8.; y = sinc(c*n);**
(j) **n =-139:1:139; c = 0.4.; y = sinc(c*n);**
(k) **n =-139:1:139; c = 0.2.; y = sinc(c*n);**
(l) **n =-139:1:139; c = 0.1.; y = sinc(c*n);**

5. Compute and plot the magnitude and phase of the DTFT of the following impulse responses x. Note the effect of the frequency parameter f and the difference in impulse responses and especially phase responses between (a)-(e) and (f)-(j).

(a) **n = -10:1:10; f = 0.05; x = ((0.9).^(abs(n))).*cos(f*pi*n)**
(b) **n = -10:1:10; f = 0.1; x = ((0.9).^(abs(n))).*cos(f*pi*n)**
(c) **n = -10:1:10; f = 0.2; x = ((0.9).^(abs(n))).*cos(f*pi*n)**
(d) **n = -10:1:10; f = 0.4; x = ((0.9).^(abs(n))).*cos(f*pi*n)**
(e) **n = -10:1:10; f = 0.8; x = ((0.9).^(abs(n))).*cos(f*pi*n)**
(f) **n = 0:1:20; f = 0.05; x = ((0.9).^(abs(n))).*cos(f*pi*n)**
(g) **n = 0:1:20; f = 0.1; x = ((0.9).^(abs(n))).*cos(f*pi*n)**
(h) **n = 0:1:20; f = 0.2; x = ((0.9).^(abs(n))).*cos(f*pi*n)**
(i) **n = 0:1:20; f = 0.4; x = ((0.9).^(abs(n))).*cos(f*pi*n)**
(j) **n = 0:1:20; f = 0.8; x = ((0.9).^(abs(n))).*cos(f*pi*n)**

6. An ideal lowpass filter should have unity gain for $|\omega| \leq \omega_c$ and zero gain for $|\omega| > \omega_c$, and a linear phase factor or constant sample delay equal to $e^{-j\omega M}$ where M represents the number of samples of delay. Use the inverse DTFT to determine the impulse response that corresponds to this frequency specification. That is to say, perform the following integration

$$x[n] = \frac{1}{2\pi} \int_{-\pi}^{\pi} X(e^{j\omega}) e^{jwn} d\omega$$

where

$$X(e^{j\omega}) = \begin{cases} 1 \cdot e^{-j\omega M} & |\omega| \leq \omega_c \\ 0 & |\omega| > \omega_c \end{cases}$$

After performing the integration and obtaining an expression for $x[n]$, use n = -20:1:20 and M = 10, and obtain four impulse responses corresponding to the following values of ω_c:

(a) 0.1π

(b) 0.25π

(c) 0.5π

(d) 0.75π

Convolve each of the four resulting impulse responses with a chirp of length 1000 samples and frequency varying from 0 to 500 Hz. For each of the four impulse responses, plot the impulse response on one subplot and the chirp response on a second subplot. On a third subplot, plot the result from numerically computing the DTFT of each impulse response, or instead use the script *LV_DTFT_Basic* or the similar short version presented in the text to evaluate the DTFT and plot it in a separate window. To display on the third subplot, create a script

$$d = LVxDTFT\_Basic(x, M, R)$$

based on *LV_DTFT_Basic*, but which delivers the DTFT as the output argument d and which does not itself create a display.

7. For each of the four impulse responses computed above, obtain a new impulse response by subtracting the given impulse response from the vector [zeros(1,20),1,zeros(1,20)] (this assumes that the vector n in the previous example ran from -20 to +20).

Determine what kind of filter the four new impulse responses form. To do this, plot the magnitude of the DTFT of each original impulse response next to the DTFT of each corresponding new impulse response.

8. Create a script in accordance with the following call syntax, which should create plots similar to that shown in Fig. 5.3; test it with the sample calls given below.

```
function LVxDTFT(x,n,M,R,FreqOpt,FigNo)
% Computes and displays the magnitude, phase, real, and
% imaginary parts of the DTFT of the sequence x having
% time indices n, evaluated over M samples.
% Pass R as 1 to evaluate over pi radians, or 2 to evaluate
% over 2*pi radians
% Use FreqOpt = 1 for symmetrical frequency evaluation
% about frequency 0 or FreqOpt = 2 for an asymmetrical
```

```
% frequency evaluation
% Sample calls:
% LVxDTFT([cos(2*pi*25*(0:1:99)/100)],[0:1:99],500,2,1,88)
% LVxDTFT([cos(2*pi*25*(-50:1:50)/100)],[-50:1:50],500,2,1,88)
% LVxDTFT([cos(2*pi*5*(0:1:20)/20)],[0:1:20],100,2,1,88)
% LVxDTFT([cos(2*pi*5*(-10:1:10)/20)],[-10:1:10],100,2,1,88)
% LVxDTFT([cos(2*pi*25*(0:1:100)/100)],[-50:1:50],500,2,1,88)
% LVxDTFT([exp(j*2*pi*25*(0:1:99)/100)],[0:1:99],500,2,1,88)
% LVxDTFT([cos(2*pi*25*(0:1:99)/100)],[0:1:99],1000,2,1,88)
```

9. Create a script that conforms to the following call syntax, which should create plots similar to that shown in Fig. 5.6; test it with the sample calls given below.

```
function LVxDTFT_MS(x,SampOffset,FreqOffsetExp,...
% M,R,TimeOpt,FreqOpt)
% Computes and displays the magnitude, phase, real, and
% imaginary parts of the DTFT of the sequence x, evaluated
% over M samples, then computes the same for a
% modified version of x that has been shifted by
% SampOffset samples and multiplied by a complex
% exponential FreqOffsetExp.
% Pass R as 1 to evaluate from 0 to pi radians, or
% Pass R as 2 to evaluate from 0 to 2*pi radians
% Pass TimeOpt as 1 to let n (the time indices for x1
% and x2) be computed as n = -(N-1)/2:1:(N-1)/2 for N odd
% or n = -N/2+1:1:N/2; for even N.
% Pass FreqOpt as 1 for symmetrical frequency computation
% and  display (-R*pi to +R*pi, for example) or as 2 for
% frequency comp.and display from 0 to R*pi
% Sample calls:
% LVxDTFT_MS([cos(2*pi*25*(0:1:100)/100)],0,...
% exp(j*2*pi*10*(0:1:100)/100),500,2,1,1)
% LVxDTFT_MS([cos(2*pi*25*(0:1:100)/100)],0,...
% exp(-j*2*pi*10*(0:1:100)/100),500,2,1,1)
% LVxDTFT_MS([cos(2*pi*25*(0:1:100)/100)],0,...
% exp(-j*pi/2),500,2,1,1) % shifts phase
% LVxDTFT_MS([exp(j*2*pi*25*(0:1:100)/100)],0,...
% exp(-j*2*pi*10*(0:1:100)/100),500,2,1,1)
% LVxDTFT_MS([cos(2*pi*25*(0:1:100)/100)],0,...
% exp(j*2*pi*12.5*(0:1:100)/100),1000,2,1,1)
% LVxDTFT_MS([1 0 1],2,1,300,2,1,1)
```

```
% LVxDTFT_MS([1 0 1],0,exp(j*2*pi*1*(0:1:2)/3),...
% 100,2,1,1)
% LVxDTFT_MS([cos(2*pi*25*(0:1:100)/100)],0,...
% exp(j*2*pi*12.5*(0:1:100)/100),200,2,1,1)
% LVxDTFT_MS([1 0 1],0,exp(j*2*pi*(0:1:2)/3),300,2,1,1)
```

10. Compute 1200 evenly distributed frequency samples of the DTFT (i.e., a numerical approximation of the DTFT) for the three following sequences:

   (a) [0.1,0.7,1,0.7,0.1]
   (b) [1,0,0,0,1]
   (c) $[0.95^n - 0.85^n]$ for $0 \leq n \leq \infty$

11. Numerically compute the inverse DTFTs of the three numeric DTFTs computed in the previous exercise (for item (c), compute the inverse DTFT for n = 0:1:2).

12. Compute the net output signal obtained by convolving the signal

$$x = 0.5\cos(2\pi n/16) + 0.3\sin(5\pi n/16)$$

with the impulse response [1,0,-1] two ways, first, by performing the time domain convolution, and second, by computing the DTFT of the impulse response to determine the magnitude and phase responses for the frequencies in the signal, and then scaling and shifting the two signal components, and finally summing the two scaled and shifted signal components to obtain the net output response. Verify that the results are the same during steady state.

CHAPTER 6

The z-Transform

6.1 OVERVIEW

In the previous chapter, we took a brief look at the Fourier and Laplace families of transforms, and a more detailed look at the DTFT, which is a member of the Fourier family which receives a discrete time sequence as input and produces an expression for the continuous frequency response of the discrete time sequence. With this chapter, we take up the z-transform, which uses correlators having magnitudes which can grow, decay, or remain constant over time. It may be characterized as a discrete-time variant of the Laplace Transform. The z-transform can not only be used to determine the frequency response of an LTI system (i.e., the LTI system's response to unity-amplitude correlators), it reveals the locations of poles and zeros of the system's transfer function, information which is essential to characterize and understand such systems. The z-transform is an indispensable transform in the discrete signal processing toolbox, and is virtually omnipresent in DSP literature. Thus it is essential that the reader gain a good understanding of it.

The z-transform mathematically characterizes the relationship between the input and output sequences of an LTI system using the generalized complex variable z, which, as we have already seen, can be used to represent signals in the form of complex exponentials. Many benefits accrue from this:

- An LTI system is conveniently and compactly represented by an algebraic expression in the variable z; this expression, in general, takes the form of the ratio of two polynomials, the numerator representing the FIR portion of the LTI system, and the denominator representing the IIR portion.
- Values of z having magnitude 1.0, which are said to "lie on the unit circle" can be used to evaluate the z-transform and provide a frequency response equivalent to the DTFT.
- Useful information about a digital system can be deduced from its z-transform, such as location of system poles and zeros.
- Difference equations representing the LTI system can be constructed directly from inspection of the z-transform.
- An LTI system's impulse response can be obtained by use of the Inverse z-transform, or by constructing a digital filter or difference equation directly from the z-transform, and processing a unit impulse.
- The z-transform of an LTI system has, in general, properties similar or analogous to various other frequency domain transforms such the DFT, Laplace Transform, etc.

By the end of this chapter, the reader will have gained a practical knowledge of the z-transform, and should be able to navigate among difference equations, direct-form, cascade, and parallel filter topologies, and the z-transform in polynomial or factored form, converting any one representation to another. Additionally, an understanding will have been acquired of the inverse z-transform, and use of the z-transform to evaluate frequency response of various LTI systems such as the FIR and the IIR.

6.2 SOFTWARE FOR USE WITH THIS BOOK

The software files needed for use with this book (consisting of m-code (.m) files, VI files (.vi), and related support files) are available for download from the following website:

http://www.morganclaypool.com/page/isen

The entire software package should be stored in a single folder on the user's computer, and the full file name of the folder must be placed on the MATLAB or LabVIEW search path in accordance with the instructions provided by the respective software vendor (in case you have encountered this notice before, which is repeated for convenience in each chapter of the book, the software download only needs to be done once, as files for the entire book are all contained in the one downloadable folder).

See Appendix A for more information.

6.3 DEFINITION & PROPERTIES

6.3.1 THE Z-TRANSFORM

The z-transform of a sequence $x[n]$ is:

$$X(z) = \sum_{n=-\infty}^{\infty} x[n]z^{-n}$$

where z represents a complex number. The transform does not converge for all values of z; the region of the complex plane in which the transform converges is called the **Region of Convergence (ROC)**, and is discussed below in detail. The sequence z^{-n} is a complex correlator generated as a power sequence of the complex number z and thus $X(z)$ is the correlation (CZL) between the signal $x[n]$ and a complex exponential the normalized frequency and magnitude variation over time of which are determined by the angle and magnitude of z.

6.3.2 THE INVERSE Z-TRANSFORM

The formal definition of the inverse z-transform is

$$x[n] = \frac{1}{2\pi j} \oint X(z)z^{n-1}dz \tag{6.1}$$

where the contour of integration is a closed counterclockwise path in the complex plane that surrounds the origin ($z = 0$) and lies in the ROC.

There are actually many methods of converting a z-transform expression into a time domain expression or sequence. These methods, including the use of Eq. (6.1), will be explored later in the chapter.

6.3.3 CONVERGENCE CRITERIA

Infinite Length Causal (Positive-time) Sequence

When $x[n]$ is infinite in length, and identically 0 for $n < 0$, the ratio of $x[n]$ to z^n (or in other words, $x[n]z^{-n}$) must generally decrease in magnitude geometrically as n increases for convergence to a finite sum to occur.

Note that if the sequence $x[n]$ is a geometrically convergent series, then the z-transform will also converge provided that

$$\left|\frac{x[n+1]}{x[n]}\right| < |z|$$

In terms of numbers, if

$$\left|\frac{x[n+1]}{x[n]}\right| = 0.9$$

for example, then it is required that $|z| > 0.9$ for convergence to occur.

If in fact $x[n]$ is a geometrically convergent series, and z is properly chosen, the sum of the infinite series of numbers consisting of $x[n]z^{-n}$ may conveniently be written in a simple algebraically closed form.

Example 6.1. Determine the z-transform for a single pole IIR with a real pole p having a magnitude less than 1.0.

The impulse response of such a filter may be written as

$$[p^0(=1), p^1, p^2, p^3, \ldots\ p^n]$$

etc., or to pick a concrete example with the pole at 0.9,

$$0.9^n = [1, 0.9, 0.81, 0.729, \ldots]$$

and the z-transform would therefore be:

$$A(z) = 1 + 0.9z^{-1} + 0.81z^{-2} + 0.729z^{-3} + \ldots\ p^n z^{-n}$$

or in generic terms as

$$A(z) = 1 + pz^{-1} + p^2z^{-2} + p^3z^{-3} + \ldots\ p^n z^{-n}$$

The summation of an infinite number of terms of the form c^n where $|c| < 1$ with

$$0 \leq n < \infty$$

is

$$\frac{1}{1-c} \tag{6.2}$$

For the single pole IIR with a pole at p, and by letting $c = pz^{-1}$ (note that $p^0 = 1$) in Eq. (6.2), we get the closed-form z-transform as

$$A(z) = \frac{1}{1-pz^{-1}} = \frac{z}{z-p} \tag{6.3}$$

The z-transform in this case is defined or has a finite value for all z with $|z| > |p|$ or

$$\left|\frac{p}{z}\right| < 1$$

Example 6.2. Plot the ROC in the z-plane for the z-transform corresponding to the sequence $0.8^n u[n]$.

The z-transform is

$$X(z) = \frac{1}{1-0.8z^{-1}}$$

which converges for $|z| > 0.8$. The shaded area of Fig. 6.1 shows the ROC.

Example 6.3. Write the z-transform for the sequence $0.7^n u[n]$, evaluate it at 500 values of z lying on the unit circle (i.e., having a magnitude of 1.0, and radian frequencies between 0 and 2π), and plot the magnitude of the result.

Values of z on the unit circle have magnitude greater than 0.7 and thus lie in the ROC for the z-transform, which is $1/(1-0.7z^{-1})$; Figure 6.2 shows the result of running the following m-code, which computes and plots the magnitude of the z-transform:

```
radFreq = [0:2*pi/499:2*pi]; z = exp(j*radFreq);
Zxform = 1./(1-0.7*z.^(-1)); plot(radFreq/pi,abs(Zxform))
```

For the positive-time or causal sequence $x[n]$, convergence of the z-transform is guaranteed for all z with magnitude greater than the pole having greatest magnitude (sometimes referred to as the dominant pole) in a transfer function.

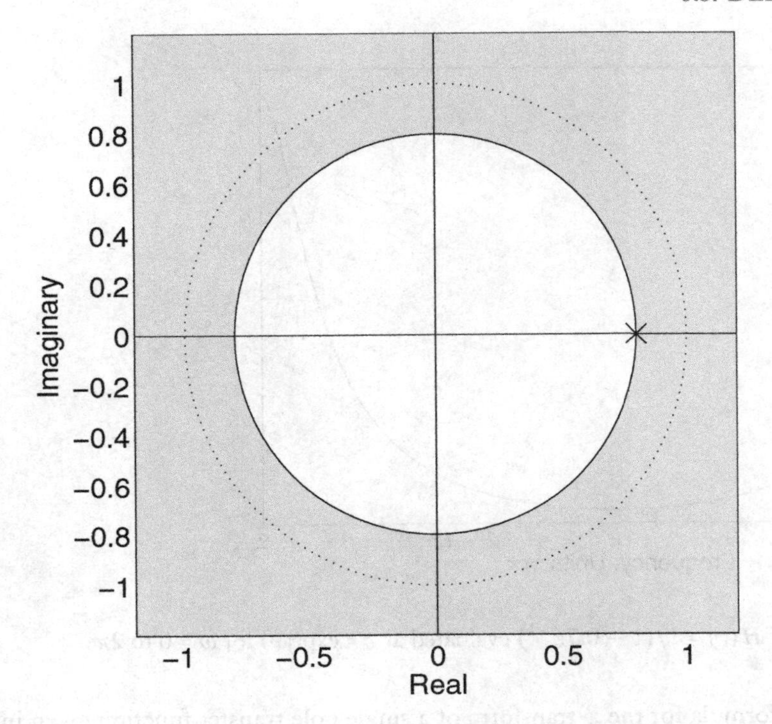

Figure 6.1: The Region of Convergence for a system having one pole at 0.8 is shown as the shaded area outside (not including) a circle of radius 0.8. The unit circle is shown as a dotted circle at radius 1.0.

For example, a transfer function having two poles at 0.9 and 0.8, respectively, would have the z-transform

$$A(z) = \frac{1}{(1 - 0.9z^{-1})(1 - 0.8z^{-1})}$$

which would converge for all z having $|z| > 0.9$.

The z-transform is undefined where it does not converge. Consider an example where $p = 0.9$, and we choose to evaluate the transform at $z = 0.8$. We write the first few terms of the definition of the z-transform as

$$A(z) = p^0 z^0 + pz^{-1} + p^2 z^{-2} + p^3 z^{-3} + \ldots p^n z^{-n}$$

which in concrete terms would be

$$A(z) = (\frac{0.9}{0.8})^0 + (\frac{0.9}{0.8})^1 + (\frac{0.81}{0.64})^2 + (\frac{0.9}{0.8})^3 + \ldots (\frac{0.9}{0.8})^n$$

which clearly diverges, and thus the transform is undefined.

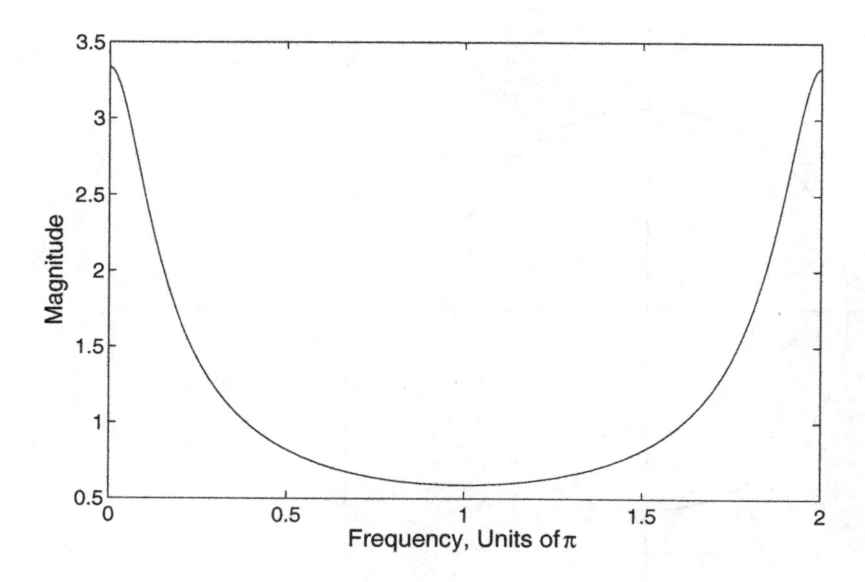

Figure 6.2: The magnitude of $H(z) = 1/(1 - 0.7z^{-1})$ evaluated at $z = \exp(j\omega)$ for $\omega = 0$ to 2π.

Note that the general formula for the z-transform of a single pole transfer function given in Eq. (6.3), if filled in with a value of z which is not in the ROC, will give an answer, but it will not be valid. Thus it is very important to know what the region of convergence is for a given problem.

Example 6.4. Determine, and verify using a numerical computation, the z-transform and ROC of the sequence defined as

$$x[n] = 0.9^n u[n-1]$$

Taking into account the delay of one sample induced by the delayed unit step function, the z-transform would be

$$X(z) = \sum_{n=1}^{\infty} x[n]z^{-n} \tag{6.4}$$

the first few terms of which are $0.9z^{-1}, 0.81z^{-2}, 0.9^3z^{-3}$, etc., which is a geometric series having its first term FT equal to $0.9z^{-1}$ and the ratio R between successive terms equal to $0.9z^{-1}$. The sum of a geometric series is $FT/(1 - R)$, which leads to the z-transform as

$$X(z) = \frac{0.9z^{-1}}{1 - 0.9z^{-1}} \tag{6.5}$$

with ROC: $|z| > 0.9$.

To verify the validity of this expression, we can first compute the summation in Eq. (6.4) to a large number of samples until convergence is (essentially) reached, using a value of z lying in the ROC. Then we will substitute the chosen value of z in Eq. (6.5) and compare results, using the following code. Note that in accordance with Eq. (6.4), we start each sequence, $x[n]$ and the negative power sequence of z, at power 1 rather 0.

```
N = 150; z = 1; zpowseq = z.^(-1:-1:-N); x = 0.9.^(1:1:N);
NumAns = sum(zpowseq.*x)
zXformAns = (0.9/z)/(1-0.9*(1/z))
```

Note that as $|z|$ gets closer to 0.9, a larger value of N is necessary for convergence. For example, if $z = 0.91$, N will need to be about 1500.

Infinite Length, Negative Time Sequence

An infinite-length sequence that is identically zero for values of n equal to or greater than zero is called a negative-time sequence. A geometric sequence defined as

$$x[n] = -b^n u[-n-1]$$

where c has a finite magnitude and n runs from -1 to -∞ has the z-transform

$$X(z) = -\sum_{n=-\infty}^{-1} b^n z^{-n} = -\sum_{n=-\infty}^{-1} (\frac{b}{z})^n = -\sum_{n=1}^{\infty} (\frac{z}{b})^n$$

which can be reduced to

$$X(z) = 1 - \sum_{n=0}^{\infty} (\frac{z}{b})^n = 1 - \frac{1}{1 - z/b} = \frac{z}{z-b} = \frac{1}{1 - bz^{-1}}$$

which converges when $|z| < |b|$.

Example 6.5. A certain negative-time sequence is defined by $x[n] = -0.9^n u[-n-1]$ where $n = -1:-1:-\infty$. Determine the z-transform and the ROC. Pick a value of z in the ROC, numerically compute the z-transform, and then compare the answer to that obtained by substituting the chosen value of z into the determined z-transform.

The z-transform is $1/(1\text{-}0.9z^{-1})$, and the ROC includes all z having magnitude less than 0.9. The value of the transform for $z = 0.6$ is 1/(1-0.9/0.6) = -2.0.

We can numerically compute the z-transform, say, for 30 points, using $z = 0.6$, which gives the answer as -2.0:

```
N = 30; b = 0.9; n = -N:1:-1; z = 0.6; sig = -(b.^n); z2n = z.^(-n);
subplot(2,1,1); stem(n,sig); xlabel('n'); ylabel('Signal Amp')
```

```
subplot(2,1,2); stem(n,z2n); xlabel('n'); ylabel('z^{-n}')
zXform = -sum( (b.^n).*(z.^(-n)) )
```

Infinite Length, Two-Sided

Sequences that have infinite extent toward both +∞ and -∞ are called two-sided and the ROC is the intersection of the ROC for the positive and negative sequences.

Example 6.6. Determine the z-transform and ROC for the following two-sided sequence:

$$x[n] = 0.8^n u[n] - 0.9^n u[-n-1]$$

The net z-transform is the sum of the z-transforms for each of the two terms, and the ROC is the intersection of the ROCs for each. Thus we get

$$X(z) = \frac{z}{z-0.8} - \frac{z}{z-0.9}$$

with the ROC for the left-hand term (the right-handed or positive-time sequence) comprising $|z| > 0.8$ and for the left-handed or negative-time sequence, the ROC comprises $|z| < 0.9$. The net ROC for the two-sided sequence is, therefore, the open annulus (ring) defined by

$$0.8 < |z| < 0.9$$

Note that it may happen that the two ROCs do not intersect, and thus the transform in such case is undefined, i.e., does not exist.

Example 6.7. Determine the z-transform for the system defined by

$$x[n] = 0.9^n u[n] - 0.8^n u[-n-1]$$

Here we see that for the positive-time sequence, the ROC includes z having magnitude greater than 0.9, whereas for the negative-time sequence, z must have magnitude less than 0.8. In other words, there are no z that can satisfy both criteria, and thus the transform is undefined.

Note that the z-transform alone cannot uniquely define the underlying time domain sequence since, for example, the positive time sequence $0.9^n u[n]$ and the negative-time sequence $-0.9^n u[-n-1]$ have the same z-transform

$$\frac{1}{1-0.9z^{-1}}$$

Therefore, it is necessary to specify both the z-transform and the ROC to uniquely define the underlying time domain sequence.

Finite Length Sequence

When the sequence $x[n]$ is finite in length and $x[n]$ is bounded in magnitude, the ROC is generally the entire complex plane, possibly excluding z = 0 and/or z = ∞.

Example 6.8. Evaluate the z-transform and state the ROC for the following three sequences:

(a) x[n] = [1,0,-1,0,1] with time indices [0,1,2,3,4].
(b) x[n] = [1,0,-1,0,1] with time indices [-5,-4,-3,-2,-1].
(c) x[n] = [1,0,-1,0,1] with time indices [-2,-1,0,1,2].

For (a) we get

$$z^0 - z^{-2} + z^{-4}$$

which has its ROC as the entire complex plane except for $z = 0$. For (b) we get

$$z^5 - z^3 + z^1$$

which has its ROC as the entire complex plane except for z = ∞. For (c) we get

$$z^2 - z^0 + z^{-2}$$

which has its ROC as the entire complex plane except for $z = 0$ and $z = \infty$.

6.3.4 SUMMARY OF ROC FACTS

- **Since convergence is determined by the magnitude of z, ROCs are bounded by circles.**
- **For finite sequences that are zero-valued for all $n < 0$, the ROC is the entire z-plane except for $z = 0$.**
- **For finite sequences that are zero-valued for all $n > 0$, the ROC is the entire z-plane except for $z = \infty$.**
- **For infinite length sequences that are causal (positive-time or right-handed), the ROC lies outside a circle having radius equal to the pole of largest magnitude.**
- **For infinite length sequences that are anti-causal (negative-time or left-handed), the ROC lies inside a circle having radius equal to the pole of smallest magnitude.**
- **For sequences that are composites of the above criteria, the sequence should be considered as the sum of a number of subsequences, and the ROC is generally the intersection of the ROCs of each subsequence.**

6.3.5 TRIVIAL POLES AND ZEROS

The generalized z-transform for an finite-length sequence may be written as

$$B(z) = b_0 + b_1 z^{-1} + b_2 z^{-2} + ... + b_{N-1} z^{-(N-1)}$$

where the sequence length is N. If the right-hand side of this equation is multiplied by z^{N-1}/z^{N-1} (i.e., 1), the result is

$$B(z) = (b_0 z^{N-1} + b_1 z^{N-2} + b_2 z^{N-3} + ... + b_{N-1})/z^{N-1}$$

which has $(N-1)$ poles at $z = 0$. These are referred to as **Trivial Poles.**

Likewise, the generalized expression for an IIR

$$A(z) = 1/(a_0 - a_1 z^{-1} - a_2 z^{-2} - ... - a_{N-1} z^{-(N-1)})$$

if multiplied on the right by z^{N-1}/z^{N-1} will yield a transfer function with N -1 **Trivial Zeros** at $z = 0$.

The trivial poles of a finite sequence (or FIR) cause the z-transform to diverge at $z = 0$ as mentioned above.

6.3.6 BASIC PROPERTIES OF THE Z-TRANSFORM

Linearity

If the z-transforms of two functions $x_1[n]$ and $x_2[n]$ are $X_1(z)$ and $X_2(z)$, respectively, then the z-transform of

$$Z(c_1 x_1[n] + c_2 x_2[n]) = c_1 X_1(z) + c_2 X_2(z); \text{ ROC: ROC}(x_1[n]) \cap \text{ROC}(x_2[n])$$

for all values of c_1 and c_2.

Shifting or Delay

If $X[z]$ is the z-transform of $x[n]$, then the z-transform of the delayed sequence $x[n-d]$ is $z^{-d}X[z]$ with the ROC the same as that for $x[n]$. This is a very useful property since it allows one to write the z-transform of difference equations by inspection.

Example 6.9. Using the shifting property, write the z-transform of the following causal system:

$$y[n] = x[n] + ay[n-1] + bx[n-1]$$

The z-transform may be written as

$$Y(z) = X(z) + az^{-1}Y(z) + bz^{-1}X(z)$$

which simplifies to

$$\frac{Y(z)}{X(z)} = \frac{1 + bz^{-1}}{1 - az^{-1}}; \text{ ROC: } |z| > |a|$$

Convolution
The z-transform of the convolution of two functions $x_1[n]$ and $x_2[n]$ is the product of the z-transforms of the individual functions. Stated mathematically,

$$Z(x_1[n] * x_2[n]) = X_1(z)X_2(z); \text{ ROC: ROC}(x_1[n]) \cap \text{ROC}(x_2[n])$$

where the symbol $*$ is used here to mean convolution.

Example 6.10. Compare the convolution of two sequences using linear convolution and the z-transform technique.

Consider the two time domain sequences [1,1,1] and [1,2,-1]. The time domain linear convolution is y = [1,3,2,1,-1]. Doing the problem using z-transforms, we get

$$Y(z) = (1 + z^{-1} + z^{-2})(1 + 2z^{-1} - z^{-2}) = [1 + 3z^{-1} + 2z^{-2} + z^{-3} - z^{-4}]$$

from which we can write the equivalent impulse response by inspection, [1,3,2,1,-1], which is identical to the time domain result.

Time Reversal or Folding
If

$$Z(x[n]) = X(z)$$

then

$$Z(x[-n]) = X(1/z)$$

and the ROC is inverted.

Example 6.11. A sequence is $x[n]$ = [1, 0, −1, 2, 1], having time indices of [2, 3, 4, 5, 6]. Determine the z-transform of $x[-n]$.

The z-transform of $x[n]$ is

$$z^{-2} + 0z^{-3} - z^{-4} + 2z^{-5} + z^{-6}; \text{ ROC: } |z| > 0$$

According to the folding property, the z-transform of $x[-n]$ should be

$$(1/z)^{-2} - (1/z)^{-4} + 2(1/z)^{-5} + (1/z)^{-6} = z^2 - z^4 + 2z^5 + z^6$$

The time reversed sequence is [1,2,-1,0,1] having time indices [-6,-5,-4,-3,-2], and computing the z-transform of $x[-n]$ directly we get

$$z^6 + 2z^5 - z^4 + 0z^3 + z^2; \text{ ROC: } |z| < \infty$$

Multiplication By a Ramp

$$Z(nx[n]) = -z\frac{dX(z)}{dz}; \text{ ROC: ROC}(x[n])$$

Example 6.12. Determine the z-transform of the sequence $nu[n]$.

We get initially

$$X(z) = -z\frac{d}{dz}\left(\frac{1}{1 - z^{-1}}\right) = -z\frac{d}{dz}([1 - z^{-1}]^{-1})$$

$$= -z(-(1 - z^{-1})^{-2}(z^{-2})) = \frac{z^{-1}}{(1 - z^{-1})^2} = \frac{0z^0 + z^{-1}}{1 - 2z^{-1} + z^{-2}}$$

We can verify this using the following code, which yields the sequence $nu[n]$ i.e., [0,1,2...].

```
y = filter([0 1],[1,-2,1],[1,zeros(1,20)])
```

6.3.7 COMMON Z-TRANSFORMS

Sequence	z-Transform	ROC
$\delta[n]$	1	$\forall z$
$u[n]$	$1/(1 - z^{-1})$	$\|z\| > 1$
$-u[-n-1]$	$1/(1 - z^{-1})$	$\|z\| < 1$
$a^n u[n]$	$1/(1 - az^{-1})$	$\|z\| > \|a\|$
$-b^n u[-n-1]$	$1/(1 - bz^{-1})$	$\|z\| < \|b\|$
$[a^n \sin \omega_o n]u[n]$	$\frac{(a \sin \omega_o)z^{-1}}{1-(2a \cos \omega_o)z^{-1}+a^2z^{-2}}$	$\|z\| > \|a\|$
$[a^n \cos \omega_o n]u[n]$	$\frac{(a \cos \omega_o)z^{-1}}{1-(2a \cos \omega_o)z^{-1}+a^2z^{-2}}$	$\|z\| > \|a\|$
$na^n u[n]$	$\frac{az^{-1}}{(1-az^{-1})^2}$	$\|z\| > \|a\|$
$-nb^n u[-n-1]$	$\frac{bz^{-1}}{(1-bz^{-1})^2}$	$\|z\| < \|b\|$

Example 6.13. Represent in the z-domain the time domain convolution of the sequence [1, -0.9] with a single-pole filter having its pole at 0.9. Use the representation to determine the time domain convolution.

The convolution in the time domain of the two sequences can be equivalently achieved in the z-domain by multiplying the z-transforms of each time domain sequence, i.e., in this case

$$\frac{(1-0.9z^{-1})}{(1-0.9z^{-1})}=1$$

Looking at the table of common z-transforms, we see that the unit impulse is the time domain signal that has as its z-transform the value 1. This can be confirmed computationally (for the first 20 samples) using the following code:

```
imp = 0.9.^(0:1:100); sig = [1,-0.9]; td = conv(imp,sig)
```

which returns the value of td as a unit impulse sequence. Note that even though the single pole generates an infinitely long impulse response, this particular input sequence results in a response which is identically zero after the first sample of output.

Example 6.14. The sequence randn(1,8) is processed by an LTI system represented by the difference equation below; specify another difference equation which will, when fed the output of the first LTI system, return the original sequence.

$$y[n]=x[n]+x[n-2]+1.2y[n-1]-0.81y[n-2]$$

We rewrite the difference equation as

$$y[n]-1.2y[n-1]+0.81y[n-2]=x[n]+x[n-2]$$

and then write the z-transform as

$$Y(z)(1-1.2z^{-1}+0.81z^{-2})=X(z)(1+z^{-2})$$

which yields

$$Y(z)/X(z)=\frac{1+z^{-2}}{1-1.2z^{-1}+0.81z^{-2}}$$

The b and a coefficients for this system are b = [1,1] and a = [1,-1.2,0.81]. To obtain the inverse system, simply exchange the values for b and a, i.e., obtain the reciprocal of the z-transform. To check the answer, we will first process the input sequence with the original difference equation, then process the result with the inverse difference equation:

```
x = randn(1,8), y = filter([1,1],[1,-1.2,0.81],x);
ans = filter([1,-1.2, 0.81],[1,1],y)
```

6.3.8 TRANSFER FUNCTIONS, POLES, AND ZEROS

LTI System Representation

If an LTI system is represented by the difference equation

$$y[n] + \sum_{k=1}^{N-1} a_k y[n-k] = \sum_{m=0}^{M-1} b_m x[n-m]$$

Taking the z-transform and using properties such as the shifting property, we get

$$Y(z)(1 + \sum_{k=1}^{N-1} a_k z^{-k}) = X(z) \sum_{m=0}^{M-1} b_m z^{-m}$$

which yields the system transfer function as

$$H(z) = \frac{\sum_{m=0}^{M-1} b_m z^{-m}}{1 + \sum_{k=1}^{N-1} a_k z^{-k}} \tag{6.6}$$

Example 6.15. Determine the system transfer function for the LTI system represented by the following difference equation:

$$y[n] = 0.2x[n] + x[n-1] + 0.2x[n-2] + 0.95y[n-1]$$

We rewrite the difference equation as

$$y[n] - 0.95y[n-1] = 0.2x[n] + x[n-1] + 0.2x[n-2]$$

and then take the z-transform (using the shifting property)

$$Y(z) - 0.95Y(z)z^{-1} = 0.2X(z) + X(z)z^{-1} + 0.2X(z)z^{-2}$$

which yields

$$Y(z)/X(z) = \frac{0.2 + z^{-1} + 0.2z^{-2}}{1 - 0.95z^{-1}}$$

Eq. (6.6) can be expressed as the ratio of products of individual zero and pole factors by converting it to an expression in positive powers of z:

$$H(z) = (\frac{z^{N-1}}{z^{M-1}})(\frac{z^{M-1}}{z^{N-1}})\frac{b_0 + b_1 z^{-1} + ...b_{(M-1)} z^{-(M-1)}}{1 + a_1 z^{-1} + ...a_{(N-1)} z^{-(N-1)}}$$

which becomes

$$b_0 z^{N-M}(\frac{z^{M-1} + (b_1/b_0)z^{M-2} + ... + (b_{(M-2)}/b_0)z^1 + (b_{(M-1)})/b_0}{z^{N-1} + a_1 z^{N-1} + ...a_{(N-2)}z^1 + a_{(N-1)}})$$

which can then be written as the ratio of the product of pole and zero factors:

$$H(z) = b_0 z^{N-M} \frac{\Pi_{m=1}^{M-1}(z - z_m)}{\Pi_{k=1}^{N-1}(z - p_k)} \tag{6.7}$$

where z_m and p_k are the zeros and poles, respectively. Each of the factors may be interpreted, for any specific value of z, as a complex number having a magnitude and angle, and thus we can write

$$H(z) = b_0 z^{N-M} \frac{\Pi_{m=1}^{M-1} M_m \angle\theta_m}{\Pi_{k=1}^{N-1} M_k \angle\theta_k} \tag{6.8}$$

(An example pertaining to Eq. (6.8) will be presented below).

FIR Zeros

The difference equation for an FIR of length M is

$$y[n] = b_0 x[n] + b_1 x[n-1] + b_2 x[n-2] + ... b_{(M-1)} x[n - (M-1)]$$

and translates (using the time-shifting property) directly into the z-domain as follows:

$$Y(z) = b_0 X(z) + b_1 X(z) z^{-1} + b_2 X(z) z^{-2} + ... b_{(M-1)} X(z) z^{-(M-1)}$$

Then

$$\frac{Y(z)}{X(z)} = b_0 + b_1 z^{-1} + b_2 z^{-2} + ... b_{(M-1)} z^{-(M-1)} \tag{6.9}$$

The roots of Eq. (6.9) are the zeros of the transfer function, i.e., values of z such that

$$b_0 + b_1 z^{-1} + b_2 z^{-2} + ...b_{(M-1)} z^{-(M-1)} = 0$$

While Eq. (6.9) is in the form of a polynomial in negative powers of z, Eq. (6.10) shows the factored form

$$\frac{Y(z)}{X(z)} = \Pi_{m=1}^{M-1}(1 - r_m z^{-1}) \tag{6.10}$$

where r_m refers to the mth root of Eq. (6.9).

Example 6.16. Obtain the z-transform and its roots for the finite impulse response [1, 0, 0, 0, 1].

A simple filter having a periodic or comb-like frequency response can be formed by adding a single delayed version of a signal with itself. The impulse response

$$[1, 0, 0, 0, 1]$$

has, by definition, the z-transform

$$Y(z)/X(z) = 1 + z^{-4} \tag{6.11}$$

with ROC being the entire z-plane less the origin ($z = 0$).

The roots of Eq. (6.11) can be found using DeMoivre's Theorem. Setting Eq. (6.11) equal to zero and multiplying both sides of the equation by z^4 we get

$$z^4 = -1$$

By expressing -1 as odd multiples of π radians (180 degrees) in the complex plane, we get

$$z^4 = 1\angle(\pi + 2\pi m) = 1\angle\pi(2m + 1)$$

where m can be 0, 1, 2, or 3. The roots are obtained, for each value of m, by taking the fourth root of the magnitude, which is 1, and one-quarter of the angle. We would thus get the four roots as $\pi/4$, $3\pi/4$, $5\pi/4$, and $7\pi/4$. Plots of the magnitude of the z-transform and the transfer function zeros are shown in Fig. 6.3. The z-plane plot shown in Fig. 6.3, plot (b), can be obtained using the function

$$zplane(z, p)$$

where z and p are column vectors of zeros and poles of the transfer function, or, alternately

$$zplane(b, a)$$

where b and a are row vectors of the z-transform numerator and denominator coefficients, respectively. The m-code used to generate Fig. 6.3 was

```
freq = -pi:0.02:pi; z = exp(j*freq);
FR = abs(1 + z.^(-4));
figure(44); subplot(211);
plot(freq/pi,FR)
xlabel('Frequency, Units of π')
ylabel('Magnitude')
subplot(212);
zplane([1 0 0 0 1],1)
xlabel('Real Part')
ylabel('Imaginary Part')
```

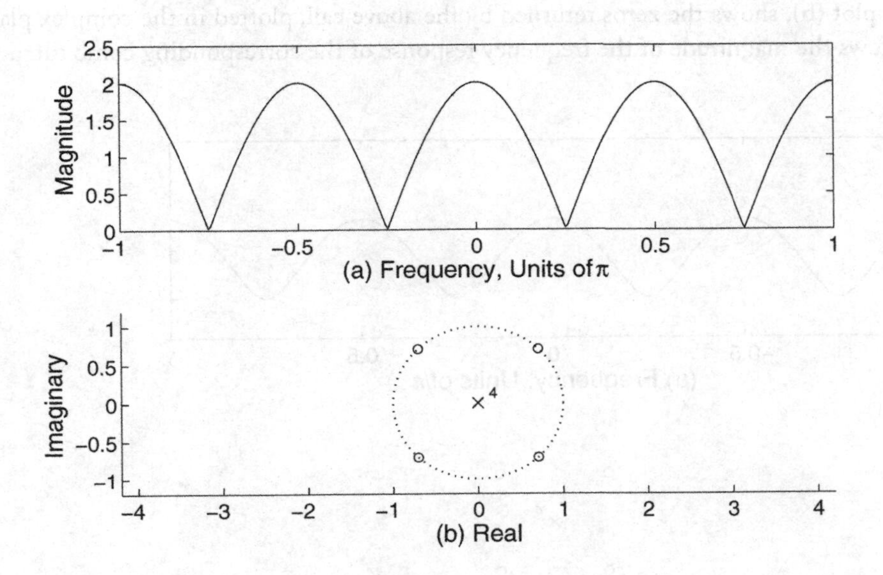

Figure 6.3: (a) Magnitude of the z-transform of a unity-additive comb filter having a delay of four samples, having the z-transform $H(z) = 1 + z^{-4}$; (b) Plot of the zeros of the z-transform of the comb filter of (a). Note the four poles at $z = 0$.

For more complicated transfer functions, the roots can be found using the function *roots*.

Example 6.17. Obtain the z-transform and its roots for the finite impulse response [1, 0, 0, 0, 0.5].

The z-transform is

$$Y(z)/X(z) = 1 + 0.5z^{-4}$$

The zeros of this transfer function do not lie on the unit circle. They can be found as

$$z = \sqrt[4]{0.5}\angle((\pi + 2\pi m)/4)$$

where m = 0,1,2, and 3, by making this m-code call

```
m = 0:1:3; angs = exp(j*(pi*(2*m +1)/4)); theZeros = 0.5 ^(0.25)*angs
```

or by making the call

```
theZeros = roots([1,0,0,0,0.5])
```

Figure 6.4, plot (b), shows the zeros returned by the above call, plotted in the complex plane, while plot (a) shows the magnitude of the frequency response of the corresponding comb filter.

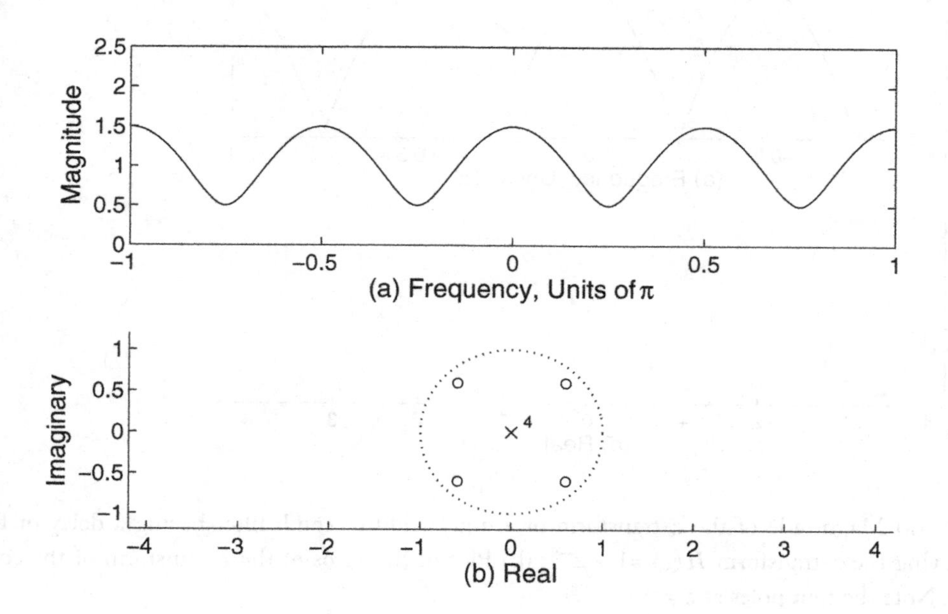

Figure 6.4: (a) Magnitude of the z-transform $H(z) = 1 + 0.5z^{-4}$ over the frequency range $-\pi$ to π radians; (b) Plot of the zeros of the z-transform of the comb filter of (a).

Example 6.18. Determine the roots (zeros) of the transfer function corresponding to the impulse response $[ones(1,6)]$.

The z-transform of the impulse response is

$$H(z) = 1 + z^{-1} + z^{-2} + z^{-3} + z^{-4} + z^{-5}$$

The transfer function length is $M + 1 = 6$, and therefore there are $M = 5$ roots, obtained by the call

```
rts = roots([1,1,1,1,1,1])
```

which yields the answer 0.5 ± 0.866j, -1, -0.5 ± 0.866j.

IIR-Poles

The z-transform for a single pole IIR is

$$\frac{Y(z)}{X(z)} = \frac{1}{1 - pz^{-1}} \tag{6.12}$$

The value of p in Eq. (6.12) is called a pole, because if z assumes the same value, the transfer function becomes infinite or undefined, since

$$\lim_{z \to p} (1/(1 - p(\frac{1}{z}))) \to \infty$$

A multiple-pole transfer function can be written in factored form as

$$\frac{Y(z)}{X(z)} = 1/\ \Pi_{n=1}^{N}(1 - p_n z^{-1})$$

where p_n are the poles and N is the total number of poles. The coefficient or polynomial form can be arrived at by performing the indicated multiplication.

Example 6.19. A certain LTI system has poles at $\pm 0.6j$. Determine the z-transform in polynomial form.

We perform the multiplication

$$H(z) = \frac{1}{(1 - (j0.6)z^{-1})(1 - (-j0.6)z^{-1})} = \frac{1}{1 + 0.36z^{-2}}$$

This multiplication of polynomials can be performed as a convolution when a vector of roots is available:

```
theRoots = [j*0.6,-j*0.6];
NetConv = [1];
for ctr = 1:1:length(theRoots)
NetConv = conv(NetConv,[1 -theRoots(ctr)]);
end
theCoeff = NetConv
```

which returns

```
theCoeff = [1 ,0, 0.36]
```

Example 6.20. Compute the z-transform coefficients of an IIR having poles at $0.9j$ and $-0.9j$; check your results using the MathScript function *poly*.

The coefficients are

conv([1, -(0 + 0.9j)],[1, -(0 - 0.9j)]) = [1,0,0.81]

and the z-transform is

$$Y(z)/X(z) = 1/(1 + 0z^{-1} + 0.81z^{-2})$$

This can be checked using the function *poly* which converts a vector of roots into polynomial coefficients:

```
a = poly([0.9*j,-0.9*j]) = [1,0,0.81]
```

Example 6.21. The poles of a certain causal LTI system are at 0.9 and $\pm 0.9j$. There are three zeros at -1. Using Eq. (6.8), determine the magnitude of this system's response at DC. Use the function poly to obtain the polynomial form of the z-transform, and evaluate it to check your work.

We note that with three zeros, $M = 4$ and with three poles, $N = 4$. To get b_0, we must obtain the polynomial coefficients for the numerator of the z-transform. Note that the ROC is for $|z| > 0.9$.

```
zzs = [-1,-1,-1]; pls = [0.9, 0.9*j,-0.9*j]; Denom = poly(pls);
N = length(pls) +1; M = length(zzs) +1; Num = poly(zzs);
b0 = Num(1); z = 1; NProd = 1; for NCtr = 1:1:M-1;
NProd = NProd*(abs(z - zzs(NCtr))); end
DProd = 1; for DCtr = 1:1:N-1;
DProd = DProd*(abs(z - pls(DCtr))); end
MagResp = abs(b0)*abs(z^(N-M))*NProd/DProd
AltMag = abs(sum(Num.*(z.^(0:-1:-(M-1))))/sum(Denom.* (z.^(0:-1:-(N-1)))))
```

6.3.9 POLE LOCATION AND STABILITY

The impulse response of a single pole IIR is [1,p,p^2,p^3...]. It can be seen that if $|p| < 1$, then a geometrically convergent series results. A bounded signal (one whose sample values are all finite) will not produce an unbounded output. If, however, $|p| > 1$ then the impulse response does not decay away, and a bounded input signal can produce an output that grows without bound. For the borderline case when $|p| = 1$, the unit impulse sequence as input produces a unit step ($u[n]$) as the corresponding output sequence. If the input signal is $u[n]$, then the output will grow without bound.

The difference equation for the single-pole IIR

$$y[n] = x[n] + py[n-1]$$

can be used to observe the relationship between pole magnitude, stability, and pole location in the complex plane by choosing different values of p and then processing a test signal $x[n]$ such as the unit impulse [1,0,0 ...] or the unit step [1,1,1 ...]. For $|p| \geq 1$ and the unit impulse as $x[n]$, the output $y[n]$ continues growing without further input.

For an LTI system to be stable, all of its poles must have magnitude less than 1.0.

Example 6.22. Determine the pole magnitudes of the LTI system represented by the z-transform

$$H(z) = \frac{1}{1 - 1.9z^{-1} + 0.95z^{-2}}$$

We run the code

```
pmags = abs(roots([1,-1.9, 0.95]))
```

which yields *pmags* = [0.9746, 0.9746], meaning that the system is stable.

In summary

- **The poles of a transfer function must all have magnitudes less than 1.0 in order for the corresponding system to have a stable, bounded response to a bounded input signal (i.e., BIBO).**
- **Stable poles, when graphed in the z-domain (i.e., the complex plane), all lie inside the unit circle.**

Example 6.23. For an IIR having a single real pole p, plot the first 45 samples of the impulse and unit step responses for the following values of p:

(a) 0.9
(b) 1.0
(c) 1.01

We can use this code to generate the impulse and step responses:

```
p = 0.9; ImpResp = filter(1,[1,-p],[1,zeros(1,44)])
StepResp = filter(1,[1,-p],[ones(1,95)])
```

Figure 6.5 shows the results for each of the three pole values. Note that for the one stable case (p = 0.9), the impulse response decays to zero, and the step response converges to a finite value.

You can experiment with pole location for single or complex conjugate pairs of poles and the corresponding/resultant system impulse response by calling the VI

DemoDragPolesImpRespVI.

Figure 6.6 shows an example of the VI *DemoDragPolesImpRespVI* using a pair of complex poles having an approximate magnitude of 0.9 and angles of approximately $\pm\pi/4$ radians.

A MATLAB script that performs a similar function is

ML_DragPoleZero

which, when called on the Command Line, opens up a GUI that allows you to select a single or complex conjugate pair of poles or zeros and move the cursor around the z-plane to select the pole(s)

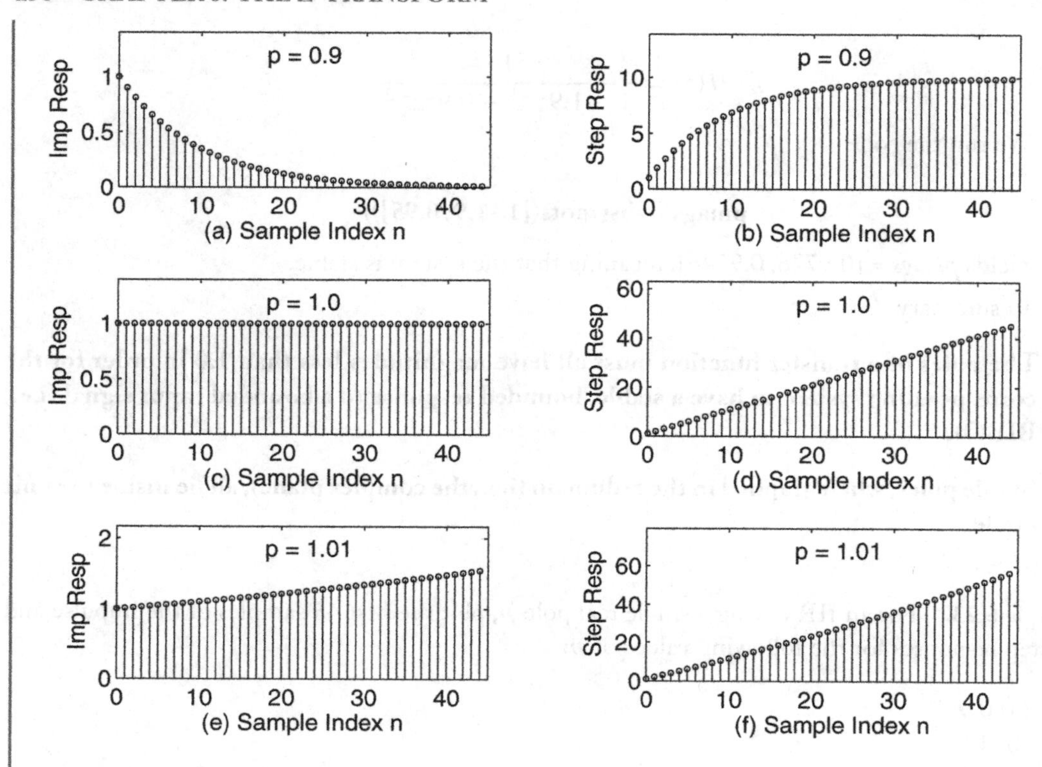

Figure 6.5: (a) Impulse response for a single pole IIR, p = 0.9; (b) Step response for same; (c) Impulse response for a single pole IIR, p = 1.0; (d) Step response for same; (e) Impulse response for a single pole IIR, p = 1.01; (f) Step response for same.

or zero(s). The magnitude and phase response of the z-transform and the real and imaginary parts of the impulse response are dynamically displayed as you move the cursor in the z-plane. For poles of magnitude 1.0 or greater, instead of using the z-transform to determine the frequency response (which is not possible since values of z along the unit circle are not in the ROC when the poles have magnitude 1.0 or greater), the DFT of a finite length of the test impulse response is computed. While this is not a true frequency response (none exists for systems having unstable poles since the DTFT does not converge), it gives an idea of how rapidly the system response to a unity-magnitude complex exponential grows as pole magnitude increases beyond 1.0.

6.4 CONVERSION FROM Z-DOMAIN TO TIME DOMAIN

There are a number of methods for computing the values of, or determining an algebraic expression for, the time domain sequence that underlies a given z-transform. Here we give a brief summary of some of these methods:

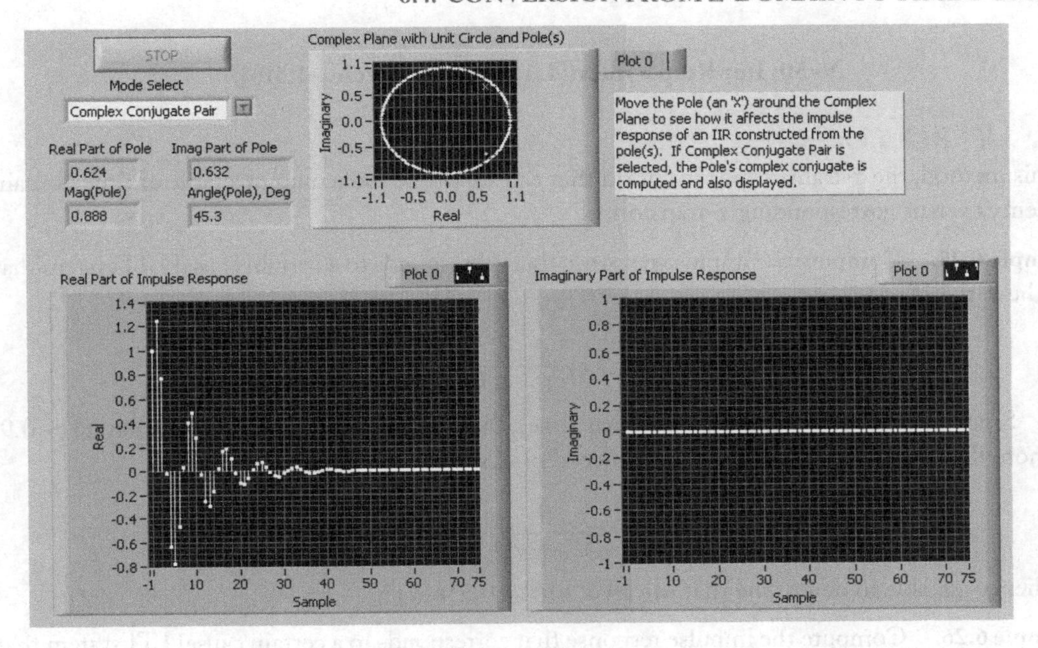

Figure 6.6: A VI allowing the user to drag a pole or pair of complex conjugate poles in the z-plane and observe the impulse response of an IIR constructed from the pole or poles. The first 75 samples of the impulse response are computed and displayed.

6.4.1 DIFFERENCE EQUATION

A simple, direct method is to write the difference equation of the system by inspection from the z-transform, and then process a unit impulse.

Example 6.24. Compute the impulse response that corresponds to a causal LTI system that has the z-transform

$$X(z) = \frac{1 + z^{-1}}{1 - 0.9z^{-1}}$$

The difference equation is

$$y[n] = x[n] + x[n - 1] + 0.9y[n - 1]$$

with $y[n] = x[n] = 0$ for $n < 0$. We can compute an arbitrary number of samples N of the impulse response using the function *filter* with b = [1,1], a = [1, -0.9], and x = [1, zeros(1,N)]. Thus we make the call

N=50; ImpResp = filter([1,1],[1,-0.9],[1,zeros(1,50)])

6.4.2 TABLE LOOKUP

In this method, the z-transform is in a form that can simply be looked up in a table of time domain sequences versus corresponding z-transforms.

Example 6.25. Compute the impulse response that corresponds to a certain causal LTI system that has the z-transform

$$X(z) = \frac{1}{1 - 0.9z^{-1}}$$

Since the system is causal (i.e., positive-time), the ROC will include all z such that $|z| > 0.9$. We note that the z-transform of the sequence $a^n u[n]$ where $|a| < 1$ is

$$\frac{1}{1 - az^{-1}}$$

and hence are able to determine that $h[n] = a^n u[n] = 0.9^n u[n]$.

Example 6.26. Compute the impulse response that corresponds to a certain causal LTI system that has the z-transform

$$X(z) = \frac{1 + z^{-1}}{1 - 0.9z^{-1}}$$

Since the system is causal (i.e., positive-time), the ROC will include all z such that $|z| > 0.9$. We can exercise a little ingenuity by breaking this z-transform into the product of the numerator and denominator, obtain the impulse response corresponding to each, and then convolve them in the time domain. We will thus convolve the sequences $0.9^n u[n]$ and $[1, 1]$. A suitable call to compute the first 30 samples is

ImpResp = conv([0.9.^(0:1:29)],[1,1])

We can check this with the call

ImpRespAlt = filter([1,1],[1,-0.9],[1,zeros(1,29)])

6.4.3 PARTIAL FRACTION EXPANSION

In this method, the z-transform is not in a form that can readily be looked up. A z-transform of the general form

$$X(z) = \frac{b_0 + b_1 z^{-1} + ... + b_M z^{-M}}{1 + a_1 z^{-1} + ... + a_N z^{-N}} \tag{6.13}$$

can be rewritten as

$$X(z) = \frac{b'_0 + b'_1 z^{-1} + ... + b'_{N-1} z^{-(N-1)}}{1 + a_1 z^{-1} + ... + a_N z^{-N}} + \sum_{k=0}^{M-N} C_k z^{-k} \tag{6.14}$$

which can be rewritten as a sum of fractions, one for each pole, plus a sum of polynomials if $M \geq N$.

$$X(z) = \sum_{n=1}^{N} \frac{R_n}{1 - p_n z^{-1}} + \sum_{k=0}^{M-N} C_k z^{-k} \tag{6.15}$$

The values R_n are called residues, p_n is the n-th pole of $X(z)$, and it is assumed that the poles are all distinct, i.e., no duplicate poles. In such a case, the residues can be computed by:

$$R_n = \frac{b'_0 + b'_1 z^{-1} + ... + b'_{N-1} z^{-(N-1)}}{1 + a_1 z^{-1} + ... + a_N z^{-N}} (1 - p_n z^{-1}) |_{z=p_n}$$

The above expressions are true for distinct p_n. If p_n consists of S duplicate poles, the partial fraction expansion is

$$\sum_{s=1}^{S} \frac{R_{n,s} z^{-(s-1)}}{(1 - p_n z^{-1})^s} = \frac{R_{n,1}}{1 - p_n z^{-1}} + \frac{R_{n,2} z^{-1}}{(1 - p_n z^{-1})^2} + ... + \frac{R_{n,S} z^{-(S-1)}}{(1 - p_n z^{-1})^S}$$

The time domain impulse response can be written from the partial fraction expansion as

$$x[n] = \sum_{n=1}^{N} R_n Z^{-1} \left[\frac{1}{1 - p_n z^{-1}} \right] + \sum_{k=0}^{M-N} C_k \delta[n-k]$$

For causal sequences, the inverse z-transform of

$$\frac{1}{1 - p_n z^{-1}}$$

is

$$p_n^k u[k]$$

The function

$$[R, p, C_k] = residuez(b, a)$$

provides the vector of residues R, the corresponding vector of poles p, and, when the order of the numerator is equal to or greater than that of the denominator, the coefficients C_k of the polynomial in z^{-1}.

Example 6.27. Construct the partial fraction expansion of the z-transform below; use the function residuez to obtain the values of R, p, and C_k. Determine the impulse response from the result.

$$X(z) = \frac{0.1 + 0.5z^{-1} + 0.1z^{-2}}{1 - 1.2z^{-1} + 0.81z^{-2}}$$

We make the call

```
[R,p,Ck] = residuez([0.1,0.5,0.1],[1,-1.2,0.81])
```

and receive results

```
R = [(-0.0117 -0.4726i),(-0.0117 + 0.4726i)]
p = [(0.6 + 0.6708i),(0.6 - 0.6708i)]
Ck = [0.1235]
```

from which we construct the expansion as

$$X(z) = \frac{R(1)}{1 - p(1)z^{-1}} + \frac{R(2)}{1 - p(2)z^{-1}} + Ck(1)z^0$$

and the impulse response is

$$x[n] = R(1)p(1)^n u[n] + R(2)p(2)^n u[n] + Ck(1)\delta[n] \tag{6.16}$$

A script that will numerically evaluate (6.16) and then compute the impulse response using *filter* with a unit impulse is

```
[R,p,Ck] = residuez([0.1,0.5,0.1],[1,-1.2,0.81])
n = 0:1:50; x = R(1)*p(1).^n + R(2)*p(2).^n; x(1) = x(1) + Ck;
altx = filter([0.1,0.5,0.1],[1,-1.2,0.81],[1 zeros(1,50)])
diff = x - altx, hold on; stem(x); stem(altx)
```

6.4.4 CONTOUR INTEGRATION IN THE COMPLEX PLANE

Numerical Method

For this method, we use the formal definition of the inverse z-transform:

$$x[n] = \frac{1}{2\pi j} \oint X(z) z^{n-1} dz \tag{6.17}$$

where the contour of integration is a closed counterclockwise path in the complex plane that surrounds the origin ($z = 0$) and lies in the ROC.

The integral at (6.17) can be evaluated directly using numerical integration. The script (see exercises below)

$$LVxNumInvZxform(NumCoefVec, DenCoefVec, M, ContourRad, nVals)$$

allows you to enter the z-transform of a causal sequence as its numerator and denominator coefficient vectors in ascending powers of z^{-1}. You also specify the number of points M along the contour to use to approximate the integral, the radius of the circle *ContourRad* that will serve as the contour, and $nVals$, the values of n (for $x[n]$) to be computed; $nVals$ must consist of nonnegative integers corresponding to the sample indices of a causal sequence $x[n]$.

The script computes the answer and generates two plots, the first showing the real part of $x[n]$ and the second showing the imaginary part of $x[n]$. If the z-transform coefficients are real, then $x[n]$ will be real, and the imaginary part of $x[n]$ should be zero. If this is the case, then it is possible to judge the accuracy of $x[n]$ (i.e., $real(x[n])$) by judging how close $imag(x[n])$ is to zero. Generally, the closer $imag(x[n])$ is to zero, the more accurate is $real(x[n])$.

A suitable call to obtain the inverse of the z-transform

$$X(z) = \frac{1}{1 - 1.2z^{-1} + 0.81z^{-2}} \qquad \text{(ROC: } |z| > 0.9\text{)}$$

is

LVxNumInvZxform([1],[1,-1.2, 0.81],10000,1,[0:1:50])

which results in Fig. 6.7. Theoretically, any counterclockwise closed contour in the ROC will result in the same answer; circular contours are particularly easy to specify and compute. You can verify that any circular contour in the ROC will give the same answer by changing the value of *ContourRad*.

Example 6.28. Estimate, using numerical contour integration, the time domain sequence corresponding to the z-transform

$$X(z) = \frac{z^{-1}}{1 - 2z^{-1} + z^{-2}} \qquad \text{(ROC: } |z| > 1.0\text{)}$$

We make the call

LVxNumInvZxform([0,1],[1,-2,1],10000,1.05,[0:1:50])

which results in Fig. 6.8.

The call immediately above yields the sequence

$$x[n] = nu[n]$$

and thus you can see how accurate the result is since the answer should be a sequence of integers; the imaginary parts should all be zero-valued. To improve accuracy, increase M; to decrease computation time, at the expense of accuracy, decrease M.

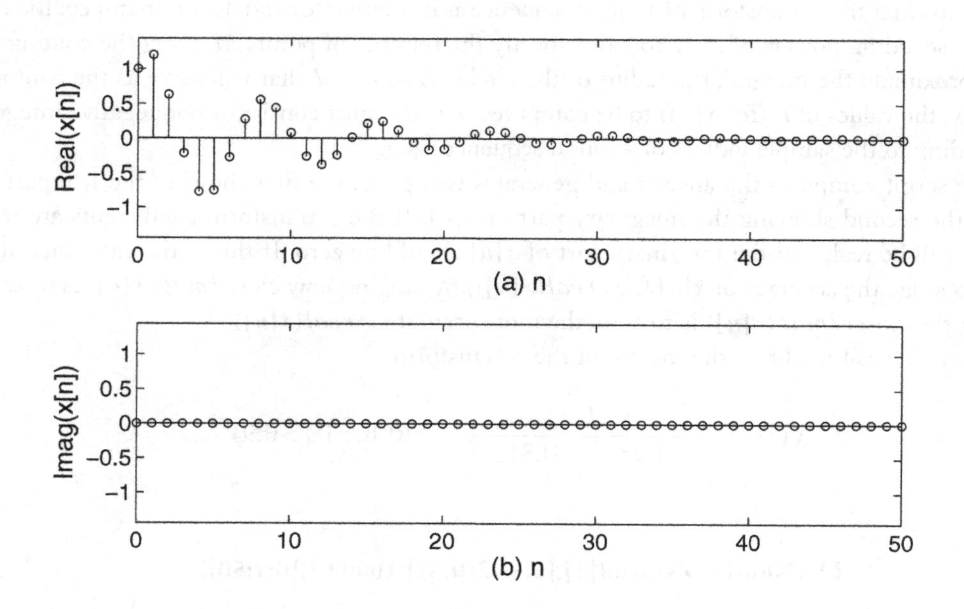

Figure 6.7: (a) The real part of the first 50 samples of the impulse response corresponding to the z-transform $1/(1 - 1.2z^{-1} + 0.81z^{-2})$, computed using numerical contour integration along a circular contour of radius 1.0 since the largest magnitude pole is 0.9; (b) Imaginary part of same.

Sum of Residues

The traditional way to evaluate (6.17) is to use a theorem (the Cauchy Residue Theorem) that states that the value of a contour integral in the z-plane over any closed counterclockwise path that encircles the origin and lies within the region of convergence is equal to the sum of the residues of the product of $X(z)$ and z^{n-1}within the chosen contour, which may be, for example, for right-handed z-transforms (corresponding to a causal sequence $x[n]$), a circle around the origin whose radius is greater than the magnitude of the largest pole of $X(z)$. For simple (nonrepeated) poles, all of which lie within the chosen contour, this statement may be stated mathematically as

$$x[n] = \frac{1}{2\pi j}\oint X(z)z^{n-1}dz = \sum_{k}(z - z_k)X(z)z^{n-1}\,|_{z=z_k}$$

Executing this method requires that the expression $X(z)z^{n-1}$ be expanded by partial fractions so that the term $(z - z_k)$ will cancel the corresponding pole in one fraction and reduce the others to zero when z_k is substituted for z.

Example 6.29. Using the sum of residues method, determine $x[n]$ for a certain causal LTI system that has poles at $\pm 0.9j$ and the corresponding z-transform

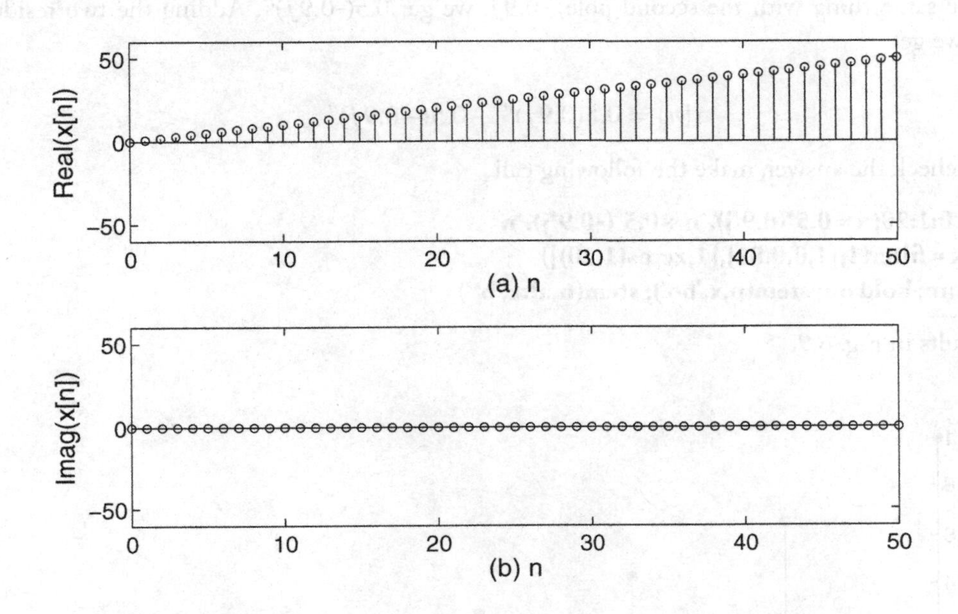

Figure 6.8: (a) The real part of the first 51 samples of the numerical approximation to the inverse z-transform of the transfer function $(z^{-1})/(1 - 2z^{-1} + z^{-2})$; (b) Imaginary part of same.

$$X(z) = \frac{1}{(1 - j0.9z^{-1})(1 + j0.9z^{-1})} = \frac{1}{1 + 0.81z^{-2}}$$

Making the call

[R,p,Ck] = residuez([1],[1,0,0.81])

allows us to write the partial fraction expansion of $X(z)$ (in positive powers of z) as

$$X(z) = \frac{0.5z}{(z - 0.9j)} + \frac{0.5z}{(z + 0.9j)}$$

and the net integrand as

$$\frac{0.5z^n}{(z - 0.9j)} + \frac{0.5z^n}{(z + 0.9j)}$$

We obtain the first residue by multiplying the expression by the first pole $(z - 0.9j)$ and then setting $z = 0.9j$ from which we get

$$0.5(0.9j)^n + \frac{0.5(0.9j)^n(0.9j - 0.9j)}{(0.9j + 0.9j)} = 0.5(0.9j)^n$$

Doing the same thing with the second pole, -0.9j, we get 0.5(-0.9j)^n. Adding the two residues together we get

$$x[n] = 0.5(0.9j)^n + 0.5(-0.9j)^n$$

To check the answer, make the following call

```
n = 0:1:20; x = 0.5*(0.9*j).^n + 0.5*(-0.9*j).^n
altx = filter(1,[1,0,0.81],[1,zeros(1,20)])
figure; hold on; stem(n,x,'bo'); stem(n,altx,'b*')
```

which results in Fig. 6.9.

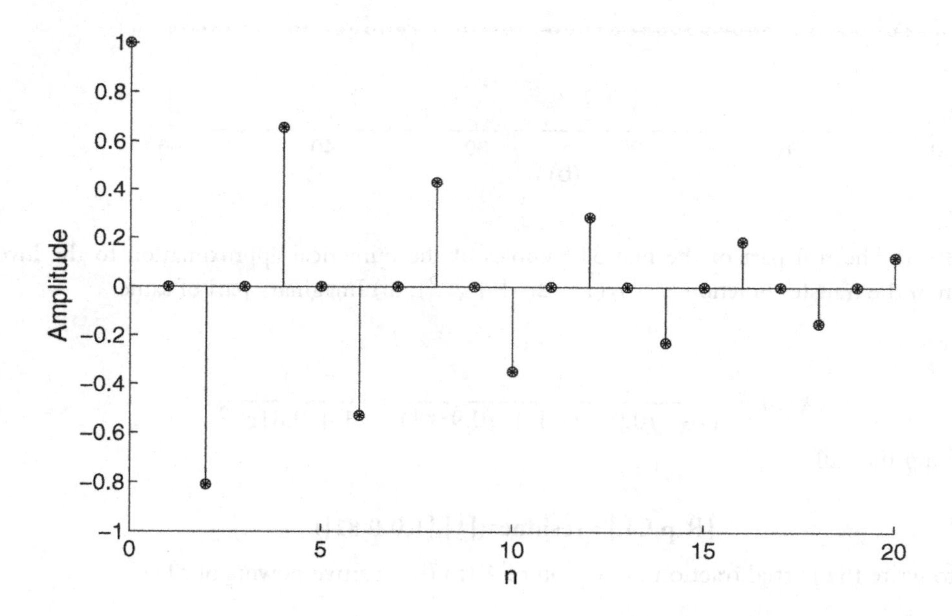

Figure 6.9: The inverse z-transform of the z-transform $1/(1 + 0.81z^{-2})$, computed using partial fraction expansion, as well as by filtering a unit impulse sequence using the b and a filter coefficients. Both results are plotted as specified by the m-code in the text.

6.5 TRANSIENT AND STEADY-STATE RESPONSES

Having discussed the Inverse-z-transform, we are now in a position to place the ideas of transient and steady-state responses on a more formal basis. To do this, we'll examine the response of a single-pole IIR to excitation by a complex exponential that begins at time zero, i.e., $n = 0$, that is to say,

$$x[n] = e^{j\omega n}u[n]$$

which has the z-transform

$$X(z) = \frac{1}{1 - e^{j\omega}z^{-1}}$$

The z-transform of a single pole LTI system represented by the difference equation

$$y[n] = b_0 x[n] + a_1 y[n-1]$$

is

$$H(z) = \frac{b_0}{1 - a_1 z^{-1}}$$

and therefore the z-transform of the system under excitation by the complex exponential is

$$Y(z) = H(z)X(z) = (\frac{b_0}{1 - a_1 z^{-1}})(\frac{1}{1 - e^{j\omega}z^{-1}})$$

which can be converted into an equivalent time domain expression using a partial fraction expansion which yields

$$y[n] = (C_S a_1^n + C_E e^{j\omega n})u[n]$$

where

$$C_S = b_0 a_1/(a_1 - e^{j\omega});\ C_E = b_0/(1 - a_1 e^{-j\omega})$$

Note that the excitation or driving signal, $e^{j\omega n}$ is a unity-magnitude complex exponential that does not decay away, whereas the component due to $H(z)$, namely $C_S a_1^n$, will decay away with increasing n if $|a| < 1$, i.e., if the system is stable.

Example 6.30. Compute and plot the transient, steady state, and total responses to the complex exponential $x[n] = exp(j\pi/6)$, and independently verify the total response, of the LTI system defined by the following difference equation:

$$y[n] = x[n] - 0.85y[n-1]$$

The following code generates Fig. 6.10:

```
w = pi/6; b0 = 1; a1 = - 0.85; Cs = b0*a1/(a1-exp(j*w));
Ce = b0/(1-a1*exp(-j*w)); n = 0:1:50; figure(11);
ys = Cs*a1.^n; yE = Ce*exp(j*w*n); yZ = ys+yE;
```

```
subplot(321); stem(real(ys)); subplot(322); stem(imag(ys));
subplot(323); stem(real(yE)); subplot(324); stem(imag(yE));
subplot(325); stem(real(yZ)); subplot(326); stem(imag(yZ));
```

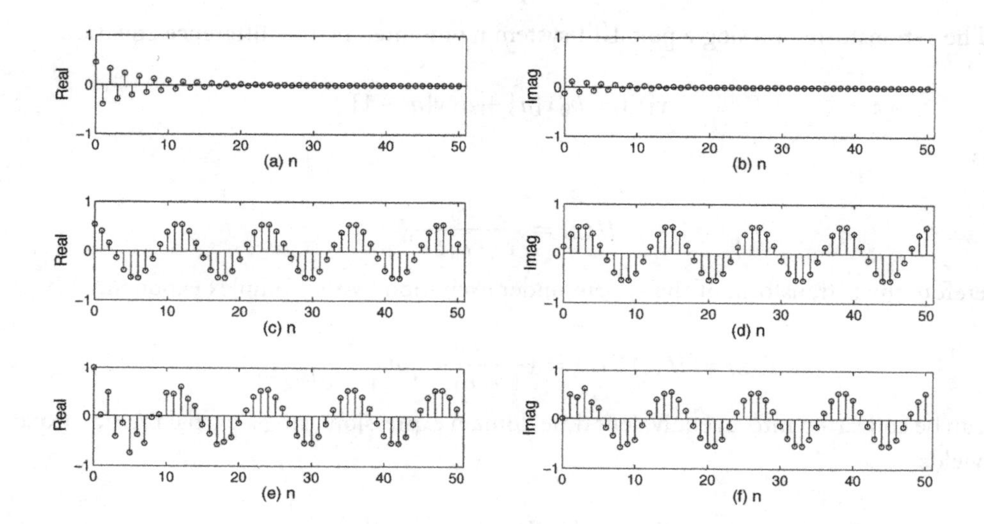

Figure 6.10: (a) Real part of transient response of IIR having a pole at -0.85 to a complex exponential at frequency pi/6; (b) Imaginary part of same; (c) Real part of steady-state response; (d) Imaginary part of steady-state response; (e) Real part of total response; (f) Imaginary part of total response.

In Fig. 6.10, we see, in plots (a) and (b), the real and imaginary parts of the transient response. Since the system pole that generates this response lies inside the unit circle, the system is stable, and this system-generated response decays away. In plots (c) and (d), we see the steady-state response due to the excitation function, a complex exponential. The total response is shown in plots (e) and (f).

We can verify the total response by filtering (say) 50 samples of the impulse response corresponding to the single pole filter having its pole at -0.85, and 50 samples of a complex exponential at radian frequency $\pi/6$. The following code results in Fig. 6.11:

```
n = [0:1:49]; y = exp(j*(pi/6)).^n;
ans = filter(1,[1,0.85],y);
figure(8); subplot(211); stem(n,real(ans));
subplot(212); stem(n,imag(ans))
```

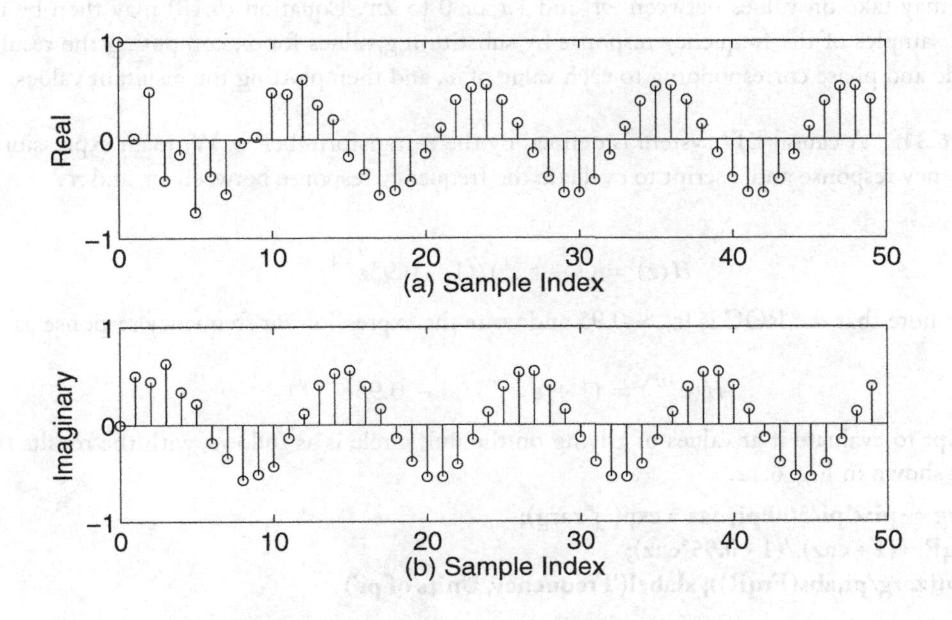

Figure 6.11: (a) Real part of complete response of an IIR having a single pole at -0.85, used to filter 50 samples of a complex exponential having radian frequency $\pi/6$; (b) Imaginary part of same.

6.6 FREQUENCY RESPONSE FROM Z-TRANSFORM

6.6.1 FOR GENERALIZED TRANSFER FUNCTION

If the Region of Convergence (ROC) of the z-transform of a digital filter includes the unit circle, the filter's frequency response can be determined by evaluating the z-transform at values of z lying on the unit circle. In other words, for each appearance of z in the z-transform, substitute a chosen value of z from the unit circle:

$$Y(e^{jw})/X(e^{jw}) = \frac{b_0 + b_1(e^{jw})^{-1} ... + b_{N-1}(e^{jw})^{-(N-1)}}{1 - a_1(e^{jw})^{-1} ... - a_{N-1}(e^{jw})^{-(N-1)}}$$

which simplifies to

$$Y(e^{jw})/X(e^{jw}) = \frac{b_0 + b_1 e^{-j\omega} ... + b_{N-1}e^{-(N-1)j\omega}}{1 - a_1 e^{-j\omega} ... - a_{N-1}e^{-(N-1)j\omega}} \qquad (6.18)$$

where ω may take on values between $-\pi$ and $+\pi$ or 0 to 2π. Equation (6.18) may then be used to obtain samples of the frequency response by substituting values for ω, computing the resulting magnitude and phase corresponding to each value of ω, and then plotting the resultant values.

Example 6.31. A causal LTI system is defined by the z-transform below. Write an expression for the frequency response and a script to evaluate the frequency response between $-\pi$ and π.

$$H(z) = (1 + z^{-1})/(1 - 0.95z^{-1})$$

We note that the ROC is $|z| > 0.95$ and write the expression for frequency response as

$$H(e^{jw}) = (1 + e^{-jw})/(1 - 0.95e^{-jw})$$

and a script to evaluate it at values of z lying on the unit circle is as follows, with the results from the script shown in Fig. 6.12.

```
zarg = -pi:2*pi/500:pi; eaz = exp(-j*zarg);
FrqR = (1 + eaz)./(1 - 0.95*eaz);
plot(zarg/pi,abs(FrqR)); xlabel('Frequency, Units of pi')
```

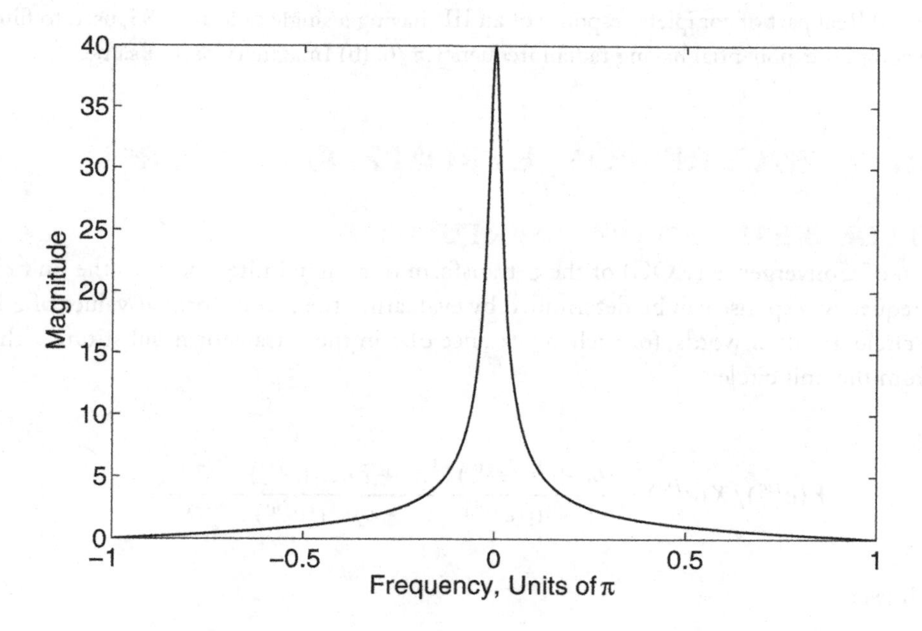

Figure 6.12: The magnitude of $H(z) = (1 + z^{-1})/(1 - 0.95\, z^{-1})$ evaluated at about 500 points along the unit circle from $-\pi$ to π radians.

6.6.2 RELATION TO DTFT

The z-transform of a sequence $x[n]$ is defined as

$$X(z) = \sum_{n=-\infty}^{\infty} x[n]z^{-n}$$

When the ROC includes the unit circle in the complex plane, we can substitute $e^{j\omega}$ for z and get

$$X(e^{j\omega}) = \sum_{n=-\infty}^{\infty} x[n]e^{-j\omega n}$$

which is the definition of the DTFT. Thus, when the unit circle is in the ROC of the z-transform of the LTI system, values of the z-transform evaluated along the unit circle are identical to values of the DTFT.

6.6.3 FINITE IMPULSE RESPONSE (FIR)

The z-transform of a finite sequence $x[n]$ is

$$x[0]z^0 + x[1]z^{-1} + x[2]z^{-2} + ... x[N-1]z^{-(N-1)}$$

For a finite causal sequence, the **ROC** (Region Of Convergence) is everywhere in the z-plane except the origin ($z = 0$). Choosing z somewhere on the unit circle, we have

$$z = e^{j\omega}$$

We can test the response of, say, an 11-sample FIR with one-cycle correlators (equivalent to those used for Bin 1 of an 11-point DFT, for example) generated by the z-power sequence shown in plot (a) of Fig. 6.13. To test the response of the same 11-sample FIR with, say, 1.1 cycle correlators, we would use $z = \exp(j2\pi k/N)$, where k =1.1.

$$b(z) = b_0 z^0 + b_1 z^{-1} + b_2 z^{-2} + ...b_{10} z^{-10}$$

- The z-transform evaluated at a value of z lying on the unit circle at a frequency corresponding to a DTFT frequency produces the same numerical result as the DTFT for that particular frequency. We can also evaluate the z-transform at many more (a theoretically unlimited number) values of z on the unit circle, corresponding to any sample of the DTFT.

Values of z not on the unit circle may also be used, provided, of course, that they lie in the **ROC**. Values of z having magnitude less than 1.0, for example, will result in a negative power sequence of z (z^0, z^{-1}, z^{-2}, etc.) that increases in magnitude, as shown in Fig. 6.13, plots (d)-(f).

Example 6.32. Evaluate the magnitude and phase response of the z-transform of an eight-point rectangular impulse response.

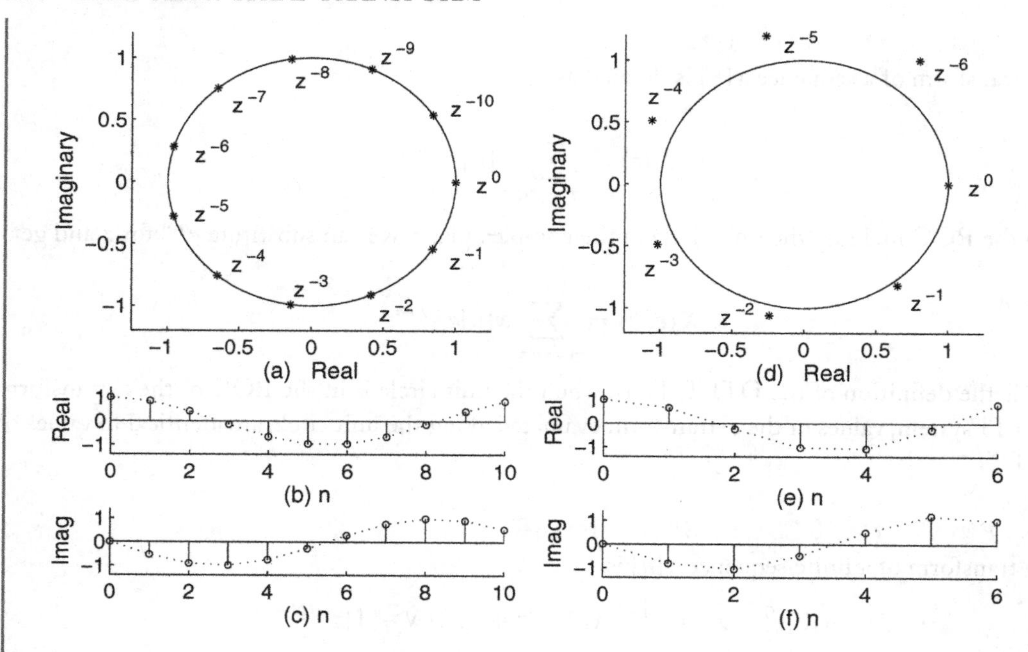

Figure 6.13: (a) Plot of complex correlator $C_1 = \exp(j2\pi/11)^n$ where n = 0:-1:-10; (b) Real part of C_1; (c) Imaginary part of C_1; (d) Plot of complex correlator $C_2 = (0.96\exp(j2\pi/7))^n$ where n = 0:-1:-6; (e) Real part of C_2; (f) Imaginary part of C_2.

The desired impulse response is

$$\textbf{Imp = [1,1,1,1,1,1,1,1]} \tag{6.19}$$

and the corresponding z-transform is

$$b(z) = 1 + z^{-1} + z^{-2} + ... + z^{-7}$$

Let's write two scripts to do the evaluation. The first method will use a loop and evaluate the z-transform for one value of z at a time; the second uses vectorized methods that avoid looping.

For the first method, we'll create a vector of values of z along the unit circle at which to evaluate the z-transform, by use of a loop:

```
SR=200; for A = 1:1:SR+1; z = exp(j*(A-1)*2*pi/SR);

B = 0 :-1: -length(Imp)+1; zvec = [z.^B];
```

and then the z-transform for index A (i.e., at radian frequency $\pi(A-1)/SR$) is obtained by multiplying $zvec$ with the impulse response (i.e., obtaining the correlation between the impulse response and the complex correlator generated by the negative power sequence of z)

```
zXform(A) = sum(Imp.*zvec); end
```

The above uses values of z all around the unit circle, i.e., positive and negative frequencies. To plot the above with a simple 2-D plot, make the call

```
plot(abs(zXform))
```

To evaluate the frequency response at only positive frequencies, use

```
z = exp(j*(A-1)*pi/SR);
```

For the second method, we create a matrix, each column of which is the z test or evaluation vector raised to a power such as 0, -1, etc. The matrix *pzMat* of arguments for the complex exponential is first created and then used as an argument for the *exp* function. Each row of the matrix *ZZMat* is a power sequence based on a given value of z; collectively, all the rows represent power sequences for all the values of z at which the z-transform is to be evaluated.

```
Imp = [1 1 1]; B=0:1:length(Imp)-1; SR = 256;
pzMat = (2*pi*(0:1:SR)/SR)'*(-B);
ZZMat = exp(j*pzMat); zXform = ZZMat*Imp'; plot(abs(zXform))
```

The script *LVxFreqRespViaZxform* (see exercises below) was used to generate Fig. 6.14 by making the call

```
LVxFreqRespViaZxform([1,1,1,1,1,1,1,1],512)
```

where the first argument is an impulse response, and the second argument is the number of z-transform and DTFT samples to compute. The figure shows, in plots (a) and (b), the magnitude and phase responses, respectively, of the z-transform of the impulse response. In the script, all values of z used to evaluate the z-transform are on the unit circle, and hence the result is identical to samples of the DTFT when evaluated at the same frequencies. The script computes the equivalent DTFT results, which are shown in plots (c) and (d) of the figure (which, for convenience shows only the positive frequency response in all plots).

Example 6.33. Efficiently evaluate the z-transform of the FIR whose z-transform is $Y(z) = 1 + 0.9z^{-n}$ where n is large and not known ahead of time for any particular computation. If, for example, n were, say, 1000, there would be 1000 terms to evaluate, as

$$Y(z) = 1 + b_1 z^{-1} + b_2 z^{-2} + ... + b_{1000} z^{-1000}$$

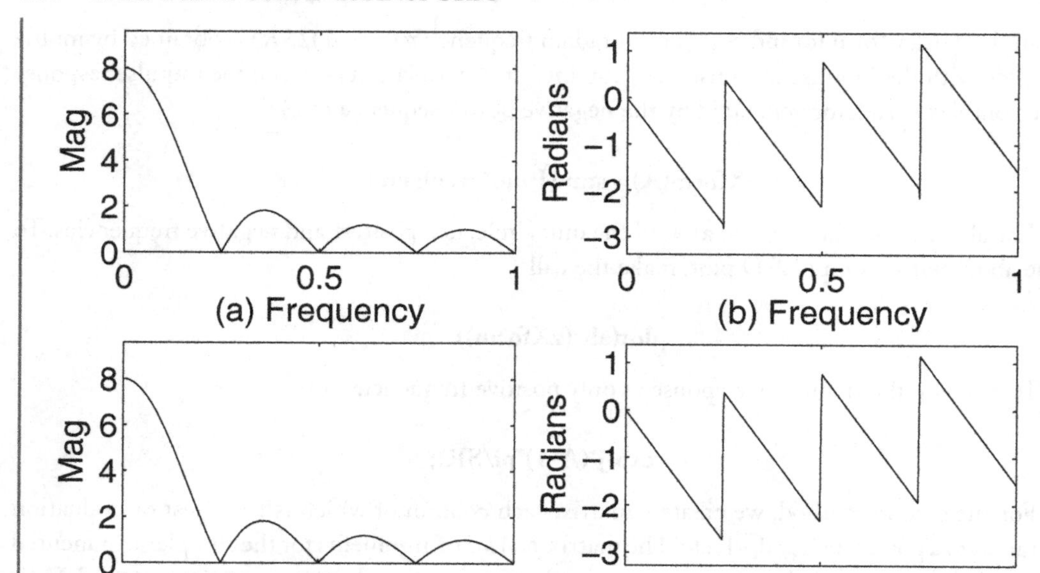

Figure 6.14: (a) Magnitude of z-Transform; (b) Phase Response of z-Transform; (c) Magnitude of DTFT; (d) Phase Response of DTFT. Note that only positive frequencies have been evaluated, i.e., only values of z lying along the upper half of the unit circle were used.

where b coefficients 1 through 999 are equal to zero.

A useful way to proceed is to assume that the vector of b coefficients will be supplied in complete form, with only a few nonzero coefficients, such as (for example)

b = [1, zeros(1,600), 0.7, zeros(1,600), 0.9]

Only nontrivial computations need to be made if we first detect the indices of the nonzero coefficients and compute only for those. The following code will work:

```
CoeffVec = [1,zeros(1,50),0.7,zeros(1,50),0.9];
SR = 256; zVec = exp(j*pi*(0:1/SR:1));
nzCoef = find(abs(CoeffVec)>0); num = zeros(1,length(zVec));
for Ctr = 1:1:length(nzCoef)
AnsThisCoeff = CoeffVec(nzCoef(Ctr))*(zVec.^(-nzCoef(Ctr)+1));
num = num + AnsThisCoeff; end
figure(888); plot(abs(num))
```

Example 6.34. Modify the code from the previous example to compute the z-transform of an IIR.

This can be accomplished simply by the computation **zXform = 1./num**. Let's take the simple IIR having z-transform

$$Y(z) = \frac{1}{1 - 1.27z^{-1} + 0.81z^{-2}}$$

having ROC: $|z| > 0.9$. We'll be using values of z lying on the unit circle, which is in the ROC. We thus use the following code (where we rename *num* as *den*):

```
CoeffVec = [1, -1.27, 0.81];
SR = 256; zVec = exp(j*pi*(0:1/SR:1));
nzCoef = find(abs(CoeffVec)>0); den = zeros(1,length(zVec));
for Ctr = 1:1:length(nzCoef)
AnsThisCoeff = CoeffVec(nzCoef(Ctr))*(zVec.^ (-nzCoef(Ctr)+1));
den = den + AnsThisCoeff; end
den = 1./den;
figure(777); plot(abs(den))
```

A script (see exercises below) which evaluates the magnitude of a generalized z-transform using the efficient code of the examples above for sparse coefficient vectors and which generates a 3-D plot is

$$LVxPlotZXformMagCoeff(NumCoeffVec,$$
$$...DenCoeffVec, rLim, Optr0, NSamps)$$

This script, which is intended for use with z-transforms having their ROCs lying outside a circle of radius equal to the magnitude of the largest pole of the z-transform, plots the magnitude of a z-transform that is supplied as a Numerator Coefficient Vector (*NumCoeffVec*), a Denominator Coefficient Vector (*DenomCoeffVec*), the desired number of circular contours to use (*rLim*), an optional initial contour radius, *Optr0* (pass as [] if not used), and *NSamps*, the number of frequency samples to use. When evaluating an FIR, pass *DenCoeffVec* as [1].

We can generate a 3-D plot of the magnitude of the z-transform of an eight-sample rectangular impulse response, as shown in Fig. 6.15 by making the call

LVxPlotZXformMagCoeff([1,1,1,1,1,1,1,1],[1],1,1,512)

Positive frequencies are represented by angles between 0 (normalized frequency 0) and 180 degrees (normalized frequency 1.0, or radian frequency π), measured counter-clockwise relative to the positive real axis. Negative frequencies are represented by angles between 0 and -180 degrees, or equivalently, between 180 and 360 degrees relative to the positive real axis.

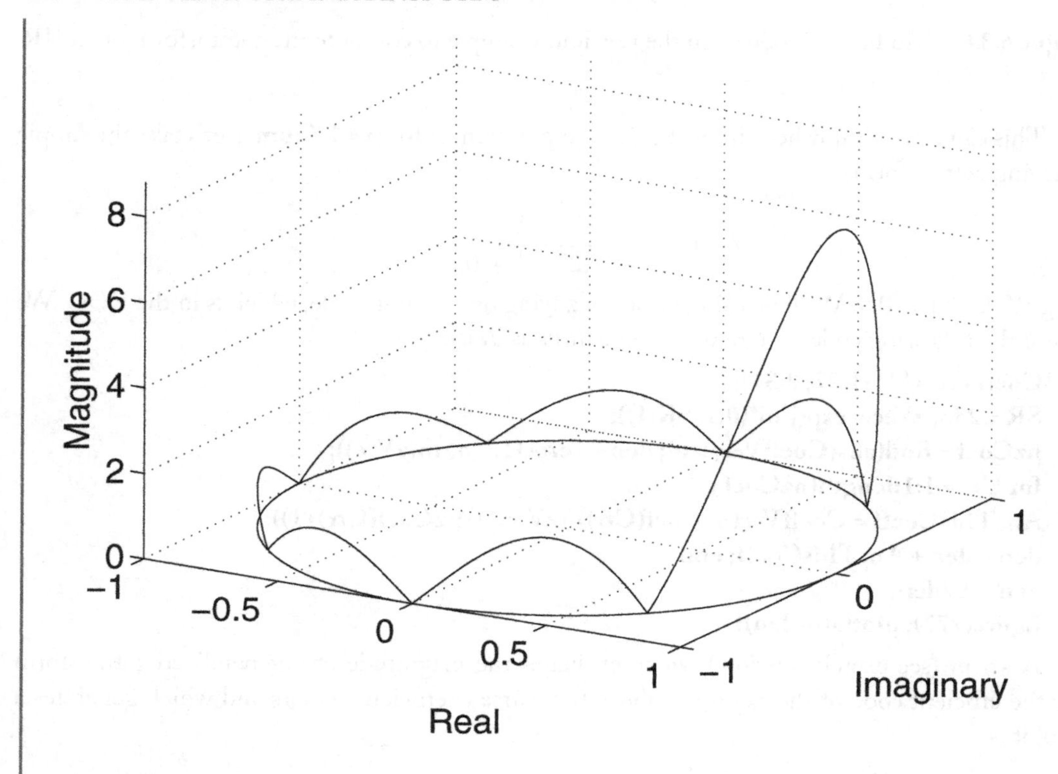

Figure 6.15: A 3-D plot of the magnitude of the z-transform of an 8-sample rectangular impulse response evaluated at many values of z along the unit circle, which is plotted in the complex plane for orientation.

Example 6.35. Evaluate the z-transform of an eight-point rectangular impulse response along contours within the unit circle.

The call

LVxPlotZXformMagCoeff([1,1,1,1,1,1,1,1],[1],30,0.95,512)

yields a figure which shows a surface formed by taking the z-transform magnitude along many circular contours in the z-plane, starting at radius 0.95 and moving outward to about radius 1.5, as shown in Fig. 6.16. A finite impulse response, which is often said to contribute only zeros to a transfer function, nonetheless has $L - 1$ trivial poles at the origin of the complex plane (L is the impulse response length), which drive the value of the z-transform to infinity at the origin, as can readily be inferred by inspection of Fig. 6.16. The plot in Fig. 6.16 was generated by evaluating the

z-transform at a limited number of points along a limited number of circular contours; as a result, fine structure, if any, between the evaluation points will be lost.

A VI that allows you to drag a zero, complex conjugate pair of zeros, or a quad of zeros (a complex conjugate pair and their reciprocals) in the z-plane, and see the magnitude and phase of the z-transform evaluated along the unit circle is

$$DemoDragZerosZxformVI$$

an example of which is shown in Fig. 6.17. This VI, when in quad mode, devolves to the minimum number of zeros necessary to maintain a linear phase filter. For example, if the main zero is given magnitude 1.0, and is complex, it and its complex conjugate are used for the FIR filter (in other words, the duplicate pair of zeros are discarded). If the main zero is given magnitude 1.0 and an angle of 0 or 180 degrees, the FIR's transfer function is built with just one zero, not four zeros. Similarly, when in Complex Conjugate mode, the number of zeros used in the FIR devolves to only one when the imaginary part of the main zero is equal to zero.

This resultant transfer function magnitude and phase responses, and the resultant impulse responses, can be compared to those obtained (in the quad case) using all four zeros in all situations if the user can run the script given below for use with MATLAB. Otherwise, the user can copy *DemoDragZerosZxformVI* to a new file name and then modify the m-code in the new VI's MathScript node to allow all four zeros to be used (see exercises below).

A script for use with MATLAB that performs functions similar to that of the VI above is

$$ML\_DragZeros$$

This script, when called, opens a GUI that allows you to select a single zero, a complex conjugate pair of zeros, or a quad of zeros (a complex conjugate pair and their reciprocals). The magnitude and phase of the z-transform as well as the real and imaginary parts of the equivalent impulse response are dynamically plotted as you move the cursor in the z-plane. This script, unlike the VI above, does not devolve to use of one or two zeros only in certain cases; four zeros are always used to create the FIR transfer function. Note that the quad of zeros always results in a linear phase characteristic and a real impulse response that is symmetrical about its midpoint. Phase linearity is of great value and FIRs can easily be designed having linear phase by ensuring that zeros of the transfer function are either single real zeros having magnitude 1.0, real pairs having reciprocal magnitudes, complex conjugate pairs having magnitude 1.0, or quads.

6.6.4 INFINITE IMPULSE RESPONSE (IIR) SINGLE POLE

The single pole IIR's z-transform

$$Y(z)/X(z) = \frac{1}{1 - pz^{-1}}$$

with ROC: $|z| > p$ (and $|p| < 1$) becomes

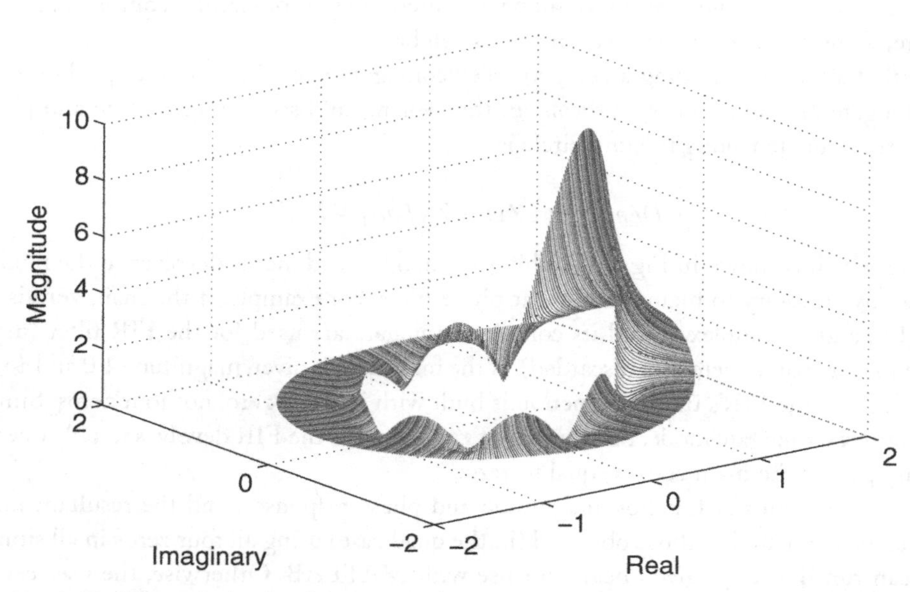

Figure 6.16: The magnitude of the z-transform of an 8-sample rectangular impulse response, evaluated over a number (30) of circular contours beginning at $|z| = 0.95$.

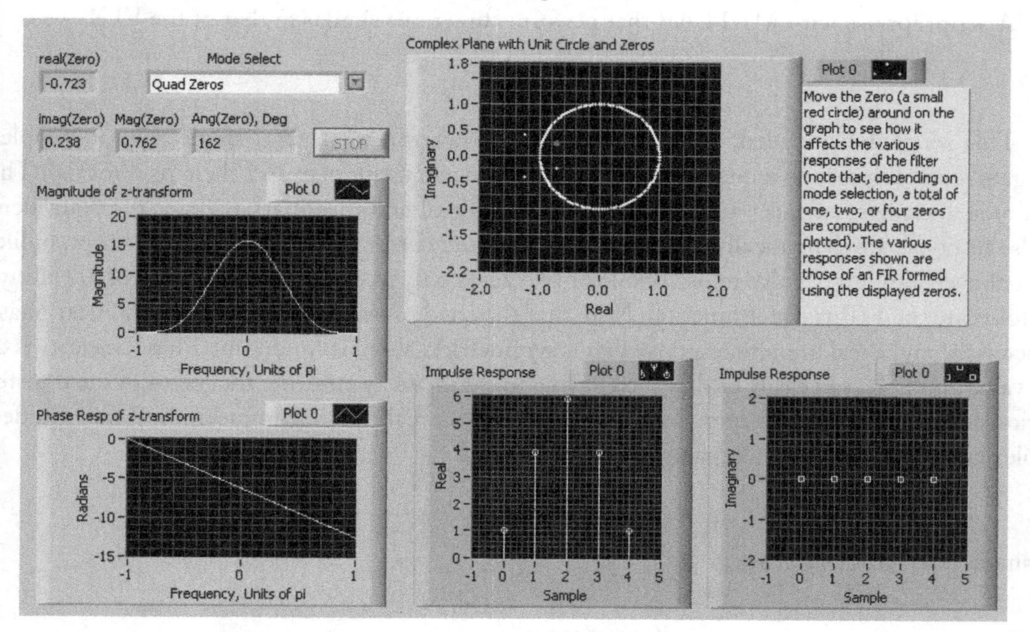

Figure 6.17: A VI that displays the magnitude and phase of the z-transform evaluated along the unit circle of an FIR filter formed from a single zero, pair of complex conjugate zeros, or a quad of zeros; the corresponding FIR impulse response is also displayed.

$$Y(z)/X(z) = \frac{1}{1 - pe^{-j\omega}}$$

after letting $z = e^{j\omega}$.

Example 6.36. Write several lines of m-code to evaluate and plot the magnitude of the z-transform of the difference equation

$$y[n] = x[n] + 0.9y[n-1]$$

at 2000 evenly-spaced frequency samples between 0 and 2π radians.

Initially, a vector of values of z at which to evaluate the z-transform must be formed, and then a vectorized expression can be written to evaluate the z-transform at all the chosen values of z. We choose (for this example), to evaluate the z-transform at 2000 values evenly spaced along the unit circle, and thus write

```
N = 2000; zarg = 0:2*pi/(N-1):2*pi;
zVec = exp(j*zarg); DenomVec = 1 - 0.9*( zVec.^(-1));
NetzXform = 1./DenomVec;
figure(10001); plot(zarg/pi,abs(NetzXform))
xlabel('Frequency, Units of pi')
```

Example 6.37. Determine the frequency response of the system defined by the difference equation shown below by use of the z-transform. Assume that $y[n] = x[n] = 0$ for $n < 0$. Note that the frequency response is the response to constant, unity-amplitude complex exponentials.

$$y[n] = x[n] + 1.1y[n-1]$$

This system is a single-pole IIR with a growing impulse response. The z-transform is only convergent for values of z having a magnitude greater than 1.1. Since the frequency response is determined by evaluating the z-transform for values of z on the unit circle (i.e., magnitudes of 1.0), we cannot evaluate the frequency response since the unit circle is not in the Region of Convergence.

6.6.5 CASCADED SINGLE-POLE FILTERS

If we were to cascade two identical single-pole IIRs, their equivalent z-transform would be the product of the two z-transforms, or

$$Y(z)/X(z) = (\frac{1}{1 - pz^{-1}})^2$$

Supposing we cascaded two IIRs with poles related as complex conjugates. We would have

$$Y(z)/X(z) = (\frac{1}{1 - pz^{-1}})(\frac{1}{1 - p_{cc}z^{-1}})$$

where

$$p_{cc}$$

denotes the complex conjugate of p. Doing the algebra, we get

$$Y(z)/X(z) = \frac{1}{1 - (p_{cc} + p)z^{-1} + (p_{cc}p)z^{-2}}$$

which reduces to

$$Y(z)/X(z) = \frac{1}{1 - (2 \cdot \text{real}(p))z^{-1} + |p|^2 z^{-2}} \tag{6.20}$$

Example 6.38. Evaluate the magnitude of the z-transform for a cascaded connection of two single-pole IIR filters each having a pole at 0.9.

Using Eq. (6.20), the z-transform would be:

$$H(z) = \frac{1}{1 - 1.8z^{-1} + 0.81z^{-2}} \tag{6.21}$$

The z-transform of Eq. (6.21), evaluated along a constant radius contour having radius 1.0 (i.e., the unit circle) is shown in Fig. 6.18. It was obtained by making the call

LVxPlotZXformMagCoeff([1],[1,-1.8,0.81],[1],[1],256)

Example 6.39. Evaluate the z-transform of a system consisting of two poles at $0.9j$ and $-0.9j$.

These poles resonate at the half-band frequency, and, using Eq. (6.20), the net z-transform is

$$\frac{1}{1 + 0.81z^{-2}}$$

Figure 6.19, which shows a peaked frequency response at the positive and negative half-band frequencies, was obtained by making the call

LVxPlotZXformMagCoeff([1],[1,0,0.81],[1],[1],256)

Example 6.40. Evaluate the magnitude of the z-transform along the unit circle of a system consisting of a pole at $(0.5 + 0.85j)$ and its complex conjugate.

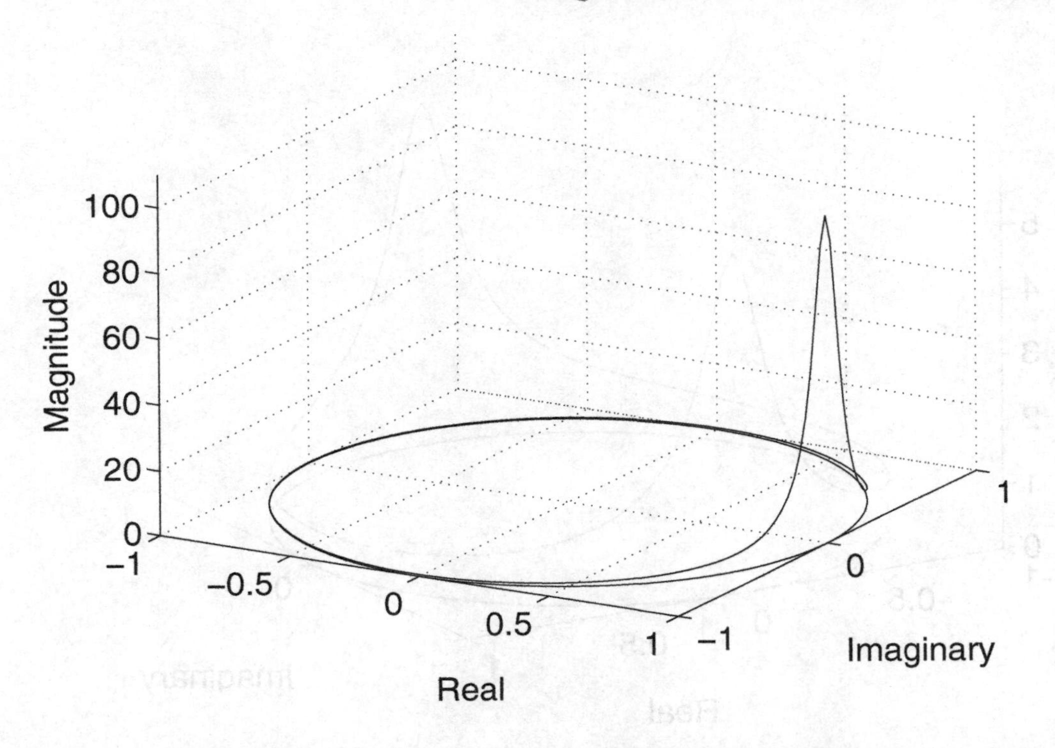

Figure 6.18: The magnitude of the z-transform evaluated along the unit circle (i.e., the Frequency Response) for the cascaded connection of two simple IIRs each having a real pole at 0.9 (the unit circle is shown in the complex plane for reference).

Again using Eq. (6.20),we get the z-transform as

$$Y(z)/X(z) = \frac{1}{1-(2\cdot 0.5)z^{-1} + (0.5^2 + 0.85^2)z^{-2}}$$

The z-transform is then evaluated as discussed in previous examples by substituting values of z on the unit circle.

A script that is equivalent to *LVxPlotZXformMagCoeff* and which works well with MATLAB (and which plots the unit circle underneath the z-transform magnitude as shown, for example, in Fig. 6.19) is

$$MLPlotZXformMagCoeff(NCoeffVec, DCoeffVec, rLim, Optr0, NSamps)$$

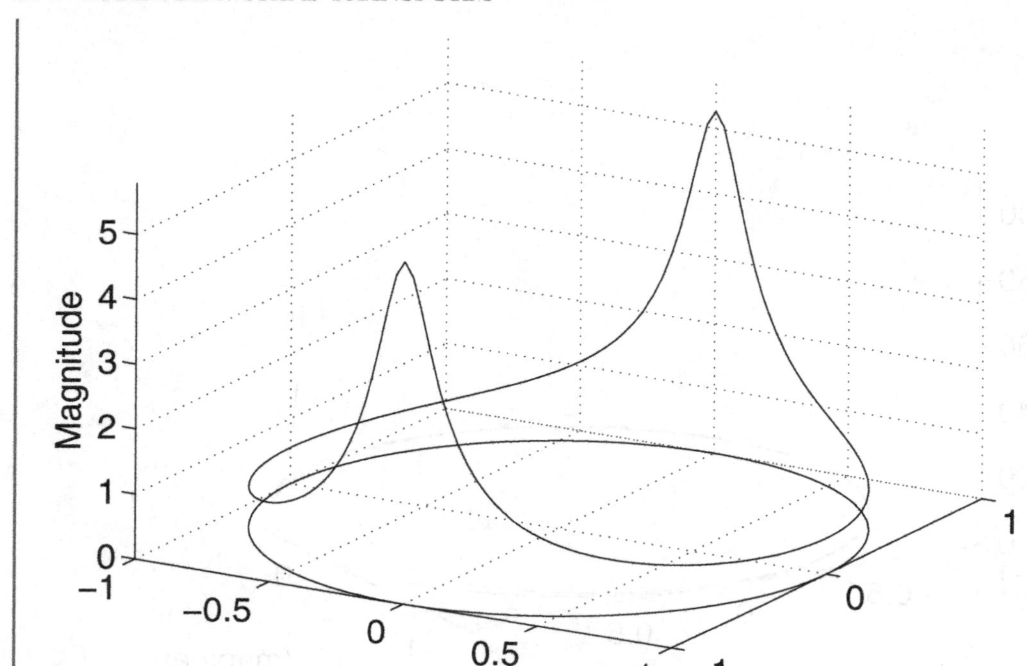

Figure 6.19: The magnitude of the z-transform evaluated along the unit circle (i.e., the Frequency Response) for two cascaded single-pole IIRs, the two poles being complex conjugates, tuned to the half-band frequency.

where $NCoeffVec$ is the numerator coefficient vector (i.e., b coefficients) and $DCoeffVec$ is the denominator coefficient vector (i.e., the a coefficients). For a description of all arguments and sample calls, type the following on the MATLAB Command Line and then press Return:

help MLPlotZXformMagCoeff

Another script

$$LVxPlotZTransformMag(ZeroVec, PoleVec, rMin, rMax, NSamps)$$

will take arguments as a row vector of zeros (*ZeroVec*), a row vector of poles (*PoleVec*), the minimum radius contour to compute (*rMin*), the maximum radius contour to compute (*rMax*), and the number of frequency samples along each contour to compute, *NSamps*. By passing both *rMin* and *rMax* as 1, the standard frequency response can be obtained.

A call to solve the current example would be

LVxPlotZTransformMag([],[(0.5 + 0.85*j),(0.5 - 0.85*j)],1,1,256)

which results in Fig. 6.20.

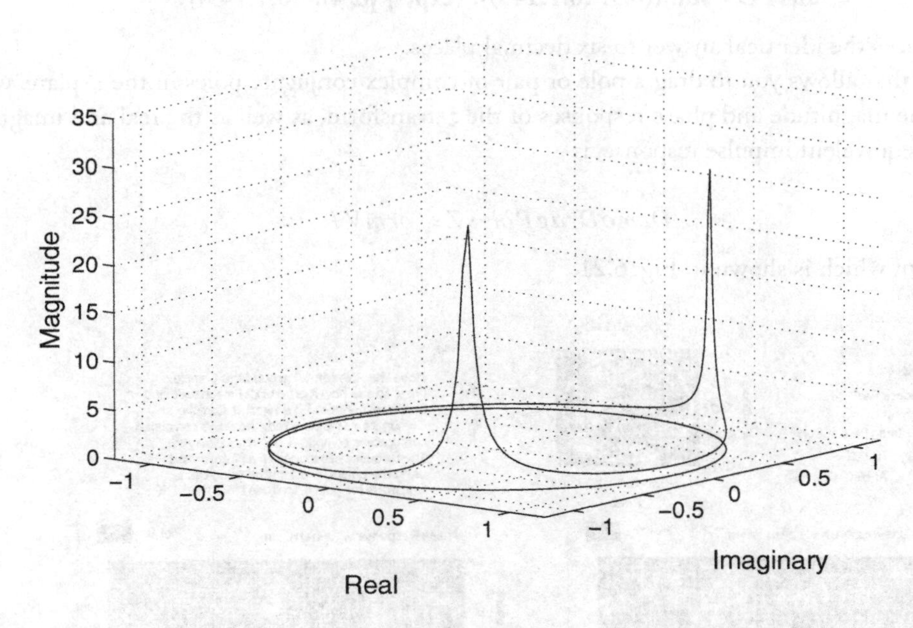

Figure 6.20: Magnitude of z-transform of the IIR formed from the poles (0.5 + 0.85j) and (0.5 - 0.85j), evaluated along the unit circle.

Example 6.41. Evaluate the frequency response of the z-transform below at radian frequency $\pi/4$, and then obtain the impulse response for the first 150 samples, correlate it with the complex exponential corresponding to the same radian frequency (i.e., compute the CZL), and compare the correlation value to the answer obtained directly from the z-transform. The z-transform is

$$X(z) = \frac{1}{1 - 0.9z^{-1}}$$

For the direct evaluation, we substitute exp(j*pi/4) for z in the z-transform and make the following call:

ansZD = 1./(1 - 0.9*exp(-j*pi/4))

which yields $ansZD$ = 0.6768 -1.1846i. To estimate the frequency response using time domain correlation, we obtain the impulse response (for the first 150 points, which should be adequate since it decays quickly), and correlate with the first 150 samples of the complex exponential power series of **exp(-j*pi/4)**

ansTD = sum((0.9.^(0:1:149)).*(exp(-j*pi/4).^(0:1:149)))

which produces the identical answer to six decimal places.

A VI that allows you to drag a pole or pair of complex conjugate poles in the z-plane while observing the magnitude and phase responses of the z-transform, as well as the real and imaginary parts of the equivalent impulse response, is

$$DemoDragPolesZxformVI$$

an example of which is shown in Fig. 6.21.

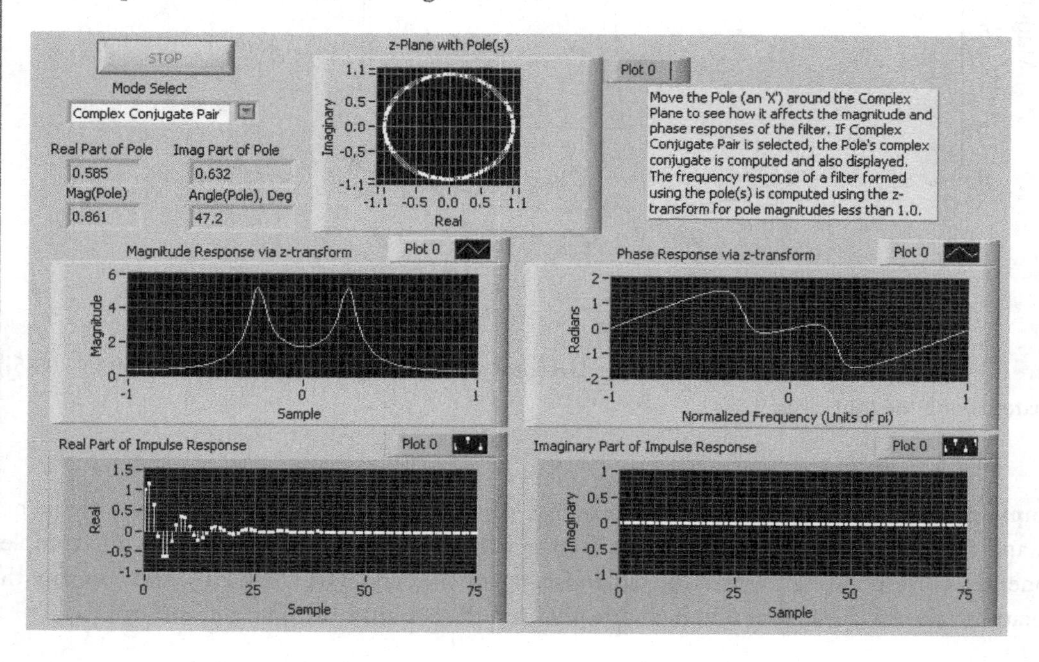

Figure 6.21: A VI that allows the user to drag a pole or pair of complex conjugate poles in the z-plane and observe the effect on impulse response and z-transform of an IIR formed from the pole(s).

6.6.6 OFF-UNIT-CIRCLE ZEROS AND DECAYING SIGNALS

When transfer function zeros are off the unit circle in the z-plane, their effect on the transfer function (for steady-state complex exponentials) is diminished. Such "off-unit-circle zeros" do not drive the output of the system to zero for a steady-state, unity-magnitude complex exponential, but rather for one which has a decaying (for zeros inside the unit circle) or growing (for zeros outside the unit circle) magnitude with time.

Example 6.42. Determine, for an FIR having the following transfer function

$$1 - 0.9z^{-1} \tag{6.22}$$

what input signal will result in a zero-valued output signal.

We can set expression (6.22) to zero, i.e.,

$$1 - 0.9z^{-1} = 0 \tag{6.23}$$

and solving for z we get $z = 0.9$.

To see this in the time domain, consider the decaying exponential series $x[n] = 1, 0.9, 0.81$, ..., i.e.,

$$x[n] = 0.9^n$$

where $0 \leq n \leq \infty$. Let's compute the first few terms of the response of the single zero system whose z-transform is shown at (6.22) and whose difference equation is

$$y[n] = x[n] - 0.9x[n-1] \tag{6.24}$$

Since Eq. (6.24) represents an FIR, the impulse response is identical to the FIR coefficients, which are [1, -0.9]. We can then convolve $x[n]$ with the system impulse response to obtain the system output $y[n]$:

```
n = 0:1:20; x = 0.9.^n;
y = conv([1, -0.9],x)
stem(y)
```

Note that the first and last samples of the above computation are not zero–why?

Example 6.43. Compute the output for an FIR with one zero at 0.9 and a constant amplitude signal having the same frequency as that of the zero, i.e., DC.

We can use the same call as above, but with the signal amplitude constant:

```
n = 0:1:20; x = 1.^n;
y = conv([1, -0.9],x)
stem(y)
```

or, alternatively,

```
stem(filter([1,-0.9],[1],[1.^(0:1:50)]))
```

The output sequence is 1 followed by the steady-state value of 0.1 and the final transient (non-steady-state) value as the end of $x[n]$ is reached.

6.7 TRANSFER FUNCTION & FILTER TOPOLOGY

6.7.1 DIRECT FORM

Consider the difference equation

$$y[n] = b_0 x[n] + b_1 x[n-1] + b_2 x[n-2] - a_1 y[n-1] - a_2 y[n-2]$$

or

$$y[n] + a_1 y[n-1] + a_2 y[n-2] = b_0 x[n] + b_1 x[n-1] + b_2 x[n-2]$$

and its z-transform:

$$Y(z)/X(z) = \frac{b_0 + b_1 z^{-1} + b_2 z^{-2}}{1 + a_1 z^{-1} + a_2 z^{-2}}$$

which can be written as a product of two terms

$$Y(z)/X(z) = (\frac{1}{1 + a_1 z^{-1} + a_2 z^{-2}})(\frac{b_0 + b_1 z^{-1} + b_2 z^{-2}}{1})$$

each of which can be realized as a separate structure, and the two structures cascaded, as shown in Fig. 6.22. This arrangement is known as a Direct Form I realization. Note that the numerator and denominator of the z-transform are separately realized and the two resultant time domain filters are cascaded, which provides the time domain equivalent (convolution) of the product of the two transfer functions.

Note that in Fig. 6.22, the two delay chains come from the same node, and hence may be replaced by a single delay chain, resulting in a Direct Form II realization as shown in Fig. 6.23. As many delay stages as needed may be used in a Direct form filter realization, as shown in Fig. 6.24.

A somewhat simpler method of illustrating a filter's topology is a signal flow diagram, an example of which is shown in Fig. 6.25, which, like Fig. 6.24, depicts an m-th order Direct Form II filter.

6.7.2 DIRECT FORM TRANSPOSED

The exact same transfer function as depicted in Fig. 6.26, for example, can be realized by following the simple procedure of: 1) reversing the directions of all signal flow arrows; 2) exchanging input and output. The result is a Direct Form Transposed structure.

6.7.3 CASCADE FORM

In addition to the Direct Form, there are several alternative structures for realizing a given transfer function. When the order is above two, it is often convenient to realize the net transfer function as a cascade of first and second order sections, each second order section being known as a **Biquad**. The z-transform is then a product of second order z-transforms with possibly an additional first order z-transform. The general form is illustrated in Fig. 6.27.

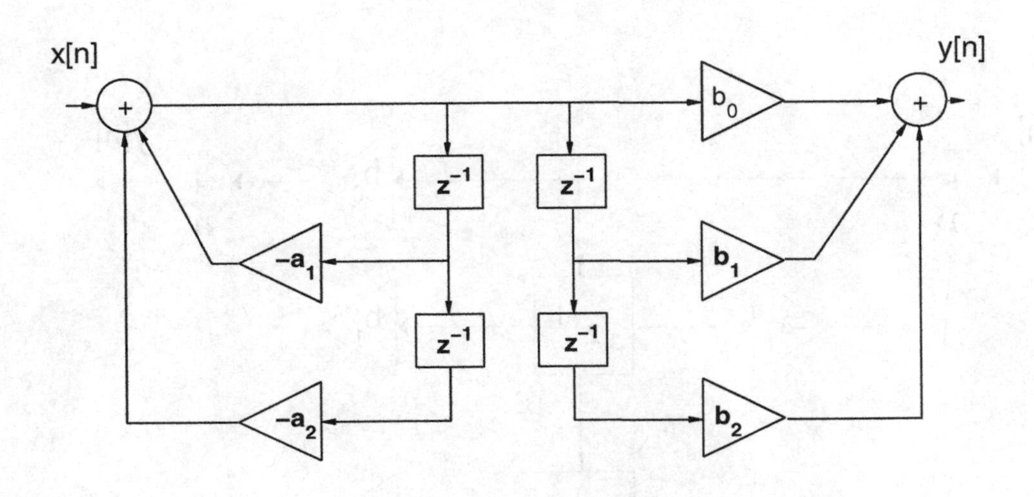

Figure 6.22: A Direct Form I realization of a second-order LTI system.

Thus the N-th order z-transform

$$H(z) = \frac{b_0 + b_1 z^{-1} + ...b_N z^{-N}}{1 + a_1 z^{-1} + ...a_N z^{-N}}$$

may be converted into the equivalent form

$$H(z) = b_0 \prod_{k=1}^{K} \frac{1 + B_{k,1} z^{-1} + B_{k,2} z^{-2}}{1 + A_{k,1} z^{-1} + A_{k,2} z^{-2}}$$

where $K = N/2$ when N is even, or $(N - 1)/2$ when N is odd, and the A and B coefficients are real numbers. When the order is even, there are only second order sections, and when it is odd there is one first order section in addition to any second order sections. Most filter designs generate poles in complex conjugate pairs, with perhaps one or more single, real poles, and therefore making each second order section from complex conjugate pairs and/or real pairs results in second order sections having all-real coefficients.

Referring to the function *LVDirToCascade,* the input arguments b and a are the numerator and denominator coefficients of a Direct Form filter; while the output arguments Ac and Bc are the biquad section coefficients, and $Gain$ is the corresponding gain, the computation of which is shown in the m-code following the next paragraph.

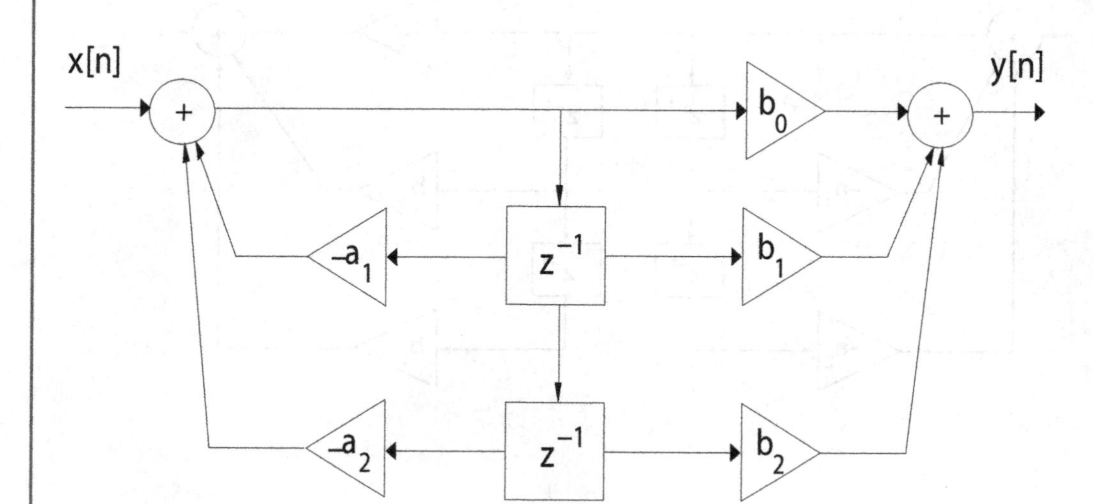

Figure 6.23: A Direct Form II realization of a second-order LTI system.

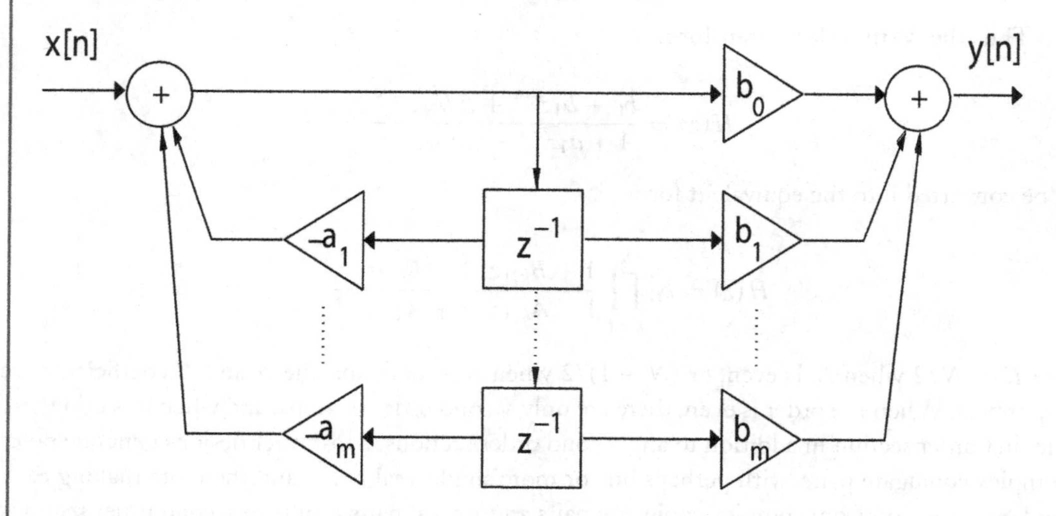

Figure 6.24: A Direct Form II realization of an m-th order LTI system.

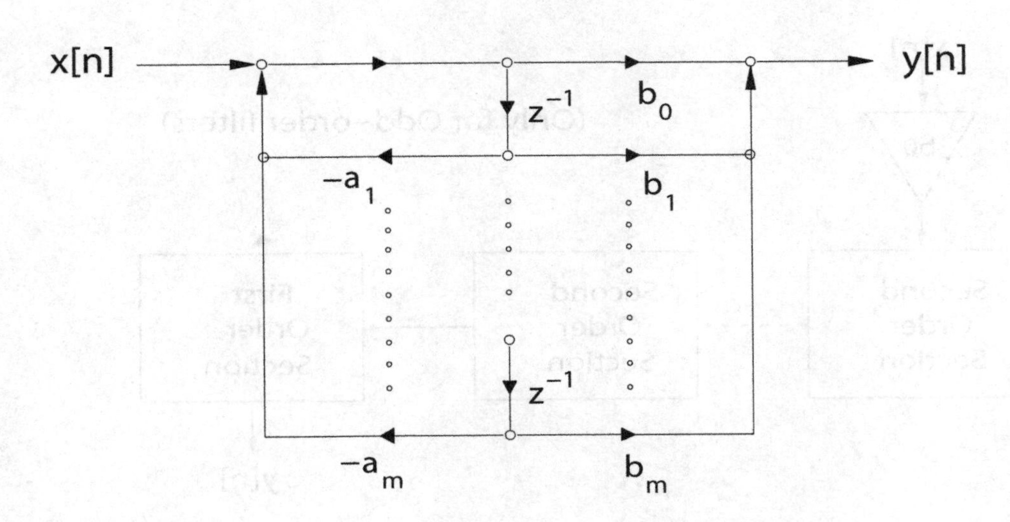

Figure 6.25: Signal Flow diagram for a generalized m-th order Direct Form II filter structure or realization. Signal flow direction is depicted by small arrows, which have unity gain unless otherwise labeled. Typical labels include a number representing a gain, a variable representing a gain, or the symbol z^{-1}, which represents a delay of one sample. The summing of signals is performed when any two signals flowing in the same direction meet at a node, represented by a small circle. The reader should compare this figure to the previous one.

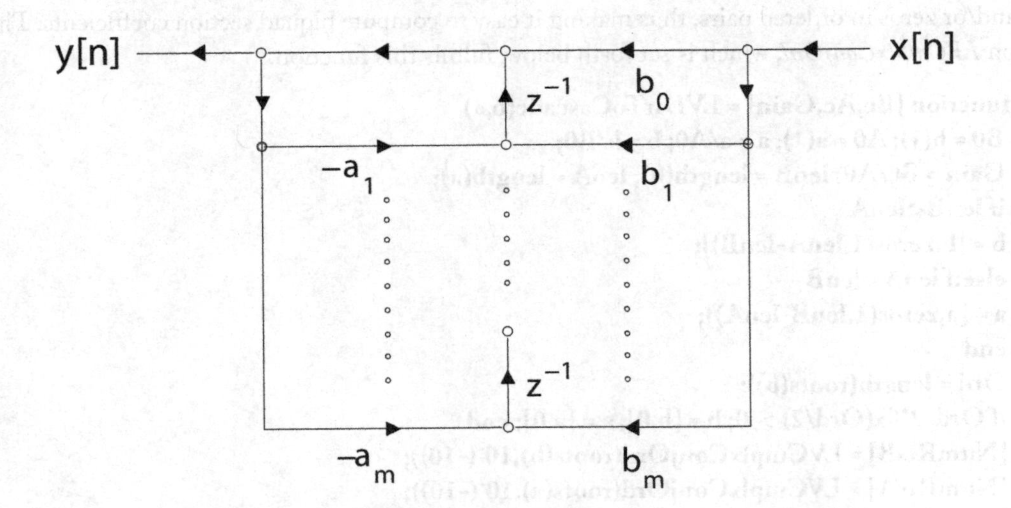

Figure 6.26: The signal flow diagram for an m-th order Direct Form II Transposed filter structure.

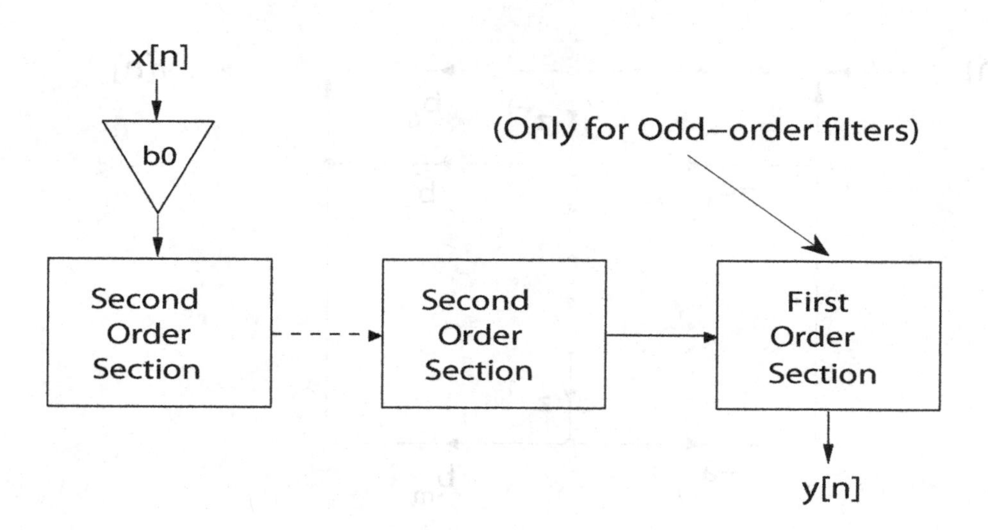

Figure 6.27: A Cascade Form filter arrangement, in which pairs of complex conjugate poles and pairs of real poles have been collected to form second order sections having real coefficients; if the order is odd, a single real pole will be left over to form a first order section.

The heart of a script to convert from Direct to Cascade form is a function to place the poles and/or zeros in ordered pairs, thus making it easy to compute biquad section coefficients. The function *LVCmplxConjOrd*, which is set forth below, fulfills this function.

```
function [Bc,Ac,Gain] = LVDirToCascade(b,a)
 B0 = b(1); A0 = a(1); a = a/A0; b = b/B0;
 Gain = B0/A0; lenB = length(b); lenA = length(a);
 if lenB<lenA
 b = [b,zeros(1,lenA-lenB)];
 elseif lenA<lenB
 a = [a,zeros(1,lenB-lenA)];
 end
 Ord = length(roots(b));
 if Ord-2*fix(Ord/2) > 0; b = [b,0]; a = [a,0]; end
 [NumRtsB] = LVCmplxConjOrd(roots(b),10^(-10));
 [NumRtsA] = LVCmplxConjOrd(roots(a),10^(-10));
 LenNumRts = length(NumRtsB); NoPrs = fix(LenNumRts/2)
 for Ctr = 1:1:NoPrs; ind = [2*Ctr-1, 2*Ctr];
```

```
Bc(Ctr,1:3) = real(poly(NumRtsB(ind)));
Ac(Ctr,1:3) = real(poly(NumRtsA(ind)));
end
```

The function *LVCmplxConjOrd,* used in the script *LVDirToCascade* above, sorts a list of randomly-ordered complex conjugate poles and real poles into pairs of complex conjugates followed by real-only poles: x is a vector of complex conjugate poles and real poles, in random (or unknown) order, tol defines how close a pole must be to its conjugate to be detected as such (tol is necessary due to roundoff error, which may prevent true conjugates from being so detected) $CCPrs$ is the output vector of pairs of complex conjugates, with real poles at the trailing end.

```
function [CCPrs] = LVCmplxConjOrd(x,tol)
Lenx = length(x); CCPrs = x;
for ctr = 1:2:Lenx-1
absdiff = abs(CCPrs(ctr)-conj(CCPrs(ctr+1:Lenx)));
y = find( absdiff < tol );
if isempty(y) % real number
temp = CCPrs(ctr); CCPrs(ctr:Lenx-1) = CCPrs(ctr+1:Lenx);
CCPrs(Lenx) = temp;
else; temp = CCPrs(ctr+1); CCPrs(ctr+1) = CCPrs(y(1)+ctr);
CCPrs(y(1)+ctr) = temp; end; end
```

To perform filtering using the Cascade Form filter coefficients, the input signal is filtered using the first cascade filter section, the output of which is used as the input to the second cascade filter section, and so on, until the signal has passed through all of the cascaded filter sections. The script

$$y = LVCascadeFormFilter(Bc, Ac, Gain, x)$$

works in this manner to filter the input vector x to produce the net filtered output y:

```
function [y] = LVCascadeFormFilter(Bc,Ac,Gain,x)
szA = size(Ac);
for ctr = 1:1:szA(1)
a = Ac(ctr,:); b = Bc(ctr,:);
x = filter(b,a,x);
end
y = Gain*x;
```

Example 6.44. Write a test script that will begin with a set of Direct Form filter coefficients, convert to Cascade Form, filter a linear chirp first with the Direct Form coefficients, then (as a separate experiment), filter a linear chirp using the Cascade Form coefficients, and plot the results from both experiments to show that the results are identical.

```
[b,a] = butter(7,0.4),
x = chirp([0:1/1023:1],0,1,512);
y1 = filter(b,a,x);
[Bc,Ac,Gain] = LVDirToCascade(b,a),
[y2] = LVCascadeFormFilter(Bc,Ac,Gain,x);
figure(10); subplot(211); plot(y1);
subplot(212); plot(y2)
```

To convert from Cascade to Direct form, a script must separately convolve each of the numerator factors and each of the denominator factors. Ac is a K by 3 matrix of biquad section denominator coefficients, K being $N/2$ for even filter order N, or $(N+1)/2$ for odd filter order, Bc is an r by 3 matrix of biquad section numerator coefficients, with r = floor($N/2$) where N is the filter order.

```
function [b,a,k] = LVCas2Dir(Bc,Ac,Gain)
k = Gain; szB = size(Bc);
b = 1; for ctr = 1:1:szB(1),
b = conv(b,Bc(ctr,:)); end;
szA = size(Ac); a = 1;
for ctr = 1:1:szA(1),
a = conv(a,Ac(ctr,:)); end
```

Example 6.45. Use the call $[b, a] = butter(4, 0.5)$ to obtain a set of Direct Form Butterworth filter coefficients, convert to Cascade Form, and then convert from Cascade back to Direct Form.

Using the scripts given above, a composite script is

```
[b,a] = butter(4,0.5)
[Bc,Ac,Gain] = LVDirToCascade(b,a)
[b,a,k] = LVCas2Dir(Bc,Ac,Gain)
```

which yields, from the first line,

b = [0.094,0.3759,0.5639,0.3759,0.094]

which can be rewritten as

0.094([1,4,6,4,1])

and

a = [1,0,0.486,0,0.0177]

The second m-code line above yields (in compact reformatted form) the Cascade Form coefficients as

Bc = [1,2,1;1,2,1]

Ac = [1,0,0.4465;1,0,0.0396]

The third m-code line above converts the Cascade Form coefficients back into Direct Form:

b = [1,4,6,4,1]

a = [1,0,0.4860,0,0.0177]

k = 0.094

6.7.4 PARALLEL FORM

In this form, illustrated in Fig. 6.28, a number of filter sections are fed the input signal in parallel, and the net output signal is formed as the sum of the output signals from each of the sections. Eq. (6.13) shows a Direct Form z-transform, possibly having the numerator order exceeding that of the denominator, and Eq. (6.14) rewrites this as a proper fraction (i.e., having numerator order equal to or less than that of the denominator) plus a polynomial in z. In the Parallel Form, Eq. (6.14) is rewritten as

$$H(z) = \frac{B_{k,0} + B_{k,1}z^{-1}}{1 + A_{k,1}z^{-1} + A_{k,2}z^{-2}} + \sum_{0}^{M-N} C_k z^{-k}$$

where M and N are the orders of the numerator and denominator of the z-transform, respectively, and, assuming N is even, $K = N/2$, and the A, B, and C coefficients are real numbers. Note that in this form, the numerator of the second order sections is first order.

To generate the Parallel Form filter coefficients Bp, Ap, and Cp, the script (see exercises below)

$$[Bp, Ap, Cp] = LVxDir2Parallel(b, a)$$

receives as input the Direct Form filter coefficients (b, a) and begins by using the function *residuez* to produce the basic vectors of residues, poles, and C coefficients, as described earlier in the chapter with respect to partial fraction expansion. It is then necessary to reorder the poles in complex conjugate pairs, followed by real poles. This is done using the function *LVCmplxConjOrd*, introduced above with respect to conversion to Cascade Form. Since it is necessary to keep a given residue with its corresponding pole, it is necessary, once the properly-ordered vector of poles has been obtained, to locate each one's index in the original pole vector returned by the call to *residuez*, and then reorder the residues obtained from that call to correspond to the new pole order obtained from the call to

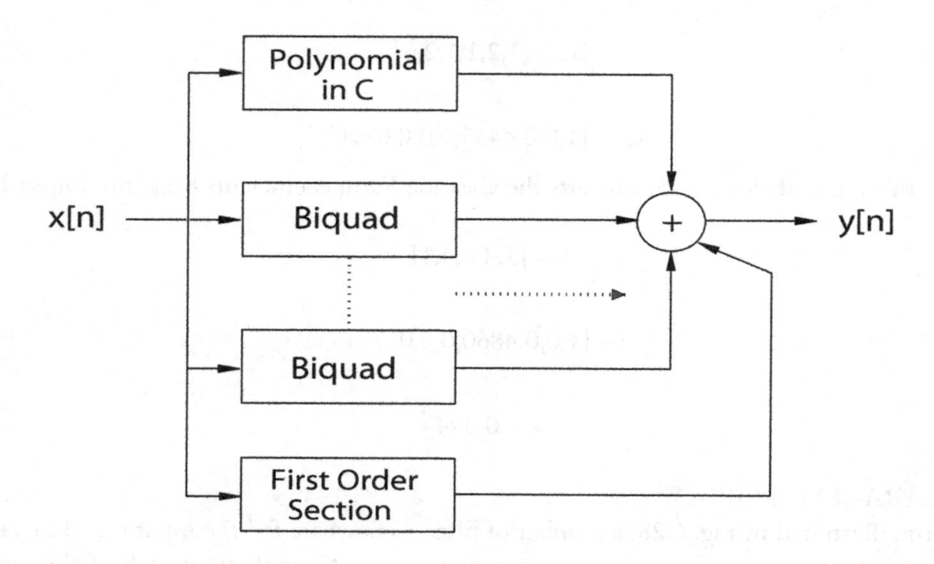

Figure 6.28: A Parallel Form Filter. Assuming even order N, there will be $K = N/2$ biquad sections. Each biquad section has a second-order denominator and a first-order numerator. If any of the C coefficients are nonzero, the section labeled "Polynomial in C" will be present. If N is odd, there will be one additional first-order section in parallel with the other sections, as shown.

LVCmplxConjOrd. Once this has been done, the poles and corresponding residues are collected two at a time and a call is made to *residuez*, which in this mode delivers a set of b and a coefficients for a single biquad section. This is done until all pole pairs have been used, and if there is one pole and residue left over, a single first order section is also created.

In order to filter using the Parallel Form, the input signal is convolved with each filter section, and the outputs of the sections are added. The script (see exercises below)

$$y = LVxFilterParallelForm(Bp, Ap, Cp, x)$$

filters the input vector x using the Parallel Form coefficients (Bp, Ap, Cp) to produce the output signal y.

Example 6.46. Write a short script that generates a basic set of Direct Form coefficients, filters a linear chirp using them with the function filter, and then converts the Direct Form coefficients into Parallel Form coefficients, filters the linear chirp using the script *LVxFilterParallelForm*, and then displays plots of the output signals from the two filtering operations.

The following code results in Fig. 6.29.

```
[b,a] = butter(5,0.2)
x = chirp([0:1/1023:1],0,1,512);
y1 = filter(b,a,x);
figure(9); subplot(211); plot(y1);
[Bp,Ap,Cp] = LVxDir2Parallel(b,a);
y2 = LVxFilterParallelForm(Bp,Ap,Cp,x);
subplot(212); plot(y2)
```

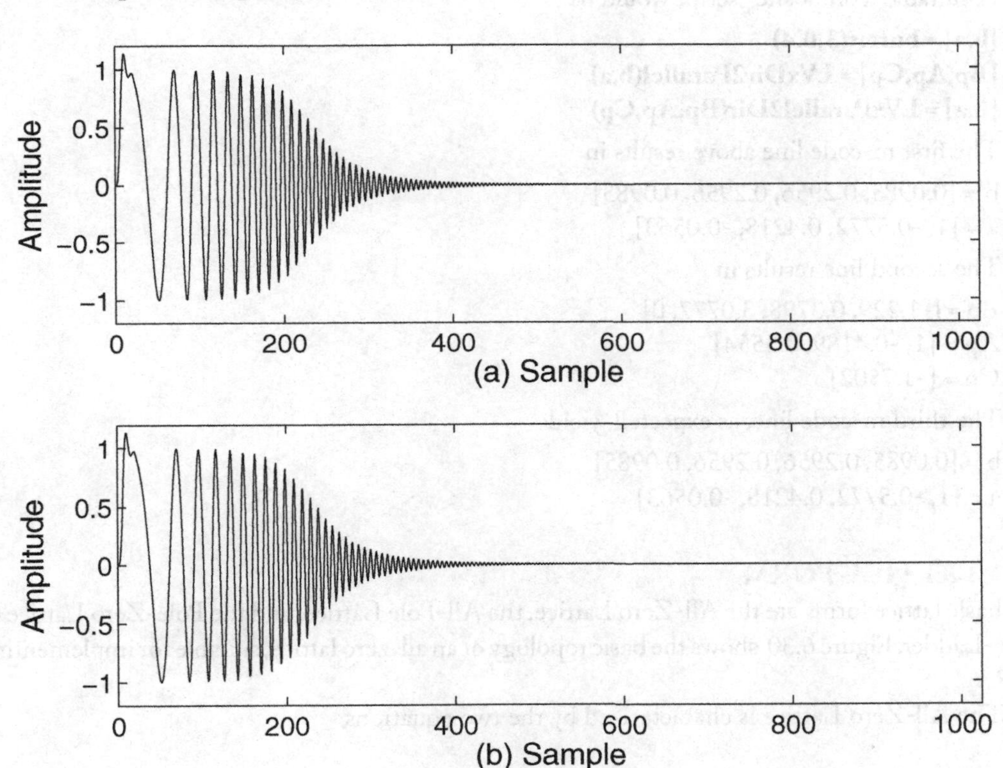

Figure 6.29: (a) A linear chirp filtered using Direct Form coefficients of a fifth order lowpass Butterworth filter; (b) The same linear chirp, filtered by the equivalent Parallel Form coefficients of the fifth order lowpass Butterworth filter.

In order to convert from Parallel Form to Direct Form, the script (see exercises below)

$$[b, a] = LVxParallel2Dir(Bp, Ap, Cp)$$

uses the coefficients for each filter section to obtain the corresponding set of poles and residues by making a call to *residuez*. This is done for all Parallel Form filter sections, and the entire accumulated set of poles and residues is used in a single "reverse" call to *residuez* to obtain the Direct Form b and a coefficients.

Example 6.47. Use the call $[b, a] = butter(3, 0.4)$ to obtain a set of coefficients (b, a) for a lowpass filter, convert to a set of Parallel Form coefficients $[Bp, Ap, Cp]$, and then convert these coefficients back into Direct Form coefficients using the scripts described above.

A suitable "composite" script would be

[b,a] = butter(3,0.4)
[Bp,Ap,Cp] = LVxDir2Parallel(b,a)
[b,a] = LVxParallel2Dir(Bp,Ap,Cp)

The first m-code line above results in

b = [0.0985, 0.2956, 0.2956, 0.0985]
a = [1, -0.5772, 0.4218, -0.0563]

The second line results in

Bp = [-1.229, 0.3798; 3.0777, 0]
Ap = [1, -0.4189, 0.3554]
Cp = [-1.7502]

The third m-code line, as expected, yields

b = [0.0985, 0.2956, 0.2956, 0.0985]
a = [1, -0.5772, 0.4218, -0.0563]

6.7.5 LATTICE FORM

Three basic lattice forms are the All-Zero Lattice, the All-Pole Lattice, and the Pole-Zero Lattice or Lattice-Ladder. Figure 6.30 shows the basic topology of an all-zero lattice, suitable for implementing an FIR.

The All-Zero Lattice is characterized by the two equations

$$g_m[n] = g_{m-1}[n] + K_m h_{m-1}[n-1],\ m = 1, 2, ...M-1$$

$$h_m[n] = K_m g_{m-1}[n] + h_{m-1}[n-1],\ m = 1, 2, ...M-1$$

where the values of K are known as the reflection coefficients. For a given FIR in Direct Form, a recursive algorithm can be solved to determine the appropriate values of K. However, if the magnitude of any value of K_m is 1, the algorithm will fail. This is the case for linear phase filters, so linear phase filters cannot be implemented using a lattice structure.

Example 6.48. For a Direct Form FIR having $b = [1, 0.5, -0.8]$, determine the equivalent K coefficients by writing the lattice transfer function and equating to b.

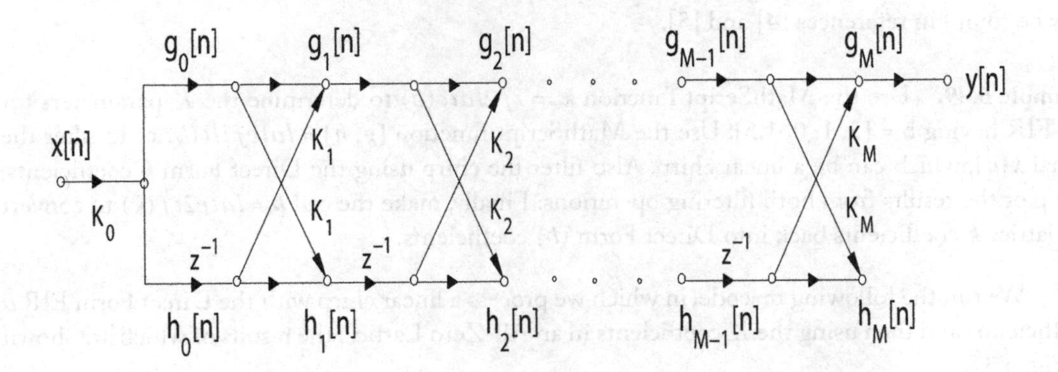

Figure 6.30: The basic structure of an All-Zero Lattice, suitable to implement an FIR.

Referring to Fig. 6.30, we would need two stages, so in this case the output would be at $g_2[n]$. By tracing the three signal paths from the input to the output at $g_2[n]$, we can write the lattice output transfer function as

$$Y(z)/X(z) = 1 + (K_1 + K_1 K_2)z^{-1} + K_2 z^{-2}$$

and since the Direct Form z-transform is

$$Y(z)/X(z) = b[0] + b[1]z^{-1} + b[2]z^{-2}$$

we get

$$(K_1 + K_1 K_2) = b[1];\ K_2 = b[2]$$

or

$$K_2 = b[2];\ K_1 = b[1]/(1 + b[2])$$

We thus get K_1 = 0.5/(1-0.8) = 2.5 and K_2 = -0.8. We can check this using the following call to the MathScript function *tf2latc*:

k = tf2latc([1,0.5,-0.8])

which yields k = 2.5000, -0.8000. Note that the above derivation assumes $b[0]$ = 1, so b should be scaled by its first element, i.e., **b = b/b[0]**.

The general recursive algorithm to compute all-zero lattice coefficients for any order of FIR may be found in references [4] and [5].

Example 6.49. Use the MathScript function $k = tf2latc(b)$ to determine the K parameters for the FIR having b = [2, 1, 0, 1.5]. Use the MathScript function $[g, h] = latcfilt(k, x)$ to filter the signal $x[n]$, which can be a linear chirp. Also filter the chirp using the Direct Form b coefficients, and plot the results from both filtering operations. Finally, make the call $b = latc2tf(k)$ to convert the lattice k coefficients back into Direct Form (b) coefficients.

We run the following m-code, in which we process a linear chirp with the Direct Form FIR b coefficients, and then using the K coefficients in an All-Zero Lattice, the results of which are shown in Fig. 6.31.

```
x = chirp([0:1/1023:1],0,1,512);
b = [2,1,0,1.5]; b1 = b(1); b = b/b1; y = filter(b,1,x);
k = tf2latc(b), [g,h] = latcfilt(k,x);
figure(8); subplot(211); plot(y); xlabel('Sample')
subplot(212); plot(g); xlabel('Sample')
b = latc2tf(k); b = b*b1
```

The topology of an All-Pole Lattice is shown in Fig. 6.32; note that the reflection coefficients are feedback rather than feedforward (see reference [4] for details). The MathScript function $tf2latc$ can be used to determine the values of K using the format $K = tf2latc(1, a)$ where the input argument a represents the Direct Form IIR coefficients. To filter a sequence x[n] using the all-pole lattice coefficients k, make the following call: $[g, h] = latcfilt(k, 1, x)$.

Example 6.50. A certain IIR has a = [1, 1.3, 0.81]. Compute the All-Pole Lattice coefficients K and filter a linear chirp using both the Direct Form and Lattice coefficients. Plot the results from each filtering operation.

The following code will suffice, the results of which are shown in Fig. 6.33.

```
x = chirp([0:1/1023:1],0,1,512);
a = [1,1.3,0.81]; y = filter(1,a,x);
k = tf2latc(1,a),
[g,h] = latcfilt(k,1,x);
figure(8); subplot(211); plot(y);
subplot(212); plot(g)
```

A filter having both IIR and FIR components has an equivalent in the Lattice-Ladder, the basic design of which is shown in Fig. 6.34. The call

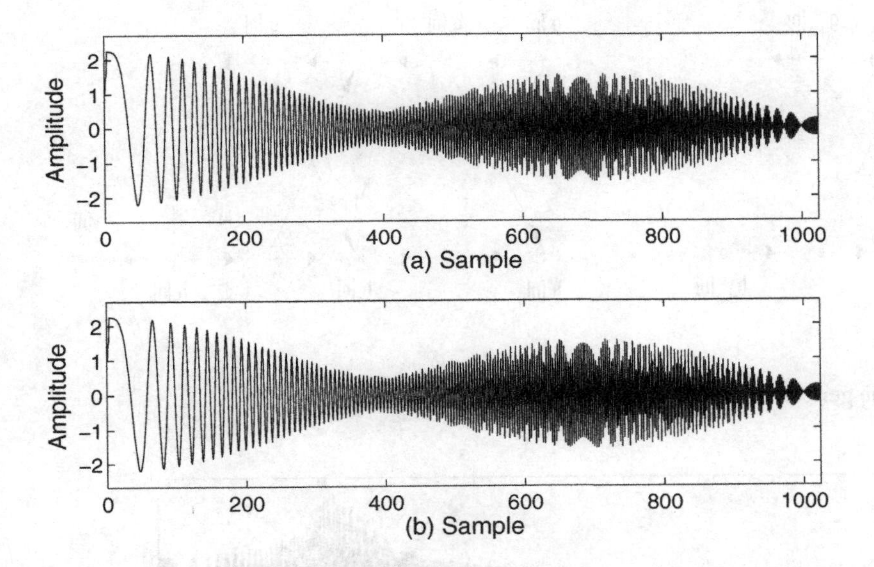

Figure 6.31: (a) A linear chirp after being filtered using the Direct Form FIR coefficients [2,1,0,1.5]; (b) A linear chirp after being filtered using the equivalent All-Zero Lattice structure.

$$[K, C] = tf2latc(b, a)$$

computes the lattice coefficients K and the ladder coefficients C that form the equivalent Lattice-Ladder to the Direct Form filter characterized by the z-transform numerator coefficients b and denominator coefficients a. To filter a signal $x[n]$ using the lattice and ladder coefficients K and C, respectively, make the call

$$[g, h] = latcfilt(K, C, x)$$

Example 6.51. A certain IIR has Direct Form coefficients b = [0.3631, 0.7263, 0.3631] and a = [1, 0.2619, 0.2767]. Compute the lattice and ladder coefficients K and C, respectively, and filter a linear chirp using both the Direct Form and Lattice-Ladder coefficients.

The results from the following code are shown in Fig. 6.35.

```
x = chirp([0:1/1023:1],0,1,512);
a = [1,0.2619,0.2767];
b = [0.3631,0.7263,0.3631];
y = filter(b,a,x);
```

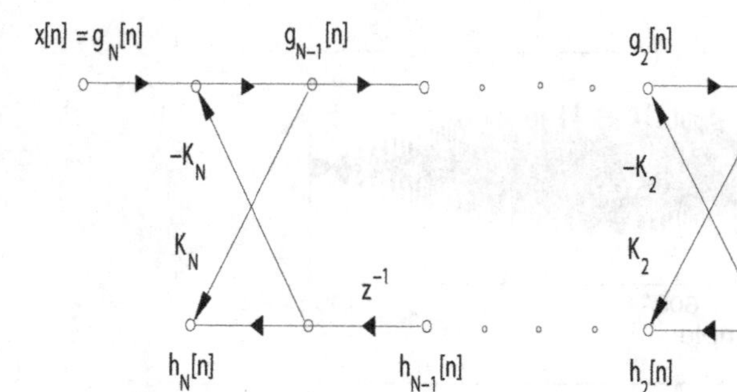

Figure 6.32: The general topology of an All-Pole Lattice filter.

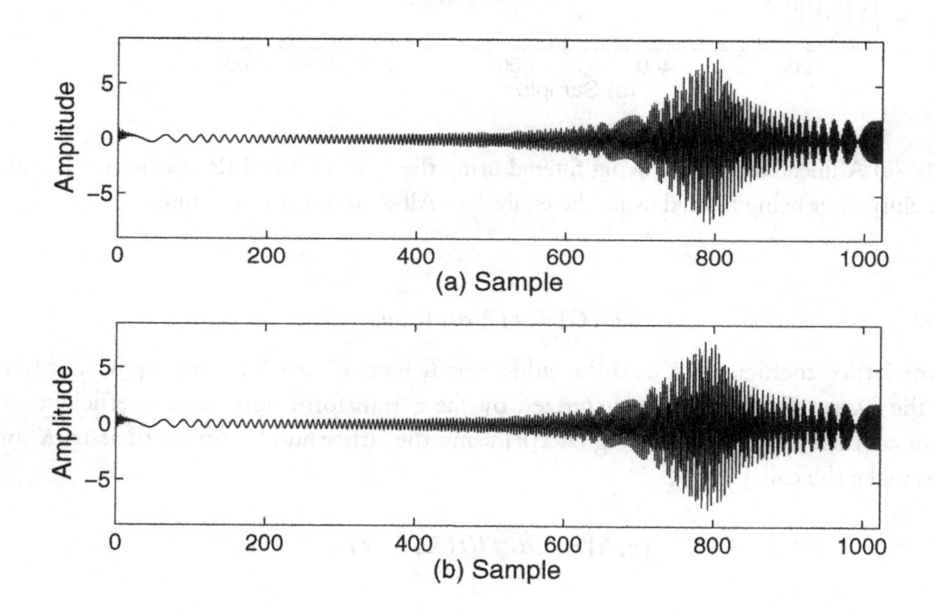

Figure 6.33: (a) A linear chirp after being filtered using the Direct Form IIR coefficients b =1 and a = [1,1.3,0.81]; (b) A linear chirp after being filtered using an All-Pole Lattice having k = [0.7182, 0.81].

```
[k,c] = tf2latc(b,a),
[g,h] = latcfilt(k,c,x);
figure(8); subplot(211); plot(y);
subplot(212); plot(g);
```

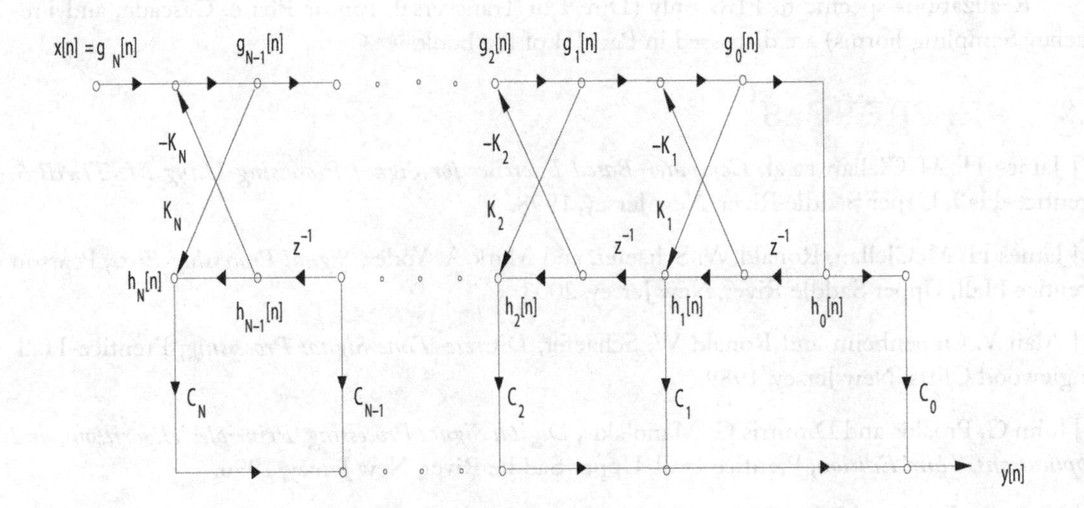

Figure 6.34: A generalized Lattice-Ladder structure.

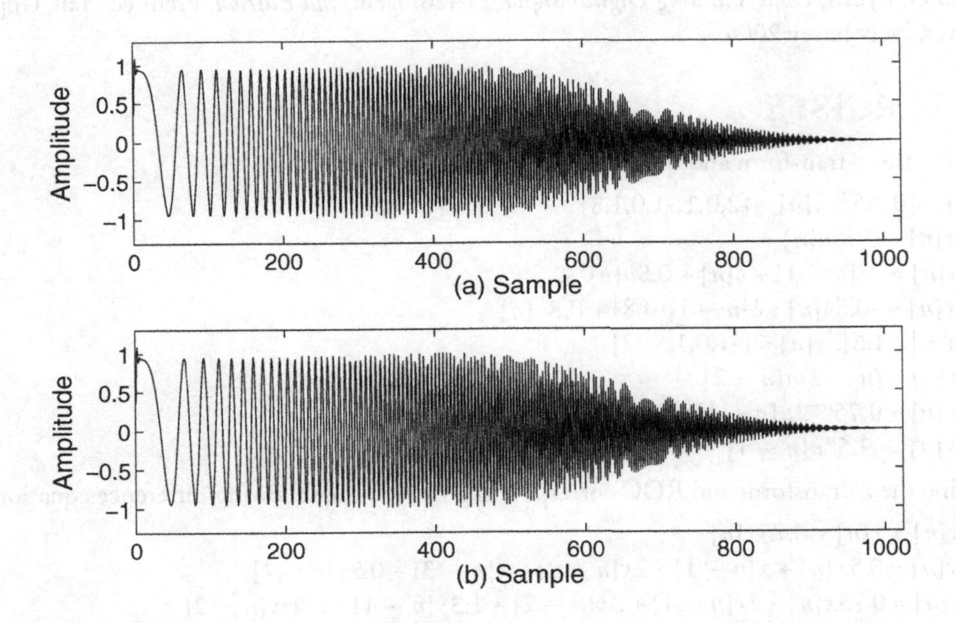

Figure 6.35: (a) A linear chirp filtered by an IIR having b = [0.3631,0.7263,0.3631] and a = [1,0.2619,0.2767]; (b) A linear chirp filtered using a Lattice-Ladder structure having k = [0.2051, 0.2767] and c = [0.1331, 0.6312, 0.3631].

Realizations specific to FIRs only (Direct or Transversal, Linear Phase, Cascade, and Frequency Sampling Forms) are discussed in Part III of the book.

6.8 REFERENCES

[1] James H. McClellan et al, *Computer-Based Exercises for Signal Processing Using MATLAB 5*, Prentice-Hall, Upper Saddle River, New Jersey, 1998.

[2] James H. McClellan, Ronald W. Schaefer, and Mark A. Yoder, *Signal Processing First*, Pearson Prentice Hall, Upper Saddle River, New Jersey, 2003.

[3] Alan V. Oppenheim and Ronald W. Schaefer, *Discrete-Time Signal Processing*, Prentice-Hall, Englewood Cliffs, New Jersey, 1989.

[4] John G. Proakis and Dimitris G. Manolakis, *Digital Signal Processing, Principles, Algorithms, and Applications, Third Edition*, Prentice Hall, Upper Saddle River, New Jersey, 1996.

[5] Vinay K. Ingle and John G. Proakis, *Digital Signal Processing Using MATLAB V.4*, PWS Publishing Company, Boston, 1997.

[6] Richard G. Lyons, *Understanding Digital Signal Processing, Second Edition*, Prentice Hall, Upper Saddle River, New Jersey 2004.

6.9 EXERCISES

1. Determine the z-transform and the ROC for the following sequences:

    (a) n = [0:1:5]; $x[n]$ = [2,0,1,-1,0,1,8]
    (b) $x[n] = 0.95u[n]$
    (c) $x[n] = 2\delta[n+1] + \delta[n] + 0.8u[n]$
    (d) $x[n] = -0.5\delta[n] - \delta[n-1](0.8) + 0.8u[n]$
    (e) n = [2:1:6]; $x[n]$ = [-1,0,1,5,-3]
    (f) $x[n] = (n-2)u[n-2]$
    (g) $x[n] = 0.75^{n-1}u[n-1] + 0.95^{n}u[-n-1]$
    (h) $x[n] = 2.5^{n}u[n-1]$

2. Determine the z-transform and ROC corresponding to each the following difference equations:

    (a) $y[n] = x[n] + 0.75x[n]$
    (b) $y[n] = 0.5x[n] + x[n-1] - 2x[n-2] + x[n-3] + 0.5x[n-4]$
    (c) $y[n] = 0.75x[n] - 2x[n-1] + 3x[n-2] + 1.3y[n-1] - 0.95y[n-2]$
    (d) $y[n] = x[n] - 1.1x[n-1] - 2.3x[n-2] - 0.98y[n-2]$

3. Since two time domain sequences may be equivalently convolved by multiplying their z-transforms, correlation can be equivalently performed by multiplying the z-transform of one sequence and the z-transform of a time-reversed version of the other sequence. The time reversal can be obtained

by substituting $1/z$ (see the time reversal property of the z-transform earlier in the chapter) for z in the z-transform of the sequence to be reversed. Perform correlation of the following pairs of sequences $x_1[n]$ and $x_2[n]$ by obtaining the time domain sequence corresponding to the product of the z-transform of $x_1[n]$ and $x_2[-n]$, then verify the answer by using time domain correlation. Don't forget to take note of the applicable ROCs.

(a) $x_1[n] = [1,2,3,4]$; $x_2[n] = [4,3,2,1]$; $n = [0,1,2,3]$
(b) $x_1[n] = [1,2,3,4]$; $x_2[n] = [1,2,3,4]$; $n = [0,1,2,3]$
(c) $x_1[n] = 0.9(u[n] - u[n-5])$; $x_2[n] = 0.8(u[n] + u[n-3])$

4. Determine the poles and zeros corresponding to each of the following z-transforms, and, using paper and pencil, plot them in the z-plane. Characterize each system's stability.

(a) $X(z) = (1 + z^{-1}) / (1 - 1.4z^{-1} + 0.9z^{-2})$
(b) $X(z) = (1 + z^{-3}) / (1 + 0.9z^{-3})$
(c) $X(z) = (0.0528 + 0.2639z^{-1} + 0.5279z^{-2} + 0.5279z^{-3} + 0.2639z^{-4} + 0.0528z^{-5}) / (1 + 0.6334z^{-2} + 0.0557z^{-4})$
(d) $X(z) = (0.0025z^{-1} + 0.0099z^{-2} + 0.0149z^{-3} + 0.0099z^{-4} + 0.0025z^{-5}) / (1 - 2.9141z^{-1} + 3.5179z^{-2} - 2.0347z^{-3} + 0.4729z^{-4})$

5. Should the roots of a polynomial with real, random coefficients be real or complex? If complex, what special characteristics should they have? Distinguish between polynomials of even length and odd length. An easy way to investigate this is to repeatedly run the code

x = roots([randn(1,N)])

where N is a conveniently small integer such as 4 or 5. Justify or prove your conclusion.

6. Determine an inverse signal that will result in the unit impulse sequence when processed by the following systems, some of which are unstable.

(a) $y[n] = x[n] + 1.05y[n-1]$
(b) $y[n] = x[n] - y[n-2]$
(c) $y[n] = x[n] + 0.5x[n-1] - 0.25x[n-2]$

7. For each of the following sets of zeros and poles, which are associated with an underlying causal LTI system, determine the equivalent z-transform (with ROC) and the difference equation. For (c), write m-code to plot the zeros in the z-plane.

(a) **zeros = [j,-1,-j]; poles = [0.9*exp(j*pi/20), 0.9*exp(-j*pi/20)]**
(b) **zeros = [-1,-1,-1]; poles = [0.5774*j, -0.5774*j,0]**
(c) **zeros = [exp(j*(pi/4:pi/20:3*pi/4)),exp(-j*(pi/4:pi/20:3*pi/4))]**

8. A certain sequence $x[n]$ has $X(z) = 1 - z^{-1} + 2z^{-2}$. ($z \neq 0$) Determine the z-transforms and corresponding ROCs for the following sequences:

(a) $x_1[n] = 2x[n] - x[n-1]$
(b) $x_2[n] = 2x[2-n] - x[n-1]$

(c) $x_3[n] = nx[1-n] - x[n-1]$

9. Using Eq. (6.8), evaluate the magnitude of response of the following systems, specified with poles (pls) and zeros (zzs), to unity-amplitude complex exponentials having the following radian frequencies: $\pi/4, \pi/2. -\pi/4, 0, \pi$.

(a) **zzs = [-1,-1]; pls = [(0.3739 + 0.3639*j), (0.3739 - 0.3639*j)]**

(b) **zzs = [-1,-1,1,1]; pls = [(-0.26 + 0.76*j), (-0.26 - 0.76*j), (0.26 + 0.76*j), (0.26 - 0.76*j)]**

10. Write a script that implements the following call syntax:

$$LVxFreqRespViaZxform(Imp, SR)$$

as described in the text, and which creates a plot of the magnitude and phase of both 1) the z-transform, and 2) the DTFT of Imp.

11. Write a script that implements the function *LVxPlotZXformMagCoeff* introduced in the text, according to the following format:

```
function LVxPlotZXformMagCoeff(b,a,rLim,Optionalr0,NSamps)
% intended for use with positive-time z-transforms, i.e., those
% whose ROC lies outside a circle of radius equal to the mag-
% nitude of the largest pole; both a and b must be passed in the
% order of: Coeff for z^0, Coeff for z^-1, Coeff for z^-2, etc.
% b is the vector of numerator coefficients; a  is the vector of
% denominator coefficients;
% rLim is the number of circular contours along which to evaluate
% the z-transform. The spacing between multiple contours is 0.02.
% Optionalr0 is the radius of the first contour–if passed as [], the
% default is 1.0 for FIRs (i.e., DenomCoeffVec=[1]) or the magni-
% tude of the largest pole plus 0.02. If the largest pole magnitude
% is > 0.98 and < 1.0, the radius of the first contour is set
% at 1.0 unless Optionalr0 is larger than 1.0. If there are no poles,
% pass a as [1]. If there are no zeros, pass b  as [1].
% NSamps is the number of frequencies at which to evaluate the
% z-transform along each contour.
% Sample calls:
% LVxPlotZXformMagCoeff([1],[1,-1.8,0.81],[1],[1],256)
% LVxPlotZXformMagCoeff([1],[1,0,0.81],[1],[1],256)
% LVxPlotZXformMagCoeff([1,1,1,1,1,1,1,1],[1],[1],[1],256)
```

12. Write a script that implements the function *LVxPlotZTransformMag* as described in the text and further defined below. The script should plot at least the magnitude of the z-transform of the transfer function defined by the input arguments *ZeroVec* and *PoleVec*.

```
function LVxPlotZTransformMag(ZVec,PVec,rMin,rMax,NSamps)
```

```
% LVxPlotZTransformMag is intended for use with right-handed
% z-transforms, i.e., those whose ROC is the area outside of a
% circle having a radius just larger than the magnitude of the
% largest pole of the z-transform. ZVec is a row vector of zeros;
% if no zeros, pass as []
% PVec is a row vector of poles; if no poles, pass as []
% rMin and rMax are the radii of the smallest and largest contours
% to compute; rMin and rMax may be equal to compute one con-
% tour. Pass rMin = rMax = 1 for standard frequency response.
% NSamps is the number of frequencies at which to evaluate the
% z-transform along each contour. rMin is set at 0.02 greater than
% the magnitude of the largest pole unless the pole magnitude is
% greater than 0.98 and less than 1.0 in which case rMin is set
% at 1.0.
% Sample calls:
% LVxPlotZTransformMag([-1,1],[0.95*exp(j*2*pi/5),...
% 0.95*exp(-j*2*pi/5)],1,1,256)
% LVxPlotZTransformMag([exp(j*2*pi/4) exp(-j*2*pi/4)],...
% [0.95*exp(j*2*pi/4),0.95*exp(-j*2*pi/4)],1,1,256)
% LVxPlotZTransformMag([-1,j,-j],[],1,1,256)
% LVxPlotZTransformMag([],[0.9],1,1,256)
% LVxPlotZTransformMag([(0.707*(1+j)),j,(0.707*(-1+j)),...
% -1,(0.707*(1-j)),-j,-(0.707*(1+j))],[],1,1,256)
```

13. Evaluate the frequency response of the z-transforms below at radian frequencies 0, $\pi/4$, and $\pi/2$, then obtain the impulse response for the first 150 samples and correlate it with the complex exponential corresponding to the same radian frequencies. Plot the impulse response and the real and imaginary parts of the exponential power series used in the correlation. Compare the correlation values to the answers obtained directly from the z-transform. The z-transforms are

a) $X(z) = [1 + z^{-1}]/(1 - 0.9z^{-1})$;ROC: $|z| > 0.9$
b) $X(z) = [z^{-1} + z^{-2}]/(1 - z^{-1} + 0.9z^{-2})$;ROC: $|z| > 0.9487$
c) $X(z) = [1 + z^{-4}]/(1 - 1.3z^{-1} + 0.95z^{-2})$;ROC: $|z| > 0.9747$

14. Write a script that can receive set of b and a coefficients of the z-transform of a causal sequence whose ROC either lies outside a circle of radius equal the magnitude of the largest pole, or includes the entire complex plane except $z = 0$, and compute the time domain sequence $x[n]$ using the method of partial fraction expansion, employing the function *residuez*. Test your script on the following z-transform coefficients:

(a) **b = [0.527,0,1.582,0,1.582,0,0.527]; a = [1,0,1.76, 0,1.182, 0,0.278]**
(b) **b = [0.518 0 1.554 0 1.554 0 0.518]; a = [1,0,1.745,0,1.172,0,0.228]**
(c) **[b,a] = butter(8,[0.4,0.6])**

(d) **[b,a] = cheby1(8,1,[0.6,0.8],'stop')**

15. Determine the frequency response of the following LTI systems, characterized by their poles and zeros, at each of the following radian frequencies: $0, \pi/4, \pi/2, 3\pi/4, \pi$.

(a) Poles = [0.9,0.9,0.9j,-0.9j]; Zeros = [0.65(1+j),0.65(1-j),-0.65(1+j),-0.65(1-j)]
(b) Poles = [(0.0054 $\pm$ 0.9137j),(0.3007 $\pm$ 0.6449j),0.551]; Zeros = [-1,-1,-1,-1,1]
(c) Poles = [(-0.5188 $\pm$ 0.7191j), $\pm$0.4363,(0.5188 $\pm$ 0.7191j)]; Zeros = [$\pm$j, $\pm$j, $\pm$j]
(d) Poles = [($\pm$0.5774j),0]; Zeros = [1,1,1]

16. Write a script that implements a numerical approximation of Eq. (6.17) according to the following function specification:

```
function LVxNumInvZxform(b,a,M,ContourRad,nVals)
%
% Intended to perform the inverse z-transform for transforms whose
% ROC lies outside a circle of radius equal to the largest pole
% magnitude of the z-transform b is the numerator coefficient vector;
% a is the denominator coefficient vector
% M is the number of samples to use along the contour
% ContourRad is the radius of the circle used as the contour
% nVals is vector of time domain indices to compute, i.e., x[nVals]
% will be computed via the inverse z-transform
% Sample Call:
% LVxNumInvZxform([1],[1,-0.9],5000,1,[0:1:20])
% LVxNumInvZxform([0 1],[1,-2,1],5000,1.1,[0:1:20])
% LVxNumInvZxform([1],[1,-1.2,0.81],60000,0.91,[0:1:100])
% LVxNumInvZxform([1],[1,-1.2,0.81],5000,1,[0:1:100])
```

A simple way to perform numerical integration of, for example, the function $y = x^2$, is to construct small boxes of width Δx and height $f(x)$ under the curve and take the limit of the sum of the areas of the boxes as $\Delta x \rightarrow 0$. Figure 6.36, plot (a) shows this approach. An approach that uses trapezoids instead of rectangles, and which converges more quickly, is shown in plot (b). It is easy to show that the area of each trapezoid is Δx multiplied by

$$0.5(f(x[n]) + f(x[n+1]))$$

where

$$\Delta x = x[n+1] - x[n]$$

This latter method, the trapezoidal, is the one you should use for your script as the faster convergence greatly reduces the number of sample points needed for acceptable results.

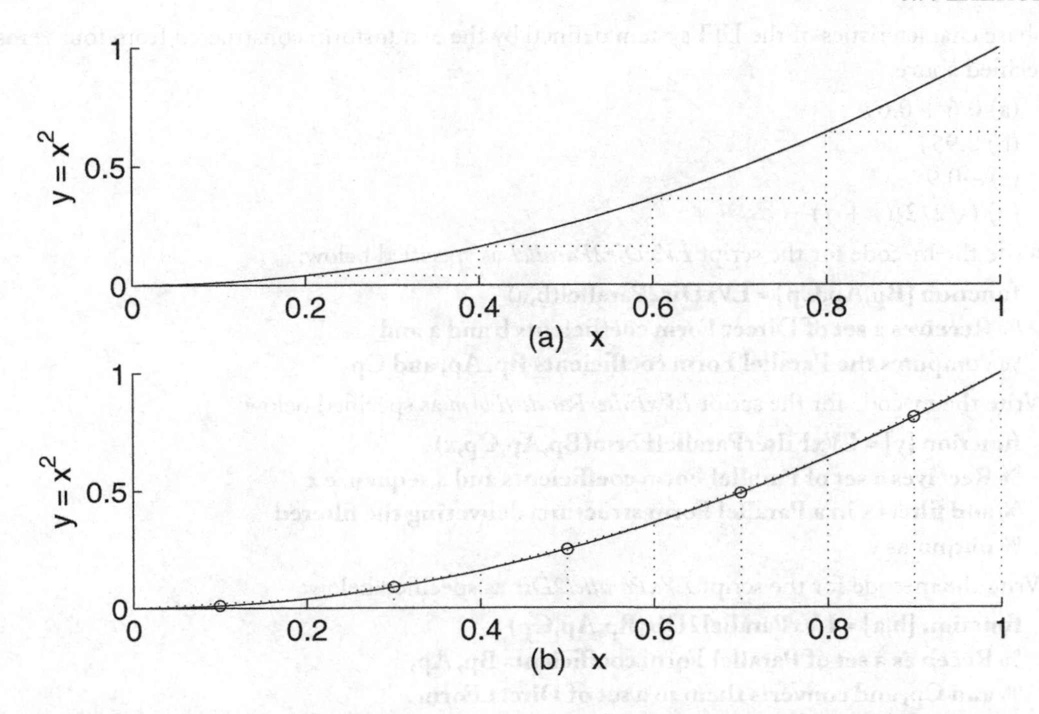

Figure 6.36: (a) Rectangular approach to numerically estimating the integral of the function $y = x^2$ over the interval 0:1; (b) Trapezoidal approach to numerically estimating the integral of the function $y = x^2$ over the interval 0:1, with the average function value on the interval marked at the mid-point of each interval with a circle.

17. Write a script that will receive the location of a zero of the form $M\exp(j\theta)$, where M and θ are real numbers, construct a z-transform consisting of the following four zeros: $M\exp(j\theta)$, $M\exp(-j\theta)$, $(1/M)\exp(j\theta)$, and $(1/M)\exp(-j\theta)$, two different ways.

First, using paper and pencil, derive an algebraic expression in M and θ for the z-transform coefficients, and second, construct the z-transform coefficients using m-code to generate the zero locations and from them generate the polynomial coefficients (using, for example, the function *poly*) for the z-transform.

Next, using the values of z given below, compute the z-transform coefficients using both methods, i.e., using your algebraic expression to compute the coefficients, and second, using the m-code method to obtain the zeros and compute the z-transform coefficients using *poly*. The two results should agree.

Next, evaluate the z-transform along the unit circle at 256 points from $-\pi$ to π, and plot the magnitude and phase of the z-transform. Test your script with the following zeros, and comment on the phase characteristics of the LTI system defined by the z-transform constructed from four zeros as specified above.

(a) $0.6 + 0.6j$

(b) $0.95j$

(c) -0.9

(d) $(\sqrt{2}/2)(1 + j)$

18. Write the m-code for the script *LVxDir2Parallel* as specified below:

function [Bp,Ap,Cp] = LVxDir2Parallel(b,a)
% Receives a set of Direct Form coefficients b and a and
% computes the Parallel Form coefficients Bp, Ap, and Cp

19. Write the m-code for the script *LVxFilterParallelForm* as specified below:

function [y] = LVxFilterParallelForm(Bp,Ap,Cp,x)
% Receives a set of Parallel Form coefficients and a sequence x
% and filters x in a Parallel Form structure, delivering the filtered
% output as y

20. Write the m-code for the script *LVxParallel2Dir* as specified below:

function [b,a] = LVxParallel2Dir(Bp,Ap,Cp)
% Receives a set of Parallel Form coefficients Bp, Ap,
% and Cp, and converts them to a set of Direct Form
% coefficients [b,a]

21. For the following LTI systems, sketch and label with coefficients the Direct Form I, Direct Form II, Cascade (using Direct Form II sections), Parallel (using Direct Form II sections), and Lattice implementations.

(a) $y[n] = \sum_{m=0}^{2}(1/3)^m + \sum_{n=1}^{3}(1/2)^n$

(b) **b = [0.0219, 0.1097, 0.2194, 0.2194, 0.1097, 0.0219], a = [1, -0.9853, 0.9738, -0.3864, 0.1112, -0.0113]**

(c) **b = [0.0109, 0.0435, 0.0653, 0.0435, 0.0109], a = [1, -2.1861, 2.3734, -1.3301, 0.3273]**

22. Convert the following sets of Cascade Form coefficients into Direct Form coefficients, and then into Parallel Form coefficients. Sketch the Cascade, Direct, and Parallel filter signal flow diagrams. To verify equivalence, filter a linear chirp using all three filter forms, and plot and compare the results from each filtering operation.

(a) **Bc = [1, 2.002, 1.002; 1, 1.9992, 0.9992; 0, 1, 0.9987]; Ac = [1, 0.5884, 0.8115; 1, -0.0714, 0.4045; 0,1, -0.3345]; Gain = 0.0665**

(b) **Bc = [1,-2.0001, 1.0001; 1,-1.9999, 0.9999; 1,2,1; 1,2,1]; Ac = [1, 0.5973, 0.9023; 1, -0.5973, 0.9023; 1, 0.2410, 0.7621; 1, -0.2410, 0.7621]; Gain = 0.0025**

23. Convert the following Parallel Form filters into equivalent Lattice-Ladder filters, filter a linear chirp using both filters, and plot and compare the results from each filtering operation to verify equivalence.

(a) **Bp = [-0.0263,-0.0846;-0.0263,0.0846;-0.6919,0.4470; -0.6919,-0.4470]; Ap = [1,-0.5661,0.9068; 1,0.5661,0.9068;1,-0.5713,0.5035; 1,0.5713,0.5035]; Cp = 1.8148**

(b) **Bp = [-0.3217,-0.0475; -0.3217,0.0475; -0.7265,0]; Ap = [1,-0.4764,0.7387; 1,0.4764,0.7387; 1,0, 0.5095]; Cp = 1.8975**

24. Assume that an impulse response that is symmetrical about its midpoint has a linear phase characteristic. Prove that the following LTI systems have linear phase characteristics:

(a) A system having only single real zeros of magnitude 1.0 (i.e., zeros equal only to 1 or -1).

(b) A system having pairs of zeros that are real only and in reciprocal-magnitude pairs, i.e., of the form $[M, (1/M)]$ where M is real.

(c) A system having pairs of zeros that are complex conjugate pairs having magnitude 1.0.

(d) A system having quads of zeros, i.e., a set of four zeros of the form

$$[M\angle\theta, M\angle(-\theta), (1/M)\angle\theta, (1/M)\angle(-\theta)]$$

25. This exercise will copy a VI and modify m-code in its MathScript node to change its behavior for comparison to the original version. The VI *DemoDragZerosZxformVI* constructs an FIR from, in general, a quad of four zeros when in quad mode. However, it uses only one or two zeros when certain conditions occur such that only the one zero or pair of zeros are necessary to create a linear phase FIR (see discussion in this chapter). In this exercise, we'll copy the VI to a new file name and modify it to use four zeros in all situations to see what the difference is in transfer functions for the same main zero location.

Open the VI *DemoDragZerosZxformVI* and use *Save As* from the *File* menu to copy the original VI to a new file name. In the new VI, go to the Window menu and select *Show Block Diagram* (or Press Control+E). You should see a large box in the Block Diagram filled with m-code; modify this code appropriately to use all four zeros for all situations, save the new VI with your m-code modifications, and run the new VI in quad mode (four zeros). If you have done your work properly, all four zeros will be used for all situations. When this is the case, the impulse response will always be five samples in length. Compare impulse response length and magnitude and phase responses to those obtained with the original VI for the following values of the main zero:

(a) z = 1.0

(b) z = -1.0

(c) z = 0.707 + 0.707j

CHAPTER 7

The DFT

7.1 OVERVIEW

In previous chapters, we have studied the DTFT and the z-transform, both of which have important places in discrete signal processing theory, but neither is a numerically computable transform. Signal processing as applied in industry and commerce, however, for such applications as audio and video compression algorithms, etc., relies on computable transforms. Unlike the DTFT and z-transform, however, the Discrete Fourier Transform is a numerically computable transform. It is a reversible frequency transform that evaluates the spectrum of a finite sequence at a finite number of frequencies. It is perhaps the best known and most widely used transform in digital signal processing. Many papers and books have been written about it. There are other numerically computable frequency transforms, such as the Discrete Cosine Transform (DCT), the Modified Discrete Cosine Transform (MDCT), the Discrete Sine Transform (DST), and the Discrete Hartley Transform (DHT), but the DFT is the most often discussed and used frequency transform. Its fast implementation, the FFT (actually an entire family of algorithms) serves as the computational basis not only for the DFT per se, but for other transforms that are related to the DFT but do not have their own fast implementations. While the end goal of much DFT use may be spectral analysis, there is a growing body of applications that use the DFT-IDFT (and relatives of it, such as the Discrete Cosine Transform) for data compression.

In this chapter we cover the DFS and its basis, the reconstruction of a sequence from samples of its z-transform, followed by the definition(s), basic properties, and computation of the DFT. We then delve into a mix of practical and theoretical matters, including the DFTs of common signals, determination of Frequency Resolution/Binwidth, the FFT or Fast Fourier Transform, the Goertzel Algorithm (a simple recursive method to compute a single DFT bin which finds utility in such things as DTMF detection and the like), implementation of linear convolution using the DFT (a method permitting efficient convolution of large blocks of data), DFT Leakage (an issue in spectral analysis and signal detection), computation of the DTFT via the DFT, and the Inverse DFT (IDFT), which we compute directly, by matrix methods, or through ingenious use of the DFT.

7.2 SOFTWARE FOR USE WITH THIS BOOK

The software files needed for use with this book (consisting of m-code (.m) files, VI files (.vi), and related support files) are available for download from the following website:

http://www.morganclaypool.com/page/isen

The entire software package should be stored in a single folder on the user's computer, and the full file name of the folder must be placed on the MATLAB or LabVIEW search path in accordance with the instructions provided by the respective software vendor (in case you have encountered this notice before, which is repeated for convenience in each chapter of the book, the software download only needs to be done once, as files for the entire book are all contained in the one downloadable folder).

See Appendix A for more information.

7.3 DISCRETE FOURIER SERIES

From our earlier introductory discussion, several chapters ago, recall that the DFS coefficients $\widetilde{X}[k]$ of a periodic sequence $x[n]$ ($-\infty < n < \infty$) are

$$DFS(x[n]) = \widetilde{X}[k] = \sum_{n=0}^{N-1} \widetilde{x}[n]e^{-j2\pi nk/N} \tag{7.1}$$

where k is an integer and $\widetilde{x}[n]$ is one period of the periodic sequence $x[n]$. Typical ranges for k are: 0:1:N-1, or -N/2+1:1:N/2 for even length sequences, or -(N-1)/2:1:(N-1)/2 for odd length sequences. The sequence $x[n]$ can be reconstructed from the DFS coefficients as follows:

$$x[n] = \frac{1}{N}\sum_{k=0}^{N-1} \widetilde{X}[k]e^{j2\pi nk/N} \tag{7.2}$$

Note that not only is $x[n]$ periodic over n, but the DFS coefficients $\widetilde{X}[k]$ are periodic over k.

Example 7.1. One period of a certain periodic sequence is $[(2 + j),-1, j, 3]$; compute the DFS coefficients, and then reconstruct one period of the sequence $x[n]$ using the coefficients.

A straightforward script using a single loop with n as a vector would be

```
x = [(2+j),-1,j,3]; N = length(x);
n = 0:1:N-1; W = exp(-j*2*pi/N);
for k = 0:1:N-1, DFS(k +1) = sum(x.*(W.^(n*k))); end
```

And reconstruction can be performed with this code:

```
N = length(DFS); k = 0:1:N-1; W = exp(j*2*pi/N);
for n = 0:1:N-1, x(n +1) = sum(DFS.*(W.^(n*k))); end
x = x/N
```

More efficient (vectorized) code to perform the analysis and synthesis, respectively, is

```
x = [(2+j),-1,j,3]; N = length(x); n = 0:1:N-1;
k = n; DFS = x*(exp(-j*2*pi*(n'*k)/N))

N = length(DFS); n = 0:1:N-1;
```

```
k = n; x = DFS*(exp(j*2*pi*(n'*k)/N))/N
```

Example 7.2. Evaluate the DFS of the sequence $[1, 0, 1]$ and plot and discuss the result, contrasting it to the DTFT of the sequence.

The following code will work for any sequence x:

```
x = [1,0,1]; N = length(x); n = 0:1:N-1; W = exp(-j*2*pi/N);
k = 0:1:N-1; DFS = x*(W.^(n'*k)),
figure(9); stem(n,abs(DFS))
xlabel('Normalized Frequency'); ylabel('Magnitude')
```

Observing the magnitude plot, it is difficult to ascertain the true frequency content of the sequence with only three frequency points. We can determine, however, where the three DFS frequencies lie on the DTFT plot. The three DFS frequencies are $2\pi k/3$ where k = 0,1, and 2. To get both the DTFT and the three DFS samples onto the same plot, the following code evaluates the DTFT from 0 to 2π radians, allowing all three DFS frequencies to be plotted:

```
x = [1 0 1]; N = length(x); n = 0:1:N-1; k = n;
W = exp(-j*2*pi/N); DFS = x*(W.^(n'*k));
w = 0:0.01:2*pi; DTFT = 1+exp(-j*2*w);
figure(8); clf; hold on; plot(w/(pi),abs(DTFT));
stem(2*[0:1:2]/3,abs(DFS),'ro');
xlabel('Norm Freq (Units of pi)'); ylabel('Magnitude')
```

Figure 7.1 illustrates the essential results from the code above.

From the above it is possible to see that the DFS coefficients of a sequence $x[n]$ (computed using one period $\widetilde{x}[n]$ of $x[n]$) are essentially equally spaced samples of the DTFT of $\widetilde{x}[n]$. Assuming that the ROC of the z-transform of $\widetilde{x}[n]$ includes the unit circle, then it is also possible to say that the DFS coefficients are essentially equally spaced samples of the z-transform of $\widetilde{x}[n]$ evaluated along the unit circle. Prior to proceeding to the DFT, we briefly investigate the effect of sampling the z-transform.

7.4 SAMPLING IN THE Z-DOMAIN

The z-transform of an absolutely summable sequence $x[n]$, which may be finite or infinite in extent, is defined as

$$X(z) = \sum_{n=-\infty}^{\infty} x[n]z^{-n}$$

Assuming that the ROC includes the unit circle, and sampling $X(z)$ at highfrequencies whose radian arguments are $2\pi k/M$ where M is an arbitrary integer and $k = 0, \pm 1, \pm 2 ...$ we define

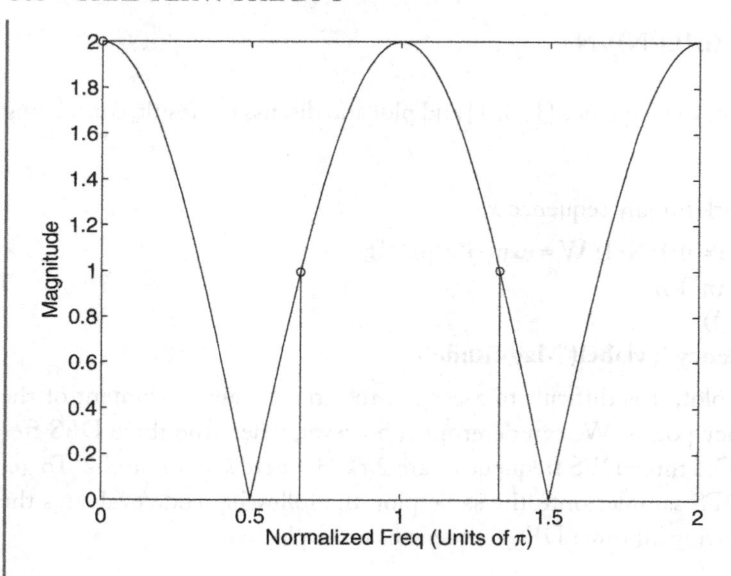

Figure 7.1: A dense grid of samples of the DTFT of the sequence [1 0 1], with a stem plot of the three DFS samples.

$$\widetilde{X}[k] = X(z)|_{z=\exp(j2\pi k/M)} = \sum_{n=-\infty}^{\infty} x[n]e^{-j2\pi kn/M} \tag{7.3}$$

which is periodic over M, and has the form of a Discrete Fourier Series. Computing the inverse DFS of $\widetilde{X}[k]$ will result in a periodic sequence $\widetilde{x}[n]$ where

$$\widetilde{x}[n] = \sum_{p=-\infty}^{\infty} x[n - pM] \tag{7.4}$$

When the number of samples M taken of $X(z)$ is smaller than N, the original length of $x[n]$, which may be infinite or finite, aliasing occurs. If M samples of $X(z)$ are obtained, the reconstructed sequence $\widetilde{x}[n]$ consists of a superposition of copies of $x[n]$ offset in time by multiples of M samples according to Eq. (7.4).

Example 7.3. Write a script that can receive a set of coefficients $[b, a]$ representing the z-transform (the ROC of which includes the unit circle) of a time domain sequence $x[n]$, a desired number *NumzSamps* of samples of the z-transform, and a desired number of time domain samples $nVals$ to reconstruct, and reconstruct $x[n]$ to demonstrate the periodicity of the reconstruction. Use $x[n] = u[n] - u[n-4]$ and test the reconstruction for values of *NumzSamps* = 8, 5, and 3, and plot the results. Compute several periods of $\widetilde{x}[n]$ as given by Eq. (7.4) and compare results to those obtained using the script below.

The script is straightforward, and the results are shown in Fig. 7.2.

```
function LV_TDReconViaSampZXform(b,a,NumzSamps,n)
kz = 0:1:NumzSamps-1;
z = exp(j*2*pi*(kz/NumzSamps));
Nm = 0; Dm = 0;
for nn = 0:-1:-length(b)+1
 Nm = Nm + (b(-nn+1))*(z.^nn); end
for d = 0:-1:-length(a)+1
 Dm = Dm + (a(-d+1))*(z.^d); end
zSamps = Nm./Dm;
figure(125);
clf; N = length(zSamps);
W = exp(j*2*pi/N); k = 0:1:N-1;
IDFS = real((W.^((n')*k))*conj(zSamps')/N);
stem(n,IDFS)
% Test calls:
% LV_TDReconViaSampZXform([ones(1,4)],[1],8,[-15:1:15])
% LV_TDReconViaSampZXform([ones(1,4)],[1],5,[-15:1:15])
% LV_TDReconViaSampZXform([ones(1,4)],[1],3,[-15:1:15])
```

To use Eq. (7.4), we can sum a few offset versions of $x[n]$ to obtain several periods of the periodic sequence. For the case of M = 3, we can sum $x[n]$, $x[n-3]$, $x[n+3]$, $x[n-6]$, $x[n+6]$, etc. This can be done by repetitively using the script $[y, nOut] = LVAddSeqs$ which was introduced in Part I of the book. The results, shown in Fig. 7.3, are valid from n = -3 to n = +6

```
x = [1,1,1,1]; n = [0,1,2,3];
[y, nOut] = LVAddSeqs(x,n,x,n+3);
[y, nOut] = LVAddSeqs(y,nOut,x,n-3);
[y, nOut] = LVAddSeqs(y,nOut,x,n+6);
[y, nOut] = LVAddSeqs(y,nOut,x,n-6);
figure(7); stem(nOut,y)
```

We have just seen that a reconstruction of $x[n]$ using samples of $X(z)$ results in a periodic version of $x[n]$. It is also possible, under certain conditions given below, to completely reconstruct $X(z)$ from $x[n]$, which implies that the original $x[n]$ (i.e., not a periodic version of it) can be reconstructed by using the inverse z-transform on the reconstructed $X(z)$.

The conditions under which $X(z)$ can be completely reconstructed are as follows: if a time domain sequence $x[n]$ is of finite extent, i.e., $x[n]$ is identically zero for $n < 0$ and $n \geq N$, and the unit circle is in the ROC, then $\widetilde{X}[k]$, which consists of N samples of $X(z)$ along the unit circle,

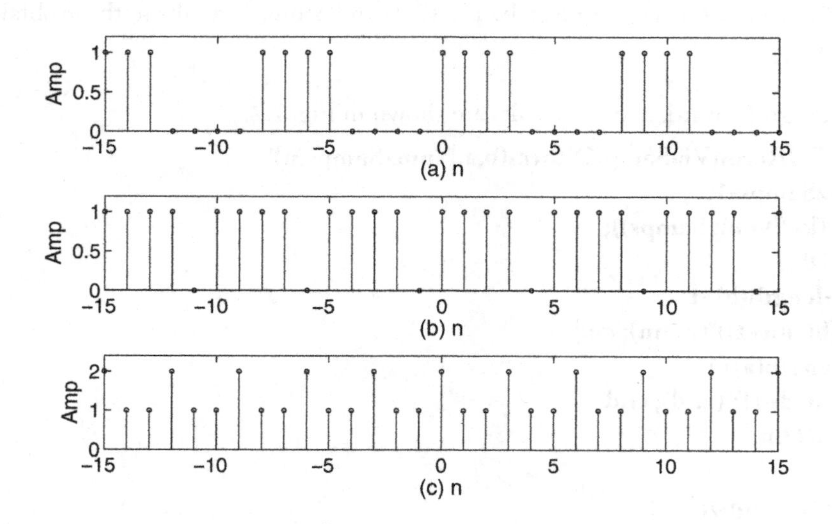

Figure 7.2: (a) Periodic reconstruction of $x[n]$ (= $u[n]$-$u[n-4]$) from eight samples of $X(z)$; (b) Periodic reconstruction from five samples of $X(z)$; (c) Periodic reconstruction from three samples of $X(z)$.

located at frequencies specified as $2\pi k/N$,where k = 0:1:N -1, can determine $X(z)$ for all z according to the following equation:

$$X(z) = \frac{1 - z^{-N}}{N} \sum_{k=0}^{N-1} \frac{\widetilde{X}[k]}{1 - e^{j2\pi k/N} z^{-1}} \tag{7.5}$$

Example 7.4. For the sequence [1, 1, 1, 1], compute and display 1000 samples of the magnitude of the z-transform directly, then compute and display 1000 samples of the magnitude of the z-transform using Eq. (7.5).

In the m-code below, we first compute 1000 points of the z-transform by writing the z-transform of the sequence and evaluating; we then extract four properly located samples from the 1000 samples and use Eq. (7.5) to reconstruct the entire 1000 samples of the z-transform. The results are shown in Fig. 7.4.

```
inc = 1/999; xvec = inc/2:inc:1; zp = 2*pi*xvec; z = exp(j*zp);
ZX1 = 1 + z.^(-1) + z.^(-2) + z.^(-3);
figure(33); subplot(211);
plot(xvec,abs(ZX1)); axis([0,1,0,5])
k = 0:1:3; Fndx = fix((k/4)*1000) + 1;
```

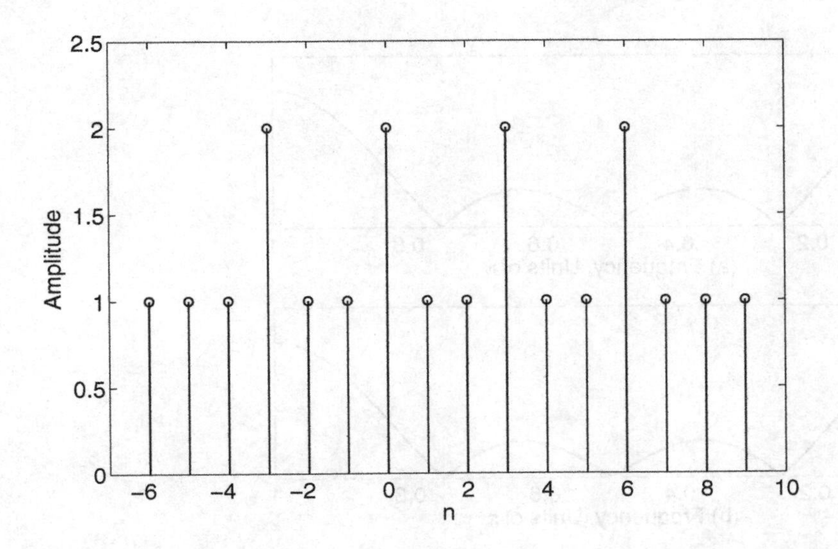

Figure 7.3: A partial reconstruction (using Eq. (7.4)) of the infinite-extent periodic sequence $\widetilde{x}[n]$ that results from reconstruction of $x[n]$ from samples of the z-transform $X(z)$, where $x[n]$ = [1,1,1,1].

```
Xtil = ZX1(Fndx); S = 0;
for k = 0:1:3
S = S + Xtil(k+1)./(1-(exp(j*2*pi*k/4))*(z.^(-1)));
end
ZX2 = S.*((1-z.^(-4))/4);
subplot(212); plot(xvec,abs(ZX2))
axis([0,1,0,5])
```

The script

$$LVxZxformFromSamps(b, a, NumzSamps, M, nVals)$$

will, for a set of coefficients $[b, a]$, compute samples of $X(z)$, then reconstruct $X(z)$, then reconstruct $x[n]$ using contour integration, and, for contrast, reconstruct a periodic version of $x[n]$ using the inverse DFS transform on the samples of $X(z)$. The call

LVxZxformFromSamps([1,1,1,1],[1],6,5000,30)

results in Fig. 7.5.

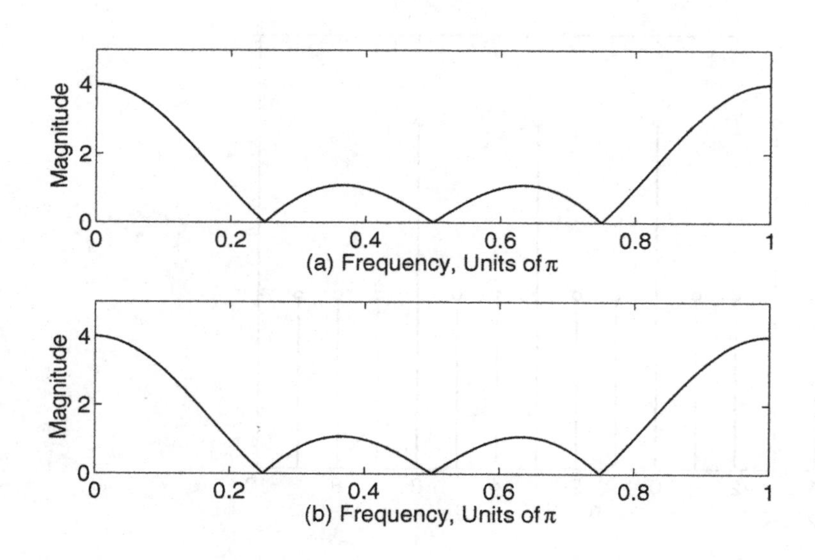

Figure 7.4: (a) 1000 samples of the z-transform $X(z)$ of the sequence [1,1,1,1], obtained by direct evaluation of the z-transform $X(z) = 1 + z^{-1} + z^{-2} + z^{-3}$;(b) 1000 samples of the z-transform of the same sequence, reconstructed from four samples of $X(z)$ using Eq. (7.5).

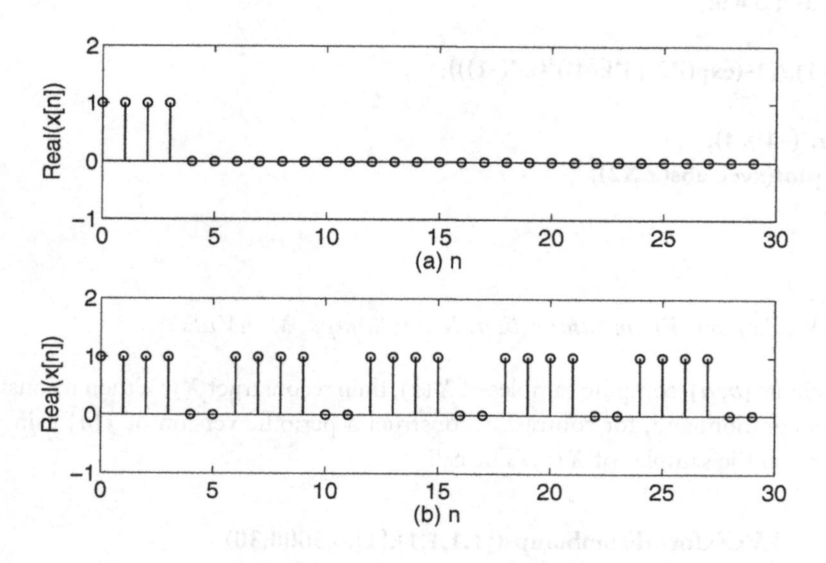

Figure 7.5: (a) A reconstruction of $x[n]$ = [1,1,1,1] from a reconstruction of $X(z)$ based on six samples of $X(z)$; (b) A reconstruction of a periodic version of $x[n]$ directly from the six samples of $X(z)$.

7.5 FROM DFS TO DFT

In the DFS, we have a computable transform; a periodic sequence $x[n]$ of periodicity N can be reconstructed from N (correlation) coefficients $X[k]$ obtained from N correlations performed between N harmonically related basis functions and a single period of $x[n]$, designated $\widetilde{x}[n]$.

In many cases, however, a sequence $y_1[n]$ $(0 \leq n \leq N-1)$ to be analyzed is not periodic, but rather aperiodic and finite. The DFT is defined for such sequences by stipulating that $y_1[n]$ is hypothetically a single period $\widetilde{y}[n]$ of an infinite length periodic sequence $y[n]$. By entertaining this line of argument, it is possible to decompose a finite sequence of length N into N coefficients, and reconstruct it using essentially the same basis functions for analysis and synthesis used for the DFS.

Thus the DFS analysis formula

$$DFS(x[n]) = \widetilde{X}[k] = \sum_{n=0}^{N-1} \widetilde{x}[n]e^{-j2\pi nk/N}$$

can be rewritten using notation for a finite sequence as

$$DFT(x[n]) = X[k] = \sum_{n=0}^{N-1} x[n]e^{-j2\pi nk/N}$$

and the DFS synthesis formula

$$x[n] = \frac{1}{N}\sum_{k=0}^{N-1} \widetilde{X}[k]e^{j2\pi nk/N}$$

can be rewritten as

$$x[n] = \frac{1}{N}\sum_{k=0}^{N-1} X[k]e^{j2\pi nk/N}$$

7.6 DFT-IDFT PAIR

7.6.1 DEFINITION-FORWARD TRANSFORM (TIME TO FREQUENCY)

The complex Discrete Fourier Transform (DFT) may be defined as:

$$X[k] = \sum_{n=0}^{N-1} x[n](\cos[2\pi nk/N] - j\sin[2\pi nk/N]) \tag{7.6}$$

or, using the complex exponential form,

$$X[k] = \sum_{n=0}^{N-1} x[n]e^{-j2\pi nk/N} \tag{7.7}$$

where n is the sample index (running from 0 to $N-1$) of an N point sequence $x[n]$. Harmonic number (also called mode, frequency, or bin) is indexed by k and runs from 0 to $N-1$.

Using Eq. (7.7) results in computation by the **Direct Method**, as opposed to much more efficient algorithms lumped under the umbrella term **FFT (Fast Fourier Transform)**.

7.6.2 DEFINITION-INVERSE TRANSFORM (FREQUENCY TO TIME)

If the DFT is defined as in Eq. (7.6), then the Inverse **Discrete Fourier Transform (IDFT)** is defined as:

$$x[n] = \frac{1}{N}\sum_{k=0}^{N-1} X[k](\cos[2\pi kn/N] + j\sin[2\pi nk/N]) \tag{7.8}$$

for $n = 0$ to $N-1$, or, in complex exponential notation

$$x[n] = \frac{1}{N}\sum_{k=0}^{N-1} X[k]e^{j2\pi nk/N} \tag{7.9}$$

7.6.3 MAGNITUDE AND PHASE

Since DFT bins are, in general, complex numbers, the magnitude and phase are computed as for any complex number, i.e.,

$$|X[k]| = \sqrt{\mathrm{Re}(X[k])^2 + \mathrm{Im}(X[k])^2}$$

and

$$\angle X[k] = \mathrm{arcTan}(\mathrm{Im}(X[k])/\mathrm{Re}(X[k]))$$

7.6.4 N, SCALING CONSTANT, AND DFT VARIANTS

The definitions above are independent of whether N is even or odd. An alternate definition of the DFT for N even makes the ranges of n and k run from $-N/2+1$ to $N/2$. An alternate definition of the DFT for N odd has the ranges of n and k running from $-(N-1)/2$ to $(N-1)/2$.

Note in Eqs. (7.8) and (7.9) above that the scaling constant $1/N$ was applied to the Inverse DFT, and no such constant was applied to the DFT. In some DFT definitions, the scaling constant $1/N$ is applied to the DFT rather than to the Inverse DFT. It is also possible to apply a scaling constant of $1/\sqrt{N}$ to both the DFT and IDFT.

There are yet other ways of defining the DFT; they all result in essentially the same information, however. The variations consist of different ways of defining the range of k or n, and whether or not the DFT or the IDFT has a scaling coefficient (such as $1/N$) applied to it. Reference [1], which is devoted entirely to the DFT, discusses a number of these variations.

7.7 MATHSCRIPT IMPLEMENTATION

Since MathScript vectors cannot use index values that are equal to or less than zero, MathScript implements the complex DFT in this way:

$$X[k] = \sum_{n=1}^{N} x[n]e^{-j2\pi(k-1)(n-1)/N} \quad (7.10)$$

where both n and k run from 1 to N. Note that the MathScript version does not scale the DFT by $1/N$–instead, the factor $1/N$ is applied in the function $ifft$, which is MathScript's inverse DFT function. When using the standard DFT definition Eq. (7.7), since the factor $1/N$ is applied in the DFT itself, it is not applied when using the IDFT.

MathScript can be used to compute the DFT of a sequence by using the function *fft*.

Example 7.5. Compute and display the magnitude and phase of the DFT of the signal sequence $[ones(1, 64)]$ using MathScript.

We make the call

```
s = [ones(1,64)]; y = fft(s);
```

and follow it with

```
subplot(2,1,1); stem(abs(y)); subplot(2,1,2); stem(unwrap(angle(y)))
```

7.8 A FEW DFT PROPERTIES

Before proceeding, it should be noted that a number of DFT properties pertain to sequences or their DFTs that have been circularly folded. A circular folding of a sequence $x[n]$ of length N is defined as

$$x[(-n)]_N = \begin{cases} x[0] & n = 0 \\ x[N-n] & 1 \le n \le N-1 \end{cases}$$

1. **Linearity**
The DFT is a linear operator, i.e., the following is true:

$$DFT(ax_1[n] + bx_2[n]) = aDFT(x_1[n]) + bDFT(x_2[n])$$

2. **Circular folding**
If the DFT of a sequence $x[n]$ is $X[k]$, the DFT of a circularly folded version of $x[n]$ is a circularly folded version of the DFT of the sequence, i.e.,

$$DFT(x[(-n)]_N) = X[(-k)]_N$$

The following code demonstrates this property.

```
n = 0:1:7; x = [0:1:7]; xret = x(mod(-n,8)+1);
subplot(3,2,1);stem(n,x); y = fft(x); yret = fft(xret);
subplot(3,2,2);stem(n,xret); subplot(3,2,3); stem(n,real(y));
subplot(3,2,4);stem(n,real(yret));subplot(3,2,5); stem(n,imag(y));
subplot(3,2,6); stem(n,imag(yret))
```

3. **Shift (circular) in Time Domain**

Since the DFT is periodic in n, shifting the sequence some number of samples m to the right can be equivalently achieved by a circular shift, and

$$DFT(x[n-m]_N) = X(k)e^{-j2\pi mk/N}$$

The following code verifies this for a short sequence. The DFT *prxshfft* is constructed according to the formula from xft, the DFT of the original sequence x.

```
x=[1 2 3 4]; xsh1 = [4 1 2 3]; xft = fft(x), xshfft = fft(xsh1),
k=0:1:length(x)-1; prxshfft = xft.*(exp(-j*2*pi*1*k/4))
```

4. **Circular Convolution of Two Time Domain Sequences**

If the circular convolution of two sequences $x_1[n]$ and $x_2[n]$ is defined as

$$x_1[n] \circledast x_2[n] = \sum_{m=0}^{N-1} x_1[m]x_2[(n-m)]_N$$

for $0 \leq n \leq N-1$, then

$$DFT(x_1[n] \circledast x_2[n]) = X_1[k]X_2[k]$$

This property and its usefulness to perform ordinary (or linear) convolution will be discussed in detail later in the chapter.

5. **Multiplication of Time Domain Sequences**

The DFT of the product of two time domain sequences is $1/N$ times the circular convolution of the DFTs of each:

$$DFT(x_1[n]x_2[n]) = \frac{1}{N}(X_1[k] \circledast X_2[k])$$

6. **Parseval's Relation**

The energy of a sequence is the sum of the squares of the absolute values of the samples, and this has an equivalent computation using DFT coefficients:

$$\sum_{n=0}^{N-1} |x[n]|^2 = \frac{1}{N} \sum_{k=0}^{N-1} |X[k]|^2$$

The following code verifies this property using a random sequence:

```
N = 9; x = randn(1,N); td = sum(abs(x).^2),
fd = (1/N)*sum(abs(fft(x)).^2)
```

7. **Conjugate Symmetry**
The DFT of a real input sequence is **Conjugate-Symmetric**, which means that

$$\mathrm{Re}(X[k]) = \mathrm{Re}(X[-k])$$

and

$$\mathrm{Im}(X[k]) = -\mathrm{Im}(X[-k])$$

This property implies that (for a real sequence $x[n]$) it is only necessary to compute the DFT for a limited number of bins, namely

$$\begin{array}{ll} k = 0{:}1{:}N/2 & N\ even \\ k = 0{:}1{:}(N-1)/2 & N\ odd \end{array}$$

Example 7.6. Demonstrate conjugate symmetry for several input sequences.

We'll use the input sequence x = [16:-1:-15]. Since the length is even, there are two bins that will not have complex conjugates, namely Bins 0 and 16 in this case, or Bins 0 and $N/2$ generally. The following code computes and displays the DFT; conjugate symmetry is shown except for the aforementioned Bins 0 and 16.

```
n=[0:1:31]; x=[16:-1:-15]; y=fft(x); figure
subplot(2,1,1); stem(n,real(y)); subplot(2,1,2); stem(n,imag(y))
```

Figure 7.6 shows the result of the above code.

If we modify the sequence length to be odd, then there is no Bin $N/2$, so conjugate symmetry is shown for all bins except Bin 0. The result from running the code below is shown in Fig. 7.7.

```
n=[0:1:32]; x=[16:-1:-16]; y=fft(x); figure
subplot(2,1,1); stem(n,real(y)); subplot(2,1,2); stem(n,imag(y))
```

8. **Even/Odd TD-Real/Imaginary DFT Parts**
The circular even and odd decompositions of a time domain sequence are defined as

$$x_{cE}[n] = \begin{cases} x[0] & n = 0 \\ (x[n] + x[N-n])/2 & 1 \leq n \leq N-1 \end{cases}$$

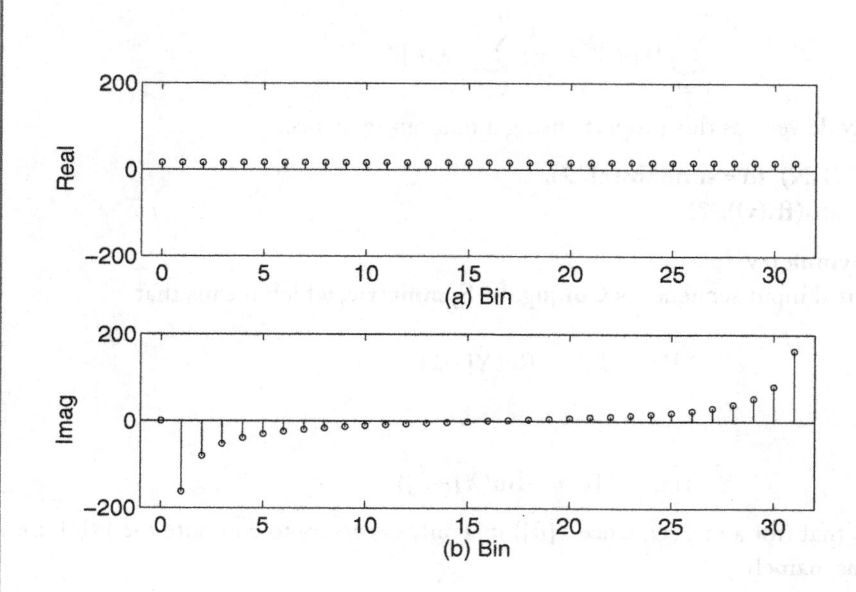

Figure 7.6: (a) Real part of DFT of the sequence $x[n]$ = [16:-1:-15]; (b) Imaginary part of same.

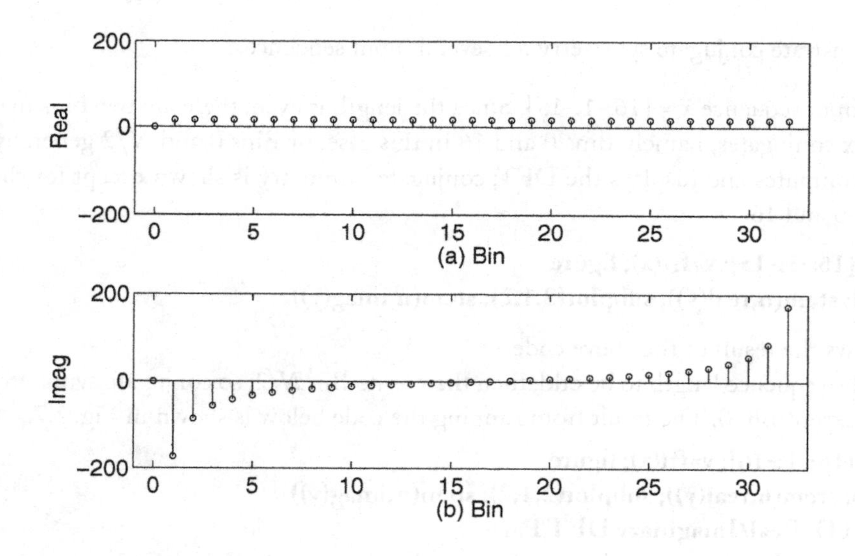

Figure 7.7: (a) Real part of DFT of the sequence $x[n]$ = [16:-1:-16]; (b) Imaginary part of same.

and

$$x_{cO}[n] = \begin{cases} 0 & n = 0 \\ (x[n] - x[N-n])/2 & 1 \leq n \leq N-1 \end{cases}$$

In such a case, it is true that

$$DFT(x_{cE}[n]) = \text{Re}(X[k]) \tag{7.11}$$

and

$$DFT(x_{cO}[n]) = \text{Im}(X[k]) \tag{7.12}$$

Example 7.7. Write a short script to verify Eqs. (7.11) and (7.12).

The following will suffice; different sequences can be substituted for x if desired.

```
x=[1:1:8]; subx = x(1,2:length(x)); xe = 0.5*(subx + fliplr(subx));
xo = 0.5*(subx - fliplr(subx)); xeven = [x(1), xe], xodd = [0, xo];
DFTevenpt = fft(xeven), ReDFTx = real(fft(x))
N=length(x); var = sum([(DFTevenpt-ReDFTx)/N].^2)
```

7.9 GENERAL CONSIDERATIONS AND OBSERVATIONS

The computation of each bin value $X[k]$ is performed by doing a CZL (correlation at the zeroth lag) between the signal sequence of length N and the complex correlator

$$\cos[2\pi kn/N] - j\sin[2\pi nk/N]$$

Computation of the DFT using (7.7) directly is possible for smaller values of N. For larger values of N, a **Fast Fourier Transform (FFT)** algorithm is usually employed. A subsequent section in this chapter discusses the FFT.

7.9.1 BIN VALUES

- If $x[n]$ is real-valued only, then $X[0]$ is real-valued only and represents the average or DC component of the sequence $x[n]$.

- If $x[n]$ is real-valued only, $X[N/2]$ is real-valued only, since the sine of the frequency $N/2$ is identically zero. For odd-length DFTs, there is no Bin $N/2$.

Example 7.8. Compute the DFT of the sequence [1, 1].

Note that for a 2-point DFT, the only frequencies are DC and 1-cycle (which is the Nyquist rate, $N/2$). Both Bins 0 and 1 are therefore real. Bin 0 = **sum(([1,1]).*cos(2*pi*(0:1)*0/2))** = 2 (not scaling the DFT by $1/N$). For Bin 1, we get Bin 1 = **sum(([1,1]).*cos(2*pi*(0:1)*1/2))** = 0 (intuitively, you can see that [1,1] is DC only, and has no one-cycle content).

Example 7.9. Demonstrate the conjugate-symmetry property of the DFT of the real input sequence

cos(2*pi*1.3*(0:1:3)/4) + sin(2*pi*(0.85)*(0:1:3)/4)

We make the call

fft([cos(2*pi*1.3*(0:1:3)/4) + sin(2*pi*(0.85)*(0:1:3)/4)])

which returns the DFT as

[1.611, (1.133 - 0.291i), 0.120, (1.133 + 0.291i)]

We note conjugate symmetry for Bins 1 and 3, where Bin 3 is an alias of Bin -1. Note that Bins 0 and 2 (i.e., $N/2$) are unique, read-only bins within a single period of k and do not have negative frequency counterparts.

7.9.2 PERIODICITY IN N AND K

For the sake of simplicity, the following discussion assumes that N is even (the DFT of an odd-length sequence differs chiefly in that there is no Bin $N/2$, and the range of k must be adjusted accordingly).

The DFT is periodic in both n and k, meaning that if n assumes, for example, the range N to $2N$ - 1, or $2N$ to $3N$ - 1, etc., the result will be the same as it would with n running from 0 to N - 1. In a similar manner, k running from N to $2N$ - 1 yields the same result as k running from 0 to N - 1. Or, using symmetrical k-indices, with k from $-N/2$ + 1 to $N/2$, k could instead run from $N/2$ + 1 to $3N/2$, or from $3N/2$ + 1 to $5N/2$, etc.

Stated formally, we have

$$x[n+N] = x[n]$$

and

$$X[k+N] = X[k]$$

These two relationships are true for all real integers n and k. This is the result of the periodicity of the complex exponential, $\exp(j2\pi nk/N)$ over 2π. Thus we have

$$\exp(j[2\pi k(n+N)/N]) = \exp(j[2\pi nk/N + 2\pi k]) = \exp(j[2\pi nk/N])$$

and likewise

$$\exp(j[2\pi n(k+N)/N]) = \exp(j[2\pi nk/N + 2\pi n]) = \exp(j[2\pi nk/N])$$

- Figure 7.8 illustrates the periodicity of the DFT. The DFT of a signal of length 32 has been computed for bins -15 to 32. A complete set of DFT bins may be had by using either k = 0 to N - 1 (standard MathScript method) or $k = -N/2 + 1$ to $N/2$. By computing 48 bins, as was done here, instead of the usual 32, both arrangements may be seen in Fig. 7.8.

- In plotting a MathScript-generated DFT, a simple command (for example) such as

```
y = ones(1,32); stem(abs(fft(y)))
```

will plot the bins with indices from 1 to N. In order to plot the bins with indices from 0 to N - 1, write

```
x = 0:1:length(y) - 1; stem(x, abs(fft(y)))
```

- For the case of k from $-N/2 + 1$ to $N/2$, $X[k]$ and $X[-k]$ show conjugate symmetry, i.e., conjugate symmetry is shown about Bin[0]. For example, at plot (a) of Fig. 7.8, Re($X[1]$) = Re($X[-1]$), showing even symmetry, and at plot (b), Im($X[k]$) = -Im($X[-k]$), showing anti-symmetry.

- For the case of k from 0 to N - 1, the negative bins are aliased as bins having values of k greater than $N/2$. Thus, for the example shown in Fig. 7.8, for $k > 16$ (i.e., $N/2$), note that $X[k-N]$ = $X[k]$. For k = 17, the equivalent bin is Bin[17-32] = Bin[-15], Bin[18] = Bin[-14],...and finally, Bin[31] = Bin[-1]. Note that the periodicity extends to positive and negative infinity for all real integer values of k. Bin[32], for example, is equivalent to Bin[0], as is Bin[64] (not shown), etc.

- For the asymmetrical bin arrangement, conjugate symmetry is had between Bin[N/2 - m] and Bin[N/2 + m]. For example, as seen in Fig. 7.8, Bin[15] is the complex conjugate of Bin[17], which is equivalent to Bin[17-32] = Bin[-15].

Example 7.10. Demonstrate the periodicity of k for the length-three DFT of the signal $[1, -1, 1]$.

An easy way to proceed is to make this call, letting k assume various values.

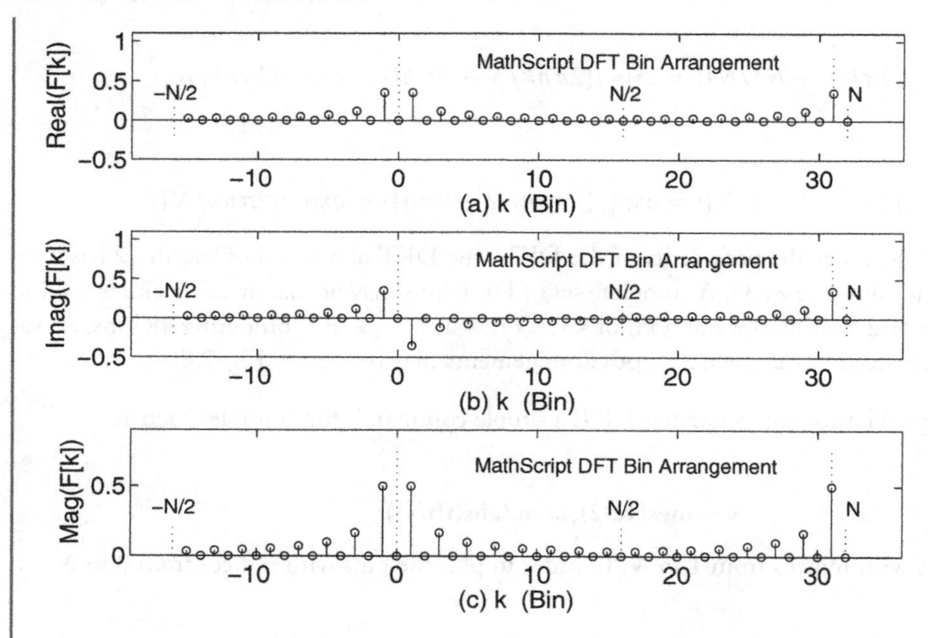

Figure 7.8: (a) The real part of the DFT of a signal, evaluated from $-N/2+1$ to N where $N = 32$; (b) The imaginary part of the DFT of the same signal; (c) The magnitude of the DFT of the signal. See the text for details of the "MathScript DFT Bin Arrangement."

```
k = 0; y = sum(exp(-j*2*pi*(0:1:2)*k/3 ).*([1,-1,1]))
```

Bin 0 is, of course, real only, and there is no Bin $N/2$ since $N/2 = 1.5$. Bin 2 should prove to be the same as Bin -1. The following answers are obtained by making the calls using the values of k shown:

k	Re($X[k]$)	Im($X[k]$)
0, 3	1	0
1, 4	1	1.7321i
2, 5	1	−1.7321i
−4, −1	1	−1.7321i

7.9.3 FREQUENCY MULTIPLICATION IN TIME DOMAIN

Suppose that the time domain sequence we have been considering is multiplied by a periodic sequence with a certain frequency k_0. Let's compare the DFT of the original (unmultiplied) sequence and the DFT of the sequence after it has been multiplied (before sampling) by some frequency k_0.

Before proceeding, let's try to see what to expect. From trigonometry, we know

$$\sin(\alpha)\sin(\beta) = \frac{1}{2}\cos(\alpha - \beta) - \frac{1}{2}\cos(\alpha + \beta)$$

or

$$\cos(\alpha)\cos(\beta) = \frac{1}{2}\cos(\alpha - \beta) + \frac{1}{2}\cos(\alpha + \beta)$$

and so on for various combinations of cosine and sine. When you multiply two frequencies in the time domain, the resultant signal contains the sum and difference of the original frequencies.

With this in mind, consider Fig. 7.9, which shows at (a) a cosine of frequency five; its DFT (real only, since the time domain signal is a cosine) at (c) shows spikes at frequencies 5 and -5 (Bin 31 is equivalent to Bin -1, Bin 30 is equivalent to Bin -2, etc.).

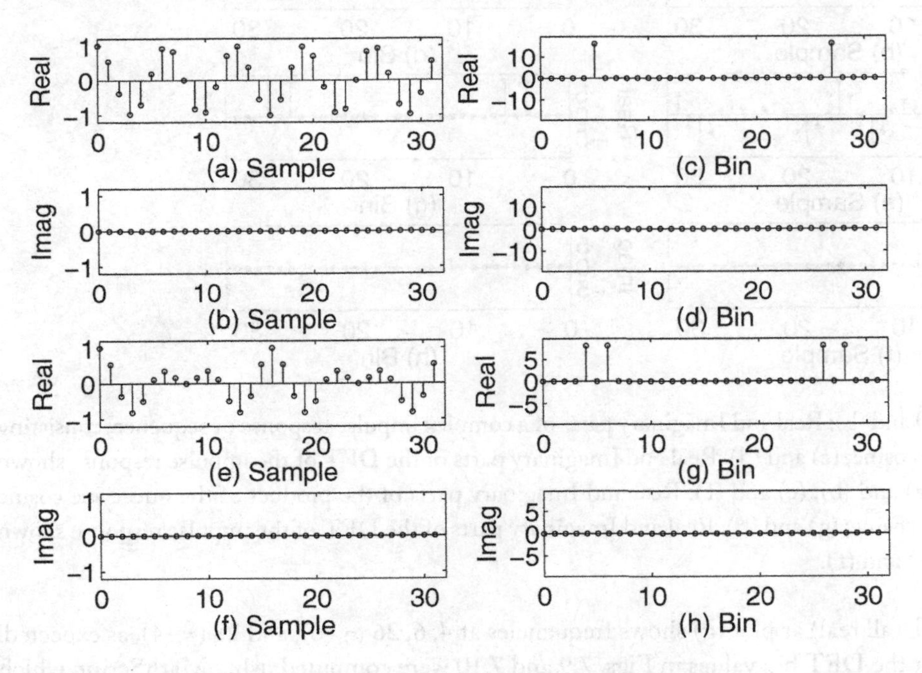

Figure 7.9: (a) and (b) Real and Imaginary parts of a complex impulse response or sequence, consisting of five cycles of a cosine; (c) and (d) Real and Imaginary parts of the DFT of the impulse response shown collectively by (a) and (b); (e) and (f) Real and Imaginary parts of the product of a five-cycle cosine with a one-cycle cosine; (g) and (h) Real and Imaginary parts of the DFT of the impulse response shown collectively by (e) and (f).

At (e), the cosine of five cycles has been multiplied by a cosine of one cycle. Thus the expected frequencies in the new time domain signal would be 5 ± 1, and indeed, we see spikes in the DFT (plot (g)) at frequencies 4, 6, and -4 and -6 (the two latter are in their aliased positions).

In Fig. 7.10, we initially show the time domain signal as a cosine of one cycle at (a); the DFT (at (c)) shows frequencies of 1 and -1, as expected. At (e), the time domain signal is now the product of a cosine of one cycle and a cosine having five cycles. We would thus expect frequencies of $1 \pm 5 = 6; -4$.

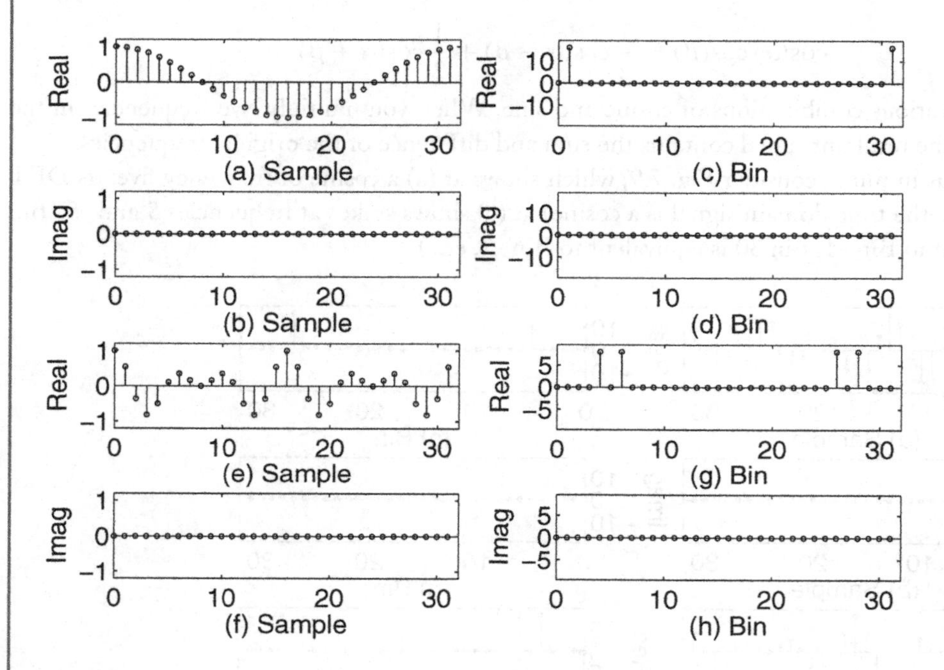

Figure 7.10: (a) and (b): Real and Imaginary parts of a complex impulse response or sequence, consisting of one cycle of a cosine; (c) and (d): Real and Imaginary parts of the DFT of the impulse response shown collectively by (a) and (b); (e) and (f): Real and Imaginary parts of the product of the one-cycle cosine with a five-cycle cosine; (g) and (h): Real and Imaginary parts of the DFT of the impulse response shown collectively by (e) and (f).

The DFT (all real) at plot (g) shows frequencies at 4, 6, 26 (= -6), and 28 (= -4), as expected.

Note that the DFT bin values in Figs. 7.9 and 7.10 were computed using MathScript, which does not scale the forward (i.e., DFT) transform by $1/N$.

7.10 COMPUTATION OF DFT VIA MATRIX

By setting up a matrix W each row of which is a DFT complex correlator, and multiplying by the signal (in column vector form), we can get the DFT.

Example 7.11. Compute the DFT of the sequence $x[n] = [1, 2, 3, 4]$ using the matrix method and check using the MathScript function *fft*.

We construct the matrix as

```
x = [1 2 3 4]; N=length(x); nkvec = 0:1:N-1;
W = exp(nkvec'*nkvec).^(-j*2*pi/N); dft = W*x',
mfft = fft(x)
```

If x is complex, care must be taken as MathScript automatically conjugates a vector when it is transposed. In such case this code, which restores the signs of the imaginary parts of x after transposing, gives proper results:

```
x = [(1+j) 2 (3+j) 4]; N=length(x); nkvec = 0:1:N-1;
W = exp(nkvec'*nkvec).^(-j*2*pi/N); dft = W*conj(x'),
mfft = fft(x)
```

7.11 DFT OF COMMON SIGNALS

We'll investigate the DFT of a number of standard signals using a series of examples. All of the scripts described in the examples below scale the DFT by $1/N$.

Example 7.12. Compute and display the DFT of a square wave synthesized from a finite number of harmonics.

Figure 7.11 shows the computation window generated by the script call

LVDFTCompute(0)

as it appears for $k = 0$. The test signal at (e) is a square wave constructed with a truncated series of odd-only harmonics as

$$W = \sum_{k=1}^{N/4} (1/(2k-1)) \sin(2\pi n(2k-1)/N)$$

where n runs from 0 to $N-1$ and N is the sequence length.

The real and imaginary components (i.e., cosine and sine) of each complex correlator are displayed for $k = 0$ up to $k = N-1$ (press any key to compute and display the next $F[k]$). Correlations at the zeroth lag (CZLs) are done between each complex correlator and the test signal. The real and imaginary parts of $F[k]$, as well as the magnitude, are plotted as each bin is computed.

The script uses $N = 32$ and hence the test signal contains the frequencies 1, 3, 5, 7, 9, 11, 13, and 15 with amplitudes inversely proportional to frequency.

The code that computes each bin value $F[k]$ is

```
F(k+1) = (1/N)*sum(Signal.*exp(-j*2*pi*t*k))
```

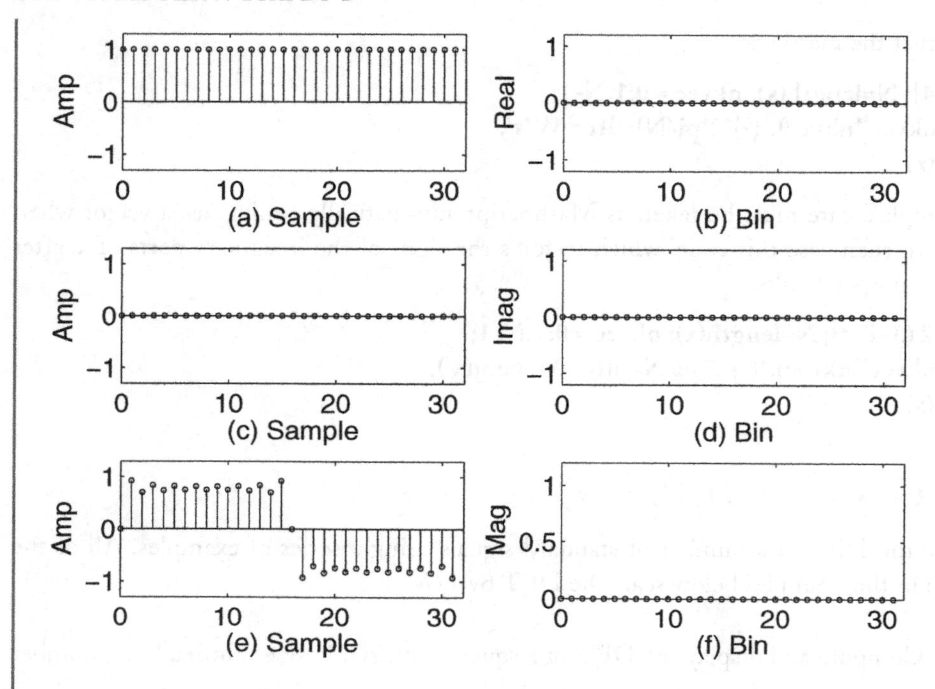

Figure 7.11: (a) and (c): Real and Imaginary correlators for Bin 0, i.e., frequency zero cosine and sine (over 32 samples), respectively; (b) and (d): Real and Imaginary parts of DFT for waveform at (a) and (c), initialized to zero with Bin 0 plotted; (e) Truncated square wave serving as test signal; (f) Magnitude of DFT, based on real and imaginary parts plotted in (b) and (d).

where n = 0:1:N -1 and $t = n/N$. The bins for the expression above are indexed from 1 to N as needed by MathScript, so the DC bin is F(1) rather than F(0), etc. This is corrected in the plot by plotting the output array against the vector [0:1:N -1], such as by the code line

plot([0:1:N-1], F(1,1:N))

When first making the call **LVDFTCompute(0)**, the initial plot displays information for Bin 0 (i.e., k = 0). The real or cosine correlator is identically one for 32 samples, and the imaginary or sine correlator is identically zero for 32 samples, and the net DFT value is zero. If you are running the script, press any key, and k will be set to 1, and the result is shown in Fig. 7.12.

Note that in the complex DFT, the positive frequencies are tested with an inverted sine wave (this is the result of the negative sign in front of the j in the formula), and the negative frequencies are tested with a noninverted sine wave. Actually, it need not be this way; the negative sign in front of the j could just as well have been positive. However, for the complex DFT, the sign in front of the j used in the Inverse DFT must be opposite to the sign used in the DFT.

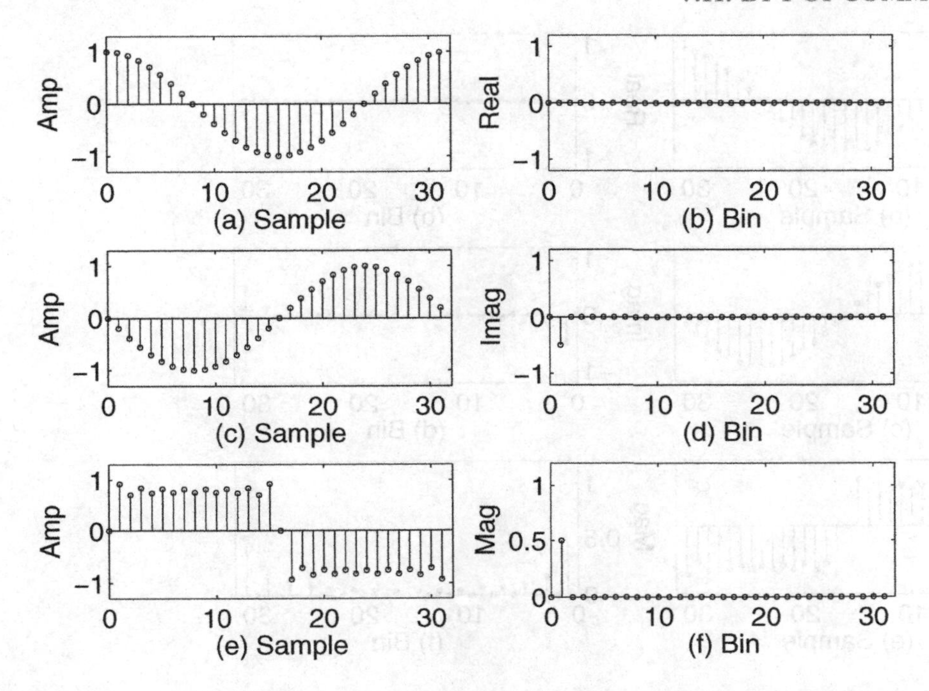

Figure 7.12: (a) and (c): Real and Imaginary correlators for Bin 1, i.e., one cycle (over 32 samples) cosine and sine, respectively; (b) and (d): Real and Imaginary parts of DFT for waveform at (a) and (c), initialized to zero with Bins 0 and 1 plotted; (e) Truncated square wave serving as test signal; (f) Magnitude of DFT, based on real and imaginary parts plotted in (b) and (d).

In Fig. 7.13, we see the final result for performing the DFT on the truncated square wave. There are a number of things to observe. First, the square wave is an odd function, and is made of a series of odd harmonics with amplitudes inversely proportional to the harmonic number. Hence, the real part of the DFT is zero.

Let's analyze the lower right plot of Fig. 7.13, which is the magnitude of the DFT.

Bin 1 (the fundamental) should have an amplitude of 1, Bin 2 should be zero (it's an even harmonic), Bin 3 should be 0.33, and so forth. Note instead that the values are half this since both positive and negative frequency bins with the same magnitude of k contribute to the reconstructed waveform when using the inverse DFT. Hence, when we add Bins 1 and 31 together, we get the required magnitude of 1 for the first harmonic ($k = 1$), 0.33 for the 3rd harmonic, 0.2 for the 5th, harmonic, etc. (later in the chapter we'll demonstrate the truth of this mathematically in a more detailed discussion of the IDFT). Recall that Bin 31 is the equivalent of Bin −1 when k runs from 0 to $N - 1$. Recall that for all Bins of the Real DFT (other than Bins 0 and $N/2$), it was necessary to double the values of the reconstructed harmonics since only positive frequencies were used in the original set of correlations. From this it can be seen that mathematically, the Real DFT is a sort of

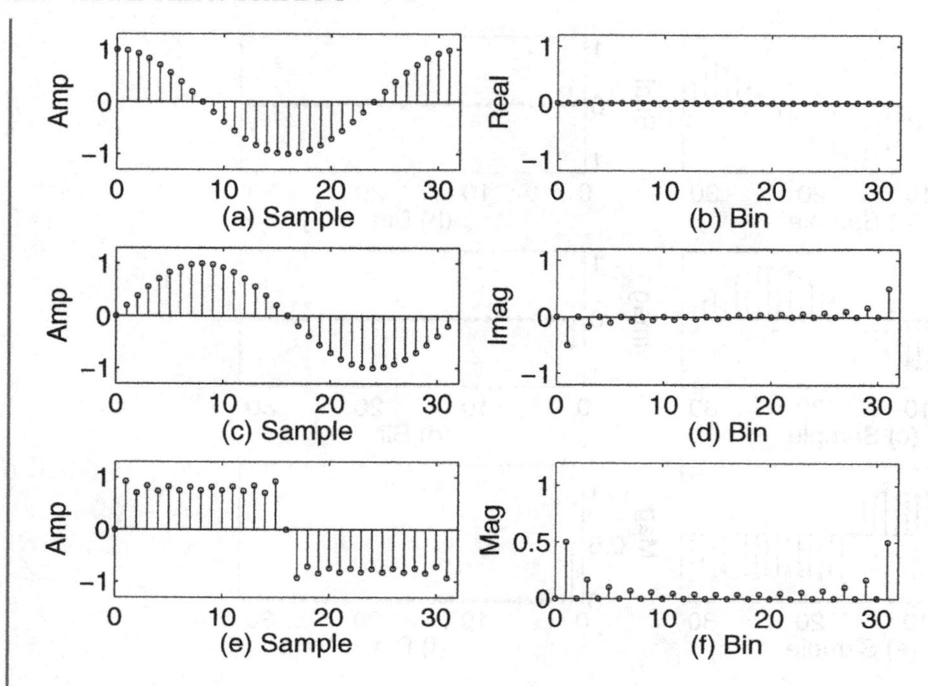

Figure 7.13: (a) and (c): Real and Imaginary correlators for Bin -1, shown after computing all bin values; (b) and (d): Real and Imaginary parts of DFT for waveform at (a) and (c), with all bin values plotted; (e) Truncated square wave serving as test signal; (f) Magnitude of DFT, based on real and imaginary parts plotted in (b) and (d).

corollary or special case of the complex DFT, which possesses a symmetry that allows the scaling value to be equal for all bins.

Example 7.13. Compute the DFT of a sawtooth wave.

The script (see exercises below)

LVxDFTComputeSawtooth

computes the DFT, step-by-step, of a truncated harmonic sawtooth which is synthesized using

$$W = \sum_{k=1}^{N/2} (1/k)\sin(2\pi nk/N)$$

where n runs from 0 to $N-1$. Note that both odd and even harmonics are included. The result from running the script through all bin values is shown in Fig. 7.14.

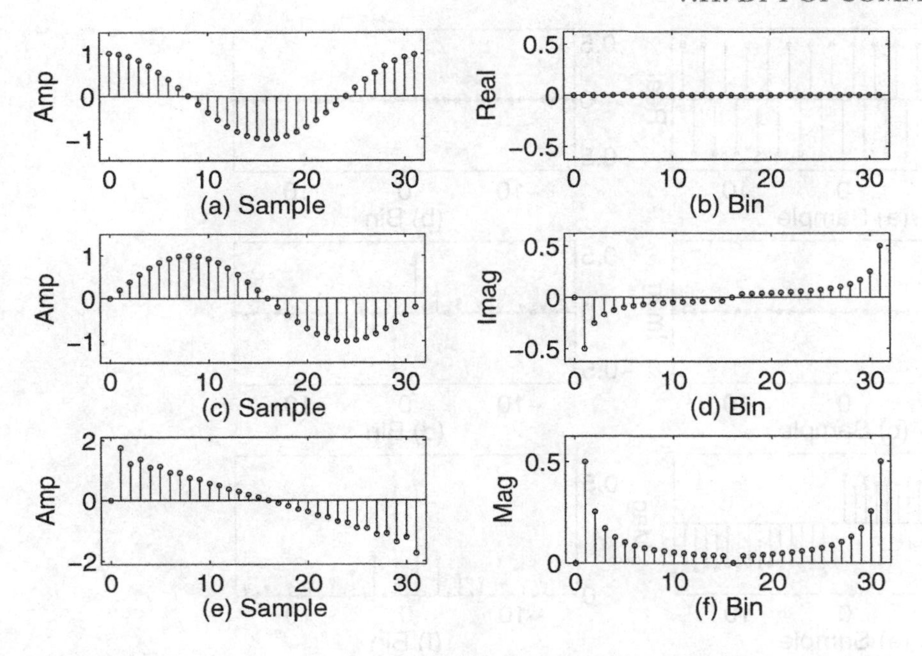

Figure 7.14: (a) Real correlator, Bin 31; (b) Real part of DFT, Bins 0-31; (c) Imaginary correlator, Bin 31; (d) Imaginary part of DFT, Bins 0-31; (e) Bandlimited sawtooth wave (input signal the DFT of which is shown in plots (b), (d), and (f)); (f) Magnitude of DFT of sawtooth shown in plot (e).

Example 7.14. Compute the DFT of a test signal using symmetric bin values.

In this case, k runs from $-N/2+1$ to $N/2$ rather than from 0 to $N-1$. The result of computation is exactly the same as for asymmetrically valued k, except for the arrangement of the output. The script (see exercises below)

LVxDFTComputeSymmIndex

computes the symmetrically-indexed DFT of a bandlimited square wave, constructed from a limited number of harmonically-related sine waves.

Figure 7.15 shows the result. In the symmetric-index DFT, negative and positive frequencies per se are used and displayed in the output as such; if N is an even number, the values of k run from -$N/2$+1 to $N/2$. Other than the arrangement of bin values in the output display, the values correspond exactly to the values obtained using the nonsymmetrical system, where k runs from 0 to $(N-1)$. Bin $(N-1)$, for example, is identical to Bin (-1) of the symmetric-index system, Bin $(N-2)$ is equivalent to Bin (-2), and so forth.

Example 7.15. Compute the DFT, and graph the results, for a unit impulse signal.

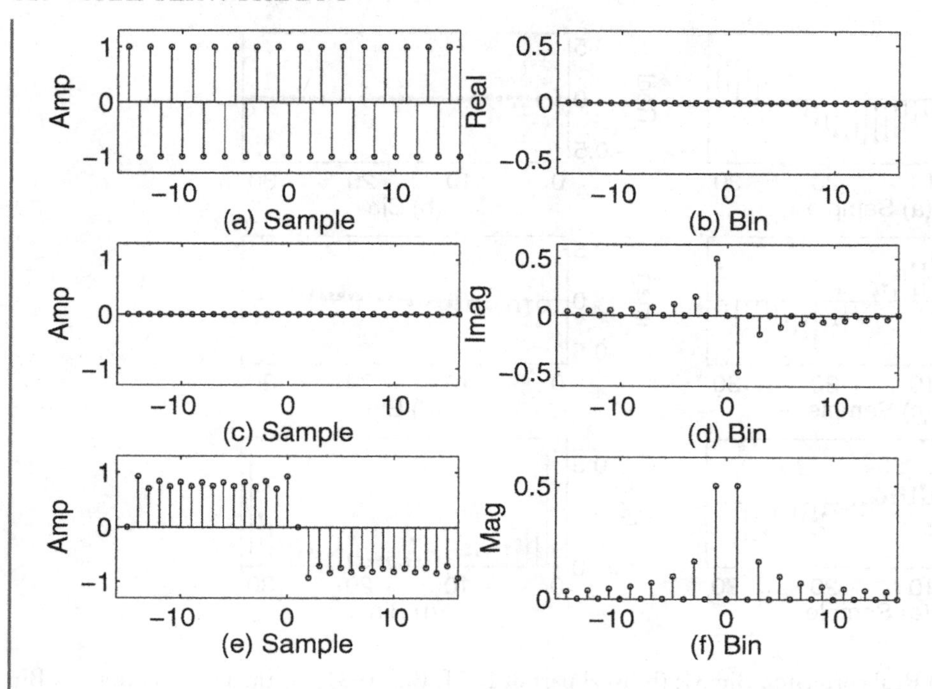

Figure 7.15: (a) and (c): Real and Imaginary correlators for Bin 16; (b) and (d) DFT with all Bin values plotted, using symmetrical Bin index; (e) Test truncated harmonic square wave; (f) Magnitude of DFT based on real and imaginary parts from (b) and (d).

In the continuous domain, the frequency spectrum of an impulse is flat, all frequencies being present with equal amplitude. In a sampled data set, we use the Unit Impulse, and all possible frequencies (positive and negative) within the sequence length are present with equal amplitudes.

The script (see exercises below)

LVxDFTComputeImpUnBal

computes and displays the DFT, one bin at a time, of a unit impulse sequence of length 32.

Figure 7.16 shows the computation completed up through bin 16. Note that the impulse is a "1" at time zero and zero thereafter, and that the imaginary correlators for all harmonics (sine waves), are all zero at time zero, thus yielding a zero-valued imaginary part for all bins. Furthermore, the real correlators for all harmonics (cosine waves) are all valued at "1" at time zero, so the DFT coefficient (bin value) for every bin is the same, and equal to (1/32)(1)(1) = 0.0313. Thus the DFT over N samples of an N-sample Unit Impulse sequence consists of an equal-amplitude harmonic series of cosines.

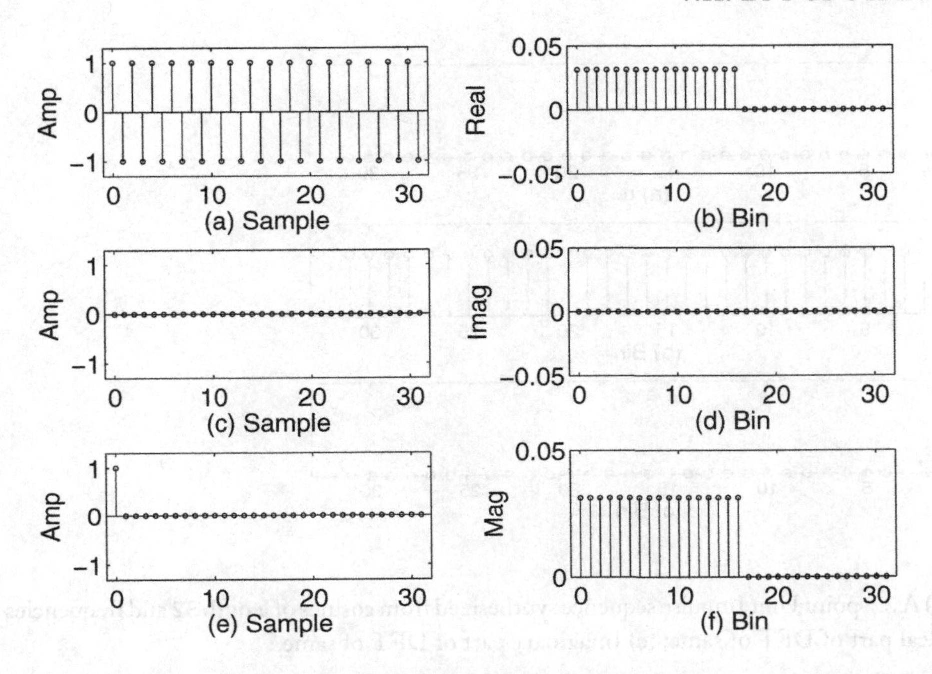

Figure 7.16: (a) and (c): Real and Imaginary correlators for Bin 16; (b) and (d): DFT with Bin values plotted up to Bin 16; (e) Unit Impulse test signal; (f) Magnitude of DFT based on real and imaginary parts from (b) and (d).

Example 7.16. Verify that, for example, a Unit Impulse sequence of length 32 is composed of a superposition of cosine waves. After synthesizing a length-32 Unit Impulse sequence, compute its DFT using the function *fft* and compare results to those of Fig. 7.16.

The following code synthesizes a unit impulse sequence of length SR and then computes its DFT using the function *fft*. The results are then plotted. Figure 7.17 shows the result. Note that the function *fft* does not scale the DFT coefficients by $1/N$; if the DFT employed had been one that scales the DFT coefficients, then the magnitude of the bin values in plot (b) would have been 0.0313 (1/32) rather than 1.0.

```
SR = 32; n = 0:1:SR-1; w = 0;
for ctr = 1:1:SR; w = w + (1/SR)*cos(2*pi*n/SR*(ctr-1)); end;
dftans = fft(w); figure(9); subplot(311); stem(w);
subplot(312); stem(real(dftans));
subplot(313); stem(imag(dftans))
```

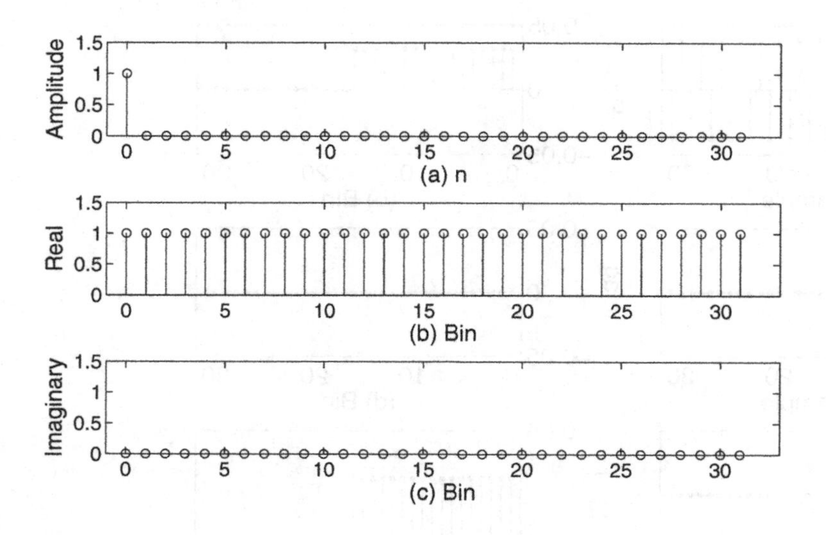

Figure 7.17: (a) A 32-point Unit Impulse sequence synthesized from cosines of length 32 and frequencies of 0 to 31; (b) Real part of DFT of same; (c) Imaginary part of DFT of same.

7.12 FREQUENCY RESOLUTION

Consider a sample sequence of length N obtained over a two second period; now correlate it with a sinusoid having one cycle over N samples. We are in effect correlating the signal sequence with a 0.5 Hz sine wave. In this particular example, the one cycle DFT correlators, sine and cosine, are equivalent to 0.5 Hz, so the next orthogonal pair (at 2 cycles) would in effect be testing the original sequence for 1 Hz, the next orthogonal pairs would be testing for 1.5 Hz, and so on.

With this concrete example in mind, we can say that the **Frequency Resolution**, or **Bin width,** or **Bin Spacing,** available from the DFT is equal to the reciprocal of the total sampling duration T. In the example above, a total sampling duration of two seconds yielded bin center frequencies of 0 Hz, 0.5 Hz, 1 Hz, 1.5 Hz, etc. It is evident that the frequency or bin spacing, in terms of original signal frequencies, is 0.5 Hz, which is the reciprocal of the total sampling duration. We'll use the term **Bin width** in this book to mean frequency resolution or bin spacing relative to the sampled signal's actual (or original) frequencies.

Figure 7.18 shows this concept in generic form for a total sampling duration of T seconds. We see at (a) the signal, at (b) the cosine and sine correlators for or Bin 0, at (c), the cosine and sine correlators for Bin 1, at (d) for Bin 2, and at (e), the correlators for Bin 3. These sets of correlators have frequencies over the sequence length N that go all the way up to $N/2$ for N even or $(N-1)/2$ for N odd. The equivalent signal frequency of Bin 1 is the reciprocal of T, and all higher bins represent harmonics thereof. Thus the equivalent bin (signal) center frequencies are 0, $1/T$, $2/T$, $3/T$, ... L/T with $L = N/2$ for N even or $(N - 1)/2$ for N odd.

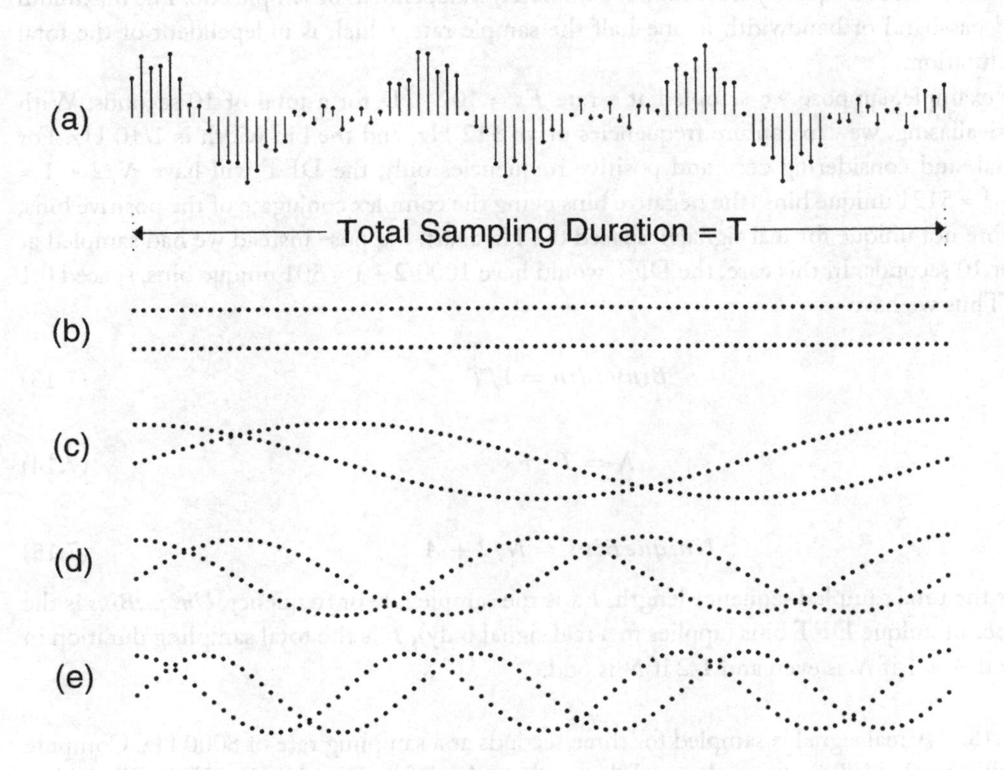

Figure 7.18: (a) Sample sequence obtained over duration T seconds; (b) Cosine and Sine correlators for frequency 0 (DC); (c) Correlators for frequency 1; (d) Correlators for frequency 2; (e) Correlators for frequency 3.

Example 7.17. A signal is sampled for 2 seconds at a sampling rate of 8 kHz, resulting in 16,000 samples. Give the equivalent signal (center) frequency of Bin 1, and give the highest bin number and its equivalent signal frequency.

The center frequency of Bin 1 in signal terms is $1/T$ = 0.5 Hz. The highest bin index is 16000/2 = 8000. The equivalent signal frequency would be 8000/T = 4000 Hz. This is consistent with the sample rate of 8 kHz, which can only represent nonaliased frequencies up to its Nyquist rate of 4 kHz.

7.13 BIN WIDTH AND SAMPLE RATE

Note that bin width or frequency resolution is completely independent of sample rate. The maximum nonaliased passband or bandwidth is one-half the sample rate, which is independent of the total sampling duration.

For example, suppose we sampled at a rate Fs = 1024 Hz for a total of 10 seconds. With proper anti-aliasing, we can capture frequencies up to 512 Hz, and the bin width is 1/10 Hz. For a real signal, and considering zero and positive frequencies only, the DFT will have $N/2 + 1$ = 10240/2 + 1 = 5121 unique bins (the negative bins being the complex conjugate of the positive bins, and therefore not unique for real signals), spaced 0.1 Hz apart. Suppose instead we had sampled at 100 Hz for 10 seconds. In this case, the DFT would have 1000/2 + 1 = 501 unique bins, spaced 0.1 Hz apart. Thus we have

$$Binwidth = 1/T \tag{7.13}$$

$$N = T \cdot Fs \tag{7.14}$$

$$UniqueBins = N/2 + A \tag{7.15}$$

where N is the total sampled sequence length, Fs is the sample rate or frequency, *UniqueBins* is the total number of unique DFT bins (applies to a real signal only), T is the total sampling duration in seconds, and A is 1 if N is even and 1/2 if N is odd.

Example 7.18. A real signal is sampled for three seconds at a sampling rate of 8000 Hz. Compute the bin width, total number of samples, and the total number of unique bins the DFT will yield.

The Nyquist limit is 8000/2 = 4000 Hz, and from Eq. (7.13), *Binwidth* = 0.333 Hz, total samples N (from Eq. (7.14)) = (3)(8,000) = 24,000, and *UniqueBins*, from Eq. (7.15), = 24,000/2 + 1 = 12,001.

Example 7.19. A total of 100 samples are obtained of a real signal, sampled at a constant rate for three seconds. What is the bin width, Nyquist limit, and number of unique bins?

The bin width is 0.333 Hz and the number of unique bins is 51. Since the sampling frequency is 100/3 = 33.33 Hz, the Nyquist rate is 33.33/2 = 16.66 Hz; the anti-aliasing filter should cutoff at this frequency.

Example 7.20. A real signal is sampled for three seconds at a rate of 100 Hz. Give the bin width, the highest nonaliased frequency, the number of unique bins, and whether the following frequencies in the signal would be on-bin or off-bin in the DFT: 2.0 Hz, 4.5 Hz, 6.666 Hz.

A total of 300 samples result. The bin width is 0.333 Hz, and the bin frequencies are therefore 0, 0.333 Hz, 0.666 Hz, 1 Hz, 1.333 Hz, etc. The (nonaliased) highest frequency based on the sampling rate and proper anti-aliasing, is 50 Hz; the number of unique bins is 300/2 + 1 = 151. The frequencies 2.0 Hz and 6.666 Hz would lie squarely on-bin (in fact, on bins 6 and 20, respectively), while 4.5 Hz is off-bin.

7.14 THE FFT

There are many specialized algorithms for efficiently computing the DFT. The direct implementation of a DFT of length N requires roughly N^2 computations. The Cooley-Tukey radix-2 FFT is perhaps the most well-known of such algorithms, and requires only $Nlog_2(N)$ operations, which makes possible real-time computation of larger DFTs.

Example 7.21. Assume that a certain computer takes one second to compute the DFT of a length-2^18 sequence using an FFT algorithm. Compute how long (approximately) it would take to compute the same DFT using the direct DFT implementation.

We evaluate the ratio

$$N^2/(N\log_2(N)) = N/\log_2(N)$$

as $2^{18}/18$ = 14,564 seconds or about 4.05 hours.

7.14.1 N-PT DFT FROM TWO N/2-PT DFTS

We'll consider the simple radix-2 **Decimation-in-Time** (**DIT**) algorithm, which is based on the idea that any sequence having a length equal to a power of two can be divided into two subsequences, comprised of the even and odd indexed members of the original sequence. A DFT can then be computed for the two shorter sequences (each of which is half the length of the original sequence), and the two DFTs can be put together to result in the DFT of the original sequence. There is a computational savings achieved by doing this. You can keep dividing each subsequence into two parts until the original sequence of length N has been divided into N sequences of length one, each of which is its own DFT. From this, $N/2$ DFTs each having two bins can be assembled. The $N/2$ two-bin DFTs are then assembled into $N/4$ four-bin DFTs, and so on until a length-N DFT has been computed.

Consider the DFT ($X[k]$) of a sequence $x[n]$ of length eight samples, which can be computed in MathScript as

X(k) = sum(x(n).*exp(-j*2*pi*(0:1:7)*k/8))

The signal $x[n]$ can be rewritten as the sum of two length-4 DFTs by dividing it into its even and odd indexed samples:

$$xE = x[0:2:7];\, xO = x[1:2:7]$$

which, in MathScript, would be, owing to the fact that MathScript array indices start with 1 rather than 0,

xE = x(1:2:8); xO = x(2:2:8)

For an 8-pt DFT, the complex correlator is

$$\exp(-j2\pi k(0:1:7)/8) = \exp(-j2\pi k(0:\frac{1}{8}:\frac{7}{8}))$$

For a 4-pt DFT, the complex correlator is

$$\exp(-j2\pi k(0:\frac{1}{4}:\frac{3}{4})) = \exp(-j2\pi k(0:\frac{2}{8}:\frac{6}{8}))$$

The latter expression is just the even indexed correlators (starting with index 0) for an 8-pt DFT. The odd indexed correlators for an 8-pt DFT can be obtained from the 4-pt DFT by a simple phase shift:

$$\exp(-j2\pi k(\frac{1}{8}:\frac{2}{8}:\frac{7}{8})) = \exp(-j2\pi k/8)\exp(-j2\pi k(0:\frac{2}{8}:\frac{6}{8}))$$

Then setting **n = 0:1:3**; and **Phi = exp(-j*2*pi*k/8)**, the 8-pt DFT, expressed as the sum of two 4-pt DFTs, would be

X(k) = sum(xE.*exp(-j*2*pi*(0:1:3)*k/4)) +...

Phi*sum(xO*exp(-j*2*pi*(0:1:3)*k/4)) (7.16)

At this point, we can simplify the notation by setting

$$W_N = \exp(-j2\pi/N)$$

Using this notation

$$W_N^{nk} = \exp(-j2\pi nk/N)$$

Equation (7.16) can be expressed symbolically as

$$X[k] = \sum_{n=0}^{3} x_E W_4^{(0:1:3)k} + W_8^k \sum_{n=0}^{3} x_O W_4^{(0:1:3)k} \tag{7.17}$$

where we have used the more standard notation x_E instead of the MathScript-suitable xE for the even subsequence, and similarly x_O for xO.

Restating Eq. (7.17) generically, we would have

$$X[k] = \sum_{n=0}^{N/2-1} x_E W_{N/2}^{nk} + W_N^k \sum_{n=0}^{N/2-1} x_O W_{N/2}^{nk}$$

which is a DFT of length N expressed as the sum of two DFTs of length $N/2$, the second of which is multiplied by a phase (or **Twiddle**) factor which adjusts the phase of its complex correlators to correspond to those of the odd-indexed values of the length-N DFT. Letting k run from 0 to $N/2$ -1, in order to account for all bins for a length N DFT, we would have

$$X[k] = X_E[k] + W_N^k X_O[k]$$

and

$$X[k + N/2] = X_E[k + N/2] + W_N^{(k+N/2)} X_O[k + N/2]$$

Note that the sequences X_E and X_O are periodic over $N/2$ samples, so $X_E[k + N/2] = X_E[k]$ and $X_O[k + N/2] = X_O[k]$ and that

$$W_N^{(k+N/2)} = W_N^k W_N^{N/2}$$

Since

$$W_N^{N/2} = \exp(-j2\pi(N/2)/N) = -1$$

we have as a result the two **Butterfly Formulas**

$$X[k] = X_E[k] + W_N^k X_O[k] \tag{7.18}$$

$$X[k + N/2] = X_E[k] - W_N^k X_O[k] \tag{7.19}$$

Figure 7.19 illustrates, in plots (a)-(c), the complex correlator and its real and imaginary parts for an 8-pt DFT, for $k = 1$. Plots (d)-(f) show the complex correlator for a 4-pt DFT, plotted as circles in all three plots. The complex values marked as stars were generated by offsetting the phase of the 4-pt DFT correlators by the phase factor exp(-j2π(1)/8).

7.14.2 DECIMATION-IN-TIME

This principle of breaking a length 2^n sequence into even and odd subsequences, and expressing the DFT of the longer sequence as the sum of two shorter DFTs, can be continued until only length-1 DFTs remain. A sequence of length 8, for example, is decimated into even and odd subsequences several times, in this manner.

Matrix (7.20) shows the steps for DIT of a length-8 sequence. It may be interpreted as follows: TD (1 X 8), for example, means that the sequence (whose original sample indices lie to the right),

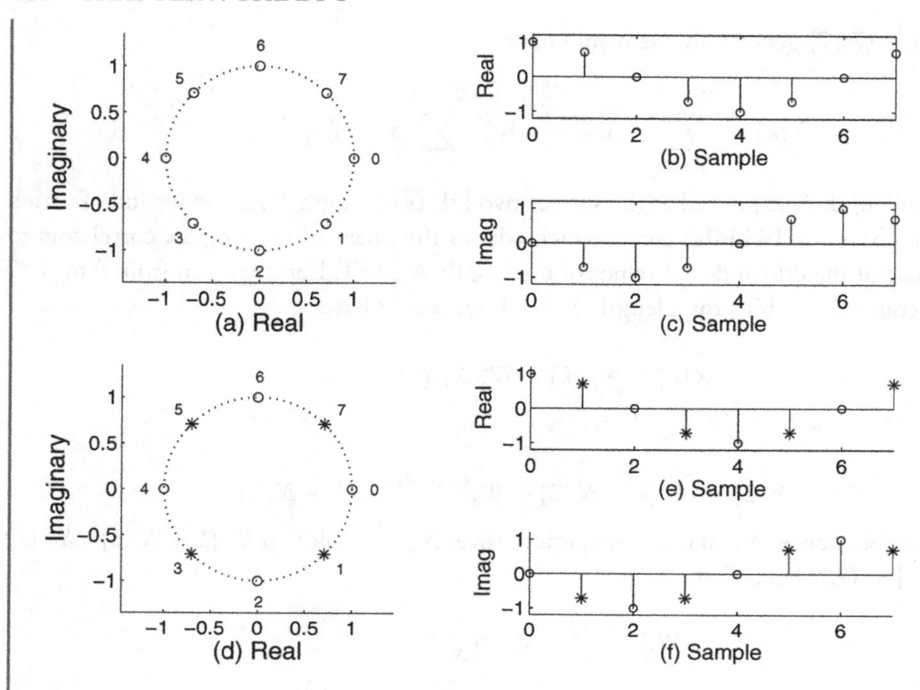

Figure 7.19: (a) The complex correlators for an 8-pt DFT, for $k = 1$; (b) Real part of the complex correlators shown in (a); (c) Imaginary part of the complex correlators shown in (a); (d) The complex correlators for an 8-pt DFT ($k = 1$), generated as the complex correlators for a 4-pt DFT (marked with circles), and the same 4-pt correlators multiplied by the phase factor exp(-j$2\pi(1)/8$), marked with stars.

is 1 sequence of length 8, whereas TD (2 X 4) means that the original sample indices to the right form 2 sequences of 4 samples each, and so on.

$$\begin{array}{lcccccccc} \text{TD}(1 \text{ X } 8) & 0 & 1 & 2 & 3 & 4 & 5 & 6 & 7 \\ \text{TD}(2 \text{ X } 4) & 0 & 2 & 4 & 6 & 1 & 3 & 5 & 7 \\ \text{TD}(4 \text{ X } 2) & 0 & 4 & 2 & 6 & 1 & 5 & 3 & 7 \\ \text{TD}(8 \text{ X } 1) & 0 & 4 & 2 & 6 & 1 & 5 & 3 & 7 \\ \text{FD}(8 \text{ X } 1) & 0 & 4 & 2 & 6 & 1 & 5 & 3 & 7 \end{array} \tag{7.20}$$

The length-8 sequence (in the first row of the matrix) is divided into two length-4 sequences by taking its even and odd indexed parts, shown in the second row. Then the two 4-sample sequences are each split into two 2-sample sequences, forming a total of four 2-sample sequences. These are then subdivided again to form eight 1-sample sequences. The result, however, is identical to the ordering of the four 2-sample sequences. Since the last three rows are identical, there is obviously no separate step needed to generate the final two rows.

The DFT of a single sample sequence is itself, so the row TD(8 X 1) in the matrix, is also a row of eight 1-sample DFTs, labeled FD(8 X 1).

7.14.3 REASSEMBLY VIA BUTTERFLY

From there, the sequences are recombined to form four 2-pt DFTs, then two 4-pt DFTs, then one 8-pt DFT. This is done according to the Butterfly formulas given above.

In practical terms, when the DIT routine arrives at 2^{N-1} subsequences of length two, the butterfly routine can be started. For the first set of butterflies, eight 1-point DFTs are converted to four two-point DFTs using the butterfly formulas with $N = 2$ and k taking on only the single value of 0. These formulas

$$X[0] = X_E[0] + W_2^0 X_O[0]$$

$$X[0+1] = X_E[0] - W_2^0 X_O[0]$$

which simplify to

$$X[0] = X_E[0] + X_O[0]$$

$$X[0+1] = X_E[0] - X_O[0] \tag{7.21}$$

are applied in turn to the four pairs of single-point DFTs to produce the four 2-point DFTs.

At this stage we now have four 2-pt DFTs which must be assembled into two 4-pt DFTs. Using the butterfly formulas, and letting k run from 0 to 1, we have

$$X[0] = X_E[0] + W_4^0 X_O[0]$$

$$X[0+2] = X_E[0] - W_4^0 X_O[0]$$

$$X[1] = X_E[1] + W_4^1 X_O[1]$$

$$X[1+2] = X_E[1] - W_4^1 X_O[1] \tag{7.22}$$

which thus produce a new four point DFT for every four bins formed by the two bins of each of X_E and X_O.

The final step is to assemble two 4-pt DFTs into one 8-pt DFT using the butterfly formulas. A summary of the entire process in schematic form, known as a **Butterfly Diagram,** is shown in Fig. 7.20. The input to this butterfly diagram or algorithm is the time-decimated signal x (shown above the topmost row of boxes) from the final stage of time domain decimation. The values $x[0]$,

$x[4]$, etc., are reinterpreted as eight 1-pt DFTs which are labeled in pairs as $X_E[0]$ and $X_O[0]$ preparatory to combining to form four 2-pt DFTs. The applicable values for k and twiddle factors W are shown to the right of each set of butterflies.

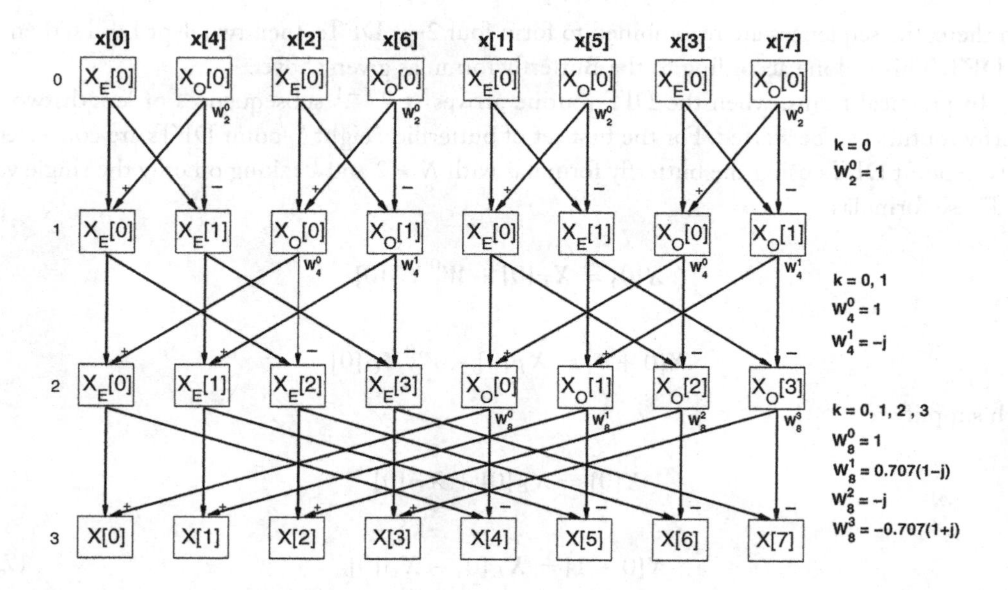

Figure 7.20: The Butterfly Diagram for a length-8 Decimation-in-Time FFT algorithm.

Example 7.22. Compute the DFT of the sequence [1 2 3 4] using the radix-2 DIT algorithm discussed above.

The first step is to decimate in time, which results in the net sequence [1,3,2,4]. Then we obtain the 2-pt DFTs of the sequences [1,3] and [2,4] which are, respectively, [4,-2] and [6,-2] by using the 2-pt butterflies (7.21). Using the 4-pt butterflies at (7.22), we combine the two 2-pt DFTs into one 4-pt DFT:

$$X[0] = 4 + 6 = 10$$

$$X[0 + 2] = 4 - 6 = -2$$

and

$$X[1] = -2 + (-j)(-2) = -2 + j$$

$$X[1+2] = -2 - (-j)(-2) = -2 - j$$

yielding a net DFT of

$$[10, (-2+j), -2, (-2-j)]$$

The MathScript call

```
y = fft([1, 2, 3, 4])
```

yields the identical result.

7.14.4 ALGORITHM EXECUTION TIME

The DIT FFT is made more efficient by making its two components, the time decimation, and the reassembly by successive butterflies, as efficient as possible. To make the decimation more efficient, it is noted that if the samples are indexed in binary notation, the entire sequence can be properly rearranged as discussed above simply by exchanging sample values that have bit-reversed or mirror-image binary addresses. The following table shows this:

Binary	Decimal	Binary	Decimal
000	0	000	0
001	1	100	4
010	2	010	2
011	3	110	6
100	4	001	1
101	5	101	5
110	6	011	3
111	7	111	7

(7.23)

The first two columns of Table (7.23) are the forward binary count sequence and its decimal equivalent, and the third and fourth columns are the bit-reversed binary count sequence and its decimal equivalent, which can be seen as forming the proper time decimation of the 8-pt sequence. It is only necessary to exchange samples when the bit-reversed counter's value exceeds the forward counter's value. An in-place bit-reversal routine is as follows:

```
J = 0;
for I = 1:1:length(Signal)-2
k = length(Signal)/2;
while (k <= J)
J = J - k;
k = k/2;
end
```

```
J = J + k;
if I < J
 temp = Signal(J+1);
 Signal(J+1) = Signal(I+1);
 Signal(I+1) = temp;
 end
end
```

The script

$$LV\_FFT(L, DecOrBitReversal)$$

graphically illustrates a test signal, its progressive decimation using either direct decimation or bit reversal, and the reassembly by butterfly. The variable *DecOrBitReversal* does the signal rearrangement by even/odd decimation if passed as 0, or by bit reversal if passed as 1. The test signal is the ramp 0:1:$L - 1$, which makes evaluation/visualisation of the decimation easy. Figure 7.21, for example, shows the result from the call

LV_FFT(8,0)

Since the test signal's sample values are the same as their respective (original) indices, plot (c)'s amplitude values are the same as the test sequence's original index values. Thus we see that the sequence 0:1:7, shown at (a), has been properly decimated into the sequence [0,4,2,6,1,5,3,7], shown at (c).

A butterfly routine, compact in terms of number of lines of code, can easily be written in m-code (see exercises below) in accordance with Eqs. (7.18) and (7.19). However, a less-efficient looking, but far more efficient in practice routine to compute the butterflies from the decimated-in-time sequence is as follows:

```
Rx = real(x); Ix = imag(x);
M = log2(length(x)); LenSig = length(x);
for L = 1:1:M
 LE = 2^L; LE2 = LE/2;
 uR = 1; uI = 0;
 sR = cos(pi/LE2); sI = -sin(pi/LE2);
for J = 1:1:LE2
Jmin1 = J-1;
 for I = Jmin1:LE:LenSig - 1
  Ip = I + LE2;
  tR = Rx(Ip+1)*uR - Ix(Ip+1)*uI;
  tI = Rx(Ip+1)*uI + Ix(Ip+1)*uR;
  Rx(Ip+1) = Rx(I+1) - tR;
```

```
        Ix(Ip+1) = Ix(I+1) - tI;
        Rx(I+1) = Rx(I+1) + tR;
        Ix(I+1) = Ix(I+1) + tI;
    end
    tR = uR;
    uR = tR*sR - uI*sI;
    uI = tR*sI + uI*sR;
    end
end
x = Rx + j*Ix;
```

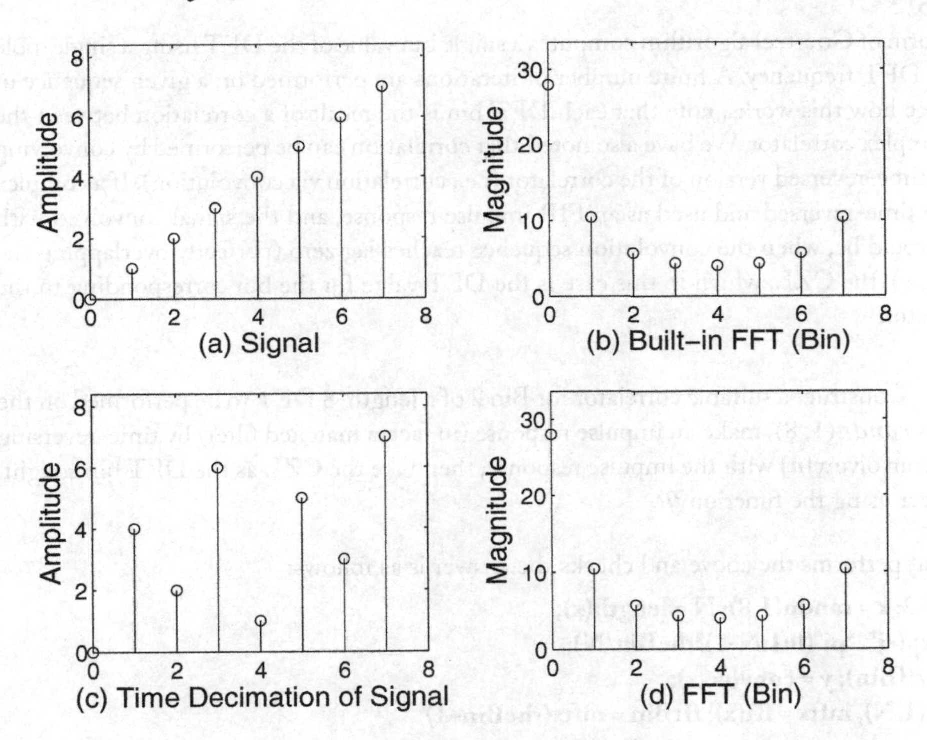

Figure 7.21: (a) Test signal, the ramp 0:1:8; (b) Built-in fft of the test signal; (c) Time decimation of the test signal; (d) FFT of test signal as performed using butterfly routines on the signal at (c).

7.14.5 OTHER ALGORITHMS

Beside the DIT algorithm, there are **Decimation-In-Frequency** (**DIF**) algorithms. For both types of these two basic algorithms, there are many variations as to whether or not the samples input to the butterfly routine are in natural or bit-reversed order. There are algorithms in which the input is presented in natural order and the output is in correct bin order, without any explicit bit-reversal-like reordering. There are also algorithms that do not require that the signal length be a power of two, such as relative prime factor algorithms, for example.

Reference [1] gives a very succinct explanation of the radix-2 decimation-in-time algorithm; [2] and [3] present many butterfly diagrams and discussions covering both DIT and DIF algorithms; [4] discusses factored, radix-2, radix-4, and split radix FFTs; [5] and [6] discuss matrix factorizations in addition to various standard FFT algorithms.

7.15 THE GOERTZEL ALGORITHM

7.15.1 VIA SINGLE-POLE

The simplest form of Goertzel algorithm computes a single bin value of the DFT using a single-pole IIR tuned to a DFT frequency. A finite number of iterations are performed on a given sequence of length N. To see how this works, note that each DFT bin is the result of a correlation between the signal and a complex correlator. We have also noted that correlation can be performed by convolving a signal with a time-reversed version of the correlator (i.e., correlation via convolution). If a complex correlator were time-reversed and used as an FIR impulse response, and the signal convolved with it, the output would be, when the convolution sequence reaches lag zero (perfectly overlapping the impulse response), the CZL, which in this case is the DFT value for the bin corresponding to the complex correlator.

Example 7.23. Construct a suitable correlator for Bin 2 of a length-8 DFT to be performed on the sequence $x[n] = randn(1, 8)$, make an impulse response (in fact, a matched filter) by time-reversing the correlator, convolve $x[n]$ with the impulse response, then take the CZL as the DFT bin sought. Verify the answer using the function *fft*.

Code that performs the above and checks the answer is as follows:

```
theBin = 2; x = randn(1,8); N = length(x);
fBin = exp(-j*2*pi*(0:1:N-1)*theBin/N);
cc = fliplr(fBin); y = conv(cc,x);
CZL = y(1,N), mftx = fft(x); fftBin = mftx(theBin+1)
```

Now imagine that instead of performing the convolution using an FIR, it is done using an IIR having an impulse response that, over a finite number of samples, matches the complex correlator needed If, for example, the bin sought is Bin k for a length-N DFT, we set

p = exp(j*2*pi*k/N)

Let's consider the case of a length-4 sequence for which we want to compute Bin 1. If the samples of the sequence are designated $s[0]$, $s[1]$, $s[2]$, and $s[3]$, and we process the sequence with an IIR having a single pole at

p = exp(j*2*pi*1/4)

then we can write the output values of the filter as

s[0]
p^1s[0] + s[1]
p^2s[0] + p^1s[1] +s[2]
p^3s[0] + p^2s[1] +p^1s[2] + s[3]
p^4s[0] + p^3s[1] +p^2s[2] + p^1s[3]

the last term of which, due to periodicity, is the same as

p^0s[0] + p^{-1}s[1] +p^{-2}s[2] + p^{-3}s[3]

which is a correlation at the zeroth lag between the signal and the complex power sequence

exp(-j*2*pi*1/4).^(0:1:3)

which is the same complex power sequence used to compute DFT Bin 1 for a length-4 DFT.

Example 7.24. Verify that, for the above example, that $p.\hat{}(4:-1:1)$ is the same as $p.\hat{}(0:-1:-3)$.

We run the following code:

```
p = exp(j*2*pi*1/4); ans1 = p.^(4:-1:1), ans2 = p.^(0:-1:-3)
```

which shows that the two expressions yield the same result.

Example 7.25. Compute Bin 3 for the 8-pt DFT of the sequence $x = [-3:1:4]$ using the Goertzel algorithm and using the function *fft*.

We run the code

```
Bin = 3; x = [-3:1:4]; N=length(x);
GBin = filter(1,[1 -(exp(j*2*pi*Bin/N))],[x 0]);
GoertzelBin = GBin(N+1), ft = fft(x); ftBin = ft(Bin+1)
```

You can change the values for Bin and x in the code above to experiment further.

7.15.2 USING COMPLEX CONJUGATE POLES

We can convert the single-complex-pole Goertzel IIR to an all-real coefficient IIR by using the standard technique for converting a fraction having a complex denominator to one having a real denominator. A complex FIR is created by this conversion, but its computation need only be done for the final or $(N+1)$-th output of the IIR.

$$G(z) = \frac{(1 - p^* z^{-1})}{(1 - pz^{-1})(1 - p^* z^{-1})} = \frac{(1 - p^* z^{-1})}{1 - 2\text{Re}(p)z^{-1} + |p|^2 z^{-2}}$$

Note that since p lies on the unit circle, its magnitude is 1. Note also that the real part of $\exp(\text{j}2\pi k/N)$ is $\cos(2\pi k/N)$. Thus the simplified z-transform is

$$G(z) = \frac{1 - \exp(-j2\pi k/N)z^{-1}}{1 - 2\cos(2\pi k/N)z^{-1} + z^{-2}}$$

To implement this filter, we perform the IIR portion and obtain the N-th and $(N+1)$-th outputs; the net output is the $(N+1)$-th IIR output minus $\exp(-j2\pi k/N)$ times the N-th IIR output.

Example 7.26. Compute Bin 3 for the 8-pt DFT of the sequence x = [−3 : 1 : 4] using the Goertzel algorithm as described above using a pair of complex conjugate poles.

```
Bin = 3; x = [-3:1:4]; N=length(x);
GB = filter(1,[1 -2*(cos(2*pi*Bin/N)) 1],[x, 0]);
GrtzlBin = GB(N+1) - exp(-j*2*pi*Bin/N)*GB(N),
ft = fft(x); ftBin = ft(Bin+1)
```

7.15.3 MAGNITUDE ONLY OUTPUT

If only the magnitude of the DFT bin is needed, the final step of the Goertzel algorithm, which uses complex arithmetic, can be replaced with read-only arithmetic, yielding an all-real algorithm. The magnitude squared of the DFT bin as computed by the expression for *GrtzlBin* in the code above can be obtained readily as

```
MagSqGBin = GB(N+1)^2 - 2*cos(2*pi*Bin/N)*GB(N+1)*GB(N) + GB(N)^2,
AbsGBin = MagSqGBin^0.5
```

7.16 LINEAR, PERIODIC, AND CIRCULAR CONVOLUTION AND THE DFT

The mathematical formula for convolution, which we now further denote, for purposes of this discussion, as **Linear Convolution**, is, for two sequences $b[n]$ and $x[n]$ and Lag Index k, defined as

$$y[k] = \sum_{n=0}^{N-1} b[n]x[k-n] \tag{7.24}$$

Figure 7.22 shows two sequences of length eight, and their linear convolution, which has a length equal to 15 (i.e., $2N$ -1).

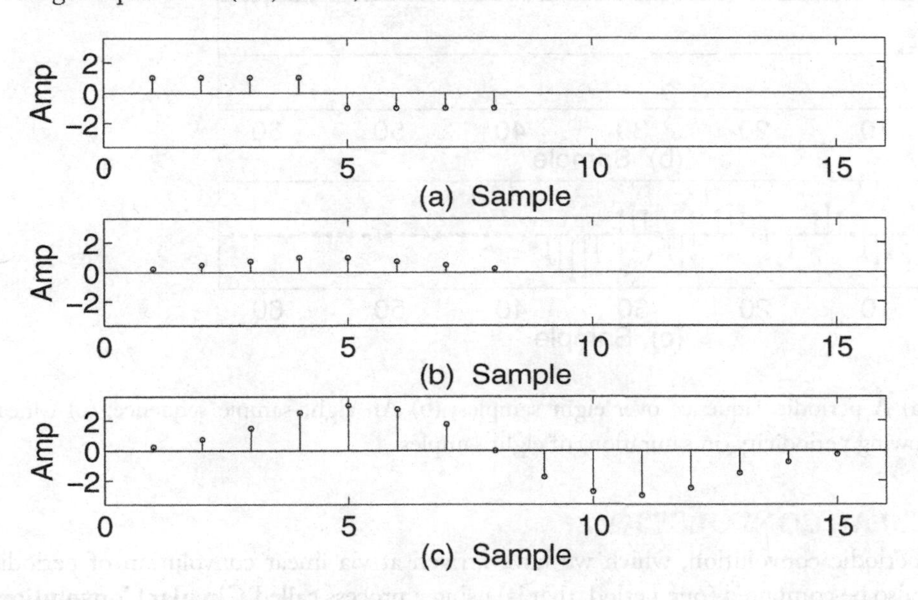

Figure 7.22: (a) Sequence 1; (b) Sequence 2; (c) Linear Convolution of Sequences 1 and 2.

7.16.1 CYCLIC/PERIODIC CONVOLUTION

Consider the case in which two sequences are both periodic over N samples. In this case, the linear convolution (when in steady-state or saturation) produces a convolution sequence that is also periodic over N samples.

Figure 7.23 shows a simple scheme wherein one sequence is just a repetition of an eight sample subsequence, and the other is a single period of an eight sample sequence. The linear convolution of these two sequences results in a periodic, or cyclic, convolution sequence when the two sequences are in saturation (steady-state response). If you look closely, you'll see that the two end (transient-response) segments of the overall convolution would actually, by themselves (removing the cyclic part of the convolution and bringing the two end segments together) constitute the linear convolution of the two eight sample sequences. Figure 7.24 shows more or less the same thing only with two periods of the second sequence. The result is that the noncyclic part of the convolution is about twice as long (nonsaturation is as long at each end as the shorter of the overall lengths of the two sequences, minus one) and the cyclic part has twice the amplitude.

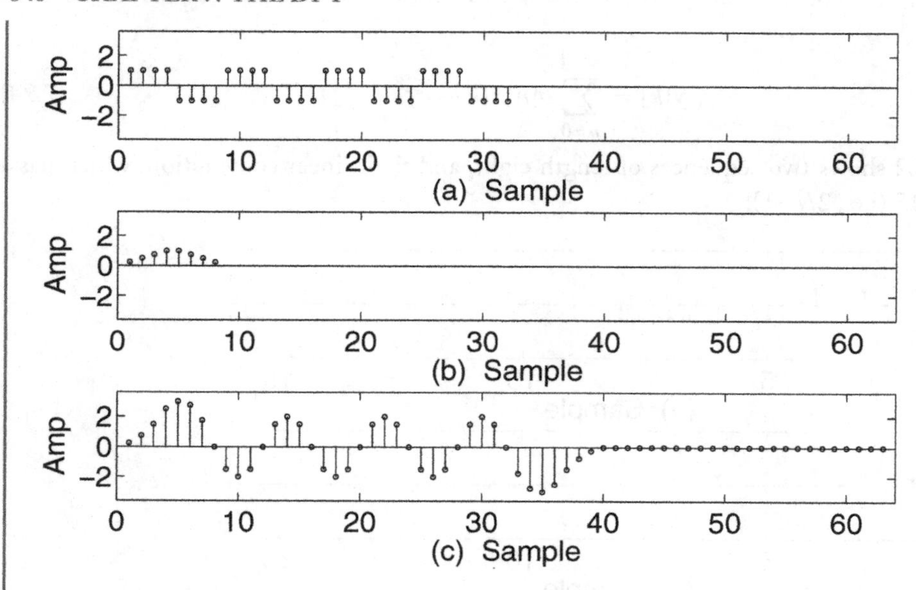

Figure 7.23: (a) A periodic sequence over eight samples; (b) An eight-sample sequence; (c) Linear convolution, showing periodicity (in saturation) of eight samples.

7.16.2 CIRCULAR CONVOLUTION

The cyclic or periodic convolution, which we have arrived at via linear convolution of periodic sequences, can also be computed (one period, that is) using a process called **Circular Convolution**, which we can define as

$$y[k] = \sum_{n=0}^{N-1} b[n](x[k-n]_N) \tag{7.25}$$

where $b[n]$ and $x[n]$ are two N-point sequences and the expression

$$x[k-n]_N$$

is evaluated modulo-N, which effectively extends the sequence $x[n]$ forward and backward so the N point convolution is a periodic convolution of $b[n]$ and $x[n]$. The term circular convolution is used since with the modulo index evaluation, a negative index effectively rotates to a positive one. For example, if $[k-n] = 0$, $[k-n]_N = 0$; if $[k-n] = -1$, $[k-n]_N = N - 1$; if $[k-n] = N$, $[k-n]_N = 0$, and so forth. Using the MathScript function $mod(x, N)$, for example, a Command Line call such as

y = mod(-2,8)

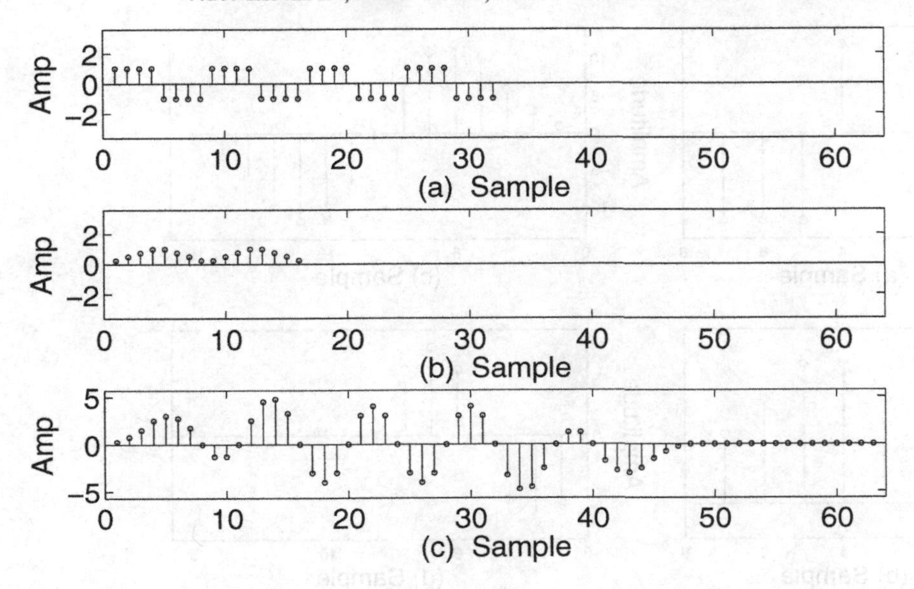

Figure 7.24: (a) A periodic sequence over eight samples; (b) Two cycles of an eight-sample sequence; (c) Linear convolution, showing periodicity (in saturation) of eight samples.

yields $y = 6$ which is N - 2. Recall that in evaluating the linear convolution formula (Eq. (7.24)), $x[k-n] = 0$ if $k-n < 0$. Note also that in evaluating $x[k-n]$ in MathScript, an offset of +1 is needed since 0 indices, like negative indices, are not permitted in MathScript. Thus, for example, to evaluate a circular convolution of two length-N sequences $b[n]$ and $x[n]$ in MathScript, you might employ the statement

y(m) = sum(b(n).*(x(mod(m - n, N) +1))); (7.26)

where n is the vector 1:1:N and m may assume values from 1 to N.

Figure 7.25 shows, at (a) and (b), two eight point sequences and, at (c), their linear convolution, having length 15 (8 + 8 - 1), and, at (d), their (8-pt) circular convolution computed using Eq. (7.25) as implemented in MathScript by Eq. (7.26). Compare this figure, plots (a), (b), and (d), to plots (a)-(c) of Fig. 7.23.

7.16.3 DFT CONVOLUTION THEOREM

If $f[n]$ and $g[n]$ are both time domain sequences that are periodic over N samples, then the DFT of the (periodic) convolution over N samples is N times the product of the DFTs of each sequence. Stated more compactly

$$DFT(f[n] \circledast_N g[n]) = NF[k]G[k]$$

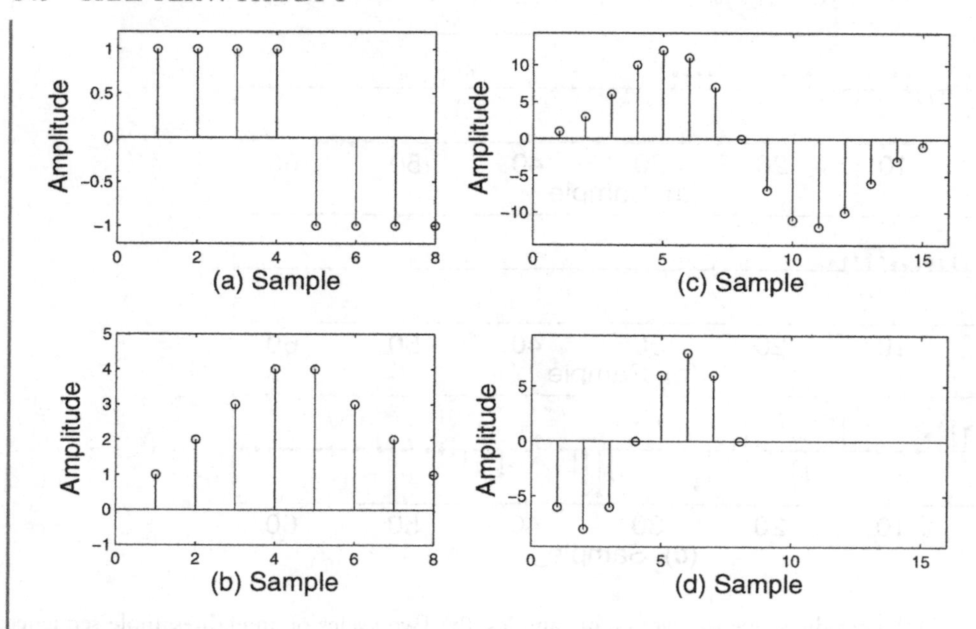

Figure 7.25: (a) First sequence; (b) Second sequence; (c) Linear convolution of sequences at (a) and (b); (d) Circular convolution of sequences at (a) and (b).

where $\circledast_N$ means a periodic (or circular) convolution over N samples. Taking the inverse DFT of each side of the equation, we get

$$f[n] \circledast_N g[n] = N \cdot DFT^{-1}(F[k]G[k]) \quad (7.27)$$

which states that the periodic convolution of two sequences of length N, is N times the inverse DFT of the product of the DFT of each sequence. This assumes that the DFTs are scaled by $1/N$. Since MathScript does not scale the DFT by $1/N$, but rather scales the IDFT by $1/N$, it follows that for MathScript

$$f[n] \circledast_N g[n] = DFT^{-1}(F[k]G[k]) \quad (7.28)$$

Example 7.27. Perform circular convolution of the sequences b = [1, 2, 2, 1] and x = [1, 0, −1, 1] using both the time domain method and the DFT Convolution Theorem, and plot the results for comparison.

The following code will suffice:

```
b = [1,2,2,1]; x = [1,0,-1,1]; N = 4; n = 1:1:N;
for m =1:1:N; CirCon(m) = sum(b(n).*(x(mod(m-n,N)+1)));
```

```
end; subplot(2,1,1), stem(CirCon)
subplot(2,1,2); y = real(ifft(fft(b).*fft(x))); stem(y)
```

7.16.4 LINEAR CONVOLUTION USING THE DFT

- A linear convolution of two sequences cannot be done using DFTs of the original sequences, but, by padding the original sequences with zeros, a problem in linear convolution can be converted to one computable as a periodic convolution, which can be computed using DFTs. For longer sequences, far greater efficiency in terms of number of computations needed can be obtained by using the DFT (implemented by FFT) technique instead of linear convolution.

Figure 7.26 illustrates how to perform a linear convolution of two length-8 sequences using circular convolution. First, pad the original sequences with zeros to a length equal to or greater than the expected length of the linear convolution. Since the ultimate goal is to use DFTs, and standard FFTs operate on sequence lengths that are powers of two, you should pick the smallest power of two that is equal to or larger than the expected linear convolution length, which is the sum of the lengths of the two sequences, less one. In this case, we pick 16 as the padded length since 15 is the required length, and 16 is the lowest power of two which equals or exceeds 15. The linear convolution of the two padded sequences is the first 15 samples of plot (c), while 16-pt circular convolutions are shown in (d) and (e), the first 15 samples of each being equivalent to the linear convolution. The circular convolution at (d) was performed using Eq. (7.25), while the circular convolution at (e) was performed using Eq. (7.28).

Example 7.28. Consider the two sequences [1, 2, 2, 1] and [1, 0, −1, 1]. Perform the linear convolution, then obtain the same result using circular convolution. Then verify the second computation using the DFT Convolution Theorem. Plot each of the three computations on a separate axis of the same figure for comparison.

We perform and plot the linear convolution with the call

```
figure(120); subplot(3,1,1); stem(conv( [1,2,2,1], [1,0,-1,1]))
```

For the circular convolution, we first pad both sequences with zeros to at least a length which is the sum of the lengths of the two (unpadded) sequences minus one. We then evaluate using Eq. (7.26). We define the sequences and their padded length, perform and display the computation, and verify this using the DFT Convolution Theorem with this script

```
b = [1,2,2,1,zeros(1,3)]; x = [1,0,-1,1,zeros(1,3)];
N = 7; n = 1:1:N;subplot(3,1,2)
for m =1:1:N; CirCon(m) = ...
sum(b(n).*(x(mod(m-n,N)+1))); end;
stem(CirCon); subplot(3,1,3);
```

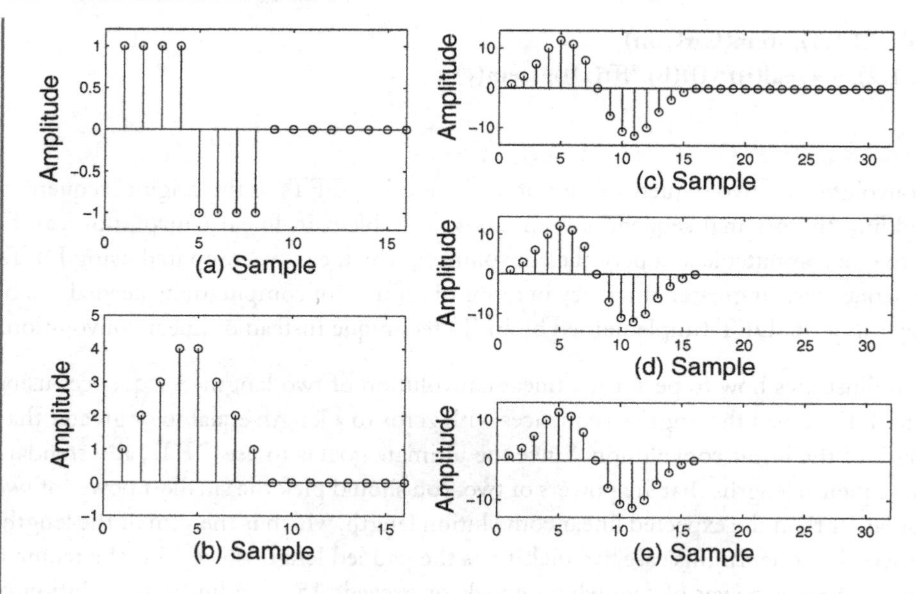

Figure 7.26: (a) First sequence, padded to length 16; (b) Second sequence, padded to length 16; (c) Linear convolution of sequences at (a) and (b); (d) Circular convolution of sequences at (a) and (b) using direct formula; (e) Circular convolution of sequences at (a) and (b) using the DFT technique.

```
y = real(ifft(fft(b,8).*fft(x,8))); stem(y(1:7))
```

Note that the DFTs have been specified as having a length equal to the smallest power of two equal to or greater than the minimum length necessary to result in a linear convolution, which is 4 + 4 -1 = 7, yielding a DFT length of eight.

7.16.5 SUMMARY OF CONVOLUTION FACTS

- Two sequences of length N when linearly convolved, yield a linear convolution of length $2N - 1$ (see Fig. 7.22).

- Two sequences, each periodic over N samples (at least one of which contains many periods) when linearly convolved, result in a convolution sequence which, in steady-state, is periodic over N samples, and is hence called a Periodic Convolution (see Fig. 7.24).

- One period of a periodic convolution can be computed using the Circular Convolution formula, Eq. (7.25).

- A linear convolution of two length-N sequences can be performed by padding each with zeros to a length at least equal to $2N - 1$, and performing circular convolution (see Fig. 7.26).

- A circular convolution in the time domain can be equivalently performed using DFTs (see Eq. (7.27) or (7.28)).

- By padding two length-N sequences with zeros up to a power of 2 which is at least equal to $2N - 1$, a linear convolution of the unpadded sequences using DFTs can be performed. This can save computation for larger sequences due to the efficiency of the FFT.

7.16.6 THE OVERLAP-ADD METHOD

Let's consider a common situation, in which we want to convolve a filter impulse response of finite length N, and a signal sequence of much greater length.

If both sequences are of finite length, the simplest approach is to add the lengths of the two sequences and then pad that number up to the next power of two, then take the real part of the *ifft* of the product of the DFTs of the two padded sequences.

Example 7.29. Perform linear convolution in the time domain of a 1024 sample linear chirp and the impulse response [1, 1, 1, 1], and then perform the same computation using the DFT Convolution Theorem.

We define the signals, compute the convolution both ways, and plot the results with this call

```
SR = 1000; ts = chirp( [0:1/(SR-1):1],0,1,SR/2);
s = [1,1,1,1]; lincon = conv(s, ts);
linconbyfft = real(ifft( fft(ts, 1024).*fft(s,1024)));
figure(14); subplot(2,1,1); plot(lincon);
subplot(2,1,2); plot(linconbyfft(1:1003))
```

Note that the linear convolution has a length equal to 1000 + 4 -1 = 1003, and thus we take the leftmost 1003 samples of the convolution-via-fft as the linear convolution.

If one of the sequences is of indefinite length (such as a sample stream from a real-time process), an approach is to size the DFTs according to the length N of the impulse response (the shorter sequence) and to break the longer sequence into shorter subsequences of length P, each of which is convolved with the impulse response, and the partial responses overlapped with the proper offset and added. The subsequence length P might also equal N, but it could be longer as well. Either way, the impulse response and the subsequence are padded with zeros to a length at least equal to the sum of their lengths less one, and then up to the next power of two.

As an example, we'll use an impulse response of length N as one period of an N-periodic sequence, and divide the signal sequence into subsequences, each of length N.

The following procedure is known as the **Overlap and Add Method**:

- 1) Pad the impulse response with another N zeros to make a sequence of $2N$. If $2N$ is not a power of 2, pad with additional zeros until a power of 2 length is reached (call the final padded length M).

- 2) Take the first N samples of the signal and pad them to length M.
- 3) Compute the inverse DFT of the product of the DFTs of the two sequences of length M.
- 4) The result from 3) above is a sequence of length M—take the leftmost $2N-1$ samples, which is the proper length of the linear convolution of the original two sequences of length N. These samples form the first $2N-1$ output samples (however, only the first N samples will remain unchanged, as the next iteration will be superposed on this first output sequence starting at sample $N+1$).
- 5) Take the second N samples of the signal sequence, pad to length M, and use with the padded impulse response to obtain the next $2N-1$ samples of the output sequence using the DFT method.
- 6) Superpose these $2N-1$ samples onto the current output sequence starting at sample $N+1$.
- 7) Keep repeating the procedure for each N signal samples, and superposing the result of the Pth computation starting at output sequence sample index $(P-1)(N)+1$. For example, our first computation ($P=1$) was added into the output sequence starting at sample 1, and our second computation (a sequence of length $2N-1$) was added in starting at sample $(2-1)(N)+1 = N+1$, and so forth until all signal samples have been processed.

Example 7.30. Perform linear convolution using the DFT and the Overlap-Add Method.

Figure 7.27 shows a time just after the beginning of the DFT convolution of an eight-sample impulse response with a much longer test signal. The test signal is divided into nonoverlapping eight sample subsequences. The impulse response, and each subsequence in its turn, is padded to length-16. Then the inverse DFT of the product of the DFTs of the impulse response and subsequence is added into the cumulative convolution, starting at the beginning sample index of the particular subsequence being computed. Figure 7.27 shows the first subsequence's contribution graphed in plot (f) as a stem plot, with the second subsequence's contribution overlaid, without yet being added.

Figure 7.28 shows the result after the second subsequence's contribution in Fig. 7.27 has been added to the composite output. Figure 7.29 shows the result after the third subsequence's contribution has been added to the composite output.

Example 7.31. Break the linear convolution of [1, 1] and [1, 2, 3, 4] into two convolutions, do each using the DFT, and then combine the results using the overlap-and-add method to achieve the final result.

For the first convolution, we pad [1,1] and [1,2] each with two zeros to form length-4 sequences, then take the real part of the inverse DFT of the product of the DFT of each, or in m-code

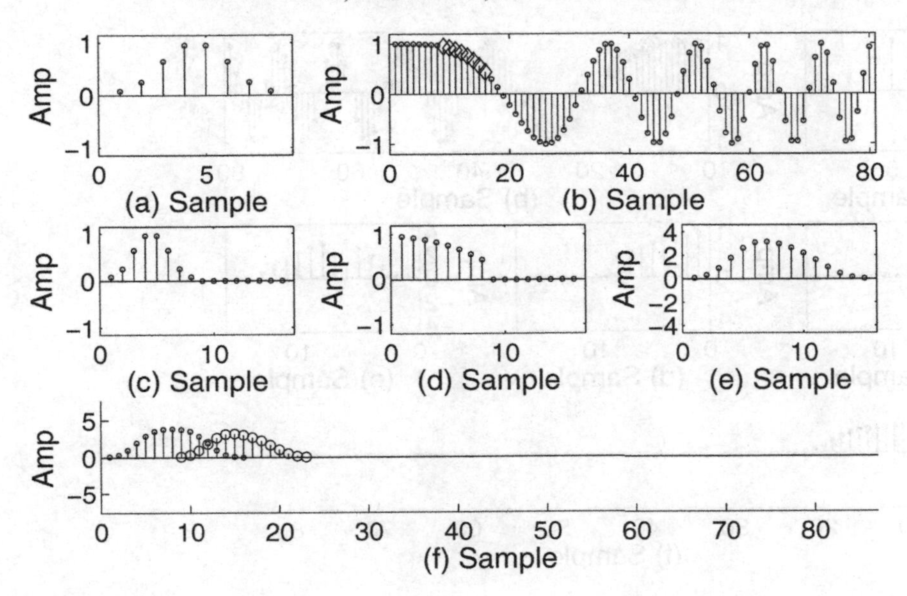

Figure 7.27: (a) An eight-sample impulse response; (b) A chirp, with second subsequence of eight-sample marked; (c) Eight-sample impulse response padded with another eight zeros; (d) Second eight-sample subsequence from (b), padded with zeros; (e) First 15 samples of circular convolution of (c) and (d), performed using DFTs; (f) Result from (e), plotted on output graph at samples 9-23, prior to being added to result from previous computation, plotted as samples 1-15.

```
FirstConv = real(ifft(fft([1,1,0,0]).*fft([1,2,0,0])))
```

which yields the sequence [1,3,2,0]. The second convolution is stated in m-code as

```
SecConv = real(ifft(fft([1,1,0,0]).*fft([3,4,0,0])))
```

which results in the sequence [3,7,4,0] which is then added to the first sequence, but shifted to the right two samples:

1	3	2	0		
		3	7	4	0
1	3	5	7	4	

This can be checked using the call

```
y = conv([1,1], [1,2,3,4])
```

which yields the identical result.

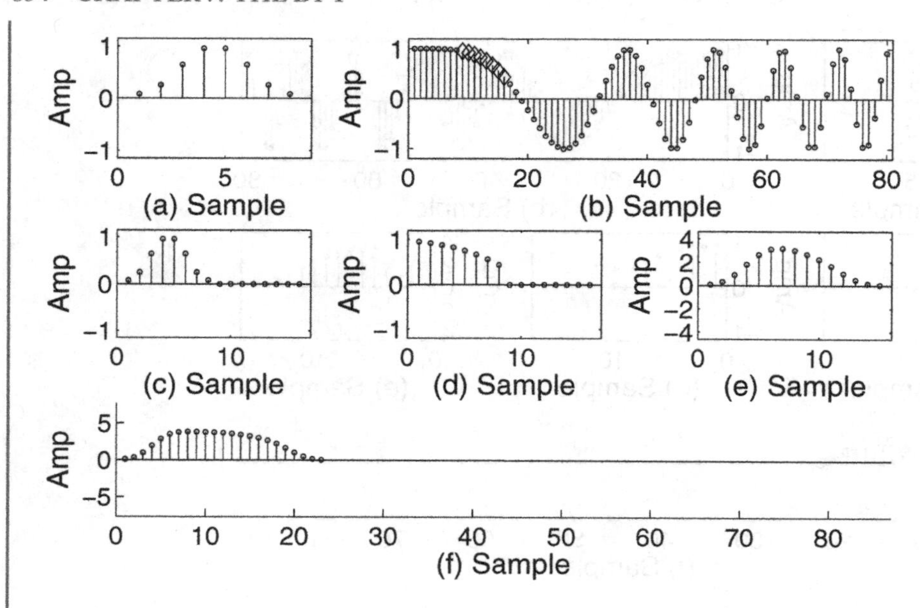

Figure 7.28: (a) Eight-sample impulse response; (b) Chirp, with second subsequence of eight samples marked; (c) Eight-sample impulse response padded with another eight zeros; (d) Second eight-sample subsequence from (b), padded with zeros; (e) First 15 samples of circular convolution of (c) and (d), performed using DFTs; (f) Result from (e), after being added to previous result, giving a valid linear convolution up to sample 16.

7.17 DFT LEAKAGE

7.17.1 ON-BIN/OFF-BIN: DFT LEAKAGE

Consider as an example a 16 sample sequence. The test correlator frequencies that the DFT will use are -7:1:8. Any signal frequency equal to one of these will have a high correlation at the same test correlator frequency, and a zero-valued correlation at all other test correlator frequencies due to orthogonality. Such integer-valued signal frequencies are described as **On-Bin** or evoking an on-bin response, which is a response confined to a single bin.

In general, noninteger signal frequencies (i.e., not equal to any of the test frequencies 0,1,2, etc) will evoke some response in most bins. This property is usually referred to as **Leakage**, i.e., **Off-Bin** signal energy "leaks" from the closest DFT bin into other bins.

Example 7.32. Demonstrate DFT Leakage by taking the DFT of two sequences, each of which is 64 samples long, the first of which contains a five-cycle cosine, and the second of which contains a 5.3-cycle cosine.

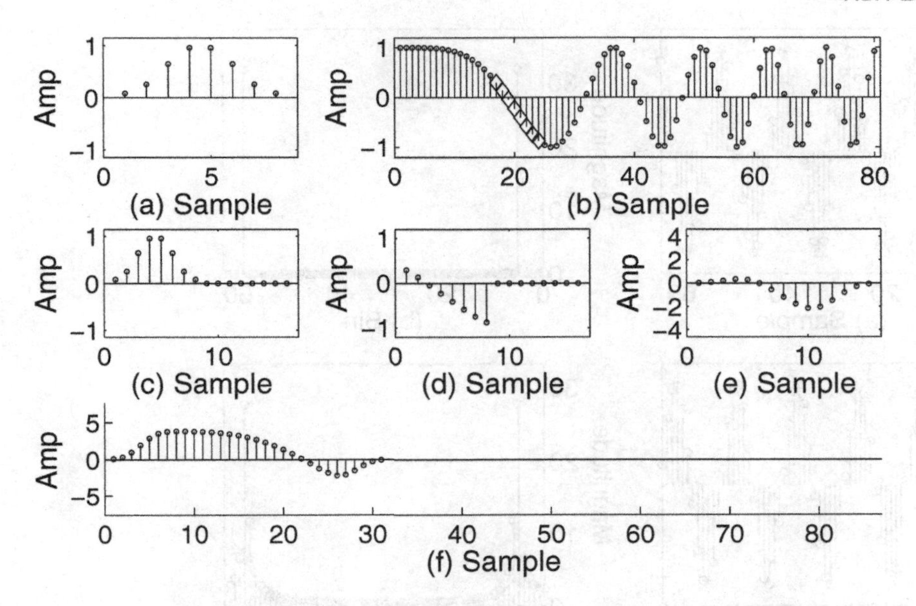

Figure 7.29: (a) Eight-sample impulse response; (b) Chirp, with third subsequence of eight-samples marked; (c) Eight-sample impulse response padded with another eight zeros; (d) Third eight-sample subsequence from (b), padded with zeros; (e) First 15 samples of circular convolution of (c) and (d), performed using DFTs; (f) Result from (e), after being added to previous result, giving a valid linear convolution up to sample 24.

Figure 7.30, in plot (a), shows the DFT of the signal containing the five-cycle signal; note that only Bins ±5 are nonzero (Bin -5 appears as Bin 59 since the DFT uses k = 0:1:N-1).

7.17.2 AVOIDING DFT LEAKAGE-WINDOWING

When the DFT of a sequence is taken, off-bin frequencies leak to some extent into all bins. The result can be, in the general case, a greatly diminished ability to distinguish discrete frequencies from each other when they are close. Leakage can, however, be reduced by multiplying the signal sequence by a smoothing function called a **Window** prior to taking the DFT.

Example 7.33. Demonstrate windowing on two 64-sample sequences, one of which contains a five-cycle cosine and the other of which contains a 5.3 cycle cosine.

Figure 7.31 shows what happens when a smoothing function, in this case a Hamming window (discussed in more detail below), is used to smooth each sequence. Note that the on-bin signal is somewhat degraded, but the off-bin signal is considerably improved. In most cases, when the frequency content of a signal is unknown or random, use of a window similar to that shown in Fig. 7.31 is advantageous. This is explored in detail immediately below.

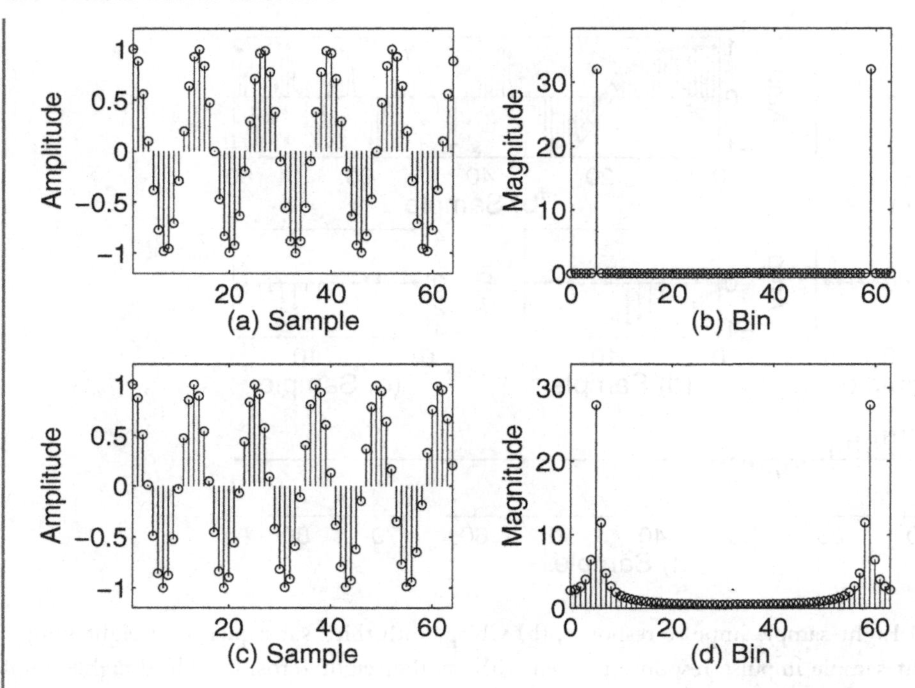

Figure 7.30: (a) Signal containing a five-cycle cosine; (b) DFT of signal in (a); (c) Signal containing a 5.3 cycle cosine; (d) DFT of signal in (c).

7.17.3 INHERENT WINDOWING BY A RECTANGULAR WINDOW

The Rectangular window of length N, $R_N[n]$, is defined as

$$R_N[n] = \begin{cases} 1 & n = 0{:}1{:}N\text{-}1 \\ 0 & \text{otherwise} \end{cases} \tag{7.29}$$

The act of obtaining a finite sequence of length N essentially or *inherently* multiplies it by a rectangular sequence of length N, and the product of two sequences in the time domain has a frequency spectrum that is the periodic convolution of the individual frequency responses of the two sequences.

Mathematically, denoting the DTFT of $x[n]$ as $X(e^{j\omega})$, the DTFT of a given window sequence $w[n]$ as $W(e^{j\omega})$, the frequency response $X_W(e^{j\omega})$ of the windowed sequence $w[n]x[n]$ is

$$X_W(e^{j\omega}) = \frac{1}{2\pi}\int_{-\pi}^{\pi} X(e^{j\phi})W(e^{j(\omega-\phi)})d\phi$$

This process is shown in Fig. 7.32, in which sequences of length 2^14 were used to generate a good approximation to the DTFT, and the periodic convolution was performed numerically via

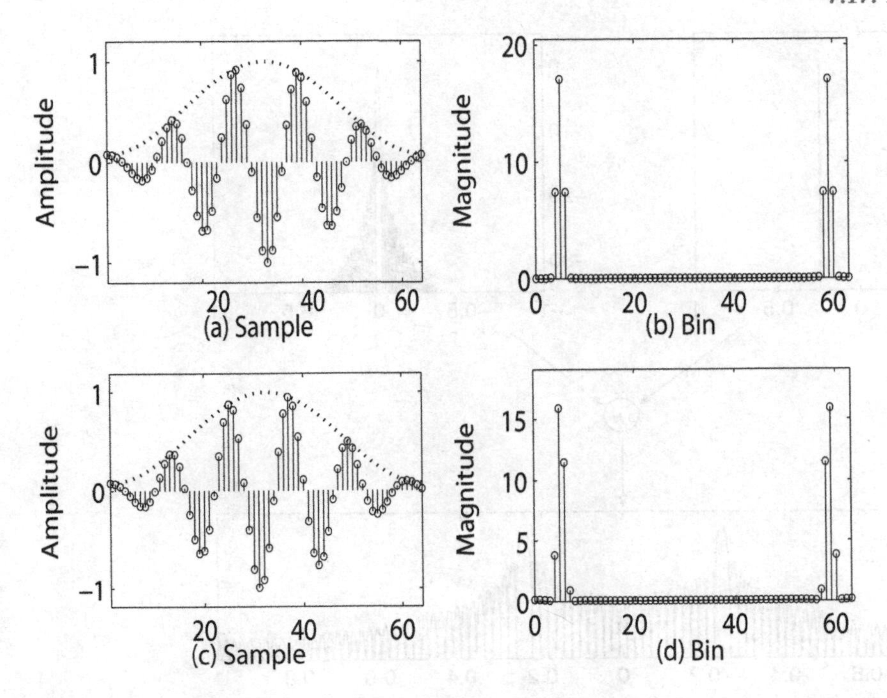

Figure 7.31: (a) Signal containing a five-cycle cosine, multiplied by a Hamming window, shown as a dashed line; (b) DFT of signal in (a); (c) Signal containing a 5.3 cycle cosine, multiplied by a Hamming window, shown as dashed line; (d) DFT of signal in (c).

circular convolution. A different look at the same signals, with a close-up of the DTFT of the window, is shown in Fig. 7.33.

We can see the effect of window length by using a rectangular window having 1001 (instead of 101) central samples valued at 1.0, the result of which is shown in Fig. 7.34. Note that with the longer window length, the frequency response of the window (plot(b)) is much narrower, and hence the net frequency response of the windowed sequence is closer to the DTFT (plot(a)) than with a shorter window. We can further emphasize this relationship with one more experiment, in which the window's central nonzero portion is shortened to 10 samples, as shown in Fig. 7.35.

As shown above, a rectangularly-windowed sequence is, due to the scalloped characteristic of its DTFT, "leaky," meaning it lets frequencies differing widely from the desired one(s) leak into the output. Fortunately, we do not have to accept this outcome. Since the net frequency response of any windowed sequence is the circular convolution in the frequency domain of the sequence's frequency response (DTFT) and the frequency response (DTFT) of the window it has been multiplied by, it should be possible to achieve a more desirable spectral response by choosing a window which, when

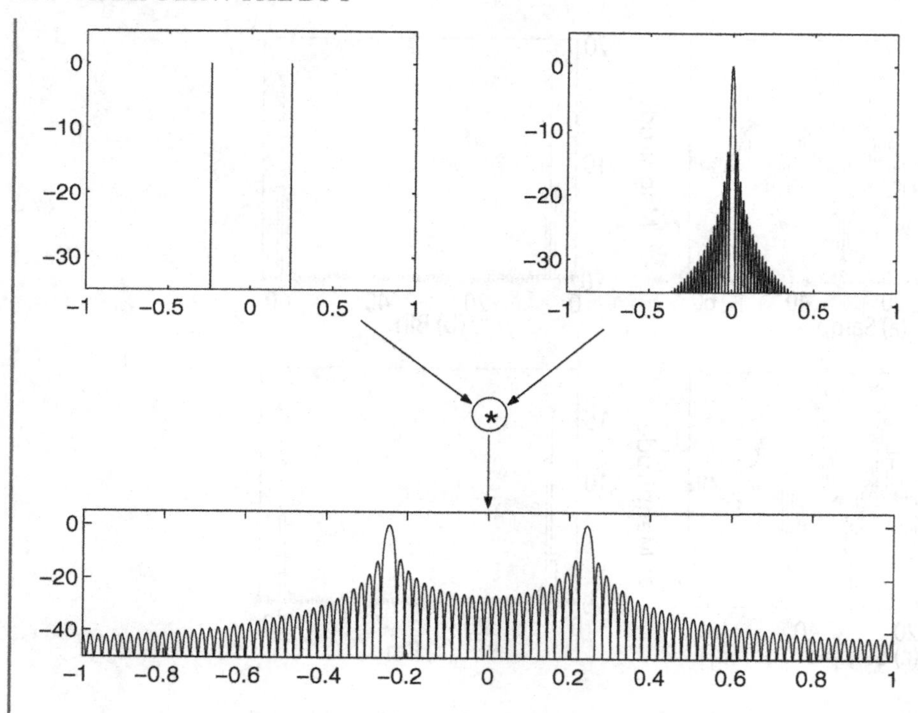

Figure 7.32: Upper left, magnitude (dB) of DTFT (via DFT of length 2ˆ14) of a cosine wave of frequency 2000; Upper right, magnitude (dB) of DTFT of Rectangular window of length 2ˆ14, with central 101 samples valued at 1.0, and all other samples valued at 0; Lower, Magnitude (dB) of the Circular Convolution of the two DTFTs shown above, giving the frequency response of the windowed cosine sequence. All horizontal axes are frequency in units of π, and all vertical axes are magnitude in dB.

convolved with the sequence's spectrum, results in less spreading out or smearing of the sequence's spectrum.

To this end, many different windows have been developed. In general, it is desirable to apply a (nonrectangular) window to a sequence prior to performing the DFT. We will explore why this is so in the following sections.

7.17.4 A FEW COMMON WINDOW TYPES

Rectangular

We defined the Rectangular window above in Eq. (7.29); Figure 7.36 shows a rectangular window having $N = 64$ which is used to window a sinusoid having 16 cycles over 64 samples; the magnitude of the DTFT is shown in plot(d). Note the heavily scalloped effect in plot(d).

Hamming

The *hamming* window of length N is computed according to the formula:

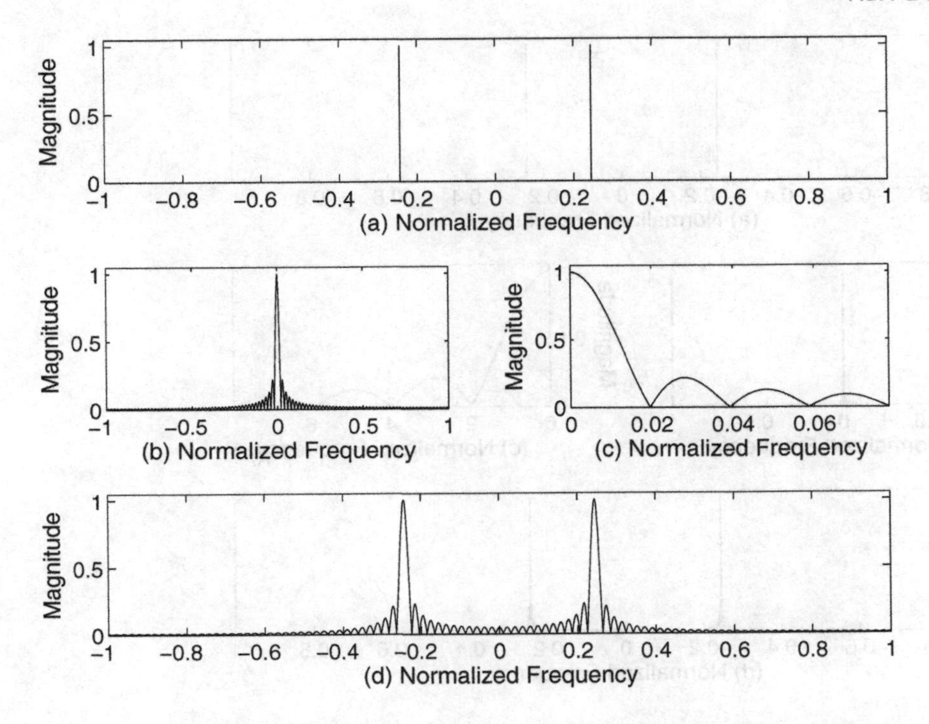

Figure 7.33: (a) DTFT magnitude of a cosine wave of length 2^14 and frequency 2000; (b) DTFT magnitude of a length-2^14 rectangular window having 101 contiguous, central values of 1.0 with all other values equal to 0; (c) A zoomed-in view of part of (b); (d) Frequency response of the windowed cosine sequence, i.e., the magnitude of the circular convolution of the DTFTs at (a) and (b).

$$w[n] = \begin{cases} 0.54 - 0.46(\cos[2\pi n/(N-1)]) & n = 0{:}1{:}N\text{-}1 \\ 0 & \text{otherwise} \end{cases}$$

Figure 7.37 shows the *hamming* window applied to a 16-cycle test sequence.

Blackman

The *blackman* window of length N is computed according to the following formula, in which $M = N - 1$:

$$w[n] = \begin{cases} 0.42 - 0.5\cos[2\pi n/M] + 0.08\cos[4\pi n/M] & n = 0{:}1{:}N\ \text{-}1 \\ 0 & \text{otherwise} \end{cases}$$

Let's take a look at the attenuation characteristic of the *blackman* window using a log plot, which shows the fine structure of the sidelobes. From Fig. 7.38, plot(d), you can readily see that the scalloped sidelobes are still there, only greatly attenuated relative to the rectangular window's

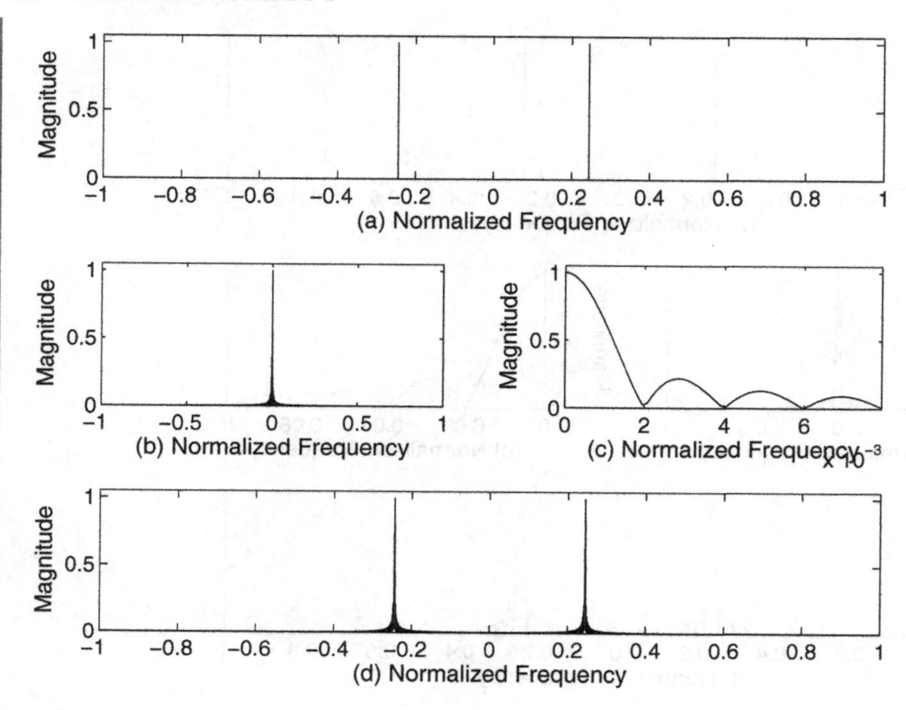

Figure 7.34: (a) DTFT magnitude of a cosine wave of length 2^14 and frequency 2000; (b) DTFT magnitude of a length-2^14 rectangular window having 1001 contiguous, central values of 1.0 with all other values equal to 0; (c) A zoomed-in view of part of (b); (d) Frequency response of the windowed cosine sequence, i.e., the magnitude of the circular convolution of the DTFTs at (a) and (b).

sidelobes. The central lobe, however, is much wider than that of either of the rectangular or hamming windows.

Kaiser
The *kaiser* window is not a single window, but a family of windows that are specified by the number of samples in the desired window and a parameter β that essentially allows you to choose the tradeoff between main lobe width and sidelobe amplitude. Figure 7.39 shows the *kaiser* window with β =5.

You can run these scripts

LVDTFTWindowsOnly(64)

LVDTFTWindowing(64)

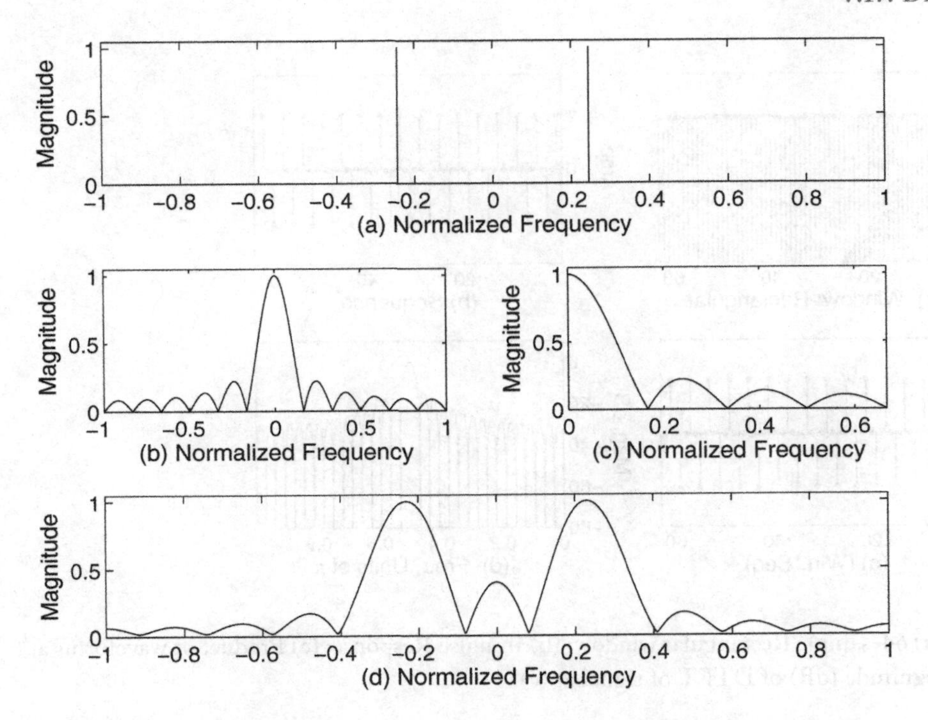

Figure 7.35: (a) DTFT magnitude of a cosine wave of length 2^14 and frequency 2000; (b) DTFT magnitude of a length-2^14 rectangular window having only 11 contiguous, central values of 1.0 with all other values equal to 0; (c) A zoomed-in view of part of (b); (d) Frequency response of the windowed cosine sequence, i.e., the magnitude of the circular convolution of the DTFTs at (a) and (b).

to view a sequence of about 15 different windows. The number in parentheses is the number of samples in the window, and you may change that number when calling the script(s).

7.17.5 DFT LEAKAGE V. WINDOW TYPE

You may run the script (see exercises below)

$$LVxWindowingDisplay(NoiseAmp, freq1, freq2)$$

which was used to generate the plots associated with this section of text. The arguments are as follows: $NoiseAmp$ is an amount of noise, $freq1$ and $freq2$ are frequencies that are immersed in the noise and which must be discriminated by performing a DFT on the composite signal (specific calls are given for the examples discussed below).

Let's do several experiments in which there are two sinusoids in white noise. In the first experiment, the sinusoids will have frequencies of 66 and 68 cycles and amplitudes of 1.0 each. These integral frequencies coincide perfectly with FFT test correlators, so such frequencies are

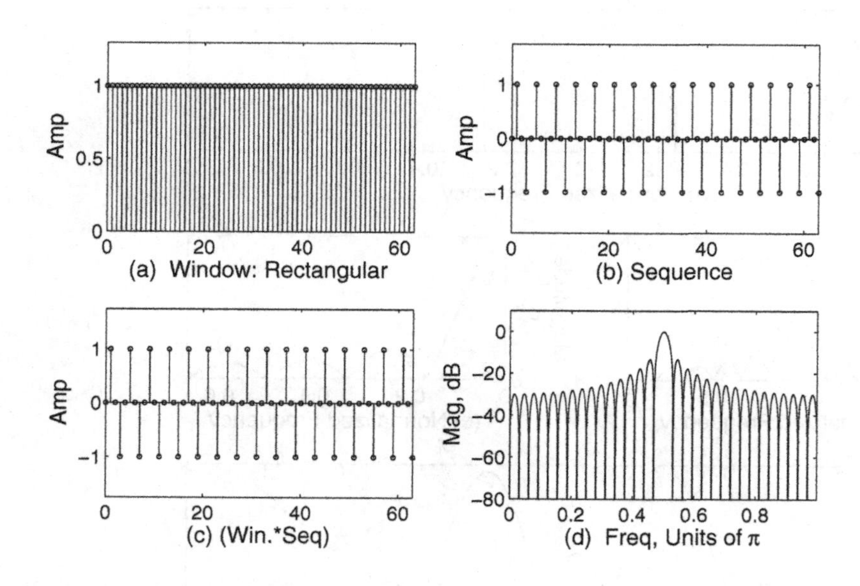

Figure 7.36: (a) 64-sample Rectangular window; (b) Impulse Response; (c) Product of waveforms at (a) and (b); (d) Magnitude (dB) of DTFT of signal at (c).

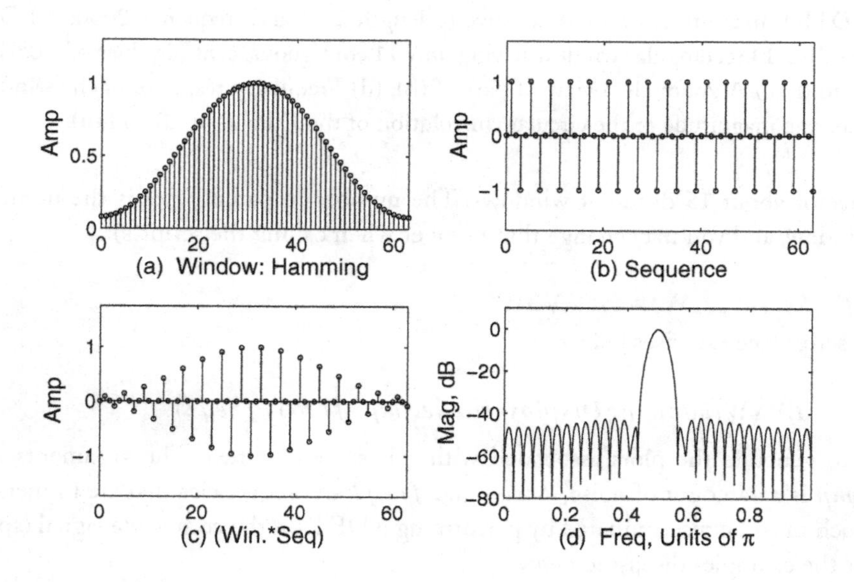

Figure 7.37: (a) 64-sample Hamming window; (b) Signal; (c) Product of waveforms at (a) and (b); (d) Magnitude (dB) of DTFT of signal at (c).

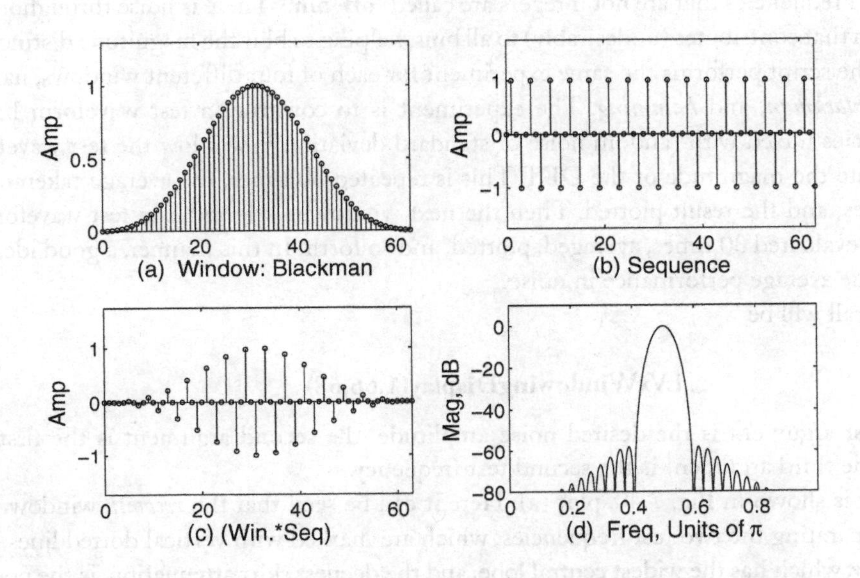

Figure 7.38: (a) 64-sample Blackman window; (b) Signal; (c) Product of waveforms at (a) and (b); (d) Magnitude (dB) of DTFT of signal at (c).

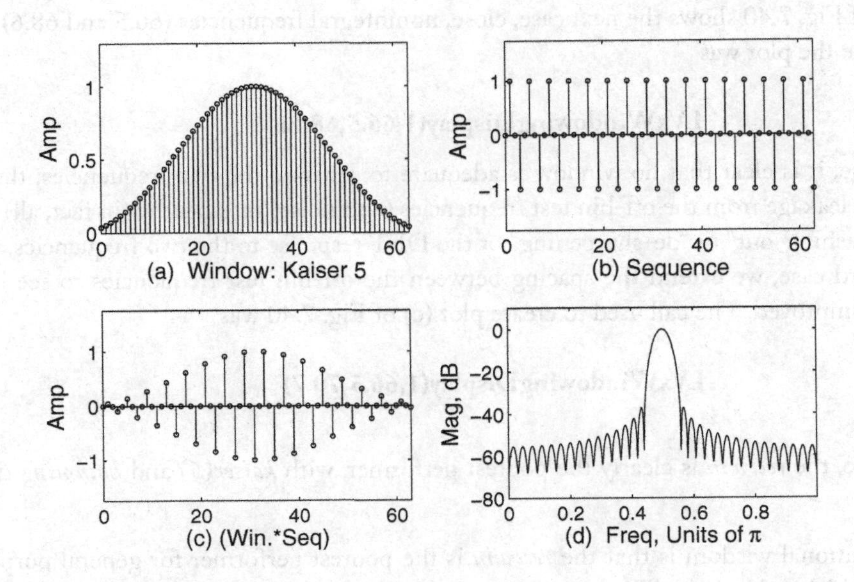

Figure 7.39: (a) 64-sample Kaiser window with $\beta = 5$; (b) Signal; (c) Product of waveforms at (a) and (b); (d) Magnitude of DTFT of signal at (c).

called "**on-bin**." Frequencies that are not integers are called "**off-bin**." There is noise throughout the signal's spectrum that contributes (undesirably) to all bins, helping to blur the magnitude distinctions between bins. The script performs the same experiment for each of four different windows, namely, *rectwin*, *kaiser*, *blackman*, and *hamming*. The experiment is to construct a test waveform having the two frequencies mixed with random noise of standard deviation k, window the test waveform, and then compute the magnitude of the DFT. This is repeated 30 times, the average taken of the DFT magnitudes, and the result plotted. Then the next window is selected, the test waveform is constructed and evaluated 30 times, averaged, plotted, and so forth. In this manner, a good idea can be obtained of the average performance in noise.

Our first call will be

LVxWindowingDisplay(1,66,68)

in which the first argument is the desired noise amplitude, the second argument is the first test frequency, and the third argument is the second test frequency.

The result is shown in Fig. 7.40, plot (a). Here it can be seen that the *rectwin* window is by far the best at separating the two test frequencies, which are marked with vertical dotted lines. The *blackman* window, which has the widest central lobe, and the deepest skirt attenuation, is the poorest in this case, with the *kaiser*(5) and *hamming* windows placing between the *blackman* and the *rectwin*. Note that the two test frequencies in this case are "on-bin," and hence are orthogonal to one another and therefore cannot influence each other's DFT response.

Plot (b) of Fig. 7.40 shows the next case, close, nonintegral frequencies (66.5 and 68.6). The call used to create the plot was

LVxWindowingDisplay(1,66.5,68.6)

In this case, it is clear that no window is adequate to separate the two frequencies; there is simply too much leakage from the off-bin test frequencies (and noise) into nearby (in fact, all) bins; the result is a "washing-out" or "de-sharpening" of the DFT response to the two frequencies.

In the third case, we extend the spacing between the off-bin test frequencies to see if the situation can be improved. The call used to create plot (c) of Fig. 7.40 was

LVxWindowingDisplay(1,66.5,70.7)

In this case, the *rectwin* is clearly the poorest performer, with *kaiser*(5) and *hamming* doing much better.

- The conventional wisdom is that the *rectwin* is the poorest performer for general purposes, due to its wideband leakage. The vast majority of frequencies in an unknown signal are likely to be off-bin (a good assumption unless the contrary is known), and hence a nonrectangular window is likely to be the better choice.

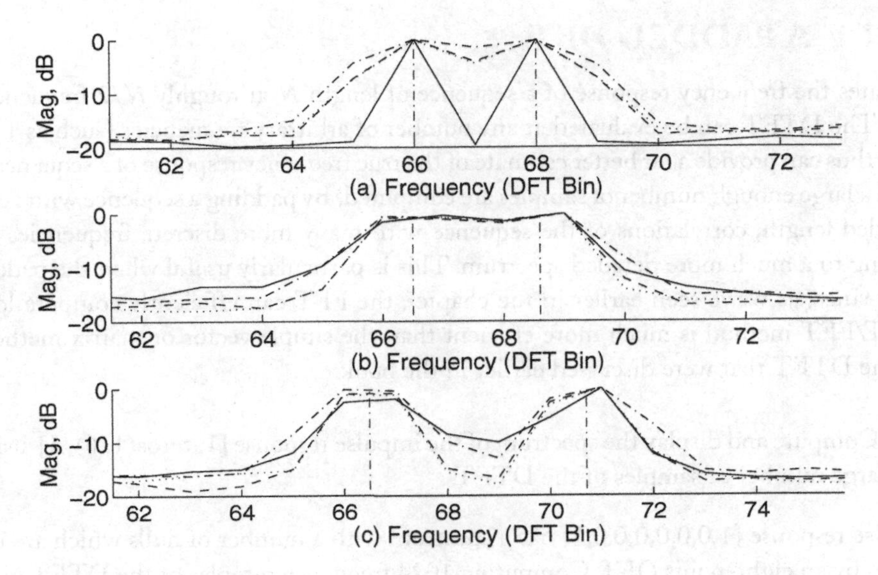

Figure 7.40: (a) Comparison of DFT response to a signal having closely-spaced on-bin frequencies with added noise for various windows–Boxcar: solid; Kaiser(5): dotted; Blackman: dash-dot; Hamming: dashed; (b) Comparison of DFT response to closely-spaced off-bin frequencies with added noise for various windows, plotted as in (a); (c) Comparison of DFT response to widely spaced, off-bin discrete frequencies with added noise, plotted as in (a). In each case the discrete test frequencies are marked with vertical lines.

- When all frequencies being detected are orthogonal to each other, and each is "on-bin" for the DFT, and noise levels are not excessively high, the rectangular window can perform well for frequency discrimination. Each frequency, in this case, will either correlate perfectly with a DFT correlator, or not at all. Hence, the problem of leakage will exist in this case only with respect to wideband noise components in the signal which, if present, will contaminate all bins, lowering the signal-to-noise ratio and blurring the distinction among bins close to each other.

7.17.6 ADDITIONAL WINDOW USE

Windows are also commonly applied to FIR impulse responses. A later chapter (found in Part III of the book) on basic FIR design will explore the benefits and tradeoffs associated with different windows used in FIR design.

7.18 DTFT VIA PADDED DFT

The DFT evaluates the frequency response of a sequence of length N at roughly $N/2$ frequencies such as 0, 1, etc. The DTFT can be evaluated at any number of arbitrary frequencies (such as 1.34, 2.66,3, etc.), and thus can provide a far better estimate of the true frequency response of a sequence or system, provided a large enough number of samples are computed. By padding a sequence with zeros to a quite extended length, correlations of the sequence with many more discrete frequencies are performed, leading to a much more detailed spectrum. This is particularly useful when the padded sequence is long since, as we've seen earlier in the chapter, the FFT can efficiently compute long DFTs. The DFT/FFT method is much more efficient than the simple vector or matrix methods for computing the DTFT that were discussed earlier in the book.

Example 7.34. Compute and display the spectrum of the impulse response $[1, \text{zeros}(1, 6), 1]$ using the DFT and a large number of samples of the DTFT.

The impulse response [1,0,0,0,0,0,0,1] has a spectrum with a number of nulls which are not adequately shown by an eight-point DFT. Computing 1024 frequency samples of the DTFT gives a more complete picture. This was achieved by padding the original impulse response with zeros to a length of 1024 samples. When using the function fft, however, you can simply specify the length of DFT to perform, and MathScript performs the zero-padding automatically. Thus the call

x = [1,0,0,0,0,0,0,1]; y = fft(x,1024); plot(abs(y))

will compute and plot the magnitude of 1024 samples of the DTFT between normalized radian frequencies 0 to 2π.

Figure 7.41 shows the difference. The DFT values are correct for the particular frequencies they test, but the number of frequencies tested is far too small to detect the "fine" structure of the actual frequency response. The script

$$LVDTFTUsingPaddedFFT(FIRXferFcn, FFTLength)$$

allows you to see the difference in spectra between a DFT having the same length as a desired impulse response (*FIRXferFcn*), and *FFTLength* samples of the DTFT. The call used to generate Fig. 7.41 was

LVDTFTUsingPaddedFFT([1,0,0,0,0,0,0,1],1024)

Example 7.35. An easy way to compute the frequency response at a high number of frequencies is to simply pad the input sequence with zeros and then compute the DFT of the resultant sequence.

The script

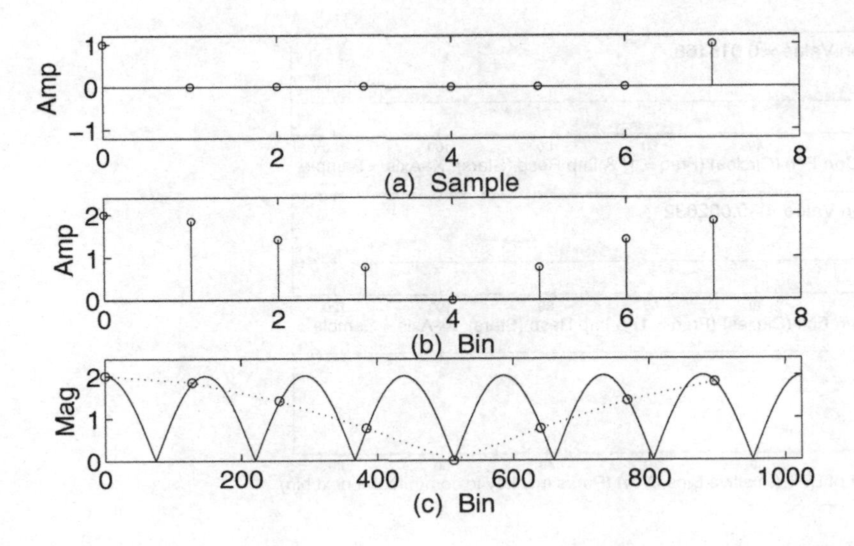

Figure 7.41: (a) Eight-sample impulse response; (b) Magnitude of eight-point DFT; (c) Magnitude of 1024-point DFT (i.e., 1024 samples of the DTFT), solid line, with eight-point DFT magnitude values plotted as stars.

$$LVPaddedDFTMovie(ImpResp, LenDTFT, CompMode)$$

allows you to experiment by inputting a desired impulse response as *ImpResp*, and a desired DFT length, as *LenDTFT*. In making the script call, pass *CompMode* as *1* for automatic computation of the next sample, or *2* if you want to press any key to compute the next sample. The call used to generate Figs. 7.42 and 7.43 was

LVPaddedDFTMovie([1,0,0,0,0,0,0,1],128,1)

Figure 7.42 shows the computation for $k = 1$. The two upper plots show the sequence or impulse response under test plotted from index numbers 0 to 7, with indices 0 and 7 being valued 1 and indices 1-6 being zero (indices 8-127 are padded with zeros to compute the 128-point DFT). The cosine and sine correlators, one cycle, are also plotted, and the correlation value is plotted at index 1 in plot (c). Figure 7.43 shows the next correlation, at $k = 2$. Since the correlators are 128 samples long, k will run from 0 to 64 to cover all unique bins. The DFT of an 8-point real sequence would test five unique frequencies, i.e., k = 0,1,2,3,and 4, while the 128-pt DFT gives us 65 frequencies, i.e., k = 0:1:64. Consequently, there is a much finer gradation of frequency with the longer DFT. The 65 unique bins (Bin 0, positive Bins, and Bin $N/2$) of the 128-pt DFT are shown in Fig. 7.44.

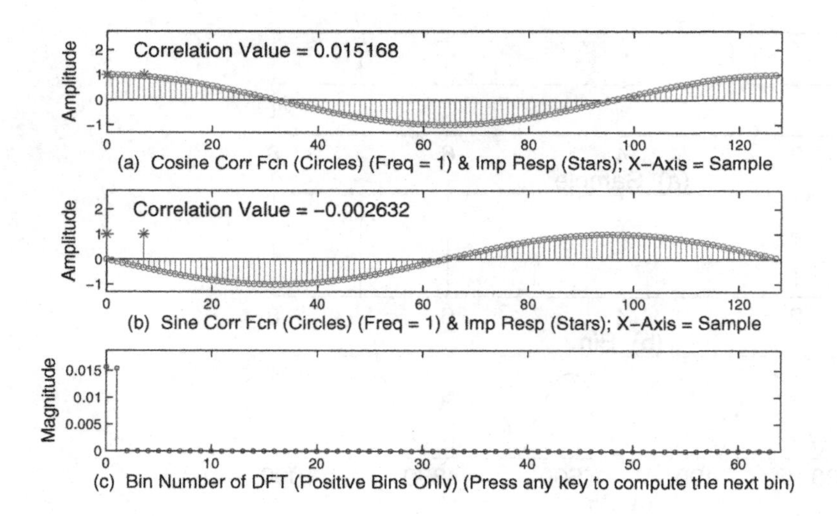

Figure 7.42: (a) Eight-sample impulse response, nonzero values plotted with stars (zero values at indices 1-6 and 8-127 not plotted), and 128-point, one-cycle cosine (real) correlator, to compute the Real part of Bin 1; (b) Eight-sample impulse response, plotted with stars (zero values at indices 1-6 and 8-127 not plotted), and 128-point, one-cycle sine (imaginary) correlator, to compute the Imaginary part of Bin 1; (c) Magnitude of DFT, plotted up to Bin 1.

Example 7.36. Estimate the harmonic content of the output of a zero-order hold DAC converting one cycle of a sine wave to an analog signal, prior to any post-conversion lowpass filtering.

Plot (a) of Fig. 7.45 shows the theoretical output of a zero-order-hold-reconstructed 1 Hz sine wave sampled at a rate of 11 Hz. Plot (b) shows a 150 sample segment of the sampled representation of the waveform at (a) (which was simulated with 1000 samples), from samples 250 to 400, which correspond to the waveform at (a) from time 0.25 second to 0.4 second (since the samples represent times 1/1000 of a second apart). A DFT is then performed on the sampled sequence, and a portion of the result (the first 60 bins) is plotted at (c). The 1000-point DFT of the sequence at (b) thus yields 1000 samples of the DTFT of the simulated 11 Hz-sampled zero-order-hold-reconstructed 1 Hz sine wave at (a). Note that the one-cycle 11-stairstep sine wave's spectrum has a high amplitude component at Bin 1, and harmonics at Bins 10 and 12, 21 and 23, 32 and 34, etc. Since the sample rate is 11 Hz, the Nyquist limit is 5.5 Hz, and this lies well below the lowest harmonics shown in plot (c) – so, with a good lowpass filter cutting off sharply at the Nyquist limit, it should be possible to eliminate all of the harmonics and have a smooth sine wave without any stairstep characteristic.

Example 7.37. Consider the sequence [−0.1, 1, 1, −0.1]. Figure 7.46 shows, in plots (a)-(c), the real part, imaginary part, and magnitude of the 4-pt DFT of the sequence, while plots (d)-(f)

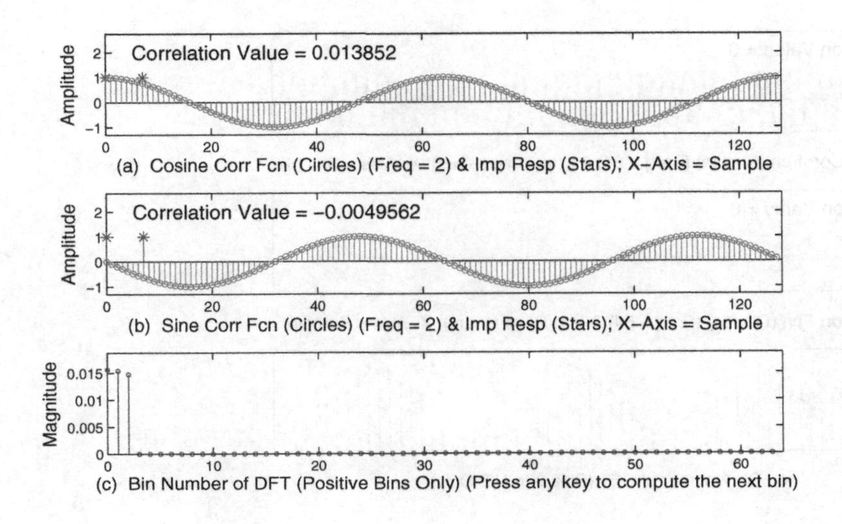

Figure 7.43: (a) Eight-sample impulse response (stars), and 128-point, two-cycle cosine (real) correlator, to compute the Real part of Bin 2; (b) Eight-sample impulse response (stars), and 128-point, two-cycle sine (imaginary) correlator, to compute the Imaginary part of Bin 2; (c) Magnitude of DFT, plotted up to Bin 2.

show the zero-padded 128-pt DFT of the sequence. Explain what each plot represents in terms of frequency components which make up the sequence and frequency response of the sequence.

The DFT of the four-sample sequence shows the frequency and amplitudes of the cosine and sine components used to make the sequence–i.e., the coefficients can be used to reconstruct the sequence exactly. Plots(d)-(f) constitute samples of the DTFT, which contain the same information of plots (a)-(c), plus much more. For example, Bin 1 of plot (a) represents one cycle over four samples, or 128/4 = 32 cycles over 128 samples. Hence, the information found for Bin 1 of plot (a) is the same as that for Bin 32 of plot (d). Using similar reasoning, Bins 0 and 2 of plot (a) would be equivalent to Bins 0 and 64 of plot (d), and analogously between plots (b) and (e), and plots (c) and (f).

Example 7.38. Estimate the true frequency response of the cascade of three filters comprising a first IIR filter having a pole at $0.9j$, a second IIR filter having a pole at $-0.9j$, and a third (FIR) filter having the impulse response $[1, 0, 0, 0, 0, 0, 1]$ by computing 1024 samples of the DTFT of the net impulse response. Verify the answer by determining the b and a coefficients of the z-transform, using them to filter a unit impulse sequence, and then computing 1024 samples of the DTFT (using the DFT) of the unit impulse sequence. Plot the positive frequency response from both results for comparison.

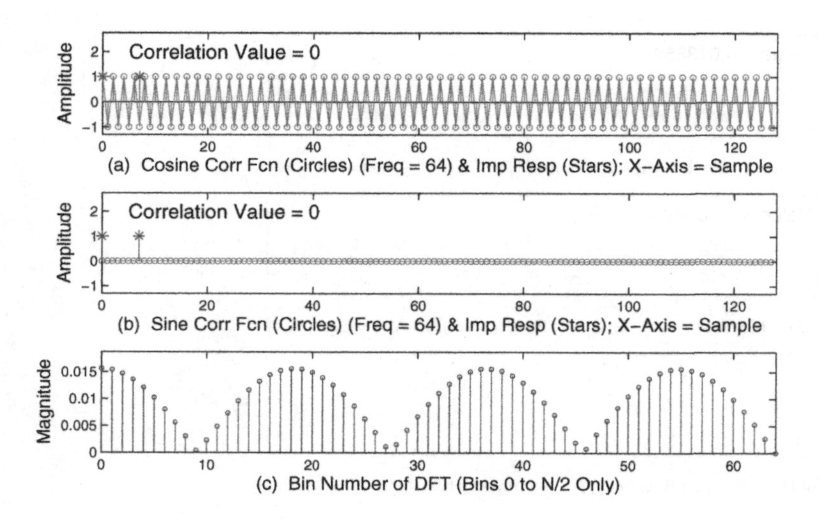

Figure 7.44: (a) Eight-sample impulse response (stars), and 128-point, 64-cycle cosine (real) correlator, to compute the Real part of Bin 64; (b) Eight-sample impulse response (stars), and 128-point, 64-cycle sine (imaginary) correlator, to compute the Imaginary part of Bin 64; (c) Magnitude of DFT, plotted up to Bin 64.

The first 100 samples of the IIR impulse responses should be adequate to represent the true impulse responses since $0.9^{99} = 3*10^{-5}$, i.e., a very small number–in other words, samples of the impulse response beyond this make a negligible contribution in practical terms to the net result. The truncated impulse responses are therefore (0.9*j).^(0:1:99) and (-0.9*j).^(0:1:99), respectively. We can obtain the composite impulse response by convolving the two IIR impulse responses and then convolving the result with the FIR impulse response; then samples of the DTFT can be computed using a zero-padded DFT. The following code performs and plots the computation, and checks it using the b and a coefficients and the MathScript function *filter*. The results are shown in Figure 3.47.

```
y = abs(fft(conv(conv((0.9*j).^(0:1:99),(-0.9*j). ^(0:1:99)),...
[1,0,0,0,0,0,1]),1024)); figure(8); subplot(211);
plot(y(1,1:513)); axis([0,513,0,inf])
a = conv([1,0.9*j],[1,-0.9*j]); b=[1,0,0,0,0,0,1]
ans = filter(b,a,[1,zeros(1,1000)]);fr=abs(fft(ans,1024));
subplot(212); plot(fr(1,1:513))
axis([0,513,0,inf])
```

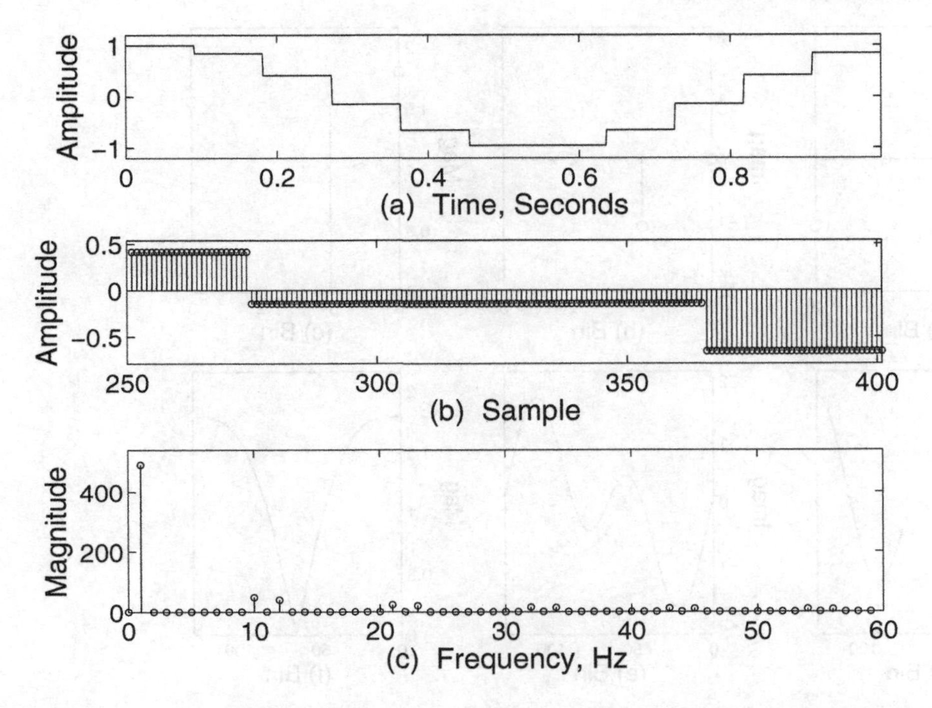

Figure 7.45: (a) A zero-order-hold-reconstructed 1 Hz sine wave prior to being lowpass filtered; (b) A 150-sample portion of the waveform at (a) simulated with 1000 samples; (c) The 1000-point DFT of the entire simulated sequence representing the waveform at (a).

7.19 THE INVERSE DFT (IDFT)

The Inverse DFT is used to convert a set of Bin values from a DFT back into a time domain sequence. If the DFT is defined as

$$X[k] = \sum_{n=0}^{N-1} x[n](\cos[2\pi kn/N - j\sin[2\pi kn/N])$$

then the corresponding Inverse Discrete Fourier Transform (IDFT) is defined as

$$x[n] = \frac{1}{N}\sum_{k=0}^{N-1} X[k](\cos[2\pi kn/N + j\sin[2\pi kn/N]) \tag{7.30}$$

where n runs from 0 to N - 1, as does k. Thus the original time domain signal is reconstructed sample-by-sample.

Example 7.39. Compute the IDFT using the DFT coefficients of the sequence [2, 1, −1, 1].

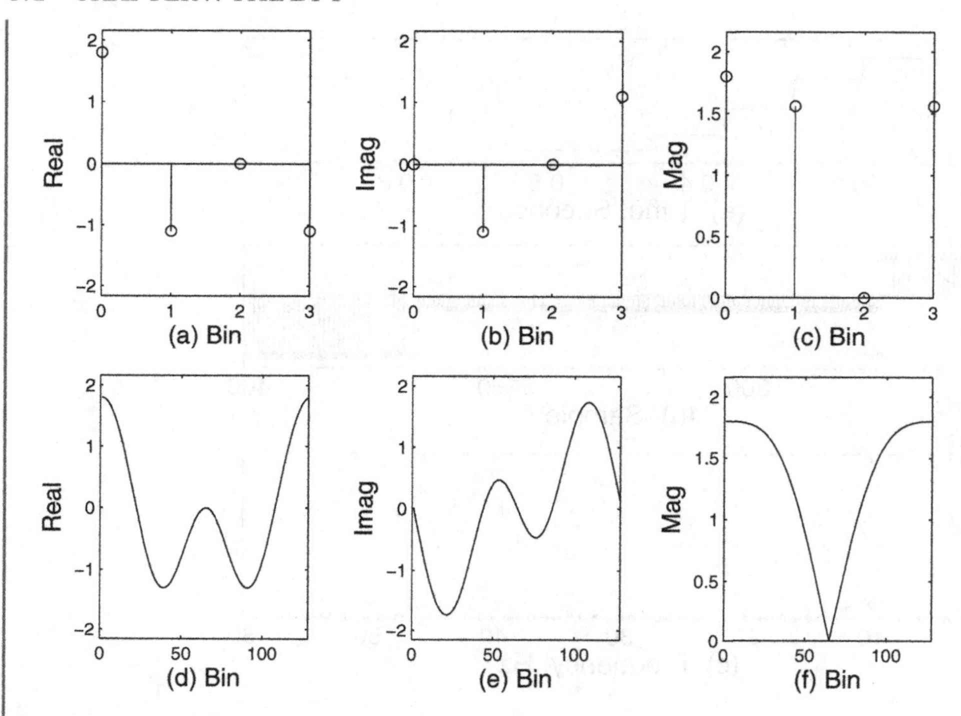

Figure 7.46: (a) Real part of 4-pt DFT; (b) Imaginary part of 4-pt DFT; (c) Magnitude of 4-pt DFT; (d) Real part of 128-pt DFT; (e) Imaginary part of 128-pt DFT; (f) Magnitude of 128-pt DFT.

The following code will obtain the DFT coefficients, initialize several values, then compute and display the answer:

```
s = [2,1,-1,1]; n = 0:1:3; F = fft(s); k = [0:1:3]'; Arg = 2*pi*k/4;
for n = 0:1:3; hold on;
stem(n, real(0.25*sum(F(k+1)*exp(j*Arg*n)))); end
```

Perhaps a more intuitive way of thinking about the IDFT is to generate it not point-by-point, but harmonic-by-harmonic. In this case,

$$x[0 : N - 1]_k = X[k]e^{j2\pi nk/N}$$

or using the rectangular notation,

$$x[0 : N - 1]_k = X[k](\cos[2\pi kn/N] + j\sin[2\pi nk/N])$$

where the vector $n = 0 : N - 1$, and the complete IDFT is the accumulation of all $X[k]$-weighted harmonic basis vectors $x[0 : N - 1]_k$

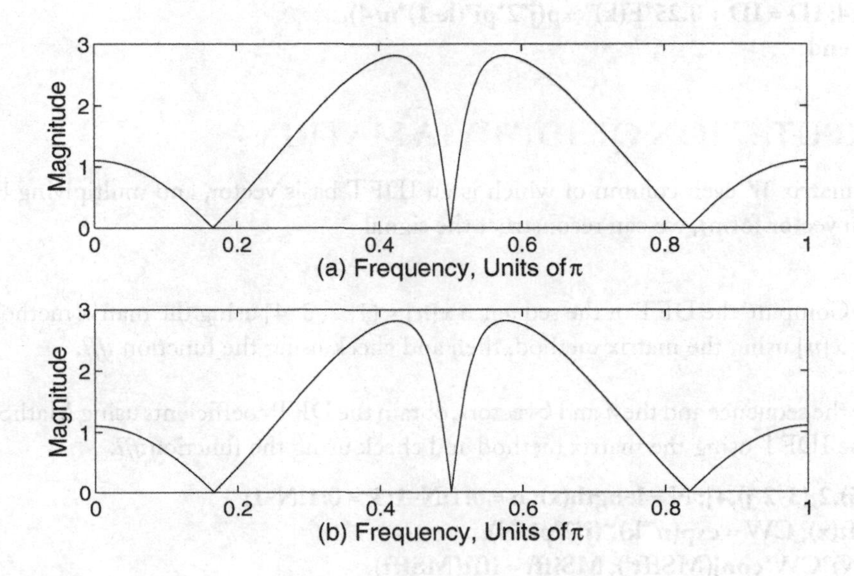

Figure 7.47: (a) Samples of the DTFT of a filter, computed as a long DFT of the convolution of two truncated IIR impulse responses and one FIR impulse response; (b) Samples of the DTFT of the same filter, estimated by computing a long DFT of the truncated impulse response of the composite filter, which was computed from the z-transform b and a coefficients.

$$x[0:N-1] = \sum_{k=0}^{N-1} x[0:N-1]_k$$

Notice how similar the IDFT is to the DFT; they differ only by using $x[n]$ in the forward transform vice $X[k]$ in the reverse transform, a positive sign in the exponential for the reverse transform as opposed to a negative sign for the forward transform, and the scaling constant $1/N$. The negative sign may be used in the exponential in either one of the forward or reverse transform, providing the positive sign is used in the other transform, and the scaling constant may be used on either the forward or reverse transform. Furthermore, the scaling constant can be balanced as $\sqrt{1/N}$ and applied to both the forward and reverse transforms.

Example 7.40. Devise a script to obtain the DFT of a 4-sample sequence and then reconstruct the original sequence from the DFT coefficients one harmonic at a time.

The following code computes the IDFT one harmonic at a time; the script will advance to the next computation after a one second pause:

```
s = [2,1,-1,1]; n = 0:1:3; F = fft(s); ID = zeros(1,4);
```

```
for k=1:1:4; ID = ID + 0.25*F(k)*exp(j*2*pi*(k-1)*n/4),
pause(1); end
```

7.20 COMPUTATION OF IDFT VIA MATRIX

By setting up a matrix W each column of which is an IDFT basis vector, and multiplying by the DFT (in column vector form), we can reconstruct the signal.

Example 7.41. Compute the DFT of the sequence $x[n] = [1, 2, 3, 4]$ using the matrix method and then reconstruct $x[n]$ using the matrix method, then and check using the function *ifft*.

We define the sequence and the n and k vectors, obtain the DFT coefficients using MathScript, then compute the IDFT using the matrix method and check using the function*ifft*.

```
x = [(1+2*j),2,(3-2*j),4]; N = length(x); n = 0:1:N-1; k = 0:1:N-1;
MSfft = fft(x); CW = exp(n'*k).^(j*2*pi/N);
idft = (1/N)*CW*conj(MSfft'), MSifft = ifft(MSfft)
```

A symbolic rendering of this would be

$$x = \frac{1}{N} CW \cdot D \tag{7.31}$$

where **CW** is the IDFT basis vector matrix

exp(n'*k).^(j*2*pi/N)

where $n = k =$ 0:1:N -1 and **D** is the DFT vector in column form. Here, note that we have scaled by $1/N$ since the DFT has not been so scaled. Note that the conjugate must be taken of the DFT coefficients $MSiff$ when they are transposed into a column vector since MathScript automatically conjugates vectors when they are transposed.

7.21 IDFT VIA DFT

The similarity between the DFT and the IDFT can be used to redefine the IDFT so that only a single algorithm, the DFT needs to be employed to compute both the DFT and IDFT.

If the DFT is defined as

$$X[k] = \frac{1}{N} \sum_{n=0}^{N-1} x[n] e^{-j2\pi nk/N} \tag{7.32}$$

then it can be shown that the Inverse DFT is

$$x[n] = \left(\sum_{n=0}^{N-1} (X[k]^*) e^{-j2\pi nk/N} \right)^* \tag{7.33}$$

where the symbol * is used to indicate complex conjugation. If the DFT is not scaled by $1/N$ (as is true for MathScript's function *fft*), then Eq. (7.33) (rather than Eq. (7.32)) should be scaled by $1/N$.

Described in words, the steps to take to perform Eq. (7.33) would be to first take the complex conjugate of the DFT coefficients, then perform the DFT on them, and then take the complex conjugate of the result.

Example 7.42. Verify the IDFT-Via-DFT concept with a four-sample complex sequence.

An easy experiment to verify Eq. (7.33) is to run the following code:

```
idft = (1/4)*conj(fft(conj(fft([1+4*j,2+3*j,3+2*j,4+j]))))
```

which returns the input complex vector (in brackets as the argument for the innermost FFT) as the value of the variable $idft$.

Example 7.43. Compute and display the inverse DFT of a 25% duty cycle rectangle using direct implementation of the DFT to obtain the DFT coefficients, followed by computation of the IDFT using the harmonic-by-harmonic method.

Here we outline the necessary computations of the script (not provided, see description immediately following and exercises below)

LVxInvDFTComputeRect25

The IDFT portion of the script involves generating cosine and sine basis signals of various frequencies k and length N, weighted and phase shifted according to the magnitude and phase of the DFT bin coefficient for the particular value of k, in this manner:

$$x[0:N-1]_k = X[k](\cos[2\pi kn/N] + j\sin[2\pi kn/N])$$

where n is a vector running from 0 to N - 1. For each value of k, a complex multiplication of the DFT bin value $X[k]$ with the complex basis vector for k is performed. Thinking in polar coordinates, we are multiplying two complex vectors, each having a magnitude and a phase angle. The product is a complex exponential consisting of two waveforms, a real (cosine) and imaginary (sine). This is done for each value of k, and the results are accumulated to obtain the output.

For k = 1, the net synthesis vector would be

$$X[1](\cos[2\pi(0:N-1)(1)/N] + j\sin[2\pi(0:N-1)(1)/N])$$

If we take the real part of the above product and plot it, we would see a cosine scaled by the magnitude and shifted by the phase of $X[1]$. Similarly, taking the imaginary part of the above sequence, and plotting it, we would see a sine wave scaled and shifted by $X[1]$.

The above process is performed for $k = 0$ to $N - 1$, or in the case of the symmetrical DFT, $k = -N/2 + 1$ to $k = N/2$ for N even, or $-(N - 1)/2$ to $(N - 1)/2$ for N odd, and we sum all N complex synthesis vectors to get the net IDFT.

If the DFT bin values display conjugate symmetry, as they do for real input signals (which is the case here), the resultant sum of waveforms will be identically zero for the imaginary part, and the original time domain sequence will be the real part of the inverse DFT. A real input signal yields a conjugate symmetric DFT and hence a conjugate symmetric DFT returns a real sequence when reverse transformed. Of course, it also returns an imaginary sequence, but with all values equal to zero.

The script *LVxInvDFTComputeRect25* computes the DFT and then uses DFT bin values to reconstruct the original signal one harmonic (or weighted basis vector) at a time; a figure is created to display the process, computing and displaying the DFT-bin-weighted IDFT basis vectors for $+k$ and $-k$ simultaneously, making it easy to see that the imaginary components cancel for each bin pair, $\pm k$.

Figure 7.48, plot (a), shows a 25% duty cycle rectangular wave as the original source wave. Plots (c) and (d) show the real and imaginary parts of the DFT-bin-weighted IDFT basis vector for $k = 1$, plots (e) and (f) show the real and imaginary parts of the DFT-bin-weighted IDFT basis vector for $k = -1$, while plot (b) shows the summation of the first two weighted basis vectors (i.e., harmonics) for $k = 0$ and $k = 1$.

Figure 7.49 shows the situation up to $k = 3$. You can already see the partial IDFT starting to look like the original signal. Of course, the partial IDFT will go through a lot of appearance changes before the original signal is completely reconstructed. In Fig. 7.50, for $k = 15$, we see essentially the original signal (reconstructed) at (b), and when the last bin ($k = 16$) is reached (see Fig. 7.51), we note no additional contribution since the Bin 16 DFT coefficient was 0.

Example 7.44. Show mathematically how the cosine and sine components shown in plots (e) and (f) of Figs. 7.48-7.51 obtain their apparent phase shift (i.e., note that the cosine components in the figures generally do not start at zero degrees phase, nor do the sine components). Show mathematically how imaginary components cancel in the Inverse DFT if the original time domain sequence being reconstructed was real only.

Since Bins 0 and $N/2$ are real, we need only consider Bins 1:1:($N/2 - 1$) and their negatives, i.e., Bins 1, -1, 2, -2, etc., for symmetrical indexing, or Bins 1 and $(N - 1)$, 2 and $(N - 2)$, etc., for asymmetrical indexing (i.e., when $k = 0{:}1{:}(N - 1)$). For a real input (time domain) signal, DFT Bins k and $-k$ (k not equal to 0 or $N/2$) are complex conjugates. If we let $DFT[k] = a + jb$ and $n = 0{:}1{:}N - 1$ then we can write

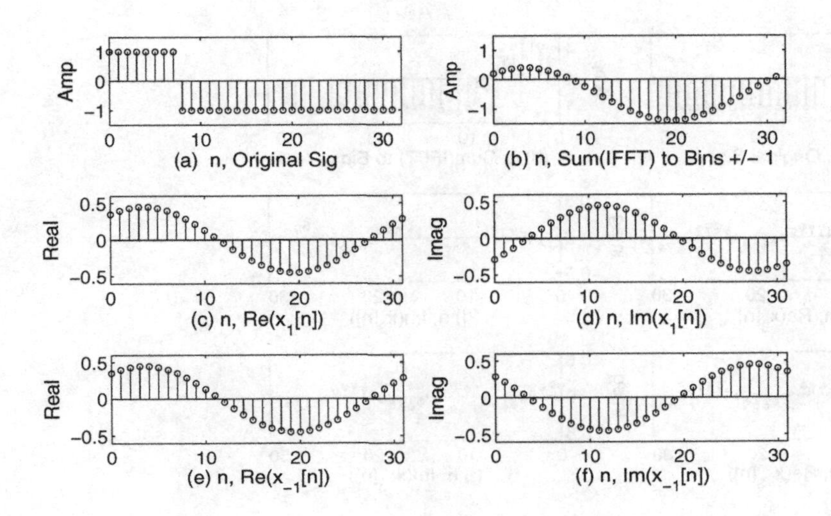

Figure 7.48: (a) Original signal; (b) Cumulative reconstructed signal, to Bin 1; (c) and (d): Real and Imaginary parts of Weighted IDFT basis vector (or harmonic) for k = 1; (e) and (f): Real and Imaginary parts of Weighted IDFT basis vector (or harmonic) for k = -1.

$$IDFT[k] = [a + jb](\cos[2\pi nk/N] + j\sin[2\pi nk/N])$$

and

$$IDFT[-k] = [a - jb](\cos[2\pi nk/N] - j\sin[2\pi nk/N])$$

where we note that $\cos(-\theta) = \cos(\theta)$ and $\sin(-\theta) = -\sin(\theta)$. Doing the complex multiplications and adding the two resultant harmonic components, we get

$$IDFT[k] + IDFT[-k] = 2a\cos[2\pi nk/N] \;-\; 2b\sin[2\pi nk/N] \tag{7.34}$$

We see that, for a real time domain signal, the sum of all complex DFT-Bin-weighted IDFT basis vectors (harmonics) is real. The contribution from Bins 0 and $N/2$ must also be real since the DFT Bin values themselves are real as are the IDFT basis vectors for k = 0 and $N/2$. It can also be seen that the reconstructed sinusoid for Bin k will have a phase determined by the values of a and b, i.e., $DFT[k]$.

A complete IDFT formula for N even using Eq. (7.34) would be

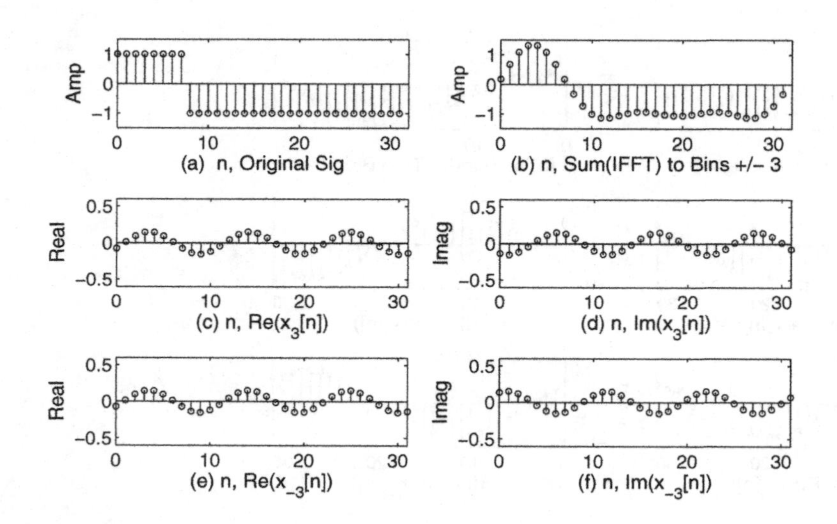

Figure 7.49: (a) Original signal; (b) Cumulative reconstructed signal, to Bin 3; (c) and (d): Real and Imaginary parts of Weighted IDFT basis vector (or harmonic) for k = 3; (e) and (f): Real and Imaginary parts of Weighted IDFT basis vector (or harmonic) for k = -3.

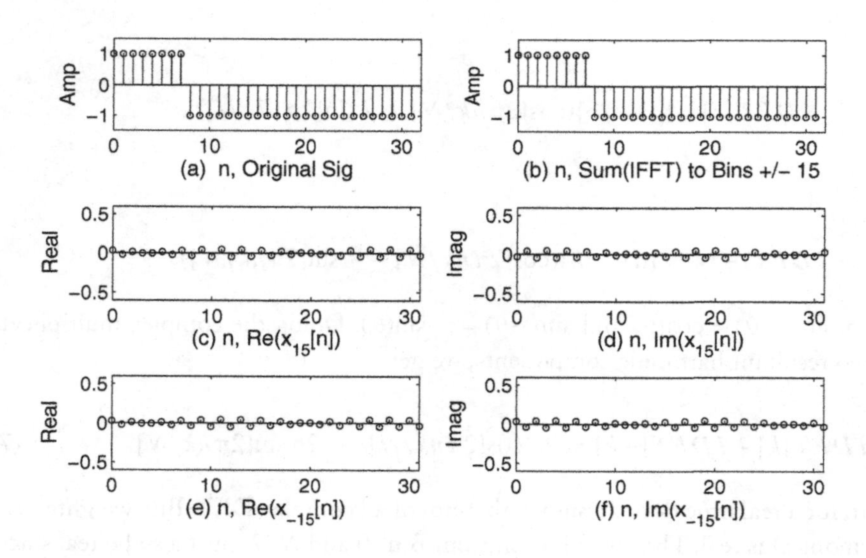

Figure 7.50: (a) Original signal; (b) Cumulative reconstructed signal, to Frequency/Bin 15; (c) and (d): Real and Imaginary parts of Weighted IDFT basis vector (or harmonic) for k = 15; (e) and (f): Real and Imaginary parts of Weighted IDFT basis vector (or harmonic) for k = -15.

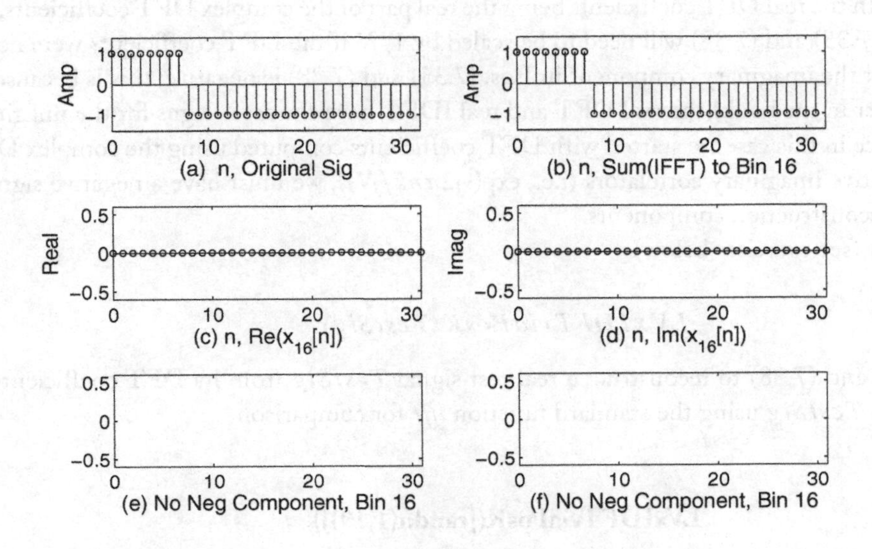

Figure 7.51: (a) Original signal; (b) Cumulative reconstructed signal, to Bin 16; (c) and (d): Real and Imaginary parts of Weighted IDFT basis vector (or harmonic) for k = 16; (e) and (f): (No plot as there is no Bin -16).

$$x[n] = Bin[0] + Bin[N/2](-1)^n + \tag{7.35}$$

$$2\sum_{k=1}^{N/2-1}(\mathrm{Re}(Bin[k])\cos[2\pi nk/N] - \tag{7.36}$$

$$\mathrm{Im}(Bin[k])\sin[2\pi nk/N]) \tag{7.37}$$

where n = 0:1:(N − 1), and for N odd,

$$x[n] = Bin[0] + 2\sum_{k=1}^{(N-1)/2}(\mathrm{Re}(Bin[k])\cos[2\pi nk/N] - \mathrm{Im}(Bin[k])\sin[2\pi nk/N]) \tag{7.38}$$

where n = 0:1:(N − 1).

Note initially that Eqs. (7.35) and (7.38) are only valid when the sequence $x[n]$ is real since their derivation was predicated on that premise. Notice also that the complex DFT bins have been decomposed into their real and imaginary parts, which are separately applied to the real and imaginary IDFT basis vectors (i.e., cosine and sine vectors). Note also that Bins 0 and $N/2$ get a weight of 1 and Bins 1 to $N/2-1$ (or $(N-1)/2$ if N is odd) get a weight of 2–this is essentially the real

DFT/IDFT, with the real DFT coefficients being the real part of the complex DFT coefficients, etc. Note that Eqs. (7.35) and (7.38) will need to be scaled by $1/N$ if the DFT coefficients were not so scaled. Note that the imaginary component in Eqs. (7.35) and (7.38) is negative; this is because (as mentioned earlier in the book) the real DFT and real IDFT have the same signs for the imaginary component. Since in this case we started with DFT coefficients computed using the complex DFT, which uses negative imaginary correlators (i.e., exp(-j$2\pi nk/N$)), we must have a negative sign for the imaginary reconstruction components.

The script (see exercises below)

$$LVxIDFTviaPosK(TestSig)$$

uses Eqs. (7.35) and (7.38) to reconstruct a real test signal $TestSig$ from its DFT coefficients. It also reconstructs $TestSig$ using the standard function *ifft* for comparison.

The call

LVxIDFTviaPosK([randn(1,19)])

results in Fig. 7.52.

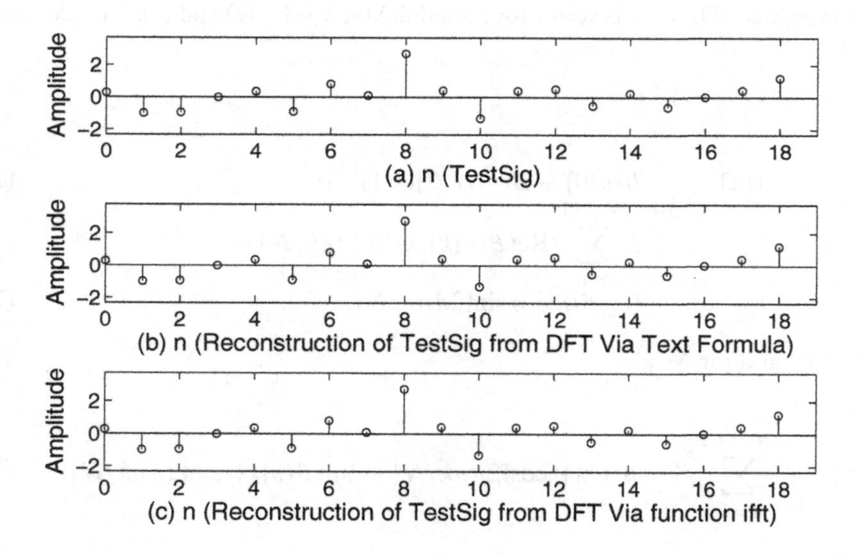

Figure 7.52: (a) Test signal; (b) Reconstruction of test signal from its DFT coefficients using Eqs. (7.35) and (7.38); (c) Reconstruction of test signal from its DFT coefficients using the standard function $ifft$.

7.22 IDFT PHASE DESCRAMBLING

7.22.1 PHASE ZEROING

An interesting and instructive use of the DFT/IDFT pair is to realign the phases of all the frequencies in a signal. Suppose we had a periodic signal with the phases randomly arranged, and we wanted them all to be aligned.

The phase angle of any bin can be set to zero by multiplying the bin value by its own complex conjugate, which yields a real number, and then taking the square root of the resultant real number to correct the magnitude to the original value.

Any bin may be represented as

$$a + jb = M\angle\theta$$

Multiplying by the complex conjugate, we get

$$(M\angle\theta)(M\angle(-\theta)) = M^2$$

Then to return to the correct magnitude, we take the square root. Thus the entire operation (per bin) is

$$NewBinVal = \sqrt{(a + jb)(a - jb)}$$

7.22.2 PHASE SHIFTING

To adjust the phase of any bin to be whatever you want, simply multiply the positive frequency bins by whatever phase angle you want to shift by, and multiply the corresponding negative frequency bins by the complex conjugate of that phase angle.

Recall that a phase angle in degrees is specified as a complex number, such as

$$e^{j2\pi(\theta/360)}$$

Let's say we wanted to shift every bin by $\pi/4$ radians (45 degrees); this can be done by multiplying each positive frequency bin value by

$$e^{j2\pi(45/360)} = e^{j\pi/4} = 0.707 + j0.707$$

and each negative bin value by

$$e^{j2\pi(45/360)} = e^{-j\pi/4} = 0.707 - j0.707$$

For bin zero, it is not necessary to multiply by a phase angle, since DC has no phase. This is automatically taken care of in the script *LVxPhaseShiftViaDFT*, which we'll discuss below, by specifying the phase factor as dependent on frequency k; hence for $k = 0$ the phase factor turns out to be 1.

A script designed to demonstrate this (see exercises below) is

LVxPhaseShiftViaDFT

which generates a waveform having the harmonic amplitudes of a square wave, but totally random phases for all harmonics present. All phases are returned to zero, which produces a cusped waveform; to correct this, the phases of all bins are then shifted 90 degrees using the same technique, which places all harmonics in just the proper phase so the waveform appears as a square wave. The script *LVxPhaseShiftViaDFT* then presents a general phase-shifting demo, in which the initial derandomized result is incrementally shifted in phase until its phase has been shifted 90 degrees from its initial value of zero degrees for all frequencies. Each time the derandomized waveform is phase-shifted, the Hilbert Transform of the resultant waveform is taken and also plotted. The result is that initially, the derandomized waveform is the cusped waveform, and its Hilbert Transform is a square wave. After phase shifting by 90 degrees, the cusped waveform becomes a square wave, while its Hilbert transform becomes a cusped waveform.

Figure 7.53 shows the initial random-phase test signal, the derandomized result (in the time domain), and then the Hilbert Transform of that.

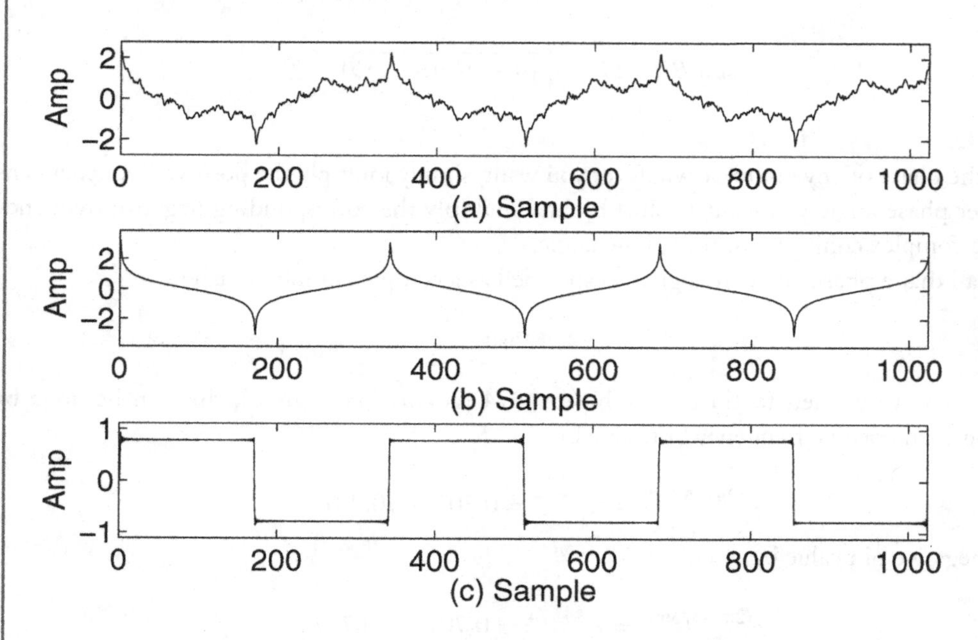

Figure 7.53: (a) Test waveform, sum of odd harmonic cosines with amplitudes inversely proportional to harmonic number and random phase; (b) Signal from (a) with all harmonics having had their phase angles set to zero; (c) Signal from (b), with all phase angles shifted 90 degrees.

7.22.3 EQUALIZATION USING THE DFT

An interesting use of the DFT/IDFT is to analyze the impulse response of a system that distorts signals and to generate an inverse or deconvolution filter to reverse the distortion induced by the system. The DFT can be used to determine the magnitude and phase characteristics of the system by taking the DFT of the system's impulse response. The reciprocal of the DFT coefficients can be used to generate a time domain filter suitable for circular convolution, or the reciprocal DFT coefficients can be multiplied by the DFT of a block of distorted signal samples, and the IDFT computed to produce the equalized (or deconvolved) signal.

The script (see exercises below)

$$LVxDFTEqualization(tstSig, p, k, SR, xplotlim)$$

creates a test signal of length 1024 samples, which is selected by input argument $tstSig$, and is one of several square wave trains or a chirp. A test digital impulse response to model the distorting analog channel or system is generated by multiplying random noise of standard deviation k by a decaying impulse response generated by a single real-pole IIR having a z-transform $H(z) = 1/(1 - pz^{-1})$. A DFT of length SR, which should be two or more times the length of the test sequence, is computed and its reciprocal is used to deconvolve or equalize the distorted test signal (which has been distorted by filtering the test signal using the generated test impulse response) two different ways. In the first method, the reciprocal DFT coefficients are multiplied by the DFT of the distorted signal, and the real part of the IDFT of this product produces the deconvolved signal. In the second method, a time domain impulse response is produced by taking the IDFT of the reciprocal DFT coefficients, and then circularly convolving the distorted (time domain) test signal with it.

The call

LVxDFTEqualization(3,0.9,0.2,2048,450)

results in (for example) Fig. 7.54.

7.23 REFERENCES

[1] William L. Briggs and Van Emden Henson, *The DFT*, SIAM, Philadelphia, 1995.

[2] Richard G. Lyons, *Understanding Digital Signal Processing*, Addison Wesley Longman, Inc., Reading, Massachusetts, 1997.

[3] Alan V. Oppenheim and Ronald W. Schaefer, *Discrete-Time Signal Processing*, Prentice-Hall, Englewood Cliffs, New Jersey, 1989.

[4] John G. Proakis and Dimitris G. Manolakis, *Digital Signal Processing, Principles, Algorithms, and Applications, Third Edition*, Prentice Hall, Upper Saddle River, New Jersey, 1996.

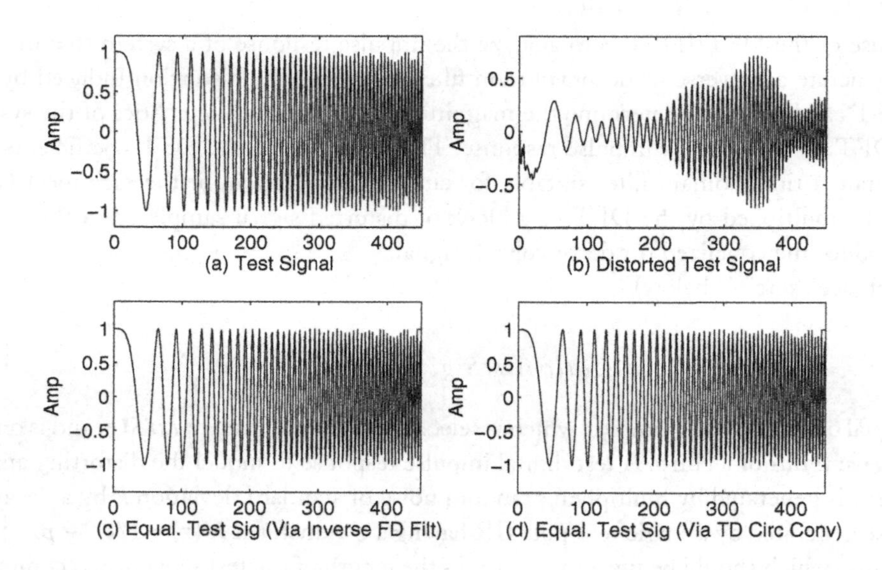

Figure 7.54: (a) Test Signal; (b) Distorted Test Signal, after being convolved with simulated signal processing channel's impulse response; (c) Deconvolved Test Signal, produced using frequency domain filtering following by the IDFT; (d) Deconvolved Test Signal, produced by using the IDFT on the reciprocal of the DFT of the channel impulse response to produce a time domain filter which is circularly convolved with the distorted test signal.

[5] E. Oran Brigham, *The Fast Fourier Transform and Its Applications*, Prentice-Hall, Inc., Upper Saddle River, New Jersey, 1988.

[6] Charles Van Loan, *Computational Frameworks for the Fast Fourier Transform*, SIAM, Philadelphia, 1992.

[7] Steven W. Smith, *The Scientist and Engineer's Guide to Digital Signal Processing*, California Technical Publishing, San Diego, 1997.

7.24 EXERCISES

1. For the following signals, state how many total bins will be generated by the DFT, what the bin width (bin spacing) is, and how many unique bins there are. The term random is used to mean that the sequences possess no particular symmetry such as being even, odd, etc.

(a) A real random sequence $x[n]$, obtained over a period of 2.5 seconds at a sample rate of 750 Hz.

(b) A real random sequence comprising 101 samples obtained at a sample rate of 4000 Hz.

(c) A complex random sequence of 499 samples obtained at a sample rate of 11025 Hz.

2. Show, using Eq. (7.7), that a) the DFT of an arbitrary real sequence $x[n]$ multiplied by a complex exponential will be shifted in frequency, and b) the DFT of an arbitrary real sequence $x[n]$ multiplied by a cosine will consist of frequency components that are the sum and difference of the original components and the frequency of the cosine.

3. (a) For a length-8 DFT, specify eight different sinusoids the DFTs of which will be real and have no leakage, i.e., the energy will be concentrated in one or two (considering both positive and negative frequencies) bins only.

(b) For a length-7 DFT, specify seven different sinusoids the DFTs of which will be imaginary-only and have no leakage, i.e., the energy will be concentrated in one or two bins only.

4. Write a script that computes the DFT of an arbitrary real input sequence $x[n]$ of length N using the definition of the DFT (either of Eqs. (7.6) or (7.7)) but which uses symmetric indexing, i.e., the range for both n and k is $-N/2+1:N/2$ for N even and $-(N-1)/2:(N-1)/2$ for N odd. On a single figure, plot $x[n]$ and the real and imaginary parts of the DFT.

Test the script with the input signals given below, and, verify results by creating a separate figure, with the same plots, but populate them by using the MathScript function fft on the various input signals. Use the function *fftshift* to bring the results into asymmetric format. For each grouping of like test signals, compare results. For the first two groupings ((a),(b) and (c),(d)), how does the length of the test sequence affect the outcome? For the last grouping, how does the number of cycles of the sawtooth affect the DFTs fundamental and harmonic series?

(a) **x = [1 zeros(1,11)]**; (b) **x = [1 zeros(1,51)]**
(c) **x = [ones(1,17)]**; (d) **x = [ones(1,73)]**
(e) **x = [-31:1:32]**
(f) **x = mod([-31:1:32],32)**; (g) **x = mod([-31:1:32],16)**
(h) **x = mod([-31:1:32],8)**; (i) **x = mod([-31:1:32],4)**

5. Write a script that will verify the DFT shift property for an arbitrary input signal $x[n]$, i.e., verify that, for an arbitrary shift of m samples,

$$DFT(x[n-m]_N) = X(k)e^{-j2\pi mk/N}$$

More specifically, the script should receive $x[n]$ and m and 1) compute the fft of $x[n-m]_N$, 2) compute the fft of $x[n]$ and multiply it by the appropriate phase factor, and 3) plot the magnitude and phase of the results from 1) and 2) above.

6. Compute and display the magnitude of the DTFT and DFS on the same plot for the sequences below.

(a) **[1,0,0,0,0,1]**
(b) **fir1(10,0.5)**
(c) **[ones(1,8)]**
(d) **[real(j.^(0:1:10))].*blackman(11)'**

7. For the periodic sequence $x[n]$ of infinite length, one period $\tilde{x}[n]$ of which is [1,2,3,4], compute the DFS coefficients, then use the inverse DFS formula to compute $x[n]$ over the range $-21 < n < 21$. Plot $\tilde{x}[n]$ and the 41 reconstructed samples of $x[n]$, with proper indices for both, on the same axis.

8. For the three scripts described in the text, namely

$$LVxDFTComputeSawtooth$$

$$LVxDFTComputeSymmIndex$$

$$LVxDFTComputeImpUnBal$$

create either a single script or a single VI that will compute the DFT of any of the following signals, all of which have user-designatable length and, in cases 1) and 2), user-designatable fundamental frequency: 1) a bandlimited sawtooth; 2) a bandlimited square wave, 3) a bandlimited triangle wave, 4) a unit step. The format of DFT display should also be user-designatable as either symmetrical or asymmetrical.

The DFTs should be computable step-by-step and displayed similarly to any of Figs. 7.14, 7.15, or 7.16.

9. If a certain 4-pt sequence is represented symbolically as [a,b,c,d], derive an expression for the DFT using decimation-in-time and the 4-pt butterfly.

10. Using paper and pencil, and subsequence lengths of two samples, compute the convolution of the sequences [-1, 7] and [2, -1, 3, 0, 2, -3, 2, 4] using the overlap and add method. Proceed in the following manner: for each pair of length-2 subsequences to be convolved, pad to length-4, obtain the 4-pt DFTs of each subsequence using the expression derived in the previous problem. Compute the product of the two 4-pt DFTs and compute the inverse DFT of the product by using the DFT-by-IDFT method and the 4-pt DFT expression derived in the previous problem. Take the resulting time domain answer and add it to the cumulative result with the proper offset, etc. Check your answer by performing the time domain linear convolution using the *conv* function.

11. If a certain computer can perform a direct-implementation DFT of length 4096 in 36 seconds, approximately how long should the same computer need to perform a DFT of length 262,144 samples using a decimation-in-time FFT?

12. Write a script that will receive a sequence and perform a radix-2 decimation-in-time FFT on the sequence. Sequences that are not a power of two in length should be automatically padded out with zeros to a power-of-two length by your script. Write the script to allow the use of two methods to perform the decimation in time: 1) a direct implementation that divides vectors into two vectors, each of half the original length, and so forth as outlined in the text, and 2) the bit reversal code given in the text.

Compare the results and execution time for the sequences below. To compare results, have the script perform the decimation, and measure the execution time, for both methods. To determine whether or not both methods arrive at the same answer, take the sum of the absolute value of the differences between the two answers, divided by the sequence length. The result should be zero or, due to roundoff error, in the vicinity of 10^-15 or less.

(a) **Signal = ones(1,32);**
(b) **Signal = ones(1,256);**
(c) **Signal = ones(1,2^10);**

13. Write the code to implement the script

```
function LVx_FFTDecnTimeAltCode(x,CmprMode,..
BitRCd,BtrFCd)
% x is the test signal; if not a power of 2 in length,
% it will be padded with zeros so that it is. Pass
% CmprMode as 0 to compare the time of execution
% for various user-written m-coded FFT's to the built-in
% fft; Pass CmprMode as 1 to compare the time of execution
% of a direct DFT to the built-in fft. When CmprMode is
% passed as 0, pass BitRCd as 0 to use a standard
% bit-reversal routine (from text) or as 1 to use direct
% decimation (user-written). Pass BtrFCd as 0 to run the
% butterfly routine from the text or as 1 to run a user-written
% butterfly routine. When CmprMode is passed as 1, you may
% pass BitRCd and BtrFCd as the empty matrix [].
% Test calls:
% LVx_FFTDecnTimeAltCode([0:1:31],1,[],[])
% LVx_FFTDecnTimeAltCode([0:1:1023],1,[],[])
% LVx_FFTDecnTimeAltCode([0:1:31],0,0,0)
% LVx_FFTDecnTimeAltCode([0:1:1023],0,0,0)
% LVx_FFTDecnTimeAltCode([0:1:31],0,1,0)
% LVx_FFTDecnTimeAltCode([0:1:1023],0,1,0)
% LVx_FFTDecnTimeAltCode([0:1:31],0,0,1)
% LVx_FFTDecnTimeAltCode([0:1:1023],0,0,1)
% LVx_FFTDecnTimeAltCode([0:1:31],0,1,1)
% LVx_FFTDecnTimeAltCode([0:1:1023],0,1,1)
```

The script described above has two general comparison modes. The first mode, specified by input argument *CmprMode* = 1, compares the execution time and numerical result from the built-in FFT (function *fft*) to the results from a direct DFT implementation written in m-code by yourself. The second comparison mode, specified by input argument *CmprMode* = 1, compares execution time and numerical result from the built-in FFT to results from an FFT routine written by you, in which you specify by the input arguments how the decimation-in-time is performed, and how the butterfly routine is performed.

The two decimation-in-time routines, selected by input argument *BitRCd*, consist of the efficient bit-reversal code given in the text, and m-code written by yourself to repeatedly subdivide the sequences into even and odd parts, as described in the text. You can probably reuse the code written for the previous exercise to implement this portion of the current project.

The two butterfly routines, selected by *BtrFCd*, consist of the efficient butterfly code given in the text, and m-code written by yourself that works in accordance with Eqs. (7.18) and (7.19).

Test the completed script by performing the FFT on the sequences below, and compare (using the sum of differences method mentioned in the previous exercise) the result to that obtained using the function *fft*. Use the *tic* and *toc* functions in the script to determine the execution time for the built-in FFT, the direct DFT, the two decimation-in-time routines, and the two butterfly routines. For each sequence given below, vary the input arguments as necessary to check the execution time and consistency of numerical results. There are a total of five different combinations possible. In most cases, the built-in FFT will far exceed the performance of any of the m-coded FFTs, and the m-coded FFTs will far exceed the direct DFT in performance.

(a) **Signal = [0:1:31];**
(b) **Signal = ones(1,128);**
(c) **Signal = ones(1,2^8);**

14. Write a script that implements the functions of the script *LVxCyclicVLinearConv*, described below, which performs convolution using the DFT Convolution theorem. The script also performs the convolution using the function *conv*, and compares the results for the test calls given below. Figure 7.55 shows the result from one possible script and corresponding set of displays for the call

LVxCyclicVLinearConv(39,571)

```
function LVxCyclicVLinearConv(ImpRespN,TestSigN)
% ImpRespN sets the length of an impulse response which
% is a cosine at normalized frequency 0.5, which is windowed with
% a Hamming window to form a smooth lowpass impulse
% response. TestSigN is the length a of chirp which is
% convolved with the lowpass impulse response using the
% DFT with padded sequences to effect linear convolution.
% Test calls:
% LVxCyclicVLinearConv(28,280)
% LVxCyclicVLinearConv(32,1024)
% LVxCyclicVLinearConv(32,1024)
% LVxCyclicVLinearConv(39,571)
```

15. The DTMF (Dual Tone, Multiple Frequency) system is used in telephone dialing; each dialed digit is encoded as two tones sounded simultaneously, sampled at 8000 Hz. Write a script that

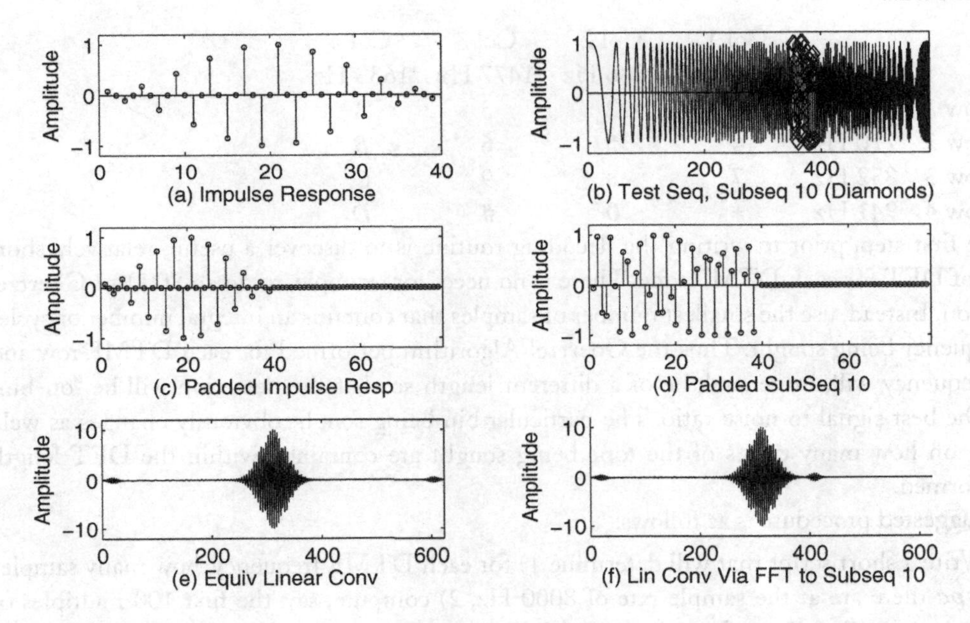

Figure 7.55: (a) Windowed impulse response; (b) Test chirp; (c) Zero-padded impulse response; (d) Subsequence 10, padded with zeros; (e) Complete linear convolution of test chirp and impulse response, performed using the function *conv*; (f) Partially complete convolution (through subsequence 10) of the convolution as performed using the DFT Convolution Theorem.

will receive a series of digits simulating a telephone number to be dialed which will then create an appropriate DTMF audio waveform. This waveform will then be input into a second script which will decode the audio waveform and provide as an output the string of dialed digits. Use the list of frequencies given below. Each portion encoding a given transmit digit should be 0.25 seconds long, and the time between successive digit-encoded waveforms should be 0.125 second. There should be a user-specified amount of white noise added to the signal. Once a transmit waveform has been created, the script should methodically decode the encoded digits using the all-real Goertzel Algorithm that provides a magnitude squared output (see function description below). DFT lengths should be chosen for each DTMF tone so that a particular DTMF frequency being sought will be "on-bin" or nearly so. The dialing digits are encoded according to their column and row, each of which is represented by a particular frequency. The particular dialed digit can be identified by methodically checking the transmitted audio waveform for each of the eight frequencies, thus identifying, for each transmitted digit a particular row and column represented by the two tones in the waveform.

		Col 1	Col 2	Col 3	Col 4
		1209 Hz	1336 Hz	1477 Hz	1633 Hz
Row 1	697 Hz	1	2	3	*A*
Row 2	770 Hz	4	5	6	*B*
Row 3	852 Hz	7	8	9	*C*
Row 4	941 Hz	*	0	#	*D*

The first step, prior to writing the decoding routine is to discover a useful, relatively short length N of DFT for each DTMF tone. There is no need, for example, to do an 8000-pt Goertzel computation. Instead, use the smallest number of samples that contains an integral number of cycles of the frequency being sought. Thus the Goertzel Algorithm performed for each DTMF row and column frequency will, in general, be of a different length so that the detection will be "on-bin" and give the best signal to noise ratio. The particular bin being sought obviously changes as well, depending on how many cycles of the tone being sought are contained within the DFT length being performed.

A suggested procedure is as follows:

a. Write a short script that will determine 1) for each DTMF frequency, how many samples per cycle spc there are at the sample rate of 8000 Hz, 2) compute, say, the first 100 multiples of spc, 3) either manually or by code, pick the multiple that is closest to being a whole number N, and note how many cycles of the DTMF frequency this is. The length of DFT to be performed when searching for this DTMF frequency using the Goertzel Algorithm will be N and the bin being sought will be equal to the number of cycles noted.

b. Write a function in the form

function [DFTGBin,MagDFTGBin] = LVxDFTViaGoertzelBin(Bin,Sig)

which returns the DFT Bin value (as *DFTGBin*) and its magnitude when called and supplied with the desired Bin (equivalent to frequency k) and a signal Sig comprised of N contiguous signal samples where N is the desired DFT length for the particular DTMF frequency being sought determined in step (a) above.

c. Write an encoding function in the format

function OutputWave = LVxEncodeDTMF(DialedDigits,NoiseVar)

where *DialedDigits* is an array of alphanumeric data to be encoded, limited to the alphanumeric data given in the table above, and *NoiseVar* is a number setting the amplitude of white noise to mix with the output audio signal. Let each digit be encoded using 2000 samples of the two tones, sampled at 8000 Hz, followed by 1000 samples of silence. Once all digits have been encoded and the complete audio waveform sequence generated, add white noise to it, weighted by *NoiseVar*.

d. Write a decoding function in the format

function [DialedDigits] = LVxDecodeDTMF(InputSignal)

where *DialedDigits* is the output string of alphanumeric data that was originally encoded, and *InputSignal* is *OutputWave* from your function *LVxEncodeDTMF*. Use the function *LVxDFTViaGoertzelBin* to search for each row and column frequency in each time period of the input audio waveform that contains an encoded alphanumeric character.

Test your scripts with several call pairs, such as:

(I). **OW = LVxEncodeDTMF([7 0 3],0.05); [DialedDigits] = LVxDecodeDTMF(OW)**

(II) **OW = LVxEncodeDTMF(['A', '0', '3'],0.05); [DialedDigits] = LVxDecodeDTMF(OW)**

Increase the value of *NoiseVar* to see how much noise is necessary to disrupt proper detection. An interesting experiment is to use Goertzel lengths that are constant, such as 200 samples, and adjust the bin value sought to be 200/800 = 1/40 times the DTMF frequency being sought. Using this scheme, you should find that a smaller amount of noise will lead to unreliable detection since the DTMF tones will generally be "off-bin" with this scheme, resulting in a lower signal to noise ratio in general.

16. Use the flow diagram for an 8-pt decimation-in-time FFT (found in Fig. 7.20) to write an expression for Bin 3 of an 8-pt DFT. Determine how many complex multiplications are involved to compute just Bin 3, and compare this to similar complex multiplication counts for a complete 8-pt DFT using direct implementation, a complete 8-pt decimation-in-time FFT, and the Goertzel Algorithm for Bin 3 of an 8-pt DFT. Which is the most economical method to use if it is only necessary to compute a single bin of an 8-pt DFT?

17. Assume that your answer for the previous problem to compute Bin 3 of an 8-pt DFT using the minimum number of operations from the flow diagram is of the form N_{CM} complex multiplications, how many complex multiplications would be required to compute Bin 3 (or any other bin in particular) of a 1024-pt DFT using the minimum number of operations from the flow diagram?

18. Assume that you have two sequences of length-N to be linearly convolved, and N is a power of 2. Determine 1) how many multiplications are necessary to perform the linear convolution, 2) how many multiplications would be necessary using the DFT method assuming the DFTs are performed by direct implementation, and 3) how many multiplications would be necessary using the DFT method implemented using decimation-in-time FFTs.

19. Use DFTs of the specified lengths to compute samples of the DTFT of the following sequences, and compare results among the different DFT lengths performed on each sequence (comparison should be made using magnitude).

(a) **[1,zeros(1,29),-1]**; use DFT lengths of 31,32,33,34,64,256,...
1024,4096

(b) **[ones(1,39)]**; use DFT lengths of 39,40,41,42,64,256,1024,4096.

(c) **[real(j.^(0:1:31))]**; use DFT lengths of 32,33,34,35,36,37,38,...

39,64,256,1024

20. Write the m-code to implement the script

$$LVxInvDFTComputeRect25$$

as described and illustrated in the text.

21. Write a script that will accept as an input a user-specified complex-valued sequence, for which the script will perform the DFT and then synthesize the IDFT harmonic-by-harmonic, displaying the real and imaginary parts of the original sequence, the real and imaginary parts of the cumulative IDFT, which is accumulated harmonic-by-harmonic (press any key for the next harmonic); provide additional subplots to display the real and imaginary parts of both the positive and negative frequencies for each harmonic. The figure and subplots created by the script *LVxInvDFTComputeRect25* as shown in the text should serve as a model (except that for your script the original signal and cumulative IDFT will both need to have real and imaginary subplots). There should be a total of eight subplots on your figure. The script should follow this format, and be tested with the calls shown.

```
function LVxInvDFTComplex(InputSignal)
% InputSignal can be real or complex
```

Test your script with the following sequences:

(a) **LVxInvDFTComplex([ones(1,4),-ones(1,4)])**
(b) **LVxInvDFTComplex([0, ones(1,3),-ones(1,3)])**
(c) **LVxInvDFTComplex([1, ones(1,3),-ones(1,3)])**
(d) **LVxInvDFTComplex([ones(1,8) +j*((-1).^(0:1:7))])**
(e) **LVxInvDFTComplex([0,j*ones(1,7),0,-j*ones(1,7)])**
(f) **LVxInvDFTComplex([ones(1,8)] +...**
[0,j*ones(1,3),0,-j*ones(1,3)])

22. Write a function that will receive as arguments 1) a sequence that is either a time domain sequence or the DFT of a time domain sequence, and 2) a flag telling the script to either perform the DFT on the sequence or the IDFT on the sequence. Write the function so that only one DFT algorithm is present, namely a direct-implementation that you write, and that when the flag specifies the IDFT, the proper steps are taken to prepare the sequence and treat the DFT output in accordance with Eq. (7.33) to result in the IDFT of the input sequence. The function should take the form

```
function [OutputSequence] = LVxDFTorIDFT(InputSequence,...
DftOrIdftFLAG)
% InputSequence is either a time domain sequence to have the
% DFT performed on it or a DFT to be turned into a time domain
% sequence via the IDFT, which is performed with the same DFT
% algorithm. Pass DftOrIdftFLAG as 0 to perform the DFT or 1
% to perform the IDFT
```

% Sample Calls:
% **[OutputSeq] = LVxDFTorIDFT([ones(1,8)],0)**
% **[OutputSeq] = LVxDFTorIDFT([ones(1,8)],1)**
% **[OutputSeq] = LVxDFTorIDFT([8, zeros(1,7)],0)**
% **[OutputSeq] = LVxDFTorIDFT([8, zeros(1,7)],1)**

23. Compute the following IDFTs using the matrix method of Eq. (7.31).

(a) [10,(-2 + 2j),-2,(-2 - 2j)]
(b) [2,(1 + 1j),0,(1 - 1j)]
(c) [-0.8557,(1.9849 - 2.2580j),(-1.4752 -1.9559j),(-1.4752 +1.9559j),(1.9849 + 2.2580j)]
(d) [1,(-0.5813 - 0.4223j),(0.0813 + 0.2501j),(0.0813 - 0.2501j),(-0.5813 + 0.4223j)]

24. For the following sequences, sampled over the duration and the rates stated, give the bin width of the resulting DFT, the total number of DFT bins and the total number of unique bins (i.e., for a real sequence, bins representing zero and positive frequencies).

(a) Sample Duration: 4 seconds; Sample Rate: 4000 Hz.
(b) Sample Duration: 0.25 seconds; Sample Rate: 8000 Hz.
(c) Sample Duration: 0.02 seconds; Sample Rate: 44,100 Hz.
(d) Sample Duration: 15 seconds; Sample Rate: 100 Hz.

25. Write a script that evaluates the frequency-discrimination performance of four different windows in noise as is done by the script *LVxWindowingDisplay*. The script constructs a signal of random noise of standard deviation *k* and adds to it two sinusoids of unity amplitude having frequencies *freq1* and *freq2*, windows the signal with one of the test windows (*rectwin*, *hamming*, *blackman*, *kaiser*(5)), then obtains the magnitude of the DFT (use DFTs of length 256 or greater). This process is repeated 30 times, and the results averaged. Each window is tested in this manner, and the composite results are plotted and a logarithmic scale. Run one of the sample calls for the script *LVxWindowingDisplay* to see the results for the four windows. Follow the format below, and test the script with the sample calls given.

function LVxWindowingDisplay(k,freq1,freq2)
% k determines the amount of random noise to be mixed with two unity-amplitude sinusoids having frequencies of freq1 and freq2.
% Test calls are:
% **LVxWindowingDisplay(1,66,68)**
% **LVxWindowingDisplay(1,66.5,68.6)**
% **LVxWindowingDisplay(1,66.5,70.7)**
% **LVxWindowingDisplay(2,66.6,70.5)**
% **LVxWindowingDisplay(2,60,70)**

26. For each of Figs. 7.33, 7.34, and 7.35, verify that the frequency responses shown in plot (d) of each of the figures are correct by creating a cosine signal of the specified length and frequency, windowing it as described for each figure, and then directly obtaining the DTFT, which may be done

using a very long DFT on the windowed cosine. Vary the exact placement of the window within the long (say 2^14 samples) cosine signal and note the variation in the DTFT.

27. Write the m-code for the script *LVxPhaseShiftViaDFT* as illustrated and described in the text and below:

```
function LVxPhaseShiftViaDFT
% Creates a signal having the harmonic spectrum of a square
% wave, but with random phases, then resets all phases to 0
% degrees initially using the DFT/IDFT, resulting in a cusped
% waveform, which is shown in the second subplot; thereafter,
% the phase of each bin is shifted a small amount via DFT every
% time any key is pressed. Eventually, all frequencies in the
% cusped waveform are shifted 90 degrees, and the cusped
% waveform is converted in a square wave. The third subplot
% is always a 90 degree phase-shifted version of the waveform
% in the second plot, and thus starts out as a square wave and is
% gradually converted into a cusped waveform as the cusped
% waveform in the second subplot is converted into
% a square wave
```

28. Write a short script that verifies or illustrates that the DFT of the product of two time domain sequences is $1/N$ times the circular convolution of the DFTs of each, i.e.,

$$DFT(x_1[n]x_2[n]) = \frac{1}{N}(X_1[k] \circledast X_2[k])$$

29. In this project, we'll write a script that can detect the frequency of an interfering sinusoid mixed with an audio signal. To do this, the audio signal is partitioned into small time windows or frames, and the DFT of each is obtained so that the spectrum is available over small durations throughout the signal. We'll use a finite signal of five seconds' duration, but this procedure (dividing a signal into short frames) would be mandated, for example, when processing an audio signal (in real time) of indefinite duration. For purposes of illustration, we'll compare the DFT of the entire signal to the spectrogram (i.e., spectrum over time) of the signal created by taking the DFT of many short duration frames of the signal. To enhance the detectability of the interfering sinusoid, we'll also analyze a subset of the frames, those having very low mean bin magnitude, which correspond to the relative periods of silence between uttered words in the audio file *'drwatsonSR8K.wav'*. We'll use several methods to attempt to detect which bin or bins are indicative of a persistent sinusoid, as might occur, for example, in a public address system or the like due to feedback.

We'll use the script created in this project to identify one or more interfering sinusoids. Later in the series, in the chapters on FIR filtering and IIR filtering, found in Part III, and LMS adaptive filtering, found in Part IV, we'll use the script created in this exercise to identify an interfering tone and then filter the signal to remove it. In this project, we'll concentrate on breaking the signal into frames arrayed as columns of a matrix, obtaining the DFT of each frame, and identifying one or more persistent frequencies. We'll also reconstruct the test signal from the DFTs of the frames. A simple modification of this technique (doubling the length of each frame using trailing zeros), will be used when we revisit this script in the filtering chapters of Part III and will allow us to not only perform the interfering or persistent tone identification analysis, but to then use frequency domain filtering (i.e., convolution via DFT multiplication) to remove the persistent tone, and then construct the filtered test signal from the filter-modified frames. This is a useful technique because real-time systems, for example, must work frame-by-frame, often having no more than a few frames of recent history available for analysis, filtering, and signal reconstruction.

If the purpose is frequency analysis only, nonoverlapping frames can be used. Overlap in framing a signal is used in filtering or compression techniques which result in reconstructed frames that are not perfect reconstructions of the original frames, such as with lossy compression techniques like MP3. When the reconstruction is perfect, nonoverlap of frames works well. When the reconstruction is lossy, 50% overlap of windowed frames is useful because it serves as a crossfading operation (i.e., a smooth transition from one frame to the next, in which the volume of one frame gradually fades while that of the other gradually increases) that hides the discontinuities at frame boundaries, which can cause audible clicks or other artifacts. When using 50% frame overlap, use of a nonrectangular window is essential to perform the crossfading operation. Use of 50% overlap also increases the number of "snapshots" of the frequency content per second, i.e., a finer-grained spectrogram.

We now present the script format, argument description, and test calls, followed by a procedure to write the m-code for it.

```
function [ToneFreq,Fnyq,BnSp] = LVx_DetectContTone(A,...
Freq,RorSS,SzWin,OvrLap,AudSig)
% Mixes a tone of amplitude A and frequency Freq with an
% audio file and attempts to identify the frequency of the
% interfering tone, which may have a steady-state amplitude
% A when RorSS is passed as 0, or an amplitude that
% linearly ramps from 0 to A over the length of the
% audio file when RorSS is passed as 1. The audio file
% is 'drwatsonSR8K.wav', 'whoknowsSR8k.wav', or white
% noise, which are selected, respectively, by passing
% AudSig as 1, 2, or 3. SzWin is the size of time window in
% samples into which the test signal (audio file plus tone) is
% partitioned for analysis. In partitioning the test signal into
% time windows or frames, an overlap of 0% or 50% of
% SzWin is performed when OvrLap is passed as 0 and 1,
% respectively. A number of figures are created, including
% a first figure displaying mean bin magnitude of all bins
```

```
% for each frame, which serves to identify periods of high
% and low energy in the signal, corresponding to active
% speech and background sound, second and third figures
% that are 3-D spectrogram plots of DFT magnitude versus
% Bin and Frame for the complete signal and a version
% based on only the background or relatively low energy
% frames. Fourth and fifth figures display the normalized
% bin derivatives versus frame for the complete signal and
% the low portion of the signal.
% The output arguments consist of a list of possible
% interfering (relatively steady-state) tones in Hz as T, the
% Nyquist rate as F, and the bin spacing in Hz as B, which
% allows another program to determine if a list of frequencies
% provided as ToneFreq contains adjacent bin frequencies
% and are therefore likely to be members of the same spectral
% component.
% Test calls
% [T,F,B] = LVx_DetectContTone(0.011,100,0,512,1,1)
% [T,F,B] = LVx_DetectContTone(0.008,100,0,512,0,1)
% [T,F,B] = LVx_DetectContTone(0.005,94,0,512,0,1)
% [T,F,B] = LVx_DetectContTone(0.07,150,0,512,1,3)
% [T,F,B] = LVx_DetectContTone(0.011,200,0,512,1,1)
% [T,F,B] = LVx_DetectContTone(0.01,200,0,512,1,2)
% [T,F,B] = LVx_DetectContTone(0.1,200,0,512,1,3)
% [T,F,B] = LVx_DetectContTone(0.01,210,0,512,1,1)
% [T,F,B] = LVx_DetectContTone(0.025,200,1,512,1,1)
% [T,F,B] = LVx_DetectContTone(0.02,200,1,512,1,2)
% [T,F,B] = LVx_DetectContTone(0.1,200,1,512,1,3)
% [T,F,B] = LVx_DetectContTone(0.08,500,0,512,1,3)
% [T,F,B] = LVx_DetectContTone(0.02,500,1,512,1,1)
% [T,F,B] = LVx_DetectContTone(0.018,500,1,512,1,2)
% [T,F,B] = LVx_DetectContTone(0.1,500,1,512,1,3)
% [T,F,B] = LVx_DetectContTone(0.005,1000,0,512,0,1)
% [T,F,B] = LVx_DetectContTone(0.004,1000,0,512,1,1)
% [T,F,B] = LVx_DetectContTone(0.002,3000,0,512,1,1)
```

Here is a suggested procedure to create the script described above:

(a) Open the audio file *'drwatsonSR8K.wav'*, scale it to have maximum magnitude of 1.0, and add a sinusoid of amplitude A and frequency $Freq$, thus forming the test signal $TstSig$.

(b) Form a matrix $TDMat$ by partitioning $TstSig$ into frames of length SR, with overlap of $SR/2$ samples (SR must be even). Each column of the matrix $TDMat$ will be a frame. For example, if the audio file were 32 samples long, and the frame size were 8 samples, then the first column of the matrix will comprise samples 1:8, the second column will be samples 5:12, the third column will be samples 9:16, and so forth, until all 32 samples have been used. If the last column is only partially filled with samples, it is completed with zeros. A good way to proceed is to make this action a function according to the following specification:

```
function OutMat = LVxVector2FramesInMatrix(Sig,SzWin,...
SampsOvrLap)
% Divides an input signal vector Sig into frames of length
% SzWin, with an amount of overlap in samples equal to
% SampsOvrLap. Each column of OutMat is one frame
% of the input Sig.
% Test calls:
% OutMat = LVxVector2FramesInMatrix([0:1:33],8,4)
% OutMat = LVxVector2FramesInMatrix([0:1:33],8,0)
% OutMat = LVxVector2FramesInMatrix([0:1:33],8,1)
% OutMat = LVxVector2FramesInMatrix([0:1:33],8,2)
```

(c) If 50% overlap has been used, window each column of $TDMat$ with a suitable window such as hamming, etc. When not using overlap, application of a window results in noticeable amplitude modulation of the output signal reconstructed from the windowed frames, so a nonrectangular window should not be applied to the columns (frames) of $TDMat$ in such a case.

(d) Compute a matrix Fty which is the DFT of $TDMat$, computed using the function fft. If no overlap was used and hence no nonrectangular window applied to the frames of $TDMat$, apply a window as the DFT is obtained (i.e., without permanently applying a window to $TDMat$ as this is not wanted in this case).

Each column of the resulting matrix Fty is the DFT of the corresponding frame (column) in $TDMat$. Resize the column length of Fty to eliminate negative bins, i.e., the new column length should include bins 0:1:$SzWin/2$.

(e) Form a matrix $Afty$ which is the absolute value of the elements of Fty, and normalize it by dividing it by the magnitude of the largest element.

(f) Determine or identify the lowest magnitude frames by computing the mean of each column of $Afty$, and computing a histogram of the vector of frame (column) mean values. Use the frames, in order of chronological occurrence, whose magnitudes are found below the lowest histogram threshold level to form a matrix $minFrameMat$. The following code will accomplish this:

```
meanframes = mean(Afty);
[Nhist,Xhist] = hist(meanframes);
histthresh = 1; % lowest histogram bin of the standard ten
% linear divisions of the range of values found in the data
```

```
[i,j] = find(meanframes < xhist(histthresh));
minFrameMat = Afty(:,j);
```

(g) Several methods should be used to attempt to identify steady-state amplitude or monotonically increasing amplitude sinusoids. The first method identifies the bin or bins that have magnitudes above a given threshold for the largest number of frames. The threshold can be a certain number of dB, for example, above the mean bin magnitude for the entire matrix. A number of candidate bins or frequencies can be selected using this method. A second method attempts to take advantage of the characteristic of a steady-state sinusoid that the derivative of bin magnitude with respect to frame (i.e., time), or an approximation thereof, the difference between bin magnitudes in adjacent frames, normalized by the mean bin magnitude for the bin over all frames (First Order Difference or FOD), is very low compared to that of bins in which no steady-state sinusoid is present. A number of candidate bins (such as those below the lowest one or two histogram threshold levels, or a certain number of standard deviations below mean) can also be selected using this method. A third method looks for bins that have the largest number of positive derivatives (i.e., FODs), indicating a steadily-rising amplitude sinusoid, which might be found, for example, in a public address system when feedback first arises. All three methods should be performed on both the full frame matrix $Afty$ and the minimum magnitude frame matrix $minFrameMat$. It will generally be found that the estimates derived from $minFrameMat$ are better than those from the full frame matrix.

(h) A voting procedure should be used, requiring that candidates appear on at least two of the three lists generated using the three methods described above, in order to be reported as a steady-state or rapidly rising sinusoid in output variable T. Alternately, certain standards, such as some multiple of standard deviations above or below mean, as the case may be, may also be used.

(i) Attempt to determine the relative degrees of reliability of the methods. Note that steady-state sinusoids are less detectable when they are close to other more transient tones in the audio signal itself. Thus the detectability of steady-state or rising-amplitude tones in the 100-300 Hz range is more difficult than in other frequency ranges. The methods discussed above can nonetheless identify a steady-state or rising-amplitude sinusoid in the 100-300 Hz range, especially using the minimum energy frame matrix method, which eliminates many of the transient, high-amplitude signal tones.

(j) Play the test signal file through the computer's audio system to determine the relative audibility of the interfering sinusoid throughout the signal. Note that at very low amplitude levels near the threshold of detection using the methods discussed above, the signal, especially when in the low frequency range (say, 100-300 Hz), may be masked at times.

(k) Reconstruct the original test signal from the matrix Fty (this is essentially the reverse of the procedure that created Fty) and play it through the computer's audio system to verify that it is correct. To perform the reconstruction, write a script having the following format:

```
function tdSig = LVxTDMat2SigVec(TDMat,SzWin,...
SampsOverLap)
% Generates a signal vector tdSig from a matrix of time
% domain frames of a signal, each frame of length SzWin,
```

```
% with SampsOverLap samples of overlap. The code
% determines which dimension of TDMat matches SzWin
% and assumes that the other dimension is the frame index.
% If neither dimension of TDMat matches SzWin, an error
% is thrown.
% Test call pair for OverLap=0
% OutMat = LVxVector2FramesInMatrix([0:1:33],8,0)
% tdSig = LVxTDMat2SigVec(OutMat,8,0)
% The following four lines of code, when run in sequence,
% return the input vector [0:1:33]:
% OutMat = LVxVector2FramesInMatrix([0:1:33],8,4);
% szO = size(OutMat);
% OutMat = OutMat.*(triang(8)*ones(1,szO(2)))
% tdSig = LVxTDMat2SigVec(OutMat,8,4)
```

The call

```
[T,F,B] = LVx_DetectContTone(0.008,1000,0,512,1,1)
```

results in the following output values (note that the two low frequencies are present in the audio file prior to addition of the 1000 Hz tone):

```
ToneFreq = [46.9, 62.5, 1000]
Fnyq = 4000
BnSp = 15.625
```

and the following figures, as described above.

30. Write the m-code for the script

```
function LVxZxformFromSamps(b,a,NumzSamps,M,nVals)
% Receives a set of coefficients [b,a] representative of the
% z-transform of a sequence x[n] for which the unit circle is
% included in the ROC. NumzSamps samples of the z-transform
% of x[n] are computed, and then the complete z-transform is
% computed from the samples. From this, nVals samples of x[n]
% are reconstructed using numerical contour integration. The
% samples of X(z) are also used to reconstruct a periodic version
% of x[n] over nVals samples. Both reconstructed sequences, x[n]
% and the periodic version are plotted for comparison. M is the
% number of dense grid z-samples to use for the contour integration
% Test call:
% LVxZxformFromSamps([1,1,1,1],[1],6,5000,10)
% LVxZxformFromSamps([1],[1,0,0.64],36,5000,100)
```

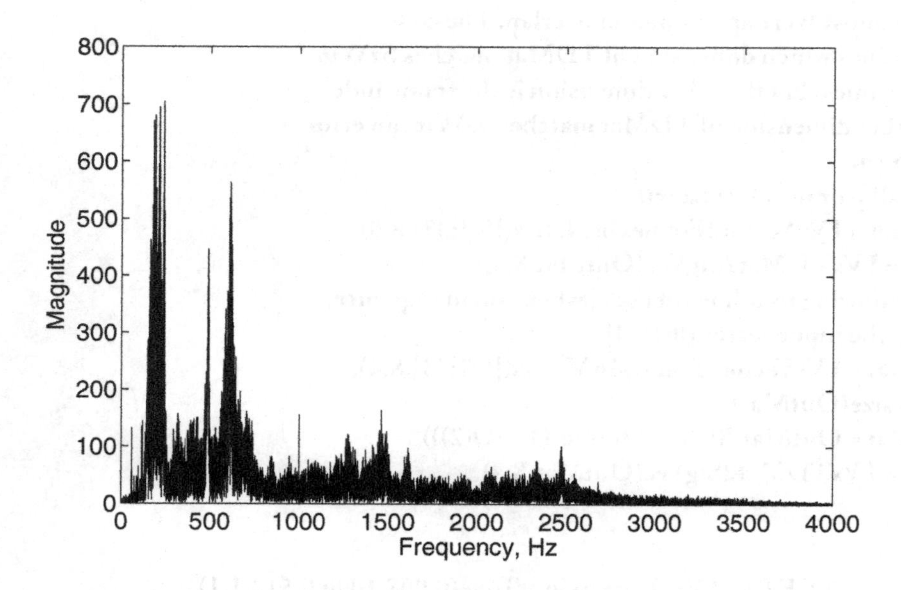

Figure 7.56: DFT (positive frequencies only) of the entire test audio signal, comprised of the audio file 'drwatsonSR8K.wav', normalized to maximum unity magnitude, to which has been added a sine wave of amplitude 0.008 and frequency of 1000 Hz.

Theoretically, $x[n]$ must be finite length, i.e., identically zero for $n < 0$ and $n \geq N - 1$. Obviously, however, many IIR impulse responses decay away to a magnitude of essentially zero in a finite number of samples, so provided that the number of z-transform samples obtained is large enough, reconstruction of $X(z)$ can be performed to a reasonable approximation. This is demonstrated by the second test call given above, the result of which is shown in Fig. 7.62.

31. Evaluate the DTFT of the original sequence, sinc-interpolated sequence, and linear-interpolated sequence created by the script *LVxInterp8Kto11025*, which was developed in Part I of the book, for the chapter on sampling and binary representation. For the interpolated sequences, evaluate the DTFT both before and after the post-interpolation lowpass filtering. For each evaluation, use 1024 signal samples, perform a DFT of length 2^15, and plot the positive frequency response only in decibels, with the maximum response for each plot being zero dB. Figures 7.63 and 7.64 show the plots that should be created, for example, in response to the call

LVxInterp8Kto11025(0,2950)

32. Write the m-code for the following function specification, the purpose of which is to verify the correctness of Eq. (7.34) in the text. Test your script with the given test calls.

function LVxIDFTviaPosK(TestSig)

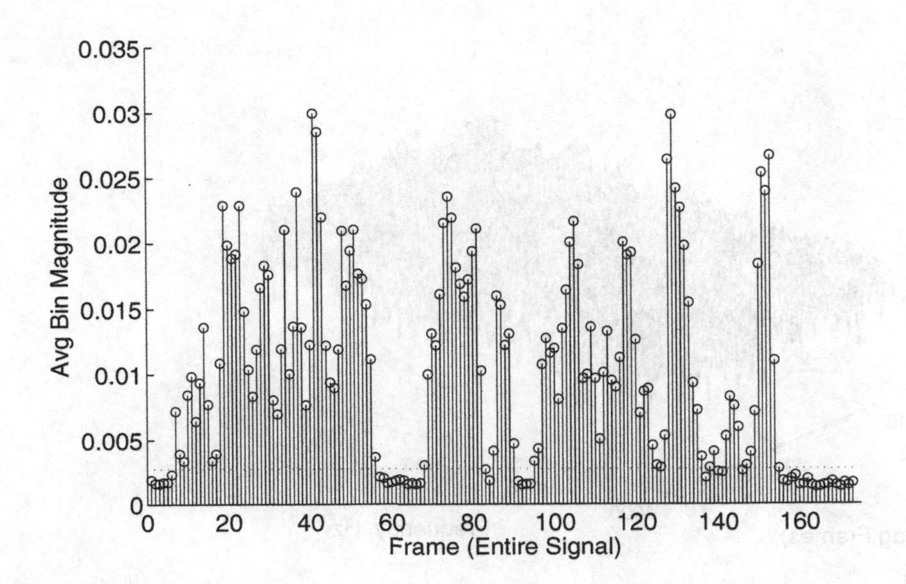

Figure 7.57: A plot of mean bin magnitude versus frame for the entire test signal. Mean bin magnitude is obtained by averaging the bin magnitudes for an entire frame.

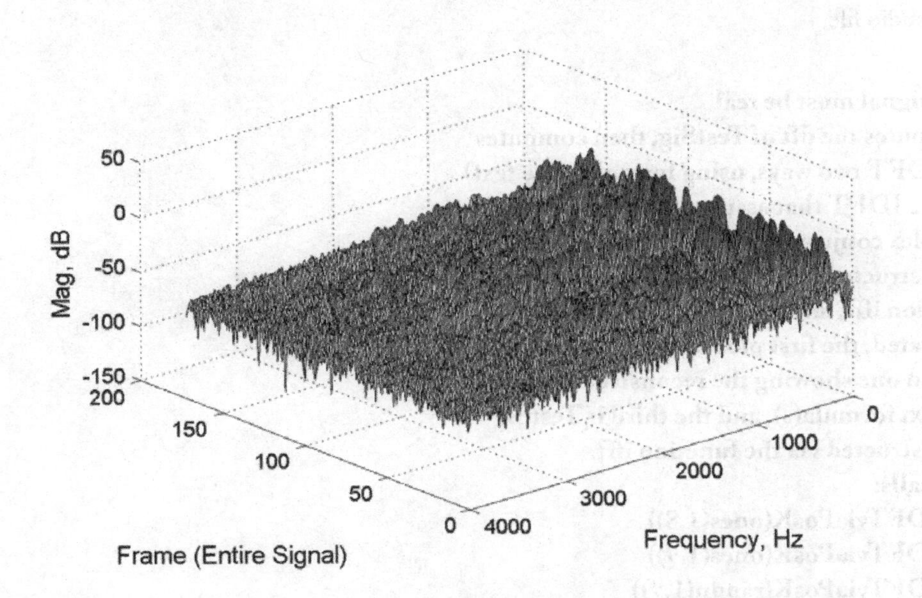

Figure 7.58: Bin magnitude versus Frequency and Frame for the entire test signal. The low-amplitude, steady-state sinusoid at 1000 Hz is barely discernible.

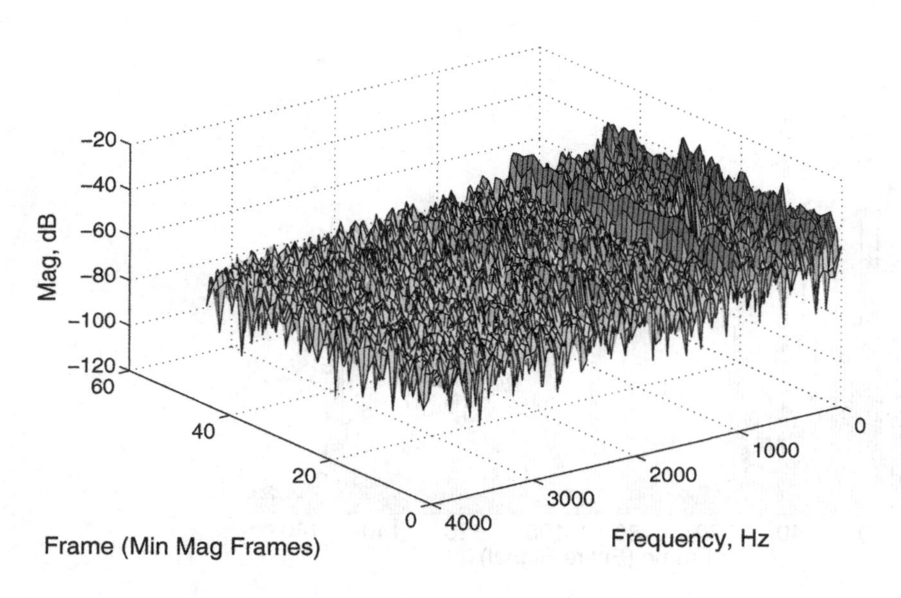

Figure 7.59: Bin magnitude versus Frequency and Frame for the low-magnitude frame matrix (*minFrameMat*). The low-amplitude, steady-state sinusoid at 1000 Hz is readily discernible in the low-energy frames of the signal. Note also an apparent steady-state sinusoid around 60 Hz; this is present in the original audio file.

```
% Test signal must be real
% Computes the dft of TestSig, then computes
% the IDFT two ways, using formulas (see text)
% for the IDFT that assume bins +k and -k are
% complex conjugates of each other. TestSig is also
% reconstructed via standard IDFT using the
% function ifft, and a figure with three plots is
% generated, the first plot showing TestSig, the
% second one showing the reconstruction via
% the text formula(s), and the third is TestSig
% reconstructed via the function ifft.
% Test calls:
% LVxIDFTviaPosK(ones(1,8))
% LVxIDFTviaPosK(ones(1,9))
% LVxIDFTviaPosK(randn(1,9))
% LVxIDFTviaPosK(j*randn(1,9))
```

33. Write the m-code for the script

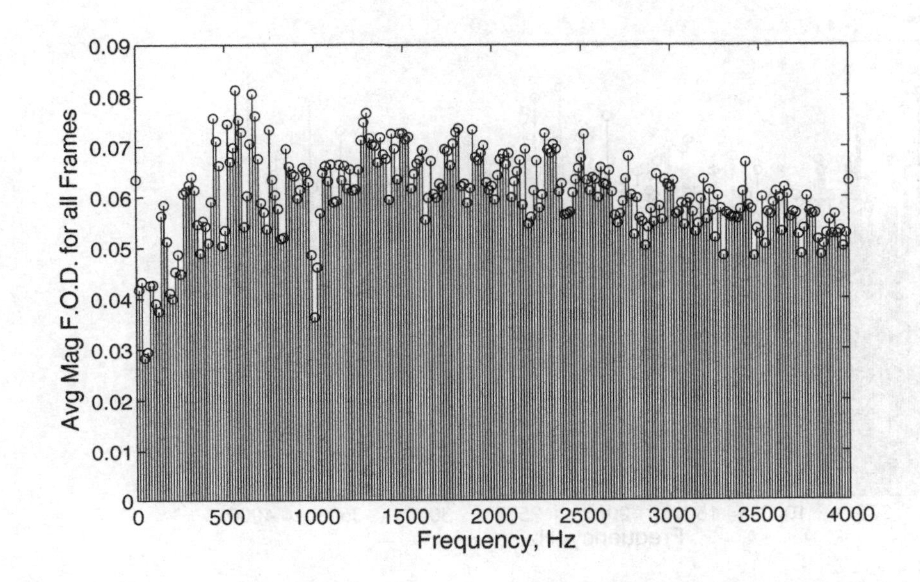

Figure 7.60: A plot of first order differences of bin magnitudes between adjacent frames. The first order differences are all normalized by the respective mean bin magnitudes over all frames in the matrix, which in this case includes all frames of the test signal.

$$LVxDFTEqualization(tstSig, p, k, SR, xplotlim)$$

as described in the text and according to the following function specification:

```
function LVxDFTEqualization(tstSig,p,k,SR,xplotlim)
% Creates a test signal; tstSig=1 gives 32 periods of a
% length-16 square wave followed by 16 samples equal to zero;
% tstSig=2 yields 64 periods of a length-16 square wave;
% tstSig = 3 yields a length-1024 chirp from 0 to 512 Hz.
% p is a single real pole to be used to generate a decaying
% magnitude profile which weights SR/4 samples of random noise
% of standard deviation k to generate the test channel
% impulse response. SR is the length of FFT to be used to
% model the channel impulse response and create the inverse
% filter.
% xplotlim is the number of samples of each of the relevant signal
% to plot; the relevant signals are the test signal, the distorted
% test signal, and two equalized versions created, respectively, by
```

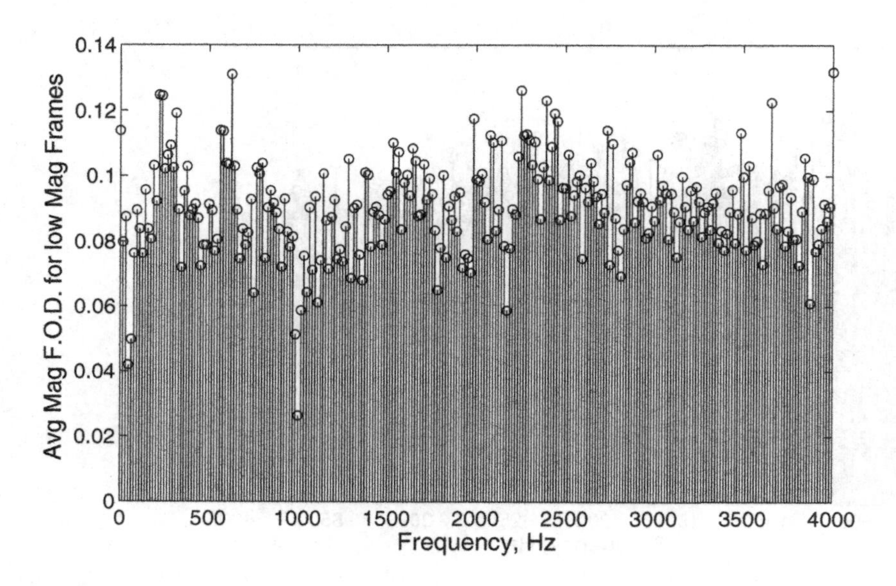

Figure 7.61: A plot of first order differences of bin magnitudes between adjacent frames. Bin indices here are converted to their Hertz equivalents so that a direct search for low FOD can be made by frequency. The first order differences are all normalized by the respective mean bin magnitudes over all frames in the matrix, which in this case includes only low-energy frames of the test signal. Note the large dip in the magnitude of FOD at 500 Hz, indicative of a relatively steady-amplitude sinusoid at that frequency.

```
% frequency domain (FD) deconvolution and time domain(TD)
% deconvolution.
% Since the impulse response is random noise with a decaying
% magnitude,every call to the function generates a completely
% different impulse response for the simulated channel, and the
% distorted test signal will have a different appearance.
% LVxDFTEqualization(1,0.9,0.5,2048,250)
% LVxDFTEqualization(2,0.9,0.5,2048,250)
% LVxDFTEqualization(3,0.9,0.5,2048,250)
```

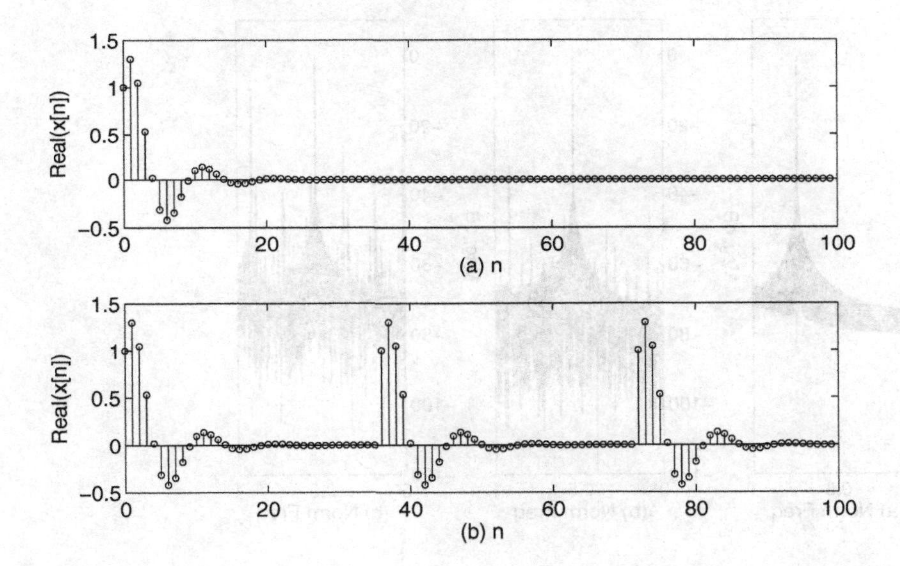

Figure 7.62: (a) A reconstruction of the first 101 samples of the impulse response of the IIR defined by its coefficients $[b, a]$ = [1],[1,-1.3,0.64], obtained by contour integration of $X(z)$, which was computed from 36 samples of $X(z)$; (b) A reconstruction of a periodic version of the same impulse response, computed as the inverse DFS of the 36 samples of $X(z)$, evaluated for n = 0:1:100.

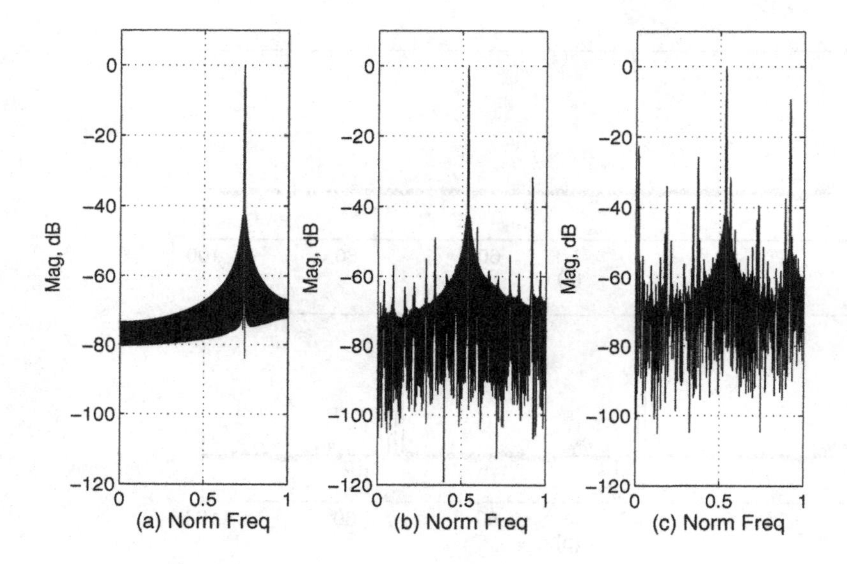

Figure 7.63: (a) DTFT of 1024 samples of original sequence of a 2950 Hz cosine sampled at 8000 Hz; (b) DTFT of 1024 samples of the 2950 Hz cosine after being resampled at 11025 Hz using sinc interpolation, without post-interpolation lowpass filtering; (c) DTFT of 1024 samples of the 2950 Hz cosine after being resampled at 11025 Hz using linear interpolation, without post-interpolation lowpass filtering.

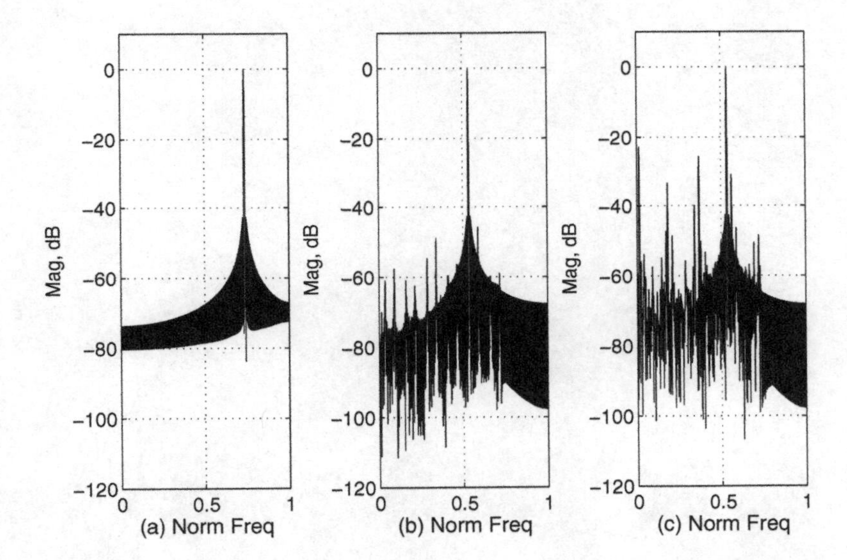

Figure 7.64: (a) DTFT of 1024 samples of original sequence of a 2950 Hz cosine sampled at 8000 Hz; (b) DTFT of 1024 samples of the 2950 Hz cosine after being resampled at 11025 Hz using sinc interpolation, and lowpass filtered with a cutoff frequency of 4 kHz; (c) DTFT of 1024 samples of the 2950 Hz cosine after being resampled at 11025 Hz using linear interpolation, and lowpass filtered with a cutoff frequency of 4 kHz.

Part III

Digital Filter Design

CHAPTER 8

Principles of FIR Design

8.1 OVERVIEW

8.1.1 IN THIS PART OF THE BOOK

In this part of the book we take up digital filter design, which addresses the question of precisely how to design filters (i.e., compute an appropriate set of filter coefficients) that meet very specific design requirements, which include band limits, maximum passband ripple, minimum stopband attenuation, steepness of roll-off, etc. There are a number of design approaches for digital filters; they require knowledge of the DFT, the DTFT, the Laplace transform (for classical IIR filters), and the z-transform for complete understanding. How each of these transforms is relevant will become apparent as we move through this and the following two chapters.

8.1.2 IN THIS CHAPTER

In this chapter we acquire additional information and tools specific to the FIR, including the effects of filter length and windowing, the relationship between impulse response symmetry and phase linearity, and the frequency response of linear-phase FIRs. We then present brief discussions of two useful FIRs that are simple to design, the Comb Filter and the Moving Average Filter. These two filters have their own distinctive uses. The Comb filter is often useful for eliminating harmonically related spectral components in a signal, and the Moving Average filter is useful for signal averaging to emphasize a coherent signal in noise. The Comb Filter, additionally, is a processing component in the Frequency Sampling Form of FIR realization, which is covered at the end of the chapter along with other filter realizations particular to the FIR.

8.2 SOFTWARE FOR USE WITH THIS BOOK

The software files needed for use with this book (consisting of m-code (.m) files, VI files (.vi), and related support files) are available for download from the following website:

http://www.morganclaypool.com/page/isen

The entire software package should be stored in a single folder on the user's computer, and the full file name of the folder must be placed on the MATLAB or LabVIEW search path in accordance with the instructions provided by the respective software vendor (in case you have encountered this notice before, which is repeated for convenience in each chapter of the book, the software download only needs to be done once, as files for the entire book are all contained in the one downloadable folder). See Appendix A for more information.

8.3 CHARACTERISTICS OF FIR FILTERS

- The impulse response of an FIR, if made symmetric or anti-symmetric, will yield a linear phase function, the result being that signals passing through the filter do not have their phases dispersed.

- Arbitrarily steep roll-offs may be had by making the impulse response correspondingly long. The cost is in computation, since linear convolution in the time domain of two sequences of length N requires about N^2 multiplications. This can often be alleviated by the use of frequency domain techniques.

- Arbitrary pass characteristics (i.e., other than standard lowpass, highpass, notch, or bandpass) may readily be generated.

- FIRs are inherently stable, meaning that a finite-valued input signal will never lead to an unbounded output signal.

8.4 EFFECT OF FILTER LENGTH

To discover several basic FIR principles, let's experiment with the length of a simple lowpass filter having impulse response $[ones(1, N)]$. By increasing N, the number of samples in the filter, we can observe the effect on frequency response, especially on steepness of roll-off. Figure 8.1 shows three such filters of increasing length in subplots (a), (c), and (e), and their frequency responses in plots (b), (d), and (f), respectively.

From Fig. 8.1 we can readily deduce that a filter can be made more selective or have a steeper roll-off by increasing its length. This may be explained by noting that as the filter length is increased, there are more distinct integral-valued frequencies (or correlators) between 0 and π radians (normalized frequencies of 0 and 1.0), that serve as potential correlation frequencies. From our studies of orthogonality (found in Part I of the book), we know that if no correlator at a given frequency is in fact present in the impulse response, the filter's response at that precise frequency must be zero, assuming the presence of at least one other integral-valued frequency component (i.e., correlator) in the impulse response. In the case of the simple lowpass filter consisting of two or more samples valued at 1.0, only DC is present, so the frequency response must go to zero at each distinct correlator frequency other than zero. For the length-3 filter [1,1,1], the only potential correlators are at normalized frequencies 0 and 0.6667 ([0,1]/1.5), and we can see in Fig. 8.1, subplot (b), that the frequency response goes to zero at frequency 0.6667. The length-7 filter shown in Fig. 8.1, subplot (c), has potential correlators at normalized frequencies of [0:1:3]/(3.5) = [0, 0.2857, 0.5714, 0.8571]. We can see from Fig. 8.1, subplot (d) that it indeed has a frequency response of zero at normalized frequencies of [0.2857, 0.5714, 0.8571], as expected.

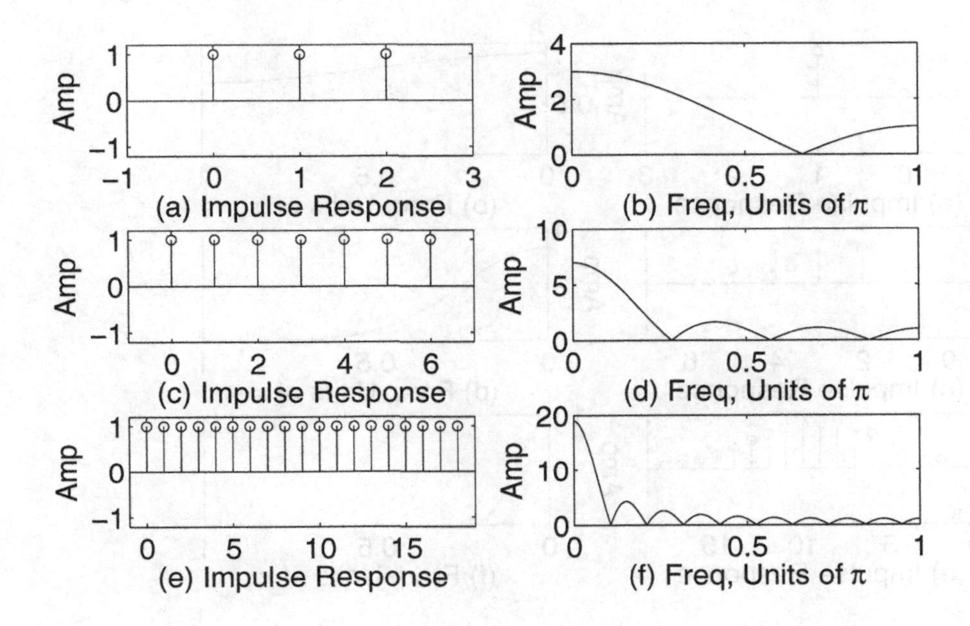

Figure 8.1: (a) 3-pt Rectangular Impulse Response; (b) Magnitude of frequency response of signal at (a); (c) 7-pt Rectangular Impulse Response; (d) Magnitude of frequency response of signal at (c); (d) 19-pt Rectangular Impulse Response; (e) Magnitude of frequency response of signal at (e).

8.5 EFFECT OF WINDOWING

It can be seen in Fig. 8.1 that longer impulse responses, although more frequency selective, suffer from leakage or scalloping in the frequency response. Fortunately, this can be alleviated by using the same kind of windows on the filter impulse response that we used on signals prior to taking the DFT. Figure 8.2 shows the same three impulse responses after smoothing with a *hamming* window, and the corresponding frequency responses. The improvement is dramatic.

In Fig. 8.3, plot (a), a multi-cycle cosine (inherently rectangularly-windowed) has been used as the impulse response. As can be seen in plot (c), the expected scalloped response results. Figure 8.3, plot (d), shows the result when a *hamming* window is applied to the impulse response (plot (b)). Note that the steepness of roll-off of the main lobe of the response is decreased with use of the *hamming* window, compared to the rectangular window.

You should notice that the rectangular window result has a main lobe width narrower than that of the *hamming* window example, as can be readily seen, but the rectangular window's side lobe amplitude is very high (i.e., the stopband attenuation is poor), making the rectangular window unacceptable for most applications.

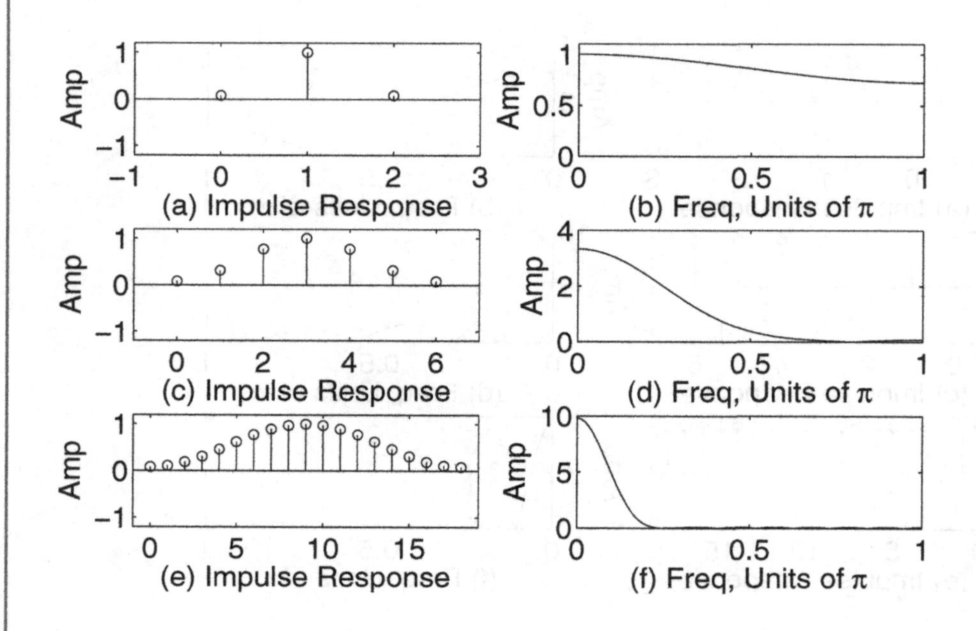

Figure 8.2: (a) 3-pt Hamming-windowed Impulse Response; (b) Magnitude of frequency response of signal at (a); (c) 7-pt Hamming-windowed Impulse Response; (d) Magnitude of frequency response of signal at (c); (d) 19-pt Hamming-windowed Impulse Response; (e) Magnitude of frequency response of signal at (e).

General Rule: For a given filter length, the greater the stopband attenuation, the shallower the main lobe roll-off must be. Stated conversely, for a given filter length, the steeper the main (or central) lobe roll-off, the poorer will be the ultimate stopband attenuation. For a given stopband attenuation, the roll-off may be improved by increasing the filter length.

8.6 LINEAR PHASE

- FIRs can be made to have a **Linear Phase** response, which means that a graph of the phase shift imparted by the filter versus frequency is a straight line, or at least piece-wise linear.
- With linear phase shift, all frequencies remain in phase, and therefore the filter acts like a bulk delay line in which all frequencies remain in time alignment. Such a filter is said to be **Non-Dispersive.**

To see why a simple delay line has a linear phase delay characteristic (and vice versa), imagine a signal simply going through a delay line that imparts a delay time of τ. A given frequency f_0 has

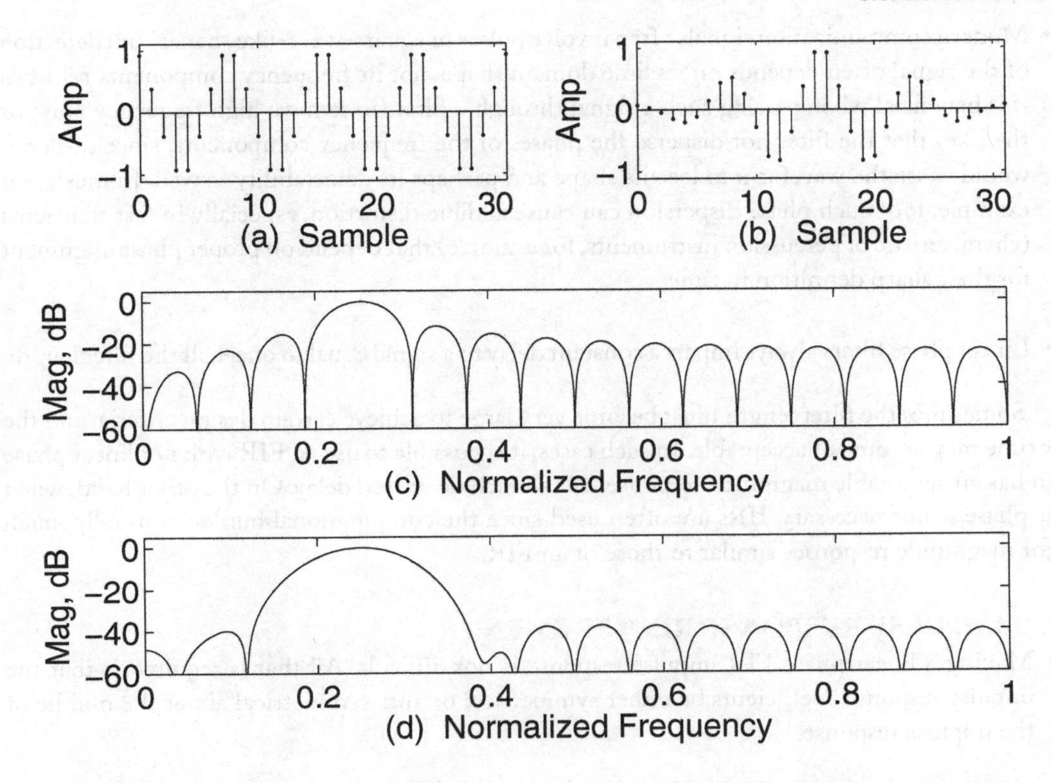

Figure 8.3: (a) Impulse response, a cosine, inherently windowed with a Rectangular window; (b) Same cosine as (a), multiplied by a Hamming window; (c) Magnitude of frequency response in dB of sequence at (a); (d) Magnitude of frequency response in dB of sequence at (b).

a period of $1/f_0$ and is therefore delayed by $\tau/(1/f_0) = f_0 \cdot \tau$ cycles, which is just a linear function of frequency If f_0 doubles, for example, so does the number of cycles of delay, and so on.

Example 8.1. Using a signal having components of 100 Hz, 200 Hz, 300 Hz, etc., show that a delay line which imparts a bulk delay of 0.01 second imparts a linear phase shift.

A 100 Hz signal has one hundred cycles in one second, or one cycle in 0.01 second, so clearly it is delayed one cycle, or 2π radians. The 200 Hz component is delayed by two cycles, since two of its cycles occur in the 0.01 second delay time. Hence, it is delayed by 4π radians, and so forth. You can see that if all the components are in phase when they go into the delay line, they will still be in phase when they come out, since the 100 Hz component will be delayed exactly one cycle, the 200 Hz component exactly two cycles, and so on, so that on exiting, they are still all beginning their cycles together, i.e., in phase with each other.

- Modern communication signals often involve pulses or square-wave-like shapes, and detection of the signal often depends on its time domain shape, not its frequency components *per se*, so it is beneficial when passing such a signal through a filter (to remove high frequency noise or the like) that the filter not disperse the phases of the frequency components, since dispersal would cause the waveform to lose its shape and perhaps its detectability as well. In music, for example, too much phase dispersion can cause audible distortion, especially in fast transients (characteristic of percussion instruments, for example) that depend on proper phase alignment for their sharp definition in time.

- Linear phase filters always impart a constant delay to a signal equal to one-half the filter length.

Sometimes the filter length must become very large to achieve certain design criteria, and the delay time may become unacceptable. In such cases, it is possible to use an FIR with nonlinear phase which has an acceptable magnitude response, but a much decreased delay. On the other hand, when linear phase is not necessary, IIRs are often used since the computational burden is usually much less for magnitude responses similar to those of an FIR.

8.6.1 IMPULSE RESPONSE REQUIREMENT

- Making a linear phase FIR impulse response is not difficult. All that is required is that the impulse response coefficients be either symmetrical or anti-symmetrical about the middle of the impulse response.

For example, a length-seven symmetrical impulse response might look like this: $[a, b, c, d, c, b, a]$, whereas a length-six filter would be $[a, b, c, c, b, a]$. The first and last coefficients are the same, the second and penultimate coefficients are the same, and so forth. An anti-symmetric impulse response of length seven might be $[a, b, c, 0, -c, -b, -a]$, whereas a length-six anti-symmetrical would be $[a, b, c, -c, -b, -a]$]. In this case, the first and last coefficients have opposite signs, and so on.

Figure 8.4 compares two impulse responses, their frequency content, and their phase responses. At (a), one cycle of a nonsymmetrical cosine has its frequency response (via DTFT) shown at (c), and it phase response at (e). Note that the phase response is slightly nonlinear at low frequencies. At (b), the same one cycle cosine has been slightly adjusted to make it symmetrical, resulting in a perfectly piece-wise linear phase characteristic at (f), having very nearly the same magnitude response (at (d)) as the nonsymmetrical impulse response.

A much more egregious example, shown in Fig. 8.5, is had by making the simple impulse response [1.5,0.5], which consists of two frequency correlators, DC and 1 cycle, weighted with 1 and 0.5, respectively, i.e., [1.5,0.5] = [1,1] + 0.5[1,-1]. The impulse response is clearly nonsymmetric, as is the phase response.

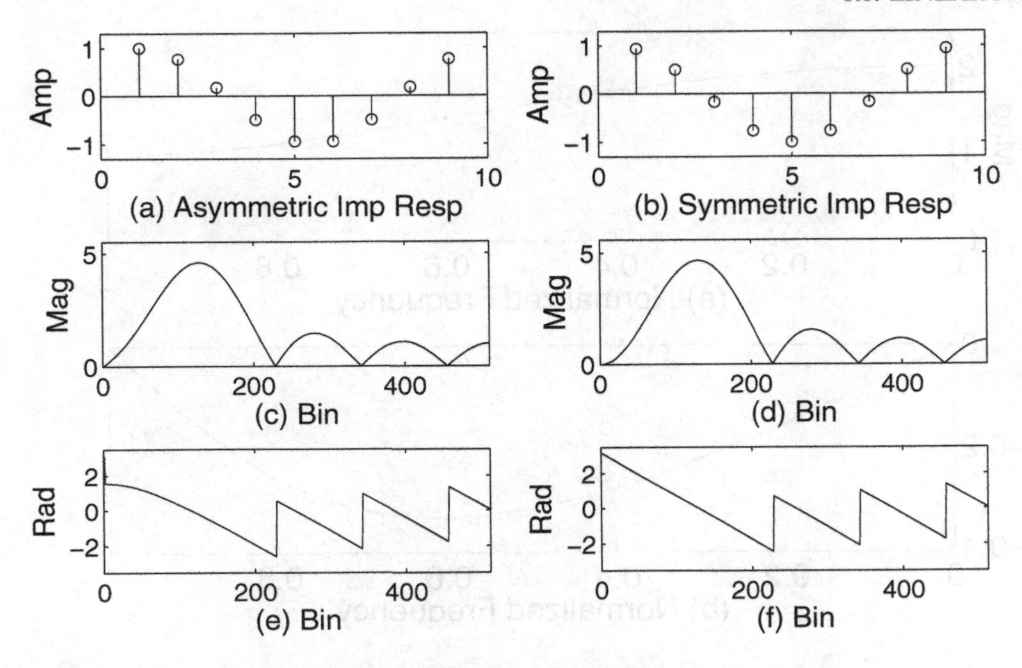

Figure 8.4: (a) Asymmetric cosine; (b) Symmetric cosine; (c) Magnitude of spectrum of signal at (a); (d) Magnitude of spectrum of signal at (b); (e) Phase response of signal at (a); (f) Phase response of signal at (b).

8.6.2 FOUR BASIC CATEGORIES OF FIR IMPULSE RESPONSE FOR LINEAR PHASE

Linear phase impulse responses must be either symmetric or anti-symmetric, and any impulse response must have either an even length (evenly divisible by two) or an odd length. There are distinct differences between even and odd length filters, as well as symmetric and anti-symmetric filters. Considering both symmetry and length, there are four basic FIR types, which are illustrated in Fig. 8.6.

Type I, Symmetric, Odd Length
This is perhaps the most used of linear phase FIR filter types because it is suitable for lowpass, highpass, bandpass, and bandstop filters. Correlator basis functions are cosines, meaning that the impulse response is generated by weighting and summing cosines of different frequencies.

Type II: Symmetric, Even Length
This filter type cannot be used as a highpass or bandstop filter since the cosine, at the Nyquist limit (necessary in the impulse response for highpass or bandstop filters), cannot be symmetrical in an even length. Lowpass and bandpass filters are possible.

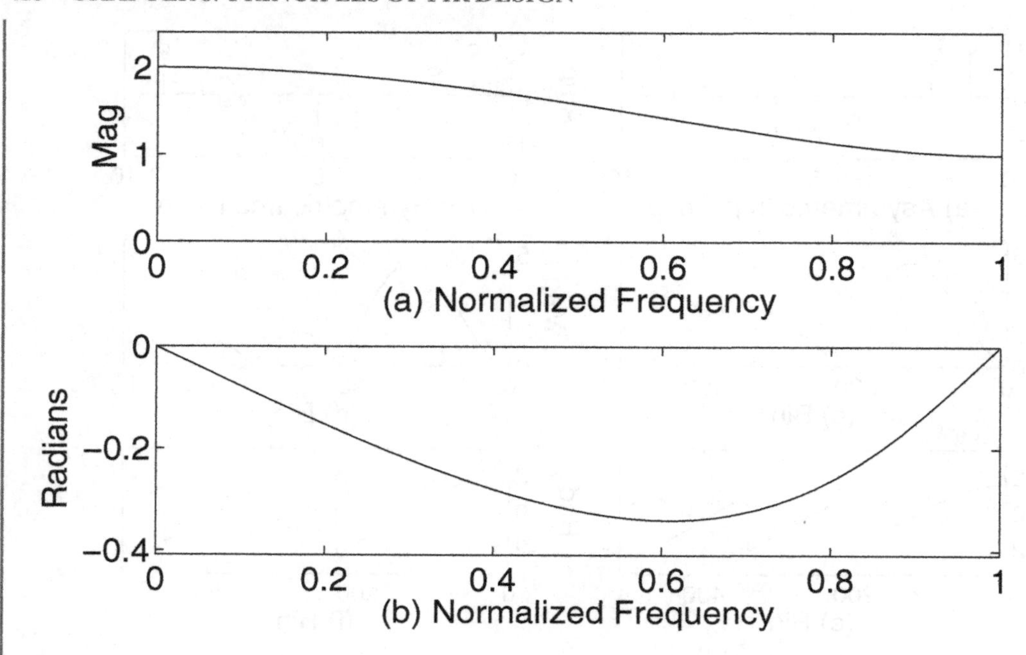

Figure 8.5: (a) Magnitude of the DTFT of the impulse response [1.5 0.5]; (b) Phase response of the same impulse response.

Type III: Anti-Symmetric, Odd Length
The correlator basis functions are sines. The basis functions are identically zero at DC and the Nyquist limit, so this type of filter cannot be used for lowpass or highpass characteristics. Bandpass filters are possible, but the primary use is in designing Hilbert transformers and differentiators. Hilbert transformers shift the phase of all frequencies in a signal by 90°, or $\pi/2$ radians. This is useful in certain communications applications, such as generating single sideband signals, demodulating quadrature modulated signals, and so forth.

Type IV: Anti-Symmetric, Even Length
Like the Type III filter, this filter uses sines as the basis correlating functions. It is not suitable for a lowpass characteristic, but is suitable for Hilbert transformers and differentiators.

8.6.3 ZERO LOCATION IN LINEAR PHASE FILTERS

A linear phase FIR's zeros conform to certain requirements. Zeros having magnitude other than 1.0 (i.e., not on the unit circle) must always be matched with a zero having the same frequency but the reciprocal magnitude. Any complex zero, of course, must also be matched with its complex conjugate to ensure real-only coefficients. The following possibilities therefore exist for zeros for linear phase FIRs:

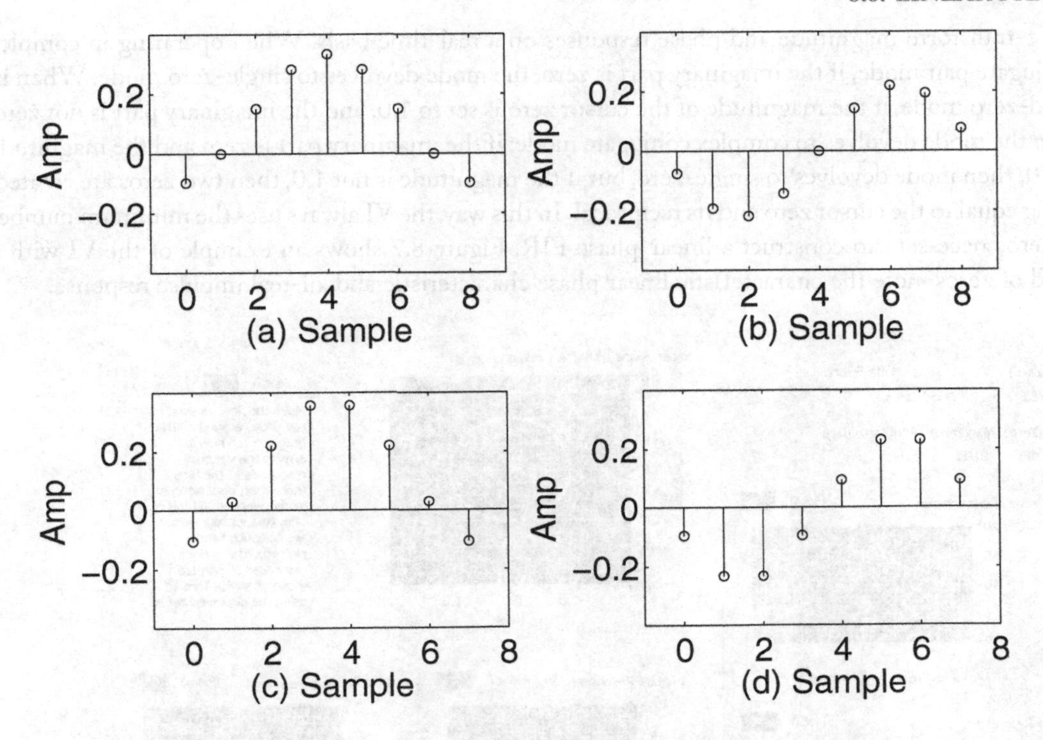

Figure 8.6: (a) Symmetric, odd-length FIR; (b) Anti-symmetric, odd-length FIR; (c) Symmetric, even-length FIR; (d) Anti-symmetric, even-length FIR.

- Zeros at 1 and -1 may exist singly since they are real and have magnitude of 1.0.
- Any zero on the real axis having magnitude other than one must be matched by a real zero having the same frequency (0 or π radians) and the reciprocal magnitude.
- Any nonreal zero on the Unit Circle must be matched with its complex conjugate. Such zeros, then, can be characterized as existing in pairs.
- Any zero that is complex and has a magnitude other than 1.0 must be matched by a zero at the same frequency and reciprocal magnitude, and both of these zeros must be matched by their complex conjugates. Therefore, this type of zero comes in sets of four, called **Quads**.

The LabVIEW VI

$$DemoDragZerosZxformVI$$

allows you to drag a single zero in the z-plane, and have the single zero, or a complex conjugate pair of zeros, or a quad of zeros, computed and displayed, along with the impulse response and the z-transform magnitude and phase responses on a real-time basis. When operating in complex conjugate pair mode, if the imaginary part is zero, the mode devolves to single-zero mode. When in quad-zero mode, if the magnitude of the cursor zero is set to 1.0, and the imaginary part is not zero, then the mode devolves to complex conjugate mode; if the imaginary part is zero and the magnitude is 1.0, then mode devolves to single-zero, but if the magnitude is not 1.0, then two zeros are created, being equal to the cursor zero and its reciprocal. In this way, the VI always uses the minimum number of zeros necessary to construct a linear-phase FIR. Figure 8.7 shows an example of the VI with a quad of zeros–note the characteristic linear phase characteristic and all-real impulse response.

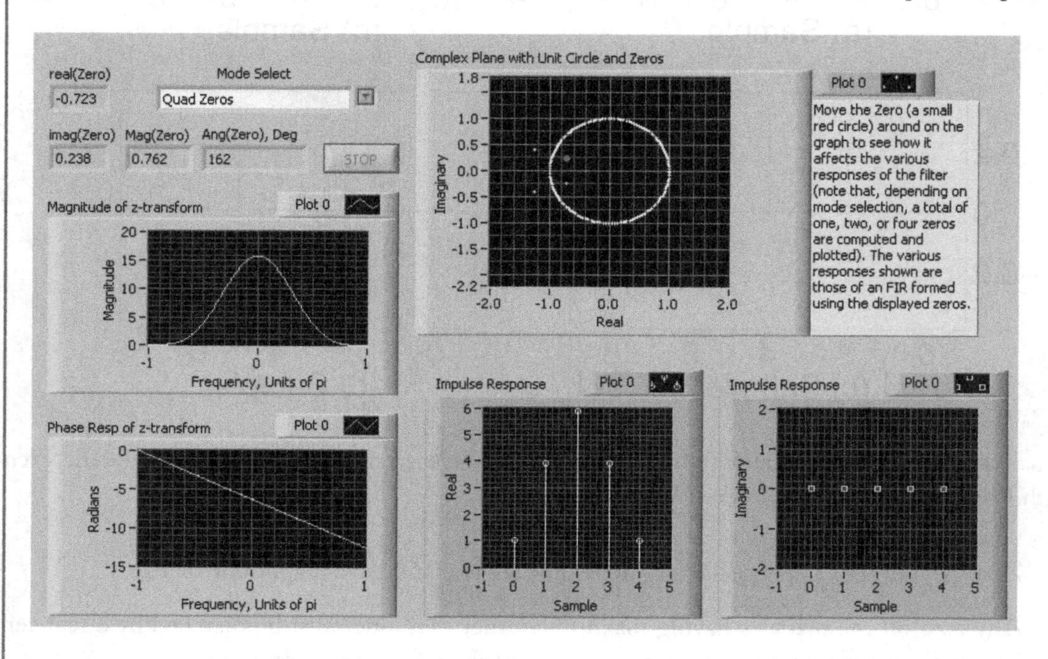

Figure 8.7: A VI that allows study of the effect of zero location and aggregation (i.e., single zero, complex conjugate pairs, or quads) on frequency, phase, and impulse responses of an FIR constructed with the computed and displayed zero(s).

A script for use with MATLAB that performs functions similar to that of the VI above is

$$ML\_DragZeros$$

This script, when called, opens a GUI that allows you to select a single zero, a complex conjugate pair of zeros, or a quad of zeros (a complex conjugate pair and their reciprocals). The magnitude and phase of the z-transform as well as the real and imaginary parts of the equivalent impulse response are dynamically plotted as you move the cursor in the z-plane. This script, unlike the VI above, does not devolve to use of one or two zeros in certain cases, nor is an attempt made to maintain phase linearity. In the quad mode, four zeros are always used to create the FIR transfer function, and in this case, of course, phase is always linear. In the complex conjugate mode, phase is only linear when the magnitude of the zeros is 1.0, and in the single zero mode, phase is only linear when the zero is real with magnitude equal to 1.0. The reader should verify these observations when running the script.

Example 8.2. Linearize the phase response of an FIR that has the transfer function

$$H(z) = 1 - 0.81z^{-2}$$

First, we compute the values of its zeros as 0.9 and -0.9. The we supply two more zeros, the first at frequency zero (DC) with magnitude 1/0.9, and the second at the Nyquist frequency, with magnitude 1/0.9. To get the new transfer function, we convolve the factors [1,-0.9]. [1,-1/0.9], [1,0.9], and [1,1/0.9] and get

$$H(z) = 1 + 0z^{-1} - 2.0446z^{-2} + 0z^{-3} + z^{-4}$$

which gives the impulse response as [1,0,-2.0446,0,1], which is a linear phase bandpass filter with the passband centered at a normalized frequency of 0.5 (one-quarter of the sampling frequency).

You can verify this by making either of the following calls:

```
LVxFreqRespViaZxform([1, 0, -2.0446, 0, 1],1024)
```

or

```
y = abs(fft([1, 0, -2.0446, 0, 1],1024)); plot(y(1,1:512))
```

Example 8.3. It is known that a certain linear phase FIR has zeros at $0.9j$, $(\sqrt{2}/2)(1 + j)$, and -1.0. Give the entire set of zeros.

The zero at 0.9j gives rise to another zero at its reciprocal magnitude, and two more zeros that are the complex conjugates of the first two zeros. The zero at $(\sqrt{2}/2)(1 + j)$ is on the Unit Circle, and thus gives rise only to its own complex conjugate. The zero at -1 is both real and of magnitude 1.0 and thus exists by itself. The entire list is therefore 0.9j, -0.9j, 1.11j, -1.11j, 0.707(1 + j), 0.707(1 - j), and -1.0.

Example 8.4. Obtain the FIR coefficients corresponding to the zeros in the above example.

We make the call

```
a = 0.9*j;b = exp(j*pi/4);Imp = poly([a,-a,(1/a),(1/-a),b,b^-1,-1.0])
```

which returns the coefficients as Imp.

8.7 LINEAR PHASE FIR FREQUENCY CONTENT AND RESPONSE

Since linear phase FIRs conform to specific forms for the impulse response, the frequency response can be written for each of the four FIR types as an expression involving cosines or sines. The **Impulse Response** for **Type I or II filters** conforms to the rule that

$$h[n] = h[L-n-1]$$

where the impulse response length is L and n = 0:1:$L-1$.

The **Frequency Response** for **Type I and II filters** conforms to the general form

$$H(\omega) = H_r(\omega)e^{-j\omega M}$$

where $M = (L-1)/2$, and $Hr(\omega)$ is a real function that can be positive or negative and is therefore called the **Amplitude Response**, while the complex exponential represents a linear phase factor.

For the Type I filter (odd length), the amplitude response is given as

$$H_r(\omega) = h[M] + 2\sum_{n=1}^{M} h[M-n]\cos(\omega\, n) \tag{8.1}$$

which is equivalent to

$$H_r(\omega) = h[M] + 2\sum_{n=0}^{M-1} h[n]\cos(\omega[M-n]) \tag{8.2}$$

Example 8.5. Consider the Type I linear phase filter having the impulse response [a, b, c, b, a]; show that Eq. (8.1) is equivalent to the DTFT of the impulse response.

The DTFT is defined as

$$X(e^{j\omega}) = \sum_{n=-\infty}^{\infty} x[n]e^{-j\omega n}$$

from which we get

$$X(e^{j\omega}) = a + be^{-j\omega} + ce^{-j\omega 2} + be^{-j\omega 3} + ae^{-j\omega 4}$$

which reduces to

$$X(e^{j\omega}) = e^{-j\omega 2}(ae^{j\omega 2} + be^{j\omega} + c + be^{-j\omega} + ae^{-j\omega 2})$$

and finally

$$X(e^{j\omega}) = e^{-j\omega 2}(c + 2b\cos(\omega) + 2a\cos(2\omega))$$

which can be written as

$$X(e^{j\omega}) = e^{-j\omega M}[x[M]\cos(0\omega) + 2(x[M-1]\cos(1\omega) + x[M-2]\cos(2\omega))]$$

where $M = (L-1)/2$ which conforms to Eq. (8.1).

For the Type II (even length) filter $H_r(\omega)$ is given as

$$H_r(\omega) = 2\sum_{n=0}^{L/2-1} h[n]\cos(\omega[M-n]) \tag{8.3}$$

which is equivalent to

$$H_r(\omega) = 2\sum_{n=1}^{L/2} h[\frac{L}{2}-n]\cos(\omega[n-\frac{1}{2}]) \tag{8.4}$$

The **Impulse Response** for **Type III and IV filters** conforms to the rule

$$h[n] = -h[L-n-1]$$

and the **Frequency Response** for **Type III and IV filters** conforms to the general form

$$H(\omega) = jH_r(\omega)e^{-j\omega M} = H_r(\omega)e^{j(\pi/2-\omega M)}$$

where $H_r(\omega)$, for Type III (odd length) is given as

$$H_r(\omega) = 2\sum_{n=0}^{M-1} h[n]\sin(\omega[M-n]) \tag{8.5}$$

and for Type IV (even length) $H_r(\omega)$ is

$$H_r(\omega) = 2\sum_{n=0}^{L/2-1} h[n]\sin(\omega[M-n]) \tag{8.6}$$

Example 8.6. Compute and plot the amplitude and phase responses for the Type I linear phase filter whose impulse response is [1, 0, 1]. Contrast the amplitude and corresponding phase plots to magnitude and phase plots obtained using the DTFT.

For the amplitude response we get $L = 3$, $M = 1$, and the formula

$$H_r(\omega) = h[M] + 2\sum_{n=0}^{M-1} h[n]\cos(\omega[M-n])$$

becomes

$$H_r(\omega) = 0 + 2\sum_{n=0}^{0} h[0]\cos(\omega[1-0]) = 2\cos(\omega)$$

and the phase is $e^{-j\omega}$. We use the following code to evaluate and plot the result at a large number of frequencies between -π and π:

```
function LVFrPhRImp101
incF = 2*pi/1024; argF = -pi+incF:incF:pi;
Hr = 2*cos(argF); Hph = -argF;
figure(8); subplot(221); plot(argF/pi, Hr)
subplot(222); plot(argF/pi,Hph)
H = fft([1 0 1],1024); Hdtft = fftshift(H); subplot(223);
xvec = [-511:1:512]/512; plot(xvec, abs(Hdtft));
subplot(224); plot(xvec, unwrap(angle(Hdtft)))
```

Figure 8.8 shows the result from running the above code. Note that the amplitude and corresponding phase functions are linear and continuous, whereas the magnitude and corresponding phase functions are not.

Example 8.7. Compute and plot the amplitude and phase responses and magnitude and phase responses for the Type II filter whose impulse response is [1, 1, 1, 1].

We use the Eq. (8.3), with L = 4 and M = 3/2, we get

$$H_r(\omega) = 2\sum_{n=0}^{1} h[n]\cos(\omega[\frac{3}{2}-n]) = 2(\cos(\frac{3}{2}\omega) + \cos(\frac{1}{2}\omega))$$

Using code similar to that given for the previous example, we get Fig. 8.9.

Example 8.8. Compute and display the amplitude and phase responses and magnitude and phase responses for the Type III filter having impulse response [1,0,-1].

Using M = 1 and Eq. (8.5) we get

$$H_r(\omega) = 2\sum_{n=0}^{0} h[0]\sin(\omega[1-0]) = 2\sin(\omega) \tag{8.7}$$

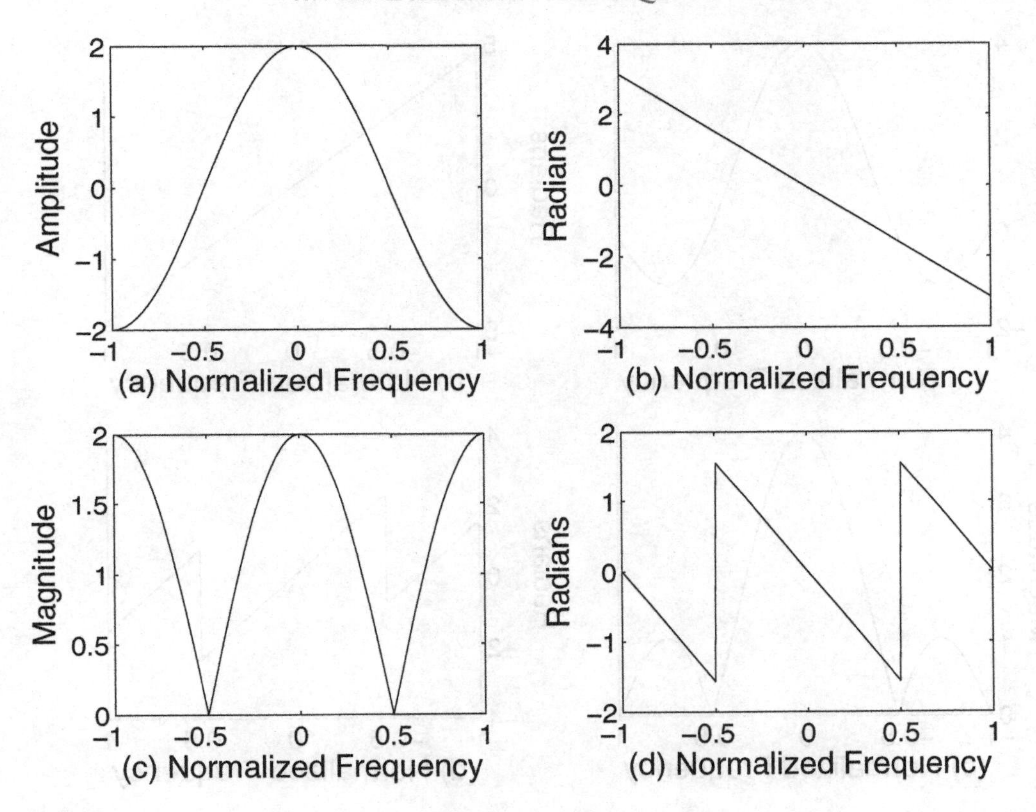

Figure 8.8: (a) Amplitude response of Type I linear phase filter whose impulse response is [1,0,1]; (b) Phase response or function of same; (c) Magnitude response of same impulse response; (d) Phase response of same.

and the corresponding frequency and phase responses are shown in Fig. 8.10.

Example 8.9. Compute and plot the amplitude and phase responses and magnitude and phase responses for the Type IV filter whose impulse response is [1, 1,-1,-1].

Using Eq. (8.6) with M = 1.5, we get

$$H_r(\omega) = 2\sum_{n=0}^{1} h[n]\sin(\omega[\frac{3}{2} - n]) = 2(\sin(\frac{3}{2}\omega) + \sin(\frac{1}{2}\omega))$$

Figure 8.11 shows the result.

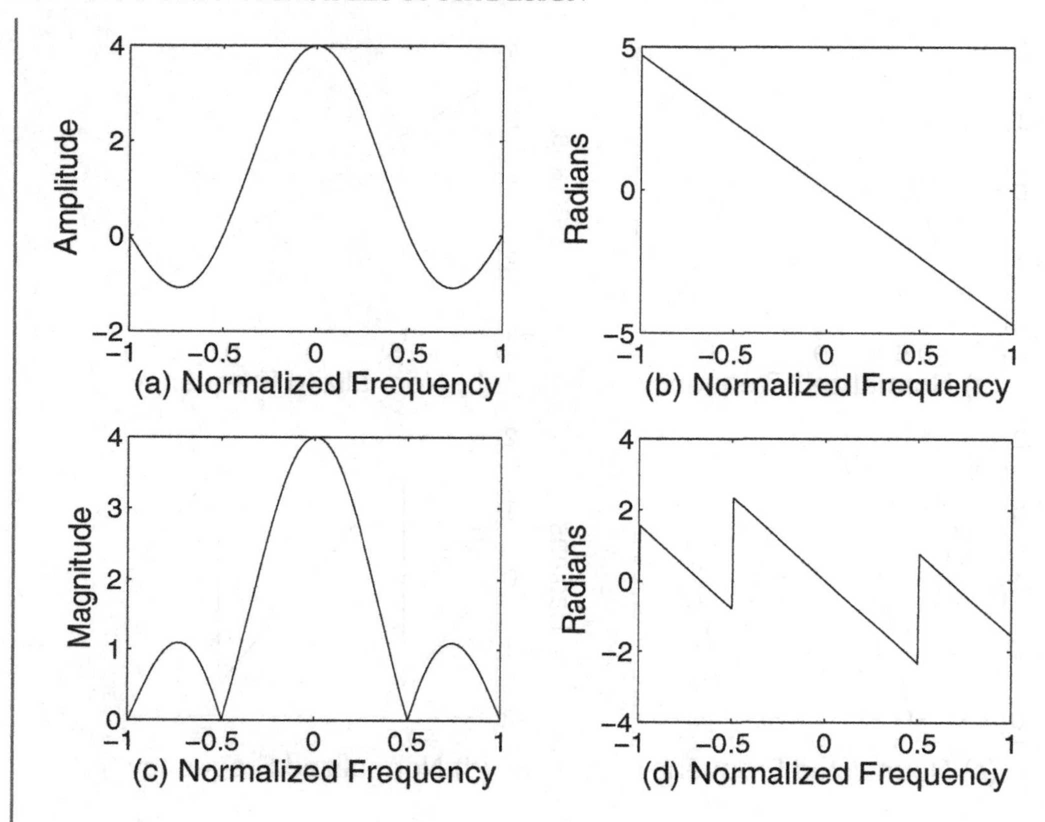

Figure 8.9: (a) Amplitude response of Type II linear phase filter whose impulse response is [1,1,1,1]; (b) Phase response or function of same; (c) Magnitude response of same impulse response; (d) Phase response of same.

8.8 DESIGN METHODS

8.8.1 BASIC SCHEME

- Designing frequency-selective filters consists of specifying **Passband(s)** (frequencies to be passed unattenuated), **Stopband(s)** (frequencies to be completely attenuated), and **Transition Band(s)**, containing frequencies lying between the passband(s) and stopband(s) which may (with some constraints) have whatever amplitudes are necessary to help optimize the responses in the passband(s) and stopband(s). Thus the entire possible frequency range from 0 to the Nyquist limit is broken into one or more passbands, stopbands, and transition bands.

- All linear-phase FIRs are composed of (or synthesized by summing) symmetrical cosines or sines (i.e., correlators) having frequencies conforming to either of two orthogonal frequency schemes, namely, whole integer frequencies such as 0, 1, 2, etc., or odd multiples of half-cycles,

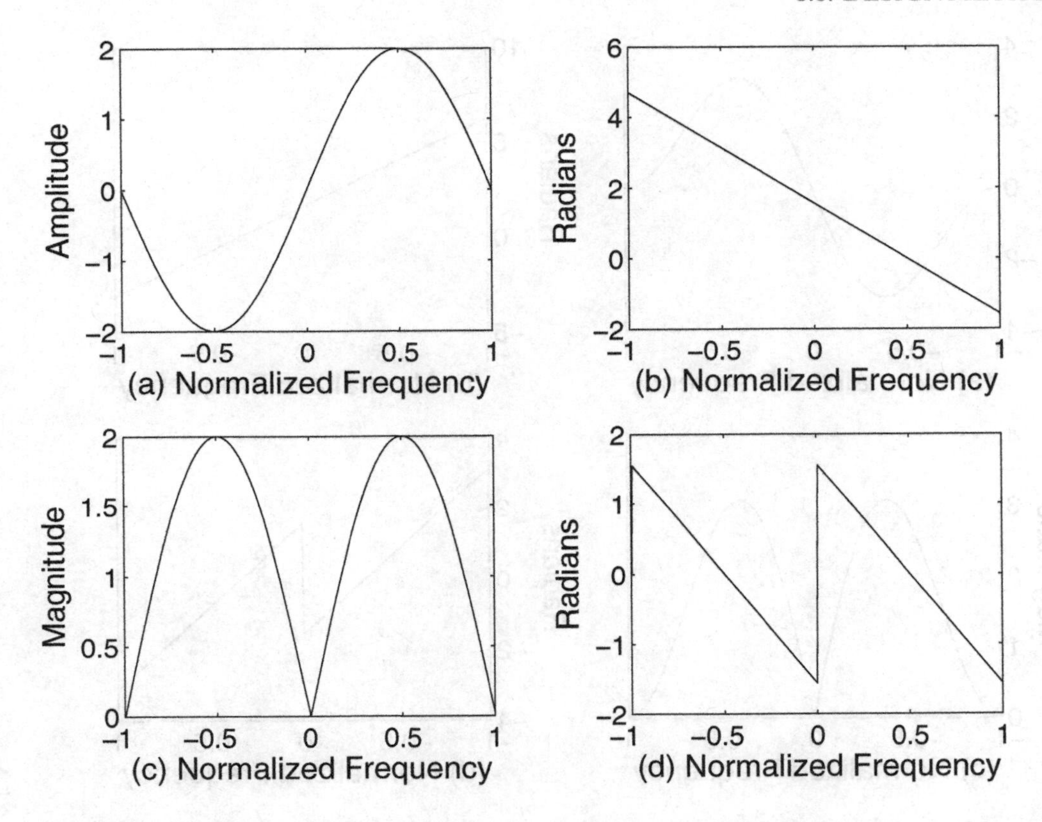

Figure 8.10: (a) Amplitude response of Type III linear phase filter whose impulse response is [1, 0, -1]; (b) Phase response or function of same; (c) Magnitude response of same impulse response; (d) Phase response of same.

yielding frequencies of 0.5, 1.5, 2.5, etc. The net frequency response is the superposition of the individual frequency responses contributed by each correlator.

8.8.2 THREE DESIGN METHODS

- One approach, called the **Window Method**, is to generate a truncated "ideal" lowpass filter and apply a window to the impulse response to achieve a certain desired stopband attenuation or reduction of passband ripple. Filter length is adjusted as necessary to achieve desired roll-off rate. Other filters such as highpass, bandpass, and notch can be generated starting with one or more lowpass filters.

- In the **Frequency Sampling Method**, the desired magnitudes of filter response at a plurality of DFT frequencies (i.e., frequencies defined as $2\pi k/L$ with k = 0:1:L-1, for example) are

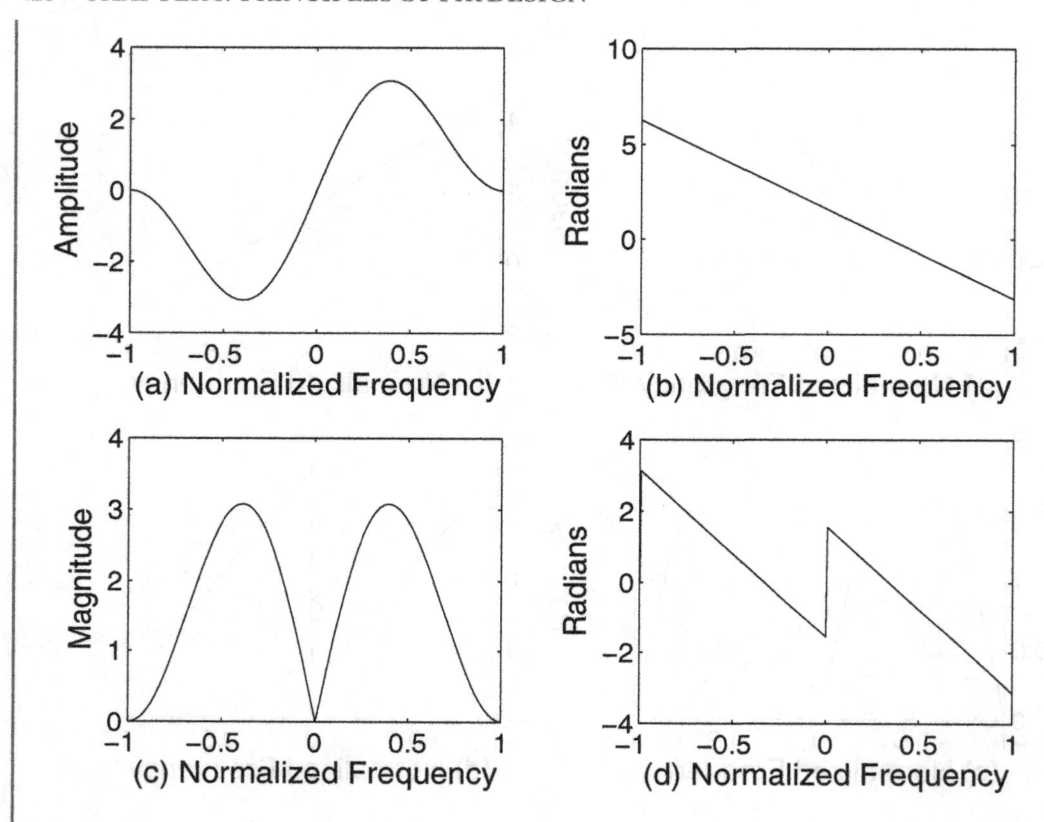

Figure 8.11: (a) Amplitude response of Type IV linear phase filter whose impulse response is [1,1,-1,-1]; (b) Phase response or function of same; (c) Magnitude response of same impulse response; (d) Phase response of same.

specified, a linear phase factor is imparted, and the inverse DFT is computed to obtain the filter's impulse response. A variant on the Frequency Sampling Method is to use, instead of the inverse DFT, simple cosine or sine summation formulas that construct an impulse response as the superposition of symmetrical cosines or sines, having frequencies conforming to one of two orthogonal systems.

- **Optimized Equiripple Method**. In this method, the approximation error is equalized in the passband(s) and stopband(s), with the maximum magnitude of error in each being user-specifiable. This method offers the greatest degree of user-control of the three methods discussed herein, and generally results in the shortest length filter that can meet a given set of design specifications.

- **A detailed discussion of these three design methods is found in the next chapter.**

8.8.3 THE COMB AND MOVING AVERAGE FILTERS

The Comb Filter

The Comb filter has only two nonzero values in its impulse response. A single delayed version of an input sequence is added to or subtracted from the undelayed version of the input sequence to form the output. For the case of a single sample of delay, the impulse response would be [1,1] when the delayed signal is added to the original, or [1,-1] when it is subtracted. For two samples of delay, it would be [1,0,1] or [1,0,-1], and so forth for different delays. The simple FIRs [1,1], [1,-1], [1,0,-1], and [1,0,1] are all comb filters which may also be characterized, respectively, as lowpass, highpass, bandpass, and notch filters. When the second non-zero value in the impulse response does not have unity magnitude, the null-depth does not go to zero-magnitude. Examples of such impulse responses would be $[1, 0.9]$, $[1, 0, -0.7]$, etc. In this book, this type of impulse response will generally be referred to as a Modified Comb Filter.

Comb filters are useful in certain types of applications. Suppose, for example, that you had an audio signal polluted with a 60 Hz fundamental wave with very high harmonic amplitudes extending into the 10^4 or higher frequency range. A comb filter is ideal for suppressing such a harmonic series, exhibiting economy and simplicity.

Figure 8.12 shows the frequency responses of two five-sample-delay comb filters, the first additive, and the second subtractive.

The script

```
function LVCombFilter(Tau)
% LVCombFilter(5)
ImpAdd = [1,zeros(1,Tau-1),1];
ImpSub = [1,zeros(1,Tau-1),-1];
DTFTLen = 1024; xplot = [1:1:DTFTLen/2+1];
xvec= (xplot-1)/(DTFTLen/2); subplot(2,1,1);
yAdd = abs(fft(ImpAdd,DTFTLen));
plot(xvec,yAdd(xplot),'b'); ylabel(['Magnitude'])
xlabel(['(a) Normalized Frequency'])
axis([0 1 0 1.2*max(yAdd)])
subplot(2,1,2); ySub = abs(fft(ImpSub,DTFTLen));
plot(xvec,ySub(xplot),'b'); ylabel(['Magnitude'])
xlabel(['(b) Normalized Frequency'])
axis([0 1 0 1.2*max(ySub)])
```

affords experimentation with additive and subtractive comb filters; the variable *Tau* is the number of samples of delay. Figure 8.12 was generated by making the call **LVCombFilter(5)**.

Example 8.10. Derive empirically, using the script *LVCombFilter*(*Tau*) with values of *Tau* such as 1, 2, 3, etc., an expression for the frequency of the first null for an additive comb filter having n samples of delay. Use the derived relationship to determine the number of samples of delay needed

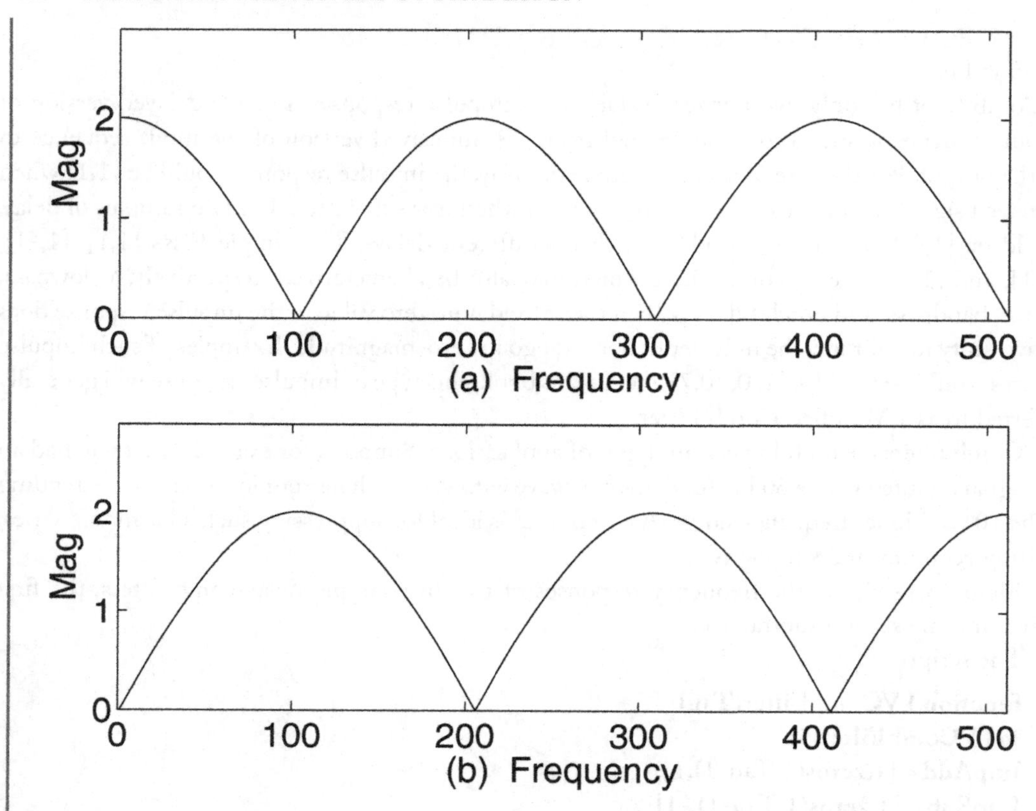

Figure 8.12: (a) Frequency response of the impulse response [1,0,0,0,0,1] (an additive comb filter having 5 samples of delay); (b) Frequency response of the impulse response [1,0,0,0,0,-1] (a subtractive comb filter having 5 samples of delay).

for an additive comb filter to have its first frequency null at 60 Hz when a signal having a sampling rate of 3000 Hz is convolved with the comb filter. Test your answer.

By making a succession of calls

$$LVCombFilter(Tau)$$

where $Tau = 1, 2, 3$, etc., and observing plot (a) of the resulting figure, we note that for one sample of delay (i.e., $Tau = 1$), the first null is at normalized frequency 1.0, for two samples delay, at 1/2 (0.5), for three samples delay, at 1/3 (0.333), and so forth, suggesting that the normalized frequency of the first null is at $1/Tau$ (this applies to an additive comb filter).

For a sampling rate of 3000 Hz, the Nyquist frequency is 1500 Hz, and therefore the normalized frequency desired for the first null is 60/1500 = 0.04, which is 1/25. We therefore need an additive comb filter with 25 samples of delay. A call which will verify this is:

```
FR = abs(fft([1,zeros(1,24),1],3000)); plot(0:1:180, FR(1,1:181))
```

Note that the first null is at 60 Hz, and the second one at 180 Hz.

The MA (Moving Average) Filter

The Moving Average filter is a single-correlator linear phase filter, the single correlator being at frequency zero (DC) The impulse response is that of a rectangle with a length of N samples, weighted by $1/N$:

$$(1/N) \cdot [1, 1....1]$$

The MA filter is useful for keeping a running average of the values of an input sequence over a certain length. An advantage of the MA impulse response is that it is possible to compute it recursively, eliminating a large amount of time-consuming convolution. To do this, let's first dispose of the scaling constant $1/N$ prior to discussing the recursion process. Note that we can either scale every input sample by $1/N$ before adding, or instead add up all samples and then scale just before delivering the sum as the next output. Let's assume the latter, so that all our operations prior to scaling by $1/N$ involve the raw, unscaled input samples.

The value we compute, then, prior to the final scaling, we will call a running sum S, which is just the output sequence generated as

$$S(n) = \sum_{i=n-N+1}^{n} x_i$$

and where N is the impulse response length, M is the length of the signal sequence x_i, which is defined as 0 for $i < 1$ or $i > M$; and n, for valid output (i.e., the impulse response saturated with input samples), runs from N to $(M - N + 1)$.

There is a very efficient way to compute the output sequence values of a running sum. Suppose that $S[8]$, which is the sum of input samples 1 through 8 (x_1 through x_8), has just been computed. To compute $S[9]$, which is the sum of x_2 through x_9, add x_9 to $S[8]$, and subtract x_1. Once $S[9]$ is in hand, of course, $S[10]$ can be computed by adding x_{10} and subtracting x_2, and so forth. Phrased mathematically, this would be:

$$S[n] = S[n-1] + x_n - x_{n-N}$$

This simple recursion formula can greatly reduce computational overhead when N gets to be very large.

Since a Moving Average filter has an impulse response which consists of samples of a cosine of frequency zero, we would expect orthogonal behavior toward signals having an integral number of cycles in the length of the MA filter. For example, a 20-point MA impulse response will yield an output identically zero (during saturation of the filter, of course) for sinusoids having exactly one cycle, two cycles, three cycles, etc. up to ten cycles in a length of 20 samples. Figure 8.13 shows a 20-pt MA filter impulse response in plot (a), and its frequency response in plot (b).

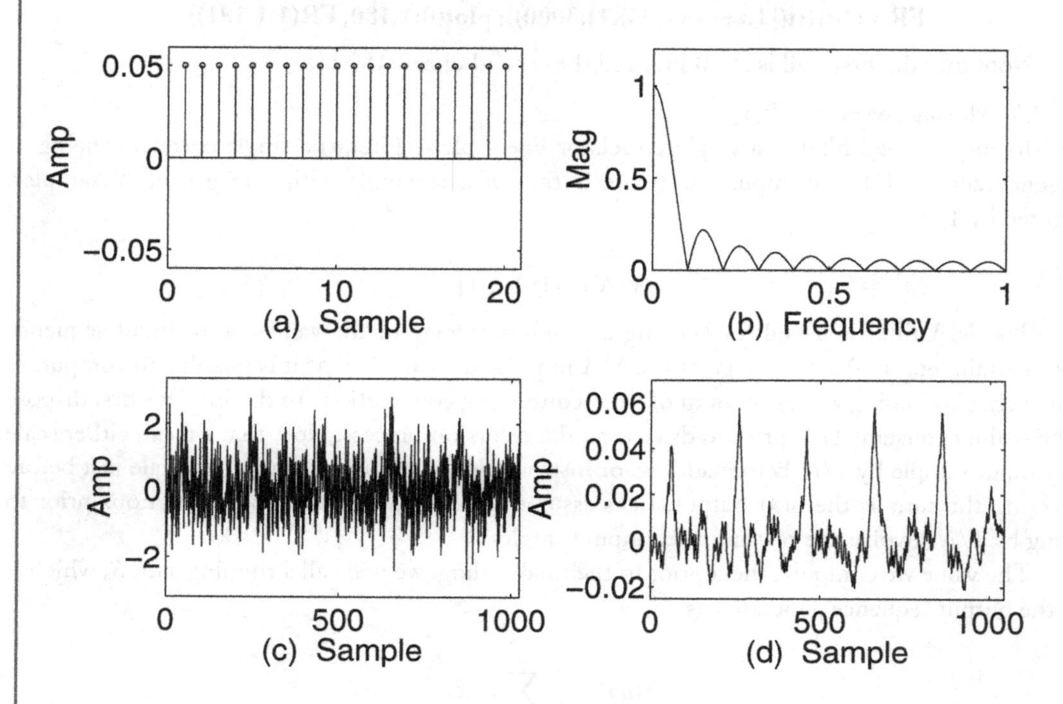

Figure 8.13: (a) 20-point Moving Average Impulse Response; (b) Normalized frequency response of impulse response in (a); (c) Test signal of noise with coherent (rectangular) signals embedded; (d) Convolution of test signal from (c) with impulse response from (a).

In some situations, it is possible to enhance the signal-to-noise ratio (SNR) of a signal by using the MA filter. This requires that the noise be generally random or *incoherent*, and that the signal be coherent, or very predictable. The easiest situation is when the signal is all of one polarity. In this case, the signal, when averaged, builds up approximately proportionately to N, the number of samples added together, while the standard deviation of the noise builds up only by $\sqrt{N}$, thus increasing the SNR by $\sqrt{N}$.

Looking again at Fig. 8.13, we see in plot (c) what appears to be nothing but noise. In reality, it is a large amount of noise to which has been added five equally spaced signals, each consisting of a sequence of 20 samples valued at 1. Plot (d) of Fig. 8.13 shows five signal peaks that have been recovered from the noise by using the 20-pt moving average filter shown in the upper left plot. The best result occurs (on the average) when the MA filter length is exactly the same length as the coherent signal, since at that point, the maximum signal gain would have been obtained; averaging in more samples would only bring in additional noise without bringing in any additional signal.

Figure 8.14 shows the result when the MA filter is only 5 samples long. The results of the averaging in plot (d) show that a number of the signal pulses in the noise have not been well-recognized or emphasized. The results, however, vary with each running since the test signal is random noise, which is different for each trial run.

8.9 FIR REALIZATION

In Part II of the book we explored the Direct, Cascade, Parallel, and Lattice Forms of realization for a generalized LTI system having both IIR and FIR components. In addition to these forms, there are several implementations that apply specifically to FIRs, including the Linear Phase FIR, which we have just introduced in this chapter. We begin with the simple Direct or Transversal form, proceed through Cascade and Linear Phase Forms, and finish with the Frequency Sampling Form, which is based on the idea of reconstructing the z-transform of an FIR from its samples, which we discussed in conjunction with the Discrete Fourier Series in Part II.

8.9.1 DIRECT FORM

The Direct Form implementation of an FIR, illustrated in Fig. 8.15, is the simple transversal arrangement, in which the signal passes down a chain of delay elements, and the output of each delay element is weighted by a respective coefficient and all weighted outputs are summed to produce the output. In this arrangement, the filter signal flow diagram can be constructed directly from the z-transform of the FIR.

8.9.2 CASCADE FORM

By collecting the FIR's zeros in complex conjugate pairs, second order real sections can be made and cascaded to implement the filter, as shown in Fig. 8.16.

The coefficients can be computed using the same scripts we used for the generalized LTI Cascade Form, namely, the scripts

$$[Bc, Ac, Gain] = LVDirToCascade(b, a)$$

$$[b, a, k] = LVCas2Dir(Bc, Ac, Gain)$$

$$[y] = LVCascadeFormFilter(Bc, Ac, Gain, x)$$

Example 8.11. For the Linear Phase filter having b = $fir1(6, 0.5)$, compute the Cascade Form coefficients, and filter a linear chip using the Direct and Cascade Form coefficients and compare the results.

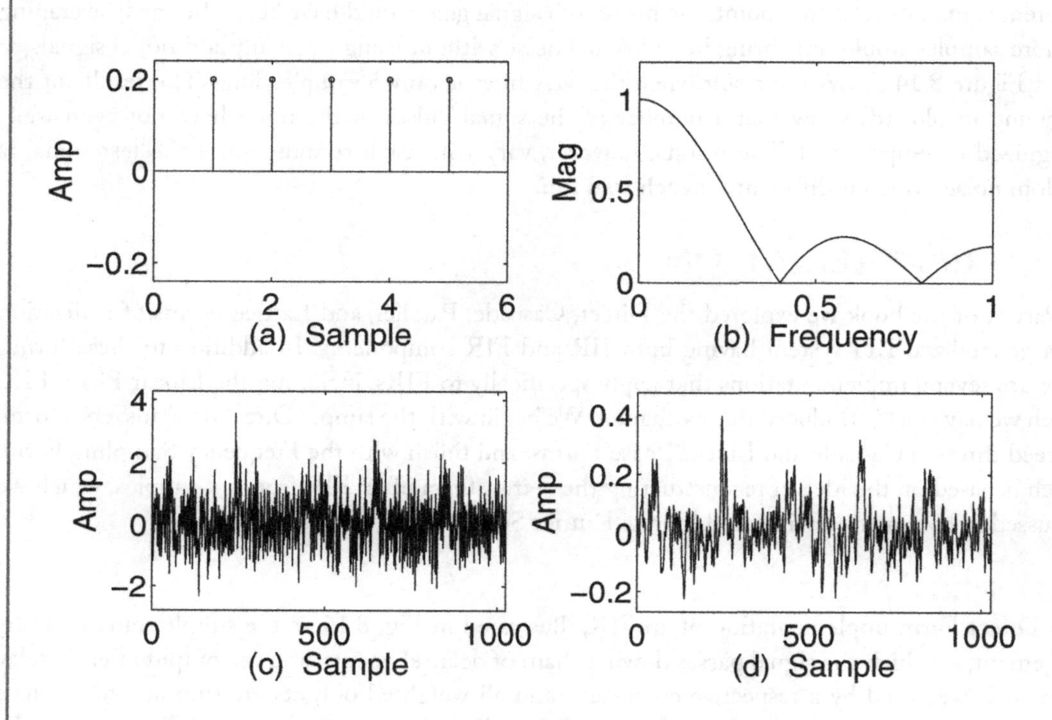

Figure 8.14: (a) 5-point Moving Average Impulse Response; (b) Normalized frequency response of impulse response in (a); (c) Test signal of noise with coherent (rectangular) signals embedded; (d) Convolution of test signal from (c) with impulse response from (a).

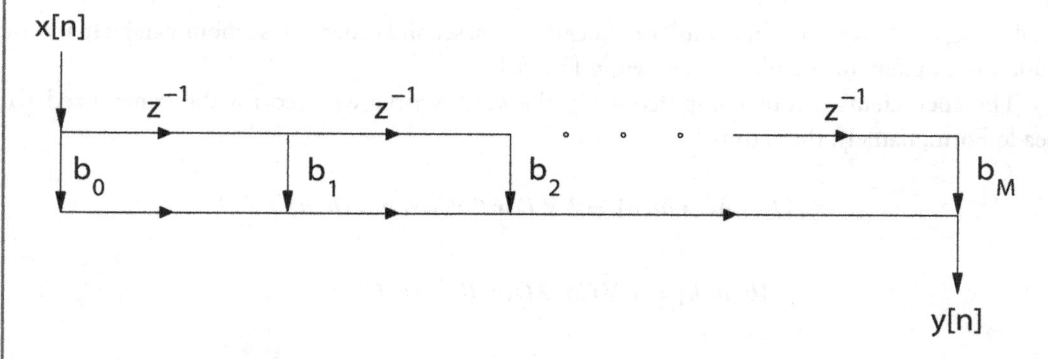

Figure 8.15: An FIR implemented in Direct Form. This arrangement is often called a Transversal filter since the signal moves across the filter. Note that the input signal $x[n]$ and the outputs of all delay elements are scaled by respective coefficients b_i and then summed to generate the output $y[n]$.

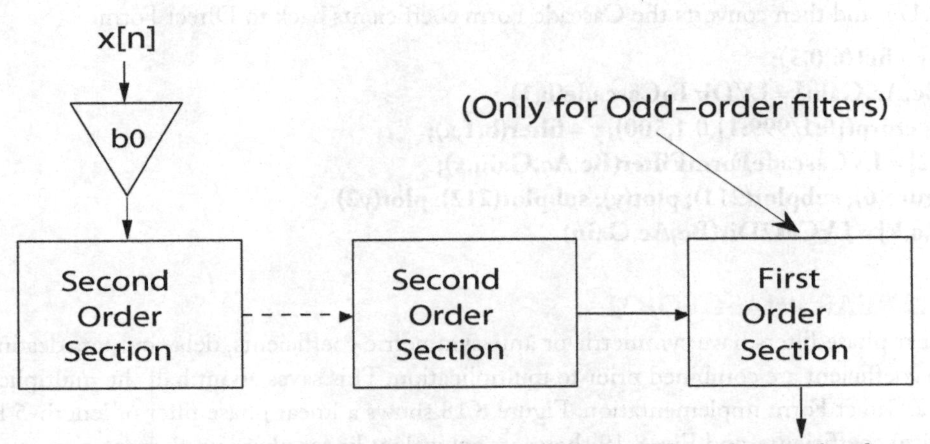

Figure 8.16: A basic cascade arrangement to implement an FIR; each second order section consists of a second order FIR implemented in Direct Form. For odd-order filters, there is one additional first order section.

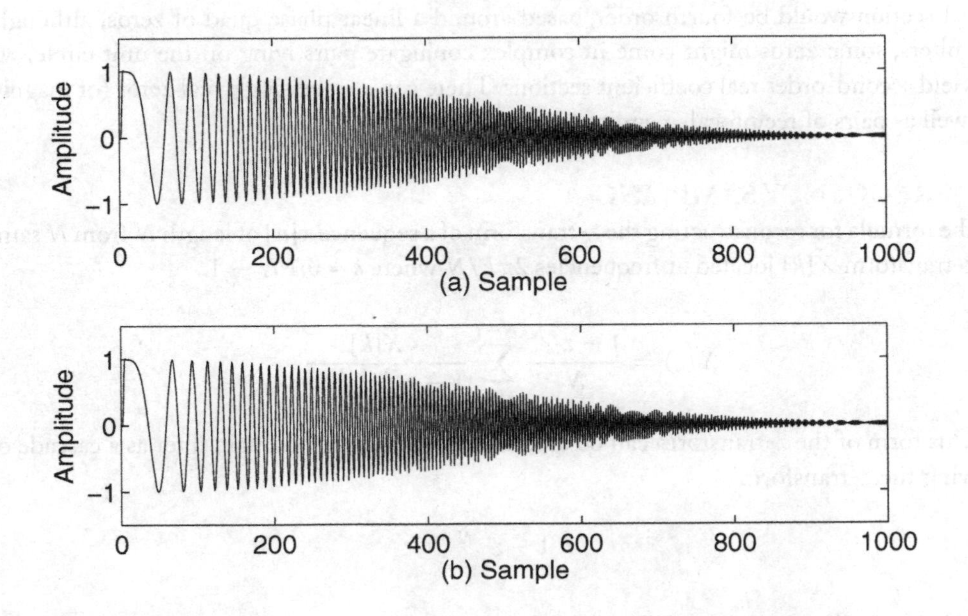

Figure 8.17: (a) A linear chirp filtered using a Direct Form lowpass filter; (b) Same, but filtered using the equivalent Cascade Form filter.

The following m-code generates a set of Direct Form coefficients for a lowpass FIR, then computes the Cascade Form coefficients, filters a test chirp using both forms, plots the results (shown in Fig. 8.17), and then converts the Cascade Form coefficients back to Direct Form.

```
[b] = fir1(6,0.5);
[Bc,Ac,Gain] = LVDirToCascade(b,1)
x = chirp([0:1/999:1],0,1,500); y = filter(b,1,x);
[y2] = LVCascadeFormFilter(Bc,Ac,Gain,x);
figure(6); subplot(211); plot(y); subplot(212); plot(y2)
[b,a,k] = LVCas2Dir(Bc,Ac,Gain)
```

8.9.3 LINEAR PHASE FORM

Since linear phase filters have symmetric or anti-symmetric coefficients, delay outputs destined for the same coefficient are combined prior to multiplication. This saves about half the multiplications need for a Direct Form implementation. Figure 8.18 shows a linear phase filter of length-5 having symmetrical coefficients, and Fig. 8.19 shows its equivalent linear phase implementation.

8.9.4 CASCADED LINEAR PHASE FORM

Another possibility for a linear phase FIR is to form a cascade of linear phase, real coefficient sections. A typical section would be fourth order, based around a linear phase quad of zeros, although for certain filters, some zeros might come in complex conjugate pairs lying on the unit circle, which would yield second-order real coefficient sections. There can also be single real zeros (of magnitude 1.0) as well as pairs of reciprocal-magnitude real zeros.

8.9.5 FREQUENCY SAMPLING

Recall the formula for reconstructing the z-transform of a sequence $x[n]$ of length N from N samples of the z-transform $\widetilde{X}[k]$ located at frequencies $2\pi k/N$ where $k = 0:1:N-1$.

$$X(z) = \frac{1 - z^{-N}}{N} \sum_{k=0}^{N-1} \frac{\widetilde{X}[k]}{1 - e^{j2\pi k/N} z^{-1}} \tag{8.8}$$

This form of the z-transform can be used to construct a time domain filter as a cascade of an FIR having the z-transform

$$\frac{1 - z^{-N}}{N}$$

followed by a parallel structure of IIRs of the form

$$\frac{\widetilde{X}[k]}{1 - e^{j2\pi k/N} z^{-1}}$$

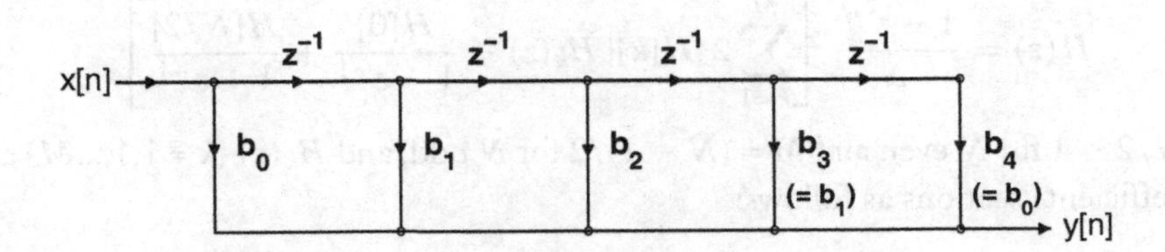

Figure 8.18: A length-5 FIR implemented using Direct Form. Note that the coefficients are symmetrical, having $b_3 = b_1$ and $b_4 = b_0$, so the signals from taps 3 and 4 can be combined with those from taps 0 and 1, respectively, and thus only two multiplications and two additions for the four coefficients b_0, b_1, b_3, and b_4 must be performed rather than the original four multiplications. This more efficient arrangement is shown in Fig. 8.19.

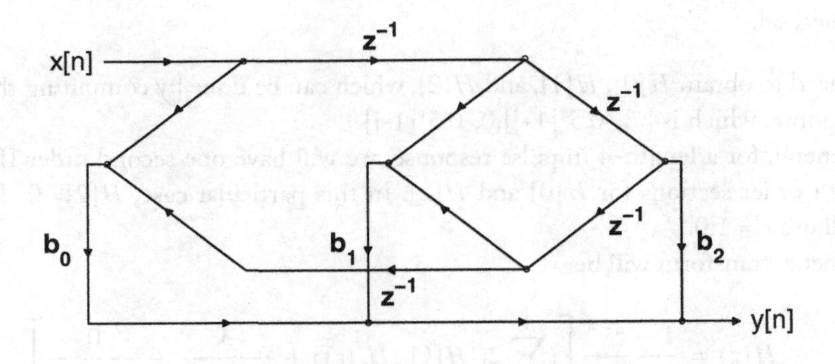

Figure 8.19: A Linear Phase Form filter arrangement for a symmetrical filter having $N = 5$. This form requires only three multiplications instead of the five required in the direct form implementation shown in the previous figure.

The z-transform samples that are not real-only form complex conjugate pairs, and hence they can be collected to make second-order real coefficient sections. The result is that the parallel structure of IIRs uses only real coefficients, which greatly simplifies implementation. A real coefficient filter impulse response $h[n]$ of length N $(0 \leq n \leq N-1)$ has a z-transform in the Frequency Sampling Form of

$$H(z) = \frac{1-z^{-N}}{N}\left[\sum_{k=1}^{M} 2\,|H[k]|\,H_k(z) + \frac{H[0]}{1-z^{-1}} + \frac{H[N/2]}{1+z^{-1}}\right] \tag{8.9}$$

where $M = N/2 - 1$ for N even and $M = (N-1)/2$ for N odd, and $H_k(z)$ (k = 1,1,...M) are second order real coefficients sections as follows:

$$H_k(z) = \frac{\cos[\angle H[k]] - \cos[\angle H[k] - (2\pi k/N)]z^{-1}}{1 - 2\cos[2\pi k/N]z^{-1} + z^{-2}} \tag{8.10}$$

In Eq. (8.9), $H[0]$ is real and if N is odd, the term $H[N/2]/(1+z^{-1})$ will not be present. Note that the values $H[k]$ are z-transform values along the unit circle at DFT frequencies and hence may be computed as the DFT of $h[n]$.

Note that each IIR has a pole of magnitude 1.0, and hence is unstable. To overcome this, the poles and zeros are given magnitude r, slightly less than 1.0, resulting in the following formula:

$$X(z) = \frac{1-r^N z^{-N}}{N}\sum_{k=0}^{N-1}\frac{\widetilde{X}[k]}{1-re^{j2\pi k/N}z^{-1}} \tag{8.11}$$

Example 8.12. Implement the FIR whose impulse response is [0.5, 1, 1, 0.5] using the Frequency Sampling method.

We need to obtain $H[0]$, $H[1]$, and $H[2]$, which can be done by computing the DFT of the impulse response, which is [3,-0.5*[1+j],0,-0.5*[1-j]].

In general, for a length-4 impulse response, we will have one second order IIR section and two real, first order sections for $H[0]$ and $H[2]$. In this particular case, $H[2]$= 0. To keep things simple, we'll use r = 1.0.

The net z-transform will be

$$H(z) = \frac{1-z^{-4}}{4}\left[\left(\sum_{k=1}^{1} 2(|H[k]|)H_k(z)\right) + \frac{3}{1-z^{-1}} + \frac{0}{1+z^{-1}}\right]$$

$$H(z) = \frac{1-z^{-4}}{4}\left[\left(\sum_{k=1}^{1} 2(0.707)H_k(z)\right) + \frac{3}{1-z^{-1}}\right]$$

with

$$H_1(z) = \frac{\cos[\angle H[1]] - \cos[\angle H[1] - (2\pi(1)/4)]z^{-1}}{1 - 2\cos[2\pi(1)/4]z^{-1} + z^{-2}}$$

$$H_1(z) = \frac{\cos[5\pi/4] - \cos[5\pi/4 - \pi/2]z^{-1}}{1 - 2\cos[\pi/2]z^{-1} + z^{-2}}$$

which reduces to

$$H_1(z) = \frac{-0.707 + 0.707z^{-1}}{1 + z^{-2}} = 0.707\left[\frac{-1 + z^{-1}}{1 + z^{-2}}\right]$$

and the final net z-transform will be

$$H(z) = \frac{1 - z^{-4}}{4}\left[\frac{-1 + z^{-1}}{1 + z^{-2}} + \frac{3}{1 - z^{-1}}\right]$$

Figure 8.20 shows the topology or layout of the filter; note that since $H[2] = 0$, the lowermost IIR would not be implemented in practice. For certain filters that have a large number of $H[k] = 0$, the Frequency Sampling Form implementation can be much more efficient than other implementations.

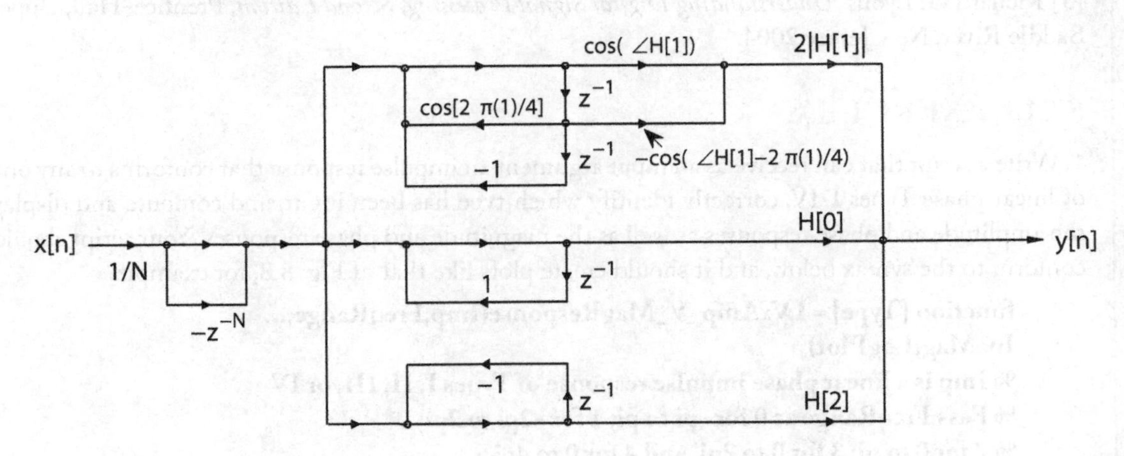

Figure 8.20: The layout of a Frequency Sampling Form equivalent for the simple impulse response [0.5,1,1,0.5]. Note that for this case, $H[2] = 0$, so the lowermost IIR does not need to be implemented.

To verify that Fig. 8.20 is correct, we can process an impulse in the FIR portion

$$\frac{1 - z^{-4}}{4}$$

and then process the result in each of the two nonzero IIRs, summing their outputs. The result should be the original impulse response [0.5,1,1,0.5].

```
unitImp = [1,zeros(1,6)]; y1 = filter([0.25*[1,0,0,0,-1]],[1],unitImp);
y2 = filter([-1,1],[1,0,1],y1); y3 = filter([3],[1,-1],y1);
y = y2 + y3
```

References [3] and [4] discuss FIR realizations in detail; [5] gives a very detailed discussion of the Frequency Sampling implementation.

8.10 REFERENCES

[1] T. W. Parks and C. S. Burrus, *Digital Filter Design*, John Wiley & Sons, New York, 1987.

[2] James H. McClellan et al, *Computer-Based Exercises for Signal Processing Using MATLAB 5*, Prentice-Hall, Upper Saddle River, New Jersey, 1998.

[3] John G. Proakis and Dimitris G. Manolakis, *Digital Signal Processing, Principles, Algorithms, and Applications, Third Edition*, Prentice-Hall, Upper Saddle River, New Jersey, 1996.

[4] Vinay K. Ingle and John G. Proakis, *Digital Signal Processing Using MATLAB V.4*, PWS Publishing Company, Boston, 1997.

[5] Richard G. Lyons, *Understanding Digital Signal Processing, Second Edition*, Prentice-Hall, Upper Saddle River, New Jersey 2004.

8.11 EXERCISES

1. Write a script that can receive as an input argument an impulse response that conforms to any one of linear phase Types I-IV, correctly identify which type has been input, and compute and display the amplitude and phase responses as well as the magnitude and phase responses. Your script should conform to the syntax below, and it should create plots like that of Fig. 8.8, for example.

```
function [Type] = LVxAmp_V_MagResponse(Imp,FreqRange,...
IncMag,LogPlot)
% Imp is a linear phase impulse response of Types I, II, III, or IV
% Pass FreqRange as 0 for -pi to pi; 1 for -2pi to 2pi;
% 2 for 0 to pi; 3 for 0 to 2pi, and 4 for 0 to 4pi
% Pass IncMag as 1 to include magnitude plots, or 0 for
% amplitude plots only. Pass LogPlot as 0 for linear
% magnitude plot or 1 for 20log10(Mag) plot;
% The output variable Type is returned as 1,2,3, or 4 for
% Types I, II, III, or IV, respectively, or 0 if Imp is not
% a linear phase impulse response.
% Test calls:
% [Type]= LVxAmp_V_MagResponse([1,0,1],0,1,0)
% [Type]= LVxAmp_V_MagResponse([1,1,1,1],0,1,0)
```

% [Type]= LVxAmp_V_MagResponse([1,0,-1],0,1,0)
% [Type]= LVxAmp_V_MagResponse([1,1,-1,-1],0,1,0)

2. Write a script that conforms to the following call syntax:

```
function LVxMovingAverageFilter(MALength,NoiseAmp)
% MALength is the length of the Moving Average filter
% NoiseAmp is the StdDev of white noise mixed with a coherent
% signal, which consists of five rectangular pulses over 1024
% samples, each coherent pulse having an amplitude of
% 1.0 and a width of 20 samples.
% Test calls:
% LVxMovingAverageFilter(5,0.8)
% LVxMovingAverageFilter(10,0.8)
% LVxMovingAverageFilter(20,0.8)
% LVxMovingAverageFilter(40,0.8)
```

3. A signal consists of a one millisecond long positive pulse followed by a one millisecond long negative pulse, immersed in random noise. The signal repeats itself every 20 milliseconds. Your receiver samples at a rate of 10 kHz, and the total signal duration is one second.

(a) Devise a suitable matched FIR filter to enhance the signal.

(b) Design a recursive algorithm to implement the filter designed in (a).

(c) Write a script that generates the test signal with a user-specified amount of noise, and filters the test signal using (1) the matched filter of (a) above using function $conv$, and (2) the recursive algorithm of (b) above. While the ideal matched filter will precisely match the sought-after signal in duration as well as shape, it is instructive to experiment with filters of the proper shape but differing durations. Thus the script should allow, using the input argument $lenMDfilt$, the number of samples in the filter to be readily changed. After filtering the test signal using both methods, plot the test signal and both filtering results for various levels of noise and various filter lengths as specified in the test calls given below in the following function specification. Figure 8.21 is an example of the result for the call

LVxMDfilter(0.75,20,2000)

```
function LVxMDfilter(k,lenMDfilt,plotlim)
% Creates a test signal sampled at 10 kHz consisting of, over
% a duration of one second, fifty equally spaced bipolar pulses
% each consisting of a one millisecond long positive pulse
% followed immediately by a one millisecond long negative
% pulse, the entire one second test signal consisting of the fifty
% bipolar pulses plus white noise having standard deviation
% equal to k. A matched FIR filter of length lenMDfilt is used
% to improve the signal-to-noise ratio, i.e., to emphasize the
```

```
% signal relative to the noise. plotlim is the number of samples
% of tstSig and the two filtered sequences to plot
% Test calls:
% LVxMDfilter(0.1,20,2000)
% LVxMDfilter(0.25,20,2000)
% LVxMDfilter(0.5,20,2000)
% LVxMDfilter(1,20,2000)
% LVxMDfilter(2,20,2000)
% LVxMDfilter(0.1,6,2000)
% LVxMDfilter(0.25,6,2000)
% LVxMDfilter(0.5,6,2000)
% LVxMDfilter(1,6,2000)
% LVxMDfilter(2,6,2000)
```

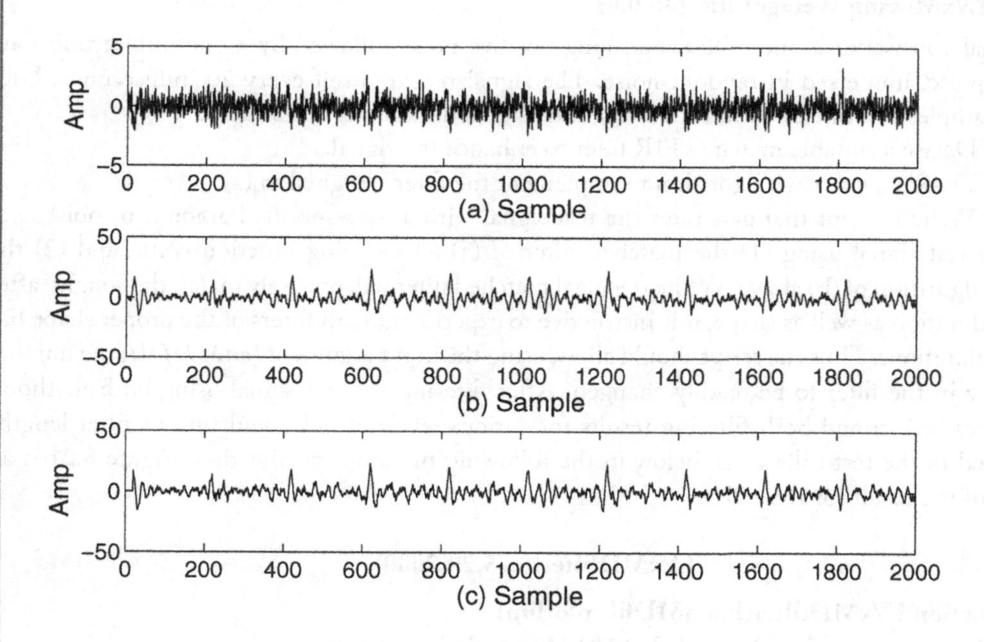

Figure 8.21: (a) Test signal, bipolar pulses in white noise (see text); (b) Test signal filtered with a matched filter using the function *conv*; (c) Test signal filtered with the matched filter using a recursive algorithm.

4. For each of the sets of conditions listed below, answer the following questions using the VI *DemoDragZeroZxformVI*:

Questions

i) Describe the resultant impulse response as real or complex

ii) Describe the phase response as linear or nonlinear
iii) Note the minimum and maximum values of the magnitude of the z-transform and the corresponding frequencies at which these occur
iv) Characterize the passband type of the filter i.e., lowpass, highpass, etc.
v) Symmetry of magnitude response about frequency zero

Conditions:

a) *Mode Select* = *Single Zero* and move the zero cursor Z to approximately (≊) $1 + 0j$
b) *Mode Select* = *Complex Conjugate Zeros*; Z ≊ $0 + j$
c) *Mode Select* = *Complex Conjugate Zeros*; Z ≊ $0 - j$
d) *Mode Select* = *Single Zero;* Z ≊ -1
e) *Mode Select* = *Complex Conjugate Zeros*; Z ≊ $0 + j$
f) *Mode Select* = *Complex Conjugate Zeros*; Z ≊ $0.9 + 0.1j$
g) *Mode Select* = *Quad Zeros*; Z ≊ $0.9 + 0.1j$
h) *Mode Select* = *Complex Conjugate Zeros*; Z ≊ $0 + 0.9j$
i) *Mode Select* = *Quad Zeros*; Z ≊ $0 + 0.9j$
j) *Mode Select* = *Complex Conjugate Zeros*; Z ≊ $-0.9 + 0.1j$
k) *Mode Select* = *Quad Zeros*; Z ≊ $-0.9 + 0.1j$
l) *Mode Select* = *Quad Zeros*; Z ≊ $0.6 + 0.6j$
m) *Mode Select* = *Complex Conjugate Zeros*; Z ≊ $-0.6 + 0.6j$
n) *Mode Select* = *Quad Zeros*; Z ≊ $0.6 + 0.6j$
o) *Mode Select* = *Complex Conjugate Zeros*; Z ≊ $-0.6 + 0.6j$
p) *Mode Select* = *Single Zero;* Z ≊ $0.707 + 0.707j$
q) *Mode Select* = *Single Zero;* Z ≊ $0.707 - 0.707j$

5. Design the shortest linear phase FIR that has [0.65 + 0.65j] as one of its zeros. Evaluate the magnitude and phase response of the resultant linear phase filter to verify its phase linearity.

6. (a) Design a comb filter having impulse response = [1,zeros(1, N),1] that is to give the maximum attenuation possible to a 60 Hz cosine wave sampled at 44,100 Hz. Determine N, then plot the log-magnitude spectrum of the signal after being filtered by your comb filter. Repeat the filtering and log-magnitude plot using values of N from three less to three more than the value you computed to verify that your value of N gives the best attenuation.

(b) Repeat part (a), this time using as the signal to be attenuated the following:

$$y = \cos(120\pi t) + \cos(360\pi t) + \cos(600\pi\ t)$$

where t = [0:1/(44,099):1]. Compare results to those of the original problem using only the 60 Hz signal.

(c) Repeat part (a), but use a subtractive comb filter of the following form: [1, zeros(1,M),-1]. You should find that the ultimate attenuation achievable is higher using this comb filter rather than that of part (a) above. Why?

7. Compute and plot the amplitude and magnitude responses of the following linear phase filters, represented by their impulse responses:

(a) [1,zeros(1,7),1]

(b) [1,zeros(1,6),1]

(c) [1,1,zeros(1,5),-1,-1]

(d) [1,1,zeros(1,4),-1,-1]

(e) [-0.0052,-0.0229,0.0968,0.4313,0.4313,0.0968,-0.0229,-0.0052]

8. Compute and plot the system (or z-transform) zeros for each of the impulse responses in the previous problem, and identify single real zeros, complex conjugate zero-pairs, and zero-quads.

9. Verify that the amplitude response for a Type-III linear phase filter, given by

$$H_r(\omega) = 2\sum_{n=0}^{M-1} h[n]\sin(\omega[M-n]) \tag{8.12}$$

(where $M = (L-1)/2$ and L is the filter length) is correct for the impulse response

$$Imp = [a, b, c, 0, -c, -b, -a]$$

by obtaining an expression for the DTFT of Imp and modifying the expression until it is in the form given by Eq. (8.12).

10. Verify that the amplitude response for a Type-IV linear phase filter, given by

$$H_r(\omega) = 2\sum_{n=0}^{L/2-1} h[n]\sin(\omega[M-n])$$

is correct for the impulse response

$$Imp = [a, b, c, -c, -b, -a]$$

11. Write a script to implement Eqs. (8.9) and (8.10), i.e., to convert a set of Direct Form FIR coefficients into a set of coefficients for a Frequency Sampling implementation, according to the following specification:

```
function [CFsB,CFsA,BFs,AFs] = LVxDirect2FreqSampFIR(Imp)
% Receives an FIR impulse response Imp and generates
% the Frequency Sampling Coefficients BFs,and AFs with
% the comb filter section coefficients as CFsB and CFsA.
% Test calls:
% [CFsB,CFsA,BFs,AFs] = LVxDirect2FreqSampFIR([1,1,1,1])
% [CFsB,CFsA,BFs,AFs] = LVxDirect2FreqSampFIR([1,-1,1,-1])
% [CFsB,CFsA,BFs,AFs] = LVxDirect2FreqSampFIR([1,0,0,1])
```

12. Write a script that receives a set of Frequency Sampling Form coefficients corresponding to an impulse response *Imp* and filters a signal *x* using both the Direct Form coefficients (i.e., *Imp* itself) and the Frequency Sampling Form coefficients. Display the results of both filtering operations. Follow the function specification below:

```
LVxFreqSampFilter(Imp,CFsB,CFsA,BFs,AFs,x)
% Receives an impulse response Imp and a signal x, filters x
% and displays the result two different ways, first, using the
% impulse response itself as a Direct Form FIR, and second,
% using Frequency Sampling implementation coefficients
% CFsB,CFsA,BFs,AFs. (created for example, by the script
% LVxDirect2FreqSampFIR).
```

The m-code

```
Imp = [1,1,1,1]; x = chirp([0:1/999:1],0,1,500);
[CFsB,CFsA,BFs,AFs] = LVxDirect2FreqSampFIR(Imp)
LVxFreqSampFilter(Imp,CFsB,CFsA,BFs,AFs,x)
```

should, for example, result in Fig. 8.22.

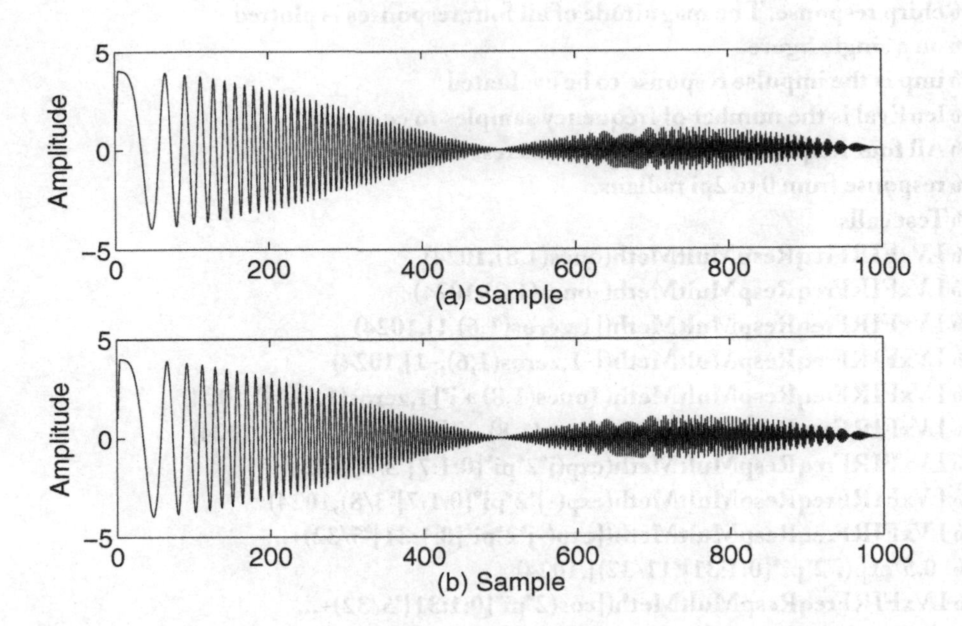

Figure 8.22: (a) Convolution of a linear chirp of length 1000 with the Direct Form coefficients [1,1,1,1]; (b) Result from filtering the linear chirp with the Frequency Sampling implementation of the Direct Form coefficients [1,1,1,1].

Additional test m-code:

```
Imp = fir1(22,0.3); x = chirp([0:1/999:1],0,1,500);
[CFsB,CFsA,BFs,AFs] = LVxDirect2FreqSampFIR(Imp)
LVxFreqSampFilter(Imp,CFsB,CFsA,BFs,AFs,x)

Imp = fir1(22,[0.3,0.5]); x = chirp([0:1/999:1],0,1,500);
[CFsB,CFsA,BFs,AFs] = LVxDirect2FreqSampFIR(Imp)
LVxFreqSampFilter(Imp,CFsB,CFsA,BFs,AFs,x)

Imp = fir1(82,[0.4,0.6],'stop'); x = chirp([0:1/999:1],0,1,500);
[CFsB,CFsA,BFs,AFs] = LVxDirect2FreqSampFIR(Imp)
LVxFreqSampFilter(Imp,CFsB,CFsA,BFs,AFs,x)
```

13. Write a script that will evaluate the frequency response of an FIR from 0 to 2π radians using the DTFT, the z-transform, a real chirp, and a complex chirp, and plot the results, in accordance with the following function specification.

```
function LVxFIRFreqRespMultMeth(imp,lenEval)
% Receives a real or complex impulse response and computes
% and displays the frequency response using four methods, namely
% the DTFT, the z-transform, real chirp response, and complex
% chirp response. The magnitude of all four responses is plotted
% on a single figure.
% imp is the impulse response to be evaluated
% lenEval is the number of frequency samples to compute
% All four frequency response methods test the frequency
% response from 0 to 2pi radians.
% Test calls:
% LVxFIRFreqRespMultMeth(ones(1,8),1024)
% LVxFIRFreqRespMultMeth(-ones(1,8),1024)
% LVxFIRFreqRespMultMeth([1,zeros(1,6),1],1024)
% LVxFIRFreqRespMultMeth([-1,zeros(1,6),-1],1024)
% LVxFIRFreqRespMultMeth( (ones(1,8) + j*[1,zeros(1,6),1]),1024)
% LVxFIRFreqRespMultMeth((ones(1,8) - j*[1,zeros(1,6),1]),1024)
% LVxFIRFreqRespMultMeth(exp(j*2*pi*[0:1:7]*3/8),1024)
% LVxFIRFreqRespMultMeth(exp(-j*2*pi*[0:1:7]*3/8),1024)
% LVxFIRFreqRespMultMeth([exp(-j*2*pi*[0:1:31]*5/32)+...
%  0.5*exp(j*2*pi*[0:1:31]*11/32)],1024)
% LVxFIRFreqRespMultMeth([cos(2*pi*[0:1:31]*5/32)+...
%  0.5*sin(2*pi*[0:1:31]*11/32)],1024)
```

CHAPTER 9

FIR Design Techniques

9.1 OVERVIEW

In the previous chapter we examined a number of general ideas or principles related to FIR design, such as the effect of filter length, the effect of windowing, requirements for linear phase, etc. Additionally, we have gained knowledge of simple filters such as the Comb and Moving Average filters, as well as simple passband filters having arbitrary band limits built by superposing two or more frequency-contiguous correlators (covered in Part I of the book).

We have at last accumulated enough knowledge to successfully undertake the design of linear phase FIRs that can meet certain user-specified filter design requirements, including particular passband and stopband boundaries, particular levels of stopband attenuation, passband ripple, etc. After presenting a brief overview of the three main design methods that will be explored in this chapter, we set forth the various standard parameters used to specify desired filter characteristics, which will allow us to not only specify the requirements for a given filter, but to evaluate and compare the performance of different filters designed to meet the same criteria. We then launch into a detailed discussion of the windowed ideal lowpass filter technique, including how to generate highpass, bandpass, and bandstop filters from lowpass filters. This is followed by an examination of the Frequency Sampling Design Method (not to be confused with the Frequency Sampling Realization Method discussed in the previous chapter with regard to the realization, rather than the design, of FIRs), which uses the inverse DFT to generate an impulse response from a user-specified set of frequency domain samples. With this design method we'll also explore the use of optimized transition band sample amplitudes or coefficients, a simple technique that can greatly improve stopband attenuation. We'll also investigate the design of certain linear phase Type III and Type IV specialty filters, the Hilbert transformer and the differentiator. The last major topic in the chapter is a very important one, equiripple FIR design; the equiripple filter design technique, although somewhat difficult to understand and implement compared to the windowed-lowpass and frequency sampling design techniques, is very popular since it results in filters that can achieve a given design with the shortest length.

9.2 SOFTWARE FOR USE WITH THIS BOOK

The software files needed for use with this book (consisting of m-code (.m) files, VI files (.vi), and related support files) are available for download from the following website:

http://www.morganclaypool.com/page/isen

The entire software package should be stored in a single folder on the user's computer, and the full file name of the folder must be placed on the MATLAB or LabVIEW search path in accordance with the instructions provided by the respective software vendor (in case you have encountered this notice before, which is repeated for convenience in each chapter of the book, the software download only needs to be done once, as files for the entire book are all contained in the one downloadable folder). See Appendix A for more information.

9.3 SUMMARY OF DESIGN METHODS

Three standard methods to design an FIR which will be covered in this chapter are:

- **The Window Method**: In this method, a truncated ideal lowpass filter having a certain bandwidth is generated, and then a chosen window is applied to achieve a certain stopband attenuation. Filter length can be adjusted to achieve a needed roll-off rate in the transition band. Filter types other than lowpass, such as highpass, bandpass, and notch, can be achieved by several techniques that start with windowed, truncated ideal lowpass filters.

- **The Frequency Sampling Method**: In this method, evenly-spaced samples of a desired frequency response are created, and the IDFT is computed to obtain an impulse response. In addition to using the IDFT to accomplish this, there are formulas that do the equivalent, i.e., superposing orthogonal cosines or sines to generate an impulse response. Both variants of this method will be explored below. Rather than creating an impulse response to be implemented either in Direct Form or Linear Phase Form, it is possible to use the frequency samples to directly realize the filter using this method's namesake, the Frequency Sampling Realization Method, which will be explored in the exercises at the end of the chapter.

- **The Equiripple Method**: This method designs an FIR having equalized ripple amplitudes in the passband, and equalized ripple in the stopband. The ripple levels may be independently controlled, allowing great flexibility. It is possible with equiripple design, for example, for a given filter length, to increase stopband attenuation by letting passband ripple increase, and vice versa. The previous two FIR design methods do not permit this degree of control.

9.4 FILTER SPECIFICATION

Figure 9.1 illustrates a typical design specification for an FIR lowpass filter. There is generally a certain amount of ripple in both passbands and stopbands. In Fig. 9.1, the maximum deviation (considered as an acceptable tolerance) from the average value in the passband is designated δ_P, while the deviation from zero in the stopband is designated as δ_S. This manner of defining the requirements for passband and stopband ripple is called an absolute specification; another manner (more common) is to specify the levels of ripple in decibels relative to the maximum magnitude of response, which is $1 + \delta_P$.

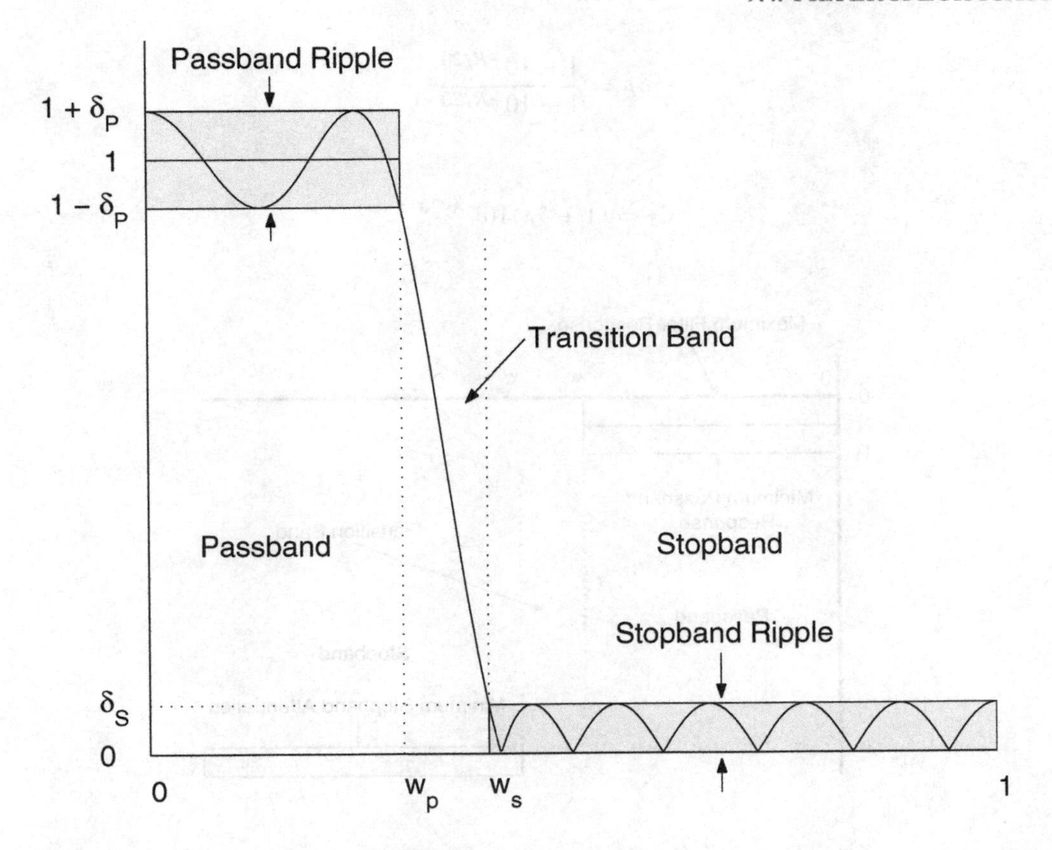

Figure 9.1: Design criteria for an FIR.

Figure 9.2 depicts another lowpass filter design specification in relative terms. The values of R and A are in decibels, and represent the passband ripple amplitude (or minimum passband response when the maximum filter/passband response is 0 db) and minimum stopband attenuation, respectively. The relationship between δ_P and δ_S and R and A are

$$R = -20 \log 10(\frac{1-\delta_P}{1+\delta_P})$$

and

$$A = -20 \log 10(\frac{\delta_S}{1+\delta_P})$$

To determine δ_P when R and A are given, use

$$\delta_P = \frac{1 - 10^{-R/20}}{1 + 10^{-R/20}}$$

and

$$\delta_S = (1 + \delta_P)10^{-A/20}$$

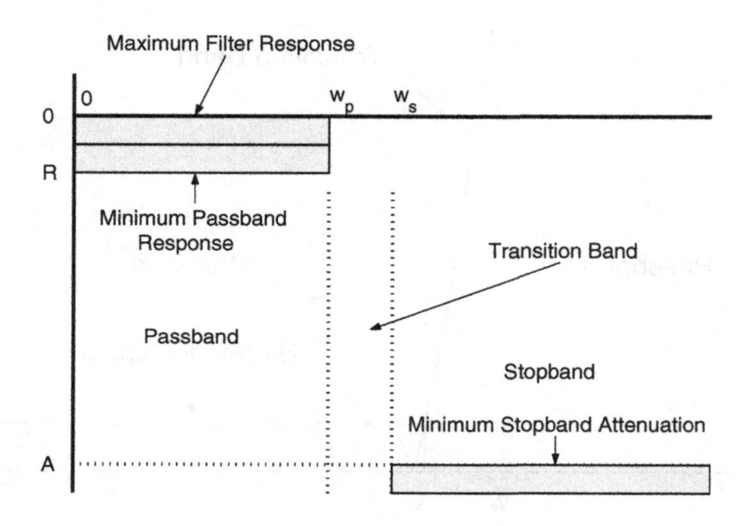

Figure 9.2: A relative filter design specification, with the (horizontal) frequency axis at the top, and (vertical) logarithmic amplitude (dB) axis at the left.

Example 9.1. A certain filter design specification is expressed in absolute terms as $\delta_P = 0.02$ and $\delta_S = 0.002$. Determine the filter design specification in relative terms.

We have

$$R = -20 \log 10(\frac{0.98}{1.02}) = 0.3475\ db$$

and

$$A = -20 \log 10(\frac{0.002}{1.02}) = 54.15\ db$$

In terms of m-code, we get

```
R = -20*log10(0.98/1.02)
A = -20*log10(0.002/1.02)
```

Example 9.2. A certain filter design specification is expressed in relative terms as $R = 0.5$ db and $A = 60$ db. Determine the absolute specification.

We get

$$\delta_P = \frac{1 - 10^{-0.5/20}}{1 + 10^{-0.5/20}} = 0.028774$$

and

$$\delta_S = (1.028774)(10^{-60/20}) = 0.001028$$

Suitable m-code to compute δ_P and δ_S is

```
function [DeltaP,DeltaS] = LVRelSpec2AbSpec(Rp,As)
% [DeltaP,DeltaS] = LVRelSpec2AbSpec(0.5,60)
Rfac= 10^(-Rp/20); DeltaP = (1-Rfac)/(1+Rfac);
DeltaS = (1+DeltaP)*10^(-As/20);
```

To check the computation, use this m-code to return to relative specification:

```
function [Rp,As] = LVAbSpec2RelSpec(DeltaP,DeltaS)
% [Rp,As] = LVAbSpec2RelSpec(DeltaP,DeltaS)
Rp = -20*log10((1-DeltaP)/(1+DeltaP));
As = -20*log10(DeltaS/(1+DeltaP));
```

9.5 FIR DESIGN VIA WINDOWED IDEAL LOWPASS FILTER

An Ideal Lowpass filter has a noncausal, infinite-length impulse response which can be determined by taking the Inverse DTFT of the frequency specification

$$X(e^{j\omega}) = \begin{cases} 1 \cdot e^{-j\omega M} & |\omega| \leq \omega_c \\ 0 & |\omega| > \omega_c \end{cases}$$

which can be written as

$$x[n] = \frac{1}{2\pi}\int_{-\omega_c}^{\omega_c} e^{-j\omega M} e^{jwn} d\omega = \frac{1}{2\pi}\int_{-\omega_c}^{\omega_c} e^{j\omega(n-M)} d\omega = \frac{\sin(\omega_c[n-M])}{\pi[n-M]} \tag{9.1}$$

To utilize such an impulse response for a linear phase FIR, it is necessary to symmetrically truncate it about M, where $M = (L-1)/2$ and L is the total length of the symmetrically-truncated impulse response. Since the impulse response is symmetrical, Type I and II linear phase filters can be designed.

Example 9.3. Write m-code that will generate a causal, symmetrically truncated impulse response of the ideal lowpass type; compute and plot for $\omega_c = 0.25$ and $L = 61$.

The following m-code generates such an impulse response; change w_c to control cutoff frequency and L to control impulse response length. Note that a true ideal lowpass filter impulse response is a continuous function of infinite duration from $t = -\infty$ to $+\infty$; in this example, we are computing a finite number of samples of such a function. For the purposes of bandlimited discrete processing and digital filtering, of course, this is acceptable.

```
function b = LVIdealLPFImpResp(wc,L)
% LVIdealLPFImpResp(0.25*pi,61)
M = (L-1)/2; n = 0:1:L-1;
b = sin(wc*(n - M + eps))./(pi*(n - M + eps));
figure(55); stem(n,b)
```

9.5.1 WINDOWS

Any finite-length (i.e., truncated) version of the ideal lowpass impulse response may be considered as the product of the infinite-length lowpass impulse response and a window function W, which has a finite number of contiguous nonzero-valued samples

$$b = (\frac{\sin(\omega_c[n-M])}{\pi[n-M]})(W_L[n-M]) \tag{9.2}$$

where the window length is L, $M = (L-1)/2$, $0 \leq n \leq L-1$, and $W_L[n]$ is generally a function $F_E[n]$ having even symmetry about M defined as

$$W_L[n] = \begin{cases} F_E[n] & n = 0{:}1{:}L-1 \\ 0 & \text{otherwise} \end{cases}$$

The right side of Eq. (9.2) is the product of an infinite-length, ideal lowpass filter with a function that is nonzero only over a finite number of contiguous samples. The result is a finite-length or truncated lowpass filter. Since W is chosen to have even symmetry around M, the window symmetrically truncates the infinite-length function (which is itself symmetrical about M), resulting in a finite-length symmetrical sequence as the net filter impulse response.

We encountered windows in Part II of the book as a tool for reducing leakage in the DFT. We discuss some of the basic information on standard windows here for convenience.

The simplest window is the rectangular window $R[n]$, which is defined as

$$R[n] = \begin{cases} 1 & 0 \leq n \leq L-1 \\ 0 & \text{otherwise} \end{cases}$$

The windowing process, using a rectangular window, is shown in Fig. 9.3. In subplot (a), a portion of an infinite-length, ideal lowpass filter is shown, centered on M; in subplot (b), a portion of the infinite-length (rectangular) window is shown, with its contiguous, nonzero-valued samples centered on M, and finally, in subplot (c), the product of the two sequences in (a) and (b), the truncated lowpass filter, is shown.

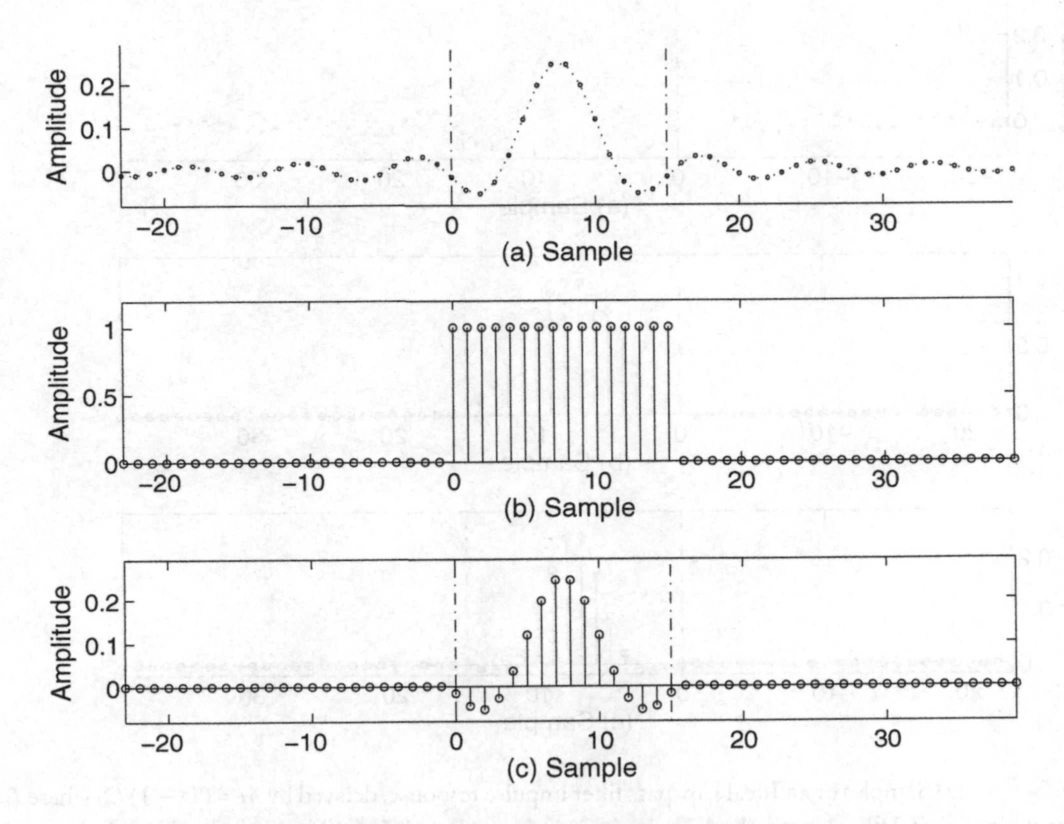

Figure 9.3: (a) Samples of an Ideal Lowpass filter impulse response, delayed by $M = (L-1)/2$, where L is the desired digital filter length; (b) A Rectangular window of length L; (c) The product of the window and the Ideal Lowpass filter, yielding, for samples n = 0:1:(L − 1), a length-L symmetrical impulse response as the lowpass digital filter finite impulse response.

The Hanning, Hamming, and Blackman windows are described by the general raised-cosine formula

$$W[n] = \begin{cases} a - b\cos(2\pi \frac{n}{L-1}) + c\cos(4\pi \frac{n}{L-1}) & n = 0:1:L-1 \\ 0 & \text{otherwise} \end{cases}$$

where L is the window length, and where, for the Hanning window, a = 0.5, b = 0.5, and c = 0; for the Hamming window, a = 0.54, b = 0.46, and c = 0; and for the Blackman window, a = 0.42, b = 0.5, and c = 0.08.

Figure 9.4 shows the windowing process using a Hamming window:

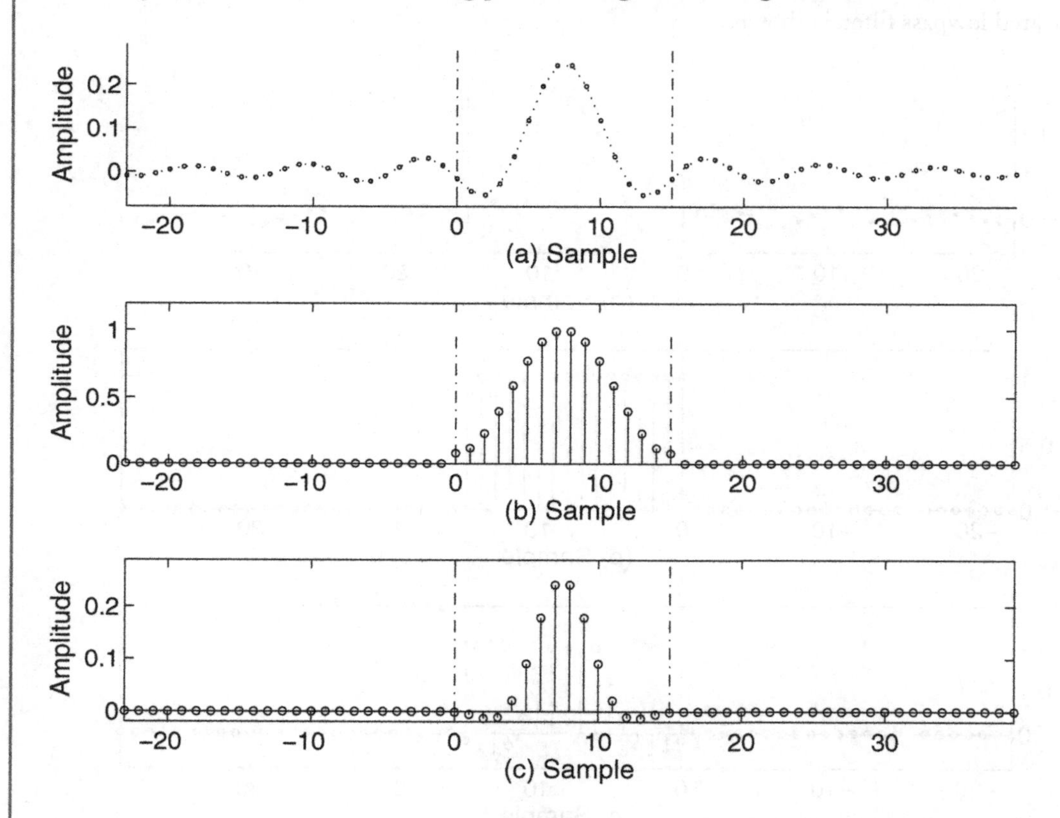

Figure 9.4: (a) Samples of an Ideal Lowpass filter impulse response, delayed by $M = (L-1)/2$, where L is the desired digital filter length; (b) A Hamming window of length L; (c) The product of the window and the Ideal Lowpass filter, yielding, for samples n = 0:1:(L − 1), a length-L symmetrical impulse response as the lowpass digital filter finite impulse response.

The Kaiser window is described by the formula

$$W[n] = \begin{cases} \frac{I_0[\beta(1-(n-M)/M]^2)^{0.5}]}{I_0(\beta)} & n = 0:1:L-1 \\ 0 & \text{otherwise} \end{cases}$$

for n = 0:1:L-1, where L is the window length, $M = (L-1)/2$, and I_0 represents the modified Bessel function of the first kind.

9.5.2 NET FREQUENCY RESPONSE

The net frequency response of the impulse response b is the circular convolution of the DTFTs of the ideal lowpass filter (an infinite-length time domain sequence) and the window (of infinite length but containing nonzero values only over the interval $0 \le n \le L-1$). This process results in a smearing or widening of the frequency response (i.e., the DTFT) of the net lowpass filter relative to that of the ideal lowpass filter, which has infinitely steep roll-off. Representing the frequency response of the ideal lowpass filter by $H_L(e^{j\omega})$, and that of the window by $W(e^{j\omega})$, the frequency response of the windowed ideal lowpass filter is

$$H(e^{j\omega}) = H_L(e^{j\omega}) \circledast W(e^{j\omega})$$

where the symbol $\circledast$ here means circular convolution.

Figures 9.5, 9.6, 9.7, and 9.8, depict this process. The generic process is shown in Fig. 9.5. The frequency domain effect of the time domain process of windowing (whereby an infinite length ideal lowpass filter impulse response is symmetrically truncated to a finite length) can be determined by numerically performing circular convolution of samples of the DTFTs of the ideal lowpass filter and the proposed window; the ripple and transition width of the window determine the ultimate frequency response of the truncated lowpass filter.

Looking at the computational process in more detail for several different windows, for each of Figs. 9.6, 9.7, and 9.8, the computation started with a good approximation of an ideal lowpass filter (an impulse response having a length of thousands of samples) and a symmetrical window of the same length consisting of the value zero everywhere except in the central portion, in which is located a symmetrical group of contiguous, nonzero samples of a desired length which form the window (Hamming, Kaiser, etc) which will truncate the ideal lowpass filter's impulse response. The DFT (i.e., samples of the DTFT) is obtained of the ideal lowpass filter impulse response (approximated by a very large number of samples) and of the window, which is of the same length as the ideal lowpass filter, and then the circular convolution of the two DFTs is obtained, which is the frequency response of the ideal lowpass filter as truncated by the window. The magnitude of all three (very long) DFTs is then plotted.

- Each type of window, when applied to a truncated ideal lowpass impulse response, results in a filter having a characteristic main lobe width, transition width, and minimum stopband attenuation, the first two of which depend strongly on filter length.

- The minimum stopband attenuation is generally quoted as a constant value. As a result, it is possible to choose a window according to what minimum level of stopband attenuation is acceptable, and then adjust L (filter length) to achieve the needed roll-off or narrowness of transition band.

- The rectangular window has the narrowest main lobe and smallest transition width, but the poorest stopband attenuation.

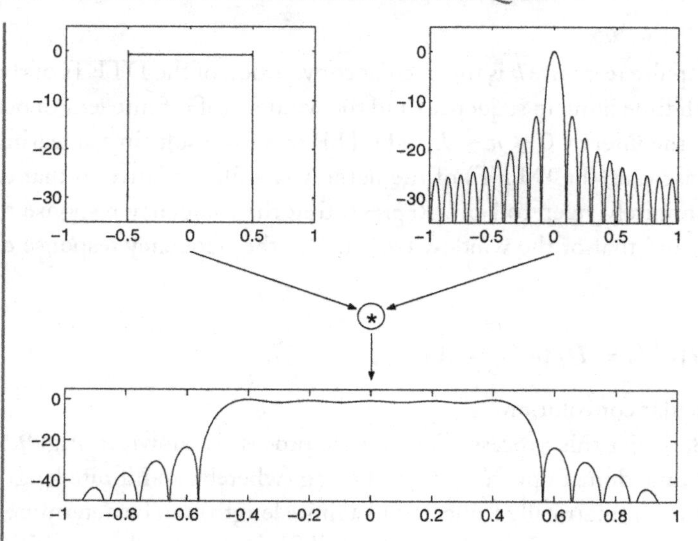

Figure 9.5: Upper left plot: samples of the DTFT (obtained via DFT) of an ideal lowpass filter having $\omega_c = 0.5\pi$ radians, showing to good approximation a very steep roll-off (theoretically infinitely steep); Upper right plot: samples of the DTFT of a rectangular window, the rectangular window having been used to symmetrically truncate the ideal lowpass filter's impulse response; Lower plot: the net frequency response of the truncated lowpass filter, computed as the circular convolution of the DTFTs of the ideal lowpass filter and the rectangular window. All horizontal axes are frequency in units of π, and all vertical axes are magnitude in dB.

- The Hanning, Hamming, and Blackman windows have broader main lobes, wider transition widths, but improved stopband attenuation.

- The Kaiser window is adjustable, allowing a chosen compromise between transition width and stopband attenuation through the choice of the parameter β.

- Approximate and exact values of the transition width as a function of window length for the standard windows have been tabulated and can be used to give a good first estimate of the needed filter length. The following table gives approximate and exact values for L in terms of the transition width, $\omega_t = w_s - w_p$ and the minimum stopband attenuation values for several standard windows. The exact values for L are better estimates, and can be used when the design target A_s is close to the window's inherent A_s. If the target A_s is much smaller than the window's inherent A_s, the filter length needed may be shorter.

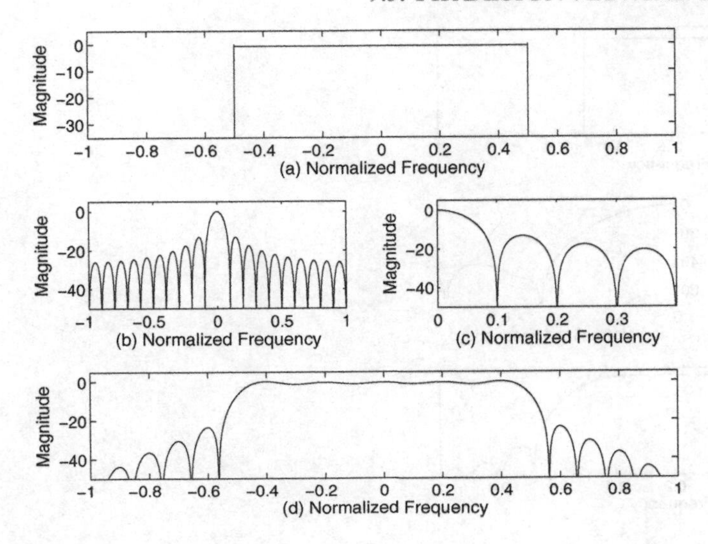

Figure 9.6: (a) Magnitude (dB) of DFT of an ideal lowpass filter ($L = 2^{15}$ samples); (b) Magnitude (dB) of DFT of a window of length 2^{15} samples, all which are zero except for the central 20 samples, which form a rectangular window; (c) Zoomed-in view of waveform at (b); (d) Net frequency response (magnitude in dB) of the truncated ideal lowpass filter, computed as the circular convolution of the DFTs of the ideal lowpass filter and the window.

Name	Approx L	Exact L	min A_s, dB
Blackman	$12\pi/\omega_t$	$11\pi/\omega_t$	74
Hamming	$8\pi/\omega_t$	$6.6\pi/\omega_t$	53
Hanning	$8\pi/\omega_t$	$6.2\pi/\omega_t$	44
Bartlett	$8\pi/\omega_t$	$6.1\pi/\omega_t$	25
Rectangular	$4\pi/\omega_t$	$1.8\pi/\omega_t$	21

The Kaiser window is adjustable according to the parameter β, and Kaiser has provided empirical formulas that allow determination of necessary values of L (filter length) and β to achieve a certain minimum stopband attenuation A_s. The needed length L for a given A_s is

$$L \simeq \frac{2\pi(A_s - 7.95)}{14.36(\omega_s - \omega_p)} + 1 \tag{9.3}$$

where ω_s and ω_p are in radians, such as 0.5π, etc., and the needed β is

$$\beta = \begin{cases} 0.1102(A_s - 8.7) & A_s \geq 50 \\ 0.5842(A_s - 21)^{0.4} + 0.07886(A_s - 21) & 21 \leq A_s < 50 \\ 0 & A_s < 21 \end{cases} \tag{9.4}$$

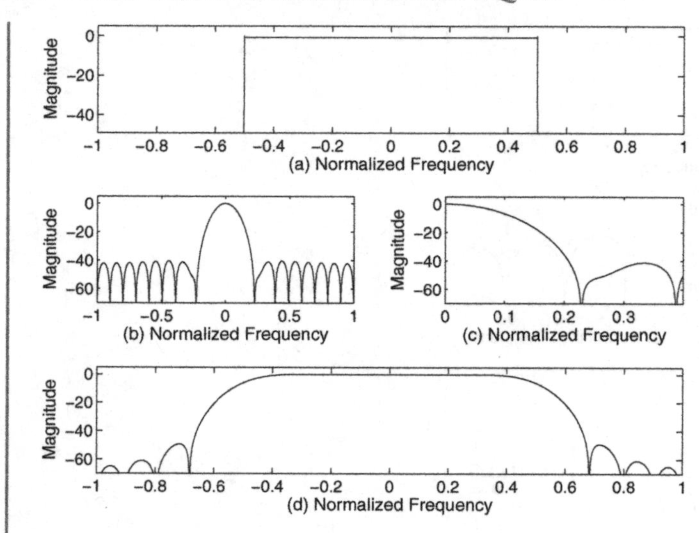

Figure 9.7: (a) Magnitude (dB) of DFT of an ideal lowpass filter ($L = 2^{15}$ samples); (b) Magnitude (dB) of DFT of a window of length 2^{15} samples, all which are zero except for the central 20 samples, which form a Hamming window; (c) Zoomed-in view of waveform at (b); (d) Net frequency response (magnitude in dB) of the truncated ideal lowpass filter, computed as the circular convolution of the DFTs of the ideal lowpass filter and the window.

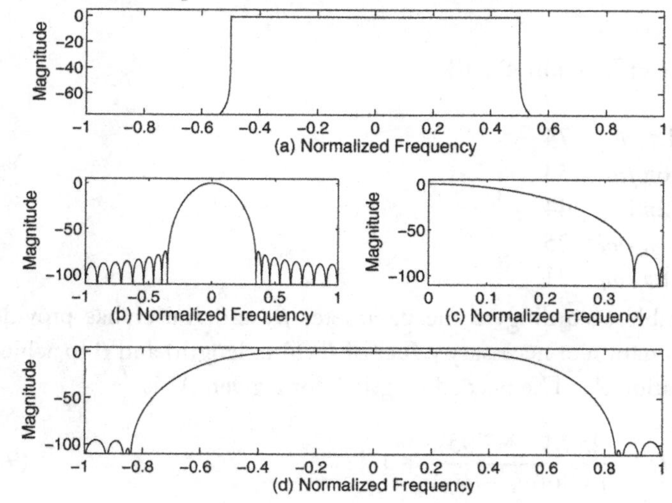

Figure 9.8: (a) Magnitude (dB) of DFT of an ideal lowpass filter ($L = 2^{15}$ samples); (b) Magnitude (dB) of DFT of a window of length 2^{15} samples, all which are zero except for the central 20 samples, which form a Kaiser(10) window; (c) Zoomed-in view of waveform at (b); (d) Net frequency response (magnitude in dB) of the truncated ideal lowpass filter, computed as the circular convolution of the DFTs of the ideal lowpass filter and the window.

9.5.3 WINDOWED LOWPASS FILTERS-PASSBAND RIPPLE AND STOPBAND ATTENUATION

We illustrate the practical application of various windows, and the effect on passband ripple and stopband attenuation to a lowpass impulse response with several examples:

Example 9.4. Design a length-17 lowpass filter having $\omega_p = 0.4\pi$ and $\omega_s = 0.5\pi$ using a rectangular window. Measure passband ripple and stopband attenuation.

The function below constructs the impulse response using an ideal lowpass impulse response with a rectangular window. The result from making the call

LVLPFViaSincRectwin(0.4*pi,0.5*pi,17)

is shown in Fig. 9.9.

```
function LVLPFViaSincRectwin(wp,ws,L)
% LVLPFViaSincRectwin(0.4*pi,0.5*pi,17)
wc = (wp + ws)/2; M = (L-1)/2; n = 0:1:L-1;
b = sin(wc*(n - M + eps))./(pi*(n - M + eps));
LenFFT = 8192; fr = abs(fft(b,LenFFT)); fr=fr(1,1:LenFFT/2+1);
Lfr = length(fr); PB = fr(1,1:round((wp/pi)*Lfr));
SB = fr(1,round((ws/pi)*Lfr):Lfr);
PBR = -20*log10(min(PB)), SBAtten = -20*log10(max(SB)),
figure(59); plot([0:1:LenFFT/2]/(LenFFT/2), 20*log10(fr+eps));
xlabel('Frequency, Units of \pi');
text(0.1,-30,['actual Rp = ',num2str(PBR,3),' dB'])
text(0.1,-45,['actual As = ',num2str(SBAtten,3),' dB'])
ylabel(['Mag, dB']); axis([0 1 -inf inf])
```

9.5.4 HIGHPASS, BANDPASS, AND BANDSTOP FILTERS FROM LOWPASS FILTERS

To create filters other than lowpass from a lowpass filter, several examples are presented that illustrate the general procedure, which can be described as spectral subtraction.

Example 9.5. Design a highpass filter using a rectangular window of length 51 having $\omega_c = 0.3\pi$ using a truncated ideal lowpass filter.

To do this, we will design a lowpass filter having the same cutoff (0.3π), and subtract it from a filter of the same length (51) that passes all frequencies from 0 to π radians. The following code illustrates this procedure; the result from making the call

LVHPFViaSincLPFRectwin(0.3*pi,51)

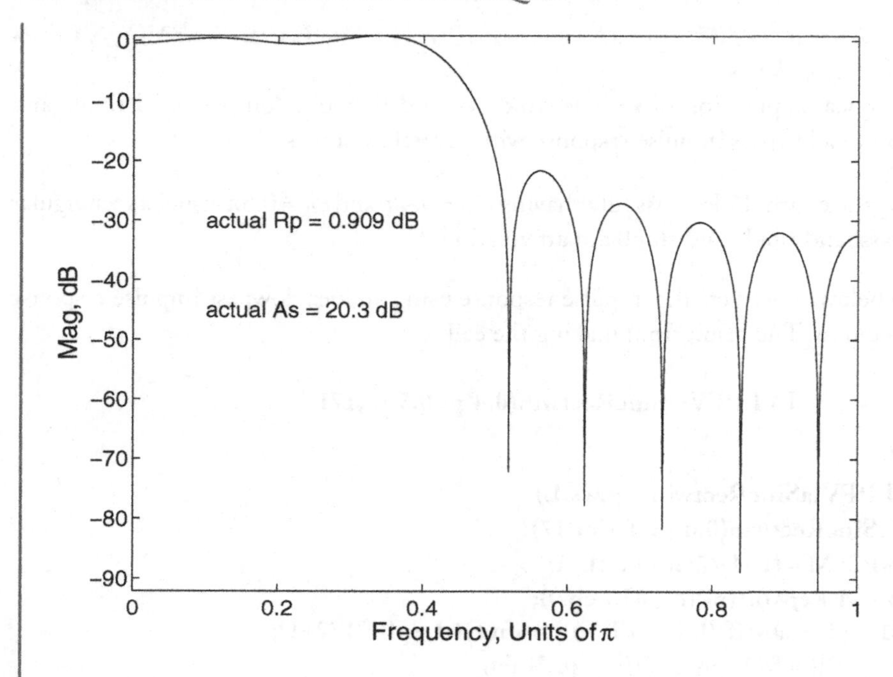

Figure 9.9: The magnitude of the frequency response of a length-17 lowpass filter having an inherent rectangular window.

is shown in Fig. 9.10.

```
function LVHPFViaSincLPFRectwin(wc,L)
% LVHPFViaSincLPFRectwin(0.3*pi,51)
M = (L-1)/2; n = 0:1:L-1;
ImpLo = sin(wc*(n - M + eps))./(pi*(n - M + eps));
ap = sin(pi*(n - M + eps))./(pi*(n - M + eps));
ImpHi = ap - ImpLo; frImpHi = abs(fft(ImpHi,1024));
frImpLo = abs(fft(ImpLo,1024)); figure(56);
set(56,'color',[1,1,1]); subplot(221); stem(n,ImpLo);
xlabel('(a) Sample'); ylabel('Amplitude')
axis([0 length(ImpLo) 1.2*min(ImpLo) 1.1*max(ImpLo)])
subplot(222); plot([0:1:512]/512, frImpLo(1,1:513));
xlabel(['(b) Freq, Units of \pi']); ylabel('Magnitude')
axis([0 1 0 1.1]); subplot(223); stem(n,ImpHi);
xlabel('(c) Sample'); ylabel('Amplitude')
axis([0 length(ImpHi) 1.2*min(ImpHi) 1.1*max(ImpHi)])
subplot(224); plot([0:1:512]/512,frImpHi(1,1:513));
```

```
xlabel(['(d) Freq, Units of \pi']); ylabel('Magnitude')
axis([0 1 0 1.1])
```

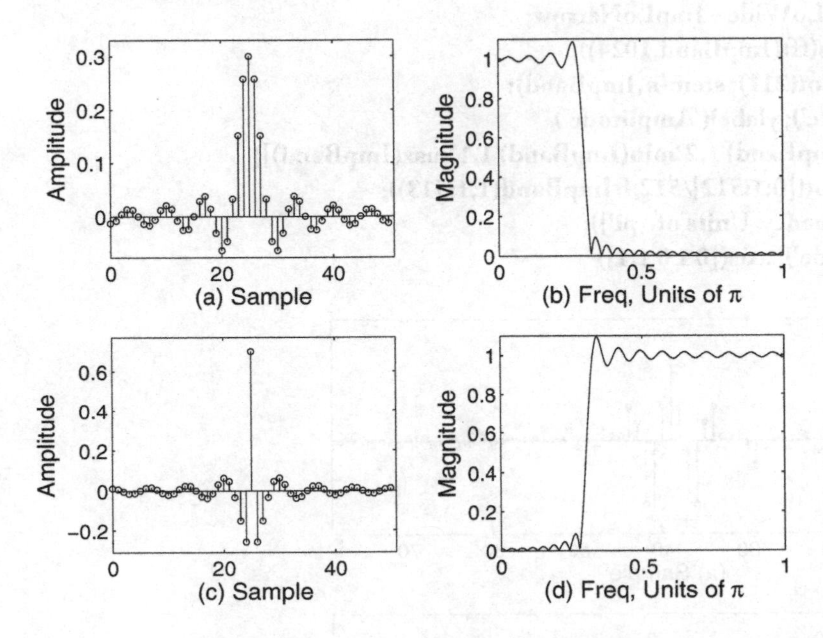

Figure 9.10: (a) Lowpass filter impulse response; (b) Magnitude of DTFT of sequence in (a); (c) Highpass impulse response; (d) Magnitude of DTFT of sequence in (c).

Example 9.6. Design a bandpass filter of length 79, using a rectangular window, having band edges [0.2, 0.3, 0.5, 0.6], where the first stopband runs from 0 to 0.2π radians, the first transition band runs from 0.2π to 0.3π radians, the second transition band runs from 0.5π to 0.6π radians, and the second stopband runs from 0.6π to π radians.

To do this, we will design a lowpass filter of length 79 having transition band from 0.5π to 0.6π ($\omega_{c2} = (0.5 + 0.6)\pi/2$), then subtract from its impulse response that of a lowpass filter of length 79 having its transition band from 0.2π to 0.3π ($\omega_{c1} = (0.2 + 0.3)\pi/2$). The following code implements this procedure; the result from making the call

LVBPFViaSincLPFRectwin(0.25*pi,0.55*pi,79)

is shown in Fig. 9.11.

```
function LVBPFViaSincLPFRectwin(wc1,wc2,L)
% LVBPFViaSincLPFRectwin(0.25*pi,0.55*pi,79)
M = (L-1)/2; n = 0:1:L-1;
```

```
ImpLoWide = sin(wc2*(n - M + eps))./(pi*(n - M + eps));
ImpLoNarrow = sin(wc1*(n - M + eps))./(pi*(n - M + eps));
ImpBand = ImpLoWide - ImpLoNarrow;
frImpBand = abs(fft(ImpBand,1024));
figure(57); subplot(211); stem(n,ImpBand);
xlabel('(a) Sample'); ylabel('Amplitude')
axis([0 length(ImpBand) 1.2*min(ImpBand) 1.1*max(ImpBand)])
subplot(212); plot([0:1:512]/512,frImpBand(1,1:513));
xlabel(['(b) Frequency, Units of \pi']);
ylabel('Magnitude'); axis([0 1 0 1.1])
```

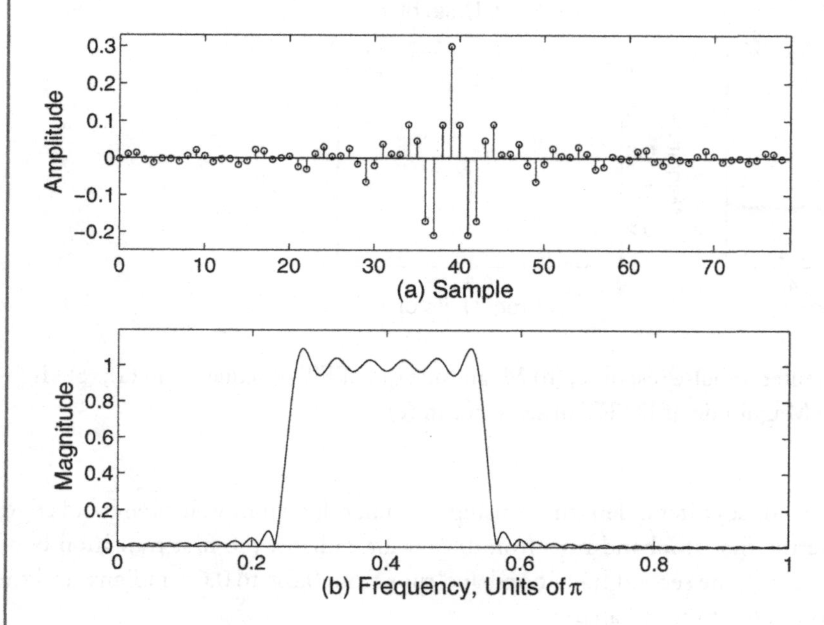

Figure 9.11: (a) Bandpass filter impulse response; (b) Magnitude of DTFT of impulse response in (a).

Example 9.7. Design a bandstop filter having band edges [0.4, 0.5, 0.7, 0.8], length 71, and using a rectangular window.

To do this, we create a bandpass filter as above having the needed band edges and length, then subtract it from a filter of the same length that passes all frequencies from 0 to π radians. The following code implements this procedure; Fig. 9.12 shows the result of making the call

LVNotchViaLPFSincRectwin(0.45*pi,0.75*pi,71)

```
function LVNotchViaLPFSincRectwin(wc1,wc2,L)
% LVNotchViaLPFSincRectwin(0.45*pi,0.75*pi,71)
M = (L-1)/2; n = 0:1:L-1;
ImpLo2 = sin(wc2*(n - M + eps))./(pi*(n - M + eps));
ImpLo1 = sin(wc1*(n - M + eps))./(pi*(n - M + eps));
ImpBand = ImpLo2 - ImpLo1;
ImpWide = sin(pi*(n - M + eps))./(pi*(n - M + eps));
ImpStop = ImpWide - ImpBand;
frImpStop = abs(fft(ImpStop,1024));
figure(58); subplot(211); stem(n,ImpStop);
xlabel('(a) Sample'); ylabel('Amplitude');
subplot(212); plot([0:1:512]/512, frImpStop(1,1:513));
xlabel(['(b) Frequency, Units of \pi']);
ylabel('Magnitude'); axis([0 1 0 1.2])
```

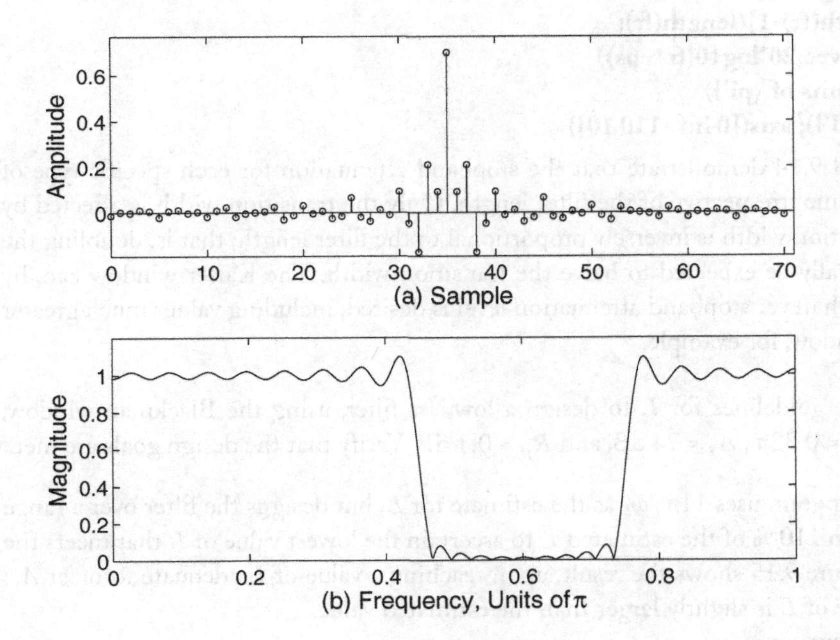

Figure 9.12: (a) Bandstop filter impulse response; (b) Magnitude of DTFT of sequence in (a).

9.5.5 IMPROVING STOPBAND ATTENUATION

The filters designed in the examples above show the high passband ripple and poor stopband attenuation associated with the rectangular window. By choosing different nonrectangular windows, the stopband attenuation can be improved at the expense of a shallower roll-off–but this can be compensated by increasing the filter length.

Example 9.8. Compare the stopband attenuation, passband ripple, and transition widths of a length-50 lowpass filter, $\omega_c = 0.5\pi$, windowed with rectangular, Hamming, Blackman, and Kaiser ($\beta = 10$) windows. Perform the experiment a second time, using $L = 101$.

We'll present code to design the lowpass filter using the Kaiser(10) window, and illustrate the other windows in Fig. 9.13.

```
function LVLPFViaSincKaiser(wc,L)
% LVLPFViaSincKaiser(0.5*pi,50)
M = (L-1)/2; n = 0:1:L-1;
Imp = sin(wc*(n - M + eps))./(pi*(n - M + eps));
win = kaiser(L,10)'; fr = abs(fft(Imp.*win, 2048));
fr = fr(1,1:fix(length(fr)/2+1));
xvec = [0:1:length(fr)-1]/length(fr);
figure(6); plot(xvec,20*log10(fr+eps))
xlabel(['Freq, Units of \pi'])
ylabel(['Mag, dB']); axis([0 inf -110 10])
```

Figures 9.13 and 9.14 demonstrate that the stopband attenuation for each specific type of window remains the same irrespective of the filter length. Only the transition width is affected by filter length. The transition width is inversely proportional to the filter length; that is, doubling the filter length may generally be expected to halve the transition width. The Kaiser window can, by changing β, be given whatever stopband attenuation level is desired, including values much greater than the Blackman window, for example.

Example 9.9. Use the guidelines for L to design a lowpass filter, using the Blackman window, having $\omega_p = 0.25\pi$, $\omega_s = 0.32\pi$, $A_s = 74$ dB, and $R_p = 0.1$ dB. Verify that the design goals are met.

The following program uses $11\pi/\omega_t$ as the estimate for L, but designs the filter over a range of L from about 90 % to 110 % of the estimated L to ascertain the lowest value of L that meets the design value of A_s. Figure 9.15 shows the result, upon reaching a value of L adequate to meet A_s; note that the final value of L is slightly larger than the estimated value.

```
function LVLPFSincBlackman(wp,ws,Rp,As)
% LVLPFSincBlackman(0.25*pi,0.32*pi,0.1,74)
wc = (wp+ws)/2; wt = ws - wp; limL = ceil(11*pi/wt),
startL = fix(0.9*limL); figure(59); LenFFT = 8192;
for L = startL:1:ceil(1.1*limL); M = (L-1)/2;
n = 0:1:L-1; b = sin(wc*(n - M + eps))./(pi*(n - M + eps));
b = b.*(blackman(L)'); fr = abs(fft(b,LenFFT));
```

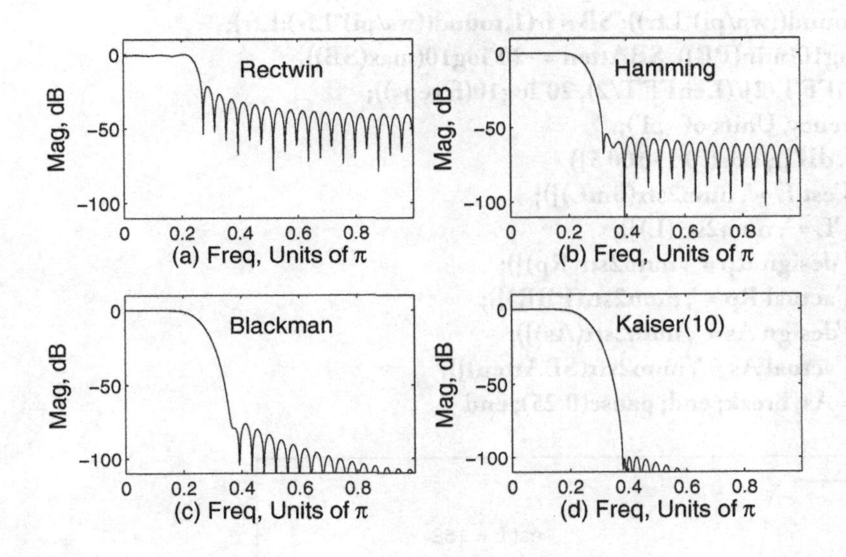

Figure 9.13: (a) DTFT of a length-50 ideal lowpass filter multiplied by a rectangular window; (b) Same, with Hamming window; (c) Same, with Blackman window; (d) Same, with Kaiser window, β = 10.

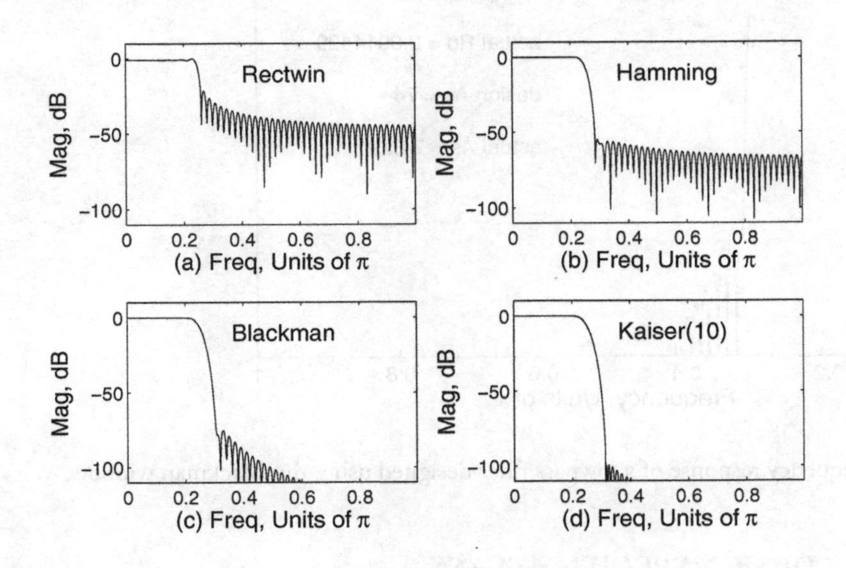

Figure 9.14: (a) DTFT of a length-101 ideal lowpass filter multiplied by a Rectangular window; (b) Same, with Hamming window; (c) Same, with Blackman window; (d) Same, with Kaiser window, β = 10.

```
fr=fr(1,1:(LenFFT/2+1)); Lfr = length(fr);
PB = fr(1,1:round((wp/pi)*Lfr)); SB = fr(1,round((ws/pi)*Lfr):Lfr);
PBR = -20*log10(min(PB)), SBAtten = -20*log10(max(SB)),
plot([0:1:LenFFT/2]/(LenFFT/2), 20*log10(fr+eps));
xlabel('Frequency, Units of \pi');
ylabel(['Mag, dB']); axis([0 1 -100 5])
text(0.6,-10,['est L = ', num2str(limL)]);
text(0.6,-20,['L = ', num2str(L)]);
text(0.6,-30,['design Rp = ', num2str(Rp)]);
text(0.6,-40,['actual Rp = ', num2str(PBR)]);
text(0.6,-50,['design As = ', num2str(As)]);
text(0.6,-60,['actual As = ', num2str(SBAtten)]);
if SBAtten>=As; break; end; pause(0.25); end
```

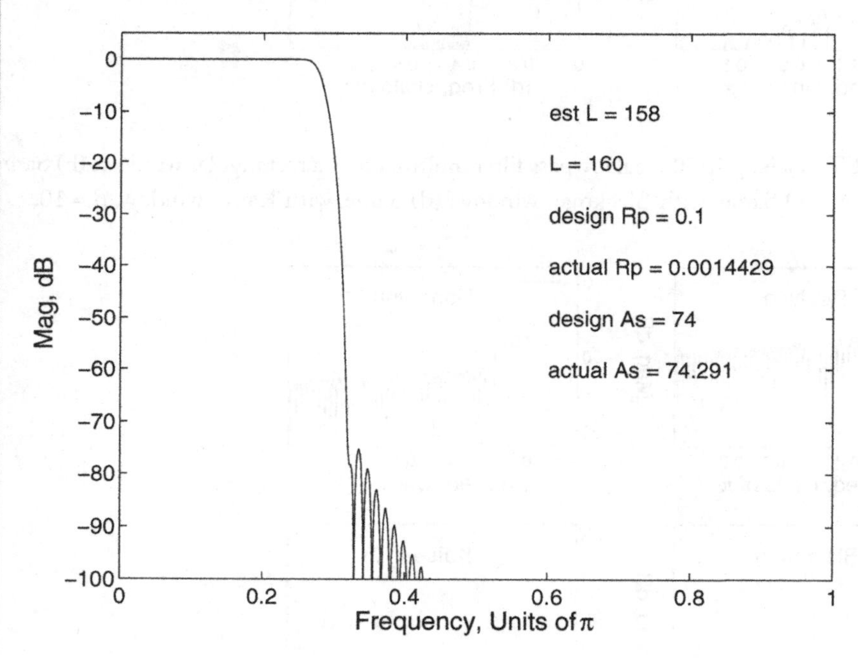

Figure 9.15: The frequency response of a lowpass filter designed using the Blackman window.

9.5.6 MEETING DESIGN SPECIFICATIONS

A basic procedure to design a filter using the windowed ideal lowpass method is as follows:

- Specify the filter design parameters, which consist of filter type, band edges, maximum desired passband ripple, and minimum permissible stopband attenuation.

- Pick a window that can produce the required minimum stopband attenuation.
- Either estimate the value of L needed, or start at a low value of L and proceed incrementally until the minimum stopband attenuation is just met. Inadequate values of L cause the window's transition band to pass into the stopband, thus producing a minimum stopband attenuation above that attainable with the given window. Compare the value of realized passband ripple to the design value.

One of the exercises at the end of this chapter is to generate a script which conforms to the following call syntax:

$$LVxFIRViaWinIdealLPF(FltType, BndEdgeVec, Win, As, Rp) \tag{9.5}$$

FltType is passed as *1* for lowpass, *2* for highpass, *3* for bandpass, and *4* for bandstop; *BndEdgeVec* is specified as ω_p, ω_s for a lowpass filter, $[\omega_s, \omega_p]$ for a highpass filter, $[\omega_{s1}, \omega_{p1}, \omega_{p2}, \omega_{s2}]$ for a bandpass filter, and $[\omega_{p1}, \omega_{s1}, \omega_{s2}, \omega_{p2}]$ for a bandstop filter, all values in fractions of π, such as 0.3, 0.5, etc, with 1.0 being the Nyquist rate, or π radians. Available window types are *rectwin*, *bartlett*, *hanning*, *hamming*, *blackman*, and *kaiser*. Minimum acceptable stopband attenuation in dB is passed as As and maximum acceptable passband ripple in dB as R_p.

The computations and figures for the following examples were generated using a script conforming to function (9.5).

Example 9.10. Design a bandpass filter having *BndEdgeVec* = [0.4, 0.5, 0.8, 0.9], *Win* = 'kaiser', and As = 78. We also require that the maximum passband ripple be 0.1 dB.

The frequency response of the resultant design is displayed along with various parameters (L, β, A_s, R_p) in Fig. 9.16. The script obtained the initial estimate for L and the value of β from the design formulas (9.3) and (9.4). We note that the actual value of passband ripple (R_p) and minimum stopband attenuation are, respectively, 0.0023 dB and 78 dB, which meet the design goals.

Example 9.11. Compare designs for a lowpass filter with the following design specifications: $A_s = 44$, $R_p = 0.1$, $\omega_p = 0.45$, $\omega_s = 0.55$, using two different windows, Hanning and Kaiser.

The Hanning design is shown in Fig. 9.17, while the Kaiser design is shown in Fig. 9.18. Note that the Kaiser window produces a much shorter filter length for the same design parameters.

Example 9.12. Compare designs for a bandstop filter having *BndEdgeVec* = [0.4, 0.45, 0.65, 0.7]. $A_s = 74$, $Rp = 0.1$, using the Blackman and Kaiser windows.

We again find that the Kaiser window needs a smaller value of L than does the other window (Blackman, in this case) for the same design parameters. The Blackman design is shown in Fig. 9.19, while the Kaiser design is shown in Fig. 9.20.

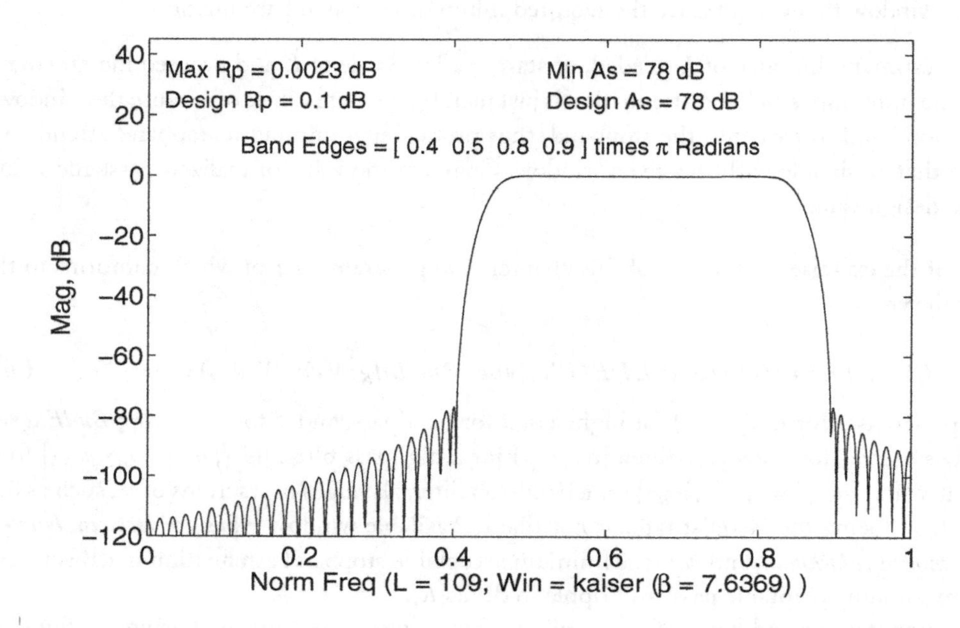

Figure 9.16: The frequency response and various design and realized parameters for a bandpass filter using a Kaiser window.

9.6 FIR DESIGN VIA FREQUENCY SAMPLING

In the previous section, we synthesized a basic lowpass filter impulse response by defining an ideal lowpass frequency response and using the inverse DTFT to determine an equivalent time domain expression which could be evaluated over any desired range of sample index n. We were further able, with some ingenuity, to synthesize highpass, bandpass, and bandstop filters from lowpass filters.

In the Frequency Sampling approach, we again start in the frequency domain with a specification, but instead of using a continuous frequency specification and the inverse DTFT, we take equally spaced samples of the frequency domain specification, and treat them as a DFT. The inverse DFT then yields an impulse response that will result in a filter whose frequency response matches that of the specification exactly at the location of the frequency samples. The filter's frequency response at values of ω lying between the frequency samples will differ from the ideal or continuous frequency response.

Once having specified the desired response, which is a set of amplitudes, one for each correlation frequency possible within the filter length, it is possible to proceed two ways:

- **Convert the specification to a correctly-formatted DFT and perform the IDFT to obtain the impulse response.**

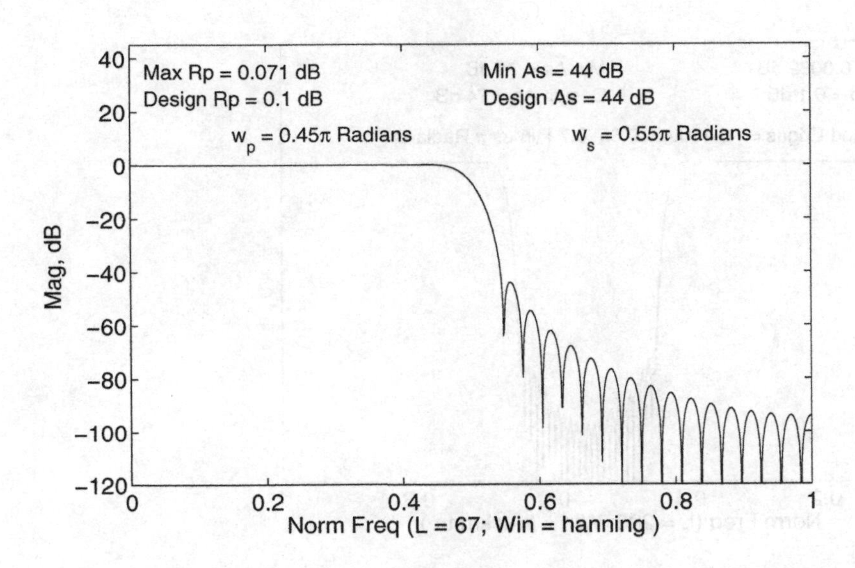

Figure 9.17: Frequency response of lowpass filter designed using a Hanning window with a target A_s of 44 dB and R_p of 0.1dB. The design specifications were met with $L = 67$.

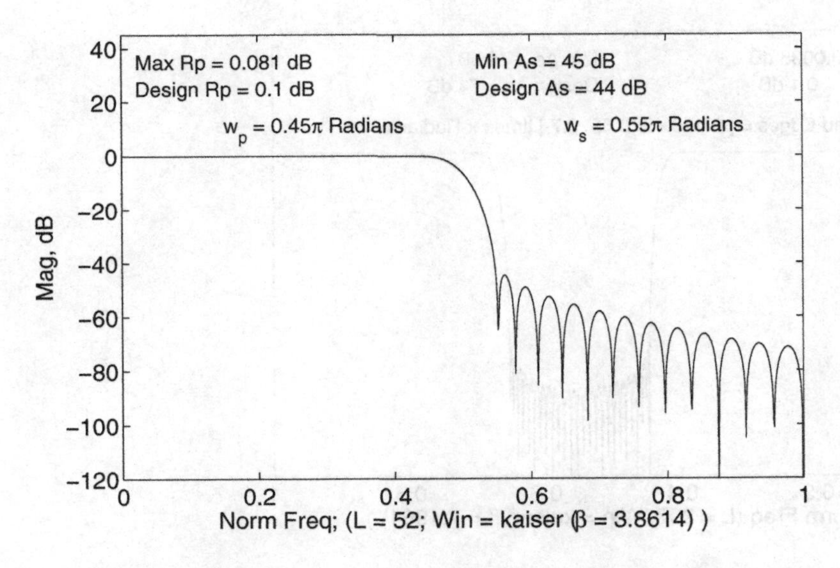

Figure 9.18: Frequency response of a lowpass filter designed using a Kaiser window (β = 3.8614) with a target A_s of 44 dB and R_p of 0.1dB. The design specifications were met with L = 52.

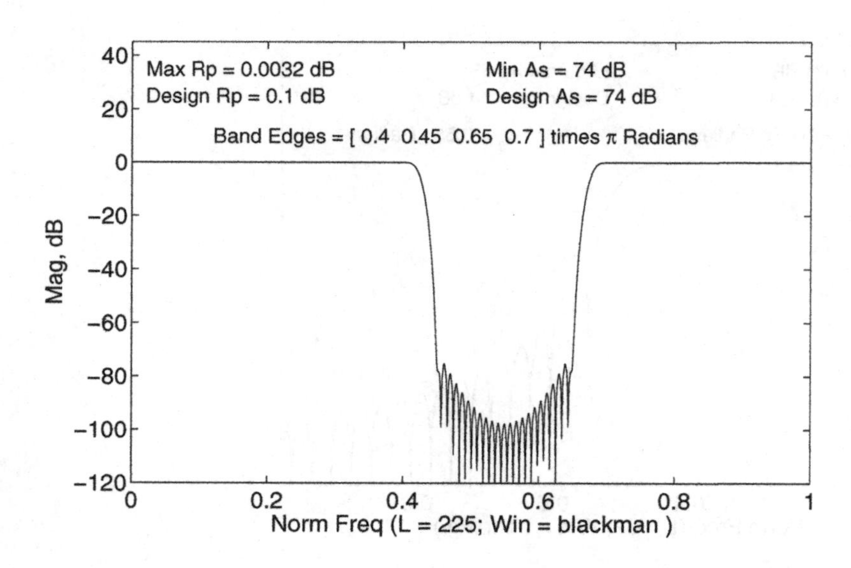

Figure 9.19: Frequency response of a bandstop filter designed using a Blackman window with a target A_s of 74 dB and R_p of 0.1dB. The design specifications were met with $L = 225$.

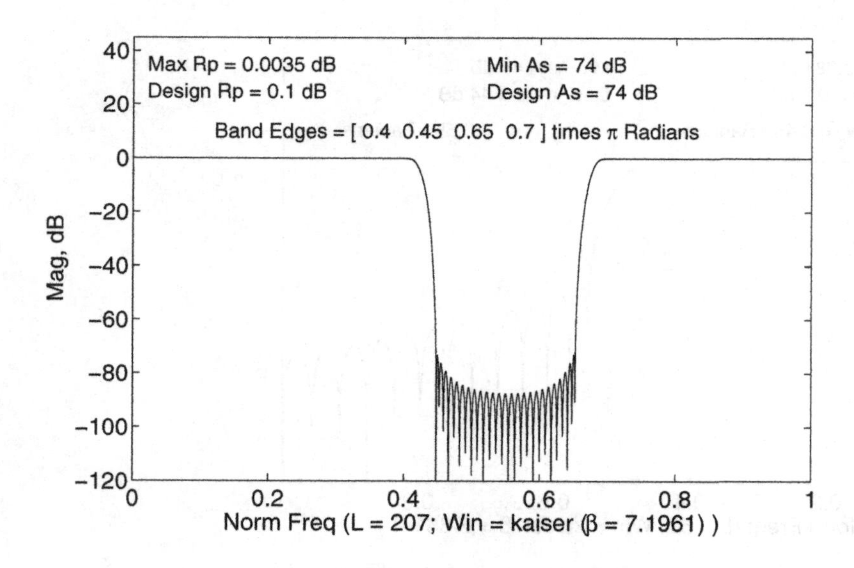

Figure 9.20: Frequency response of a bandstop filter designed using a Kaiser window ($\beta = 7.1961$) with a target A_s of 74 dB and R_p of 0.1dB. The design specifications were met with $L = 207$.

- **Use the desired response values directly in cosine or sine summation formulas to obtain the impulse response. These are formulas that produce the same net result as would be obtained by taking the IDFT. However, no complex arithmetic is involved in using the summation formulas.**

Figure 9.21 shows a filter specification for a length-40 filter. The passband and stopband desired responses are shown as solid lines, and the frequency samples, located at normalized frequencies of $2k/40$ ($k = 0{:}1{:}19$), are marked with circles. The actual frequency response is marked with a dotted line, and the borders of the transition band are marked with vertical dashed lines.

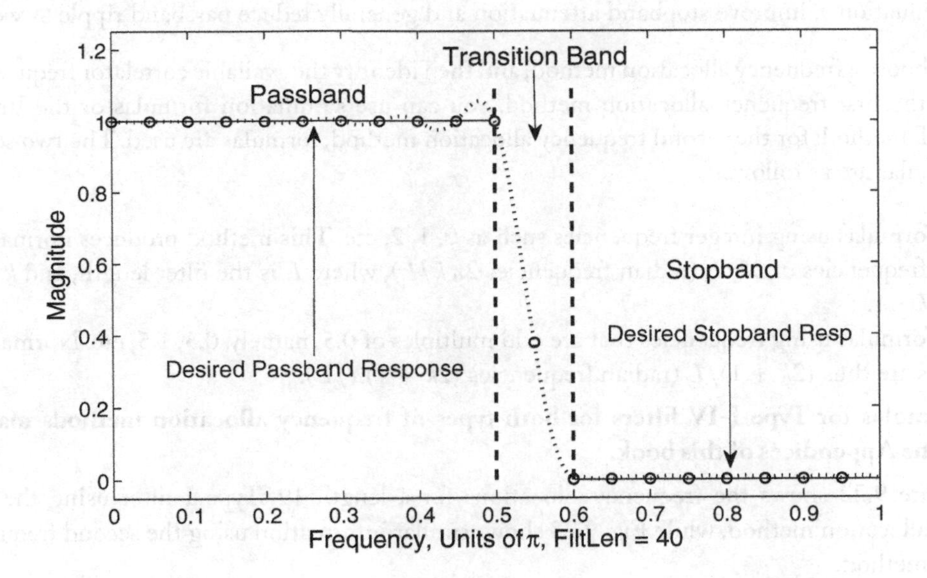

Figure 9.21: A filter frequency specification, with desired passband and stopband responses shown as solid lines. The filter's actual frequency response, shown as a dotted line that passes through all frequency samples (marked with circles), was obtained by performing the DTFT on the filter's impulse response, which was itself obtained by performing the inverse DFT on the frequency samples.

A systematic procedure can be followed to design any standard passband type using the frequency sampling method:

- 1. Define required pass, stop, and transition bands by normalized frequency (1.0 equals Nyquist rate, half the sampling rate).

- 2. Choose a filter type. Available filter types are I, II, III, and IV.

  a. Type I filters can be lowpass, highpass, bandpass, or bandstop.

b. Type II filters can be lowpass or bandpass; highpass and bandstop are prohibited.

c. Type III filters can be bandpass only; lowpass, highpass, and bandstop are prohibited. Hilbert transformers are possible.

d. Type IV filters can be bandpass or highpass; lowpass and bandstop are prohibited. Hilbert transformers and differentiators are possible.

- 3. Estimate the needed filter length L. The longer the filter, the steeper the roll-off. If uncertain, pick a length, design a filter, and, if passband ripple and stopband attenuation are inadequate, increase L. This generally provides additional samples in the transition bands, the values of which, when properly chosen (see discussion below entitled "Improving Stopband Attenuation"), improve stopband attenuation and generally reduce passband ripple as well.

- 4. Choose a frequency allocation method, and then identify the available correlator frequencies. For the first frequency allocation method, you can use summation formulas or the Inverse DFT method; for the second frequency allocation method, formulas are used. The two sets of formulas are as follows:

a) Formulas using integer frequencies such as 0, 1, 2, etc. This method produces normalized correlator frequencies of $2k/L$ (radian frequencies $2\pi k/L$), where L is the filter length, and k runs from 0 to L - 1.

b) Formulas using frequencies that are odd multiples of 0.5, namely, 0.5, 1.5, etc. Normalized frequencies are thus $(2k + 1)/L$ (radian frequencies $(2k + 1)\pi/L$).

Formulas for Type I-IV filters for both types of frequency allocation methods may be found in the Appendices of this book.

Figure 9.22 shows the frequency allocations for a length 19, Type I filter using the first frequency allocation method, while Fig. 9.23 shows similar information using the second frequency allocation method.

- 5. Choose the design formula based on the choices made in Steps 2 and 4 above. In choosing L and frequency allocation method, the goal should generally be to place frequency samples at or very near band edges, with one or more samples in the transition band. It may be necessary to change L, change frequency allocation method, or change band edge specifications to achieve this.

- 6. Assign amplitudes for each correlator frequency, based on the passband, stopband, and transition band assignments made in Step 1 above. Typically, an amplitude of 1.0 is used for passband correlators, 0 for stopband correlators, and if there are transition bands (which there should be in order to achieve good stopband attenuation), values intermediate 0 and 1. There are optimum values for transition band correlators; these have been tabulated in Reference [4] by filter length, frequency allocation method, number of transition samples, and number of consecutive bands having amplitude 1.0 that border on the transition band in question.

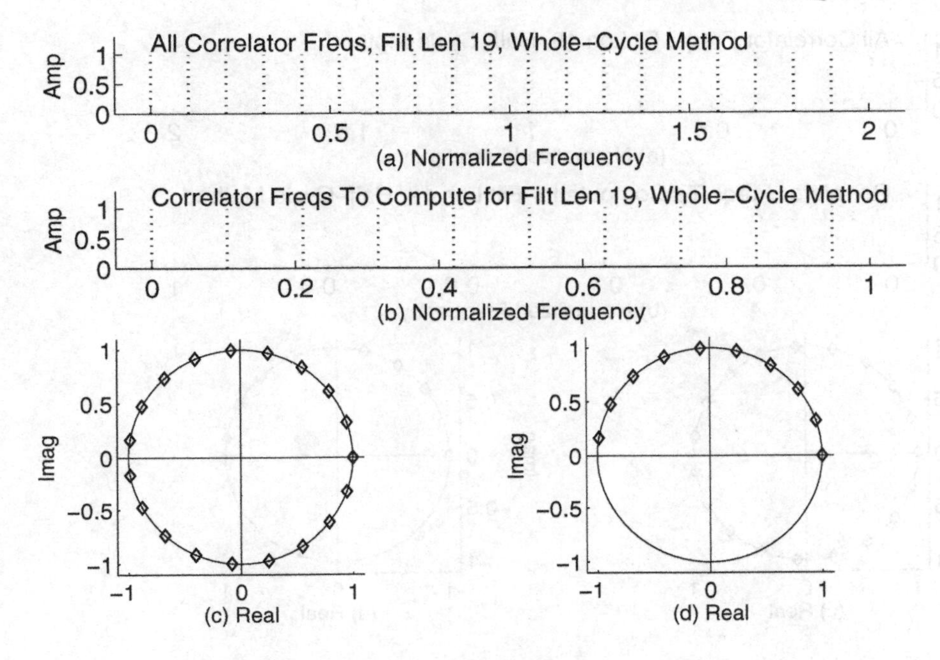

Figure 9.22: (a) All correlator frequencies for Type I filter, length 19, whole-cycle method; (b) The frequencies that must be computed to generate the filter impulse response; (c) Same as (a), but plotted in the complex plane; (d) Same as (b), but plotted in the complex plane.

- 7. Compute the impulse response using the chosen formula and parameters. If the first frequency allocation method is being used, it is straightforward to alternatively use the Inverse DFT method to obtain the impulse response. Often, a window is applied; this tends to reduce ripple and increase stopband attenuation at the expense of roll-off rate. L must be increased accordingly. Note that when a window (other than rectangular) is used, the frequency response will no longer be equal to the specified frequency sample values at the corresponding correlator frequencies.

- 8. Evaluate the frequency response of the resultant filter using the DTFT (usually plotted using a logarithmic scale to show stopband detail). Determine actual values for A_s and R_p.

- 9. Alter filter length L as necessary to attempt to bring the filter's frequency response closer to the design specification.

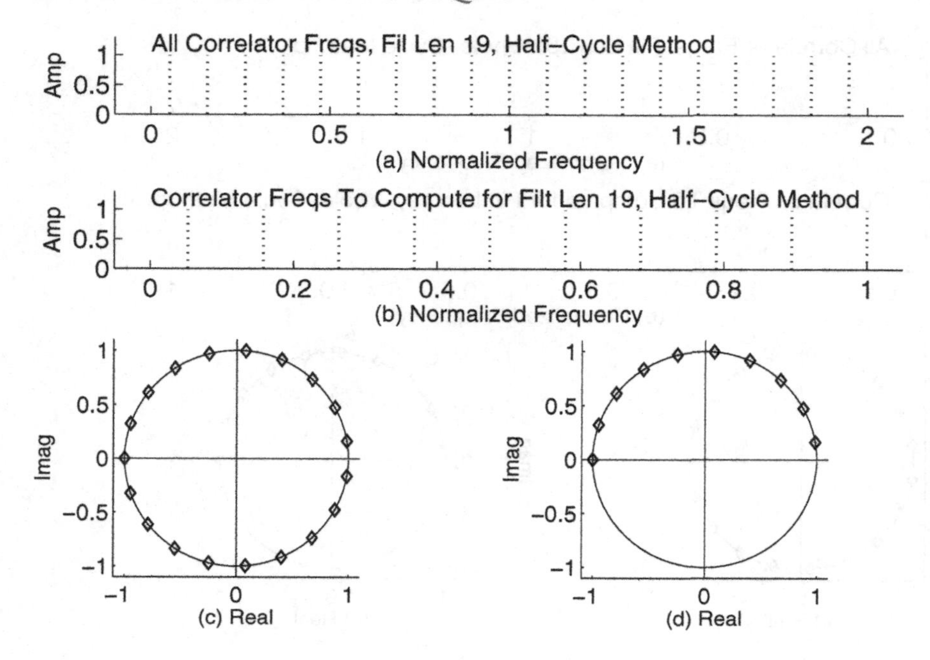

Figure 9.23: (a) All correlator frequencies for Type I filter, length 19, half-cycle method; (b) The frequencies that must be computed to generate the filter impulse response; (c) Same as (a), but plotted in the complex plane; (d) Same as (b), but plotted in the complex plane.

9.6.1 USING THE INVERSE DFT

Setting Bin Amplitudes

We begin by determining the amplitudes for the available correlators. For odd length filters, this amounts to specifying the amplitudes for Bin (or correlator or frequency) 0 and a set of positive bins. Once having this, the negative bin amplitudes may be set to the same value as for corresponding positive bins, i.e., the amplitude for Bin(-1) is the same as for Bin(1).

This produces, for **symmetrically indexed DFTs**, and **odd length** filters, the vector of amplitudes

$$A_k = [\mathrm{Bin}[\text{-}(L\text{-}1)/2],...\mathrm{Bin}[\text{-}2], \mathrm{Bin}[\text{-}1], \mathrm{Bin}[0], \mathrm{Bin}[1], \mathrm{Bin}[2],...\mathrm{Bin}[(L\text{-}1)/2]]$$

For **even length** filters, there is also a Bin($L/2$):

$$A_k = [\mathrm{Bin}[\text{-}L/2 + 1],...\mathrm{Bin}[\text{-}2], \mathrm{Bin}[\text{-}1], \mathrm{Bin}[0], \mathrm{Bin}[1], \mathrm{Bin}[2],...\mathrm{Bin}[L/2]]$$

For **asymmetrically indexed DFTs**, the arrangements are, for **odd length** filters

$$A_k = [\text{Bin}[0], \text{Bin}[1], \text{Bin}[2], ...\text{Bin}[(L-1)/2], \text{Bin}[\text{-}(L\text{-}1)/2], ..\text{Bin}[\text{-}2], \text{Bin}[\text{-}1]]$$

which can be reindexed as

$$A_k = [\text{Bin}[0], \text{Bin}[1], \text{Bin}[2], ...\text{Bin}[(L\text{-}1)/2], \text{Bin}[L\text{-}(L\text{-}1)/2], ..\text{Bin}[L\text{-}2], \text{Bin}[L\text{-}1]]$$

and for **even length** filters

$$A_k = [\text{Bin}[0], \text{Bin}[1], \text{Bin}[2], ..\text{Bin}[L/2], \text{Bin}[\text{-}L/2 + 1], ..\text{Bin}[\text{-}2], \text{Bin}[\text{-}1]]$$

which can be reindexed as

$$A_k = [\text{Bin}[0], \text{Bin}[1], \text{Bin}[2], ..\text{Bin}[L/2], \text{Bin}[L\text{-}L/2 + 1], ..\text{Bin}[L\text{-}2], \text{Bin}[L\text{-}1]]$$

Example 9.13. Write the vector of Bin indices for a length-8 (Type II) filter using symmetric and asymmetric indices. Do the same for a length-9 (Type I) filter. For the asymmetric indices, also write out the vector of actual frequencies.

For symmetric indices and length-8, we get $-L/2 + 1{:}1{:}L/2$ = -3:1:4. For asymmetric indices, we get $0{:}1{:}(L\text{-}1)$ = 0:1:7, which in terms of frequencies is [0,1,2,3,4,-3,-2,-1].

For the length-9 filter, symmetric indices, we get -4:1:4, and for asymmetric indices we get 0:1:8, which in terms of frequencies is [0,1,2,3,4,-4,-3,-2,-1].

Example 9.14. Specify the Bin amplitudes for length-8 and length-9 filters that have only Bin[0] and Bin[1] amplitudes set to 1.0 with all other bins set to amplitude 0.

For a length-8, symmetrical DFT, we have indices -3:1:4, and set Bins -1, 0, and 1 to amplitude 1, which yields the amplitude vector [0,0,1,1,1,0,0,0]. Using asymmetrical indices, we get [1,1,0,0,0,0,0,1]. For a length-9 filter, symmetrical DFT, we have indices -4:1:4, and therefore we get [0,0,0,1,1,1,0,0,0]. Using asymmetrical indices, this becomes [1,1,0,0,0,0,0,0,1].

Setting Bin Phase

This vector of amplitudes must then be multiplied by a phase factor that produces the correct linear phase angles for the various bins. We note that, for **asymmetrically indexed DFTs**, having bins from 0 to $L-1$, the phase angles in radians are given as, for Type I and II filters,

$$\angle H[k] = \begin{cases} -2\pi kM/L & k = 0, 1...M \\ 2\pi(L-k)M/L & k = M+1, M+2, ...L-1 \end{cases}$$

where $\angle H[k]$ is the phase at Bin k and $M = (L-1)/2$. For Type III and IV filters, the phase is

$$\angle H[k] = \begin{cases} (\pm\pi/2) - 2\pi k M/L & k = 0, 1...M \\ (\pm\pi/2) + 2\pi(L-k)M/L & k = M+1, M+2, ...L-1 \end{cases}$$

For **symmetrically indexed DFTs**, having k from $-M$ to M for odd filters and $-L/2+1$ to $L/2$ for even filters, a single expression for Types I and II is

$$\angle H[k] = -2\pi k((L-1)/2)/L) = -\pi k((L-1)/L)$$

and for Types III and IV

$$\angle H[k] = \pm\pi/2 - 2\pi k((L-1)/2)/L) = \pm\pi/2 - \pi k((L-1)/L)$$

By using $\angle H[k]$ as the argument for the complex exponential, the actual complex-numbered phase angles $H_{ph}[k]$ for the DFT (Types I and II) can be generated:

$$H_{ph}[k] = e^{-j2\pi k((L-1)/2)/L)} = e^{-j\pi k((L-1)/L)} \tag{9.6}$$

Example 9.15. Design a length-9 (Type I) filter with cutoff approximately equal to 0.5π using the Inverse-DFT method. Compute the impulse response and the frequency response from 0 to π radians. On the frequency response plot, also plot the frequency samples.

We note that the Nyquist limit is 9/2 = 4.5. By setting correlators 0, 1, and 2 (located at normalized frequencies of 0, 0.222, and 0.444) to amplitude 1, and correlators 3 and 4 (located at normalized frequencies of 0.666 and 0.888) to 0, the cutoff will be between band (or correlators) 2 and 3, i.e., at about sample 2.5, which is 2.5/4.5 = 0.55. We therefore have the vector of sample amplitudes as

$$A_k = [0,0,1,1,1,1,1,0,0]$$

having bin indices [-4:1:4]. We obtain the net DFT by multiplying by

$$H = e^{-j\pi[-4:1:4]((9-1)/9)}$$

Thus far we have formatted the DFT using symmetric indices. It is necessary to adjust bin location when using DFT/IDFT routines that expect the asymmetric bin arrangement, in which k runs from 0 to L-1 rather than from $-(L-1)/2$ to $(L-1)/2$ for odd length filters, or from $-L/2$ +1 to $L/2$ for even length filters, it is necessary to shift the negative bins to the right side of the DFT; this is done in the following script:

```
function LVLPFViaSymm2AsymmIDFT(Ak)
% LVLPFViaSymm2AsymmIDFT([0,0,1,1,1,1,1,0,0]); % odd
```

```
% LVLPFViaSymm2AsymmIDFT([0,0,1,1,1,1,1,0,0,0]) % even
L = length(Ak); if ~(rem(L,2)==0) % odd length filter
M = (L-1)/2; symmDFT = Ak.*exp(-j*pi*[-M:1:M]*(2*M)/L);
LenNegBins = M; NegBins = symmDFT(1,1:LenNegBins);
ZeroPosBins = symmDFT(1,LenNegBins+1:length(symmDFT));
NetDFT = [ZeroPosBins NegBins]; Imp = real(ifft(NetDFT));
k = 0:1:(L-1)/2; else % even length filter
symmDFT = Ak.*exp(-j*pi*[-L/2+1:1:L/2]*(L-1)/L);
LenNegBins = L/2-1; NegBins = symmDFT(1,1:LenNegBins);
ZeroPosBins = symmDFT(1,LenNegBins+1:length(symmDFT));
NetDFT = [ZeroPosBins NegBins]; Imp = real(ifft(NetDFT));
k = 0:1:L/2; end
figure(3); clf; subplot(211); stem([0:1:length(Imp)-1],Imp);
xlabel('Sample'); ylabel('Amplitude')
fr = abs(fft(Imp,1024));subplot(212);
plot([0:1:512]/512,fr(1,1:513));
xlabel('Frequency, Units of \pi')
ylabel('Magnitude'); hold on; for ctr = k;
plot([(2*ctr/L),(2*ctr/L)],[0 1],'b:');
plot([2*ctr/L],abs(ZeroPosBins(1,ctr+1)),'ko');
axis([0 1 -inf inf]); end
```

The result from making the call

LVLPFViaSymm2AsymmIDFT([0,0,1,1,1,1,1,0,0])

is shown in Fig. 9.24

Example 9.16. Design a length-8 (Type I) filter with cutoff approximately equal to 0.5π using the Inverse-DFT method. Compute the impulse response and the frequency response from 0 to π radians. On the frequency response plot, also plot the frequency samples.

We note that the Nyquist limit is 8/2 = 4. By setting correlators 0, 1, and 2 to amplitude 1, and correlator 3 to 0, the cutoff will be between band (or correlators) 2 and 3, i.e., with the cutoff frequency at about 2.5/4 = 0.625. The DFT amplitudes will be

$$A_k = [0,1,1,1,1,1,0,0]$$

having bin indices [-3:1:4]. We obtain the net DFT by multiplying by

$$H_{ph} = e^{-j\pi[-3:1:4](7/8)}$$

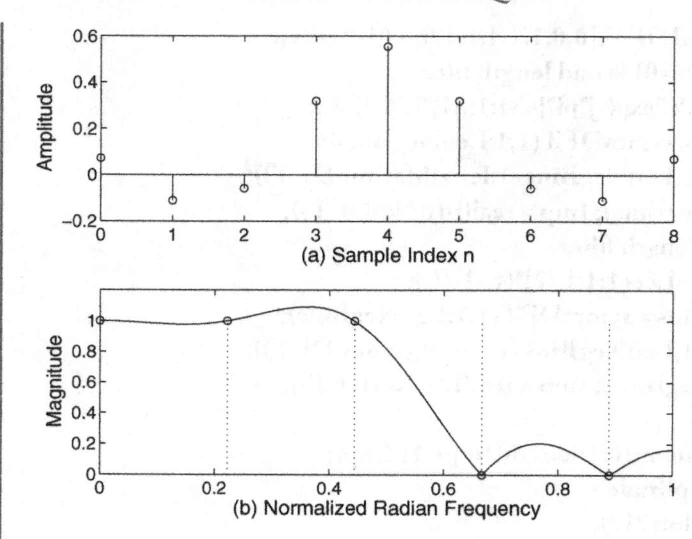

Figure 9.24: (a) Impulse response; (b) Magnitude of DTFT of impulse response in (a), with the frequency samples from the design specification marked with circles.

The computation can be done by the previously presented script, with the result from making the call

LVLPFViaSymm2AsymmIDFT([0,1,1,1,1,1,0,0])

shown in Fig. 9.25. Note that this filter is even in length, and must therefore have a frequency response of zero at the Nyquist limit. To verify this, note that the rightmost value in A_k in the code below is the $L/2$ bin; try changing its amplitude to a nonzero number. You will find that the response at π radians (normalized frequency 1.0) is still zero.

9.6.2 USING COSINE/SINE SUMMATION FORMULAS

Prior to beginning a detailed discussion, the reader should note that Cosine and Sine summation formulas for Types I, II, III, and IV filters can be found in the Appendices to this book.

To start, we will design a Type-II lowpass filter in the following example:

Example 9.17. Design a length-20 (Type II) filter meeting the band requirements shown in Fig. 9.26, where $\omega_p = 0.4\pi$, $\omega_s = 0.5\pi$, $A_s = 40$ dB, and $R_p = 0.5$ dB. The available correlator frequencies are marked for a length 20 (Type II) filter, first frequency allocation method.

The design formula for a Type I or II filter (first frequency allocation method, frequency samples at $2\pi k/L$) is

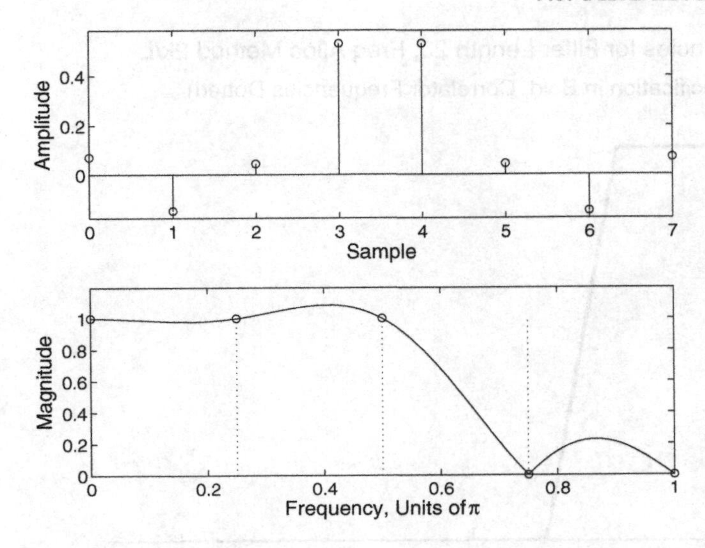

Figure 9.25: (a) Impulse response; (b) Magnitude of DTFT of impulse response in (a), with the frequency samples from the design specification marked with circles.

$$h[n] = \frac{1}{L}\left[A_0 + \sum_{k=1}^{K} 2A_k \cos(2\pi(n-M)k/L) \right]$$

where $K = (L-1)/2$ for Type I filters and $K = L/2 - 1$ for Type II filters.

For each of the available correlator frequencies ($2\pi k/20$), we'll specify the desired amplitude of response of the filter. We thus would have for the design amplitude vector

$$A_k = [1, 1, 1, 1, 1, 0, 0, 0, 0, 0]$$

which has k indices 0:1:9 (these would be the same as Bins[0:1:9] if we were using the IDFT method described above).

For both of the following guidelines, *PosBins* does not include Bin 0.

- When the filter is to be odd in length, the filter will have L = 2*length(PosBins) +1. When the filter is to be even in length, then L = 2*length(PosBins) + 2.
- When starting with a given value of L, for even length filters, length(PosBins) = $(L-2)/2$, and for odd length filters, length(PosBins) = $(L-1)/2$.

The following script (see exercises below)

WF = LVxFilterViaCosineFormula(Type,Bin0,PosBins)

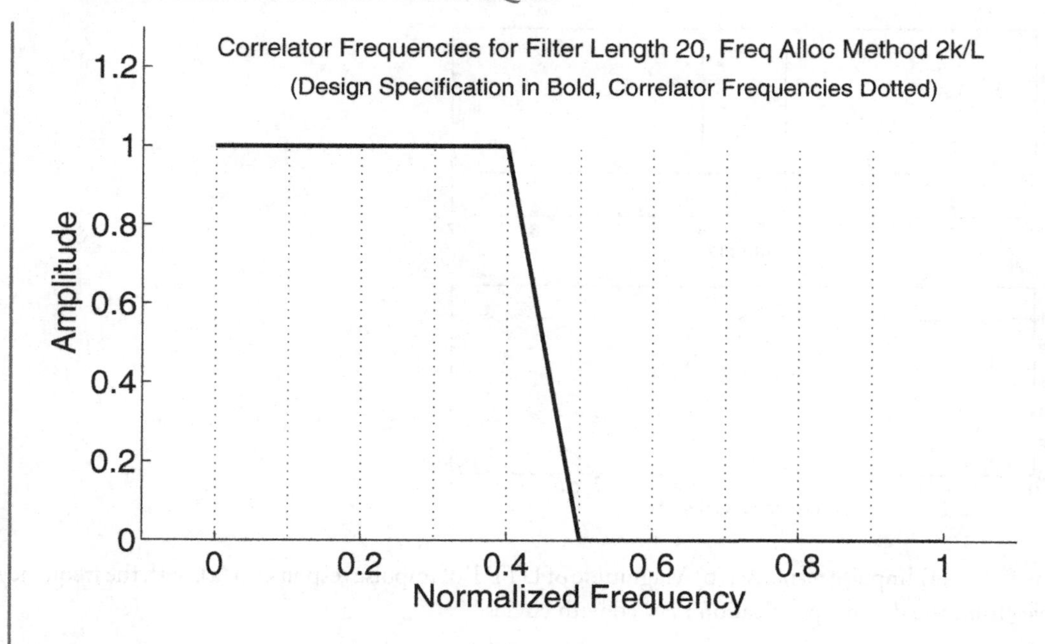

Figure 9.26: Design specification for a lowpass filter.

will compute the impulse and frequency responses for a Type I or II filter, and plot the magnitude of frequency response.

Figure 9.27 shows the result from making the call

WF = LVxFilterViaCosineFormula(2,[1],[1,1,1,1,0,0,0,0,0])

Note that the results from this simple approach are poor in terms of stopband attenuation and passband ripple.

Example 9.18. Use the script *LVxFilterViaCosineFormula* to design a length-20 bandpass filter having its passband centered around 0.5π radians, and having a passband width of about 0.2π radians.

The length of *PosBins* will be (20-2)/2 = 9. *Bin0* will have an amplitude of 0. The Nyquist limit is 20/2 = 10, and therefore samples 4-6 (*Bin0* being indexed as sample 0) represent normalized frequencies of 0.4, 0.5, and 0.6. The script call will be

WF = LVxFilterViaCosineFormula(2,[0],[0,0,0,1,1,1,0,0,0])

The result of the call is shown in Fig. 9.28.

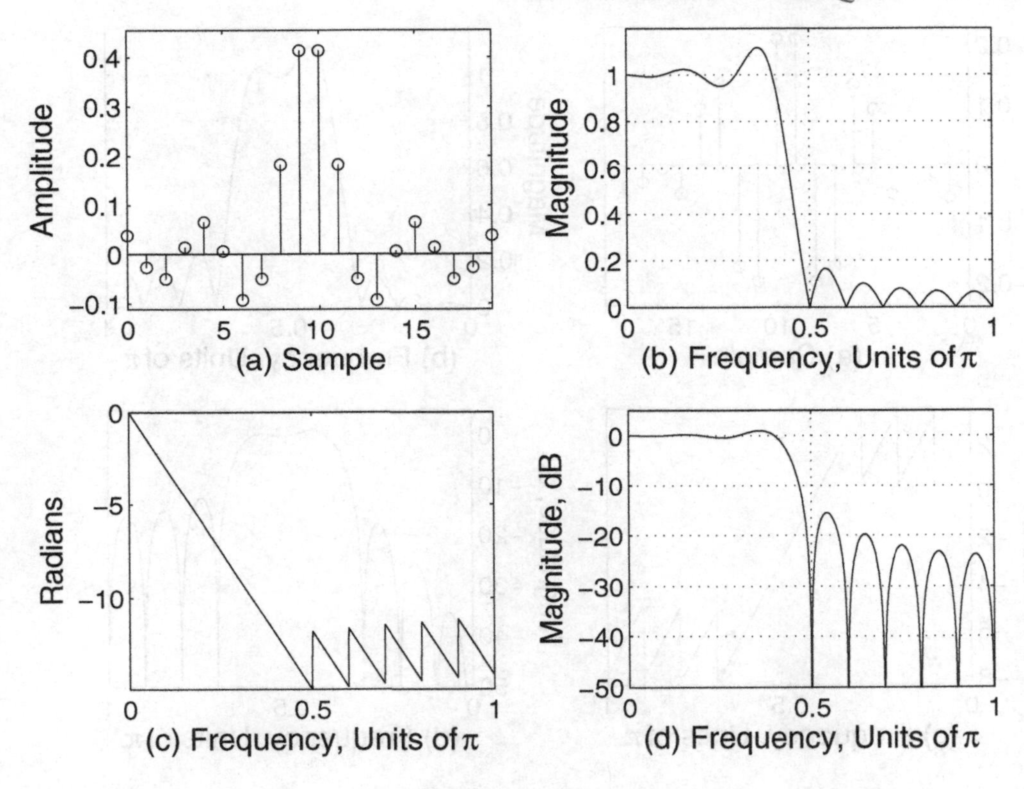

Figure 9.27: (a) Impulse response of a length-20 lowpass filter designed using the cosine summation formula; (b) Magnitude of frequency response of filter at (a); (c) Phase response of same; (d) Magnitude of frequency response in dB.

9.6.3 IMPROVING STOPBAND ATTENUATION

In order to improve stopband attenuation, it is necessary to specify a transition band having within it one or more frequency samples the amplitudes of which may be set to any necessary value to optimize stopband attenuation. This principle is illustrated in the following example.

Example 9.19. Design a filter having $\omega_p = 0.5\pi$, $\omega_p = 0.6\pi$ and having one frequency sample in the transition band. Programmatically vary the amplitude of the transition band frequency sample and observe the change in stopband attenuation.

Frequency samples at normalized frequencies of 0.5, 0.55, and 0.6 will define the limits of passband and stopband, with a single sample in the transition band as desired. The normalized sample frequencies are, for an odd length filter

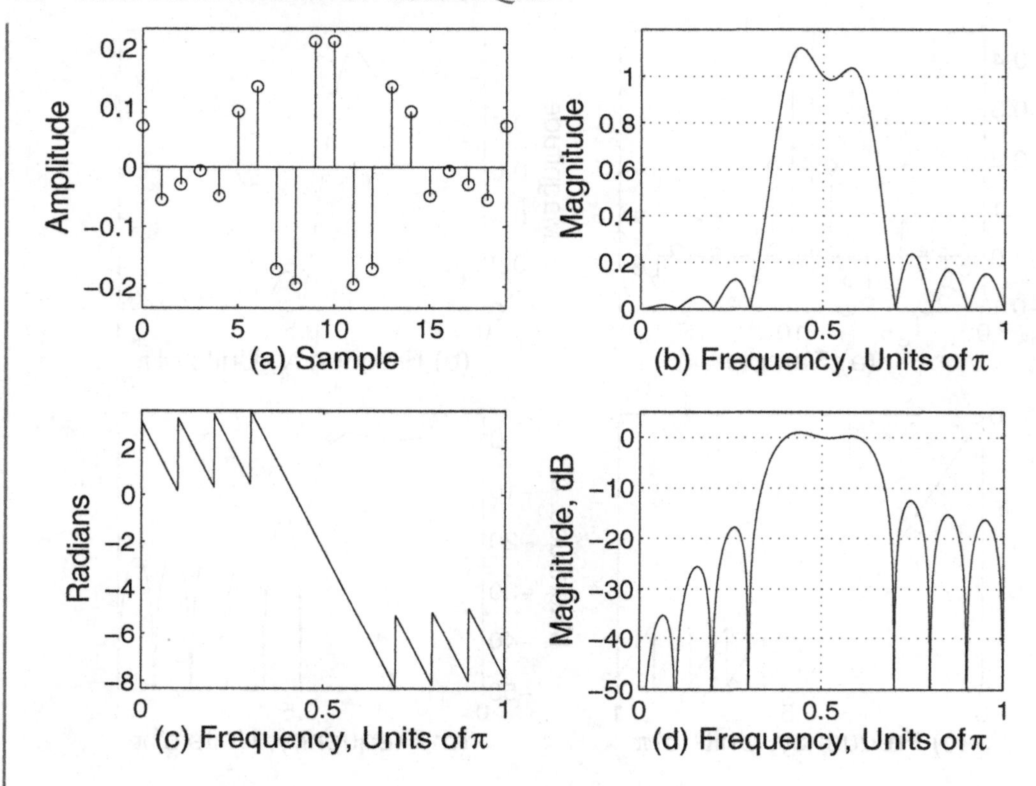

Figure 9.28: (a) The impulse response of a length-20 bandpass filter designed using the cosine summation formula; (b) Magnitude of frequency response of filter at (a); (c) Phase Response of filter at (a); (d) Magnitude of frequency response in dB.

2([0:1:M])/L

where M = (L-1)/2, or

2([0:1:(L/2-1])/L

for an even length filter. The sample spacing in radian frequency is $2\pi/L$, or just $2/L$ for normalized frequency.

Since the samples must be equally spaced at normalized frequency interval $2/L$, we get a filter length of L = 2/0.05 = 40. The frequency specification must cover Bins 0 to $L/2$-1 = 19 (for an even length filter, Bin $L/2$ is always 0 and is thus not specified), and Bins -1 to -19 having the same amplitude as Bins 1 through 19. The passband for positive Bins runs from Bin 0 to Bin 10. This is followed by the variable T, representing the amplitude of the transition band sample, and a number of zeros to fill out the stopband to sample $L/2-1$. The code below uses the cosine summation formula to obtain the impulse response of the filter for each *Ak*. For an initial test run, the limits of T (the transition band frequency sample) should be set to give a first estimate of the best value of T by stepping through all the possible values between 1.0 and 0 with a decrement of 0.01. You should find that 0.39 or thereabouts appears to be the best value. Note also that the best value of stopband attenuation appears to occur when the amplitude of stopband ripple is equalized as much as possible (later in the chapter we'll discuss equiripple FIR design). You can then rerun the code with a much reduced range of possible values for T and a smaller decrement to obtain a more accurate estimate. For the second run, try **Thigh = 0.4, Tlow = 0.38, and Dec = 0.0005,** for example.

```
function LVOptCoeffLPF(L,wp,ws,THi,TLo,Dec)
% LVOptCoeffLPF(40,0.5,0.6,0.5,0.35,0.01)
% LVOptCoeffLPF(40,0.5,0.6,0.4,0.38,0.0005)
Dec = -abs(Dec); noComp = ceil((THi-TLo)/abs(Dec)); ctr = 0;
TSB = zeros(noComp,3); LenFFT = 4096; if rem(L,2)==0;
limK = L/2-1; else; limK = (L-1)/2; end;
n = 0:1:L-1; M=(L-1)/2;
for T = THi:Dec:TLo; Ak = [ones(1,11), T, zeros(1,8)];
LA=length(Ak);
WF=(cos(((n-M)')*[0:1:LA-1]*2*pi/L))*([Ak(1),2*Ak(2:LA)]');
WF = WF'; Imp = WF/L; figure(3);
subplot(211); stem([0:1:length(Imp)-1],Imp);
xlabel('n'); ylabel('Amplitude');
fr = abs(fft(Imp,LenFFT));
fr = fr(1,1:LenFFT/2+1); fr = fr/(max(abs(fr)));
Lfr = length(fr); SB = fr(1,round(ws*(Lfr-1)):(Lfr-1));
PB = fr(1,1:round((wp/pi)*(Lfr-1))); subplot(212);
SBAt = -20*log10(max(SB) +eps);
PBR = -20*log10(min(PB)); ctr = ctr + 1;
TSB(ctr,1)=T; TSB(ctr,2)=SBAt; TSB(ctr,3)=PBR;
plot([0:1:Lfr-1]/(Lfr-1),20*log10(fr+eps));
strSBA = num2str(SBAt);
xlabel(['Norm.Freq.(T = ',num2str(T),' and As = ',strSBA,')'])
ylabel('Magnitude'); axis([0 1 -70 10]); pause(0.01);
end; sc = TSB(:,2); bestSB = max(sc),
bestSBind = find(sc==bestSB); bestT = TSB(bestSBind,1),
finalPBRipple = TSB(bestSBind,3)
```

For the example above, we would make the call

LVxFilterViaCosineFormula(2,1,...
[ones(1,10), 0.387, zeros(1,8)])

which results in Fig. 9.29.

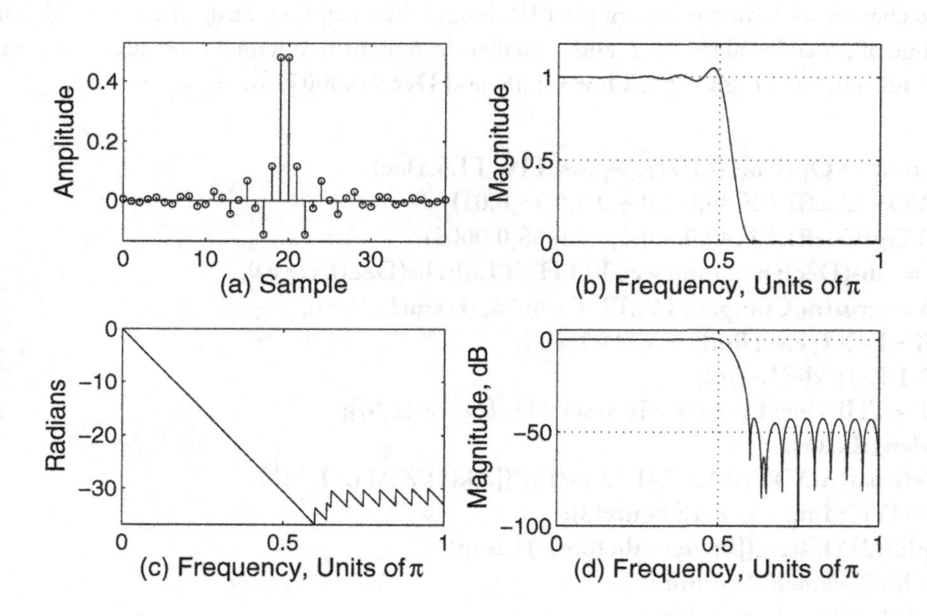

Figure 9.29: (a) Impulse response of a length-40 lowpass filter having a single optimized transition band sample; (b) Magnitude of frequency response of same; (c) Phase response of same; (d) Magnitude of frequency response, in dB.

To improve stopband attenuation further, the transition band can include additional samples, each of which will have an optimum value. Optimum values have been computed for a number of different filter types and lengths, and are found in Reference [4]. While precise determination of optimum values can require optimization algorithms, it is possible to obtain reasonable estimates using simple empirical search techniques.

Example 9.20. Redesign the filter for the previous example, which was a lowpass filter having ω_p = 0.5, ω_s = 0.6. For the redesign, use two optimized transition band samples, and then perform a second redesign using three optimized transition band samples.

The new filter length for two transition band samples can be determined by noting that the samples in or bordering on the transition band should be located at normalized frequencies of 0.5,

0.533, 0.566, and 0.6 which suggests the minimum spacing as $2\pi/L = 0.0333\pi$, which implies that $L = 2/0.0333 = 60$, with

$$A_k = [\textbf{ones(1,16)},T_1,T_2,\ \textbf{zeros(1,12)}]$$

The values $T[1] = 0.592$ and $T[2] = 0.109$, which were found by a simple search technique, provide acceptable optimization of the two transition band sample values; we thus make the call

LVxFilterViaCosineFormula(2,1,...
[ones(1,15),0.592,0.109,zeros(1,12)])

which results in Fig. 9.30.

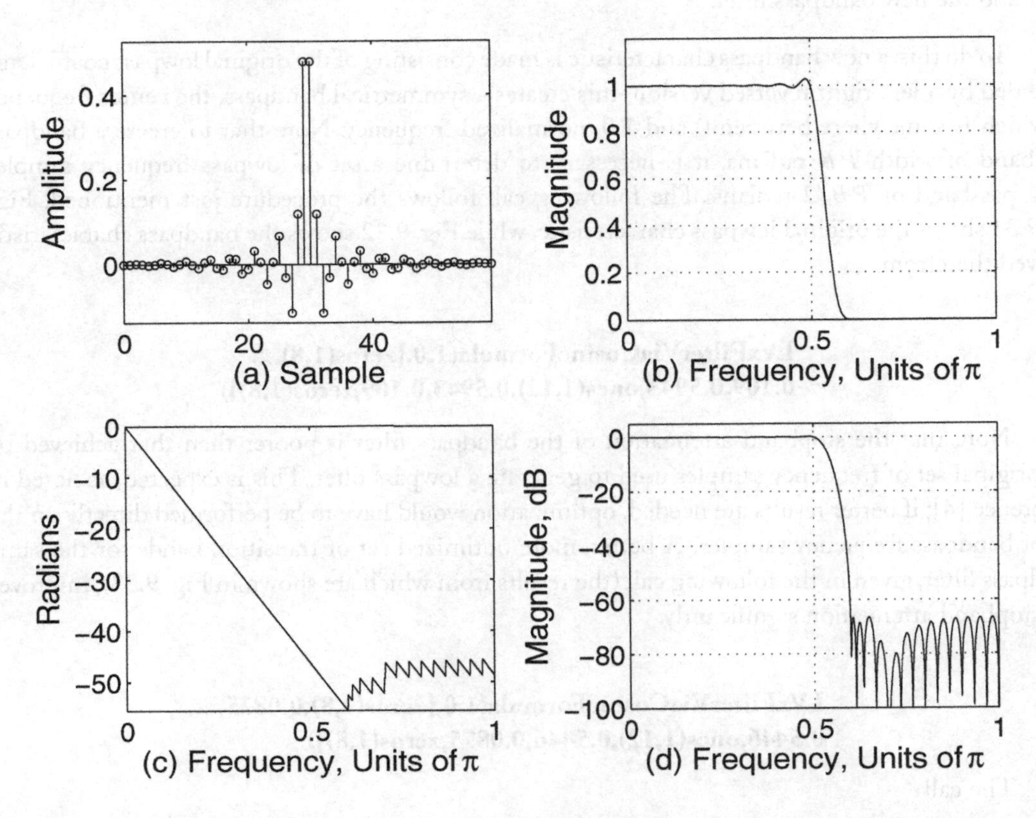

Figure 9.30: Frequency response of a length-60 LPF with $\omega_p = 0.5\pi$ and $\omega_s = 0.6\pi$, using two transition coefficients with approximately optimum values to achieve the maximum stopband attenuation.

9.6.4 FILTERS OTHER THAN LOWPASS

Recall that in designing filters using the window technique, we started with an Ideal LPF, truncated it to a desired length, and picked a window to achieve a certain stopband attenuation. Filters other than lowpass were synthesized by manipulating one or more lowpass types to generate a different type such as highpass, bandpass, and bandstop. Designing high-quality transition-band-optimized filters via Frequency Sampling, for filter types other than lowpass, in general, requires an optimization program or algorithm. Reference [4] does, however, describe creating a bandpass filter by rotating the A_k samples, optimized for a lowpass response, to a new center frequency, and duplicating the lowpass passband and transition band samples symmetrically on both sides of the new frequency, leaving stopbands of appropriate length on each side of the new pass- and transition bands.

Example 9.21. Convert the lowpass frequency sample vector Ak = $[ones(1, 6), 0.5943, 0.109, zeros(1, 25)]$ to one suitable for a bandpass filter. Display a magnitude plot for the original lowpass filter and the new bandpass filter.

To do this, a new bandpass characteristic is made consisting of the original lowpass coefficients preceded by a left-right reversed version–this creates a symmetrical bandpass, the center frequency of which is somewhere between 0 and 1.0, normalized frequency. Note that to create a bandpass passband of width PB radians, it is necessary to determine a set of lowpass frequency samples for a passband of $PB/2$ radians. The following call follows the procedure just mentioned. Figure 9.31 shows the original lowpass characteristic, while Fig. 9.32 shows the bandpass characteristic derived therefrom.

**LVxFilterViaCosineFormula(1,0,[zeros(1,8),...
0.109,0.5943,ones(1,12),0.5943,0.109,zeros(1,8)])**

Note that the stopband attenuation of the bandpass filter is poorer than that achieved by the original set of frequency samples used to generate a lowpass filter. This is expected, as noted in Reference [4]; if better results are needed, optimization would have to be performed directly on the set of bandpass frequency samples. A better, more optimized set of transition bands for the same bandpass filter, given in the following call (the results from which are shown in Fig. 9.33), improves the stopband attenuation significantly.

**LVxFilterViaCosineFormula(1,0,[zeros(1,8),0.0875,...
0.5446,ones(1,12),0.5446,0.0875,zeros(1,8)])**

The call

**LVxFilterViaCosineFormula(1,0,[zeros(1,15),...
0.0165,0.2042,0.6765,ones(1,15)])**

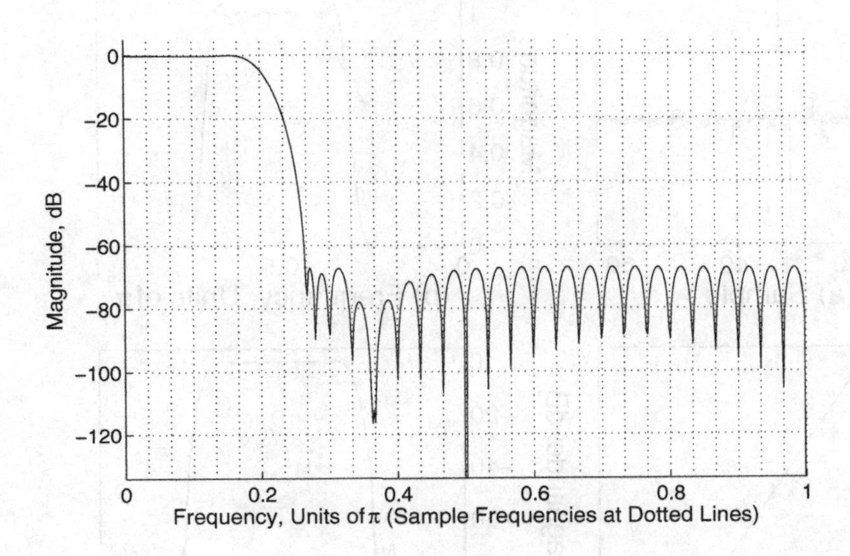

Figure 9.31: Frequency response of a lowpass filter created from the frequency sample vector [ones(1,6), 0.5943, 0.109, zeros(1,25)]. Note minimum A_s of about 67 dB.

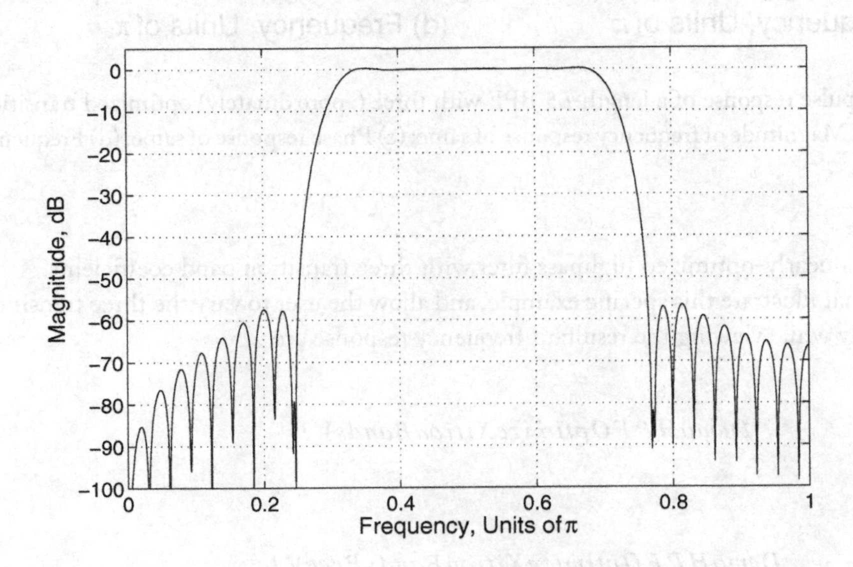

Figure 9.32: Frequency response of a bandpass filter created from the lowpass frequency sample vector [ones(1,6), 0.5943, 0.109, zeros(1,25)]. Note minimum A_s of about 57 dB.

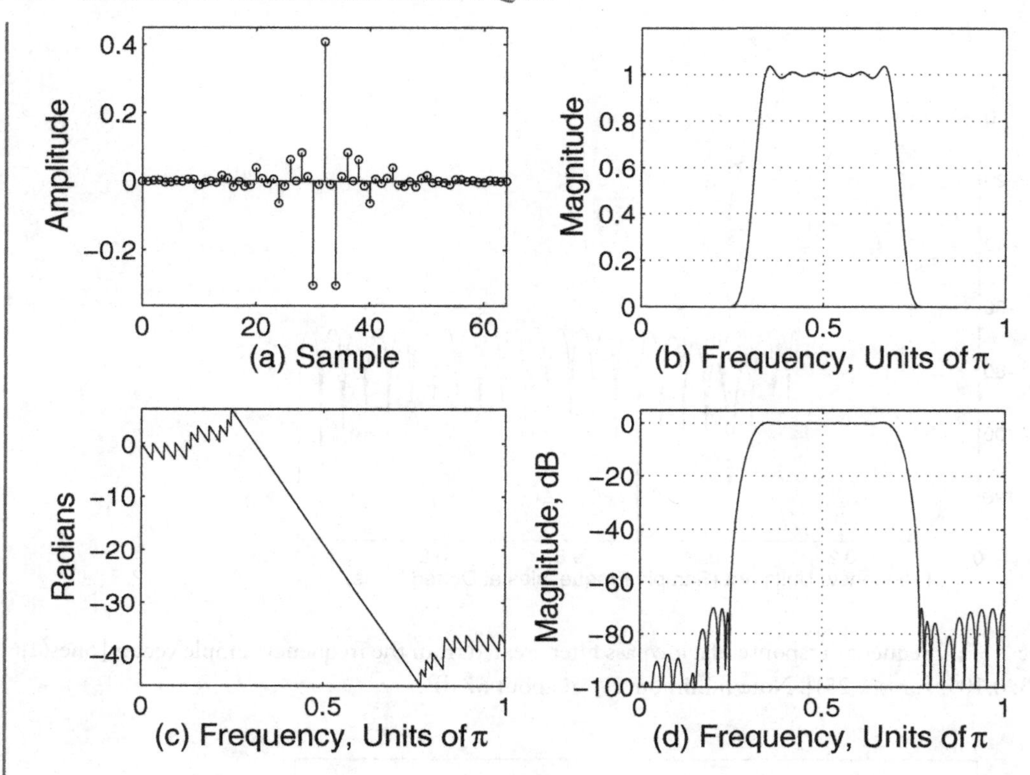

Figure 9.33: (a) Impulse response of a length-65 BPF with three (approximately) optimized transition band coefficients; (b) Magnitude of frequency response of same; (c) Phase response of same; (d) Frequency response in dB.

produces Fig. 9.34, a nearly-optimized highpass filter with three transition band coefficients.

Several VIs that illustrate this specific example, and allow the user to vary the three transition coefficients manually while viewing the resultant frequency response are

$$DemoHPFOptimizeXitionBandsVI$$

$$DemoHPFOptimizeXitionBandsPrecVI$$

The former VI allows coefficient entry using virtual slider controls, while the latter allows precise numerical entry. Figure 9.35 shows an example of the latter VI.

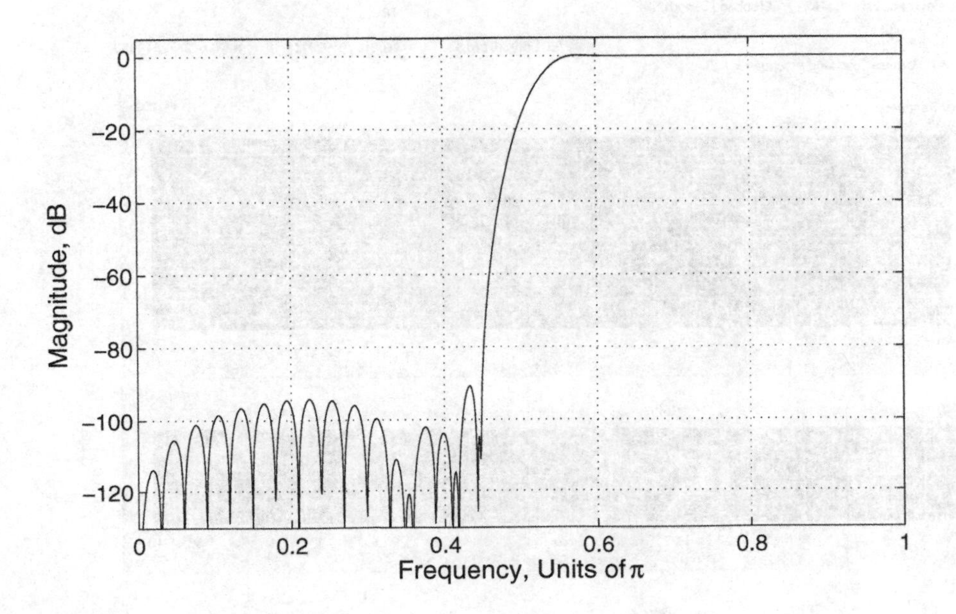

Figure 9.34: A highpass filter having three transition samples with approximate values of T = [0.0165, 0.2042, 0.6765] and stopband attenuation of more than 90 dB.

9.6.5 HILBERT TRANSFORMERS

The Hilbert transformer is a special type of filter the purpose of which is to shift the phase of all the frequencies (except DC and the Nyquist rate) in a signal by 90 degrees ($\pi/2$ radians), leaving the amplitudes untouched. Thus the ideal Hilbert Transformer has a frequency response

$$H(\omega) = \begin{cases} -j & 0 < \omega < \pi \\ +j & -\pi < \omega < 0 \end{cases} \tag{9.7}$$

The Hilbert transform is useful in communications, for example, to create single sideband signals and the like.

Two approaches to Hilbert-transforming a test waveform are:

1. Frequency Domain: Compute the DFT of the test sequence, multiply it by a frequency domain representation or mask of a Hilbert Transformer, then compute the Inverse DFT to obtain the Hilbert Transform.

2. Time Domain: Design a Hilbert Transform impulse response, and convolve the time domain test signal with it, as with any time domain filter.

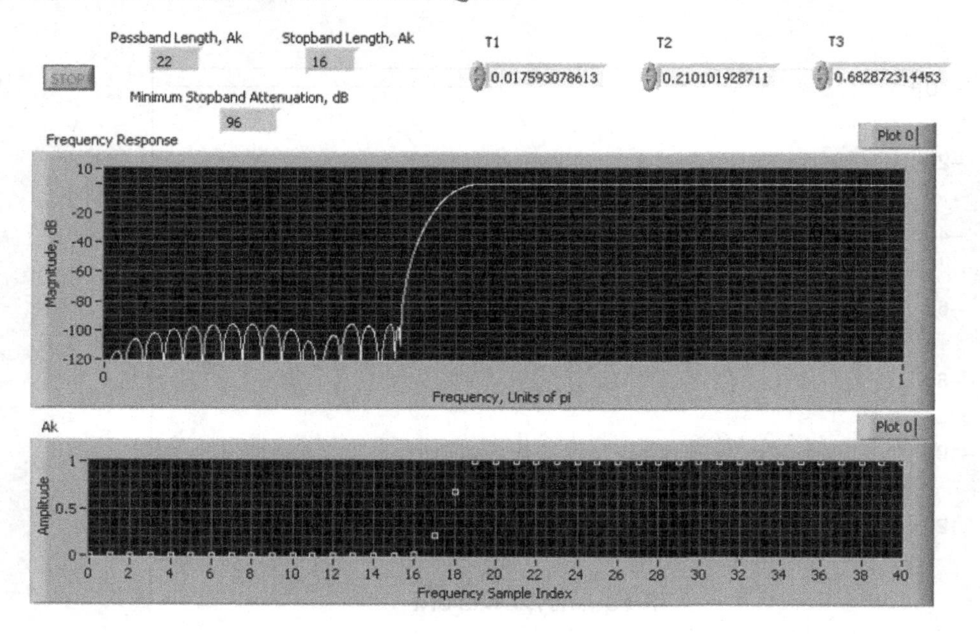

Figure 9.35: A VI allowing precise entry of transition band coefficients T1-T3 for a length-81 FIR HPF having Ak = [zeros(1,16),T1,T2,T3,ones(1,22)], showing good estimates for the three coefficients, resulting in about 96 dB maximum stopband attenuation.

In Frequency Domain

Several methods exist to generate a frequency domain representation of a Hilbert Transformer which will yield the Hilbert Transform of a signal when the DFT of the signal is multiplied by the frequency domain Hilbert mask and the product used to compute the inverse DFT.

- A first method is to generate a frequency domain mask consisting of the value 0 for bins 0 and $N/2$, -j for the remaining positive bins, and +j for negative bins. This method is referred to hereafter as the *All-Imaginary Hilbert Mask* (or Method). The mask is multiplied by the signal's DFT, the inverse DFT is taken, and the real part of the result is the Hilbert Transform.

- A second method is to generate a mask consisting of the value 1 for bins 0 and N/2, 2 for the remaining positive bins, and 0 for all negative bins. This method is referred to hereafter as the *All-Real Hilbert Mask*. The mask is multiplied by the signal's DFT, and the imaginary part of the inverse DFT of the product is the Hilbert Transform, while the real part is the original signal.

  The script

$$LVHilbertPhaseShift(DestWF, UserWF, SR)$$

allows you to experiment with the Hilbert transform as computed using frequency domain techniques. Both the All-Real Hilbert Mask and the All-Imaginary Hilbert Mask are demonstrated in the function *LVHilbertPhaseShift*.

The test signal used by the script is, for $DestWF = 1$, a waveform generated as the superposition of a harmonic series of cosines inversely weighted by harmonic number, or, for $DestWF = 2$, a waveform generated as the superposition of an odd-harmonic-only series of cosines inversely weighted by harmonic number, or, for $DestWF = 3$, a user-supplied test signal. The first two test signals, had sine waves been used instead of cosine waves as the basis functions, would have been, respectively, a sawtooth wave and a square wave. The script shifts the phases of the constituent frequencies by 90 degrees and displays the result, using both frequency domain mask methods.

Figure 9.36 shows the result of the call

LVHilbertPhaseShift(2,[],512)

while Fig. 9.37 shows the result from making the call

LVHilbertPhaseShift(3,[cos(2*pi*16*[0:1:63]/64)],[])

which computes the Hilbert transform of a cosine at the half-band frequency, and finally, Fig. 9.38 shows the result of the call

LVHilbertPhaseShift(3,[cos(2*pi*32*[0:1:63]/64)],[])

which computes the Hilbert transform of a cosine at the Nyquist rate.

Example 9.22. Compute the Hilbert Transform of a one-cycle, four-sample cosine wave using an All-Real FD Mask, using the functions fft and ifft.

We make the following call (note that cos(2*pi*(0:1:3)/4) = [1,0,-1,0]):

s = [1,0,-1,0]; h = imag(ifft(fft(s).*[1,2,1,0]))

which yields [0,1,0,-1], which is a four-sample sine wave.

We can use a recursive procedure which starts with the original signal and conducts as many 90-degree phase shifts as desired. The following Command Line call returns the original signal after four 90 degree phase shifts; the original signal is introduced as h = [1,0,-1,0] for injection into the recursive loop. Each intermediate result is printed on the Command Line for reference. The outputs will be seen to form the succession of cosine, sine, -cosine, -sine, and cosine again after the fourth 90 degree phase shift.

h = [1,0,-1,0], for ctr = 1:1:4;s = h;
h = imag(ifft(fft(s).*[1,2,1,0])),end

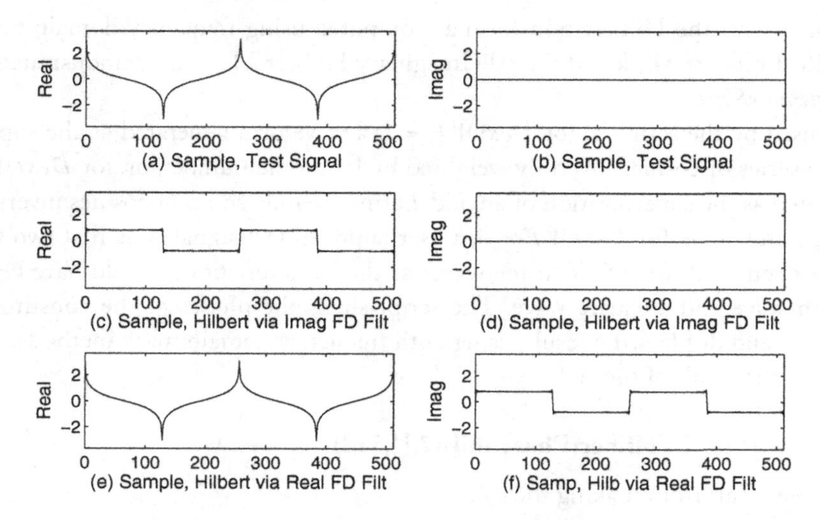

Figure 9.36: (a) Real part of test waveform, sum of odd harmonic cosines with amplitudes inversely proportional to harmonic number; (b) Imaginary part of same; (c) and (d) The real and imaginary parts of the IDFT of the product of the DFT of the waveform at (a) & (b) and an All-Imaginary FD Hilbert mask; (e) and (f) The real and imaginary parts of the IDFT of the product of the DFT of the waveform at (a) & (b) and an All-Real FD Hilbert mask.

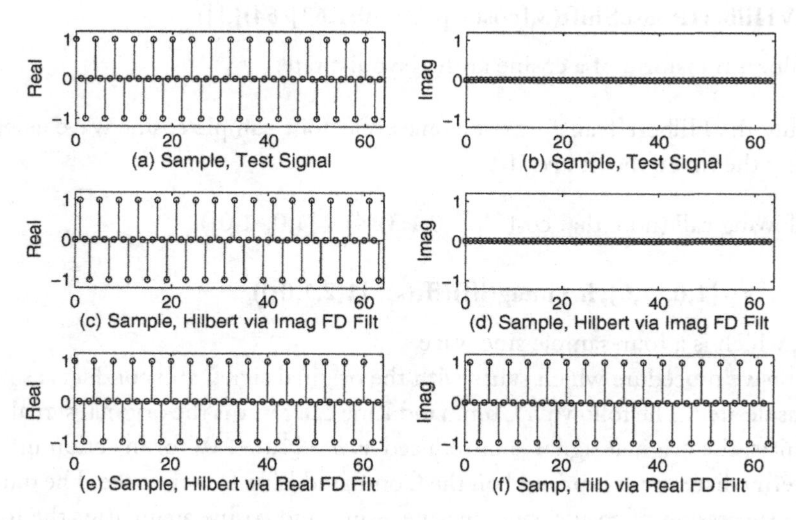

Figure 9.37: (a) Real part of test waveform, a cosine at the halfband frequency ($\pi/2$ radians); (b) Imaginary part of same; (c) and (d) The real and imaginary parts of the IDFT of the product of the DFT of the waveform at (a) & (b) and an All-Imaginary FD Hilbert mask; (e) and (f) The real and imaginary parts of the IDFT of the product of the DFT of the waveform at (a) & (b) and an All-Real FD Hilbert mask.

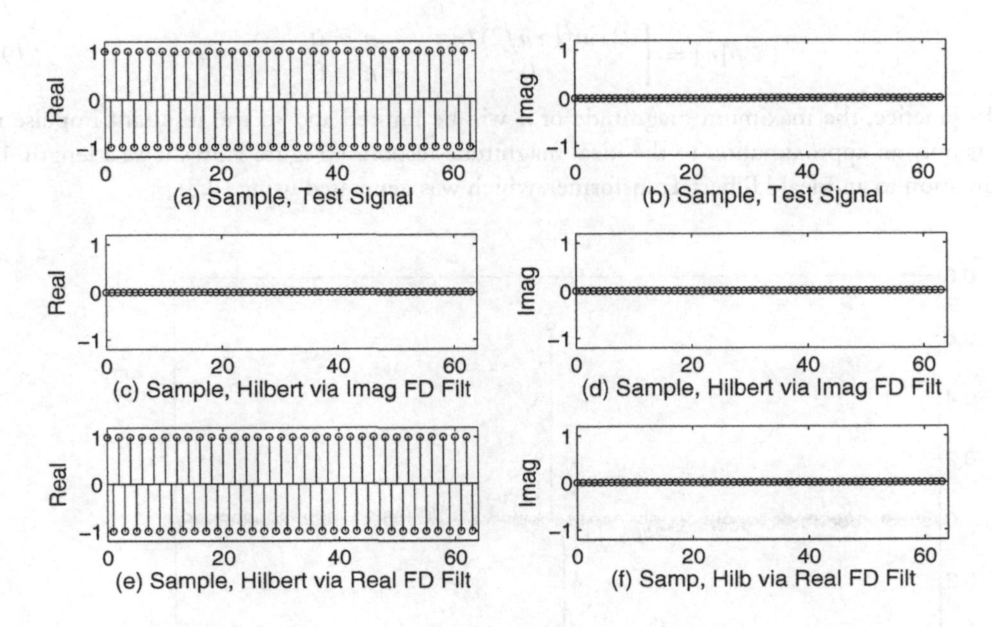

Figure 9.38: (a) Real part of test waveform, a cosine at the Nyquist rate; (b) Imaginary part of same; (c) and (d) The real and imaginary parts of the IDFT of the product of the DFT of the waveform at (a) & (b) and an All-Imaginary FD Hilbert mask; (e) and (f) The real and imaginary parts of the IDFT of the product of the DFT of the waveform at (a) & (b) and an All-Real FD Hilbert mask.

A call using the function *hilbert* is:

```
h = [1,0,-1,0], for ctr = 1:1:4; s = h; h = imag(hilbert(s)), end
```

In Time Domain
There are two approaches to employing time domain methods to implement a Hilbert Transformer.

- The first method (probably more of academic rather than practical interest) is to use the inverse DFT on either of the FD masks discussed above. This produces an impulse response suitable for circular convolution (not linear convolution as employed by typical time domain filters) with a signal of the same length to yield the Hilbert Transform.
- The second method is to generate an impulse response suitable for use as a Hilbert Transformer in linear convolution. This can be done using frequency sampling methods or explicit formulas that define the values of the impulse response. One example of the latter is the formula

$$h[n] = \begin{cases} 2\sin^2(\pi n/2)/n\pi & n \neq 0 \\ 0 & n = 0 \end{cases} \tag{9.8}$$

In practice, the maximum magnitude of n will be limited and so the resultant impulse response is only an approximation to the ideal magnitude response. Figure 9.39 shows a length-127 approximation to an ideal Hilbert Transformer, which was generated using (9.8).

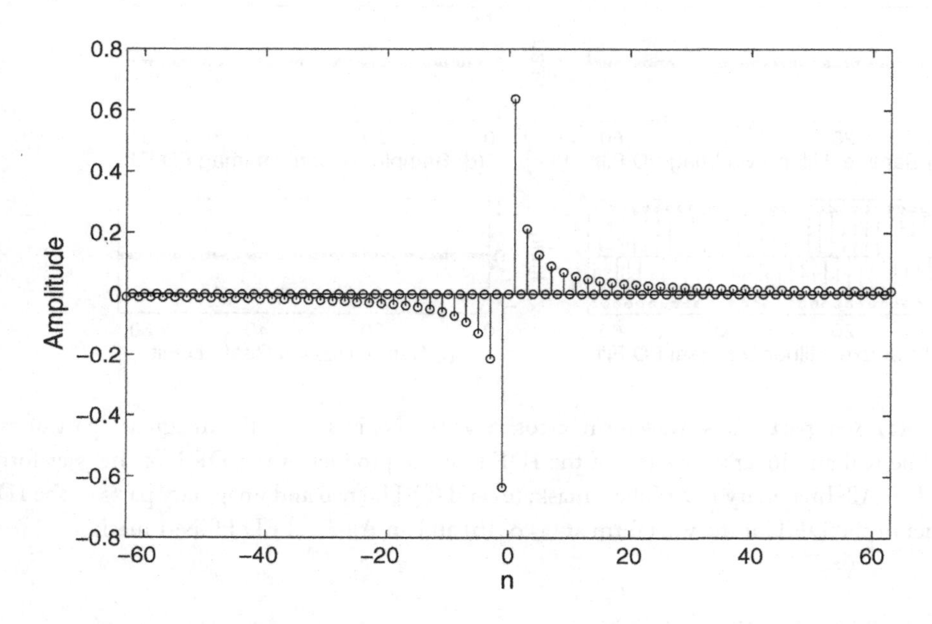

Figure 9.39: A time domain Hilbert Transformer suitable for linear convolution with a signal.

- Hilbert Transformer impulse responses can be generated using the frequency sampling method, either via IDFT or sine summation formulas.

The script (see exercises below)

$$WF = LVxFIRViaWholeSines(AkPos,AkLOver2,L)$$

provides the impulse response, frequency and phase responses, zero plot, etc. for a Type III or IV filter specified by the arguments for *Ak*, which are partitioned into *AkPos*, the positive frequencies (starting at frequency 1), and *AkLOver2*, which is passed as 0 or the empty matrix for a Type III filter, or a desired amplitude of response for a Type IV filter. The desired filter length is L, and the impulse response is output as WF. The call (in which certain frequency sample amplitudes have been set to values less than 1.0 to optimize the magnitude of the response)

WF = LVxFIRViaWholeSines([0.75,ones(1,23),0.55],[],51)

results in Fig. 9.40.

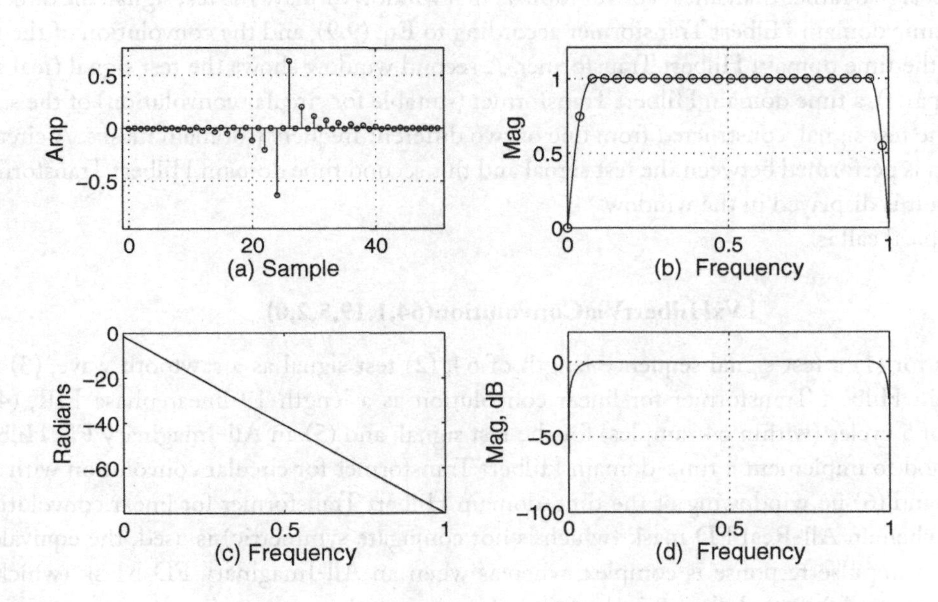

Figure 9.40: (a) Length-51 Hilbert transformer designed by the Frequency-Sampling-IDFT method; (b) Magnitude response of the Hilbert transformer, with DFT samples plotted as circles; (c) Phase response of Hilbert transformer; (d) Magnitude response in dB.

- A direct time domain formula for constructing a Hilbert transformer is as follows:

$$h[n] = \begin{cases} 2\sin^2(\pi(n-M)/2)/n\pi & n-M \neq 0 \\ 0 & n-M = 0 \end{cases} \tag{9.9}$$

where n = 0:1:L -1 and M = (L -1)/2.

The script (see exercises below)

$$LVxHilbertViaConvolution(TestSeqLen, ...$$
$$TestWaveType, FilterLen, TestSigFreq, FDMethod)$$

generates a waveform from cosines having the amplitude structure of a user-selectable sawtooth or square wave, and then phase shifts it by convolving the waveform with a time domain Hilbert

Transformer made using (1) a direct-synthesis time domain formula according to Eq. (9.9) and (2) the Inverse DFT. The first time domain impulse response has length *FilterLen* and is used in linear convolution with the test signal. The second time domain impulse response is generated by the IDFT, has the same length as the test signal, and is suitable for, and used in, circular convolution with the test signal rather than linear convolution. A first window displays the test signal, the directly-generated time domain Hilbert Transformer according to Eq. (9.9), and the convolution of the test signal and the time domain Hilbert Transformer. A second window shows the test signal (real and imaginary parts), a time domain Hilbert Transformer (suitable for circular convolution) of the same length as the test signal, constructed from one of two different frequency domain masks. A circular convolution is performed between the test signal and this second time domain Hilbert Transformer, and the result is displayed in the window.

A typical call is:

LVxHilbertViaConvolution(64,1,19,5,2,0)

which calls for (1) a test signal sequence length of 64, (2) test signal as a sawtooth wave, (3) the time-domain Hilbert Transformer for linear convolution as a length 19 linear-phase FIR, (4) a frequency of 5 cycles (within 64 samples) for the test signal, and (5) an All-Imaginary FD Hilbert Mask method to implement a time-domain Hilbert Transformer for circular convolution with the test signal, and (6) no windowing of the time-domain Hilbert Transformer for linear convolution. Note that when an All-Real FD mask (which is not conjugate symmetric) is used, the equivalent time domain impulse response is complex, whereas when an All-Imaginary FD Mask (which is conjugate symmetric) is used, the equivalent time domain impulse response is real.

The above call results in Figs. 9.41 and 9.42.

Example 9.23. Use Hilbert transforms to create a single sideband signal.

Modulating a carrier wave C with another (usually much lower frequency) frequency S_M creates sum and difference frequencies called sidebands. It is possible, using Hilbert transforms, to eliminate one of these sidebands. Since only one sideband is necessary to transmit all of the intelligence, the effective power increases dramatically over double sideband transmission, in which power goes into both sidebands. The Single Sideband (SSB) mode of radio transmission is popular for this reason.

Multiplying a carrier wave C by a modulating signal S_M results in a double sideband signal:

$$S_{DSB} = CS_M$$

The following expression will eliminate one of the sidebands.

$$S_{SSB} = CS_M \pm (hilbert(C))(hilbert(S_M)) \tag{9.10}$$

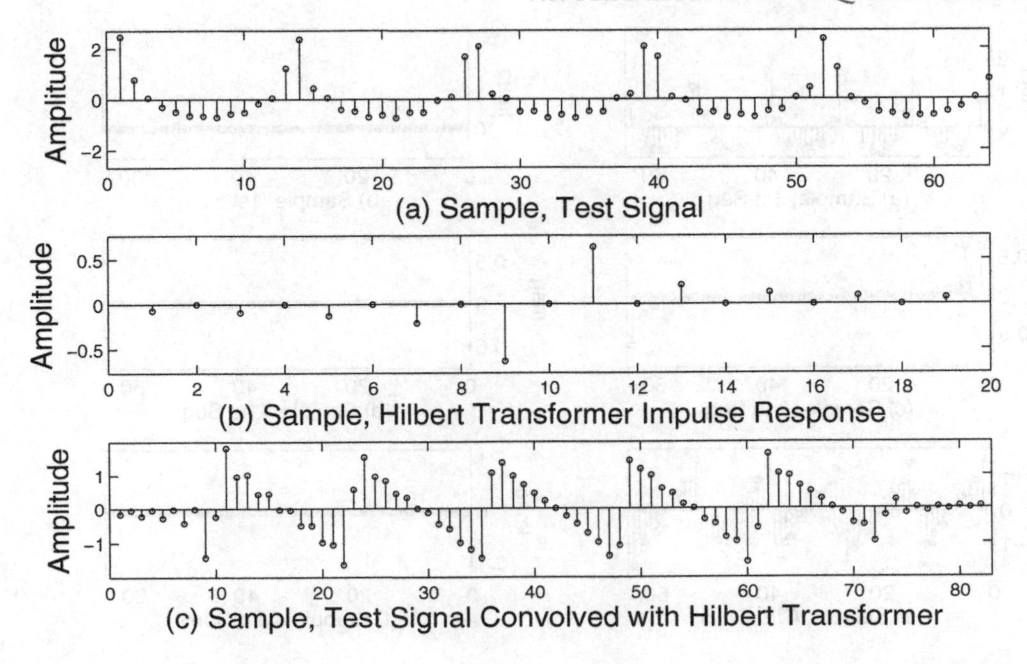

Figure 9.41: (a) Test signal; (b) Hilbert Transformer suitable for linear convolution; (c) Convolution of signals at (a) and (b).

This can be demonstrated, without loss of generality (since F_C, F_S, θ and ϕ are arbitrary), by couching it in a more specific manner as

$$S_{SSB} = \cos(2\pi t F_C + \theta)\cos(2\pi t F_M + \phi) \pm \sin(2\pi t F_C + \theta)\sin(2\pi t F_M + \phi)$$

and then substituting the Euler identity for the trigonometric terms, where, for example, in general

$$\cos(\alpha) = (\exp(j\alpha) + \exp(-j\alpha))/2$$

and

$$\sin(\alpha) = (\exp(j\alpha) - \exp(-j\alpha))/(2j)$$

After substituting the identities for all terms and simplifying, one of the sidebands vanishes.

To demonstrate this, we generate a double sideband signal, then follow that with the single sideband signal and plot the spectra of both. Using *Sb* = 0 leaves the lower sideband, using *Sb* =1 leaves the upper sideband. Here we have chosen the phase angle for the carrier to be $\pi/2$, and that of the signal as $\pi/6$. You can arbitrarily change the phase angles to verify the generality.

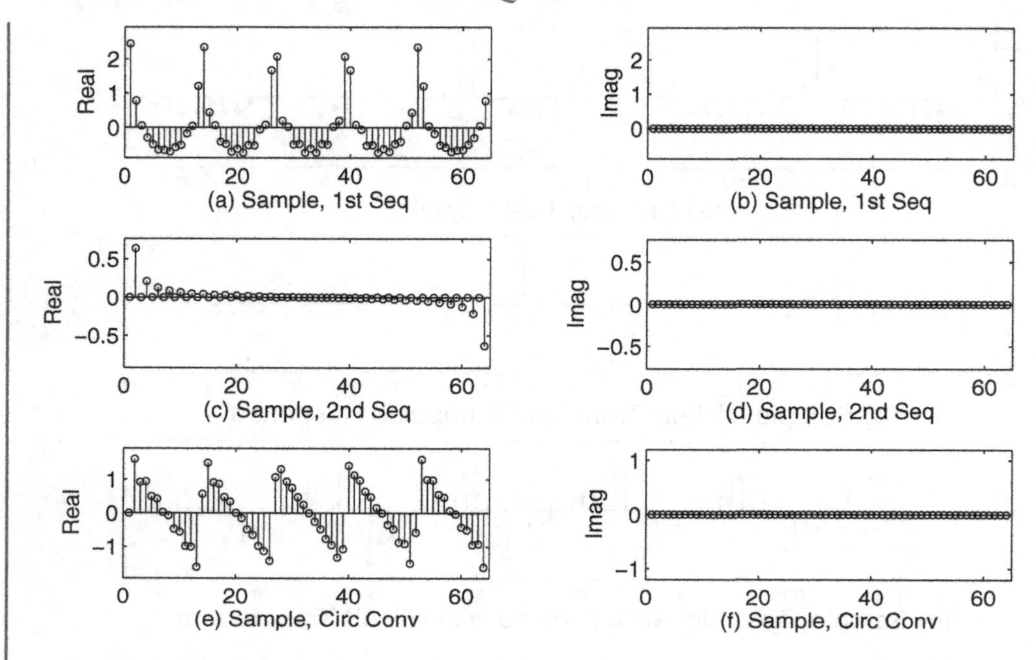

Figure 9.42: (a) and (b): Real and imaginary parts of the test signal; (c) and (d): Real and imaginary parts of TD Hilbert Mask, obtained as the IDFT of an All-Real FD Hilbert Mask; (e) and (f): Real and imaginary parts of circular convolution of test signal and TD Hilbert Mask.

```
function LVSsbDsb(Fc,Fa,Phi1,Phi2,Sb,SR)
% LVSsbDsb(100,20,pi/2,pi/6,0,1000)
phi1 = pi/2; phi2 = pi/6; t = 0:1/SR:1-1/SR;
argsC = 2*pi*t + phi1; argsS = 2*pi*t + phi2;
dsb = cos(argsC*Fc).*cos(argsS*Fa);
figure(95); subplot(2,1,1);
plot(2*t,abs(fft(dsb))); xlabel('Freq, Units of \pi')
ylabel('Mag, DSB Signal')
if Sb==0; ssb = dsb + sin(argsC*Fc).*sin(argsS*Fa);
else; ssb = dsb - sin(argsC*Fc).*sin(argsS*Fa); end
subplot(2,1,2); plot(2*t,abs(fft(ssb)))
xlabel('Freq, Units of \pi'); ylabel('Mag, SSB Signal')
```

For actual signals, which may occupy a band of frequencies, it necessary to use Eq. (9.10) to generate a single sideband signal since there would be many frequencies of arbitrary phase and amplitude present.

9.6.6 DIFFERENTIATORS

A filter that produces as its output the derivative of the input signal is called a differentiator. The frequency response of a differentiator increases linearly with frequency; its sampled response is

$$jHr[k] = \begin{cases} j2\pi k/L & k = 0:1:M \\ -j2\pi(L-k)/L & k = M+1:L-1 \end{cases}$$

where L is the filter length and $M = (L-1)/2$. To obtain a linear phase filter, the phase angles of the samples must conform to

$$\angle H[k] = \begin{cases} -\pi k(L-1)/L & k = 0:1:M \\ \pi(L-k)(L-1)/L & k = M+1:L-1 \end{cases}$$

with

$$H[k] = jHr[k]e^{j\angle H[k]}$$

Example 9.24. Design a differentiator of length 24. Use it to obtain the derivative of a triangle wave.

A Type IV filter is suitable for a differentiator since the required response at the Nyquist limit is nonzero.

```
function LVDifferentiatorLen24
Bins = pi*( [ 0,j*(1:1:11)*2/24,j*12*2/24,j*(-11:1:-1)*2/24 ] )
L = length(Bins); M = (L-1)/2;
kZeroPosBns = 0:1:ceil(M); kNegBns = ceil(M)+1:1:L-1;
angVec = [-2*pi*kZeroPosBns*M/L,2*pi*(L-kNegBns)*M/L];
PhaseFac = exp(j*angVec); NetBins = Bins.*PhaseFac;
Imp = real(ifft(NetBins))
figure(98); stem(Imp); xlabel('Sample'); ylabel('Amplitude')
```

Code that uses the sine summation formulas to construct a length-24 impulse response is

```
function LVDifferL24ViaSineSumm
AkPos = -[1:1:11]*pi/12; AkLOver2 = [-12]*pi/12;
L = 24; M = (L-1)/2; Imp = zeros(1,L); n = 0:1:L-1;
Ak = [AkPos,AkLOver2]; limK = L/2;
for k = 1:1:limK
if (k==limK); C = 1; else; C = 2; end;
Imp = Imp + C*Ak(k)*sin(2*pi*(n-M)*k/L); end;
Imp = Imp/L; figure(109); stem(Imp);
xlabel('Sample'); ylabel('Amplitude')
```

The derivative of a triangle wave is a squarewave, as shown by Fig. 9.43, in which a test triangle waveform appears in plot (a), followed in plot (b) by the convolution of the test waveform with a differentiator. The writing of a script to generate a suitable test wave and differentiator and convolve them to result in a squarewave output is covered in the exercises below.

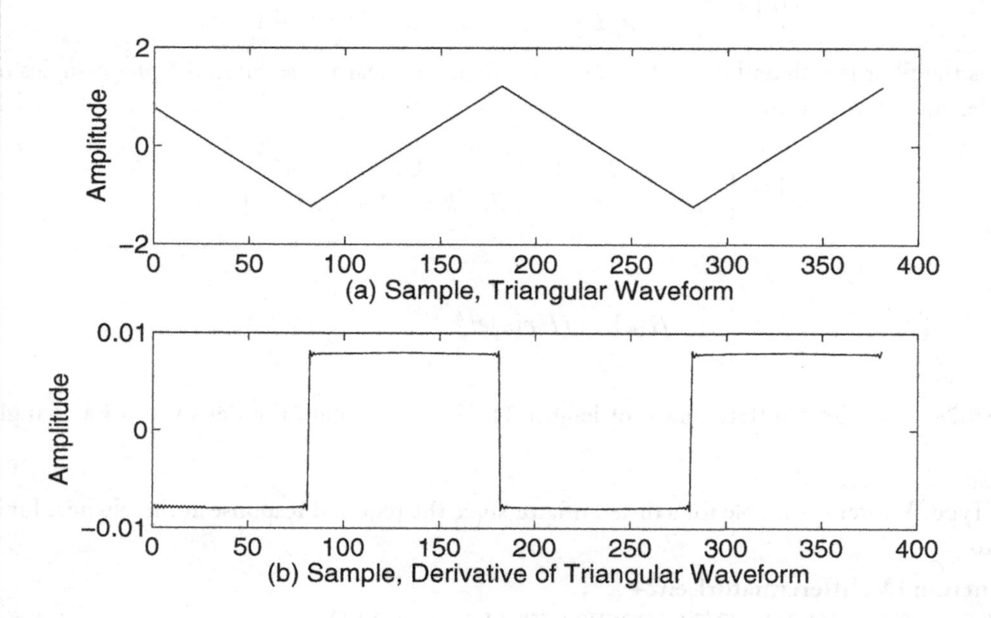

Figure 9.43: (a) Input signal, a Triangle waveform; (b) Derivative of signal in (a), delayed/offset by the filter delay.

9.7 OPTIMIZED FILTER DESIGN

Earlier in the chapter, we designed linear phase filters using the windowing and frequency sampling methods. These methods are straightforward and generally allow direct control over only stopband attenuation. Another approach, more advanced, is to design the filter such that the amplitudes of ripples in the frequency response are equalized. This results in a filter of the lowest order (shortest length) to achieve a given set of criteria; both passband ripple and stopband attenuation can be incorporated into the design.

9.7.1 EQUIRIPPLE DESIGN

The windowed-ideal-response and frequency-sampling methods of FIR design, while effective, do not allow precise control over ripple in both the passband and stopband; nor, in general, is the shortest possible filter length achieved for a given design. When the ripple in the frequency response is equalized, the minimum filter length required can be achieved. An interesting experiment that suggests this can be conducted by running the script

LVFrqSmpLPFOptOneCoeff(0.5,0.6,40,0.40,0.37,-0.001,[]);

Press any key for the next computation. The best stopband attenuation will occur when the transition band coefficient T equals 0.387. Note that the stopband ripple makes its best approach to being equalized at this point. Figure 9.44 shows the result for values of T = to 0.5, 0.387 (the ideal value), and 0.33. Passband ripple for T = 0.387 was 0.67 dB. Figure 9.45 shows the result for the same bandlimits, but with a true equiripple design which does not exceed passband ripple of 0.69 dB. In this case, the stopband attenuation is in excess of 50 dB.

For the FIR design techniques we have seen thus far, the frequency response typically contains ripples through the entire spectrum. Such ripples are not equally spaced, but rather tend to be more closely spaced the closer they are to the edges of a transition band. The number of ripples is limited and generally equal to $R + 2$, where

$$R = \begin{cases} (L-1)/2 & \text{Type-I} \\ (L-3)/2 & \text{Type-III} \\ L/2-1 & \text{Type-II, IV} \end{cases}$$

Some filters, known as extra ripple filters, can have $R + 3$ ripples. Figure 9.46 shows an equiripple lowpass filter of length 9 having $\omega_p = 0.4\pi$ and $\omega_s = 0.6\pi$. The number of extremal frequencies in the frequency response is (9-1)/2 + 2 = 6. In the case shown in the figure, the extremal frequencies are at normalized frequencies 0, 0.2625, 0.4, 0.6, 0.7375, and 1.0.

9.7.2 DESIGN GOAL

Earlier in the chapter, we noted that, for a general FIR design (for a lowpass filter), the passband frequency response design specification is

$$1 - \delta_1 \leq H_r(\omega) \leq 1 + \delta_1$$

for $\omega < \omega_p$, and for the stopband

$$-\delta_2 \leq H_r(\omega) \leq \delta_2$$

for $\omega > \omega_s$. Thus the maximum passband ripple is δ_1 and the maximum stopband ripple is δ_2.

If we define the FIR's actual response as $P(\omega)$, then the error signal $E(\omega)$ is

$$E(\omega) = H_{dr}(\omega) - P(\omega)$$

The design goal is to determine filter coefficients that will minimize the maximum magnitude of $E(\omega)$ over a filter's passband(s) and stopband(s). This is called a **minimax** problem, that is, minimizing the maximum magnitude of a quantity such as $E(\omega)$.

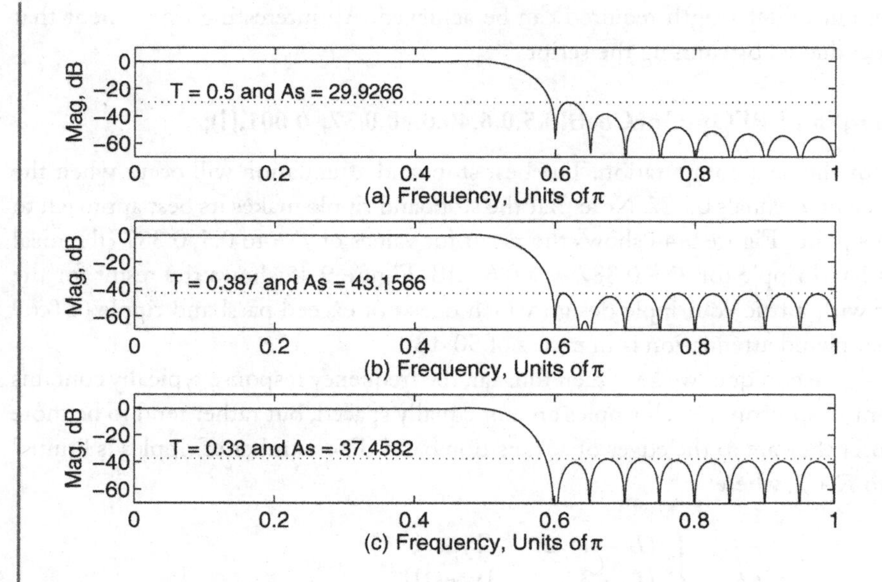

Figure 9.44: (a) A length-40 FIR having $\omega_P = 0.5\pi$, $\omega_s = 0.6\pi$, designed by the frequency sampling method, with one transition coefficient T valued at 0.5; (b) Same as (a), but with the ideal-valued T of 0.387; (c) same as (a), but with $T = 0.33$. R_P was 0.67 dB for $T = 0.387$.

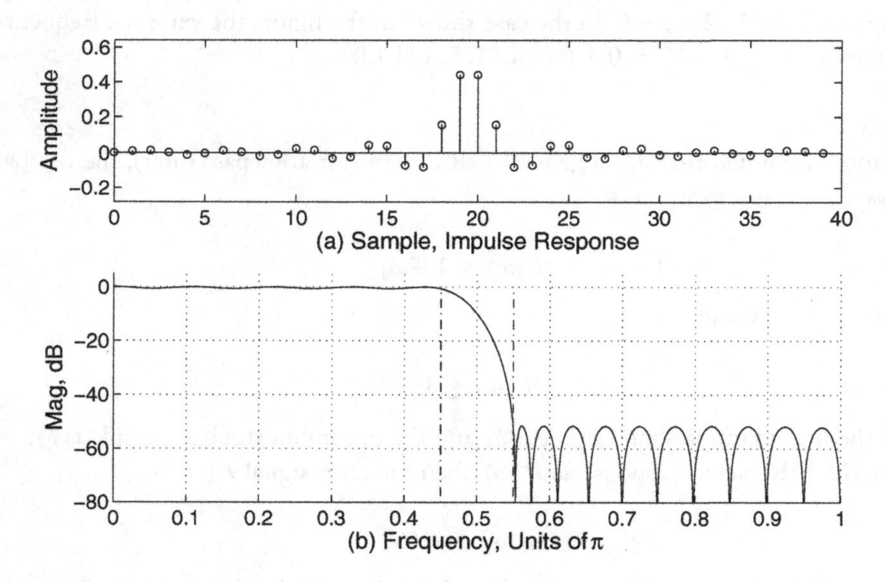

Figure 9.45: A length-40 equiripple FIR having $\omega_P = 0.5\pi$, $\omega_s = 0.6\pi$, $R_P = 0.69$ dB, and $A_S = 50.95$ dB.

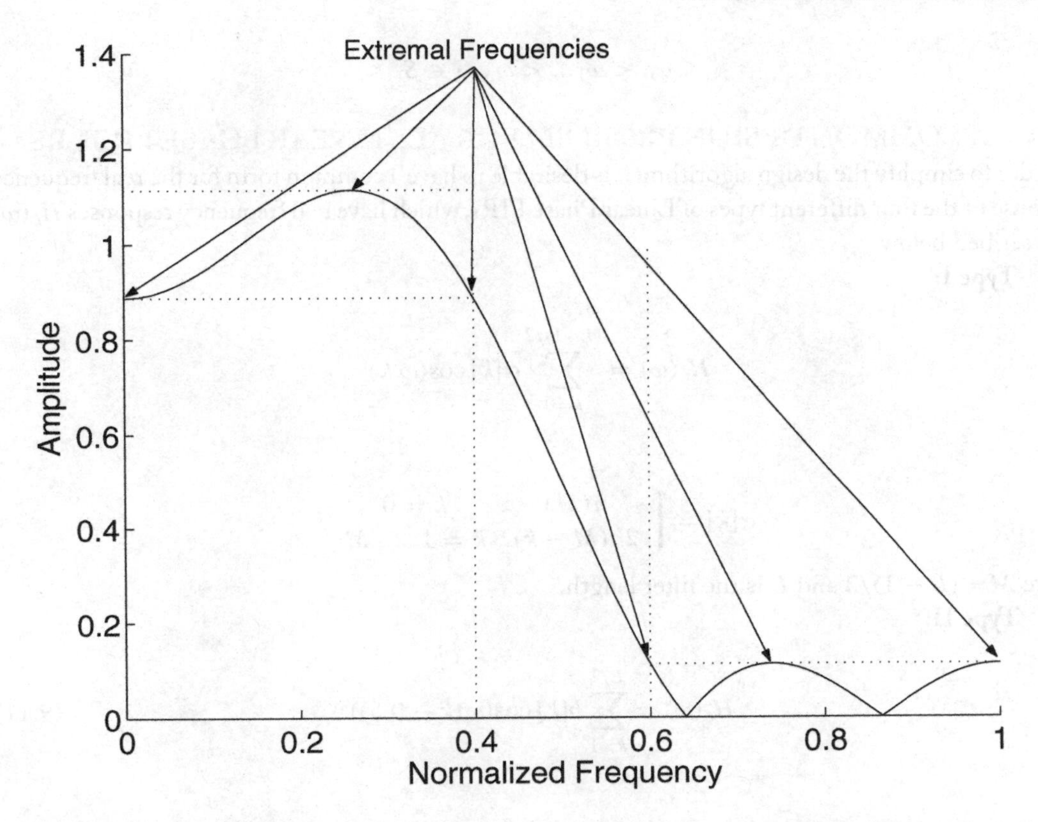

Figure 9.46: An equiripple lowpass filter having wp = 0.4 and ws = 0.6. Extremal frequencies are located at normalized frequencies 0, 0.2625, 0.4, 0.6, 0.7375, 1.0.

9.7.3 ALTERNATION THEOREM

An important theorem regarding the frequency response of an FIR is the Alternation Theorem, which states essentially that a set of filter coefficients exist which will result in a frequency response having equal ripple amplitudes, and that in fact the deviation amplitude for all ripples is limited to some value δ, with the sign of δ alternating between adjacent extremal frequencies.

Formal Statement of Alternation Theorem

If S is any closed subset of the closed frequency interval $[0, \pi]$, $P(\omega)$ is the unique minimax approximation to $H_{dr}(\omega)$ over S if $E(\omega)$ exhibits at least $(R + 2)$ unique extremal frequencies $[\omega_0, \omega_1, ...\omega_{R+1}]$ and

$$E(\omega_{i+1}) = -E(\omega_i) = \pm\delta = \pm \max_S |E(\omega)|$$

with

$$\omega_0 < \omega_1 ... < \omega_{R+1} \in S$$

9.7.4 A COMMON DESIGN PROBLEM FOR ALL LINEAR PHASE FILTERS

In order to simplify the design algorithm, it is desirable to have a common form for the real frequency response of the four different types of Linear Phase FIRs, which have real frequency responses $H_r(\omega)$ as described below.

Type I:

$$H_r(\omega) = \sum_{k=0}^{(L-1)/2} \alpha[k] \cos(\omega\, k)$$

with

$$\alpha[k] = \begin{cases} h(M) & k = 0 \\ 2h(M-k) & k = 1, 2...M \end{cases}$$

where $M = (L-1)/2$ and L is the filter length.

Type II:

$$H_r(\omega) = \sum_{k=1}^{L/2} b[k] \cos(\omega(k-0.5)) \tag{9.11}$$

with

$$b[k] = 2h(L/2 - k)$$

where k =1, 2...$L/2$. Eq. (9.11) can further be rewritten as

$$H_r(\omega) = \cos(\omega/2) \sum_{k=1}^{L/2} b'[k] \cos(\omega\, k) \tag{9.12}$$

where

$$\begin{aligned} b'[0] &= b[1]/2 \\ b'[L/2-1] &= 2b[L/2] \end{aligned}$$

and

$$b'[k] = 2b[k] - b'[k-1]; \quad k = 1, 2, ...(L/2) - 2$$

Type III:

$$H_r(\omega) = \sum_{k=1}^{M} c[k] \sin(\omega k) \tag{9.13}$$

where

$$c[k] = 2h[M-k]; \quad k = 1, 2, ...M$$

and $M = (L-1)/2$. The above can be rewritten as

$$H_r(\omega) = \sin(\omega) \sum_{k=0}^{(L-3)/2} c'[k] \cos(\omega k) \tag{9.14}$$

where $c'[n]$ is linearly related to $c[n]$ (the relationship may be found in [5]).

Type IV:

$$H_r(\omega) = \sum_{k=1}^{L/2} d[k] \sin(\omega(k-0.5)) \tag{9.15}$$

where

$$d[k] = 2h[L/2 - k]; \quad k = 1, 2, ...L/2$$

The above can be rewritten as

$$H_r(\omega) = \sin(\omega/2) \sum_{k=0}^{L/2-1} d'[k] \cos(\omega k) \tag{9.16}$$

where $d'[n]$ is linearly related to $d[n]$ (the relationship may be found in [5]).

The above real frequency responses can be rewritten as the product of two functions $Q(\omega)$ and $P(\omega)$ according to the following table:

$$\begin{array}{cccc} \text{Type} & Q(\omega) & P(\omega) & K \\ \text{I} & 1 & \sum_0^K \alpha[n]\cos(\omega n) & \frac{L-1}{2} \\ \text{II} & \cos(\omega/2) & \sum_0^K b'[n]\cos(\omega n) & \frac{L}{2}-1 \\ \text{III} & \sin(\omega) & \sum_0^K c'[n]\cos(\omega n) & \frac{L-3}{2} \\ \text{IV} & \sin(\omega/2) & \sum_0^K d'[n]\cos(\omega n) & \frac{L}{2}-1 \end{array} \tag{9.17}$$

9.7.5 WEIGHTED ERROR FUNCTION

It is possible to obtain a filter that has different ripple magnitudes in the pass- and stop- bands. If the maximum magnitude of passband ripple is δ_1 and the maximum magnitude of stopband ripple is δ_2,then the needed weight vector $W(\omega)$ is defined as

$$W(\omega) = \begin{cases} \delta_2/\delta_1 & \omega \text{ in passband} \\ 1 & \omega \text{ in stopband} \end{cases}$$

The weighted approximation error $E(\omega)$ is then defined as

$$E(\omega) = W(\omega)[H_{dr}(\omega) - H_r(\omega)]$$

which becomes

$$E(\omega) = W(\omega)[H_{dr}(\omega) - Q(\omega)P(\omega)]$$

This expression can be modified to

$$E(\omega) = W(\omega)Q(\omega)[H_{dr}(\omega)/Q(\omega) - P(\omega)]$$

and, setting

$$\hat{W}(\omega) = W(\omega)Q(\omega)$$

and

$$\hat{H}_{dr}(\omega) = \frac{H_{dr}(\omega)}{Q(\omega)}$$

we obtain as the net weighted error expression

$$E(\omega) = \hat{W}(\omega)[\hat{H}_{dr}(\omega) - P(\omega)]$$

When the proper extremal frequencies ω_n are known for a certain equiripple FIR design, it is true that

$$E(\omega_n) = \hat{W}(\omega_n)[\hat{H}_{dr}(\omega_n) - P(\omega_n)] = (-1)^n\delta$$

for $n = 0{:}1{:}K + 1$ where K is as given in Table (9.17). The set of $K + 2$ equations can be rewritten as

$$P(\omega_n) + (-1)^n\delta/\hat{W}(\omega_n) = \hat{H}_{dr}(\omega_n)$$

Replacing $P(\omega_n)$ with the equivalent cosine summation expression, we get

$$\sum_{k=0}^{K} \alpha_k \cos(\omega_n k) + \frac{(-1)^n\delta}{\hat{W}(\omega_n)} = \hat{H}_{dr}(\omega_n)$$

where n = 0:1:K +1, and K is as given in Table (9.17).

In expanded matrix form this would appear as

$$\begin{bmatrix} 1 & \cos(\omega_0) & \cdots & \cos(K\omega_0) & \frac{1}{\hat{W}(\omega_0)} \\ 1 & \cos(\omega_1) & \cdots & \cos(K\omega_1) & \frac{-1}{\hat{W}(\omega_1)} \\ \vdots & \vdots & \vdots & \vdots & \vdots \\ 1 & \cos(\omega_{K+1}) & \cdots & \cos(K\omega_{K+1}) & \frac{(-1)^{K+1}}{\hat{W}(\omega_{K+1})} \end{bmatrix} \begin{bmatrix} \alpha[0] \\ \alpha[1] \\ \vdots \\ \alpha[K] \\ \delta \end{bmatrix} = \begin{bmatrix} \hat{H}_{dr}(\omega_0) \\ \hat{H}_{dr}(\omega_1) \\ \vdots \\ \hat{H}_{dr}(\omega_{K+1}) \end{bmatrix} \tag{9.18}$$

Written in more compact form this would be

$$[WP][A] = [H]$$

where WP is the cosine-and-reciprocal-weights matrix, A is the vector of α values and δ, and H is the $\hat{H}_{dr}$ vector.

9.7.6 REMEZ EXCHANGE ALGORITHM

Unfortunately, the values for α, δ, and ω_n in matrix equation (9.18) are unknown. However, by making a guess or assumption for the initial values of the ω_n, it is possible to solve for the vector comprising the α values and δ. When this is done, the frequency response can be computed using the values of α, and a search of the frequency response is conducted for the next estimate of the extremal values. This procedure is known as the Remez Exchange algorithm, discussed and demonstrated immediately below. There are several ways to solve for the values of α and δ. One direct, but computationally expensive way is to compute the inverse (actually, the pseudo-inverse) of matrix WP:

$$[WP]^{-1}[WP][A] = [WP]^{-1}[H]$$

which reduces to

$$[A] = [WP]^{-1}[H]$$

The well-known Parks-McClellan algorithm analytically computes the value for δ, generates a set of extremal points using δ and the assumed or current-estimated extremal frequencies, interpolates on a fine grid between the extremal points to produce a frequency response, computes the error function, and then conducts a search of the error function for the next estimates of the extremal frequencies. The manner of interpolating between the extremal points is known as Barycentric Lagrangian interpolation. We will discuss the MathScript implementation of the Parks-McClellan algorithm below, after further exploration of the basic Remez Exchange concept.

Example 9.25. Write a program that will receive a set of band limits for a lowpass filter, a filter length, and a set of extremal frequencies, which will solve for delta and the alpha coefficients, and display the corresponding $P(\omega)$ in a way that allows visual selection of the next set of extremal frequencies.

The following program has been written for one specific filter design; the exercises at the end of the chapter will direct the reader in various projects to expand and generalize the code. In order to keep things conceptually simple, the values of α and δ are determined using the matrix equation method, which is acceptable for shorter filter lengths and modern, relatively fast computers.

```
function LVManualLPFRemezExch(L,LenGrid,wp,ws,curXFrqs)
% LVManualLPFRemezExch(9,145,0.45,0.55,[0,0.225,0.45,...
0.55,0.775,1])
NormFrGrid = [0:1:LenGrid-1]/(LenGrid-1); wc = (wp+ws)/2;
FrGrid = pi*NormFrGrid; WtVec = ones(1,LenGrid);
XFrindOnFG = round(curXFrqs*(LenGrid-1)+1);
Hdr = ones(1,round(wc*(LenGrid-1)));
Hdr = [Hdr, zeros(1,LenGrid-length(Hdr))];
Q = 1; kLim = (L-1)/2; WMat(1:kLim+2,1) = 1; Hdr = Hdr./Q;
WtVec = WtVec.*Q; k = 0:1:kLim+1;
Num = ((-1).^k); Denom = WtVec(1,XFrindOnFG);
WMat(k+1,2:kLim+1) = cos(pi*curXFrqs(1,k+1)'*([1:kLim]));
WMat(k+1,kLim+2) = (Num./Denom)';
HdrVec = Hdr(round((LenGrid-1)*[curXFrqs]+1));
AlDelVec = pinv(WMat)*(HdrVec)';
delta = AlDelVec(length(AlDelVec)),
P = 0; for pCtr = 1:length(AlDelVec)-1;
P = P + AlDelVec(pCtr)*(cos(FrGrid*(pCtr-1)) ); end;
E = WtVec.*([Hdr - P]); figure(8); clf;
subplot(211); hold on; ad = abs(delta);
loBndX = [0:1:round((LenGrid-1)*wp)]/(LenGrid-1);
hiBndX = [round((LenGrid-1)*ws):1:LenGrid-1]/(LenGrid-1);
plot(loBndX,E(1,1:round(wp*(LenGrid-1))+1));
line([0,max(loBndX)],[ad, ad]); line([0,max(loBndX)],[-ad, -ad]);
xlabel('Norm Freq'); ylabel('Err (Passband)');
subplot(212); hold on;
plot(hiBndX,E(1,round(ws*(LenGrid-1))+1:LenGrid));
line([ws,1],[ad, ad]); line([ws,1],[-ad, -ad]);
xlabel('Norm Freq'); ylabel('Err (Stopband)');
Hr = Q.*P; figure(9); LenHr = length(Hr);
plot([0:1:LenHr-1]/LenHr,Hr);
xlabel('Norm Freq'); ylabel('Amp')
```

The code above has for *curXFrqs* (the current set of extremal frequencies) an initial guess of linearly spaced normalized frequencies that include the two transition band edges as well as 0 and 1.0; the remaining two needed values of a total of six needed values ((9-1)/2 + 2 = 6) were spaced equally within the pass- and stop- bands.

A script which can receive several arguments to perform the same functions as the code above for Type I and II lowpass and highpass filter types is

LVDemoRemez(wp,ws,As,Rp,L,PassbandType,NormXFr,EqWt)

To demonstrate the principles of the Remez Exchange algorithm, we can start with the following call, which uses extremal frequencies that are linearly-spaced in the pass- and stop- bands for the initial guess:

LVDemoRemez(0.45,0.55,50,0.2,9,1,[0,0.225,0.45,0.55,0.775,1],1)

the result of which is shown in Fig. 9.47.

The initial guess of extremals generally includes the normalized frequencies of 0 and 1, and all transition band edges. In subsequent extremal searches, transition band edges are usually included since the design expects extremals at the transition band edges in the final solution. Extremals at normalized frequencies 0 and 1 may or may not be present after the initial guess. To be systematic and accurate, an extremal candidate list should be made which contains at least the frequency and magnitude of each candidate extremal. Another useful parameter is whether each candidate extremal is a local maximum ("positive") or minimum ("negative"), since a proper set of extremals must alternate between being positive and negative. When the list is complete, the candidates having the highest (*kLim* + 2) magnitudes (and which properly alternate in sign) are chosen as the next set of extremals. A detailed discussion of extremal frequency choice can be found in [6].

Continuing with the present example, for our next call, we use extremal frequencies estimated by inspecting plots (a) and (b) in Fig. 9.47. These appear to be 0, 0.27, 0.45, 0.55, 0.675, and 1.0. We thus make the call

LVDemoRemez(0.45,0.55,50,0.2,9,1,[0,0.27,0.45,0.55,0.675,1],1)

which results in Fig. 9.48, which shows all extremals nearly within the bounds of $\pm\delta$; one or two more iterations with very small changes should bring the design to complete convergence.

MathScript has the functions *remez* and *firpm* (which may be used interchangeably), both of which implement the Parks-McClellan algorithm. The basic format is as follows:

$$b = firpm(N, F, A, W)$$

where b is the desired impulse response, N is the order (one less than the desired FIR length), F is a vector of band edges in normalized frequencies (from 0 to 1), A is a vector (half the length of F) that gives desired response amplitude for each band (i.e., passband and stopband) defined by the

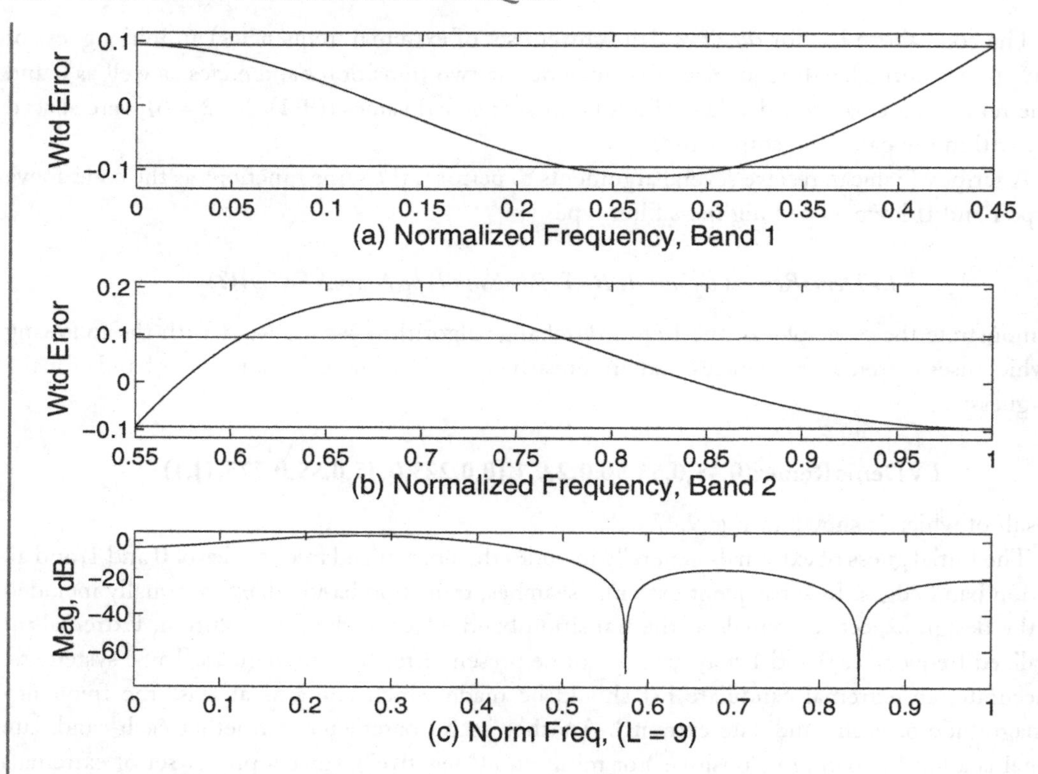

Figure 9.47: (a) Error signal in passband of a length-9 lowpass FIR; (b) Error signal in stopband of same; (c) Frequency response of filter based on the extremals [0,0.225,0.45,0.55,0.775,1]; Note the new extremal frequencies (estimated by visual inspection of the plots) at 0, 0.27, 0.45, 0.55, 0.675, and 1.0.

pairs of band edges given in vector F. W is a weight vector telling what weights to give the error in each band; W has the same length as A. The following are a few example calls:

lowpass **b = firpm(18,[0,0.5,0.6,1],[1,1,0,0])**
highpass **b = firpm(18,[0,0.5,0.6,1],[0,0,1,1])**
bandpass **b = firpm(18,[0,0.35,0.4,0.55,0.6,1],[0,0,1,1,0,0])**
bandstop **b = firpm(18,[0,0.35,0.4,0.55,0.6,1],[1,1,0,0,1,1])**

The approximate necessary filter length (one greater than the filter order, which is specified in the above calls) to meet certain desired values of A_s and R_p,can be computed according to a formula by J. F. Kaiser, which is

$$L = \frac{-20\log_{10}(\sqrt{\delta_1\delta_2}) - 13}{14.6\Delta f} \tag{9.19}$$

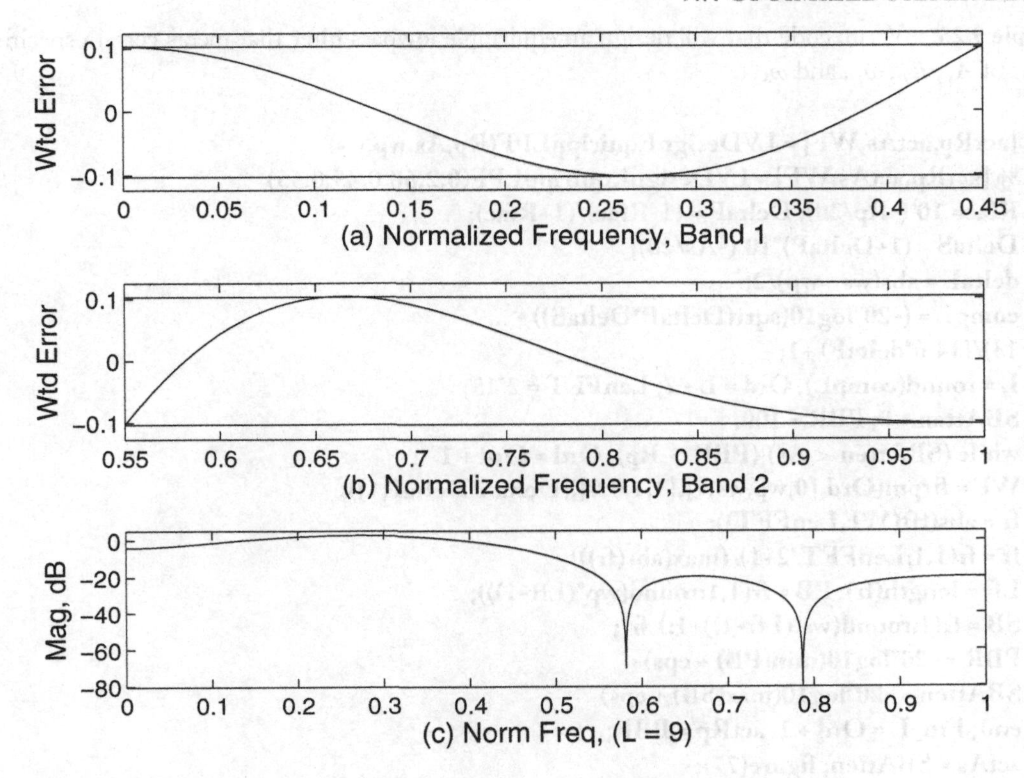

Figure 9.48: (a) Error signal in passband of a length-9 lowpass FIR; (b) Error signal in stopband of same; (c) Frequency response of filter based on the extremals [0,0.27,0.45,0.55,0.675,1], which shows the weighted error signal nearly within the limits of $\pm\delta$; one or two more iterations with minor changes in extremal estimates should be adequate to achieve complete convergence.

where $\Delta f = (\omega_p - \omega_s)/2\pi$, where ω_p and ω_s are radian frequencies, i.e., normalized frequencies multiplied by π.

To design a Hilbert transformer or a differentiator, the input argument *ftype* is necessary. Here are several sample calls:

```
b = remez(18,[0.1,0.9],[1,1],'Hilbert')
b = remez(17,[0:(1/7):1],[0:(1/7):1],'differentiator')
```

The value for L given by Eq. (9.19) is only approximate, and the needed value is often somewhat greater. To design a filter to meet certain specifications, it is necessary to run the design several times with increasing values of L, testing the result after each design run. This can be done programmatically as shown in the following example:

Example 9.26. Write code that will design an equiripple lowpass filter that meets certain specifications of A_s, R_p, ω_p, and ω_s.

```
[actRp,actAs,WF] = LVDesignEquirippLPF(Rp,As,wp,ws)
% [actRp,actAs,WF] = LVDesignEquirippLPF(0.2,60,0.45,0.55)
Rfac= 10^(-Rp/20); DeltaP = (1-Rfac)/(1+Rfac);
DeltaS = (1+DeltaP)*10^(-As/20);
deltaF = abs(ws - wp)/2;
compL = (-20*log10(sqrt(DeltaP*DeltaS)) - ...
13)/(14.6*deltaF) +1;
L = round(compL), Ord = L - 2; LenFFT = 2^15;
SBAtten = 0; PBR = 100;
while (SBAtten < As)|(PBR > Rp); Ord = Ord + 1
WF = firpm(Ord,[0,wp,ws,1],[1,1,0,0],[DeltaS/DeltaP,1]);
fr = abs(fft(WF,LenFFT));
fr = fr(1,1:LenFFT/2+1)/(max(abs(fr)));
Lfr = length(fr); PB = fr(1,1:round(wp*(Lfr-1)));
SB = fr(1,round(ws*(Lfr-1))+1:Lfr);
PBR = -20*log10(min(PB) + eps)
SBAtten = -20*log10(max(SB) + eps)
end; Fin_L = Ord + 1, actRp = PBR;
actAs = SBAtten; figure(77);
plot([0:1:LenFFT/2]/(LenFFT/2), 20*log10(fr+eps));
xlabel('Normalized Frequency (Units of \pi)')
ylabel('Magnitude, dB'), axis([0,1,-(As+20),5])
```

The result from running the above is shown in Fig. 9.49.

Example 9.27. Design a bandpass filter having band edges at [0, 0.4, 0.45, 0.65, 0.7, 1] with corresponding amplitudes of [0, 0, 1, 1, 0, 0], $As = 60$ dB, and $Rp = 0.2$ dB.

The following code is similar to that of the previous example except that the stopband evaluation is done for both stopbands.

```
function LVDesignEquirippBPF(Rp,As,ws1,wp1,wp2,ws2)
% LVDesignEquirippBPF(0.2,60,0.4,0.45,0.65,0.7)
BndLm = [0,ws1,wp1,wp2,ws2,1];
Rfac= 10^(-Rp/20); DeltaP = (1-Rfac)/(1+Rfac);
DeltaS = (1+DeltaP)*10^(-As/20);
deltaF = abs(wp1 - ws1)/2;
compL = (-20*log10(sqrt(DeltaP*DeltaS)) - ...
```

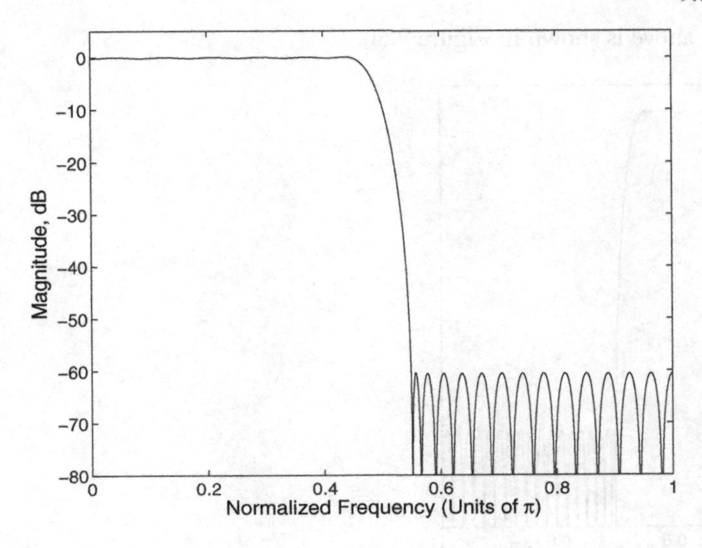

Figure 9.49: A lowpass filter designed to have the minimum length needed (53) to meet an A_s of 60 dB, R_p = 0.2 dB, with ω_p = 0.45 and ω_s = 0.55.

```
13)/(14.6*deltaF) +1;
L = round(compL), Ord = L - 2; LenFFT = 2^15;
SBAtten = 0; PBR = 100;
while (SBAtten < As)|(PBR > Rp); Ord = Ord + 1;
WF = firpm(Ord,BndLm,[0,0,1,1,0,0],[1,DeltaS/DeltaP,1]);
fr = abs(fft(WF,LenFFT));
fr = fr(1,1:LenFFT/2+1)/max(fr);
Lfr = length(fr);
PB = fr(1,round(wp1*Lfr):round(wp2*Lfr));
PBR = -20*log10(min(PB)+eps),
SB1 = fr(1,1:round(ws1*Lfr));
SB2 = fr(1,round(ws2*Lfr):Lfr);
SBAtten1 = -20*log10(max(SB1)+eps);
SBAtten2 = -20*log10(max(SB2)+eps);
SBAtten = min([SBAtten1,SBAtten2]),
end; Fin_L = Ord + 1, Fin_Rp = PBR,
Fin_As = SBAtten, figure(8); clf;
plot([0:1:LenFFT/2]/(LenFFT/2), 20*log10(fr+eps));
xlabel('Normalized Frequency (Units of \pi)')
ylabel('Magnitude, dB'); axis([0,1,-(As+20),5])
```

The result from running the above is shown in Figure 9.50.

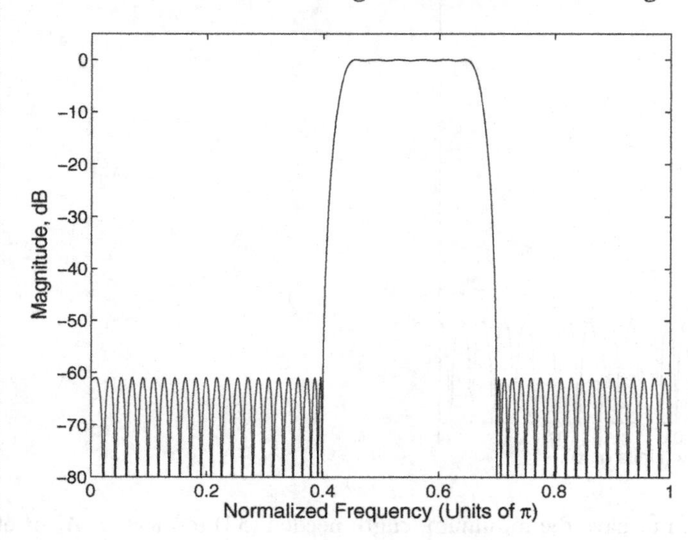

Figure 9.50: A bandpass filter designed to have the minimum length needed (110) to meet an A_s of 60 dB, R_p = 0.2 dB, with ω_{s1} = 0.4, ω_{p1} = 0.45, ω_{p2} = 0.65, and ω_{s2} = 0.7.

Example 9.28. Design a Hilbert Transformer having its passband edges at 0.1 and 0.9, with $Rp =$ 0.2 dB

The following code is straightforward, incrementing the order by two (to maintain symmetry) until the passband ripple is below the value specified for R_P.

```
function LVDesignEquirippHilbert(Rp,wp1,wp2)
% LVDesignEquirippHilbert(0.2,0.1,0.9)
PBR = 100; Ord = 10; LenFFT = 2^13;
figure(99); while (PBR > Rp); Ord = Ord + 2;
b = remez(Ord,[0.1,0.9],[1,1],'Hilbert');
y = abs(fft(b,LenFFT));
y = y(1,1:LenFFT/2+1)/(max(abs(y)));
LenGrid = LenFFT/2;
PB = y(1,round(wp1*LenGrid)+1:round(wp2*LenGrid)+1);
PBR = -20*log10(min(PB)+eps);
plot([0:1:LenGrid]/LenGrid,20*log10(y+eps))
xlabel(['Normalized Frequency (Units of \pi)'])
ylabel('Magnitude, dB'); L = Ord + 1,
axis([0,1,-40,5]); pause(0.3); end
Final_Rp = PBR, Final_L = Ord + 1
```

9.8 REFERENCES

[1] Alan V. Oppenheim and Ronald W. Schaefer, *Discrete-Time Signal Processing*, Prentice-Hall, Englewood Cliffs, New Jersey 07632, 1989.

[2] T. W. Parks and C. S. Burrus, *Digital Filter Design*, John Wiley & Sons, New York, 1987.

[3] James H. McClellan et al, *Computer-Based Exercises for Signal Processing Using MATLAB5*, Prentice-Hall, Upper Saddle River, New Jersey, 1998.

[4] Lawrence R. Rabiner, Bernard Gold, and C. A. McGonegal, *"An Approach to the Approximation Problem for Nonrecursive Digital Filters,"* IEEE Transactions on Audio and Electroacoustics, Vol. AU-18, pp. 83-106, June 1970.

[5] Lawrence R. Rabiner and Bernard Gold, *Theory and Application of Digital Signal Processing*, Prentice-Hall, Inc. Englewood Cliffs, New Jersey, 1975.

[6] Vinay K. Ingle and John G. Proakis, *Digital Signal Processing Using MATLAB V.4*, PWS Publishing Company, Boston, 1997.

[7] John G. Proakis and Demitris G. Manolakis, *Digital Signal Processing, Principles, Algorithms, and Applications (Third Edition)*, Prentice Hall, Upper Saddle River, New Jersey 07458, 1996.

[8] Andreas Antoniou, *Digital Filters, Analysis, Design, and Applications (Second Edition)*, McGraw-Hill, New York, 1993.

9.9 EXERCISES

1. Devise a script conforming to the following call syntax:

LVxFRviaCirConDFTs(LL,LenWin,wintype)

where LL is the length of the simulated ideal lowpass filter impulse response, $LenWin$ is the length of the window used to truncate the ideal lowpass filter impulse response, and $wintype$ determines the type of window used to truncate the ideal filter: pass $wintype$ as 1 for rectangular, 2 for Hamming, 3 for Blackman, and 4 for Kaiser with $\beta = 10$. The script should create an ideal lowpass filter impulse response of length $LL = L$, obtain its DFT, create a window of the same length consisting of zeros except for the central samples, which comprise the window selected by $wintype$ having length $LenWin$, obtain its DFT, and obtain the net truncated filter response as the circular convolution of the DFTs of the ideal lowpass filter and the window. A figure having the same plots shown, for example, in any of Figs. 9.6, 9.7, and 9.8, plots (a), (b), and (d), should be created.

Once the script has been written and tested, pick a value of LL of about 4096, *wintype*= 1, and run the script for values of *LenWin*=10, 50, 200, and 500. Compare results, noting the transition width for each case and the minimum stopband attenuation. Repeat this for the Hamming, Blackman, and Kaiser(10) window by specifying $wintype$ as 2, 3, and 4, respectively.

Test Calls:

(a) **LVxFRviaCirConDFTs(4096,15,1)**
(b) **LVxFRviaCirConDFTs(4096,50,1)**
(c) **LVxFRviaCirConDFTs(4096,200,1)**
(d) **LVxFRviaCirConDFTs(4096,500,1)**

2. Create and test a script conforming to the call syntax below, and as described in the text, which outlines a procedure for creating the script; the reader may use the suggested procedure or devise one functionally equivalent.

```
function LVxFIRViaWinIdealLPF(FiltType, BandEdgeVec, ...
typeWin, As, Rp)
% FiltType: 1 = Lowpass, 2 = Highpass, 3 = Bandpass, 4 = Notch
% BandEdgeVec must have 2 or 4 values; it must have length 2
% when defining ws and wp for an LPF or an HPF; it must have
% length 4 for a BPF or Notch filter. typeWin is one of the
% following functions in string format, i.e., surrounded by
% single quotes such as 'rectwin', 'kaiser', 'blackman',
% 'hanning', 'hamming', or 'bartlett'
% As is minimum desired stopband attenuation in dB, as a
% positive number such as 51, 67, etc.
% Test calls:
% Lowpass filters
% LVxFIRViaWinIdealLPF(1, [0.2,0.3], 'kaiser', 80, 0.1)
% LVxFIRViaWinIdealLPF(1, [0.2,0.3], 'boxcar', 21, 0.1)
% LVxFIRViaWinIdealLPF(1, [0.2,0.3], 'hanning', 44, 0.1)
% LVxFIRViaWinIdealLPF(1, [0.2,0.3], 'hamming', 53, 0.1)
% LVxFIRViaWinIdealLPF(1, [0.2,0.3], 'blackman', 74, 0.1)
% Highpass filters
% LVxFIRViaWinIdealLPF(2, [0.5,0.6], 'kaiser', 40, 0.1)
% LVxFIRViaWinIdealLPF(2, [0.5,0.6], 'kaiser', 60, 0.1)
% LVxFIRViaWinIdealLPF(2, [0.5,0.6], 'kaiser', 80, 0.1)
% Bandpass filters
% LVxFIRViaWinIdealLPF(3, [0.2,0.3,0.5,0.6], 'kaiser', 40, 0.1)
% LVxFIRViaWinIdealLPF(3, [0.2,0.3,0.5,0.6], 'kaiser', 60, 0.1)
% LVxFIRViaWinIdealLPF(3, [0.2,0.3,0.5,0.6], 'kaiser', 80, 0.1)
% Bandstop filters
% LVxFIRViaWinIdealLPF(4, [0.2,0.3,0.5,0.6], 'kaiser', 40, 0.1)
% LVxFIRViaWinIdealLPF(4, [0.2,0.3,0.5,0.6], 'kaiser', 60, 0.1)
% LVxFIRViaWinIdealLPF(4, [0.2,0.3,0.5,0.6], 'kaiser', 80, 0.1)
```

Here is a possible design procedure for the script *LVxFIRViaWinIdealLPF*:

a) Write a function *LVxTrunIdealLowpass* that receives a cutoff frequency ω_c in radians and a desired length L and returns to a calling function a truncated ideal lowpass filter impulse response of the given length L and cutoff frequency ω_c.

b) Write a function *LVxLPFViaWindowedSincND* that receives as arguments values for ω_p, ω_s, L, and a desired window type to use, *typeWin* (including the parameter β if *typeWin* is Kaiser). Using these parameters, the script must then obtain a truncated ideal lowpass filter impulse response of length L by calling the function *LVxTrunIdealLowpass*, apply the chosen window, evaluate the frequency response of the filter, and compute and return to a calling function the impulse response, and the actual passband and stopband ripple values.

c) Write a function *LVxDesignLPFViaWindowedSincND* that receives the arguments ω_p, ω_s, $typeWin$, and the desired A_s, estimates an initial value for L (and computes β if *typeWin* is Kaiser), and repeatedly calls *LVxLPFViaWindowedSincND*, gradually increasing L to the lowest value that can produce the desired value of A_s. This script can be used directly to design a windowed lowpass filter. Other filter types (highpass, bandpass, bandstop) must be derived using various combinations of windowed lowpass filters, which is performed by the next listed script, *LVxFIRViaWinIdealLPF*.

d) Write a script *LVxFIRViaWinIdealLPF* that receives arguments consisting of the desired filter type (lowpass, highpass, bandpass, bandstop), a vector of band edges of length two for lowpass and highpass filters, and length four for bandpass and bandstop filters; a desired value of A_s, and a desired window type. A desired maximum passband ripple may also be specified, and the script will return the actual value of passband ripple for comparison to the design goal.

3. Write the m-code for the script *LVxFilterViaCosineFormula*, as described and illustrated in the text, and test it with the given sample calls.

```
function WF = LVxFilterViaCosineFormula(Type,Bin0,PosBins)
% WF is the output impulse response designed by the script
% A Type I filter is obtained by passing Type as 1
% A Type II filter is obtained by passing Type as 2
% Bin0 is the sample amplitude for frequency zero.
% PosBins are sample amplitudes for frequencies 1 to (L-1)/2 for
% odd length, or 1 to L/2-1 for even length.
% Sample calls
% WF = LVxFilterViaCosineFormula(2,[1],[1,1,1,1,0,0,0,0,0])
% WF = LVxFilterViaCosineFormula(2,[0],[0,0,0,1,1,1,0,0,0])
% WF = LVxFilterViaCosineFormula(2,[0],[0,0,0,0,1,1,1,1,1])
```

4. Write the m-code for the script *LVxFIRViaWholeSines* as illustrated and described in the text and below, and test it using the sample calls given. Sine summation formulas for Type-II and Type-IV filters can be found in the Appendices (use the formulas for whole sine correlators rather than half-sine correlators)

```
function WF = LVxFIRViaWholeSines(AkPos,AkLOver2,L)
% Returns an impulse response WF for a Type III or Type IV
```

```
% linear phase FIR.
% AkPos are sample amplitudes for frequencies 1 to (L-1)/2 for
% odd length, or 1 to L/2-1 for even length. AkLOver2 is passed
% as 0 or the empty matrix [] for odd values of L, and as a desired
% amplitude for even values of L, which is the desired filter length
% Sample calls:
% WF = LVxFIRViaWholeSines([ones(1,38)],[],77); % Hilb
% WF = LVxFIRViaWholeSines([0.8,ones(1,36),0.55],[],77); transition
% values approximately optimized to reduce ripple.
% Type-IV differentiators
% WF = LVxFIRViaWholeSines([1:1:11]*pi/12,[12]*pi/12,24);
% WF = LVxFIRViaWholeSines([1:1:38]*pi/39,[39]*pi/39,78);
% Type-III differentiator
% WF = LVxFIRViaWholeSines([1:1:38]*pi/39,[],77);
```

5. Convert the script/function given in the text

$$LVxDifferentiatorLen24$$

into a generic script/function to generate the impulse response for a Type-IV differentiator of arbitrary even length, having the following call syntax:

$$ImpResp = LVxDifferentiatorTypeIV(L)$$

Your function should compute and display the impulse response ($ImpResp$) and corresponding frequency response for any given even filter length L.

6. Write and test a script that will create a triangle wave as a test signal, and convolve it with a differentiator, which the script creates, to generate a square wave. The script should conform to the following call syntax:

$$LVxTriang2SquareViaDiff(SR, HiHrm, Fo, Ldiff)$$

where SR is a sample rate that determines how many samples the test signal will have, *HiHrm* (an odd integer) is the highest harmonic to add in to synthesize the test waveform having a fundamental frequency of *Fo*, according to the formula for generating a triangle waveform, which is

$$WF = \sum_{n=1}^{(HiHrm+1)/2} (1/(2n-1)^2)\cos(2\pi n F_o t)$$

where t = [0:1:SR]/SR. $Ldiff$ is the length of the differentiator which is to be convolved with the test signal WF to yield a squarewave.

The call

LVxTriang2SquareViaDiff(4000,181,20,94)

should result in substantially the same plot as shown in Fig. 9.43.

7. Write the m-code for the following function, which is described and illustrated in the text with syntax below:

```
function LVxHilbertViaConvolution(TestSeqLen,TestWaveType,...
FilterLen, TestSigFreq,FDMethod,UseWin)
% TestSeqLength is the desired test sequence length;
% TestWaveType = '1' for sawtooth, and '2' for squarewave;
% FilterLen = length of the Hilbert trans. imp resp made directly
% from time domain formula;
% TestSigFreq = fundam. freq of (truncated) sawtooth or square
% wave used as the test signal;
% FDMethod = 1 for an All-Real FD Mask, or 2 for All-Imaginary
% mask the same length as.the test signal, and is converted using
% the ifft to a TD Hilbert transformer which is convolved with the
% test signal.
% UseWin = 0 to use the raw TD Hilbert impulse response, or 1 to
% window it with a Kaiser window, Beta = 5
% Test call:
% LVxHilbertViaConvolution(64, 1, 19, 5, 2,0)
```

8. Write a script that designs an equiripple notch filter to meet certain specifications as given in the following call, and test it with the given calls, noting, for each call, the length of the resulting filter.

```
function [R,A,b] = LVxDesignEquirippNotch(Rp,As,...
wp1,ws1,ws2,wp2)
% Rp is the maximum design passband ripple in positive dB
% and As is the minimum design stopband attenuation in dB,
% wp1, ws1,ws2, and wp2 specify the 1st passband, 1st
% stopband, 2nd stopband, & 2nd passband frequencies,
% respectively. R and A are the realized values of Rp
% and As, respectively, and b is the vector of designed filter
% coefficients.
% Test calls:
% [R,A,b] = LVxDesignEquirippNotch(0.2,40,0.2,0.3,0.5,0.6)
% [R,A,b] =LVxDesignEquirippNotch(0.2,55,0.2,0.3,0.5,0.6)
% [R,A,b] =LVxDesignEquirippNotch(0.2,70,0.2,0.3,0.5,0.6)
% [R,A,b] =LVxDesignEquirippNotch(0.2,40,0.2,0.25,0.4,0.45)
% [R,A,b] =LVxDesignEquirippNotch(0.2,55,0.2,0.25,0.4,0.45)
```

% [R,A,b] =LVxDesignEquirippNotch(0.2,70,0.2,0.25,0.4,0.45)

9. For each of the equiripple notch filters designed in the previous exercise, design a minimum-length notch filter meeting the same specifications using the Windowed-Ideal LPF technique with a Kaiser window. Which technique results in the shorter filter in general? Compare the filter lengths of the equiripple and Kaiser filters for the A_S = 40 dB case, the A_S = 55 dB case, and the A_S = 70 dB case, and note the relative (or percentage) length difference between the Kaiser and equiripple designs, then compare the two lengths for the A_S = 55 dB and A_S = 70 dB cases and note again the relative or percentage length differences. You should find that at higher values of A_S, the efficiency of the equiripple design compared to the Kaiser windowed design is greater than for smaller values of A_S.

10. Using the script *LVDemoRemez*, for each given call, iteratively manually estimate successive sets of extremals, attempting to achieve convergence, i.e., an equiripple filter.

(a) **LVDemoRemez(0.45,0.55,40,0.5,33,1,[],1)**
(b) **LVDemoRemez(0.45,0.55,40,0.5,9,1,[0,0.2875,0.45,0.55,...
0.7125,1],1)**
(c) **LVDemoRemez(0.65,0.75,40,0.5,19,1,[0,0.135,0.265,0.385,...
0.5,0.6,0.65,0.75,0.8,0.895,1],1)**
(d) **LVDemoRemez(0.65,0.75,40,0.5,20,1,[0,0.106,0.208,0.315,...
0.41,0.519,0.6062,0.65,0.75,0.8,0.9063],1)**
(e) **LVDemoRemez(0.75,0.65,60,0.5,19,2,[0,0.125,0.27,0.385,...
0.5,0.6,0.65,0.75,0.795,0.892,1],1)**
(f) **LVDemoRemez(0.75,0.65,40,0.5,19,2,[0,0.1,0.22,0.35,0.46,...
0.56,0.61,0.75,0.79,0.88,1],0)**
(g) **LVDemoRemez(0.75,0.65,40,0.5,19,2,[0,0.138,0.269,...
0.3875,0.5,0.6,0.65,0.75,0.8,0.894,1],1)**

11. Attempt to achieve the three optimum transition band values $T1$, $T2$, and $T3$ in either of the LabVIEW VIs

$$DemoHPFOptimizeXitionBandsVI$$

$$DemoHPFOptimizeXitionBandsPrecVI$$

Use the following procedure: Set $T1$, $T2$, and $T3$ to 1.0, then gradually reduce $T1$ until the maximum stopband attenuation is achieved. Then reduce $T2$ by 0.05, and adjust $T1$ until a new minimum is achieved. Keep reducing $T2$ in steps of 0.05 and reoptimizing $T1$ until the best stopband attenuation has been achieved using only $T1$ and $T2$. Then reduce $T3$ by 0.05, then perform again the entire $T2$, $T1$ optimization procedure. Repeat the entire general procedure until the best stopband attenuation has been achieved. In an actual computer search algorithm, of course, the step size would need to be much smaller than 0.05, but the procedure outlined above illustrates one way to determine approximate optimum values.

12. Write a script that designs an equiripple highpass filter according to the following function description:

```
function [actRp,actAs,WF] = LVxDesignEquirippHPF(Rp,As,...
ws,wp)
% Rp and As are the maximum passband ripple and minimum
% stopband attenuation, respectively.
% ws and wp are the stopband and passband edges, specified
% in normalized frequency, i.e., units of pi.
% actRp and actAs are the realized values of Rp and As from
% the filter design.
% Test calls:
% LVxDesignEquirippHPF(0.2,60,0.45,0.55)
% LVxDesignEquirippHPF(0.02,60,0.4,0.55)
% LVxDesignEquirippHPF(0.5,70,0.15,0.25)
```

13. In this project, we'll modify and use the script

$$[T, F, B] = LV\_DetectContTone(A, Freq, RorSS, SzWin, OvrLap, AudSig)$$

to not only identify an interfering steady-state or rising-amplitude sinusoid, but to remove it from the test signal using an equiripple lowpass, highpass, or notch filter designed automatically in response to the list of candidate interfering frequencies generated through analysis of the spectrogram matrix.

Once the filter is designed, filtering is performed using two methods. The first method is to filter the entire test signal in the time domain using the filter coefficients $[b, a]$ and the function $filter$. The second method is to implement convolution in the frequency domain on each frame of the signal, the output signal then being constructed from the frames. To do this, the test signal is first divided into frames with a certain overlap, such as 50%, to form the matrix $TDMat$, and then each frame (i.e., column) of $TDMat$ is filled out with zeros to double its length. The DFT of all frames (columns) is then obtained, forming the matrix $lgFty$, and the analysis proceeds using the nonnegative bins of $lgFty$. Once the interfering frequency or frequencies have been identified and suitable filter coefficients have been determined, an equivalent impulse response, truncated to the same length as the augmented columns of $lgFty$, is generated and its DFT is obtained. Each frame of $lgFty$ is then multiplied by this DFT, and then the real part of the inverse DFT of $lgFty$ is obtained to form a matrix *lgTD*. A matrix $newTD$, consisting of the upper half of the columns of *lgTD*, is then formed. To perform the proper Overlap and Add routine for the frequency domain convolution, the lower half of *lgTD* is shifted one column to the right and added to the upper half of $newTD$. The frames of $newTD$ are then concatenated with the proper overlap to generate the filtered output signal. DFTs of the test signal and both versions of the output signal (i.e., filtered in the time domain and via the frequency domain) are computed and plotted for comparison. The test signal and both filtered versions are also played through the computer's audio system for audible comparison.

The actual filter design involves using a lowpas filter when the interfering frequency or frequencies lie at or above 0.95π radians, a highpass filter when the interfering frequency or frequencies lie at or below 0.035π radians, or a notch filter otherwise. The necessary design programs can be supplied using the supplied scripts *LVDesignEquirippLPF* and *LVDesignEquirippBPF*, and two scripts written for exercises previous to this one, *LVxDesignEquirippNotch*, and *LVxDesignEquirippHPF*.

```
function [TF,Fnyq,BnSp] = LVx_DetnFiltFIRContTone(A,Freq,...
RorSS,SzWin,OvrLap,AudSig,Rp,As)
% Creates a test signal comprising an audio signal and an
% interfering sinusoid, and attempts to identify the interfering
% tone. The output arguments and the first six input arguments
% are identical to those of the script LVx_DetectContTone
% After analyzing the signal, the script then evaluates the list of
% candidate interfering tones TF to determine if filtering should
% be performed. Frequencies below 80 Hz are eliminated from the
% list as the two audio files contain prominent 60 Hz components,
% and the program is only designed to filter out one main
% spectral component which lies above 80 Hz. Also, the ear's
% sensitivity at low frequencies is very low, and the audibility of
% such tones is often well-masked by other signal components.
% The candidate frequencies above 80 Hz comprise either a single
% frequency or a number of contiguous frequencies(i.e.,lying in
% adjacent bins of the DFT). When a number of contiguous
% candidate frequencies exist, the lower and upper frequency
% bounds are established. If only a single frequency candidate
% exists, upper and lower bounds surrounding it are created so
% that a notch filter of reasonable width can be designed. When the
% interfering frequency or frequencies lie at or below 0.035pi
% radians (about 140 Hz for this exercise), a highpass filter is
% used; when the interfering frequencies lie above 0.95pi
% radians, a lowpass filter is used. Otherwise, a notch filter is used.
% The filter (an equiripple FIR) is designed to have maximum
% passband ripple of Rp dB and minimum stopband attenuation
% of As dB.
% Filtering is performed two ways, first, by time domain
% convolution of the entire test signal with the designed filter,
% and secondly, by frequency domain convolution
% that uses the original analysis matrix, which is doubled in
```

```
% column length (by addition of zeros to each column) to
% accommodate both the analysis and frequency domain
% convolution. After the columns of this special matrix have
% been multiplied by the DFT of the designed filter's impulse
% response, the IDFT is obtained to form filtered time
% domain signal frames, and the output signal vector is
% reconstructed by concatenating frames with the proper
% overlap, if any.
% The test signal, the time domain filtered version, and the
% frequency domain filtered version are all played out
% through the computer's audio system for comparison.
% Several figures are created in addition to those associated
% with frequency detection per se, including the filter impulse
% response, the filter frequency response, and the test signal
% frequency content after being filtered as one signal vector,
% and after being filtered using the DFT Convolution
% method, reconstructing the signal vector from frames
%
% [T,F,B] = LVx_DetnFiltFIRContTone(0.015,200,0,512,1,1,1,30)
% [T,F,B] = LVx_DetnFiltFIRContTone(0.01,210,0,512,1,1,1,30)
% [T,F,B] = LVx_DetnFiltFIRContTone(0.01,200,0,512,1,2,1,30)
% [T,F,B] = LVx_DetnFiltFIRContTone(0.1,200,0,512,1,3,1,30)
% [T,F,B] = LVx_DetnFiltFIRContTone(0.01,500,0,512,1,1,1,30)
% [T,F,B] = LVx_DetnFiltFIRContTone(0.01,500,0,512,1,2,1,30)
% [T,F,B] = LVx_DetnFiltFIRContTone(0.08,500,0,512,1,3,1,30)
%
% [T,F,B] = LVx_DetnFiltFIRContTone(0.025,200,1,512,1,1,1,30)
% [T,F,B] = LVx_DetnFiltFIRContTone(0.02,200,1,512,1,2,1,30)
% [T,F,B] = LVx_DetnFiltFIRContTone(0.1,200,1,512,1,3,1,30)
% [T,F,B] = LVx_DetnFiltFIRContTone(0.02,500,1,512,1,1,1,30)
% [T,F,B] = LVx_DetnFiltFIRContTone(0.018,500,1,512,1,2,1,30)
% [T,F,B] = LVx_DetnFiltFIRContTone(0.1,500,1,512,1,3,1,30)
% [T,F,B] = LVx_DetnFiltFIRContTone(0.07,150,0,512,1,3,1,30)
%
% [T,F,B] = LVx_DetnFiltFIRContTone(0.005,1000,0,512,0,1,1,30)
% [T,F,B] = LVx_DetnFiltFIRContTone(0.005,1000,0,512,1,1,1,30)
```

The call

```
[T,F,B] = LVx_DetnFiltFIRContTone(0.02,100,0,512,0,1,1,40)
```

results in Figures 9.51–9.53, in addition to those figures created as part of the frequency detection process per se.

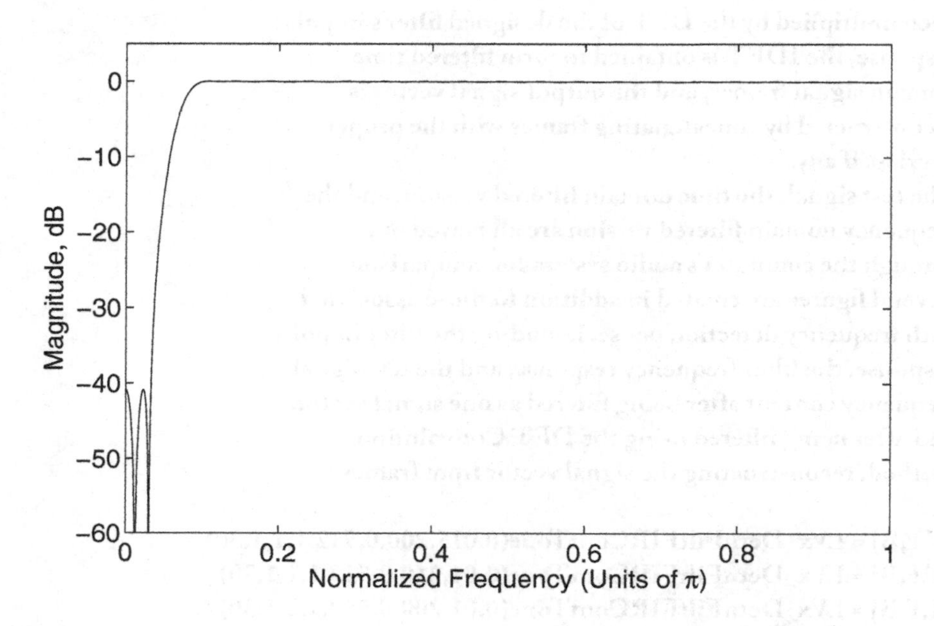

Figure 9.51: Frequency response of the filter designed by the script to remove a 100 Hz tone from the test signal.

The call

[T,F,B] = LVx_DetnFiltFIRContTone(0.01,500,0,512,1,1,1,50)

results in Figures 9.54 and 9.55, in addition to those figures created as part of the frequency detection process per se.

14. Write a script, based on the work done for the script in the previous exercise, that meets the following format and description:

```
function [T] = LVxAnalyzeModWavFile(strWavFile,SzWin,...
OvrLap,FrqRg,A,Freq,Rp,As,Act,strOutWavFile)
% strWavFile is a .wav file in the target folder (or a folder on
% the MathScript search path to be read and processed
% according the value of Act).
% Act = 1 means analyze only and
% report candidate interfering frequencies as output variable T
% Act = 2 means analyze and filter candidate frequencies if they
```

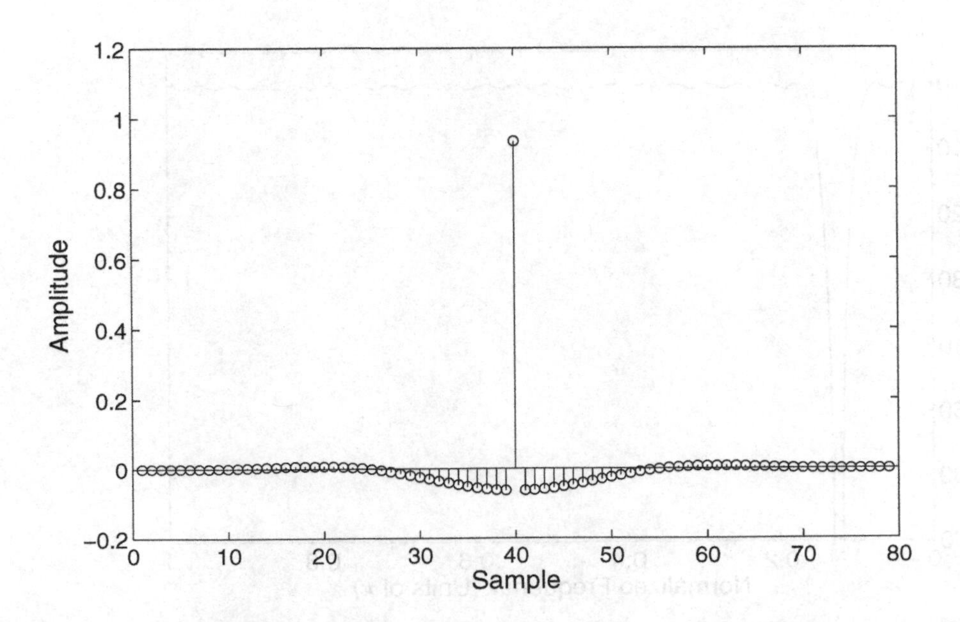

Figure 9.52: Impulse response of the filter designed by the script to remove a 100 Hz interfering tone.

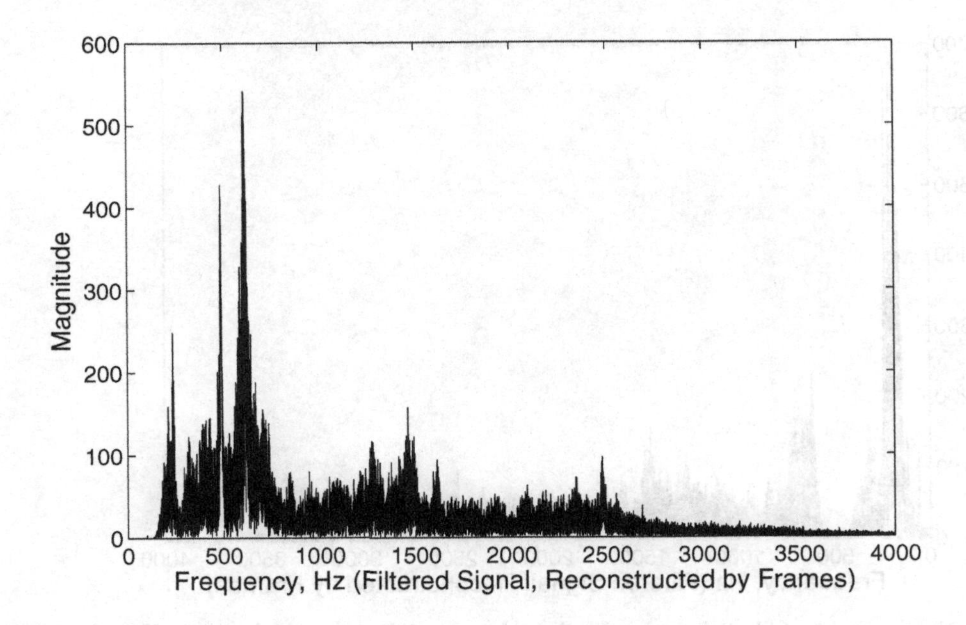

Figure 9.53: Spectrum of filtered output signal reconstructed from frames that were filtered using DFT convolution. Note the attenuated band of frequencies below about 100 Hz.

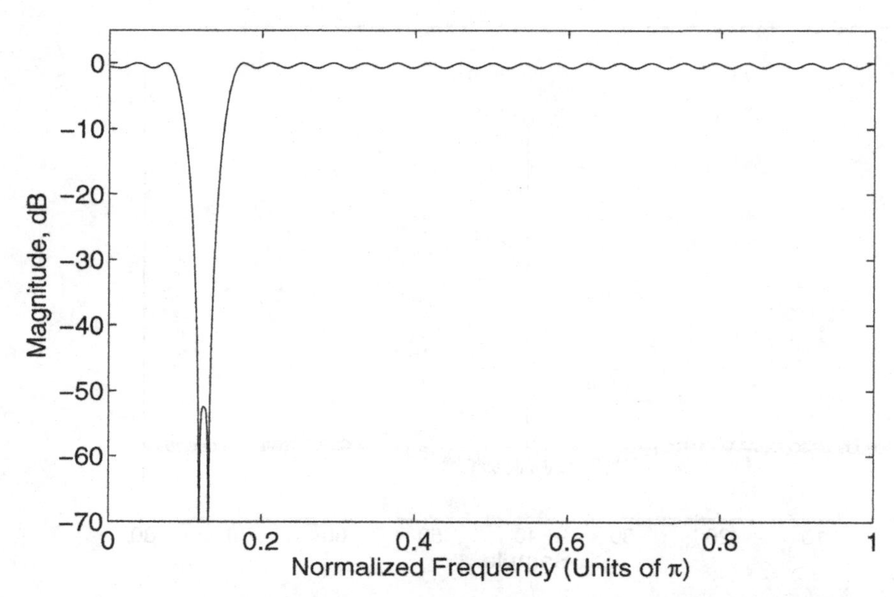

Figure 9.54: Spectrum of 500 Hz notch filtered designed automatically by the script to remove the detected 500 Hz tone.

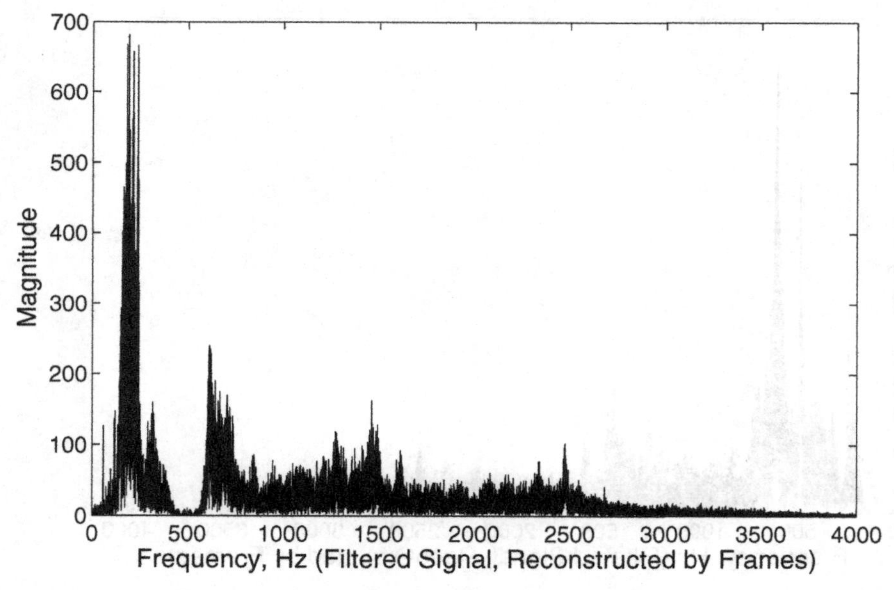

Figure 9.55: Spectrum of filtered output signal reconstructed from frames that were filtered using DFT convolution. Note the notch surrounding 500 Hz.

```
% lie within the frequency range designated by FrqRg (see below)
% Act = 3 means analyze, filter, and save the filtered output as
% a .wav file under the name strOutWavFile
% Act = 4 means do not analyze or filter, but make a .wav file
% from the test signal, which might have an added component if
% A has magnitude greater than zero. The action may be used to
% generate .wav files having specific frequencies of interfering
% sinusoid.strOutWavFile must be different from strWavFile
% TonFreq, SzWin, OvrLap, A, and Freq have the same meaning
% and use as in the scripts LVx_DetectContTone,
% LVx_DetnFiltFIRContTone, and LVx_DetnFiltIIRContTone
% FrqRg is a vector of two frequencies in Hz that define a range
% over which to consider filtering. This is useful when there are
% multiple interfering frequencies since only simple filters
% (LPF, HPF, and Notch) can be used per call to this script.
% Successive calls specifying different frequency ranges
% will allow removal, for example, of multiple interfering
% frequencies lying at a distance from one another.
%
% Rp and As are the desired maximum passband ripple in dB
% and minimum stopband attenuation in dB for the filter.
% Generally, the minimum value of As should be experimentally
% determined and used; due to masking caused by transient
% signal tones, the audibility of a low level persistent tone can
% often be eliminated using As = 20 dB or 30 dB. Very strong
% persistent tones may require correspondingly higher
% values of As.
%
% Test calls:
%
% [T] = LVx_AnalyzeModWavFile('drwatsonSR8K.wav',512,0,...
[],0,0,1,20,1,[])
%
% [T] = LVx_AnalyzeModWavFile('drwatsonSR8K.wav',512,0,...
[40,80],0,0,1,20,2,[])
%
% [T] = LVx_AnalyzeModWavFile('drwatsonSR8K.wav',512,0,...
[40,80],0,0,1,20,3,'drw8Kless60Hz.wav')
%
```

```
% [T] = LVx_AnalyzeModWavFile('drw8Kless60Hz.wav',512,0,...
[40,80],0,0,1,20,1,[])
%
% [T] = LVx_AnalyzeModWavFile('drwatsonSR8K.wav',512,0,...
[40,80],0.02,400,1,20,1,[])
%
% [T] = LVx_AnalyzeModWavFile('drwatsonSR8K.wav',512,0,...
[40,80],0,0,1,20,2,[])
%
% [T] = LVx_AnalyzeModWavFile('drwatsonSR8K.wav',512,0,...
[],0.025,400,1,20,4,'drw8Kplus400Hz.wav')
%
% [T] = LVx_AnalyzeModWavFile('drw8Kplus400Hz.wav',512,0,...
[200,600],0.025,400,1,20,2,[])
```

After making the call

$$[T] = LVx\_AnalyzeModWavFile('drwatsonSR8K.wav', 512, 0, [], 0.025, 400, 1, 20, 4, 'drw8Kplus400Hz.wav')$$

which adds a 400 Hz tone to the audio file *'drwatsonSR8K.wav'* and stores the result as *'drw8Kplus400Hz.wav'*, the call

$$[T] = LVx\_AnalyzeModWavFile('drw8Kplus400Hz.wav', ... 512, 0, [], 0.025, 400, 1, 20, 2, [])$$

is made, which displays the minimum energy frame spectrogram and the several post-filtering spectral plots as well as the filter's frequency response (see example Figures 9.56–9.58).

15. In this project, we explore the computational requirements to meet certain FIR design requirements. The goal of the project is to show that under some circumstances, the shortest filter is not the one that is most computationally efficient. Under certain conditions, the Frequency Sampling Realization Method can be employed to greatly reduce computational requirements. To explore this, we'll design a lowpass filter of low passband width ($\omega_p < 0.1\pi$) using the Frequency Sampling Design technique, determine the realized values of R_P, and A_S, implement the filter using the Frequency Sampling Realization Method, and then determine the total computation necessary to generate each sample of output. We'll then design an equivalent equiripple lowpass filter using as the design criteria the values of ω_p, ω_s and the realized values of R_P and A_S determined above. Then we will repeat the exercise for a lowpass filter having $\omega_p = 0.4\pi$.

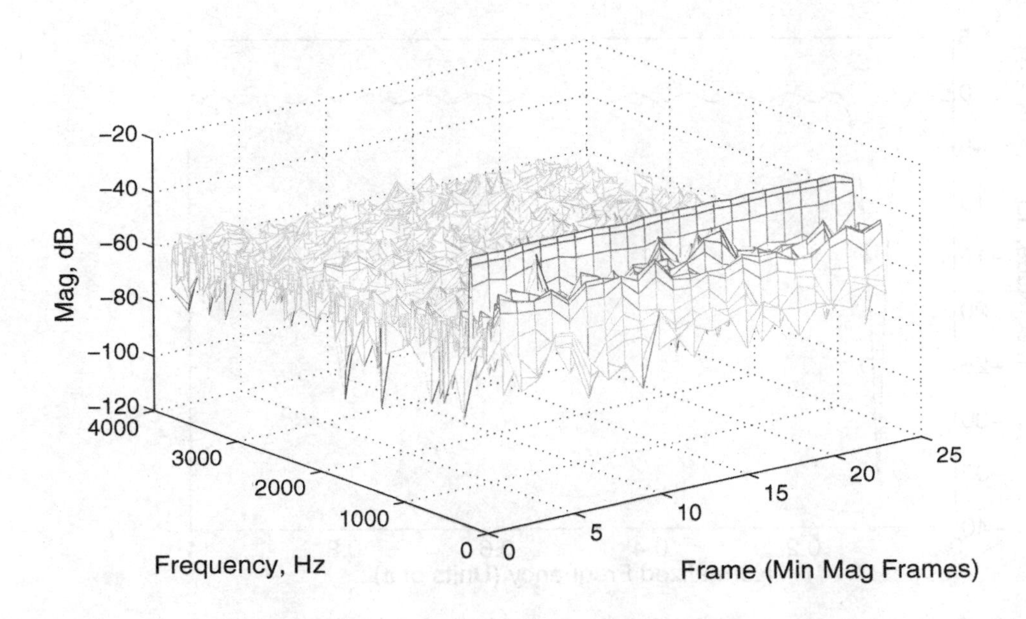

Figure 9.56: Minimum-magnitude-frame spectrogram of signal '*drwatson8Kplus400Hz.wav*'.

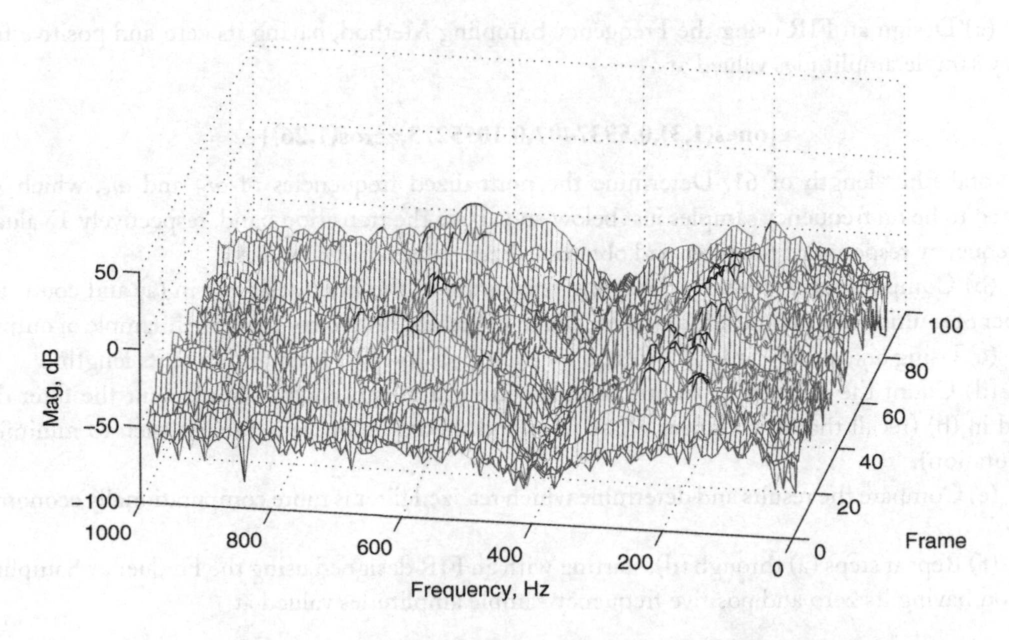

Figure 9.57: Spectrogram (to 1000 Hz, for all frames) of signal '*drwatson8Kplus400Hz.wav*' after being automatically filtered to remove the persistent 400 Hz tone.

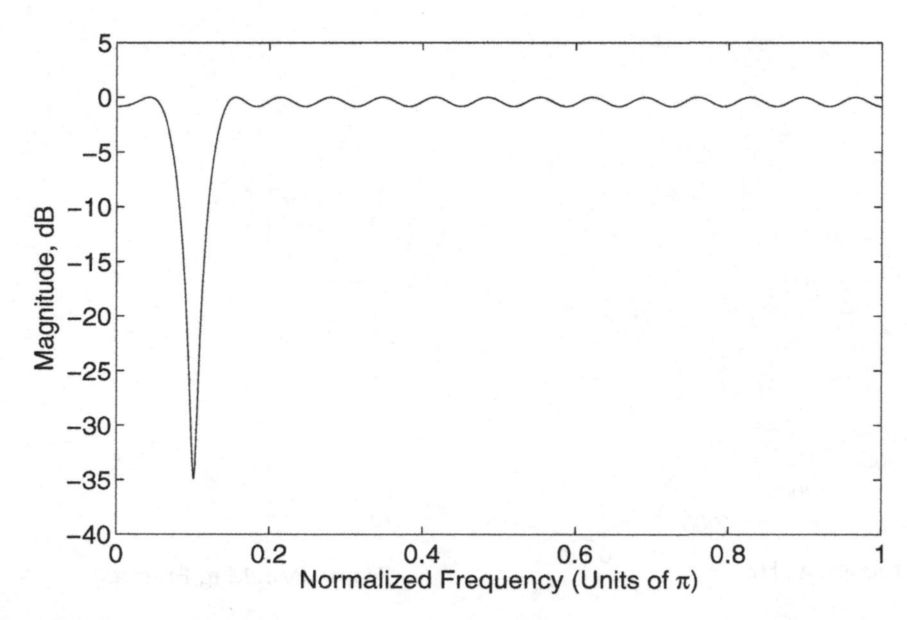

Figure 9.58: Frequency response of equiripple notch filter automatically designed to remove the persistent 400 Hz tone in the sound file 'drwatson8Kplus400Hz.wav'.

(a) Design an FIR using the Frequency Sampling Method, having its zero and positive frequency sample amplitudes valued at

[ones(1,3),0.5937402,0.1055273,zeros(1,26)]

for a total filter length of 61. Determine the normalized frequencies of ω_p and ω_s, which are assumed to lie on frequency samples just below and above the transition band, respectively. Evaluate the frequency response of the filter and obtain realized values of R_P, and A_S.

(b) Compute the Frequency Sampling realization of the filter designed in (a) and count the number of multiplications and additions needed in the realization to compute each sample of output.

(c) Using ω_p, ω_s, R_P, and A_S, design an equiripple lowpass filter and note its length.

(d) Count the number of additions and multiplications needed to implement the filter designed in (b) (recall that it is a linear phase filter and should be implemented as such to minimize computation).

(e) Compare the results and determine which realized filter is more computationally economic to use.

(f) Repeat steps (a) through (d), starting with an FIR designed using the Frequency Sampling Method having its zero and positive frequency sample amplitudes valued at

[ones(1,13),0.592998,0.1095556,zeros(1,15)]

CHAPTER 10

Classical IIR Design

10.1 OVERVIEW

In the previous chapter, we found that it was possible to design filters having linear phase with desired passband ripple, stopband attenuation, and transition widths. Achieving high stopband attenuation along with narrow transition bands (i.e., "steep roll-off"), however, comes at the price of long FIR filters having high computational cost. In this chapter, we examine the design of four types of infinite impulse response filters known as **Classical IIRs,** which were originally developed in the continuous domain using the Laplace transform. Classical IIR design theory permits ready design of recursive digital filters for all the common passband types (lowpass, highpass, bandpass, and notch). Such filters generally do not have linear phase characteristics, but steep roll-offs and high stopband attenuation can be achieved with far less computational overhead than with FIRs. Accordingly, these filters constitute yet another valuable asset in digital signal processing.

In this chapter, we briefly describe the Laplace transform and domain, followed by a description of prototype analog lowpass filters of the four standard Classical filter types (Butterworth, Chebyshev Types I and II, and Elliptic). We then see how the basic lowpass prototype filters can be converted into other filter types, such as highpass, etc. Having thus established the means to design any of the standard passband filter types (lowpass, highpass, bandpass, bandstop) in the analog domain, we then investigate two methods, Impulse Invariance and the Bilinear Transform, to convert an analog filter into an equivalent digital filter. We then describe MathScript's functions for designing Classical IIRs, and conclude the chapter with a brief mention of IIR optimization algorithms with a pointer to further reading.

10.2 SOFTWARE FOR USE WITH THIS BOOK

The software files needed for use with this book (consisting of m-code (.m) files, VI files (.vi), and related support files) are available for download from the following website:

http://www.morganclaypool.com/page/isen

The entire software package should be stored in a single folder on the user's computer, and the full file name of the folder must be placed on the MATLAB or LabVIEW search path in accordance with the instructions provided by the respective software vendor (in case you have encountered this notice before, which is repeated for convenience in each chapter of the book, the software download only needs to be done once, as files for the entire book are all contained in the one downloadable folder).

See Appendix A for more information.

10.3 LAPLACE TRANSFORM

10.3.1 DEFINITION

Just as the DTFT

$$DTFT(x[n]) = X(e^{j\omega}) = \sum_{n=-\infty}^{\infty} x[n]e^{-j\omega n}$$

is a discrete case or version of the continuous domain Fourier transform

$$F(\omega) = \int_{-\infty}^{\infty} x(t)e^{-j\omega t}dt$$

the z-transform, which we have already studied,

$$X(z) = \sum_{n=-\infty}^{\infty} x[n]z^{-n} \tag{10.1}$$

is actually a discrete form of the Laplace transform, which is defined as

$$\pounds\ (s) = \int_{-\infty}^{\infty} x(t)e^{-st}dt \tag{10.2}$$

with

$$s = \sigma + j\omega$$

where σ, a real number, is a damping factor, and $\omega = 2\pi f$.

Recall that most sequences we deal with in the real world can be characterized as right-handed, or identically zero for values of time less than zero, which simplifies the z-transform to

$$X(z) = \sum_{n=0}^{\infty} x[n]z^{-n} \tag{10.3}$$

and analogously for right-handed, continuous time signals, the Laplace transform becomes

$$\pounds\ (s) = \int_{0}^{\infty} x(t)e^{-st}dt \tag{10.4}$$

10.3.2 CONVERGENCE

When the integral in Eq. (10.2) converges (which it does for certain cases) the Laplace Transform is said to exist or to be defined; when the integral does not converge the transform is said to be undefined.

Many signals of interest take exponential form, such as real exponentials, complex exponentials, sine, cosine, etc., and the Laplace transforms for such signals are generally defined.

Example 10.1. Determine the Laplace transform and convergence criteria for the signal

$$f(t) = e^{at}u(t)$$

with a being, in general, a complex number, and $u(t)$ has the value 0 for $t < 0$ and 1 for all other t. We can determine the Laplace transform as

$$\pounds\,(f(t)) = \int_0^\infty f(t)e^{-st}dt = \int_0^\infty e^{at}e^{-st}dt \tag{10.5}$$

$$= \int_0^\infty e^{-(s-a)t}dt = \frac{-1}{s-a}\left(e^{-(s-a)t}\,|_0^\infty\right)$$

The net result is

$$\pounds\,(e^{at}u(t)) = \frac{1}{s-a} \tag{10.6}$$

where the result can also be obtained from a standard table of Laplace Transforms which can be found in many books, such as Reference [2]. Note that for the integral at (10.5) to converge, the real part of s (i.e, namely σ), must be greater than Re(a). For example, if $a = -6 + 2j$, then $\sigma > -6$ for (10.5) to converge. Stated differently, but more generally, the real part of s must lie to the right of the real part of the rightmost pole (as graphed in the complex plane) of the system or function $f(t)$ for the Laplace Transform to converge and thus be defined.

Example 10.2. Determine the Laplace transform and convergence criteria for

$$f(t) = \cos(\omega_0 t)$$

An easy approach is to use Euler's identity and construct the Laplace integral as

$$\pounds\,(f(t)) = \int_0^\infty f(t)e^{-st}dt = \int_0^\infty \frac{1}{2}(e^{j\omega_0 t} + e^{-j\omega_0 t})e^{-st}dt$$

which results in

$$\frac{1}{2}\left(\frac{1}{s-j\omega_0} + \frac{1}{s+j\omega_0}\right) \tag{10.7}$$

which results in

$$\frac{1}{2}\left(\frac{(s+j\omega_0)+(s-j\omega_0)}{(s-j\omega_0)(s+j\omega_0)}\right)=\frac{s}{s^2+\omega_0^2}$$

for $\sigma > 0$.

10.3.3 RELATION TO FOURIER TRANSFORM

It can be seen that the waveform that is correlated with the signal $x(t)$ is a complex exponential which may have an amplitude which grows, shrinks, or remains the same over time, depending on the value of σ. When $\sigma = 0, e^{-\sigma t}$ =1, and the resultant integral is identical to the Fourier Transform.

The Laplace Transform is thus a transform that includes all the information of the Fourier Transform, plus a good deal more. Not only are correlations done with constant amplitude orthogonal pairs, correlations are also done with decaying and increasing amplitude orthogonal pairs. This kind of transform uncovers more information about a system than can be obtained with the Fourier Transform. Like the z-transform, the Laplace Transform allows the poles and zeros of a system to be identified.

10.3.4 RELATION TO z-TRANSFORM

Note that $x[n]$ (a sampled sequence), used as an input to the z-transform, as shown in Eq. (10.1), is just a discrete or sampled version of the continuous function $x(t)$ used as an input to the Laplace Transform (as shown in Eq. (10.2)). The complex correlator in the z-transform,

$$z^{-n}=(Me^{j2\pi k/N})^{-n}=M^{-n}(e^{-j2\pi k/N})^n=(e^{-j2\pi k/N}/M)^n$$

where M (magnitude) is a real number, k is frequency (a continuous-valued real number), N is sequence length, and n is the sample index, is a discrete or sampled form of the complex exponential $e^{-\sigma t}e^{-j\omega t}$. This is true since by choosing k and M properly, the sequence z^{-n} will form a sampled version of the continuous Laplace correlator e^{-st}.

10.3.5 TIME DOMAIN RESPONSE GENERATED BY POLES

The time domain response generated by a single pole is

$$y(t)=e^{(\sigma_p+j\omega_p)t}$$

where $(\sigma_p + j\omega_p)$ is the value of the pole and t is time, with σ_p a real number and ω_p the radian frequency, $2\pi f_p$.

Figure 10.1 shows three pairs of complex conjugate poles, a first pair (shown in plot (a)), located in the left half-plane, a second pair on the j-Ω axis (plot (c)), and a third pair in the right half-plane (plot (e)). The corresponding time domain responses, all-real, are shown in plots (b), (d), and (f), respectively.

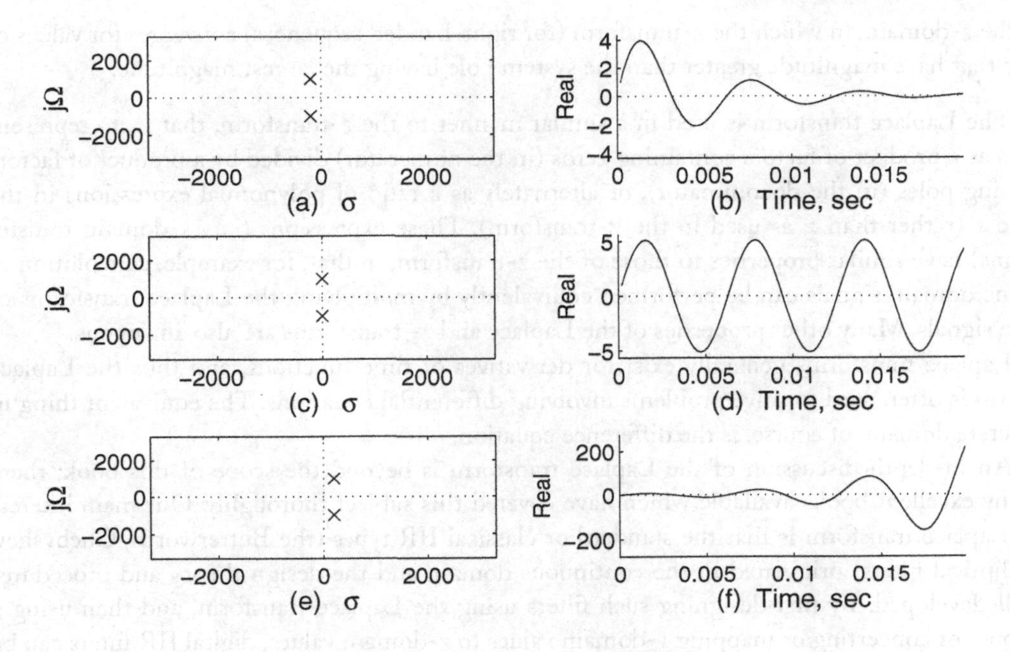

Figure 10.1: (a) A pair of complex conjugate poles in the left half-plane; (b) The time domain response arising from the poles shown in (a); (c) A pair of complex conjugate poles on the imaginary axis; (d) The time domain response arising from the poles shown in (c); (e) A pair of complex conjugate poles in the right half-plane; (f) The time domain response arising from the poles shown in (e).

10.3.6 GENERAL OBSERVATIONS

Values of the Laplace Transform are, in general, complex, and hence they are graphed in the complex plane. The real part of the Laplace Transform represents the damping factor σ, values of which are found along the real (horizontal) axis, and the imaginary part represents frequency, values of which are found along the imaginary (vertical) axis. Both axes range from negative to positive infinity.

- A pole in the left half-plane (i.e., having its real part less than zero) of the s-domain is stable, and corresponds to (or generates) a decaying exponential.
- A pole lying on the j-Ω axis (i.e., having its real part equal to zero) generates a constant, unity-amplitude exponential in response to an impulse and is borderline unstable.
- A pole lying in the right half-plane (i.e., having its real part greater than zero) of the s-Domain represents an exponential that grows with time and hence corresponds to an unstable transfer function.
- The Laplace Transform converges (and is thus defined) for values of s lying to the right of the right-most system pole as graphed in the s-domain. This is analogous to the situation in

the z-domain, in which the z-transform (for right-handed sequences) converges for values of z that have magnitude greater than the system pole having the largest magnitude.

The Laplace transform is used in a similar manner to the z-transform, that is, to represent systems as a product of factors containing zeros (in the numerator) divided by a product of factors containing poles (in the denominator), or alternately as a ratio of polynomial expressions in the variable s (rather than z as used in the z-transform). These expressions (i.e., s-domain transfer functions) have similar properties to those of the z-transform, in that, for example, convolution of two time domain signals can be performed equivalently by multiplying the Laplace transforms of the two signals. Many other properties of the Laplace and z- transforms are also analogous.

Laplace transforms generally exist for derivatives of time functions, and thus the Laplace transform is often used to solve problems involving differential equations. The equivalent thing in the discrete domain, of course, is the difference equation.

An in-depth discussion of the Laplace transform is beyond the scope of this book; there are many excellent books available which have covered this subject thoroughly. Our main interest in the Laplace transform is that the standard or classical IIR types (the Butterworth, Chebyshev, and Elliptical filters) originated in the continuous domain and the design theory and procedures are well-developed. By first designing such filters using the Laplace transform, and then using a technique for converting or mapping s-domain values to z-domain values, digital IIR filters can be efficiently designed.

10.4 PROTOTYPE ANALOG FILTERS

10.4.1 NOTATION

The specifications for analog filters differ somewhat from those of the FIRs we have studied so far; the specifications for analog filters are generally given relative to the magnitude squared of the transfer function. Passband ripple is specified by a ripple parameter ϵ, and stopband attenuation by A. Cutoff frequencies are notated using Ω_p and Ω_s (both in radians per second). For a lowpass filter, for example, the specifications would be given as

$$\begin{array}{ll} 1/(1+\epsilon^2) \le |H(j\Omega)|^2 \le 1 & |\Omega| \le \Omega_p \\ 0 \le |H(j\Omega)|^2 \le 1/A^2 & \Omega_s \le |\Omega| \end{array}$$

Figure 10.2 illustrates a typical analog filter specification.

The parameters A and ϵ are related to A_s and R_p according to the following formulas:

$$R_p = -10\log_{10}(1/(1+\epsilon^2))$$

$$\epsilon = \sqrt{10^{R_p/10} - 1}$$

$$A_s = -10\log_{10}(1/A^2)$$

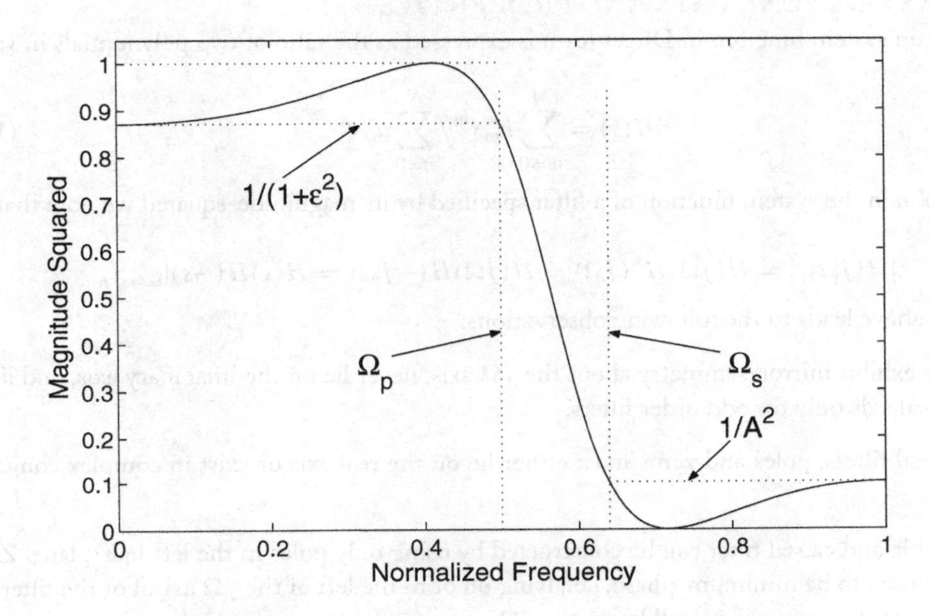

Figure 10.2: The specifications for an analog lowpass filter.

$$A = 10^{A_s/20}$$

In regard to the magnitudes of passband ripple δ_1 and stopband attenuation δ_2, we note that

$$\frac{1-\delta_1}{1+\delta_1} = \sqrt{\frac{1}{1+\epsilon^2}}$$

and therefore

$$\epsilon = \frac{2\sqrt{\delta_1}}{1-\delta_1}$$

Similarly,

$$\frac{\delta_2}{1+\delta_1} = \frac{1}{A}$$

and therefore

$$A = \frac{1+\delta_1}{\delta_2}$$

10.4.2 SYSTEM FUNCTION AND PROPERTIES

An s-domain system function in Direct form is expressed as the ratio of two polynomials in s:

$$H(s) = \sum_{m=0}^{M} b_m s^m / \sum_{n=0}^{N} a_n s^n \tag{10.8}$$

To obtain the system function of a filter specified by its magnitude-squared we note that

$$|H(j\Omega)|^2 = H(j\Omega)H^*(j\Omega) = H(j\Omega)H(-j\Omega) = H(s)H(-s)|_{\Omega=s/j}$$

The above leads to the following observations:

- Poles exhibit mirror-symmetry about the $j\Omega$ axis, never lie on the imaginary axis, and lie on the real axis only for odd order filters.
- For real filters, poles and zeros must either lie on the real axis or exist in complex conjugate pairs.
- A stable and causal filter can be constructed by using only poles in the left half-plane. Zeros are chosen to be minimum-phase, i.e., lying on or to the left of the $j\Omega$ axis. For the filters we will study, the zeros, if any, all lie on the $j\Omega$ axis.
- $H(s)$ can then be constructed from the chosen left half-plane poles and zeros.

Since the Classical IIRs we will study produce pairs of complex conjugate poles, plus a single real pole when the order is odd, a convenient way to express the system function is as the product of biquad sections (and one first order section when the order is odd), each of which is the ratio of two quadratic polynomials is s, having real-only coefficients. The resulting system function is of the form, for even order N

$$H(s) = \Pi_{i=1}^{N/2} \frac{B_{2,i}s^2 + B_{1,i}s + B_{0,i}}{A_{2,i}s^2 + A_{1,i}s + A_{0,i}}$$

where the coefficients B and A are real coefficients for the i-th biquad section. For odd order N, the system function is of the form

$$H(s) = (\Pi_{i=1}^{(N-1)/2} \frac{B_{2,i}s^2 + B_{1,i}s + B_{0,i}}{A_{2,i}s^2 + A_{1,i}s + A_{0,i}})(\frac{1}{s + A_{0,(N+1)/2}})$$

Example 10.3. A certain filter has three poles, namely $[-0.2273 \pm j0.9766, -0.5237]$ and two zeros: $[\pm j2.7584]$; write the system function using biquad sections as well as in Direct Form.

To get Direct Form, we multiply the various zero and pole factors (this can be done using the function *poly*, for example)

$$H(s) = \frac{(s - j2.7584)(s + j2.7584)}{(s + 0.2273 - j0.9766)(s + 0.2273 + j0.9766)(s + 0.5237)}$$

which yields

$$H(s) = \frac{s^2 + 7.6088}{s^3 + 0.9783s^2 + 1.2435s + 0.5265}$$

and thus we have b = [1,0,7.6088] and a = [1,0.9783,1.2435,0.5265].

To get the cascaded-biquad form, we combine the two complex conjugate poles into one factor, as well as the two complex conjugate zeros and get a single biquad section, with one first order section:

$$H(s) = (0.0692)(\frac{s^2 + 7.6089}{s^2 + 0.4546s + 1.0054})(\frac{1}{s + 0.5237}) \tag{10.9}$$

To convert an s-domain transfer function in Direct Form to one in Cascade Form, use the script

$$[Bbq, Abq, Gain] = LVDirToCascadeClassIIR(b, a, gain)$$

which is similar to the script

$$[Bbq, Abq, Gain] = LVDirToCascade(b, a)$$

presented in the discussion on filter topology in the chapter on the z-transform (found in Part II of the book), except that, in *LVDirToCascadeClassIIR*, the additional input parameter $gain$ is used to scale the output parameter $Gain$.

We can verify the correctness of Eq. (10.9), for example, by running the following m-code:

```
b = [1,0,7.6088]; a = [1,0.9783,1.2435,0.5265];
[Bbq,Abq,Gain]=LVDirToCascadeClassIIR(b,a,1)
```

which yields

Bbq = [1, 0, 7.6088]; Abq = [1,0.4546,1.0054; 0,1,0.5237]; Gain = 1

To convert from Cascade to Direct Form, use the script

$$[b, a, k] = LVCas2DirClassIIR(Bbq, Abq, Gain)$$

10.4.3 COMPUTED FREQUENCY RESPONSE

To obtain the frequency response of an analog filter, we can use the function $polyval(p, x)$, which evaluates the value of a polynomial, the coefficients of which are p, at frequency x. By making x a vector of desired frequencies, and evaluating both the b and a polynomials of a filter, and taking the ratio, we obtain the frequency response. In the function

$$H = LVsFreqResp(b, a, HiFreqLim, FigNo)$$

the argument $HiFreqLim$ is the high limit frequency at which to evaluate the response of the filter defined by its numerator and denominator polynomial coefficients b and a, respectively, and $FigNo$ is the number to assign to the figure created for the plots.

```
function H = LVsFreqResp(b,a,HiFreqLim,FigNo)
FR = (0:HiFreqLim/5000:HiFreqLim); s = j*FR;
H = polyval(b,s)./polyval(a,s);
figure(FigNo); subplot(311); yplot = 20*log10(abs(H)+eps);
plot(FR,yplot); xlabel('(a) Freq, Rad/s'); ylabel('Mag, dB')
axis([0 HiFreqLim -100 5])
subplot(312); plot(FR,abs(H)); xlabel('(b) Freq, Rad/s');
ylabel('Mag'); axis([0 HiFreqLim 0 1.05]);
subplot(313); plot(FR,unwrap(angle(H)))
xlabel('(c) Freq, Rad/s'); ylabel('Radians')
axis([0 HiFreqLim -inf inf])
```

Example 10.4. Evaluate the frequency response of a fifth-order analog lowpass Butterworth filter having a frequency cutoff of π radians.

We'll study Butterworth filters in detail shortly, but for the moment we can obtain a lowpass analog design using the m-code shown, which is followed with a call to *LVsFreqResp* to plot the net frequency response, which results in Fig. 10.3.

```
[b,a] = butter(5,pi,'s')
H = LVsFreqResp(b,a,2*pi,7)
```

Another useful function is one to compute, for an analog lowpass filter, the actual or realized values of R_P and A_S. The input arguments required are the frequency response vector H computed, for example, by the script *LVsFreqResp,* the high frequency limit *HiFreqLim* used to compute H, and the desired or user-specified band limits, Ω_P and Ω_L. As we proceed through the chapter, we will develop scripts that design lowpass filters with user-specified band limits, and hence this script will prove useful for evaluating the actual realized filter performance using the given band limits. For filters other than lowpass, we will, as we proceed through the chapter, write scripts specific to the passband type to evaluate actual filter performance.

```
function [NetRp,NetAs] = LVsRealizedFiltParamLPF(H,OmgP,...
OmgS,HiFreqLim)
Lfr = length(H); mFr = HiFreqLim;
Lenpb = round(OmgP/mFr*Lfr); pb = H(1:Lenpb);
mnpb = min(abs(pb)); NetRp = 20*log10(1/mnpb);
sbStrt = (OmgS/mFr); sb = H(round(sbStrt*Lfr):Lfr);
```

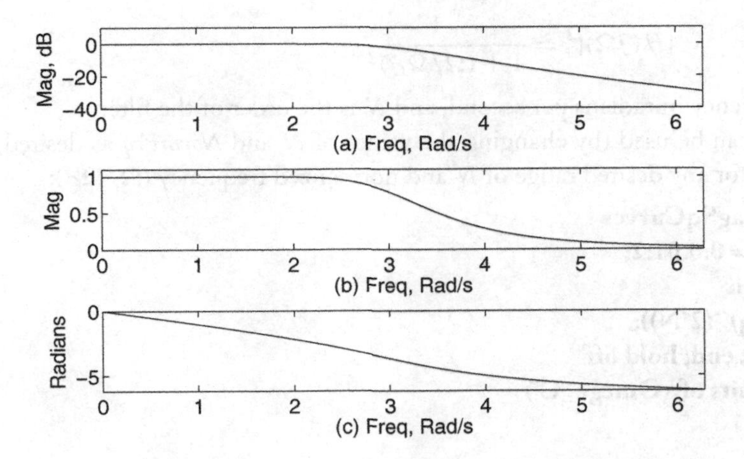

Figure 10.3: (a) Frequency response in dB to frequency 2π radians (1 Hertz) for a fifth order Butterworth lowpass filter; (b) Same, with linear scale rather than dB; (c) Phase response.

```
maxsb = max(abs(sb)); NetAs = 20*log10(1/maxsb);
```

10.4.4 GENERAL PROCEDURE FOR ANALOG/DIGITAL FILTER DESIGN

Our general procedure will be to design prototype lowpass filters based on the well-known analog filters discussed in detail below. Other analog filter types such as highpass, bandpass, and bandstop can be designed by first designing a prototype lowpass analog filter and then using a variable transform or substitution which will convert a lowpass filter transfer function to a different filter type such as highpass, etc. To obtain the desired digital filter, an analog-to-digital transform (such as the Bilinear transform, for example) can then be used to convert the design to the digital domain. An alternate method is to design a lowpass analog filter, convert it into a digital filter, and then use a digital variable-substitution method to convert a lowpass prototype digital filter into another digital passband type, such as highpass, etc. Our discussion below will concentrate on the former method, designing all passband types in the analog domain and then converting to the digital domain.

10.5 ANALOG LOWPASS BUTTERWORTH FILTERS

10.5.1 DESIGN BY ORDER AND CUTOFF FREQUENCY

The Butterworth filter characteristic is one of maximal flatness in the passband at frequency zero, and maximum flatness in the stopband at infinite frequency, which is a desirable trait in certain applications, such as audio amplifiers. The transition band roll-off rate, however, is very shallow for a given filter order N.

The desired magnitude-squared frequency response of a Butterworth filter is

$$|H(j\Omega)|^2 = \frac{1}{1 + (\Omega/\Omega_C)^{2N}}$$

where Ω_C is the cutoff frequency in radians per second, and N is the order of the filter.

The following script can be used (by changing the values of N and *NormFrq* as desired) to plot the frequency response for any desired range of N and normalized frequency (Ω/Ω_C):

```
function LVButterMagSqCurves
figure(79); NormFrq = 0:0.01:2;
for N = 1:1:12; hold on
H = 1./(1 + (NormFrq).^(2*N));
plot(NormFrq,H,'–'); end; hold off
xlabel('Frequency, Units of \Omega_C')
ylabel('Magnitude')
```

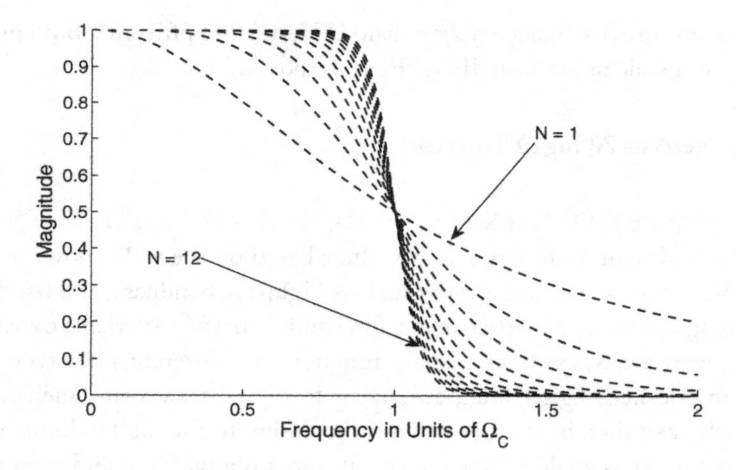

Figure 10.4: The frequency response curves for Butterworth filters of orders 1 to 12.

The system function will be

$$H(s)H(-s) = |H(j\Omega)|^2 = \frac{1}{1 + (s/j\Omega_C)^{2N}} = \frac{(j\Omega_C)^{2N}}{s^{2N} + (j\Omega_C)^{2N}}$$

The roots of the denominator (i.e., the poles of the transfer function) may be found using the expression

$$p_k = (\Omega_C)e^{j(\pi/2N)(2k+N+1)} \tag{10.10}$$

where k = 0:1:2N-1 and the system function can be written as

$$H(s) = \frac{(\Omega_C)^N}{\Pi(s - p_k)} \tag{10.11}$$

where only the p_k lying in the left half-plane are used. The numerator of Eq. (10.11) normalizes the filter gain to 1.0 at frequency 0.0 rad/sec.

Example 10.5. Determine the system function for a Butterworth filter having $\Omega_C = 2$ and $N = 3$.

The following code produces all the possible poles for the system function; however, to ensure stability, only those poles lying in the left half-plane of the s-domain are used. The code plots the poles chosen from the left half-plane and the circle upon which they lie. The system function can then be generated using Eq. (10.11) and the poles obtained as *NetP* in the code below.

```
function LVButterPoles(N,OmegaC)
% LVButterPoles(3,2)
k = 0:1:2*N-1;
P = OmegaC*exp(j*(pi/(2*N))*(2*k+N+1))
NetP = P(find(real(P)<0))
figure(90); clf; hold on; args = 0:0.02:2*pi;
plot(real(NetP),imag(NetP),'bx');
xlabel('Real'); cnums = OmegaC*exp(j*args);
plot(real(cnums),imag(cnums),':')
ylabel('Imaginary')
```

Figure 10.5 is a plot of the three poles computed by the code above, accompanied by the circle upon which they lie in the s-plane.

- The poles of a Butterworth filter all lie on a circle of radius Ω_C.
- None of the poles lie on the imaginary axis.
- All complex poles are accompanied by their complex conjugates.
- To ensure stability, only poles to the left of the imaginary axis in the s-plane are used.

Example 10.6. Evaluate the frequency response of the system function of a Butterworth lowpass filter having three poles and a cutoff frequency of 1.0 rad/s by direct evaluation of the transfer function, factor by factor. Check the results using the script *LVsFreqResp*.

The following code computes $NetP$ and then obtains the product of the factors of the transfer function over a range of normalized test frequencies, each factor magnitude being of the form

$$\frac{1}{s - p_k}$$

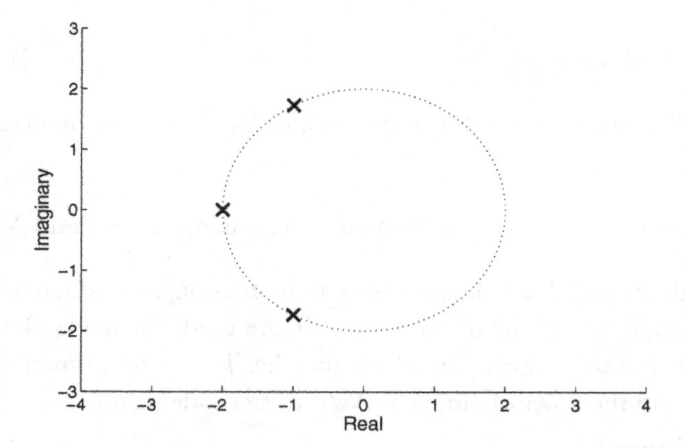

Figure 10.5: The three poles for a third order Butterworth filter. To ensure stability, only the left half-plane poles of the possible six poles are used (right half-plane poles are not shown).

The frequency response is normalized by Ω^N so that the response at normalized $\Omega = 0$ is 1.0.

```
function LVButterFR(N,OmegaC)
% LVButterFR(3,1)
k = 0:1:2*N-1;
P = OmegaC*exp(j*(pi/(2*N))*(2*k+N+1));
NetP = P(find(real(P)<0)); Freq = (0:0.01:4*OmegaC);
s = j*Freq; Resp = OmegaC^N; for Ctr = 1:1:length(NetP)
Resp = Resp.*(1./(s - NetP(Ctr))); end; figure(97);
subplot(311); plot(Freq,20*log10(abs(Resp)));
xlabel('Freq, Rad/s'); ylabel('Mag, dB')
subplot(312); plot(Freq,abs(Resp)); xlabel('Freq, Rad/s');
ylabel('Mag'); subplot(313); plot(Freq,unwrap(angle(Resp)))
xlabel('Freq, Rad/s'); ylabel('Radians')
```

The result from running the code above is shown in Fig. 10.6. Note that the phase response of the Butterworth filter is reasonably linear within the passband.

We can check the work above using the following script, *LVButterFRViaPoly*, which uses the function *LVsFreqResp* (introduced earlier) to obtain the frequency response; the results are shown in Fig. 10.7.

```
function LVButterFRViaPoly(N,OmegaC,HiFreqLim)
% LVButterFRViaPoly(3,1,4)
k = 0:1:2*N-1;
P = OmegaC*exp(j*(pi/(2*N))*(2*k+N+1))
```

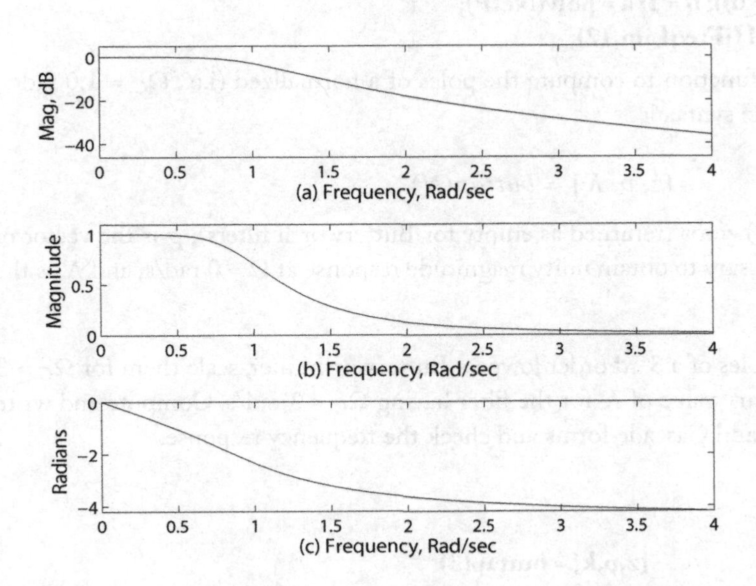

Figure 10.6: (a) Frequency response in dB of a Butterworth filter of order 3, up to a frequency equal to four times the cutoff frequency of 1 rad/s; (b) Same in linear units; (c) Phase response of same.

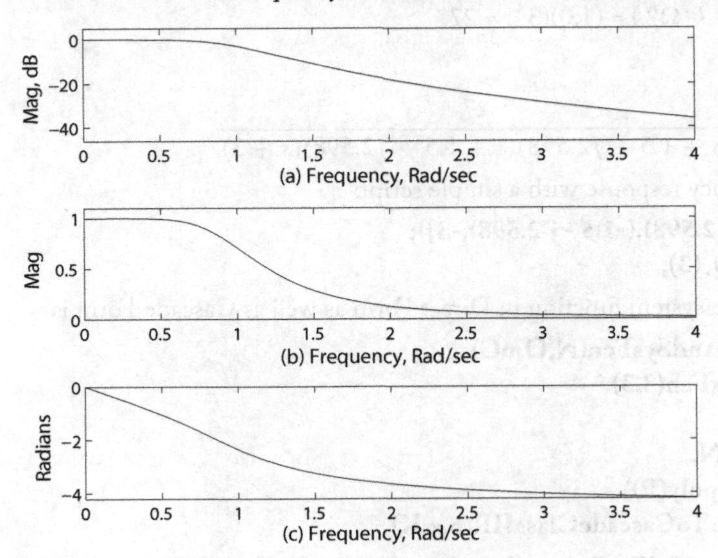

Figure 10.7: (a) The magnitude in dB of the frequency response of a 3-rd order Butterworth filter having $\Omega_C = 1$; (b) Same in linear units; (c) Phase response of same.

```
NetP = P(find(real(P)<0)); b = 1; a = poly(NetP);
H = LVsFreqResp(b,a,HiFreqLim,12);
```

MathScript provides a function to compute the poles of a normalized (i.e., Ω_C = 1.0 rad/s) Butterworth lowpass filter. The syntax is

$$[z, p, K] = buttap(N)$$

where z is the vector of (finite) zeros (returned as empty for Butterworth filters), p is the vector of poles, K is the gain (1.0) necessary to obtain unity magnitude response at Ω = 0 rad/s, and N is the desired order.

Example 10.7. Obtain the poles of a 3-rd order lowpass Butterworth filter, scale them for Ω_C = 3 rad/s, and compute the necessary value of K for the filter having Ω_C = 3 rad/s. Compute and write the system function in Direct and Cascade forms and check the frequency response.

We make the call

```
[z,p,k] = buttap(3)
```

to obtain the normalized poles p = [-0.5 $\pm$ j0.866, -1.0], which are then multiplied by the desired Ω_C (3 rad/s) to yield the poles for Ω_C = 3 rad/s as P = [-1.5 $\pm$ j2.598, -3.0]; from k, returned as 1.0, we obtain the new value of K as $k(\Omega_C^N)$ = (1.0)(3^3) = 27.

The system function is

$$H(s) = \frac{27}{(s + 1.5 + j2.598)(s + 1.5 - j2.598)(s + 3)}$$

We can check the frequency response with a simple script:

```
b = 27; a = poly([(-1.5 + j*2.598),(-1.5 - j*2.598),-3]);
H = LVsFreqResp(b,a,20,13);
```

A script that produces the system function in Direct Form as well as Cascade Form is

```
function LVButterPolesAndSysFcn(N,OmC)
% LVButterPolesAndSysFcn(3,3)
[z,p,k] = buttap(N),
P = OmC*p, K=k*OmC^N,
b = real(poly(z)), a = real(poly(P)),
[Bbq,Abq,Gain] = LVDirToCascadeClassIIR(b,a,K)
```

which yields b = 1, a = [1,6,18,27], B_bq = 1, A_bq = [1,3,9; 0,1,3], and $Gain$ = 27, and the Direct and Cascade system functions as

$$H(s) = \frac{1}{s^3 + 6s^2 + 18s + 27} = \frac{1}{(s^2 + 3s + 9)(s + 3)}$$

Note that letting $s = 0$ (i.e., evaluating the transfer function at DC or frequency zero) yields at output of $1/27$, so to achieve unity gain at DC, include *Gain* with the system transfer function:

$$H(s) = \frac{27}{s^3 + 6s^2 + 18s + 27} = \frac{27}{(s^2 + 3s + 9)(s + 3)}$$

10.5.2 DESIGN BY STANDARD PARAMETERS

To design a Butterworth lowpass filter to meet the standard specifications of Ω_P, Ω_S, R_P, and A_S, it is necessary to determine the required value of N. We start by noting that at Ω_P, the magnitude of response should be R_P, and at Ω_S, the magnitude of response should be A_S:

$$-10\log_{10}(1/(1 + (\Omega_P/\Omega_C)^{2N})) = R_P \tag{10.12}$$

and

$$-10\log_{10}(1/(1 + (\Omega_S/\Omega_C)^{2N})) = A_S \tag{10.13}$$

After solving for N, we get

$$N = \frac{\log_{10}[(10^{R_P/10} - 1)/(10^{A_S/10} - 1)]}{2\log_{10}(\Omega_P/\Omega_S)} \tag{10.14}$$

where N (usually not an integer as computed above) must be rounded up to the next integer to ensure that the specifications are met.

Once N has been obtained, it can be used in either of Eqs. (10.12) or (10.13) to solve for values of Ω_C to satisfy either R_P exactly at Ω_P, or A_S exactly at Ω_S. The resultant formulas, respectively, are

$$\Omega_{C1} = \frac{\Omega_P}{\sqrt[2N]{10^{R_P/10} - 1}} \tag{10.15}$$

$$\Omega_{C2} = \frac{\Omega_S}{\sqrt[2N]{10^{A_S/10} - 1}} \tag{10.16}$$

The range of acceptable values for Ω_C is obtained by solving for Ω_C in both of Eqs. (10.15) and (10.16) and choosing Ω_C between Ω_{C1} and Ω_{C2}.

Example 10.8. Design a Butterworth Filter having Ω_P = 0.4 Rad/s, Ω_S = 0.7 Rad/s, R_P = 0.1 dB, and A_S = 40 dB.

The following code computes N, Ω_C for both cases, the average of the two values of Ω_C, and the poles for the Butterworth filter, and then plots the frequency and phase responses. The result is shown in Fig. 10.8.

```
function [Z,P,K] = LVDesignButterworth(OmgP,OmgS,Rp,As)
```

```
% [Z,P,K] = LVDesignButterworth(0.4,0.7,0.2,40)
num = 10^(Rp/10)-1; denom = 10^(As/10)-1;
N = ceil(log10(num/denom)/(2*log10(OmgP/OmgS)))
OmcP = OmgP/((10^(Rp/10)-1)^(0.5/N))
OmcS = OmgS/((10^(As/10)-1)^(0.5/N))
OmgC = (OmcP + OmcS)/2; k = 0:1:2*N-1;
P = OmgC*exp(j*(pi/(2*N))*(2*k+N+1));
P = P(find(real(P)<0)); Z = [];
K = OmgC^N;
```

The following code will obtain the Butterworth poles and zeros, obtain and plot the frequency response, the realized values of R_P and A_S, and the cascade coefficients.

```
Rp = 0.2; As = 40; OmgP = 0.4; OmgS = 0.7;
[Z,P,K] = LVDesignButterworth(OmgP,OmgS,Rp,As)
H = LVsFreqResp(K*poly(Z),poly(P),4*OmgS,14)
[NetRp,NetAs] = LVsRealizedFiltParamLPF(H,OmgP,...
OmgS,4*OmgS)
[Bbq,Abq,Gain] = LVDirToCascadeClassIIR(poly(Z),poly(P),K)
```

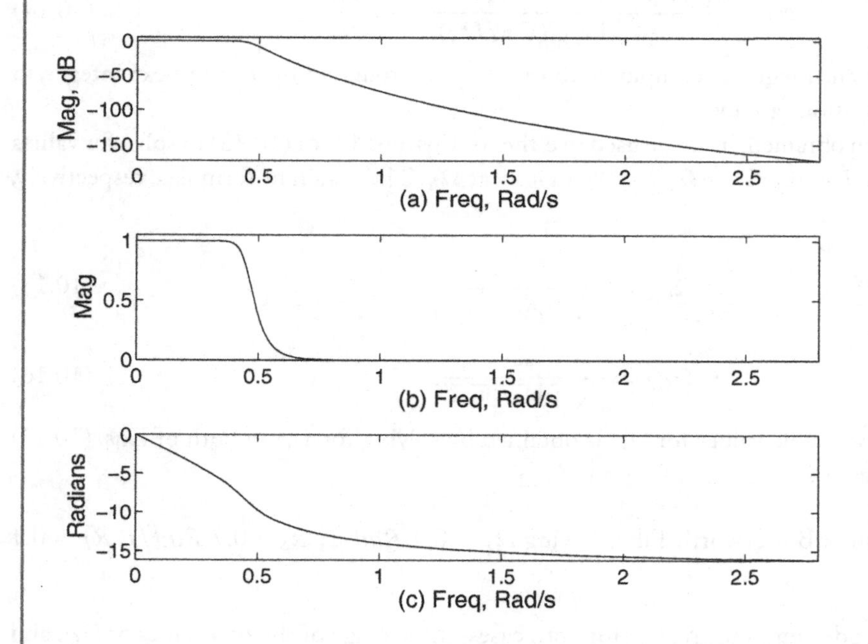

Figure 10.8: (a) Magnitude in dB of frequency response of a Butterworth filter having Ω_P = 0.4 rad/s, Ω_S = 0.7 rad/s, R_P = 0.2 dB, and A_S = 40 dB. The necessary value of N was computed to be 11; (b) Same, but in linear scale; (c) Phase response.

10.6 LOWPASS ANALOG CHEBYSHEV TYPE-I FILTERS

Chebyshev filters are based on the Chebyshev polynomials, which are defined as

$$T_N(x) = \begin{cases} \cos(N\cos^{-1}(x)) & |x| \leq 1 \\ \cosh(N\cosh^{-1}(x)) & |x| > 1 \end{cases}$$

The Chebyshev Type-I filter has an equiripple characteristic in the passband, and decreases monotonically in the stopband; the Chebyshev Type-II filter is monotonic in the passband and equiripple in the stopband.

10.6.1 DESIGN BY ORDER, CUTOFF FREQUENCY, AND EPSILON

The magnitude squared function for a **Chebyshev Type I** filter is defined as:

$$|H(\Omega)|^2 = \frac{1}{1+\epsilon^2(T_N(x))^2}$$

where $x = \Omega/\Omega_P$. The parameter ϵ determines, for a given N, the tradeoff between passband ripple and transition band steepness.

Example 10.9. Compute and display the magnitude squared function for a Type-I Chebyshev filter having ϵ = 0.4 and N = 5.

Note in the following code that the normalized frequency range is broken into two subranges to accommodate the two functions (cosine and hyberbolic cosine).

```
function LVCheby1MagSquared(Ep,N)
% LVCheby1MagSquared(0.4,5)
inc = 0.01; xLo = 0:inc:1;
xHi = 1+inc:inc:3; TnLo = cos(N*acos(xLo));
TnHi = cosh(N*acosh(xHi)); T = [TnLo,TnHi];
MagHSq = 1./(1 + Ep^2*(T.^2));
figure; xplot = [xLo, xHi]; plot(xplot,MagHSq);
xlabel('Norm Freq (\Omega/\Omegac)'); ylabel('Mag Squared')
```

Figure 10.9 shows the magnitude squared of a Chebyshev Type I filter having N = 4 for several values of ϵ. As ϵ decreases, the passband ripple decreases but the transition band becomes wider, i.e., the roll-off is less steep.

The ripple characteristics for Type-I Chebyshev filters differ according to whether N is even or odd. This is illustrated in Fig. 10.10.

$$\begin{array}{lll} |H(j0)|^2 = 1 & |H(j1)|^2 = 1/(1+\epsilon^2) & N \text{ odd} \\ |H(j0)|^2 = 1/(1+\epsilon^2) & |H(j1)|^2 = 1/(1+\epsilon^2) & N \text{ even} \end{array}$$

The poles of the system function are the roots of

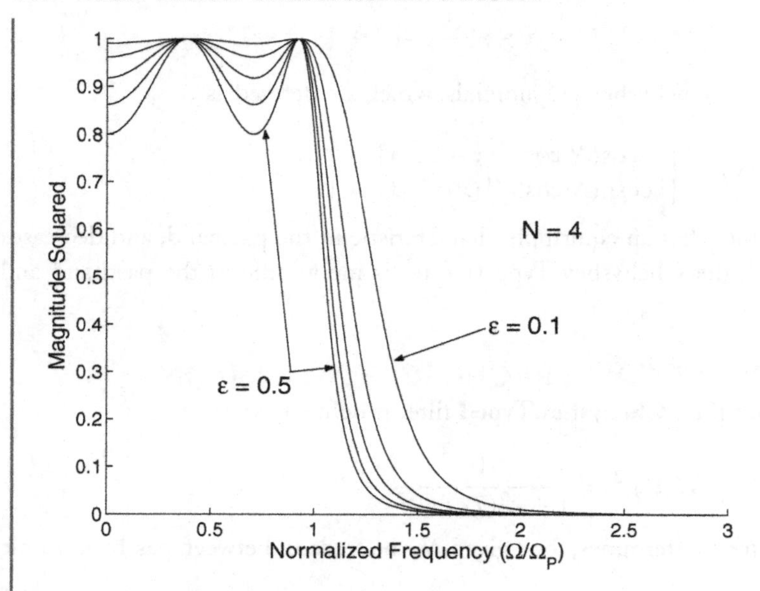

Figure 10.9: The magnitude squared function for a Chebyshev Type I filter having $N = 4$ for values of ϵ equal to 0.1, 0.2, 0.3, 0.4, and 0.5. With $\epsilon = 0.5$, the passband ripple is the largest, but the transition band is the narrowest.

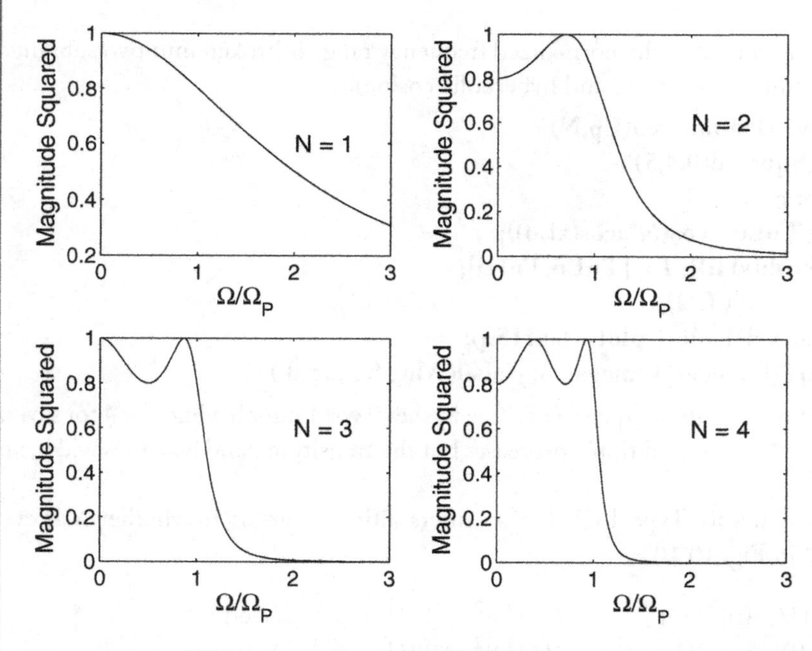

Figure 10.10: The magnitude squared characteristics of a Chebyshev Type-I lowpass filter having $\epsilon = 0.5$ for several even and odd values of N.

$$1+\epsilon^2(T_N(s/j\Omega_C))^2$$

and the left half-plane poles $p_k = \sigma_k + j\Omega_k$ can be computed as

$$\sigma_k = (A\Omega_C)\cos[\pi/2 + \pi(2k+1)/2N]$$

$$\Omega_k = (B\Omega_C)\sin[\pi/2 + \pi(2k+1)/2N]$$

where k = 0:1:N-1 and

$$A = (\sqrt[N]{\alpha} - \sqrt[N]{1/\alpha})/2$$

$$B = (\sqrt[N]{\alpha} + \sqrt[N]{1/\alpha})/2$$

with

$$\alpha = 1/\epsilon + \sqrt{1 + 1/\epsilon^2}$$

The system function is

$$H(s) = \frac{K}{\Pi(s - p_k)}$$

where K is chosen so that the magnitude function at $\Omega = 0$ is 1 for N odd or $1/\sqrt{1+\epsilon^2}$ for N even.

Example 10.10. Write a script that computes the poles, K, and the magnitude and phase responses for a Type-I Chebyshev filter having $N = 5$, $\epsilon = 0.5$, and $\Omega_C = 1$ rad/sec.

A straightforward application of the various formulas given above results in the following code, the result of which is shown in Fig. 10.11.

```
function LVCheby1(N,OmC,Epsilon)
% LVCheby1(5,1,0.5)
Alpha = 1/Epsilon + sqrt(1 + 1/(Epsilon^2));
B = 0.5*( Alpha^(1/N) + (1/Alpha)^(1/N) );
A = 0.5*( Alpha^(1/N) - (1/Alpha)^(1/N) );
k = 0:1:N-1; arg = pi/2 + pi*(2*k + 1)/(2*N);
SigK = A*cos(arg); OmK = B*sin(arg);
P = OmC*(SigK + j*OmK); K = prod(abs(P));
if rem(N,2)==0 % N even
K = 1/sqrt(1 + Epsilon^2)*K; end;
H = LVsFreqResp(K,poly(P),2*OmC,8);
```

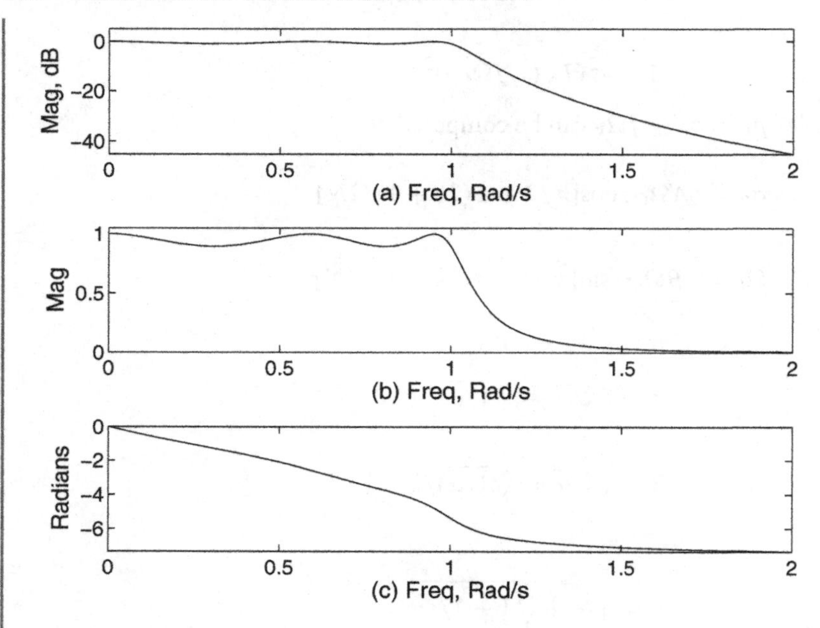

Figure 10.11: (a) Magnitude response in dB of a Chebyshev Type-I filter having N =5, ϵ = 0.5, and Ω_C = 1 rad/sec.; (b) Magnitude response (linear) of same; (c) Phase response of same.

An alternate computational method for the poles of a Type-I Chebyshev filter, which produces the same result as the previously described method, is as follows. Compute

$$\upsilon_0 = \sinh^{-1}(1/\epsilon)/N \tag{10.17}$$

and

$$k = -(N-1):2:(N-1)$$

and then

$$P_K = -\sinh(\upsilon_0)\cos(k\pi/2N) + \cosh(\upsilon_0)\sin(k\pi/2N) \tag{10.18}$$

This method assumes Ω_C = 1, so the poles P_K must be scaled by the actual value of Ω_C.

Example 10.11. For the example above, in which ϵ = 0.5, Ω_C = 1, and N = 5, compute the poles using the alternate method just described.

To compare results with the previous method, run the code from the previous example to display the pole values, and then run the code below.

```
function p = LVCheby1Poles2ndMethod(N,OmC,Epsilon)
% p = LVCheby1Poles2ndMethod(5,2,0.5)
k = -(N-1):2:N-1; r = k*pi/(2*N);
v0 = asinh(1/Epsilon)/N;
p = OmC*(-sinh(v0)*cos(r) + j*cosh(v0)*sin(r));
```

MathScript provides a function to compute the poles of a normalized (i.e., Ω_C = 1.0 rad/s) Chebyshev Type-I lowpass filter. The syntax is

$$[z, p, k] = cheb1ap(N, Rp)$$

where z is the vector of (finite) zeros (returned as empty for Chebyshev Type-I filters), p is the vector of poles, k is the gain necessary to obtain unity magnitude response at Ω = 0 rad/s, N is the desired order, and Rp is the passband ripple in dB.

Example 10.12. Obtain the poles of a second order lowpass Chebyshev Type-I filter having R_P = 0.2 dB, scale them for Ω_C = 4 rad/s, and compute the necessary value of K for the filter having Ω_C = 4 rad/s. Write the system function. Compute the Cascade Form coefficients, and convert them back to Direct Form.

We make the call

```
[z,p,k] = cheb1ap(2,0.2)
```

to obtain the normalized poles p = [-0.9635 $\pm$ j1.1952], which are then multiplied by the desired Ω_C (4 rad/s) to yield the poles for Ω_C = 4 rad/s as P = [-3.8542 $\pm$ j4.7807]; from the normalized value of k (returned from the call above as 2.3032), the new value of K is obtained as $k(\Omega_C^N)$ = $(2.3032)(4^2)$ = 36.85.

A script to perform these computations and produce the system function is

```
function LVCheby1PolesAndSysFcn(N,OmC,Rp)
% LVCheby1PolesAndSysFcn(2,4,0.2)
[z,p,k] = cheb1ap(N,Rp),
P = OmC*p, K = k*OmC^N,
b = real(poly(z)), a = real(poly(P)),
H = LVsFreqResp(K,poly(P),2*OmC,19);
[Bbq,Abq,Gain] = LVDirToCascadeClassIIR(b,a,K)
[b,a,k]=LVCas2DirClassIIR(Bbq,Abq,Gain)
```

The call

```
LVCheby1PolesAndSysFcn(2,4,0.2)
```

yields b = 1 and a = [1,7.7083,37.709], which are the same as Bbq and Abq since there are only two poles.

The system function is

$$H(s) = \frac{36.85}{s^2 + 7.7083s + 37.7093}$$

10.6.2 DESIGN BY STANDARD PARAMETERS

The parameters N, ϵ, and Ω_C are used to design a Chebyshev Type-I filter. From the input parameters Ω_P and Ω_S we can immediately compute ϵ and A (stopband attenuation in dB) as

$$\epsilon = \sqrt{10^{R_P/10} - 1}$$

and

$$A = 10^{A_S/20}$$

We note that

$$\Omega_C = \Omega_P$$

and we define

$$\Omega_T = \Omega_S/\Omega_P$$

Then

$$N = \left(\frac{\log_{10}(g + \sqrt{g^2 - 1})}{\log_{10}(\Omega_T + \sqrt{\Omega_T^2 - 1})}\right)$$

with

$$g = \sqrt{(A^2 - 1)/\epsilon^2}$$

Note that N will in general not be an integer and must be rounded up to the next integer.

Example 10.13. Design a Type-I Chebyshev filter having $R_P = 0.5$ dB, $A_S = 40$ dB, $\Omega_P = 0.5\ rad/s$, and $\Omega_S = 0.65\ rad/s$.

Using the above equations, the code is straightforward to write. Code has also been included to compute the realized or net values of R_P and A_S. Figure 10.12 shows the result.

```
function [Z,P,K] = LVDesignCheby1Filter(Rp,As,OmgP,OmgS)
% [Z,P,K] = LVDesignCheby1Filter(0.5,40,0.5,0.65)
E = sqrt(10^(Rp/10)-1); A = 10^(As/20); OmgC = OmgP;
OmgT = OmgS/OmgP; g = sqrt((A^2-1)/(E^2));
```

```
N = ceil(log10(g + sqrt(g^2 - 1))/log10(OmgT +...
sqrt(OmgT^2-1)))
k = -(N-1):2:N-1; r = k*pi/(2*N); v0 = asinh(1/E)/N;
P = OmgC*(-sinh(v0)*cos(r) + j*cosh(v0)*sin(r)),
K = prod(abs(P)); Z = []; if rem(N,2)==0 % N even
K = 1/sqrt(1 + E^2)*K; end
```

The following code will obtain the Chebyshev poles and zeros, compute and plot the frequency response, and compute the realized values of R_P and A_S as well as the cascade coefficients:

```
Rp = 0.5; As = 40; OmgP = 0.5; OmgS = 0.65;
[Z,P,K] = LVDesignCheby1Filter(Rp,As,OmgP,OmgS)
H = LVsFreqResp(K*poly(Z),poly(P),2*OmgS,15);
[NetRp,NetAs] = LVsRealizedFiltParamLPF(H,OmgP,...
OmgS,2*OmgS)
[Bbq,Abq,Gain] = LVDirToCascadeClassIIR(poly(Z),poly(P),K)
```

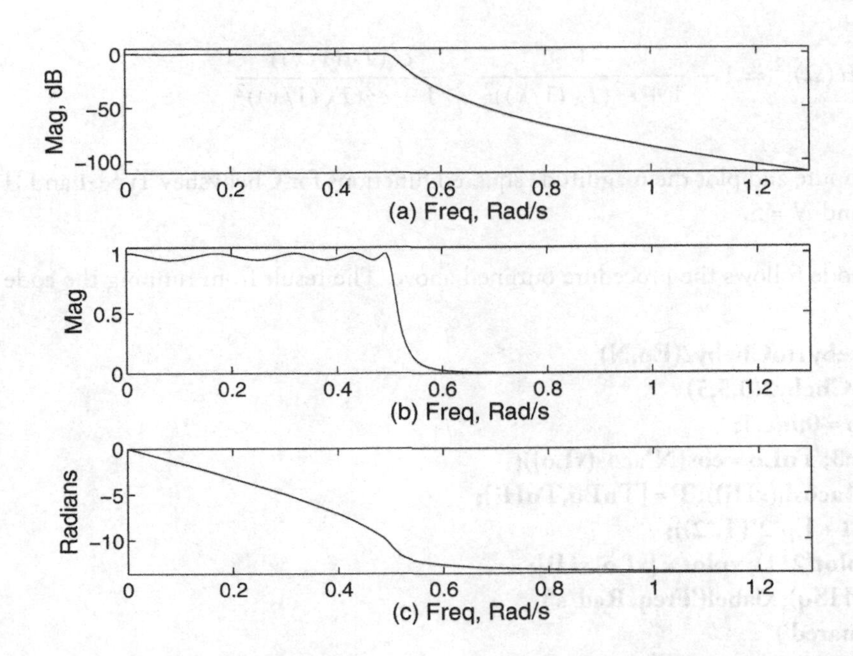

Figure 10.12: (a) Magnitude of response in dB of a Chebyshev Type-I filter designed to have A_S = 40 dB, R_P = 0.5 dB, Ω_P = 0.5 rad/s, and Ω_S = 0.65 rad/s; (b) Magnitude of response (linear) of same; (c) Phase response of same.

10.7 LOWPASS ANALOG CHEBYSHEV TYPE-II FILTERS

10.7.1 DESIGN BY ORDER, CUTOFF FREQUENCY, AND EPSILON

A Chebyshev Type-II filter is monotonic in the passband and equiripple in the stopband. One approach to deriving the magnitude squared characteristic of such a filter is to design a Chebyshev Type-I filter, substitute $1/\Omega$ for Ω, which converts the filter to highpass, and then subtract the result from 1 to convert it back to a lowpass filter.

We thus start with the magnitude squared response of a Type-I Chebyshev filter, which is

$$|H(\Omega)|^2 = \frac{1}{1+\epsilon^2(T_N(x))^2}$$

where $T_N(x)$ is the N-th order Chebyshev polynomial and $x = \Omega/\Omega_P$. We convert to a highpass filter by substituting $1/x$ for x

$$|H(\Omega)|^2 = \frac{1}{1+\epsilon^2(T_N(1/x))^2}$$

and then subtract from 1:

$$|H(\Omega)|^2 = 1 - \frac{1}{1+\epsilon^2(T_N(1/x))^2} = \frac{\epsilon^2(T_N(1/x))^2}{1+\epsilon^2(T_N(1/x))^2}$$

Example 10.14. Compute and plot the magnitude squared functions for Chebyshev Type-I and II filters having $\epsilon = 0.5$ and $N = 5$.

The following code follows the procedure outlined above. The result from running the code is shown in Fig. 10.13.

```
function LVCheby1toCheby2(Ep,N)
% LVCheby1toCheby2(0.5,5)
inc = 0.005; xLo = 0:inc:1;
xHi = 1+inc:inc:3; TnLo = cos(N*acos(xLo));
TnHi = cosh(N*acosh(xHi)); T = [TnLo,TnHi];
MagHSq = 1./(1 + Ep^2*(T.^2));
figure(33); subplot(211); xplot = [xLo, xHi];
plot(xplot,MagHSq); xlabel('Freq, Rad/s')
ylabel('Mag Squared')
% Convert to Type-II
xLo = xLo(find(~(xLo==0))); TnLo = cos(N*acos(1./xLo));
xHi = xHi(find(~(xHi==0))); TnHi = cosh(N*acosh(1./xHi));
T = [TnLo,TnHi]; MagHSq = 1 - 1./(1 + Ep^2*(T.^2));
subplot(212); xplot = [xLo, xHi];
```

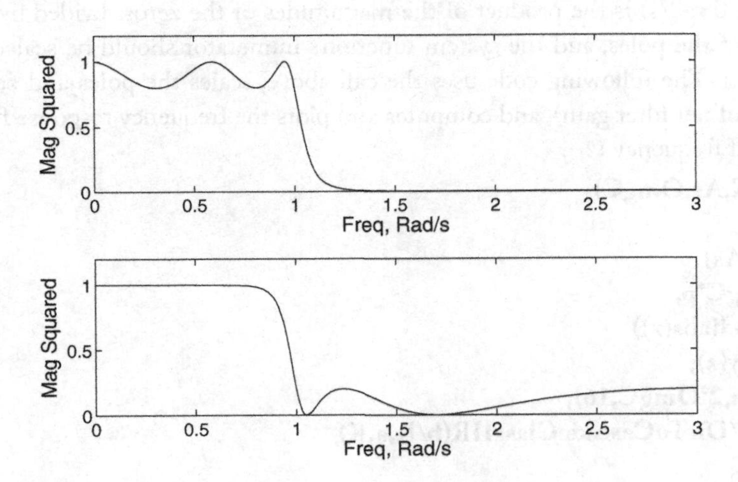

Figure 10.13: (a) Magnitude squared function for a Type-I Chebyshev filter having ϵ = 0.5 and N = 5; (b) Magnitude squared function for a Type-II Chebyshev filter derived by substituting $1/\Omega$ for Ω and subtracting from 1, as described in the text.

plot(xplot,MagHSq); xlabel('Freq, Rad/s')
ylabel('Mag Squared')

MathScript provides a function to compute the poles and zeros of a normalized (i.e., Ω_C = 1.0 rad/s) Chebyshev Type-II lowpass filter. The syntax is

$$[z, p, k] = cheb2ap(N, As)$$

where z is the vector of (finite) zeros, p is the vector of poles, k is the gain necessary to obtain unity magnitude response at Ω = 0 rad/s, N is the desired order, and A_S is the stopband ripple (the minimum stopband attenuation) in dB.

Example 10.15. Obtain the poles of a 3-rd order lowpass Chebyshev Type-II filter having A_S = 40 dB, scale them for Ω_C = 2 rad/s, and compute the necessary value of k for the filter having Ω_C = 2 rad/s. Write the system function. Plot the magnitude of frequency response.

We make the call

[z,p,k] = cheb2ap(3,40)

to obtain the normalized poles p = [-0.1611 $\pm$ j0.2959, -0.3523], which are then multiplied by the desired Ω_C (2 rad/s) to yield the poles for Ω_C = 2 rad/s as P = [-0.3222 $\pm$ j0.5918, -0.7046]. The zeros z are returned as $\pm$j1.1547, which when scaled by Ω_C yield Z = [$\pm$j2.3094].

The net filter gain (at 0 rad/s) is the product of the magnitudes of the zeros divided by the product of the magnitudes of the poles, and the system function's numerator should be scaled by the reciprocal of this number. The following code uses the call above, scales the poles and zeros, computes K (the reciprocal of net filter gain), and computes and plots the frequency response from 0 rad/s up to twice the cutoff frequency Ω_C.

```
function LVCheb2(N,As,OmgC)
% LVCheb2(3,40,2)
[z,p,k] = cheb2ap(N,As)
z = OmgC*z, p = OmgC*p,
K = prod(abs(p))./prod(abs(z))
a = poly(p); b = K*poly(z);
H = LVsFreqResp(b,a,2*OmgC,16);
[Bbq,Abq,Gain] = LVDirToCascadeClassIIR(b/K,a,K)
```

The system function is

$$H(s) = \frac{(0.06)(s^2 + 5.333)}{(s^2 + 0.6446s + 0.4542)(s + 0.7046)}$$

10.7.2 DESIGN BY STANDARD PARAMETERS

Analogously to the procedure just described, Chebyshev Type-II filters are most easily derived by first computing the pole locations for a Type-I filter having the same specifications. The pole locations for the Type-I filter are the reciprocal of the pole locations for the Type-I filter. The Type-II also has finite zeros, which are located on the imaginary axis and computed as

$$\frac{j}{\sin(k\pi/2N)} \tag{10.19}$$

where

$$k = -(N-1) : 2 : (N-1)$$

Note that when $\sin(k\pi/2N) = 0$, the frequency of the corresponding transfer function zero is infinite. Thus some of the zeros for the Chebyshev Type-II filter may be infinite, while most will be finite.

The procedure to compute the reciprocal of the Type-I poles is often given as follows: if the Type-II poles being sought are represented as $\sigma_k' + j\Omega_k'$, and the Type-I poles as $\sigma_k + j\Omega_k$, then

$$\sigma_k' = \frac{\sigma_k}{\sigma_k^2 + \Omega_k^2}$$

and

$$\Omega_k' = \frac{\Omega_k}{\sigma_k^2 + \Omega_k^2}$$

The above procedure works since the set of poles comprise complex conjugate pairs; it actually computes the reciprocal of each pole's complex conjugate. Note that

$$1/(\sigma_k + j\Omega_k) = (\sigma_k - j\Omega_k)/(\sigma_k^2 + \Omega_k^2)$$

and

$$1/(\sigma_k - j\Omega_k) = (\sigma_k + j\Omega_k)/(\sigma_k^2 + \Omega_k^2)$$

A procedure to design a Chebyshev Type-II filter is as follows:

1) Specify the allowable passband ripple which is the minimum allowed response in the passband. This can be specified in positive dB, such as the typical parameter R_P.

2) Specify the maximum allowable response in the stopband. This can be done using A_S. Calculate ϵ from this value.

3) Calculate υ_0 according to Eq. (10.17) and the poles for a Type-I filter according to Eq. (10.18).

4) Obtain the Type-II poles as the reciprocal of the Type-I poles.

5) Compute the Type-II zero locations according to Eq. (10.19).

Example 10.16. Compute the poles and zeros of a Chebyshev Type-II filter having Ω_P = 0.9 rad/s, Ω_S = 1.0 rad/s, R_P = 0.2 dB, A_S = 40 dB.

The following code scales the desired frequency limits and computes a normalized Chebyshev Type-I filter. From this, the Type-II poles are obtained, the zeros separately computed, and then both poles and zeros are scaled to reflect the actual desired values of Ω_P and Ω_S. The result from running the code below is shown in Fig. 10.14.

```
function [Z,P,K] = LVDesignCheb2(Rp,As,OmgP,OmgS)
% [Z,P,K] = LVDesignCheb2(0.2,40,0.9,1)
OmgC = 1; OmgP = OmgP/OmgS;
E = 1/sqrt(10^(As/10)-1); G = 10^(-Rp/20);
N = acosh(G/(E*sqrt(1-G^2)))/acosh(1/OmgP);
N = ceil(abs(N)); V0 = asinh(1/E)/N;
k = -(N-1):2:(N-1); r = k*pi/(2*N);
Ch1P = -sinh(V0)*cos(r) + j*cosh(V0)*sin(r);
P = OmgS*(1./Ch1P); div = sin(k*pi/(2*N));
Zdiv = find(div==0); NZerDiv = div(find(~(div==0)));
Z = OmgS*(j./NZerDiv); s=j*[0:0.001:2*OmgS];
Hs = abs(polyval(poly(Z),s)./polyval(poly(P),s));
K = 1/max(abs(Hs));
```

The following code obtains the Chebyshev Type-II poles and zeros, computes and plots the frequency response, and computes the realized values of R_P and A_S as well as the cascade coefficients:

```
Rp = 0.2; As = 40; OmgP = 0.9; OmgS = 1;
[Z,P,K] = LVDesignCheb2(Rp,As,OmgP,OmgS)
H = LVsFreqResp(K*poly(Z),poly(P),2*OmgS,17);
[NetRp,NetAs] = LVsRealizedFiltParamLPF(H,OmgP,OmgS,2*OmgS)
[Bbq,Abq,Gain] = LVDirToCascadeClassIIR(poly(Z),poly(P),K)
```

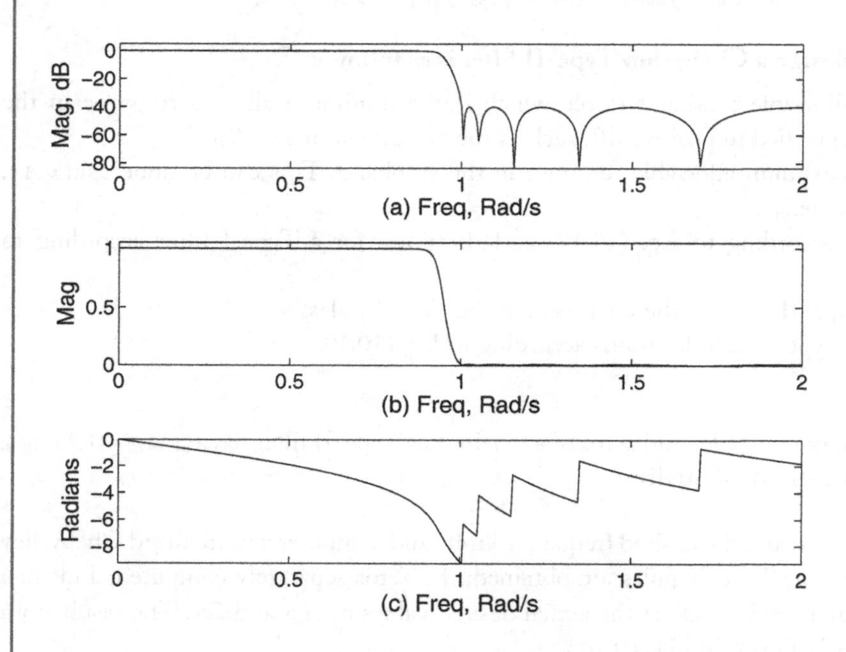

Figure 10.14: (a) Magnitude function in dB of a Type-II Chebyshev filter having Ω_P = 0.9 rad/s, Ω_S = 1.0 rad/s, R_P = 0.2 dB, and A_S = 40 dB; (b) Magnitude function (linear) of same; (c) Phase response of same.

10.8 ANALOG LOWPASS ELLIPTIC FILTERS

The Elliptic, or Cauer filter, has a steep roll-off with equiripple in both passband and stopband. Analogously to the equiripple FIR, it is possible to achieve the lowest order for a given set of specifications using an elliptic design.

The magnitude squared function of an Elliptic filter is

$$|H(j\Omega)|^2 = \frac{1}{1 + \epsilon^2 G_N^2(\Omega/\Omega_C)}$$

where ϵ is the passband ripple (related to R_P), and $G(\Omega/\Omega_C)$ is the N-th order Jacobian Elliptic function, the analysis of which is beyond the scope of this book. MathScript provides a function to compute the poles and zeros of a normalized (i.e., Ω_C = 1.0 rad/s) Elliptic lowpass filter. The syntax is

$$[z, p, k] = ellipap(N, Rp, As)$$

where z is the vector of (finite) zeros, p is the vector of poles, k is the gain necessary to obtain unity magnitude response at Ω = 0 rad/s, N is the desired order, R_P is the desired passband ripple in dB, and A_S is the passband ripple in dB.

The following code calls the *ellipap* function, scales the resultant poles and zeros, computes the value K necessary to achieve a maximum of unity gain in the passband, and computes and plots the frequency response from 0 rad/s up to three times the cutoff frequency Ω_C. The result of computation is shown in Fig. 10.15.

```
function LVellip(N,Rp,As,OmgC)
% LVellip(5,0.2,40,2)
[z,p,k] = ellipap(N,Rp,As);
Z = OmgC*z, P = OmgC*p,
nrmGainFrZero = prod(abs(z))/prod(abs(p));
nrmK = k*nrmGainFrZero;
UnrmGnFrZero = prod(abs(Z))/prod(abs(P));
K = nrmK/UnrmGnFrZero;
H = LVsFreqResp(K*poly(Z),poly(P),3*OmgC,18);
```

The system function is

$$H(s) = \frac{0.1119(s^2 + 16.6703)(s^2 + 7.8158)}{(s^2 + 1.4092s + 2.8782)(s^2 + 0.3548s + 4.343)(s + 1.1663)}$$

10.8.1 DESIGN BY STANDARD PARAMETERS

The order N needed for an Elliptic filter to meet certain specifications can be computed from the following formula:

$$N = \frac{K(k)K(\sqrt{1 - k_1^2})}{K(k_1)K(\sqrt{1 - k^2})} \tag{10.20}$$

where

$$k = \frac{\Omega_P}{\Omega_S} \text{ and } k_1 = \frac{\epsilon}{\sqrt{A^2 - 1}} \tag{10.21}$$

and

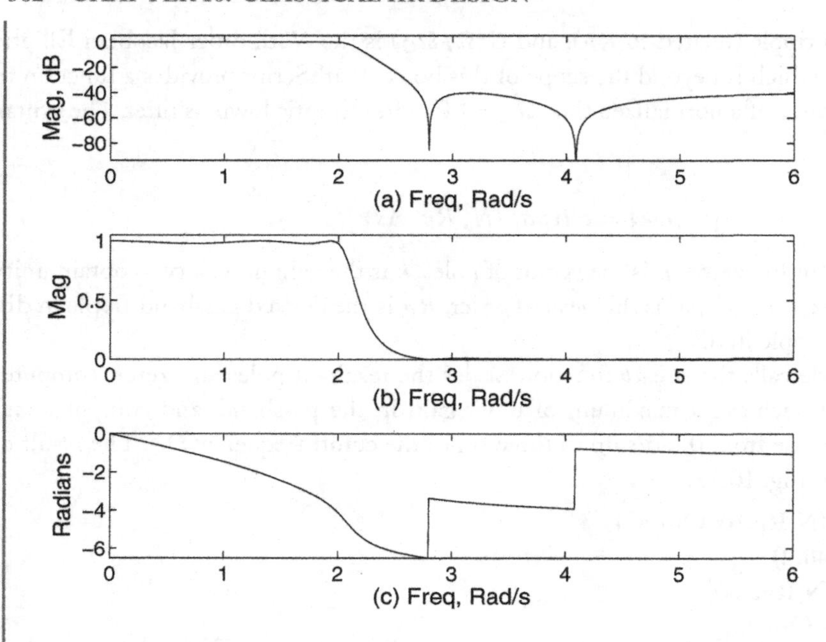

Figure 10.15: (a) Magnitude of frequency response in dB of an elliptic filter having Ω_C = 2 rad/s, N = 5, R_P = 0.2 dB, A_S = 40 dB. (b) Magnitude of frequency response (linear) of same; (c) Phase response of same.

$$K(x) = \int_0^{\pi/2} \frac{d\theta}{\sqrt{1 - x^2 \sin^2 \theta}} \tag{10.22}$$

Eq. (10.22), which defines the **Complete Elliptic Integral of the First Kind** can be evaluated using the function

$$[k, e] = ellipke(m)$$

where k is the complete elliptic integral of the first kind, e is the complete elliptic integral of the second kind, and the modulus m would correspond to x^2 in Eq. (10.22).

Example 10.17. Compute K(x) (the complete elliptic integral of the first kind) for x = 0.5.

We make the call

[k,e] = ellipke(0.5^2)

which yields $k = 1.6858$.

Example 10.18. Write a script that receives the usual input specifications for a filter (R_P,A_S,W_P,W_S) and computes N, the poles, the zeros, and the gain for an elliptic filter meeting the specifications. Compute and display the frequency response of the resulting filter as well as the realized values of R_P and A_S. Test the script using $R_P = 1.25$ dB, $A_S = 50$ dB, $W_P = 0.5$ rad/s, and $W_S = 0.6$ rad/s.

The code below computes the necessary order N from the specifications according to Eq. (10.20), and then completes the design by calling *LVellip*. The result from running the code is shown in Fig. 10.16 (the code for axis labels, etc., has been omitted for brevity).

```
function [Z,P,K] = LVDesignEllip(Rp,As,OmgP,OmgS)
% [Z,P,K] = LVDesignEllip(1.25,50,0.5,0.6)
E = (10^(Rp/10)-1)^0.5;
A=10^(As/20); OmgC = OmgP;
k = OmgP/OmgS; k1 = E/sqrt(A^2-1);
[K1k, K2k] = ellipke([k, (1-k^2)].^2);
[K1k1, K2k1] = ellipke([k1, (1-k1^2)].^2);
[Z,P,K] = LVellip(N,Rp,As,OmgC)
N = ceil(K1k(1)*K1k1(2)/(K1k1(1)*K1k(2)));
```

The following code obtains the Elliptic filter poles and zeros, computes and plots the frequency response, and computes the realized values of R_P and A_S as well as the Cascade Form coefficients:

```
Rp = 1.25; As = 50; OmgP = 0.5; OmgS = 0.6;
[Z,P,K] = LVDesignEllip(Rp,As,OmgP,OmgS)
H = LVsFreqResp(K*poly(Z),poly(P),3*OmgS,19);
[NetRp,NetAs] = LVsRealizedFiltParamLPF(H,OmgP,...
OmgS,3*OmgS)
[Bbq,Abq,Gain] = LVDirToCascadeClassIIR(poly(Z),poly(P),K)
```

10.9 FREQUENCY TRANSFORMATIONS IN THE ANALOG DOMAIN

It is possible to convert a prototype lowpass filter into a highpass, bandpass, and bandstop filter by substituting an expression in s for all values of s in the system function of the lowpass filter.

10.9.1 LOWPASS TO LOWPASS

To convert a prototype lowpass filter having passband cutoff Ω_P, replace each instance of s in the prototype lowpass system function with $s(\Omega_P/\Omega'_P)$, *i.e.*,

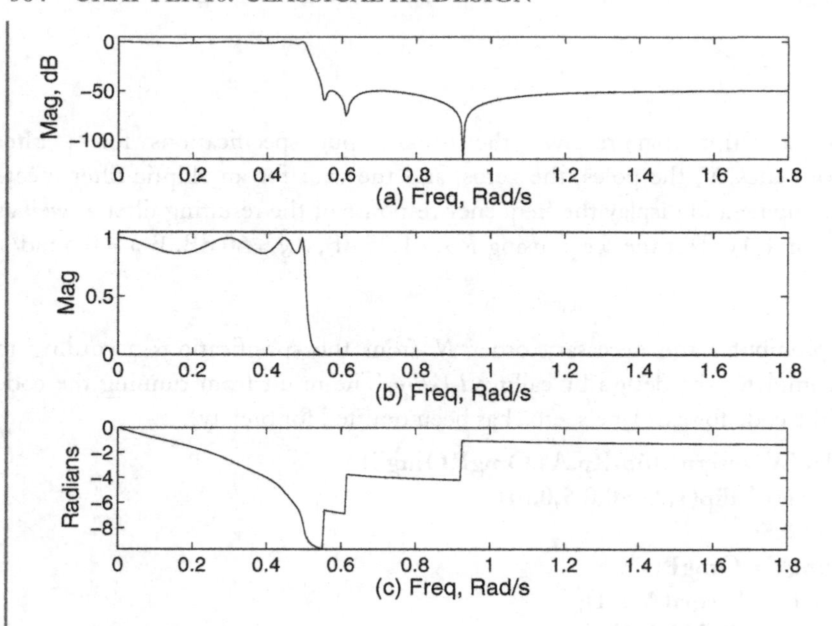

Figure 10.16: (a) Frequency response in dB of an elliptic filter having R_P = 1.25 dB, A_S = 50 dB, Ω_P = 0.5 rad/s, and Ω_S = 0.6 rad/s; (b) Frequency response (linear) of same; (c) Phase response of same.

$$s \rightarrow s\frac{\Omega_P}{\Omega'_P}$$

Example 10.19. Determine the system function for a Butterworth lowpass filter having Ω_C = 3 rad/s using the system function for a prototype Butterworth lowpass filter having Ω_C = 2 rad/s and N = 2.

To obtain the prototype lowpass filter we make the call

[z,p,k] = buttap(2)

to obtain the normalized poles p = [-0.7071 $\pm$ j0.7071], which are then multiplied by the desired Ω_C (2 rad/s) to yield the poles for Ω_C = 2 rad/s as P = [-1.414 $\pm$ j1.414]; from k, returned as 1.0, we obtain the new value of K as $k(\Omega_C^N)$ = (1.0)(2[2]) = 4. The system function for the prototype lowpass filter is therefore

$$H(s) = \frac{4}{(s + 1.414 + j1.414)(s + 1.414 - j1.414)}$$

To convert to a lowpass filter having Ω_C = 3 rad/s, we make the substitution

$$s \rightarrow \frac{\Omega_P}{\Omega'_P} s = \frac{2}{3} s$$

Thus the new lowpass filter's system function is

$$H(s) = \frac{4}{(\frac{2}{3}s + 1.414 + j1.414)(\frac{2}{3}s + 1.414 - j1.414)}$$

which reduces to

$$H(s) = \frac{9}{(s + 2.121 + j2.121)(s + 2.121 - j2.121)}$$

The following code computes and plots the frequency response for both the prototype and new filter; the result is shown in Fig. 10.17.

```
[z,p,k] = buttap(2); p = 2*p; b1=4;
a1 = poly(p); p = [2.121*(-1 + j), 2.121*(-1 - j)];
b2=9; a2 = poly(p);
LVsFreqRespDouble(b1,a1,6,22,b2,a2)
```

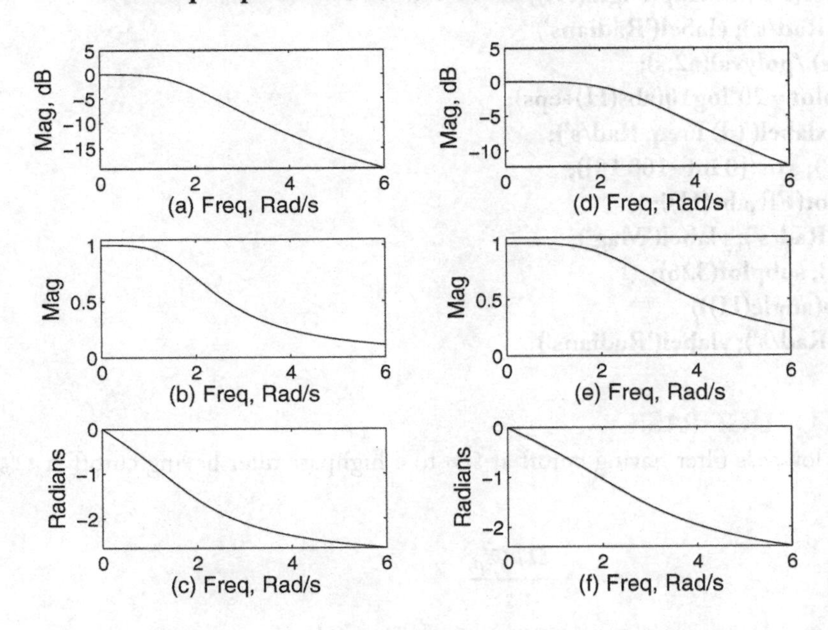

Figure 10.17: (a), (b), and (c): Magnitude (dB) response, Magnitude (linear) response, and phase response of prototype lowpass analog filter; respectively; (d), (e), (f): Magnitude (dB) response, Magnitude (linear) response, and phase response of new lowpass filter created from prototype lowpass analog filter; respectively.

The script

$$LVsFreqRespDouble(b1, a1, HiFreqLim, FigNo, b2, a2)$$

which was used in the code above, receives two sets of s-domain system function coefficients $b1, a1$, $b2, a2$, a high frequency limit for evaluation $HiFreqLim$, and a desired figure number $FigNo$ on which to plot the results:

```
function LVsFreqRespDouble(b1,a1,HiFreqLim,FigNo,b2,a2)
FR = (0:0.005:HiFreqLim); s = j*FR;
H = polyval(b1,s)./polyval(a1,s);
figure(FigNo); clf; subplot(321);
yplot = 20*log10(abs(H)+eps); plot(FR,yplot);
xlabel('(a) Freq, Rad/s'); ylabel('Mag, dB');
axis([0 inf -100 10]);
subplot(323); plot(FR,abs(H));
xlabel('(b) Freq, Rad/s'); ylabel('Mag');
axis([0 inf 0 1.1]);
subplot(325); plot(FR,unwrap(angle(H)))
xlabel('(c) Freq, Rad/s'); ylabel('Radians')
H = polyval(b2,s)./polyval(a2,s);
subplot(322); yplot = 20*log10(abs(H)+eps);
plot(FR,yplot); xlabel('(d) Freq, Rad/s');
ylabel('Mag, dB'); axis([0 inf -100 10]);
subplot(324); plot(FR,abs(H));
xlabel('(e) Freq, Rad/s'); ylabel('Mag');
axis([0 inf 0 1.1]); subplot(326);
plot(FR,unwrap(angle(H)))
xlabel('(f) Freq, Rad/s'); ylabel('Radians')
```

10.9.2 LOWPASS TO HIGHPASS

To convert a prototype lowpass filter having cutoff at Ω_P to a highpass filter having cutoff at Ω'_P, make the substitution

$$s \to \frac{\Omega_P \Omega'_P}{s}$$

Example 10.20. Determine the system function for a Butterworth highpass filter having $\Omega_C = 3$ rad/s using the system function for a prototype Butterworth lowpass filter having $\Omega_C = 2$ rad/s and $N = 2$.

The system function for the prototype lowpass filter (as determined above and converting to coefficient form) is

$$H(s) = \frac{4}{(s^2 + 2.8284s + 4)}$$

To convert to a highpass filter having Ω_C = 3 rad/s, we make the substitution

$$s \rightarrow \frac{\Omega_P \Omega'_P}{s} = \frac{(2)(3)}{s} = \frac{6}{s}$$

Thus the new highpass filter's system function is

$$H(s) = \frac{4}{(36/s^2 + 2.8284(6/s) + 4)} = \frac{4s^2}{36 + 16.97s + 4s^2}$$

which reduces to

$$H(s) = \frac{s^2}{s^2 + 4.2426s + 9}$$

The following code computes and plots the frequency response for both the prototype and new filter: The result is shown in Fig. 10.18.

```
[z,p,k] = buttap(2); p = 2*p; b1=4;
a1 = poly(p); P = roots([1,4.2426,9]);
z = [0 0]; b2 = poly(z); a2 = poly(P);
LVsFreqRespDouble(b1,a1,6,25,b2,a2)
```

10.9.3 TRANSFORMATION VIA CONVOLUTION

At this point, we have seen how substitution of an expression in s for each instance of s in a system function can transform the system function into that of a different type of filter. Manual substitution, however, is time consuming and error-prone, so it is desirable, before proceeding to the even more complex substitutions for bandpass and bandstop filters, to develop a technique to automate the filter transformation.

The product of two polynomials can be obtained by convolution of the coefficients representing the polynomials. For example, if the polynomial

$$s^2 + 2s + 1$$

is represented as

$$[1, 2, 1]$$

then its square (for example), that is,

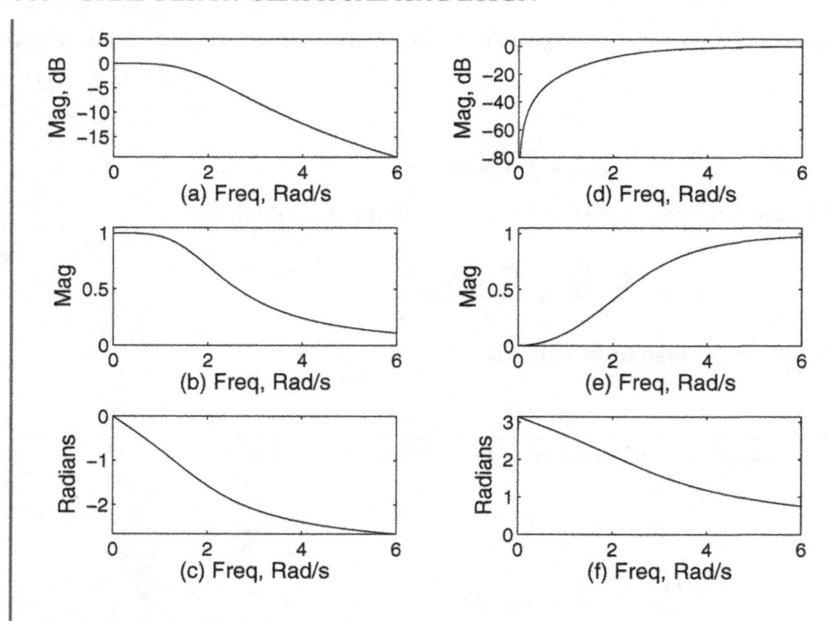

Figure 10.18: (a), (b), and (c): Magnitude (dB) response, Magnitude (linear) response, and phase response of prototype lowpass analog filter; respectively; (d), (e), (f): Magnitude (dB) response, Magnitude (linear) response, and phase response of new highpass filter created from prototype lowpass analog filter; respectively.

$$(s^2 + 2s + 1)^2$$

can be obtained as this convolution:

$$[1, 2, 1] \circledast [1, 2, 1] = [1, 4, 6, 4, 1] = s^4 + 4s^3 + 6s^2 + 4s + 1$$

where the symbol $\circledast$ here means linear convolution. Consider the generalized system function of an analog filter in factored form

$$H(s) = \Pi_{i=1}^{M}(s - z_m)/\Pi_{k=1}^{K}(s - p_k) \tag{10.23}$$

and consider a generalized form of a ratio of polynomials in s to be substituted for each instance of s in Eq. (10.23) as

$$\frac{b_2 s^2 + b_1 s + b_0}{a_2 s^2 + a_1 s + a_0}$$

which can be represented by coefficients only as

$$\frac{[b_2, b_1, b_0]}{[a_2, a_1, a_0]} = \frac{[N]}{[D]}$$

Each factor in Eq. (10.23), after making the substitution for s is of the form

$$(\frac{[N]}{[D]} - pz)$$

where pz represents a pole or zero of the system function.

Then the system function of the target filter $H(S)$ can be represented as the convolution of each factor represented in polynomial coefficient form

$$\{H(S)\} = \circledast_{i=1}^{M}(\frac{[N]}{[D]} - z_m) / \circledast_{k=1}^{K}(\frac{[N]}{[D]} - p_k)$$

which reduces to

$$\{H(S)\} = \circledast_{i=1}^{M}(\frac{[N] - z_m[D]}{[D]}) / \circledast_{k=1}^{K}(\frac{[N] - p_k[D]}{[D]})$$

where $\{H(S)\}$ means "the polynomial coefficient representation of the target system function" and the symbols $\circledast_{i=1}^{M}$ and $\circledast_{k=1}^{K}$ mean the linear convolution of the factors in the numerator and denominator, respectively.

Several cases exist in the relationship of M to K, i.e., the number of zeros compared to the number of poles.

For Butterworth and Chebyshev Type-I filters, $M = 0$, that is, there are no finite zeros. Thus the target system function in polynomial coefficient form is

$$\{H(S)\} = G / \circledast_{k=1}^{K}(\frac{[N] - p_k[D]}{[D]})$$

which becomes

$$\{H(S)\} = \frac{G \circledast_{k=1}^{K} [D]}{\circledast_{k=1}^{K}([N] - p_k[D])} \tag{10.24}$$

where G is the gain of the prototype lowpass filter and

$$\circledast_{k=1}^{K}[D]$$

means the convolution of K factors $[D]$, such as, for example,

$$\circledast_{k=1}^{3}[D] = [D] \circledast [D] \circledast [D]$$

Chebyshev Type-II and Elliptic filters have zeros in their system functions. For N even, the number of zeros is equal to the number of poles ($M = K$), but for N odd, the number of zeros is one less than the number of poles.

When $M = K$, we get

$$\{H(S)\} = \frac{\circledast_{i=1}^{M}([N] - z_m[D])}{\circledast_{k=1}^{K}([N] - p_k[D])} \tag{10.25}$$

and when $M = K - 1$ we get

$$\{H(S)\} = \frac{[\circledast_{i=1}^{M}([N] - z_m[D])] \circledast [D]}{\circledast_{k=1}^{K}([N] - p_k[D])} \tag{10.26}$$

Example 10.21. Write a script that can transform a prototype Chebyshev Type-I lowpass filter of arbitrary order N into a highpass filter using polynomial convolution.

For a Chebyshev Type-I filter, $M = 0$ (there are no finite zeros), and for a highpass filter, the transformation is

$$s \to \frac{\Omega_P \Omega'_P}{s}$$

which can be represented in polynomial coefficient form as

$$\frac{[0, \Omega_P \Omega'_P]}{[1, 0]}$$

Each factor in the denominator of the lowpass filter's system function is

$$(s - p_k)$$

and each transformed factor can be represented as

$$(\frac{[N]}{[D]} - pk) = \frac{N - p_k[D]}{[D]} = \frac{[0, \Omega_P \Omega'_P] - p_k[1, 0]}{[1, 0]} \tag{10.27}$$

The following code first obtains the poles and gain (z is empty for a Chebyshev Type-I filter) for a normalized Chebyshev Type-I filter, scales p by Ω_P, computes the new lowpass filter gain G, establishes N and D in accordance with Eq. (10.27), convolves each of the pole factors, scales the coefficients so that the coefficient of the highest power in the denominator is 1, and finally computes and displays the frequency response. For purposes of testing, the values for N, $WpLP$ (the prototype lowpass filter Ω_P), $WpHP$ (the highpass filter Ω_P), and R_P have been chosen as 7, 3 rad/s, 3 rad/s, and 1 dB, respectively, but may be varied as desired. The result from running the code is shown in Fig. 10.19.

```
function [Z,P,K] = LVCheb1Lpf2Hpf(N,Rp,WpLP,WpHP)
% [Z,P,K] = LVCheb1Lpf2Hpf(7,1,3,3)
[z,p,k] = cheb1ap(N,Rp);
```

```
P = WpLP*p; G = k*WpLP^(length(P));
b1 = G; a1 = poly(P);
Num = 1; Den = 1; D = [1,0];
c = WpLP*WpHP; N = [0,c];
for Ctr = 1:1:length(P)
Num = conv(Num,N-P(Ctr)*D);
Den = conv(Den,D); end
a = real(Num); Scale = 1/a(1);
b2 = real(Den); a2 = Scale*a;
K = Scale*G; Z = roots(b2); P = roots(a2);
LVsFreqRespDouble(b1,a1,2*WpHP,29,K*b2,a2)
```

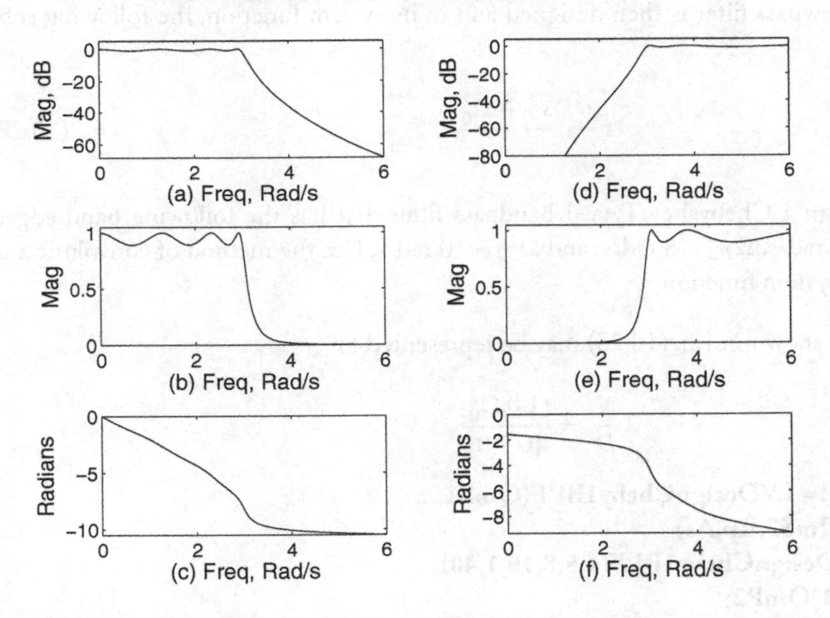

Figure 10.19: (a) Frequency response (dB) of a prototype Chebyshev Type-I lowpass filter having Ω_P = 2 rad/s; (b) Frequency response (linear) of same; (c) Phase response of same; (d) Frequency response (dB) of a highpass filter having Ω_P = 6 rad/s that was created by frequency domain transformation using the convolution method described in the text; (e) Frequency response (linear) of same; (f) Phase response of same.

10.9.4 LOWPASS TO BANDPASS

This method for transforming a prototype lowpass filter into a bandpass filter allows you to specify the new bandpass filter's four band edge frequencies, namely, Ω_{S1}, Ω_{P1}, Ω_{P2}, Ω_{S2}, and from these you compute the parameter Ω_0 (the "center" frequency) for the new bandpass filter

$$\Omega_0 = \sqrt{\Omega_{P1}\Omega_{P2}}$$

as well as the necessary values of Ω_P and Ω_S for the prototype lowpass filter:

$$\Omega_P = \frac{\Omega_{P2}^2 - \Omega_0^2}{\Omega_{P2}}$$

and

$$\Omega_S = \min\left\{\frac{\Omega_{S2}^2 - \Omega_0^2}{\Omega_{S2}}, \frac{\Omega_0^2 - \Omega_{S1}^2}{\Omega_{S1}}\right\}$$

The prototype lowpass filter is then designed and in its system function, the following substitution is made:

$$s \to \frac{s^2 + \Omega_0^2}{s} \tag{10.28}$$

Example 10.22. Design a Chebyshev Type-I bandpass filter that has the following band edges: Ω_{S1} = 4 rad/s, Ω_{P1} = 5 rad/s, Ω_{P2} = 8 rad/s, and Ω_{S2} = 10 rad/s. Use the method of convolution to obtain the new filter's system function.

The substitution shown in Eq. (10.28) may be represented as

$$\frac{N}{D} = \frac{[1,0,\Omega_0^2]}{[0,1,0]}$$

```
function [Z,P,K] = LVDesignCheby1BPF(OmS1,...
OmP1,OmP2,OmS2,Rp,As)
% [Z,P,K] = LVDesignCheby1BPF(4,5,8,10,1,40)
Om0Sq = OmP1*OmP2;
OmPlp = (OmP2^2 - Om0Sq)/OmP2;
OmSlp1 = (OmS2^2 - Om0Sq)/OmS2;
OmSlp2 = (Om0Sq - OmS1^2)/OmS1;
OmSlp = min([OmSlp1,OmSlp2]);
[Z1,P1,K1] = LVDesignChebyIFilter(Rp,As,OmPlp,OmSlp);
Num = 1; Den = 1; N = [1,0,Om0Sq]; D = [0,1,0];
for Ctr = 1:1:length(P1)
Num = conv(Num,[N-P1(Ctr)*D]);
Den = conv(Den,D); end; a = real(Num); b = real(Den);
s=j*[0:0.001:2*OmS2]; Hs = abs(polyval(b,s)./polyval(a,s));
K = 1/max(Hs); Z = roots(b); P = roots(a);
```

```
LVsFreqRespDouble(K1*poly(Z1),poly(P1),2*OmS2,30,K*b,a)
```

The following code, which uses the function above, obtains the new bandpass filter poles and zeros, computes and displays the frequency response of both the prototype lowpass filter and the new bandpass filter, computes the cascade coefficients, and computes the realized values of R_P and A_S for the design band edges using the script *LVxRealizedFiltParamBPF* (see exercises below).

```
OmS1=4; OmP1=5; OmP2=8; OmS2=10; Rp=1; As=40;
[Z,P,K] = LVDesignCheby1BPF(OmS1,OmP1,OmP2,OmS2,Rp,As)
[Bbq,Abq,Gain] = LVDirToCascadeClassIIR(poly(Z),poly(P),K)
H = LVsFreqResp(K*poly(Z),poly(P),3*OmS2,19);
[NetRp,NetAs1,NetAs2] = LVxRealizedFiltParamBPF(H,...
OmS1,OmP1,OmP2,OmS2,3*OmS2)
```

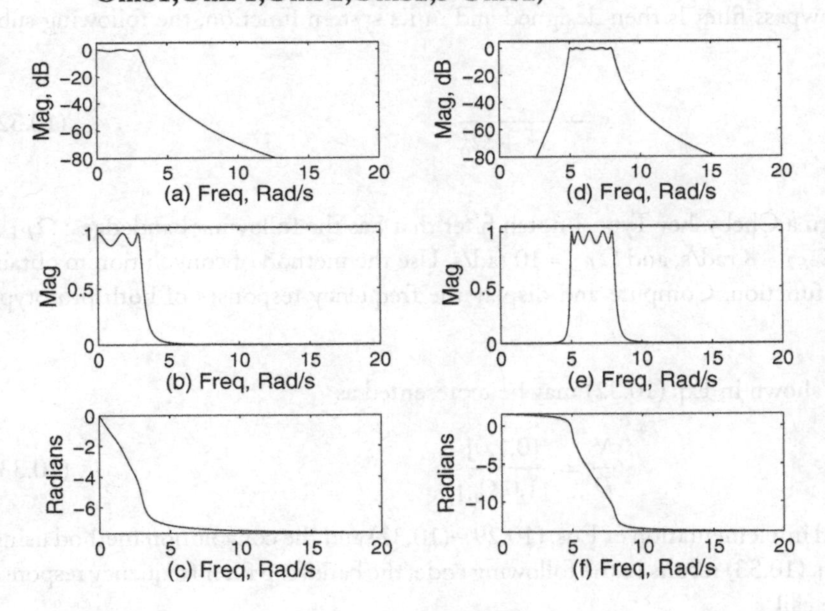

Figure 10.20: (a) Magnitude response (dB) of prototype lowpass filter having its Ω_P and Ω_S computed to result in a new bandpass filter meeting certain specifications after transformation via variable substitution implemented via convolution; (b) Magnitude response (linear) of same; (c) Phase response of same; (d) Magnitude response (dB) of the new bandpass filter; (e) Magnitude response (linear) of same; (f) Phase response of same.

10.9.5 LOWPASS TO BANDSTOP (NOTCH)

This method for transforming a prototype lowpass filter into a bandstop (or notch) filter allows you to specify the new notch filter's four band edge frequencies, namely, Ω_{P1}, Ω_{S1}, Ω_{S2}, Ω_{P2}, and from these you compute the parameter Ω_0 (the "center" frequency) for the new notch filter as

$$\Omega_0 = \sqrt{\Omega_{S1}\Omega_{S2}} \tag{10.29}$$

as well as the necessary values of Ω_P and Ω_S for the prototype lowpass filter:

$$\Omega_P = \frac{\Omega_{P1}}{\Omega_0^2 - \Omega_{P1}^2} \tag{10.30}$$

and

$$\Omega_S = \min\left\{\frac{\Omega_{S2}}{\Omega_{S2}^2 - \Omega_0^2}, \frac{\Omega_{S1}}{\Omega_0^2 - \Omega_{S1}^2}\right\} \tag{10.31}$$

The prototype lowpass filter is then designed and in its system function, the following substitution is made:

$$s \rightarrow \frac{s}{s^2 + \Omega_0^2} \tag{10.32}$$

Example 10.23. Design a Chebyshev Type-I notch filter that has the following bandedges: Ω_{P1} = 4 rad/s, Ω_{S1} = 5 rad/s, Ω_{S2} = 8 rad/s, and Ω_{P2} = 10 rad/s. Use the method of convolution to obtain the new filter's system function. Compute and display the frequency responses of both prototype and notch filters.

The substitution shown in Eq. (10.32) may be represented as

$$\frac{N}{D} = \frac{[0,1,0]}{[1,0,\Omega_0^2]} \tag{10.33}$$

A straightforward implementation of Eqs. (10.29)–(10.32) and the convolution method using N and D as shown in Eq. (10.33) results in the following code; the bandstop filter frequency response obtained by making the call

[Z,P,K] = LVDesignCheby1Notch(4,5,8,10,1,40)

is shown in Fig. 10.21. The last code line of the function *LVDesignCheby1Notch* calls another new function, *LVxRealizedFiltParamNotch* (see exercises below), which, for a notch filter, determines the realized values of R_P and A_S for the design band edges Ω_{P1}, Ω_{S1}, Ω_{S2}, Ω_{P2}.

```
function [Z,P,K] = LVDesignCheby1Notch(OmP1,...
OmS1,OmS2,OmP2,Rp,As)
% [Z,P,K] = LVDesignCheby1Notch(4,5,8,10,1,40)
Om0Sq = OmP1*OmP2; OmPlp = OmP1/(Om0Sq-OmP1^2);
OmSlp1 = OmS2/(OmS2^2 - Om0Sq);
```

```
OmSlp2 = OmS1/(Om0Sq - OmS1^2);
OmSlp = min([OmSlp1,OmSlp2]);
[Z1,P1,K1] = LVDesignCheby1Filter(Rp,As,OmPlp,OmSlp);
H = LVsFreqResp(K1*poly(Z1),poly(P1),2*OmSlp,19);
Num = 1; Den = 1; D = [1,0,Om0Sq]; N = [0,1,0];
for Ctr = 1:1:length(P1)
Num = conv(Num,[N-P1(Ctr)*D]);
Den = conv(Den,D); end;
a = real(Num); b = real(Den); s = j*[0:0.001:2*OmS2];
Hs = abs(polyval(b,s)./polyval(a,s)); K = 1/max(Hs);
Z = roots(b); P = roots(a); H = LVsFreqResp(K*b,a,2*OmS2,31);
[NetAs,NetRp1,NetRp2] = LVxRealizedFiltParamNotch(H,OmP1,...
OmS1,OmS2,OmP2,2*OmS2)
```

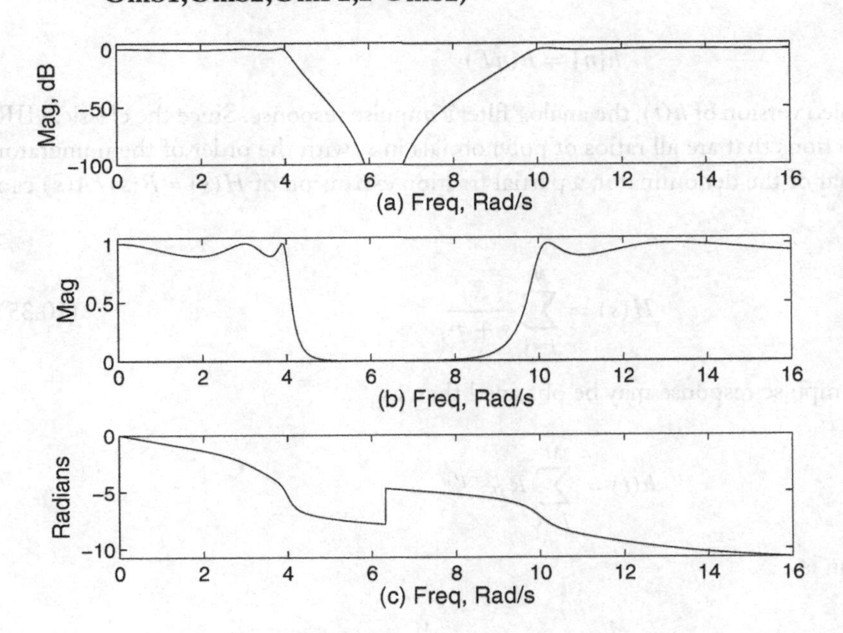

Figure 10.21: (a) Magnitude of response (dB) of bandstop (notch) filter designed by transforming a prototype Chebyshev Type-I lowpass filter according to the second method described in the text; (b) Magnitude of response (linear) of same; (c) Phase response of same.

10.10 ANALOG TO DIGITAL FILTER TRANSFORMATION

Having designed a filter in the analog domain, it is then necessary, to obtain a digital filter, to transform the poles and zeros into the z-domain in a way that preserves desirable attributes of the analog filter. Different transform methods exist to preserve different analog filter properties. The two most popular methods of doing this are the **Impulse Invariance Method** and the **Bilinear Transform.**

10.10.1 IMPULSE INVARIANCE

The Impulse Invariance Method attempts to preserve the analog filter's impulse response by sampling it at the time interval T. Since the digital filter's impulse response is a sampled version of the analog filter's impulse response, we can say that

$$H(z) = \sum_{n=0}^{\infty} h[n]z^{-n} \tag{10.34}$$

where

$$h[n] = h(nT)$$

where $h[n]$ is the sampled version of $h(t)$, the analog filter's impulse response. Since the classical IIR filters have system functions that are all ratios of polynomials in s, with the order of the numerator equal to or less than that of the denominator, a partial fraction expansion of $H(s) = B(s)/A(s)$ can be made:

$$H(s) = \sum_{j=1}^{N} \frac{R_j}{s + p_j} \tag{10.35}$$

The equivalent impulse response may be obtained then as

$$h(t) = \sum_{j=1}^{M} R_j e^{-p_j t}$$

and the sampled version is

$$h[n] = h[nT] = \sum_{j=1}^{M} R_j e^{-p_j nT} = \sum_{j=1}^{M} R_j (e^{-p_j T})^n$$

Substituting into Eq. (10.34) we get

$$H(z) = \sum_{n=0}^{\infty} \left[\sum_{j=1}^{M} R_j (e^{-p_j T})^n \right] z^{-n} \tag{10.36}$$

which results in

$$H(z) = \sum_{j=1}^{M} \frac{R_j z}{z - e^{-p_j T}} = \sum_{j=1}^{M} \frac{R_j}{1 - (e^{-p_j T})z^{-1}} \tag{10.37}$$

Since a sampling operation is involved, it follows that aliasing of the analog filter's frequency response must occur. The digital filter's transfer function is

$$H(z) = F_S \sum_{k=-\infty}^{\infty} H_a(s - j2\pi k F_S)$$

Thus frequency strips of width $2\pi k F_S$ (= $2\pi k/T$) are folded or aliased into the frequency range -πF_S to πF_S. Since no analog lowpass filter's transfer function is identically zero above any given finite frequency, it follows that degradation of the digital filter's transfer function relative to that of the analog filter must occur to a greater or lesser extent. This mapping of each horizontal strip of height $2\pi/T$ lying to the left of the s-plane imaginary axis into the interior of the unit circle in the z-plane is shown in Fig. 10.22.

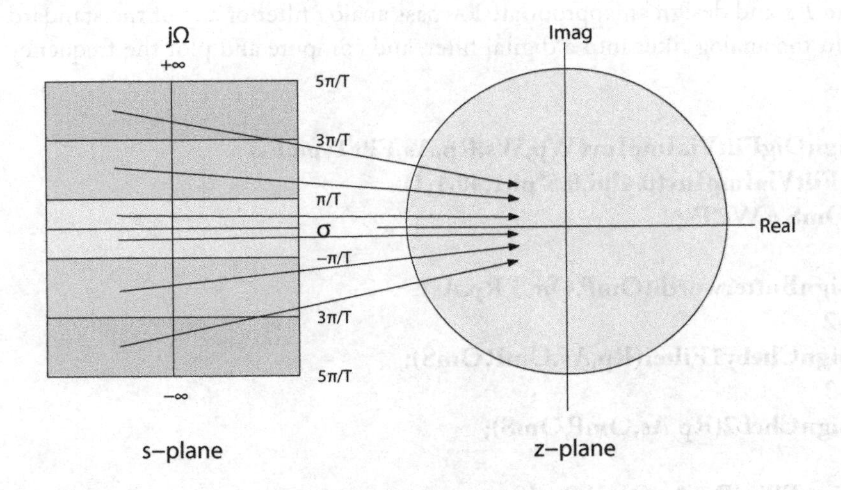

Figure 10.22: The Impulse Invariance mapping from the s-plane to the z-plane, due to aliasing, maps all horizontal strips of height $2\pi/T$ to the same place in the z-plane. The mapping is stable since values of s lying to the left of the s-plane imaginary axis map into the interior of the unit circle.

A procedure to perform the Impulse Invariance transformation is as follows:

- Set the desired digital filter values of ω_P and ω_S, pick a value for T_S (=$1/F_S$) and then compute the analog filter design frequencies Ω_P and Ω_S as

$$\Omega_P = \omega_P / T_S = \omega_P F_S$$

$$\Omega_S = \omega_S / T_S = \omega_S F_S$$

- Design a desired analog filter prototype using Ω_P and Ω_S, as well as desired values of R_P and A_S, and obtain its numerator and denominator system function coefficients, b and a.
- Perform a partial fraction expansion on b and a of the form given in Eq. (10.35).
- Convert analog poles into digital poles in accordance with Eq. (10.37).

The above steps are performed by the MathScript function

$$[Bz, Az] = impinvar(b, a, Fs)$$

which receives the analog filter b and a coefficients and a desired sample rate (= $1/T$) and delivers the z-transform coefficients of the digital filter.

Example 10.24. Devise a script that can receive desired digital filter cutoff frequencies ω_P and ω_S and a desired sample rate Fs and design an appropriate lowpass analog filter of any of the standard four types, and transform the analog filter into a digital filter, and compute and plot the frequency response of both filters.

```
function LVDesignDigFiltViaImpInv(Wp,Ws,Rp,As,FiltType,Fs)
% LVDesignDigFiltViaImpInv(0.4*pi,0.5*pi,1,40,1,1)
OmP = Wp*Fs; OmS = Ws*Fs;
if FiltType==1
[Z,P,K] = LVDesignButterworth(OmP,OmS,Rp,As);
elseif FiltType==2
[Z,P,K] = LVDesignCheby1Filter(Rp,As,OmP,OmS);
elseif FiltType==3
[Z,P,K] = LVDesignCheb2(Rp,As,OmP,OmS);
else
[Z,P,K] = LVDesignEllip(Rp,As,OmP,OmS);
end; b = K*poly(Z); a = poly(P);
[BZ,AZ] = impinvar(b,a,Fs);
LVsFRzFrLog(b,a,OmP,BZ,AZ,97)
```

We have introduced a new frequency response script, *LVsFRzFRLog*, as follows:

```
function LVsFRzFrLog(sB,sA,OmegaC,zB,zA,FigNo)
LnLm = 8*OmegaC;
Sargs = 0:0.01:LnLm;
Sargs = [Sargs]; s = j*Sargs;
Hs = polyval(sB,s)./polyval(sA,s);
```

```
Zargs = 0:0.01:pi; z = exp(j*Zargs);
Hz = polyval(zB,z)./polyval(zA,z);
figure(FigNo); subplot(211); plot(Sargs,20*log10(abs(Hs+eps)))
xlabel('(a) Freq, Radians/s'); ylabel('Magnitude, dB');
axis([0,LnLm,-100,5]); subplot(212);
ploty = 20*log10(abs(Hz)+eps);
plot(Zargs/pi,ploty); grid on;
xlabel('(b) Freq, Units of \pi'); ylabel('Magnitude, dB');
axis([10^(-2),inf,-100,(max(ploty)+10)])
```

Example 10.25. Design a Butterworth digital filter having $\omega_P = 0.5\pi$, $\omega_S = 0.7\pi$, $R_P = 0.5$ dB, and $A_S = 40$ dB using the script written for the previous example. Repeat with $\omega_P = 0.5\pi$ and $\omega_S = 0.65\pi$.

Letting $F_S = 1$ we make the call

```
LVDesignDigFiltViaImpInv(0.5*pi,0.7*pi,0.5,40,1,1)
```

and observe the results in Fig. 10.23.

Making the call

```
LVDesignDigFiltViaImpInv(0.5*pi,0.65*pi,0.5,40,1,1)
```

we obtain the results shown in Fig. 10.24, which show a definite degradation in the stopband. However, this occurs at an attenuation level exceeding the required 40 dB.

Carrying the Butterworth experiment one step further, we further narrow the transition band with the following call

```
LVDesignDigFiltViaImpInv(0.5*pi,0.62*pi,0.5,40,1,1)
```

which results in a total degradation of the filter as shown in Fig. 10.25.

Example 10.26. Design a Chebyshev Type-I digital filter having $\omega_P = 0.5\pi$, $\omega_S = 0.7\pi$, $R_P = 0.5$ dB, and $A_S = 40$ dB. Repeat the design using $\omega_P = 0.5\pi$, $\omega_S = 0.525\pi$, $R_P = 0.5$ dB, and $A_S = 40$ dB.

We make the following calls

```
LVDesignDigFiltViaImpInv(0.5*pi,0.7*pi,0.5,40,2,1)
```

```
LVDesignDigFiltViaImpInv(0.5*pi,0.525*pi,0.5,40,2,1)
```

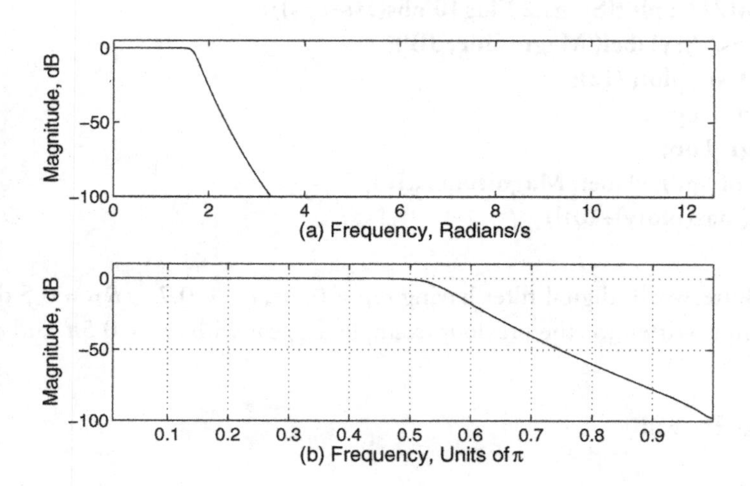

Figure 10.23: (a) Magnitude (dB) of response of prototype analog Butterworth filter; (b) Magnitude (dB) of response of digital Butterworth filter designed by Impulse Invariance.

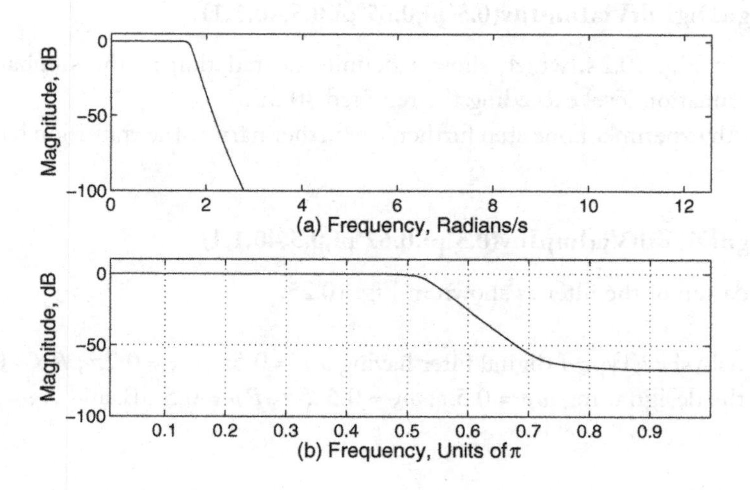

Figure 10.24: (a) Magnitude (dB) of response of prototype analog Butterworth filter; (b) Magnitude (dB) of response of digital Butterworth filter designed by Impulse Invariance.

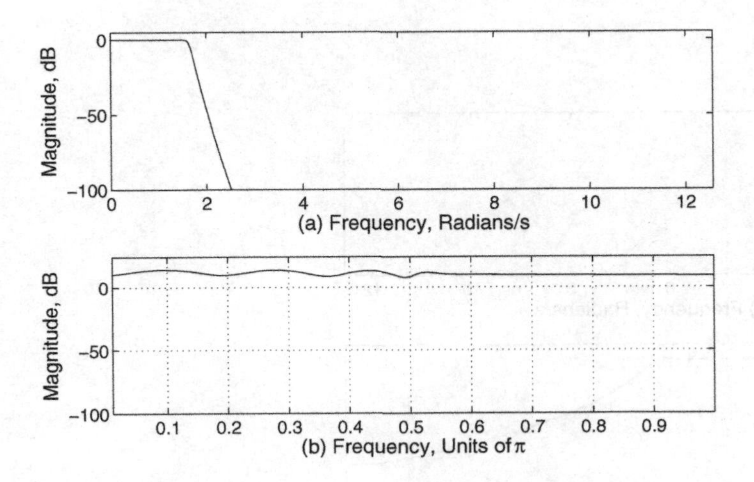

Figure 10.25: (a) Magnitude (dB) of response of prototype analog Butterworth filter; (b) Magnitude (dB) of response of digital Butterworth filter designed by Impulse Invariance.

which result in Figs. 10.26 and 10.27.

Example 10.27. Design a Chebyshev Type-II digital filter having $\omega_P = 0.5\pi$, $\omega_S = 0.7\pi$, $R_P =$ 0.5 dB, and $A_S = 40$ dB.

We make the call

LVDesignDigFiltViaImpInv(0.5*pi,0.7*pi,0.5,40,3,1)

the result from which is shown in Fig. 10.28. We note an unsatisfactory result. Note that for the Chebyshev Type-II analog filter, the frequency response does not go to zero as frequency goes to infinity, and the aliasing inherent in the Impulse Invariance technique takes its toll.

Example 10.28. Design an Elliptic digital filter having $\omega_P = 0.5\pi$, $\omega_S = 0.7\pi$, $R_P = 0.5$ dB, and $A_S = 40$ dB.

We make the call

LVDesignDigFiltViaImpInv(0.5*pi,0.7*pi,0.5,40,4,1)

the result from which is shown in Fig. 10.29. Again we note an unsatisfactory result. Note that like the Chebyshev Type-II analog filter, the frequency response of an Elliptic analog filter does not go to zero as frequency goes to infinity, and once again the aliasing inherent in the Impulse Invariance technique leads to a poor digital filter.

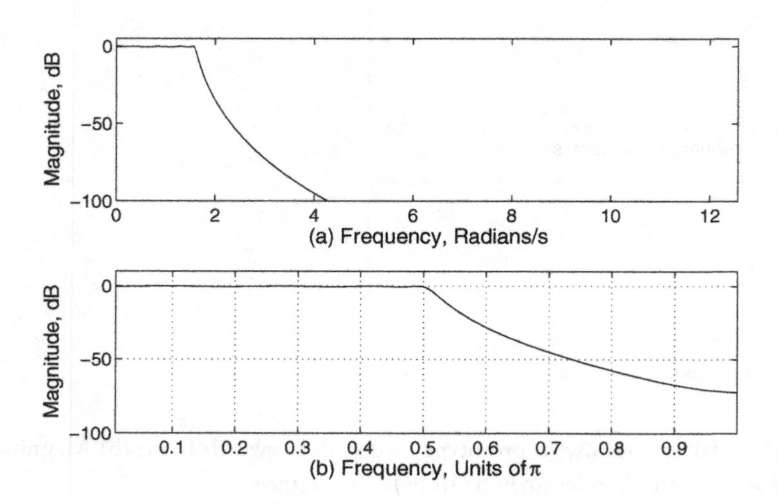

Figure 10.26: (a) Magnitude (dB) of response of prototype analog Chebyshev Type-I filter; (b) Magnitude (dB) of response of digital Chebyshev Type-I filter designed by Impulse Invariance.

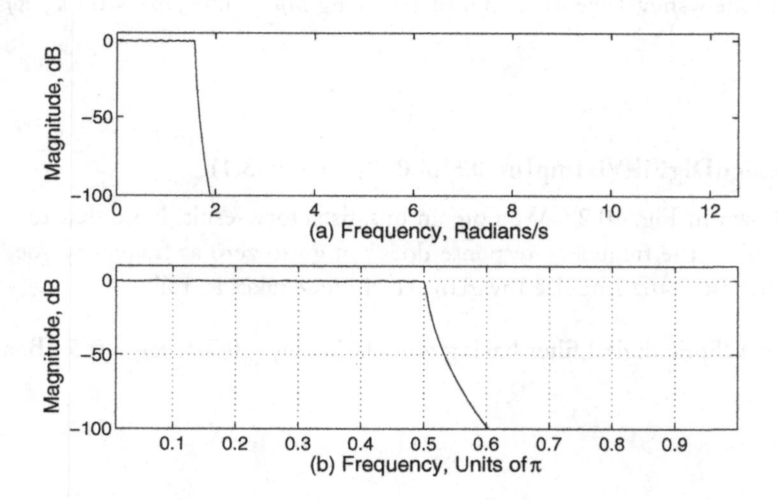

Figure 10.27: (a) Magnitude (dB) of response of prototype analog Chebyshev Type-I filter; (b) Magnitude (dB) of response of digital Chebyshev Type-I filter designed by Impulse Invariance.

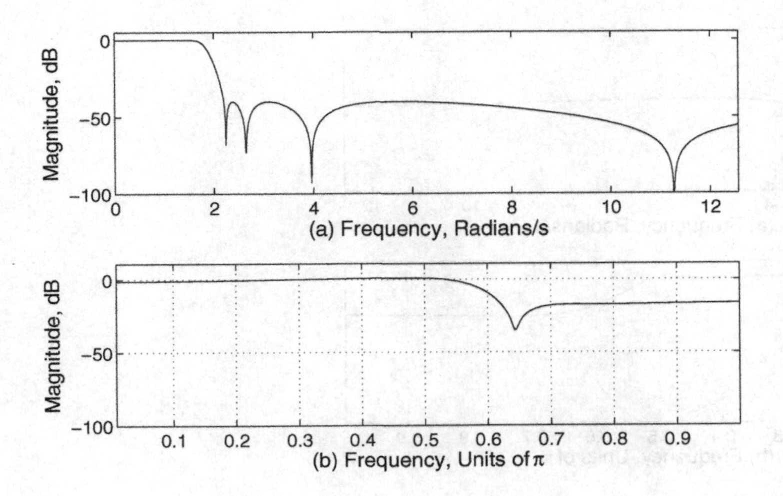

Figure 10.28: (a) Magnitude (dB) of response of prototype analog Chebyshev Type-II filter; (b) Magnitude (dB) of response of digital Chebyshev Type-II filter designed by Impulse Invariance.

10.10.2 THE BILINEAR TRANSFORM

A pole (or zero) in the Laplace domain can be mapped to a pole or zero in the z-domain using Eq. (10.38), which is known as the Bilinear transform.

$$z = \frac{1 + sT_s/2}{1 - sT_s/2} \tag{10.38}$$

where T_s is the sampling period of the digital system. To convert from a system function in the variable s, make the substitution

$$s = \frac{2}{T_s}\frac{1 - z^{-1}}{1 + z^{-1}}$$

Example 10.29. Convert the Laplace domain system function $1/(s + 1)$ to the z-domain using T_s = 1.

This is a matter of algebraic substitution and fractional simplification:

$$\begin{aligned} H(z) &= 1/(\frac{2}{T}\frac{(1 - z^{-1})}{(1 + z^{-1})} + 1) = \\ & 1/(\frac{2(1 - z^{-1}) + (1 + z^{-1})}{(1 + z^{-1})}) \end{aligned}$$

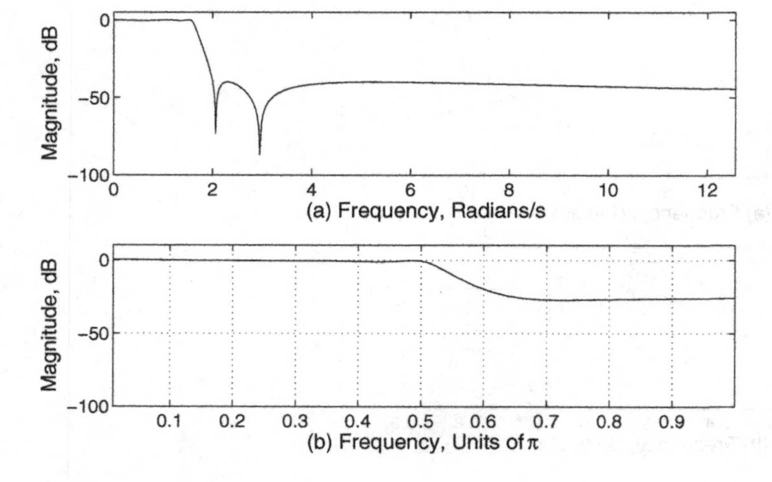

Figure 10.29: (a) Magnitude (dB) of response of prototype analog Elliptic filter; (b) Magnitude (dB) of response of digital Elliptic filter designed by Impulse Invariance.

which reduces to

$$H(z) = \frac{1 + z^{-1}}{3 - z^{-1}} = \frac{0.333(1 + z^{-1})}{1 - 0.333z^{-1}}$$

Note that the Laplace system function had one pole and one zero at infinite frequency. This has transformed into a z-domain system function having one pole and one zero at z = -1.

We can examine the Bilinear transform by letting $s = \sigma + j\omega$ and $T_s = 1/F_s$ in Eq. (10.38), which yields Eq. (10.39).

$$z = \frac{(2F_s + \sigma) + j\Omega}{(2F_s - \sigma) - j\Omega} \tag{10.39}$$

and the magnitude of z is

$$|z| = \frac{\sqrt{(2F_s + \sigma)^2 + \Omega^2}}{\sqrt{(2F_s - \sigma)^2 + \Omega^2}} \tag{10.40}$$

It can be seen from inspection of Eq. (10.40) that when $\sigma < 0$, Laplace poles are in the left half-plane (and are stable since $e^{\sigma t}e^{j\Omega t}$ decays to zero as $t \to \infty$ when $\sigma < 0$); these poles map to the interior of the unit circle i.e., $|z| < 1$) in the z-plane, which defines the stable region for z-plane poles. Similarly, Laplace poles on the imaginary axis (i.e., $\sigma = 0$), which generate constant, unity-amplitude time domain responses, map to the unit circle in the z-plane; z-plane poles lying on the unit circle also generate constant, unity-amplitude time domain responses. Figure 10.30 illustrates a few of the features of the Bilinear mapping.

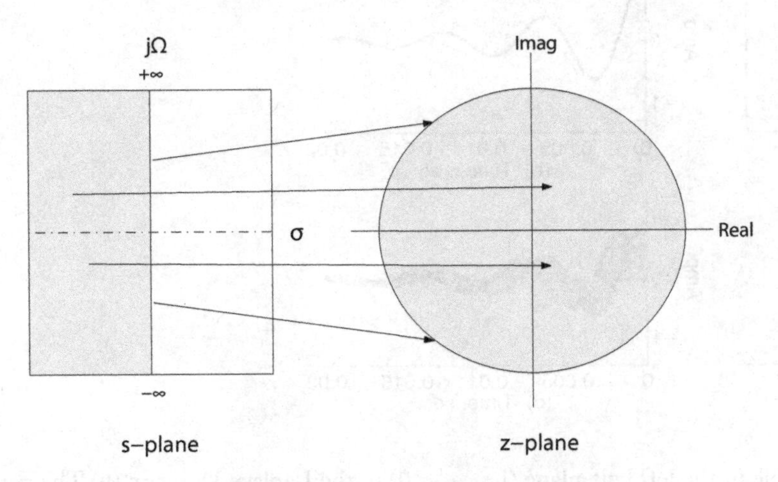

Figure 10.30: Mapping of the variable s in the Laplace (analog) plane to the variable z in the z- or digital domain. Note that the positive $j\Omega$ axis of the s-plane maps to the upper half of the unit circle in the z-plane, and the negative $j\Omega$ axis maps to the lower half of the unit circle. The upper (positive frequency) left half-plane of the Laplace domain maps to the upper interior of the unit circle, and the lower (negative frequency) half-plane maps to the lower interior of the unit circle. Both positive and negative infinity on the $j\Omega$ axis map to z = -1, and $s = 0 + j0$ maps to z = +1.

When a pole or pair of complex conjugate poles lies in the left half-plane, they map to the interior of the unit circle and both s- and z- impulse responses decay to zero. Figures 10.31 and 10.32 show the Laplace and z- domain poles and corresponding impulse responses for a single pole and a pair of complex conjugate poles, respectively.

When a pole lies on the imaginary axis in the s-domain (i.e., $\sigma = 0$), the impulse response is a constant, unity-amplitude sinusoid; this pole location is equivalent to a pole on the unit circle in the z-domain. This situation is depicted in Fig. 10.33. Poles in this location lead to marginally stable filters–for some signals, the filter will appear to be stable, while for other signals, it will be clearly unstable.

A pole to the right of the imaginary axis in the s-plane leads to an impulse response which is a complex exponential of continually increasing amplitude. This pole location corresponds to a pole outside the unit circle in the z-plane. Figure 10.34 depicts this situation.

System Transformation Via Convolution

The algebraic manipulations necessary to transform a Laplace system function into a z-domain system function are, in general, arduous and error-prone, so it behooves us to design a method for performing the operation by computer. The Laplace system function can be written as

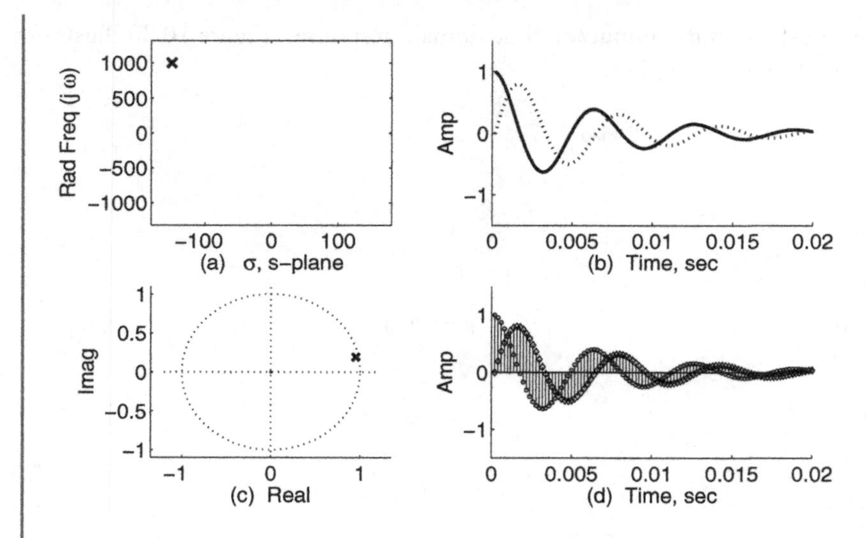

Figure 10.31: (a) A single pole in the left half-plane (i.e., $\sigma < 0$) in the Laplace Domain; (b) The real and imaginary parts of the impulse response corresponding to the pole plotted in (a); (c) A pole in the z-Plane, lying inside the Unit Circle, obtained using the Bilinear transform; (d) Real and imaginary parts of the impulse response corresponding to the pole in (c).

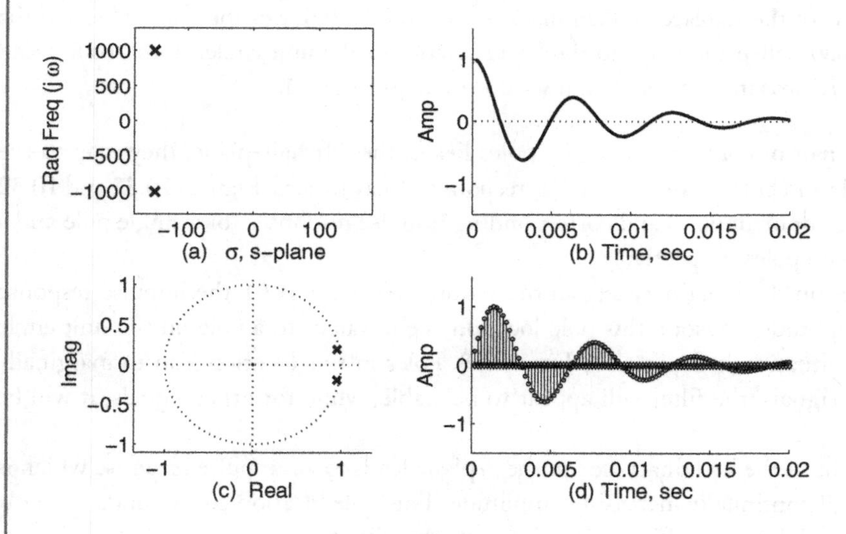

Figure 10.32: (a) A pair of complex conjugate poles in the left half-plane (i.e., $\sigma < 0$) in the Laplace Domain; (b) The (real-only) impulse response corresponding to the poles plotted in (a); (c) A pair of poles in the z-Plane, lying inside the Unit Circle, obtained using the Bilinear transform; (d) The (real-only) impulse response corresponding to the poles in (c).

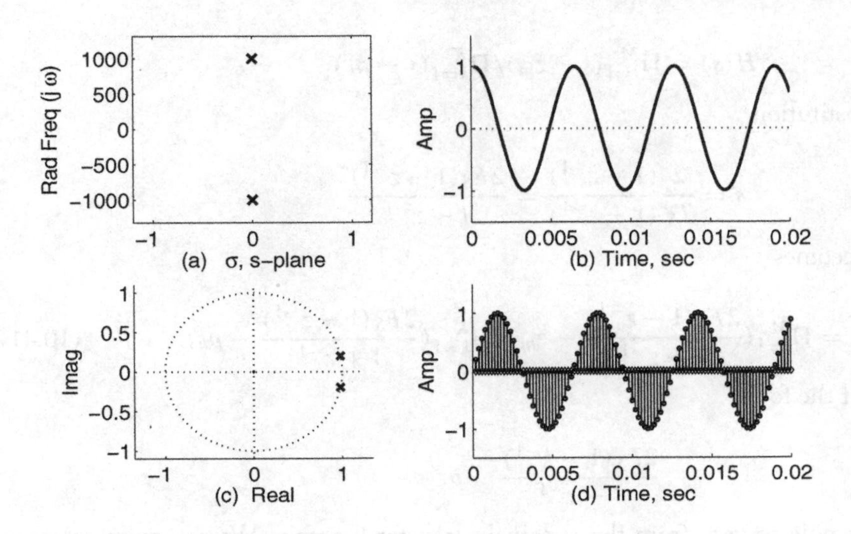

Figure 10.33: (a) A pair of complex conjugate poles on the Imaginary Axis (i.e., Damping = 0) in the Laplace Domain; (b) The real, undamped waveform generated by the poles plotted in (a); (c) The Bilinear-transformed pair of poles in the z-Plane, lying on the Unit Circle; (d) Real waveform in z-Plane generated by the poles in (c).

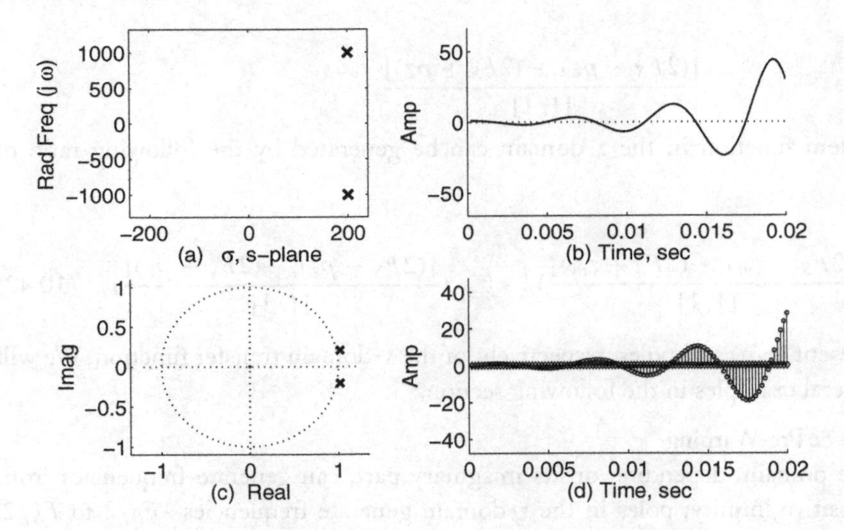

Figure 10.34: (a) A pair of complex conjugate poles to the right the Imaginary Axis (i.e., having gain greater than unity) in the Laplace Domain; (b) The real waveform generated by the poles plotted in (a); (c) The Bilinear-transformed pair of poles in the z-Plane, lying outside the Unit Circle; (d) Real waveform in z-Plane generated by the poles in (c).

$$H(s) = \Pi_{i=1}^{M}(s - z_m)/\Pi_{k=1}^{K}(s - p_k)$$

Making the substitution

$$s = \frac{2}{T_S}\frac{(1 - z^{-1})}{(1 + z^{-1})} = \frac{2F_S(1 - z^{-1})}{1 + z^{-1}}$$

the system function becomes

$$H(z) = \Pi_{i=1}^{M}(\frac{2F_S(1 - z^{-1})}{1 + z^{-1}} - z_m)/\Pi_{k=1}^{K}(\frac{2F_S(1 - z^{-1})}{1 + z^{-1}} - p_k) \tag{10.41}$$

Each factor is of the form

$$\frac{2F_S(1 - z^{-1})}{1 + z^{-1}} - pz$$

where pz represents a pole or zero from the s-domain transfer function. We can represent each factor by equivalent coefficient arrays.

$$\frac{2F_S[1, -1]}{[1, 1]} - pz = \frac{2F_S[1, -1]}{[1, 1]} - \frac{pz[1, 1]}{[1, 1]}$$

which reduces to

$$\frac{[(2F_S - pz), -(2F_S + pz)]}{[1, 1]}$$

and the symbolic system function in the z-domain can be generated by the following ratio of convolutions:

$$\{H(z)\} = \circledast_{m=1}^{M}(\frac{[(2F_S - z_m), -(2F_S + z_m)]}{[1, 1]})/ \circledast_{k=1}^{K}(\frac{[(2F_S - p_k), -(2F_S + p_k)]}{[1, 1]}) \tag{10.42}$$

where z_m and p_k represent zeros and poles, respectively, of the s-domain transfer function. We will use this method in several examples in the following section.

Frequency Relationship & Pre-Warping

A pole in the Laplace domain, depending on its imaginary part, can generate frequencies from negative infinity to positive infinity; poles in the z-domain generate frequencies -$F_S/2$ to $F_S/2$. For frequencies that are at or below about F_S (the sampling rate used in the transform), the time domain response generated in the z-plane is similar to that generated in the Laplace plane, i.e., there is a close relationship between the Laplace and z-domain frequencies. As the Laplace frequency increases relative to F_S, the relationship becomes more and more nonlinear. Figure 10.35 depicts this relationship, with the Laplace frequency normalized to multiples of F_S.

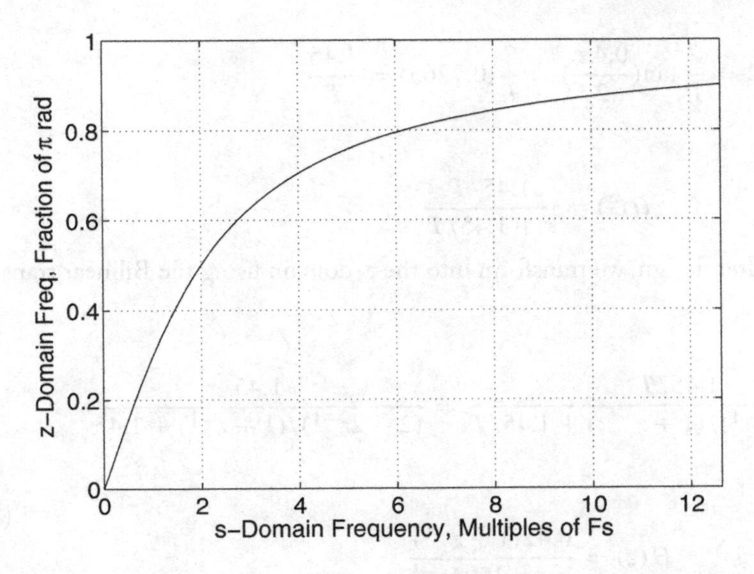

Figure 10.35: The highly nonlinear relationship between s-Domain frequency and z-Domain frequency.

As a result of the nonlinear frequency relationship between the s- and z- domains, it is necessary to pre-warp the s-domain poles so that after the Bilinear transform is performed they will be in desired locations in the z-domain.

The frequency relationships between the s- and z- domains are

$$\Omega = \frac{2}{T}\tan(\frac{\omega}{2}) \tag{10.43}$$

and

$$\omega = 2\tan^{-1}(\frac{\Omega T}{2})$$

The following examples explore the concept of **Pre-Warping,** that is, choosing analog design frequencies that must be realized in the analog design so that after the Bilinear transform is performed, the digital domain frequencies are the ones specified in the digital design.

Example 10.30. Consider the single-pole analog filter having a 3 dB bandwidth of Ω and $H(s) = \Omega/(s+\Omega)$. It is desired to transform this system function into the z-domain with $\omega_c = 0.4\pi$ radians.

We solve for the value of Ω necessary to result, after the Bilinear transform is performed, in the desired ω_c.

$$\Omega = \frac{2}{T}\tan(\frac{0.4\pi}{2}) = \frac{2}{T}(0.7265) = \frac{1.45}{T}$$

and thus

$$H(s) = \frac{1.45/T}{s + 1.45/T}$$

From this, the analog filter design, we transform into the z-domain using the Bilinear transform:

$$H(z) = \frac{1.45/T}{2/T((1 - z^{-1})/(1 + z^{-1})) + 1.45/T} = \frac{1.45}{(2 - 2z^{-1})/(1 + z^{-1}) + 1.45}$$

which reduces to

$$H(z) = \frac{0.42(1 + z^{-1})}{1 - 0.1594z^{-1}}$$

Example 10.31. Write a script suitable to convert the Laplace domain system function of a Butterworth or Chebyshev Type-I filter into a z-domain system function according to a given sample rate F_S. Test it by obtaining poles and zeros from the functions *buttap* and *cheb1ap* and converting them to the z-domain using a chosen value for Fs. Compute and plot the magnitude of frequency response of both the analog and digital filters.

Note that Ω_C = 1 for the *buttap* and *cheb1ap* functions. You can enter your own value for Ω_C as *OmegaC* in the function *LVs2zViaBilinearEx* introduced below. With *OmegaC* = 1, try several values of F_S, such as 0.1, 0.5, 1, 2, 4, and 8, and plot ω_c(the z-domain cutoff frequency of the digital filter, observed on the plot resulting from running the code below) against the ratio of Ω_C to F_S. Since it is the ratio of a particular frequency in the analog domain, such as Ω_C, to F_S, rather than the actual values of these parameters, either of Ω_C or F_S may be arbitrarily chosen and held constant while the other is arbitrarily specified to achieve a given ω_c for the digital filter. After holding *OmegaC* at 1 and varying F_S, set F_S = 1 and vary Ω_C; you should be able to produce any desired ω_c for the digital filter using either method.

In general, necessary values for $\Omega_C, \Omega_P, \Omega_S$, etc., to achieve given frequencies in the z-domain while using a given (arbitrary) value for F_S can be computed using Eq. (10.43).

```
function LVs2zViaBilinearEx(N,Rp,OmegaC,Fs,CheborButter)
% LVs2zViaBilinearEx(3,1,1,1,1)
% LVs2zViaBilinearEx(3,1,1,5,1)
% LVs2zViaBilinearEx(3,1,5,5,1)
% LVs2zViaBilinearEx(3,1,5,25,1)
```

```
% LVs2zViaBilinearEx(4,[],4,8,0)
if CheborButter==1
[Z,P,K] = cheb1ap(N,Rp);
else; [Z,P,K] = buttap(N); end
K = K*OmegaC^N; P = OmegaC*P;
sB = K; sA = poly(P);
Num = 1; Den = 1; D = [1,1];
for Ctr = 1:1:length(P)
 N = [(2*Fs - P(Ctr)), -(2*Fs + P(Ctr))];
 Num = conv(Num,N); Den = conv(Den,D);
end; zB = real(Den); zA = real(Num);
LVsFRzFr(sB,sA,OmegaC,K*zB,zA,37)
```

In the script above, a new script was introduced which can receive a set of s-domain system coefficients sB and sA, a set of z-domain system coefficients zB and zA, a cutoff frequency $OmegaC$, and a desired figure number $FigNo$ to plot the frequency responses for both filters:

```
function LVsFRzFr(sB,sA,OmegaC,zB,zA,FigNo)
LnLm = 8*OmegaC;
rootrat = ((10^6)/(LnLm))^(0.001);
Sargs = 0:LnLm/1000:LnLm-LnLm/1000;
yy = (LnLm)*(rootrat.^[1:1:1000]);
Sargs = [Sargs yy]; s = j*Sargs;
Hs = polyval(sB,s)./polyval(sA,s);
Zargs = 0:0.01:pi; z = exp(j*Zargs);
Hz = polyval(zB,z)./polyval(zA,z);
figure(FigNo); subplot(211); semilogx(Sargs,abs(Hs))
xlabel('(a) Freq, Radians/s'); ylabel('Magnitude')
subplot(212); plot(Zargs/pi,abs(Hz))
xlabel('(b) Freq, Units of \pi'); ylabel('Magnitude')
```

The call

LVs2zViaBilinearEx(3,1,1,5,1)

for example, results in Fig. 10.36.

Example 10.32. Write a script that can receive a design specification for a lowpass digital Chebyshev Type-I filter as R_P, A_S, ω_p, and ω_s, design an appropriate analog filter prototype, convert the prototype into a digital filter using the Bilinear transform implemented by convolution, and compute and display the frequency responses of both the analog prototype and digital filters.

The following script establishes F_S and then pre-warps the digital frequencies, i.e., computes the analog frequencies Ω_P and Ω_S (OmP and OmS in the script) necessary to result in the specified

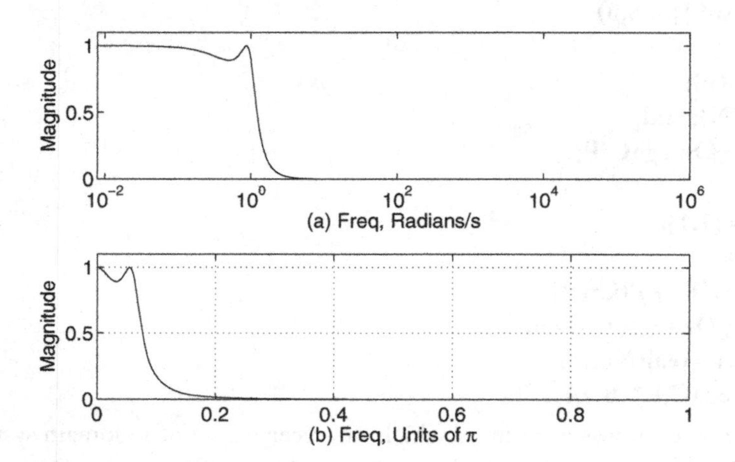

Figure 10.36: (a) A Chebyshev Type-I analog lowpass filter; (b) A digital version of the same, designed using the Bilinear Transform, with Ω_C = 1 rad/s and F_S = 1. The lowpass cutoff frequency can be controlled in the digital domain by adjusting the ratio of Ω_C to F_S.

digital frequencies ω_p, and ω_s (wp and ws in the script) after the Bilinear transform is performed. The analog filter is then designed using R_P, A_S, Ω_P, and Ω_S, and then the analog frequency response is computed and displayed. The Bilinear transform is then computed, using a simplified convolution formula since the Chebyshev Type-I filter has no finite zeros (this script will also work with a Butterworth filter since it, too, lacks finite zeros in the analog domain). The digital filter's frequency response is then computed from its b and a coefficients (*SczNum* and *SczDen* in the script).

```
function LVsCheby1LPF2zCheby1LPF(Rp,As,wp,ws)
% LVsCheby1LPF2zCheby1LPF(1,40,0.5*pi,0.6*pi)
Fs = 1; T = 1/Fs; OmP = (2/T)*tan(wp/2);
OmS = (2/T)*tan(ws/2); E = sqrt(10^(Rp/10)-1);
A = 10^(As/20); OmC = OmP; OmT = OmS/OmP;
g = sqrt((A^2-1)/(E^2));
NN = ceil(log10(g + sqrt(g^2 - 1))/log10(OmT +...
 sqrt(OmT^2-1)))
v0 = asinh(1/E)/NN; k = -(NN-1):2:(NN-1);
P = OmC*(-sinh(v0)*cos(k*pi/(2*NN)) + ...
 j*cosh(v0)*sin(k*pi/(2*NN)));
NetK = prod(abs(P)); if rem(NN,2)==0
NetK = NetK/sqrt(1 + E^2); end
sB = NetK; sA = poly(P); Num = NetK; Den = 1;
```

```
for Ctr = 1:1:length(P); D = [1,1];
N = [(2*Fs-P(Ctr)), -(2*Fs+P(Ctr))];
Num = conv(Num,N); Den = conv(Den,D);
end; zB = real(Den); zA = real(Num);
b0 = 1/zB(1); a0 = 1/zA(1);
zA = zA*a0; zB = zB*b0;
G = abs(polyval(zA,exp(j*0))./polyval(zB,exp(j*0)));
if rem(NN,2)==0; G = G/sqrt(1 + E^2); end
zB = G*zB;
LVsFRzFr(sB,sA,OmC,zB,zA,44); Hz = LVzFr(zB,zA,1000,45);
[NetRp,NetAs] = LVsRealizedFiltParamLPF(Hz,wp,ws,pi)
```

The above script introduced the script *LVzFr*, the code for which is given below. This script allows you to specify a figure number on which to plot the results, and the number of frequency points to compute.

```
function Hz = LVzFr(zB,zA,NumPts,FigNo)
% Computes and displays on Figure FigNo the magnitude and
% phase response of a digital filter having numerator
% coefficients zB and denominator coefficients zA, for NumPts
% frequency points. The complex frequency response vector Hz is
% returned by the function and can be supplied to another script if
% desired to compute realized filter parameters.
Zargs = 0:pi/NumPts:pi; z = exp(j*Zargs);
Hz = polyval(zB,z)./polyval(zA,z);
figure(FigNo); subplot(211)
plot(Zargs/pi,20*log10(abs(Hz)+eps)); grid on
xlabel('(a) Freq, Units of \pi Radians');
ylabel('Magnitude, dB'); axis([0,1,-100,10])
subplot(212); plot(Zargs/pi,unwrap(angle(Hz))); grid on
xlabel('(b) Freq, Units of \pi Radians'); ylabel('Radians')
axis([0,1,-inf,inf])
```

10.11 MATHSCRIPT FILTER DESIGN FUNCTIONS

A number of functions exist in MathScript that design digital IIR filters using the Bilinear transform and analog prototype filters as described above. The following calls return the numerator and denominator coefficients for a lowpass digital filter having ω_C equal to the input argument OmC (supplied as a normalized frequency, $0 < \omega_C < 1.0$, where 1.0 is half the sampling rate), while N, R_P, and A_S have the usual meanings. In any of the calls given below, to obtain the output as zeros, poles, and gain, replace the output argument list $[b, a]$ with the list $[z, p, k]$.

[b,a] = butter(N,OmC)
[b,a] = cheby1(N,Rp,OmC)
[b,a] = cheby2(N,As,OmC)
[b,a] = ellip(N,Rp,As,OmC)

To design an analog filter, supply the trailing argument 's', and specify the desired Ω_C as any desired frequency in rad/s.

[b,a] = butter(N,OmC,'s')
[b,a] = cheby1(N,Rp,OmC,'s')
[b,a] = cheby2(N,As,OmC,'s')
[b,a] = ellip(N,Rp,As,OmC,'s')

A highpass digital filter can be designed with any one of the following calls; analog filters may be designed by supplying the trailing argument 's' and specifying the desired band limit in rad/s.

[b,a] = butter(N,OmC,'high')
[b,a] = cheby1(N,Rp,OmC,'high')
[b,a] = cheby2(N,As,OmC,'high')
[b,a] = ellip(N,Rp,As,OmC,'high')

By specifying two band limit frequencies for OmC, a digital bandpass filter can be designed with any one of the following calls, and analog bandpass filters may be designed by supplying the trailing argument 's' and specifying desired band limits in rad/s.

[b,a] = butter(N,[OmLo,OmHi])
[b,a] = cheby1(N,Rp,[OmLo,OmHi])
[b,a] = cheby2(N,As,[OmLo,OmHi])
[b,a] = ellip(N,Rp,As,[OmLo,OmHi])

Example 10.33. Design an elliptic digital lowpass filter having $\Omega_L = 0.3\pi$ rad and $\Omega_H = 0.5\pi$ rad, $N = 7$, $R_P = 0.5$ dB, and $A_S = 50$ dB.

The following simple script designs the filter and plot its frequency response, which is shown in Fig. 10.37.

[b,a] = ellip(7,0.5,50,[0.3,0.5])
Hz = LVzFr(b,a,1000,21);

By specifying two band limit frequencies for OmC as well as the label 'stop', a bandstop or notch filter can be designed with any one of the following calls; analog filters may be designed using the same method described above for the preceding filter types.

[b,a] = butter(N,[OmLo,OmHi],'stop')
[b,a] = cheby1(N,Rp,[OmLo,OmHi],'stop')
[b,a] = cheby2(N,As,[OmLo,OmHi],'stop')
[b,a] = ellip(N,Rp,As,[OmLo,OmHi],'stop')

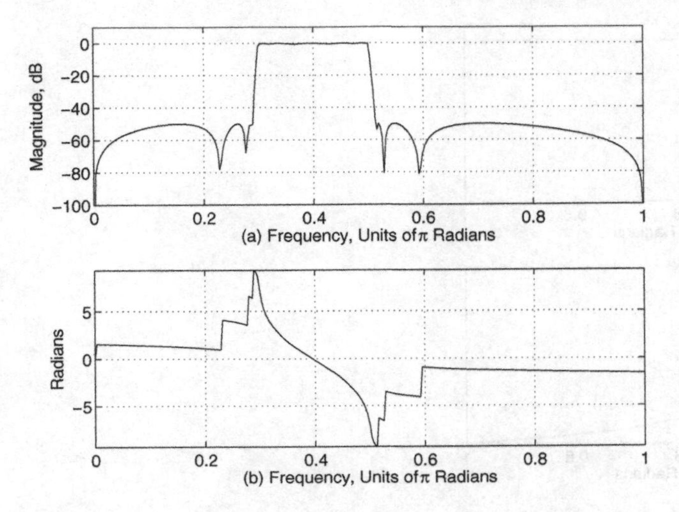

Figure 10.37: (a) Magnitude (dB) of an elliptic digital bandpass filter designed using the function *ellip*; (b) Phase response of same.

Example 10.34. Design a Chebyshev Type-I digital notch filter having $\Omega_L = 0.3\pi$ rad and $\Omega_H = 0.5\pi$ rad, $N = 7$, $R_P = 0.25$ dB, and compute the realized value of R_P, and the four band edge frequencies for the realized value of R_P and a value of $A_S = 50$ dB.

The following simple script yields Fig. 10.38. Note that the MathScript design call allows you to specify the maximum value of R_P and the two passband edges; it does not allow you to specify the value of A_S in the design. The script *LVxBW4DigitalCheb1Notch* (see exercises below) locates the stopband edges for a user-specified value of A_S. The maximum value of ripple at the specified passband edges is also measured. The following numerical results were obtained using the script below: $\Omega_{S1} = 0.3369\pi$ rad, $\Omega_{S2} = 0.4568\pi$ rad for stopband attenuation of 50 dB, and the maximum value of ripple as $Rp = 0.25$ dB.

```
[b,a] = cheby1(7,0.25,[0.3,0.5],'stop')
Hz = LVzFr(b,a,1500,18);
[rS1,rS2,Rp1,Rp2] = LVxBW4DigitalCheb1Notch(Hz,0.3,0.5,50)
```

Example 10.35. Design an elliptic digital bandstop filter having $\Omega_L = 3$ rad/s and $\Omega_H = 5$ rad/s, $N = 6$, $R_P = 0.25$ dB, and $A_S = 50$ dB. Obtain the output as poles, zeros, and gain.

The following simple script designs the filter and plot its frequency response, which is shown in Fig. 10.39.

```
[z,p,k] = ellip(6,0.25,50,[0.33,0.5],'stop')
Hz = LVzFr(k*poly(z),poly(p),1000,17);
```

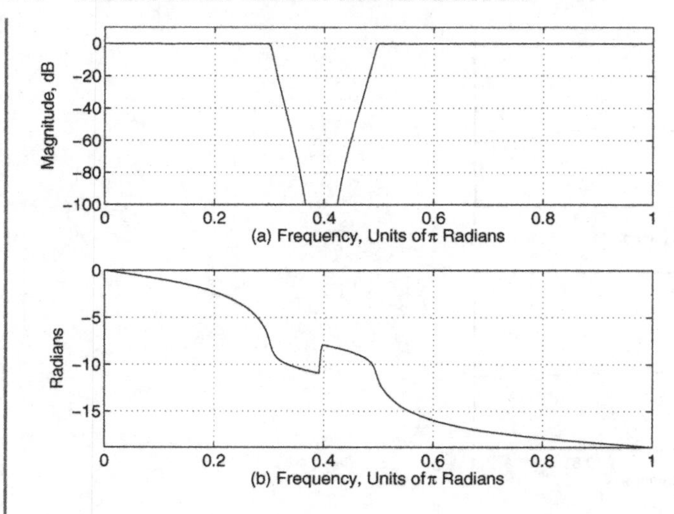

Figure 10.38: (a) Magnitude (dB) of a Chebyshev Type-I digital notch filter designed using the function *cheby*1; (b) Phase response of same.

10.12 PRONY'S METHOD

Usually, designing a filter entails starting with a desired frequency response, and computing either an impulse response in the case of an FIR, or poles and zeros in the case of an IIR.

Prony's method models an impulse response with a specified number of poles and zeros.

For the case of an IIR, if we have the impulse response $h[n]$ (or rather, a finite or truncated version of it), we can solve the IIR difference equation sequentially. The excitation sequence $x[n]$ is just the unit impulse, which is 1 for $n = 0$ and 0 otherwise. A general set of difference equations can be represented as

$$y[n] = \sum_{m=0}^{M} b_m x[n-m] - \sum_{p=1}^{N} a_p y[n-p]$$

and the first several equations generated thereby would be, for $N = 2$ and $M = 1$

$$y[0] = b_0 x[0] + b_1 x[-1] - a_1 y[-1] - a_2 y[-2]$$

$$y[1] = b_0 x[1] + b_1 x[0] - a_1 y[0] - a_2 y[-1]$$

$$y[2] = b_0 x[2] + b_1 x[1] - a_1 y[1] - a_2 y[0]$$

$$y[3] = b_0 x[3] + b_1 x[2] - a_1 y[2] - a_2 y[1]$$

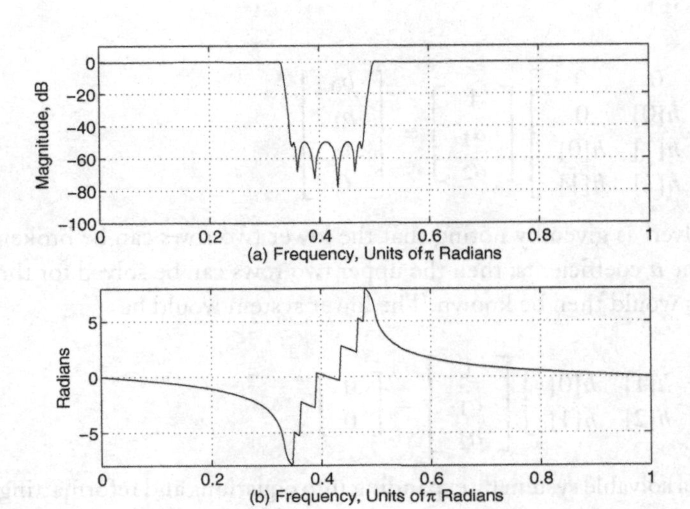

Figure 10.39: (a) Magnitude (dB) of an elliptic digital notch filter designed using the function *ellip*; (b) Phase response of same.

$$y[4] = b_0x[4] + b_1x[3] - a_1y[3] - a_2y[2]$$

Substituting in the known values of $x[n]$, and noting that $y[n] = h[n]$, the impulse response (since $x[n]$ is the unit impulse sequence), we get

$$h[0] = b_0 1 + b_1 0 - a_1 0 - a_2 0$$

$$h[1] = b_0 0 + b_1 1 - a_1 h[0] - a_2 0$$

$$h[2] = b_0 0 + b_1 0 - a_1 h[1] - a_2 h[0]$$

$$h[3] = b_0 0 + b_1 0 - a_1 h[2] - a_2 h[1]$$

$$h[4] = b_0 0 + b_1 0 - a_1 h[3] - a_2 h[2]$$

Moving all $h[n]$ terms to the left side, and writing the system in matrix form we get

$$Ha = b$$

where a and b are column vectors of the a and b coefficients; note that it is assumed that all coefficients were normalized by dividing by a_0, which is itself then equal to 1. The result, for $M = 2$ and $N = 1$, is

$$\begin{bmatrix} h[0] & 0 & 0 \\ h[1] & h[0] & 0 \\ h[2] & h[1] & h[0] \\ h[3] & h[2] & h[1] \end{bmatrix} \begin{bmatrix} 1 \\ a_1 \\ a_2 \end{bmatrix} = \begin{bmatrix} b_0 \\ b_1 \\ 0 \\ 0 \end{bmatrix}$$

This system can be exactly solved as given by noting that the lower two rows can be broken away and solved by themselves for the a coefficients; then the upper two rows can be solved for the b coefficients since the a coefficients would then be known. The lower system would be

$$\begin{bmatrix} h[2] & h[1] & h[0] \\ h[3] & h[2] & h[1] \end{bmatrix} \begin{bmatrix} 1 \\ a_1 \\ a_2 \end{bmatrix} = \begin{bmatrix} 0 \\ 0 \end{bmatrix}$$

We reconfigure the above into a solvable system by expanding into equations and reformatting into a matrix equation:

$$a_1 h[1] + a_2 h[0] = -h[2]$$

$$a_1 h[2] + a_2 h[1] = -h[3]$$

$$\begin{bmatrix} h[1] & h[0] \\ h[2] & h[1] \end{bmatrix} \begin{bmatrix} a_1 \\ a_2 \end{bmatrix} = \begin{bmatrix} -h[2] \\ -h[3] \end{bmatrix} \tag{10.44}$$

We get

$$H_{lp} a_{g0} = h_{lp}$$

where a_{g0} means the a coefficients with index greater than zero, h_{lp} means the impulse response, starting at an index one greater than the length of the desired b coefficient vector, and H_{lp} means the H matrix

$$\begin{bmatrix} h[0] & 0 & 0 \\ h[1] & h[0] & 0 \\ h[2] & h[1] & h[0] \\ h[3] & h[2] & h[1] \end{bmatrix}$$

starting with the second column and row index one greater than the number of desired b coefficients. Assuming that H_{lp} is nonsingular, we can obtain H_{lp}^{-1}, which is the pseudoinverse of H_{lp}, and solve for a_{g0} using standard matrix techniques:

$$H_{lp}^{-1} H_{lp} a_{go} = H_{lp}^{-1} h_{lp}$$

$$a_{go} = H_{lp}^{-1} h_{lp}$$

where a_{go} is returned as a column vector, and the net column vector of a coefficients is therefore

$$a = \begin{bmatrix} 1 \\ a_{go} \end{bmatrix}$$

The upper system of equations then allows us to solve the following matrix system for the b coefficients with a simple matrix multiplication:

$$\begin{bmatrix} h[0] & 0 & 0 \\ h[1] & h[0] & 0 \end{bmatrix} \begin{bmatrix} 1 \\ a_1 \\ a_2 \end{bmatrix} = \begin{bmatrix} b_0 \\ b_1 \end{bmatrix}$$

Consider the impulse response, the first few terms of which are

$$[1, 2.2, 2.05, 0.883, -0.5126]$$

and which we wish to model as an IIR having $a_0 = 1$, with a_1, a_2, b_0, and b_1 to be determined. This impulse response was generated by an IIR having b = [1,0.9] and a = [1,-1.3,0.81]. We get, from Eq. (10.44),

$$\begin{bmatrix} 2.2 & 1 \\ 2.05 & 2.2 \end{bmatrix} \begin{bmatrix} a_1 \\ a_2 \end{bmatrix} = \begin{bmatrix} -2.05 \\ -0.883 \end{bmatrix}$$

Setting this up in m-code, we get

```
Hlp = [2.2,1; 2.05,2.2]; hveclp = [-2.05,-0.883]'
a = pinv(Hlp)*hveclp
a = [1,a']'
```

which returns a = [1,-1.3,0.81]. Solving for b, we get

```
Hup = [1,0,0;2.2,1,0];
b = Hup*a
```

which returns b = [1,0.9].

When a much longer portion of the impulse response is to be modeled, which is usually the case, we must resort to a least squares fit since the matrices involved are no longer square. Consider again the previous problem, but now we wish to model a long impulse response with the same few coefficients. The first few rows would be

$$\begin{bmatrix} h[0] & 0 & 0 \\ h[1] & h[0] & 0 \\ h[2] & h[1] & h[0] \\ h[3] & h[2] & h[1] \\ h[4] & h[3] & h[2] \\ \vdots & \vdots & \vdots \\ h[N] & h[N-1] & h[N-2] \end{bmatrix} \begin{bmatrix} 1 \\ a_1 \\ a_2 \end{bmatrix} = \begin{bmatrix} b_0 \\ b_1 \\ 0 \\ 0 \\ 0 \\ \vdots \\ 0 \end{bmatrix}$$

We must solve the lower set of equations, starting at row index one greater than the that of the b coefficients, using a least squares fit. Thus notating the lower matrix system for this case as

$$H_{lp}a = b_{lp} \tag{10.45}$$

a least squares fit is defined as the set of a coefficients that minimize the sum of the squares of the differences between the elements of b_{lp} and $H_{lp}a$, i.e.,

$$\min \left| \sum (b_{lp} - H_{lp}a)^2 \right|$$

In order to obtain the values of a that obtain the minimum or least squared error case, we can solve Eq. (10.45) using the MathScript backslash operator. We illustrate this with an example.

Suppose we have the following set of equations in matrix form:

$$\begin{bmatrix} 1 & 1 & 1 \\ 1 & 2 & 4 \\ 1 & 3 & 9 \\ 1 & 4 & 16 \\ 1 & 5 & 25 \end{bmatrix} \begin{bmatrix} a_0 \\ a_1 \\ a_2 \end{bmatrix} = \begin{bmatrix} 1 \\ 4 \\ 9 \\ 16 \\ 25 \end{bmatrix} \tag{10.46}$$

which represent the algebraic system

$$y = a_0x^0 + a_1x^1 + a_2x^2$$

for x = [1:1:5]. and y = [1,4,9,16,25], i.e., the system of simultaneous equations

$$a_0 1^0 + a_1 1^1 + a_2 1^2 = 1$$

$$a_0 2^0 + a_1 2^1 + a_2 2^2 = 4$$

$$a_0 3^0 + a_1 3^1 + a_2 3^2 = 9$$

$$a_0 4^0 + a_1 4^1 + a_2 4^2 = 16$$

$$a_0 5^0 + a_1 5^1 + a_2 5^2 = 25$$

which can be written in matrix form as shown in Eq. (10.46) and written symbolically as

$$Ha = b$$

The system of equations above can be used to model the data defined by the x and y values with a second order polynomial, the coefficients of which are $[a_0, a_1, a_2]$. We can obtain a least squares-based estimate of a by making the MathScript call

$$a = b\backslash H$$

The values of a thus produced will minimize the squared error, i.e., we will obtain

$$\min \left| \sum (b - Ha)^2 \right|$$

Formulated in m-code, we have

```
h = [1,2,3,4,5]; b = [1,4,9,16,25]';
H = [h(1)^0, h(1)^1, h(1)^2; h(2)^0, h(2)^1, h(2)^2; ...
h(3)^0, h(3)^1, h(3)^2; h(4)^0, h(4)^1, h(4)^2; ...
h(5)^0,h(5)^1,h(5)^2];
a = b\H
```

This yields the coefficient estimate as [0.056, 0.23,1]. The actual coefficient values for this example are [0,0,1]. As the number of data points for this example increases, the coefficient estimate improves.

A simple program to construct a much larger matrix for the same problem might be as follows. Note that the variable *Lim* must be even for the indexing scheme to work.

```
Lim = 50; x = -Lim/2:1:Lim/2; for RowCtr = -Lim/2:1:Lim/2;
XX(RowCtr + Lim/2 + 1,1:1:3) = [1 x(RowCtr + Lim/2 + 1) ...
x(RowCtr+Lim/2+1)^2]; end;
Y = ([-Lim/2:1:Lim/2].^2)'; a = Y\XX
```

This yields a = [0.0026,0,1], which is much closer to the ideal than when using just five data points. The reader will now be equipped to undertake, in the exercises below, the writing of the m-code for the script

$$LVxProny(Imp, NumA, NumB)$$

which receives an impulse response and a desired number of a coefficients $NumA$ and b coefficients $NumB$ to model the impulse response using a least squares fit.

Example 10.36. Use Prony's method to find the numerator and denominator z-transform coefficients of an IIR that will produce the impulse response $(0.9)^n$, where n = 0:1:50.

We make the call

[b, a] = LVxProny([(0.9).^(0:1:50)],2,2)

which yields $b = [1,0]$ and $a = [1,-0.9]$.

MathScript also has a built-in function that performs Prony's Method, as follows:

$$[b, a] = prony(h, nNC, nDC)$$

where h is an impulse response to be modeled as an IIR with z-transform coefficients of b and a, nNC and nDC are the desired orders (one less than lengths) of the numerator and denominator of the z-transform, respectively.

10.13 IIR OPTIMIZATION PROGRAMS

In addition to IIR design using s-domain prototype filters, IIRs can be designed using programs that optimize one or more parameters to give a best fit to a given design specification. Reference [3] covers several basic methods to design IIR filters using frequency sampling. Reference [4] covers many different IIR optimization techniques in great detail.

10.14 REFERENCES

[1] Aram Budak, *Passive and Active Network Analysis and Synthesis*, Houghton Mifflin Company, Boston, Massachusetts, 1974.

[2] Richard G. Lyons, *Understanding Digital Signal Processing*, Addison Wesley Longman, Inc., Reading, Massachusetts, 1997.

[3] T. W. Parks and C. S. Burrus, *Digital Filter Design*, John Wiley & Sons, New York, 1987.

[4] Miroslav D. Lutovac, Dejan V. Tosic, and Brian L. Evans, *Filter Design For Signal Processing*, Prentice-Hall, Upper Saddle River, New Jersey 07458, 1996.

[5] William H. Beyer, editor, *CRC Standard Mathematical Tables, 26th Edition*, CRC Press, Inc., Boca Raton, Florida.

[6] John G. Proakis and Demitris G. Manolakis, *Digital Signal Processing, Principles, Algorithms, and Applications (Third Edition)*, Prentice Hall, Upper Saddle River, New Jersey 07458, 1996.

[7] Vinay K. Ingle and John G. Proakis, *Digital Signal Processing Using MATLAB V.4*, PWS Publishing Company, Boston, 1997.

10.15 EXERCISES

1. Compute the needed filter order for Butterworth, Chebyshev Type-I, Chebyshev Type-II, and Elliptic embodiments of an analog lowpass anti-aliasing filter. It must pass audio frequencies up to

20 kHz with no more than 1 dB of passband ripple, and the response at 22.05 kHz must be at least 60 dB below the peak passband response.

2. Compute the needed filter order for Butterworth, Chebyshev Type-I, Chebyshev Type-II, and Elliptic embodiments of an analog lowpass anti-aliasing filter for a system in which the sampling rate is 96 kHz; the passband ends at 20 kHz and must have no more than 1 dB or ripple, and the magnitude response at 48 kHz must fall to 60 dB below peak passband response.

3. Write a script that designs an analog Chebyshev Type-II bandpass filter according to the following function specification, and test it with the given test calls.

```
function [Z,P,K,NetRp,NetAs] = ...
LVxDesignAnalogCheby2BPF(OmS1,OmP1,OmP2,OmS2,Rp,As)
% Receives analog BPF band edges in radians/sec,
% plus desired maximum Rp and minimum As and returns
% the zeros (Z) and poles (P) of an analog Chebyshev
% Type-II BPF along with filter gain K and the realized
% values of Rp (NetRp) and As (NetAs).
% Test calls:
% [Z,P,K,NetRp,NetAs] = LVxDesignAnalogCheby2BPF(2,3,...
% 6,8,1,45)
% [Z,P,K,NetRp,NetAs] = LVxDesignAnalogCheby2BPF(1,...
% 1.2,2.8,3.9,1,60)
% [Z,P,K,NetRp,NetAs] = LVxDesignAnalogCheby2BPF(1,...
% 1.2,4,5,1,75)
% [Z,P,K,NetRp,NetAs] = LVxDesignAnalogCheby2BPF(4,...
% 5,8,10,1,40)
% [Z,P,K,NetRp,NetAs] = LVxDesignAnalogCheby2BPF(0.4*pi,...
% 0.45*pi,0.6*pi,0.65*pi,0.5,50)
% [Z,P,K,NetRp,NetAs] = LVxDesignAnalogCheby2BPF(19,...
% 21,27,30,0.3,60)
```

Use the lowpass-to-bandpass analog filter transformation method, implemented with the appropriate convolution transform method, and the Chebyshev Type-II prototype lowpass filter function *LVDesignCheb2,* all given in the text.

As part-and-parcel of this script, write the m-code for the script *LVxRealizedFiltParamBPF*, which is necessary to provide the realized values of R_P and A_S (considering both stopbands).

```
function [NetRp,NetAs1,NetAs2] = ...
LVxRealizedFiltParamBPF(H,S1,P1,P2,S2,HiFreqLim)
% Receives the four band edge design frequencies for a bandpass
% filter, S1,P1,P2,and S2, the computed complex frequency
% response H from frequency 0 up to HiFreqLim, and computes
% the realized value of Rp (NetRp) for the passband and the
```

```
% realized values of As returns (NetAs1 and NetAs2) for each
% stopband.
```

4. Write a script that designs an analog Elliptic bandstop filter according to the syntax below. Use the lowpass-to-bandstop analog filter transformation method, with the appropriate convolution transform method, and the Elliptic prototype lowpass filter function *LVDesignEllip,* all given in the text. The script should plot the magnitude and phase responses for the filter, and return the zeros, poles, and gain of the filter as well as the realized values of R_P and A_S. Test the script with the given test calls.

```
function [Z,P,K,NetRp,NetAs] = ...
LVxDesignAnalogEllipBPF(OmS1,OmP1,OmP2,OmS2,Rp,As)
% Receives analog BPF band edges in radians/sec, plus desired
% maximum Rp and minimum As and returns the zeros (Z) and
% poles (P) of an analog Elliptic BPF along with filter gain K and
% the realized values of Rp (NetRp) and As (NetAs), both in
% positive dB. The magnitude (in dB) and phase responses are
% also plotted.
% Test calls:
% [Z,P,K,NetRp,NetAs] = LVxDesignAnalogEllipBPF(2,3,...
% 6,8,1,45)
% [Z,P,K,NetRp,NetAs] = LVxDesignAnalogEllipBPF(1,1.2,...
% 2.8,3.9,1,60)
% [Z,P,K,NetRp,NetAs] = LVxDesignAnalogEllipBPF(1,1.2,...
% 4,5,1,75)
% [Z,P,K,NetRp,NetAs] = LVxDesignAnalogEllipBPF(4,5,...
% 8,10,1,40)
```

5. Write a script that can receive numerator and denominator coefficients of a classical IIR filter and a sampling rate and perform the Bilinear transform using the convolution method, returning the numerator and denominator coefficients of a digital filter. The script should conform to the following call; the script will be tested by using it to design digital filters in several following problems. The script should be able to deal with classical analog IIR filters, i.e., filters having either numerator and denominator coefficients of equal order, or having numerator coefficients of order one less than the denominator.

```
function [b,a] = LVxBilinearZPK(Z,P,K,Fs)
% Receives the zeros Z, poles P, and gain K of a classical IIR
% filter and performs the Bilinear transform using Fs as the
% sampling rate.
```

6. A digital signal sampled at 96 kHz will be digitally lowpass filtered to have its upper passband limit at 20 kHz, with no more than 1 dB or ripple, and the response at 22.05 kHz must have fallen by 60 dB below peak passband response. Compute the needed filter order for Butterworth, Chebyshev Type-I,

Chebyshev Type-II embodiments for an analog lowpass filter for this latter filtering operation, and then pick the lowest order filter type, pre-warp the digital band limits to obtain design band limits for an analog prototype filter, design the prototype filter and then convert it into a digital filter using the Bilinear transform as implemented by the function *LVxBilinearZPK* (see previous exercise). Compute the realized values of R_P and A_S for the digital filter.

7. Write a script that designs a digital elliptic bandpass filter according to the following syntax:

```
function [b,a,G,NetRp,NetAs] = LVxDesignDigEllipBPF(Rp,As,...
ws1,wp1,wp2,ws2)
% Receives digital BPF band edges in normalized frequency
% (units of pi), plus desired maximum Rp and minimum As
% and returns the b (numerator) and a (denominator)
% coefficients of a digital Elliptic BPF along with filter gain G
% and the realized values of Rp (NetRp) and As (NetAs). The
% magnitude (in dB) and phase responses are also plotted.
% Sample calls:
% [b,a,G,NetRp,NetAs] = LVxDesignDigEllipBPF(1,45,...
% 0.4,0.475,0.65,0.775)
% [b,a,G,NetRp,NetAs] = LVxDesignDigEllipBPF(1,60,...
% 0.45,0.55,0.7,0.83)
% [b,a,G,NetRp,NetAs] = LVxDesignDigEllipBPF(1,75,...
% 0.4,0.475,0.65,0.775)
% [b,a,G,NetRp,NetAs] = LVxDesignDigEllipBPF(1,75,...
% 0.1,0.12,0.4,0.5)
```

Prewarp the digital filter specifications to obtain the analog prototype frequencies, and use the previously script *LVxDesignAnalogEllipBPF* for the previous exercise to design the analog protype. Use the Bilinear transform (implemented using the function *LVxBilinearZPK*) to obtain the digital filter coefficients from the analog prototype filter. Test your script with the given sample calls, and plot the magnitude and phase responses.

8. Modify the scripts *LVxDesignAnalogEllipBPF* and *LVxDesignDigEllipBPF* previously written into the following two scripts that perform the same functions for Chebyshev Type-II analog and digital filters. Test the scripts using the given test calls and plot the magnitude and phase responses of the designed filters.

```
function [Z,P,K,NetRp,NetAs] = ...
LVxDesignAnalogCheby2BPF(OmS1,OmP1,OmP2,OmS2,Rp,As)
% Receives analog BPF band edges in radians/sec, plus desired
% maximum Rp and minimum As and returns the zeros (Z) and
% poles (P) of an analog Chebyshev Type-II BPF along with filter
% gain K and the realized values of Rp (NetRp) and As (NetAs).
% Also plots the magnitude (in dB) and phase responses of both
```

```
% the bandpass filter and lowpass prototype filter used in
% the design.
% Test calls:
% [Z,P,K,NetRp,NetAs] = LVxDesignAnalogCheby2BPF(2,3,...
% 6,8,1,45)
% [Z,P,K,NetRp,NetAs] = LVxDesignAnalogCheby2BPF(1,1.2,...
% 2.8,3.9,1,60)
% [Z,P,K,NetRp,NetAs] = LVxDesignAnalogCheby2BPF(1,1.2,...
% 4,5,1,75)
% [Z,P,K,NetRp,NetAs] = LVxDesignAnalogCheby2BPF(4,5,...
% 8,10,1,40)

function [b,a,G,NetRp,NetAs] = LVxDesignDigCheby2BPF(Rp,...
As,ws1,wp1,wp2,ws2)
% Receives digital BPF band edges in normalized frequency (units
% of pi), plus desired maximum Rp and minimum As and returns
% the b (numerator) and a (denominator) coefficients of a digital
% Cheby2 BPF along with filter gain G and the realized values
% of Rp (NetRp) and As (NetAs). Also plots the magnitude (in
% dB) and phase responses of the digital bandpass filter.
% Sample calls:
% [b,a,G,NetRp,NetAs] = LVxDesignDigCheby2BPF(1,45,...
% 0.4,0.475,0.65,0.775)
% [b,a,G,NetRp,NetAs] = LVxDesignDigCheby2BPF(1,60,...
% 0.45,0.55,0.7,0.83)
% [b,a,G,NetRp,NetAs] = LVxDesignDigCheby2BPF(1,75,...
% 0.4,0.475,0.65,0.775)
% [b,a,G,NetRp,NetAs] = LVxDesignDigCheby2BPF(1,75,...
% 0.1,0.12,0.4,0.5)
```

9. Write a script, as specified below, that designs an analog Chebyshev Type-II notch filter, and test the script with the given sample calls. Use the convolution method to derive the notch system function from an analog prototype lowpass Chebyshev Type-II filter.

```
function [Z,P,K,NetRp,NetAs] = ...
LVxDesignAnalogCheby2Notch(OmP1,OmS1,OmS2,OmP2,Rp,As)
% Receives analog Notch band edges in radians/sec,
% plus desired maximum Rp and minimum As and returns the
% zeros (Z) and poles (P) of an analog Chebyshev Type-II
% Notch along with filter gain K and the realized values of
% Rp (NetRp) and As (NetAs). Also plots the magnitude
% (in dB) and phase responses of both the notch filter and its
```

```
% lowpass analog prototype filter.
% Sample calls:
% [Z,P,K,NetRp,NetAs] = LVxDesignAnalogCheby2Notch(2,...
% 3,6,8,1,45)
% [Z,P,K,NetRp,NetAs] = LVxDesignAnalogCheby2Notch(1,...
% 1.2,2.8,3.9,1,60)
% [Z,P,K,NetRp,NetAs] = LVxDesignAnalogCheby2Notch(1,...
% 1.2,4,5,1,75)
% [Z,P,K,NetRp,NetAs] = LVxDesignAnalogCheby2Notch(4,...
% 5,8,10,1,40)
```

As part and parcel of writing the script *LVxDesignAnalogCheby2Notch*, write the m-code for the following script, which is necessary to compute the realized values of R_P and A_S for a notch filter:

```
function [NetAs,NetRp1,NetRp2] = ...
LVxRealizedFiltParamNotch(H,P1,S1,S2,P2,HiFreqLim)
% Receives the four band edge design frequencies for a notch
% filter, P1,S1,S2,and P2, the computed complex frequency
% response H from frequency 0 up to HiFreqLim, and returns the
% realized values of Rp (NetRp1 and NetRp2) for each passband
% and the realized value of As (NetAs) for the stopband.
```

10. Write a script, as specified below, that designs a digital Chebyshev Type-II notch filter. Design the prototype analog notch filter by prewarping the digital frequencies and calling the function written for the previous exercise (*LVxDesignAnalogCheby2Notch*), then use the Bilinear method (implemented using *LVxBilinearZPK*) to obtain the digital filter coefficients.

```
function [b,a,G,NetRp,NetAs] = ...
LVxDesignDigCheby2Notch(Rp,As,wp1,ws1,ws2,wp2)
% Receives digital Notch filter band edges in normalized frequency...
% (units of pi), plus desired maximum Rp and minimum As and
% returns the b (numerator) and a (denominator) coefficients
% of a digital Cheby2 Notch filter along with filter gain G and
% the realized values of Rp (NetRp) and As (NetAs). The
% magnitude (in dB) and phase responses of both the notch
% filter and its analog prototype filter are also plotted.
% Sample calls:
% [b,a,G,NetRp,NetAs] = LVxDesignDigCheby2Notch(1,45,...
% 0.4,0.475,0.65,0.775)
% [b,a,G,NetRp,NetAs] = LVxDesignDigCheby2Notch(1,60,...
% 0.45,0.55,0.7,0.83)
% [b,a,G,NetRp,NetAs] = LVxDesignDigCheby2Notch(1,75,...
```

```
% 0.4,0.475,0.65,0.775)
% [b,a,G,NetRp,NetAs] = LVxDesignDigCheby2Notch(1,75,...
% 0.1,0.12,0.4,0.5)
```

11. Write a script that designs a Butterworth lowpass digital filter meeting the following specifications: ω_P, ω_S, R_P, and A_S in accordance with the call syntax below. Use the Bilinear transform method, implemented by the script *LVxBilinearZPK*, on a prototype analog lowpass Butterworth filter designed by prewarping the digital frequencies to obtain the necessary analog prototype frequencies. Plot the magnitude and phase responses of the digital filter, and return the output arguments as shown in the call syntax. Test the script with the given sample calls.

```
function [zB,zA,G,NetRp,NetAs] = ...
LVxDesignDigitalButterLPF(Rp,As,wp,ws)
% function [zB,zA,G,NetRp,NetAs] = ...
% LVxDesignDigitalButterLPF(Rp,As,wp,ws)
% Designs a digital Butterworth LPF having no more than Rp
% dB of ripple at passband edge wp, and at least As dB
% attenuation at stopband edge ws. The filter coefficients are
% returned as numerator coefficients zB, denominator
% coefficients zA, and gain G. The script also plots the
% magnitude (in dB) and phase responses of both the
% digital filter, and the analog prototype used to compute
% the digital filter coefficients using the Bilinear transform.
% The realized values of Rp and As are also returned.
% Sample calls:
% [zB,zA,G,NetRp,NetAs] = LVxDesignDigitalButterLPF(1,...
% 40,0.5*pi,0.6*pi)
% [zB,zA,G,NetRp,NetAs] = LVxDesignDigitalButterLPF(0.3,...
% 60,0.25*pi,0.325*pi)
% [zB,zA,G,NetRp,NetAs] = LVxDesignDigitalButterLPF(0.1,...
% 55,0.25*pi,0.35*pi)
```

12. Write the m-code for the script *LVxBW4DigitalCheb1Notch* as specified below, and test it with the given calls.

```
function [rS1,rS2,Rp1,Rp2] = LVxBW4DigitalCheb1Notch(H,...
OmP1,OmP2,As)
% H is the complex frequency response vector (from frequency 0
%  to pi radians) of a Chebyshev Type I digital notch filter
% designed using the function cheby1; OmP1 and OmP2 are the
% passband edges, in normalized frequency (i.e., in multiples of
% pi radians) of a notch filter designed by the function cheby1,
% rS1 and rS2 are stopband edges at which the desired value of
```

% stopband attenuation As is realized. Rp1 and Rp2 are the
% realized or actual frequency responses (in positive dB) at the
% passband edges OmP1 and OmP2.

Test calls:

(a)

[b,a] = cheby1(6,0.35,[0.45, 0.7],'stop')
Hz = LVzFr(b,a,2000,29);
[rS1,rS2,Rp1,Rp2] = LVxBW4DigitalCheb1Notch(Hz,0.45,0.7,60)

(b)

[b,a] = cheby1(5,0.15,[0.25, 0.75],'stop')
Hz = LVzFr(b,a,2000,29);
[rS1,rS2,Rp1,Rp2] = LVxBW4DigitalCheb1Notch(Hz,0.25,0.75,55)

13. Write a script that conforms to the following syntax:

function LVxDigCheby1LPFViaMS(Rp,As,wp,ws)
% Performs the same function as
% LVsCheby1LPF2zCheby1LPF(Rp,As,wp,ws) but in addition it
% uses the MathScript function [b,a]=cheby1(N,Rp,wp) to compute
% the digital filter coefficients using the computed value for analog
% filter order and the desired passband edge wp. The magnitude
% (in dB) and phases responses of the digital filter computed both
% ways are displayed, and the realized filter parameters are
% computed for comparison.
% LVxDigCheby1LPFViaMS(1,40,0.5*pi,0.6*pi)
% LVxDigCheby1LPFViaMS(1,60,0.45*pi,0.65*pi)
% LVxDigCheby1LPFViaMS(1,68,0.5*pi,0.55*pi)

You should find that the results from both digital filter computation methods are essentially identical for the calls given above.

14. Write a script that compares the performance of the script

$$LVxDesignDigEllipBPF(Rp, As, ws1, wp1, wp2, ws2)$$

to the function *ellip*. Use one-half the order of the digital elliptic bandpass filter designed by *LVxDesignDigEllipBPF* as the order for the call to *ellip*, and use $wp1$ and $wp2$ (the passband edges) for the needed frequency vector. Plot the magnitude (dB) and phase responses of the filters designed by both scripts, and compute the realized values of R_P and A_S for both filters. You should be able to produce nearly identical results.

Your script to perform the above comparison should conform to the following syntax:

function LVxDigEllipBPFViaMS(Rp,As,ws1,wp1,wp2,ws2)
% Receives the values for maximum passband ripple Rp, minimum

```
% stopband attenuation As, and bandpass filter band edges ws1,
% wp1,wp2, and ws2, and designs digital bandpass filters using 1)
% the script LVxDesignDigEllipBPF and 2) the built-in function
% ellip. Plots the magnitude and phase responses of both designed
% filters as well as the realized values for Rp and As for both
% filters, allowing for visual and numerical comparison of the
% two designs.
% Test calls:
% LVxDigEllipBPFViaMS(1,75,0.4,0.475,0.65,0.775)
% LVxDigEllipBPFViaMS(1,45,0.4,0.475,0.65,0.775)
% LVxDigEllipBPFViaMS(1,60,0.45,0.55,0.7,0.83)
% LVxDigEllipBPFViaMS(1,75,0.4,0.475,0.65,0.775)
% LVxDigEllipBPFViaMS(1,75,0.1,0.12,0.4,0.5)
```

15. Write a script that will design a Chebyshev Type-I digital bandpass filter using the impulse invariance method; the script should conform to the call below. Test the script with the given test calls.

```
function LVxDesignDigCheby1BPFViaImpInv(Ws1,Wp1,...
Wp2,Ws2,Rp,As,Fs)
% Designs a digital bandpass Chebyshev Type I filter using
% impulse invariance. The four bandpass filter band limits
% are Ws1,Wp1,Wp2,and Ws2. The maximum allowable
% passband ripple in positive dB is Rp and the minimum
% stopband attenuation in positive dB is As.
% Test calls:
% LVxDesignDigCheby1BPFViaImpInv(0.2*pi,0.3*pi,...
% 0.6*pi,0.8*pi,0.5,40,0.05)
% LVxDesignDigCheby1BPFViaImpInv(0.3*pi,0.38*pi,...
% 0.6*pi,0.72*pi,0.5,50,0.05)
% LVxDesignDigCheby1BPFViaImpInv(0.2*pi,...
% 0.35*pi,0.6*pi,0.9*pi,1,60,0.1)
```

16. In this project, we'll modify and use the script

$$[T, F, B] = LVx\_DetnFiltFIRContTone(A, Freq, RorSS, SzWin, OvrLap, AudSig)$$

to identify an interfering steady-state or rising-amplitude sinusoid and remove it from the test signal using a Chebyshev Type-I filter designed automatically in response to the list of candidate interfering frequencies generated through analysis of the spectrogram matrix.

The previous script

LVx_DetnFiltFIRContTone

is modified only in that the filter to be used is a Chebyshev Type-I IIR rather than an FIR. The format of the new script is as follows:

```
function [ToneFreq,F,B] = LVx_DetnFiltIIRContTone(A,Freq,...
RorSS,SzWin,OvrLap,AudSig,Rp,FltOrd)
% This program evaluates the list of candidate interfering tones
% in the same manner as the script LVx_DetnFiltFIRContTone
% with the exception that filtering is accomplished using a
% Chebyshev Type-I filter (lowpass, notch, or highpass,
% depending on the location of the interfering frequency,
% with maximum passband ripple of Rp dB and order
% FltOrd). The first six input arguments and the output
% arguments of this script are identical to those
% of LVx_DetnFiltFIRContTone.
% Test calls:
% [T,F,B] = LVx_DetnFiltIIRContTone(0.015,200,0,512,1,1,1,2)
% [T,F,B] = LVx_DetnFiltIIRContTone(0.01,210,0,512,1,1,1,2)
% [T,F,B] = LVx_DetnFiltIIRContTone(0.01,200,0,512,1,2,1,2)
% [T,F,B] = LVx_DetnFiltIIRContTone(0.1,200,0,512,1,3,1,2)
% [T,F,B] = LVx_DetnFiltIIRContTone(0.01,500,0,512,1,1,1,2)
% [T,F,B] = LVx_DetnFiltIIRContTone(0.01,500,0,512,1,2,1,2)
% [T,F,B] = LVx_DetnFiltIIRContTone(0.08,500,0,512,1,3,1,2)
%
% [T,F,B] = LVx_DetnFiltIIRContTone(0.025,200,1,512,1,1,1,2)
% [T,F,B] = LVx_DetnFiltIIRContTone(0.02,200,1,512,1,2,1,2)
% [T,F,B] = LVx_DetnFiltIIRContTone(0.1,200,1,512,1,3,1,2)
% [T,F,B] = LVx_DetnFiltIIRContTone(0.02,500,1,512,1,1,1,2)
% [T,F,B] = LVx_DetnFiltIIRContTone(0.018,500,1,512,1,2,1,2)
% [T,F,B] = LVx_DetnFiltIIRContTone(0.1,500,1,512,1,3,1,2)
% [T,F,B] = LVx_DetnFiltIIRContTone(0.07,150,0,512,1,3,1,2)
%
% [T,F,B] = LVx_DetnFiltIIRContTone(0.005,1000,0,512,0,1,1,2)
% [T,F,B] = LVx_DetnFiltIIRContTone(0.008,1000,0,512,1,1,1,2)
% [T,F,B] = LVx_DetnFiltIIRContTone(0.005,1000,0,512,1,1,1,2)
```

The call

```
[T,F,B] = LVx_DetnFiltIIRContTone(0.015,120,0,512,0,1,1,2)
```

results in Figs. 10.40 and 10.41, in addition to the basic figures found in *LVx_DetectContTone* and *LVx_DetnFiltFIRContTone*.

The call

[T,F,B] = LVx_DetnFiltIIRContTone(0.015,500,0,512,0,1,1,2)

results in Figs. 10.42 and 10.43.

16. Write the m-code for the script

$$LVxProny(Imp, NumA, NumB)$$

as described in the text and as defined below, and test it with the given sample calls.

```
% function [b,a] = LVxProny(Imp,NumA,NumB)
% Uses Prony's Method to model an impulse response as an
% IIR having z-transform with NumB numerator coefficients
% b and NumA denominator coefficients a.
% Sample calls:
% [b, a] = LVxProny([(0.9).^( 0:1:50 )],2,2)
% [b,a] = cheby1(2,0.5,0.5), Imp = filter(b,a,[1,zeros(1,75)]);...
% [b,a]=LVxProny(Imp,3,3)
% [b,a] = cheby1(2,0.5,0.5), Imp = filter(b,a,[1,zeros(1,25)]);,,,
% [b,a]=LVxProny(Imp,3,3)
% [b,a] = cheby1(2,0.5,0.5), Imp = filter(b,a,[1,zeros(1,10)]);...
% [b,a]=LVxProny(Imp,3,3)
% [b,a] = cheby1(2,0.5,0.5), Imp = filter(b,a,[1,zeros(1,4)]);...
% [b,a]=LVxProny(Imp,3,3)
% [b,a] = cheby1(2,0.5,0.5), Imp = filter(b,a,[1,zeros(1,3)]);...
% [b,a]=LVxProny(Imp,3,3)
```

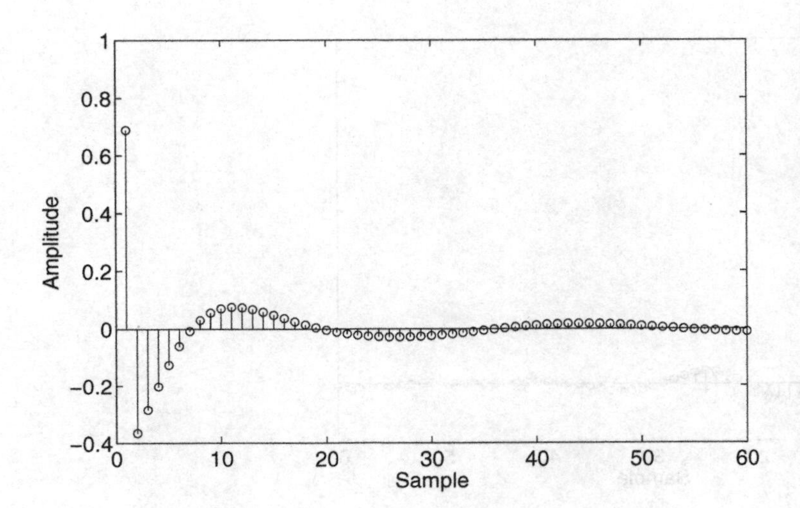

Figure 10.40: Impulse response (truncated) of HPF automatically designed by the script to eliminate an interfering tone at 120 Hz.

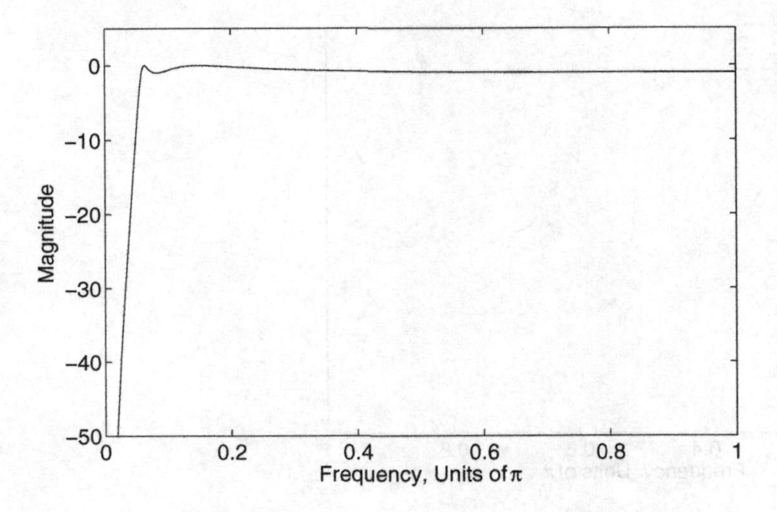

Figure 10.41: Frequency response of the 120 Hz HPF impulse response shown in the previous figure.

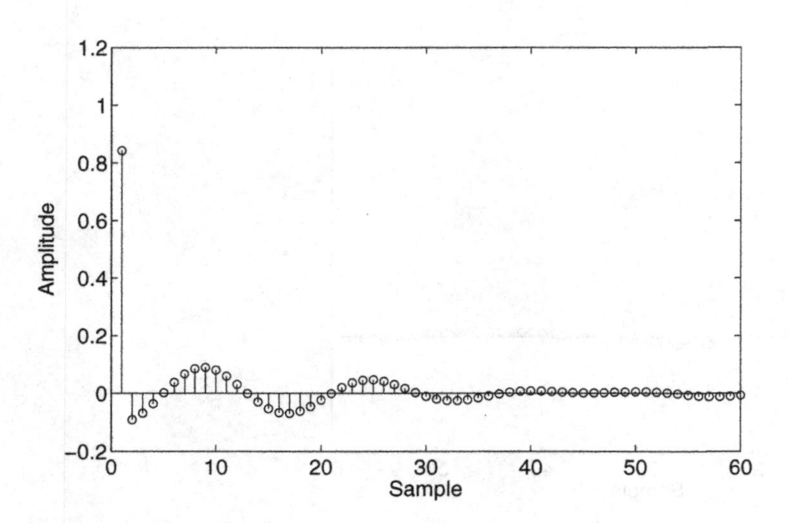

Figure 10.42: Impulse response (truncated) of a notch filter automatically designed by the script to eliminate an interfering tone at 500 Hz.

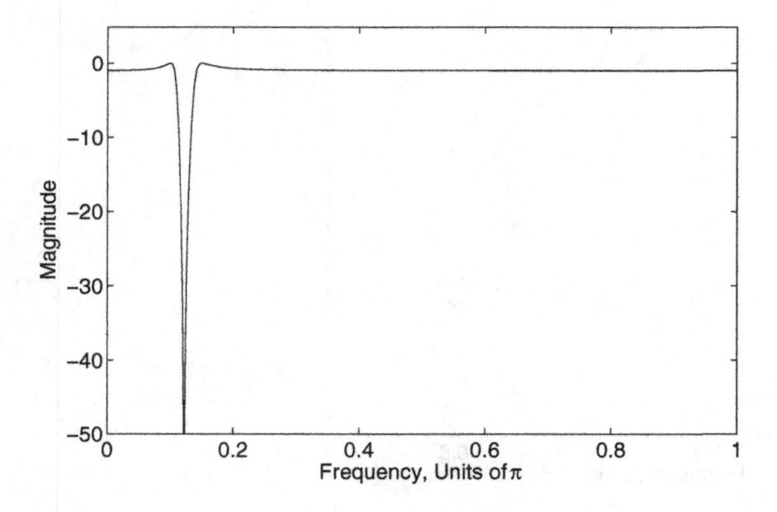

Figure 10.43: Frequency response of the notch filter designed to eliminate a 500 Hz tone.

Part IV

LMS Adaptive Filtering

CHAPTER 11

Introduction To LMS Adaptive Filtering

11.1 OVERVIEW

11.1.1 IN THIS PART OF THE BOOK

There are many situations in which needed filter characteristics change rapidly, or in which the needed characteristics are not precisely known in advance. Adaptive filtering fills these needs by automatically changing filter coefficients to achieve some particular goal. In this book, we take up the study of filters that can adapt their characteristics according to some criteria. In particular, our study will center on LMS Adaptive filtering, which is popular because of its low computational overhead. It has found applications in Active Noise Cancellation (ANC), echo cancellation in telephone systems, beamforming, narrowband signal attenuation or enhancement, equalizers, etc. In this book, consisting of two chapters, we first examine several fundamental ideas which culminate in derivation of the LMS Algorithm. In the second chapter, we examine a number of common filter topologies or applications of LMS Adaptive filtering.

11.1.2 IN THIS CHAPTER

We begin our study of adaptive filtering by first investigating cost functions, performance surfaces, and the gradient, using a simple technique called **Coefficient Perturbation**. We'll consider several examples, including one- and two- independent-variable quadratic cost functions, including the problem of fitting a line to a set of points in a plane. The simplicity of coefficient perturbation will enable the reader to readily understand fundamental concepts and issues relevant to gradient-search-based adaptive filtering. We then derive a more efficient way of estimating the gradient, the simple, elegant, and efficient algorithm known as the **LMS (Least Mean Squared)** algorithm, which, with its various derivative algorithms, accounts for the majority of adaptive algorithms in commercial use. We'll derive the algorithm for a length-2 FIR, which readily generalizes to a length-N FIR. We then explore one of the weaknesses of the LMS algorithm, its need for a wide-bandwidth signal for proper algorithm function in certain topologies, such as ANC.

By the end of the chapter, the reader will have gained fundamental knowledge of the LMS algorithm and should be able (and, in the exercises at the end of the chapter, is expected) to write scripts to implement simple LMS adaptive FIR filters. The reader will then be prepared for the next chapter in which we explore the LMS algorithm in various useful applications, such as active noise cancellation, signal enhancement, echo cancellation, etc.

11.2 SOFTWARE FOR USE WITH THIS BOOK

The software files needed for use with this book (consisting of m-code (.m) files, VI files (.vi), and related support files) are available for download from the following website:

http://www.morganclaypool.com/page/isen

The entire software package should be stored in a single folder on the user's computer, and the full file name of the folder must be placed on the MATLAB or LabVIEW search path in accordance with the instructions provided by the respective software vendor (in case you have encountered this notice before, which is repeated for convenience in each chapter of the book, the software download only needs to be done once, as files for the entire book are all contained in the one downloadable folder). See Appendix A for more information.

11.3 COST FUNCTION

A common problem is the need to fit a line to a set of test points. The test points, for example, might be data taken from an experiment or the like. We'll start by investigating how a "best fit" curve (or, in the example at hand, straight line) can be determined mathematically, i.e., what mathematical criterion can be used for an algorithm to determine when a candidate curve or line is better than another?

In Fig. 11.1, five test points (circles) are to have a horizontal line fitted in the "best" manner possible using purely mathematical criteria. For purposes of simplicity, the line is allowed to vary only in the vertical direction. In a real problem, the points would likely be irregularly placed and the line would have to assume different slopes as well as different vertical positions (we will examine line-modeling in just this way later in the chapter).

In Fig. 11.1, if we define the vertical location of the line as y; the distance from each upper point to the line is $(2 - y)$, and the distance from each of the two lower points to the line is y. We'll construct three alternative functions of y and compare their performance in determining the best value of y to place the line in an intuitively, satisfying location.

For the first function, we'll use a first order statistic, that is to say, we'll sum the various distances and normalize by the number of distances (i.e., divide the sum by the number of distances), in accordance with Eq. (11.1):

$$F_1(y) = \frac{1}{5}(3(2 - y) + 2y) \tag{11.1}$$

For the second function, we'll also use a first order statistic, but we'll sum the magnitudes of the various distances and normalize. This results in Eq. (11.2):

$$F_2(y) = \frac{1}{5}(3(|2 - y|) + 2|y|) \tag{11.2}$$

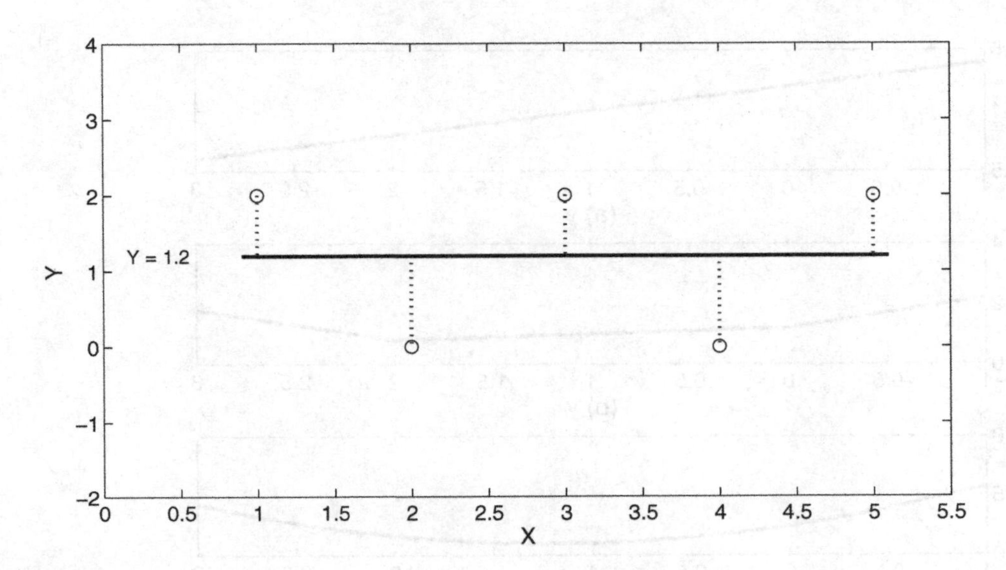

Figure 11.1: Five points in a plane to be modeled by a line $y = a$, where a is a constant.

For the third function, we'll use second-order statistics, i.e., the square of distances from the line to the various points to be modeled. To do this, we square each distance mentioned above, sum all squared distances, and normalize. We'll call this the MSE (mean squared error) method.

Using the MSE method, the function of distance looks like this:

$$F_3(y) = \frac{1}{5}(3(2-y)^2 + 2y^2) \tag{11.3}$$

We can evaluate Eqs. (11.1) through (11.3) with a test range for y to see how well each function generates an intuitively satisfying location for the line between the two rows of test points. Letting the three functions given above be represented respectively by $F1$, $F2$, and $F3$, the results from running the following m-code are shown in Fig. 11.2:

```
y = -1:0.05:3;
F1 = (1/5)*(3*(2-y) + 2*(y));
F2 = (1/5)*(3*abs(2-y) + 2*abs(y));
F3 = (1/5)*(3*(2-y).^2 + 2*(y.^2));
```

We can refer to F_1, F_2, and F_3 as **Cost Functions** since there is a "cost" or value associated with placing the line at certain positions. As in purchases using money, the goal is to minimize the cost.

Intuitively, the best location for the line is somewhere between the three upper points and the two lower points, and probably a little closer to the three upper points.

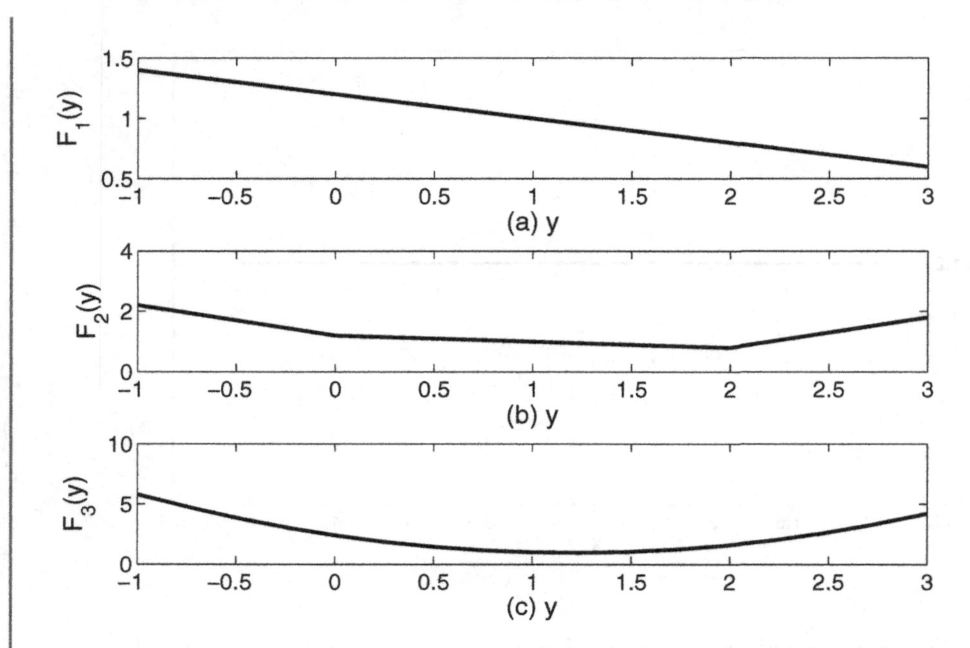

Figure 11.2: (a) The function $F_1(y)$ (see text) evaluated over the range y = -1:0.05:3; (b) Same, for function $F_2(y)$; (c) Same, for function $F_3(y)$.

Neither first order cost function (plots (a) and (b) of Fig. 11.2) does well. The first function does not provide any useful guidance; the second one does give guidance, but its minimum occurs when the line passes through the three upper points (when $y = 2$). This does not suit our intuitive notion of best placement.

The second order, or MSE cost function (plot (c) of Fig. 11.2) shows its minimum cost at $y = 1.2$, which seems a reasonable placement of the line. Notice also that there is a single, distinct minimum for the second order cost function. It is clear that the best cost function for this simple example is MSE or second order. This is true in general for many problems, and it is the cost function that we will be using in this and the following chapter in the development and use of various adaptive algorithms.

- Second order cost functions (or MSE functions), have single ("global") minima that produce an intuitively satisfying "best" result for many line and curve fitting problems.

11.4 PERFORMANCE SURFACE

Any of plots (a)-(c) of Fig. 11.2 may be described as **Performance Curves**. Each is a graph of an independent variable versus the corresponding value of a cost function. In situations where there are two or more independent variables, the cost function, graphed as a function of its independent variables, takes on three or more dimensions and hence is a surface rather than a simple curve. Hence, it is called a **Performance Surface**. We will see examples of the performance surface a little later, but first, we'll take a look at a couple of general methods for "reaching the bottom" of a performance curve algorithmically, that it to say, using a computer program.

For this kind of problem, in general, we will have one or more independent variables which when varied will cause the cost function to vary, and our goal is for a computer program or algorithm to change the values of the one or more independent variables in such a manner that eventually the cost function is minimized.

11.5 COEFFICIENT PERTURBATION

An intuitive way of determining how to adjust each independent variable to minimize the cost function is to let all of the independent variables assume certain fixed values (i.e., assign values to the variables) and then evaluate the cost function. Next, change one of the variables by a very small amount (this is called perturbation) while holding all the other variable values constant, and then evaluate the cost function. If the change to the particular variable in question was in the positive direction, for example, and the cost function increased in value, then logically, changing the value of that variable in the negative direction would tend to decrease the value of the cost function.

A mathematical way of describing this, for a cost function that depends on the single independent variable x, would be

$$CF = \frac{\Delta(CostFcn)}{\Delta x} = \frac{CostFcn(x + \Delta x) - CostFcn(x)}{\Delta x} \tag{11.4}$$

The ratio in Eq. (11.4) tells how much and in what direction the cost function varies for a unit change in the independent variable x. The limit of the ratio in Eq. (11.4) as Δx goes to zero is the derivative of the cost function CF with respect to x:

$$CF'(x) = \frac{d(CF)}{dx} = \lim_{\Delta x \to 0}\left(\frac{CF(x + \Delta x) - CF(x)}{\Delta x}\right)$$

This concept is easily extended to a cost function that is dependent upon two or more independent variables. Then if

$$y = f(x_1, x_2, \ldots x_N)$$

the partial derivative of y with respect to a particular independent variable x_i is

$$\frac{\partial(y)}{\partial(x_i)} = \lim_{\Delta x_i \to 0} \frac{f(x_1, x_2, \ldots (x_i + \Delta x_i), \ldots x_N) - f(x_1, x_2, \ldots x_N)}{\Delta x_i}$$

In computer work, there is a limit to precision, so making Δx_i smaller and smaller does not, beyond a certain point, improve the estimate of the partial derivative. Thus it is not possible (referring to the single-independent-variable example at hand) to obtain

$$\lim_{\Delta x \to 0} \frac{CF(x + \Delta x) - CF(x)}{\Delta x} \tag{11.5}$$

since, due to roundoff error, the true limit will in general not be monotonically approached or achieved. For most normal purposes, however, this is not a serious problem; a useful estimate of the partial derivative can be obtained. One method that can increase the accuracy from the very first estimate is to straddle the value of x. This is accomplished by dividing Δx by two and distributing the two parts about x as follows:

$$CF'(x) \simeq \frac{CF(x + \Delta x/2) - CF(x - \Delta x/2)}{\Delta x} \tag{11.6}$$

In Eq. (11.6), we have adopted the symbol $\simeq$ to represent the estimate or approximation of the partial derivative of $CF(x)$ with respect to x. Note that the two values of x at which the cost function is computed surround the actual value of x, and provide a better estimate of the slope of the cost function at x than would be provided by evaluating the cost function at x and $x + \Delta x$. This method is referred to as taking **Central Differences**, as described in Chapter 5 of Reference [1].

The **Gradient** of a function (such as our cost function CF) for N independent variables x_i is denoted by the **del** (∇) operator and is defined as

$$\nabla(CF) = \sum_{i=1}^{N} (\partial(CF)/\partial(x_i))\hat{u}_i \tag{11.7}$$

where $\hat{u}_i$ is a unit vector along a dimension corresponding to the independent variable x_i. The purpose of the unit vectors is to maintain each partial derivative as a distinct quantity. In the programs and examples below involving cost functions dependent on two or more independent variables, each partial derivative is simply maintained as a distinct variable.

Once all the partial derivatives have been estimated using perturbation, it is possible to update the current best estimate for each of the independent variables x_i using the following update equation:

$$x_i[k+1] = x_i[k] - \mu CF'(x_i, k) \tag{11.8}$$

where $CF'(x_i, k)$ means the estimate of the partial derivative of CF with respect to independent variable x_i at sample time k. Equation (11.8) may be described verbally thus: the estimate of the variable x_i at iteration $(k + 1)$ is its estimate at iteration k minus the product of the partial derivative of the cost function with respect to x_i at iteration k and a small value μ, which helps regulate the overall size of the update term, $\mu CF'(x_i, k)$.

Too large a value of μ will cause the algorithm to become unstable and diverge (i.e., move away from the minimum of the cost function) while too small a value will result in protracted times for convergence. New estimates of the coefficients are computed at each sample time and hopefully the coefficient estimates will converge to the ideal values, which are those that minimize the cost function. Convergence to the true minimum of the performance surface, however, can only be reliably achieved if the performance surface is unimodal, i.e., has one global minimum. Some performance surfaces may possess local minima to which the gradient may lead, and the algorithm can fail to converge to the true minimum of the cost function. Fortunately, the problems we'll investigate do have unimodal performance surfaces. As we work through various examples, we'll see how various parameters and situations affect algorithm convergence.

11.6 METHOD OF STEEPEST DESCENT

Using the negative of the gradient to move down the performance surface to the minimum value of the cost function is called the **Method of Steepest Descent;** this is because the gradient itself points in the direction of steepest *ascent*, while the negative of the gradient, which we use in coefficient update Eq. (11.8), moves in precisely the opposite direction–the direction with the steepest slope down the performance surface.

The following discussion uses a simple quadratic cost function to explore a number of important concepts involved in adaptive processes, such as gradient (slope), step size, convergence speed, accuracy of gradient estimate, etc. In the several plots of Fig. 11.3, the cost function $CF(x)$ is

$$CF(x) = x^2 + 1$$

and the ultimate goal is to determine the value of x at which the $CF(x)$ is minimized, i.e., reduced to 1.0.

Plot (a) of Fig. 11.3 shows the first step in navigating a Performance Surface to its minimum using the method of steepest descent, in which an initial value of x was chosen arbitrarily to be 4.0. From this, the value of the gradient must be estimated and the estimate used to compute a new value of x which is (hopefully) closer to the value of x which minimizes the cost function.

In any of the plots of Fig. 11.3, values of the cost function versus x are plotted as small circles; the method of steepest descent may be understood by imagining the circles as being marbles, and the performance curve as a surface down which the marbles can roll under the pull of gravity. In a more complex problem with the performance surface being a (three-dimensional) paraboloid-of-revolution rather than a simple two-dimensional curve, such a "marble" would roll straight down along the inside of the paraboloid (a bowl-shaped surface) toward the bottom.

The script

LVxGradientViaCP(FirstValX, StepSize,...

NoiseAmp, UseWgtingFcn, deltaX, NoIts)

(see exercises below) allows experimentation with the method of steepest descent using a second-order cost function having a single independent variable. A typical call would be:

LVxGradientViaCP(4,0.05,0.01,0,0.0001,50)

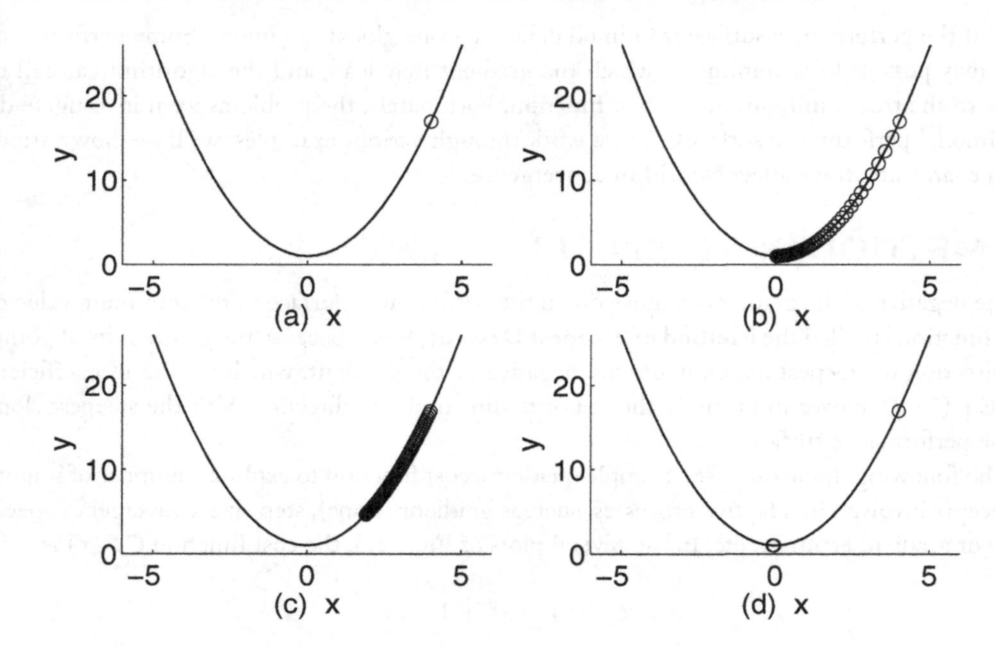

Figure 11.3: (a) Graph of cost function and first value of parameter x and the corresponding value of the cost function; (b) Successive estimated values of x and respective values of the cost function for 75 iterations; (c) Seventy-five iterations, attempting to reach the minimum value of the cost function using too small a value of StepSize (μ); (d) Three iterations using a weighting function for μ.

which results in Fig. 11.3, plot (a), for Iteration 1; the script is designed so that successive iterations of the algorithm are performed by pressing any key.

The first argument, *FirstValX*, is the starting value (the initial guess) for the only independent variable, x. *StepSize* is equivalent to μ, the parameter that regulates the size of the update term. *NoiseAmp* is the standard deviation of white noise that is added to the current estimate of x to more realistically simulate a real (i.e., noisy) situation (set it to zero if desired). The fifth argument, *deltaX* (Δx in the pseudocode below), is the amount by which to perturb the coefficients when estimating the gradient. The sixth argument, *NoIts*, is the number of iterations to do in seeking the value of x that minimizes the cost function. The variable *UseWgtingFcn* (UWF in the pseudocode below) will be discussed extensively below.

The heart of the algorithm, in pseudocode terms, is

$for\ n = 1 : 1 : NoIts$
$CF[n] = (x[n] - \Delta x/2)^2 + 1$
$TstCF = (x[n] + \Delta x/2)^2 + 1$
$\partial(CF)/\partial(x[n]) = (TstCF - CF[n])/\Delta x$

$if\ UWF = 1$
$x[n+1] = x[n] - \mu CF[n](\partial(CF)/\partial(x[n]))$
$else$
$x[n+1] = x[n] - \mu(\partial(CF)/\partial(x[n]))$
end
end

The pseudocode above parallels the preceding discussion exactly, i.e., implementing an algorithm that performs gradient estimation via central-differences-based coefficient perturbation. The product of $\partial(CF)/\partial(x[n])$ and μ (*StepSize*) is the amount by which the value of x is to be changed for each iteration. This quantity may be further modified (as shown above) by multiplying it by the current value of *CostFcn* ($CF[n]$) in order to cause the entire term to be large when the error (and hence *CF*) is large, and small when *CF* becomes small. This helps speed convergence by allowing the algorithm to take larger steps when it is far from convergence and smaller steps when it is close to convergence. The fourth input argument to the function call, *UseWeightingFcn* (UWF in the pseudocode above), determines whether or not the update term uses $CF[n]$ as a weighting function or not.

The coefficient update term, the product of $\partial(CF)/\partial(x[n])$, μ, and $CF[n]$ if applied, has great significance for convergence and stability. If the update term is too large, the algorithm diverges; if it is too small, convergence speed is too slow. In the following examples, we'll observe convergence rate and stability with the update term computed with and without the weighting function.

Example 11.1. Use the script *LVxGradientViaCP* to demonstrate stable convergence.

We call the script with no weighting function and *StepSize* = 0.03:

LVxGradientViaCP(4,0.03,0.01,0,0.0001,75)

Figure 11.3, plot (b), shows the result. Convergence proceeds in a slow, stable manner.

- A useful line of experimentation is to hold all input arguments except *StepSize* (μ) constant, and to gradually increase μ while noticing how the algorithm behaves. For this set of input parameters (i.e., with no weighting function), the reader should note that the value of μ at which the algorithm diverges is about 1.0. If using a weighting function, the algorithm will behave much differently for a given value of μ. The following examples point out different significant aspects of algorithm performance based (chiefly) on the parameters of μ and *UseWgtingFcn.*

Example 11.2. Demonstrate inadequate convergence rate due to too small a value of *StepSize.*

The result of the use of the same call as for the preceding example, but with μ = 0.005, is shown in Fig. 11.3, plot (c), in which 75 iterations are insufficient to reach the minimum value of the cost function.

Example 11.3. Demonstrate stable overshoot by use of too large a value of *StepSize*, such as 0.9, with no weighting function.

An appropriate call is

LVxGradientViaCP(4,0.9,0.01,0,0.0001,50)

which results in overshooting the minimum, but eventually descending to it. Figure 11.4 shows what happens–the successive estimates of x oscillate around the ideal value of x, but the amplitude of oscillation gradually decays, leading to convergence to the minimum of the cost function.

Example 11.4. Experimentally determine the threshold value of *StepSize* above which instability results, without the use of the weighting function.

By letting μ vary upward in successive calls, it will be found that the following call, with μ = 1, shows a general borderline stability; values of μ above 1.0 (with no weighting function) will clearly diverge. Figure 11.5 shows this condition, in which successive estimates of x lie at approximately ± 4.

LVxGradientViaCP(4,1,0.01,0,0.0001,50)

Example 11.5. Demonstrate instability using the script *LVxGradientViaCP* with a value of *StepSize* greater than about 0.0588 and a weighting function.

The call

LVxGradientViaCP(4,0.0589,0.01,1,0.0001,20)

will show that the algorithm rapidly diverges when μ is greater than about 0.058, and the fourth argument calls for *CostFcn* to be applied as a weighting function. The result is shown in Fig. 11.6.

Example 11.6. Demonstrate rapid, stable convergence by use of the weighting function and a moderate value of *StepSize*, 0.03.

A call meeting the stated requirement is

LVxGradientViaCP(4,0.03,0.01,1,0.0001,20)

and the results are shown in Fig. 11.3, plot (d).

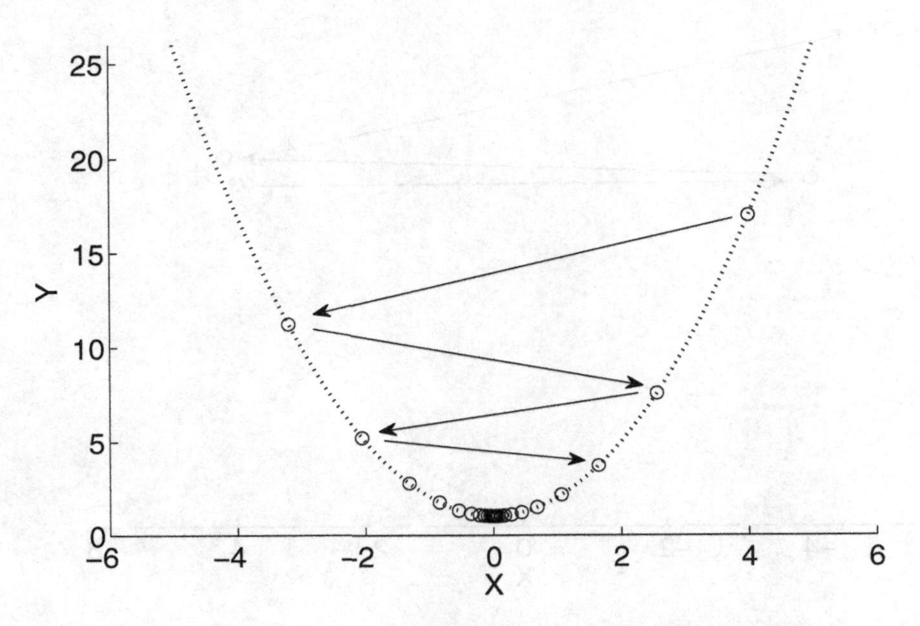

Figure 11.4: A stable overshoot (or underdamped) condition, in which, despite the fact that step size (μ) is over-large, convergence nonetheless occurs since μ is not quite large enough to lead to divergence.

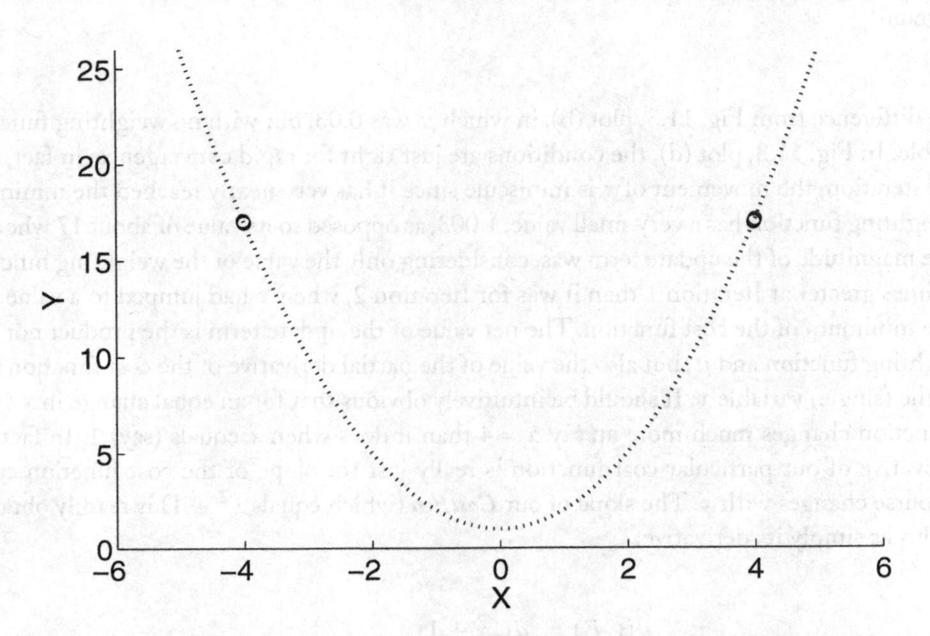

Figure 11.5: A condition in which μ is such that x neither converges nor diverges, but oscillates between two values.

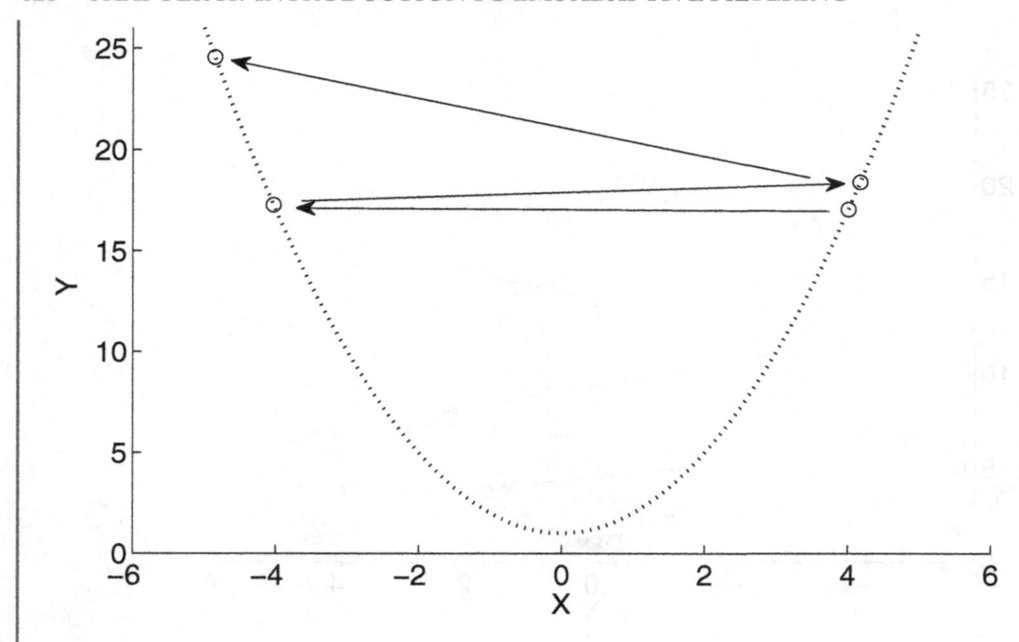

Figure 11.6: A condition in which a value of μ of 0.0589 is sufficient to make the algorithm rapidly diverge. The successive values are connected by arrows; the fifth and successive estimates of x no longer fit on the graph.

The difference from Fig. 11.3, plot (b), in which μ was 0.03, but with no weighting function, is remarkable. In Fig. 11.3, plot (d), the conditions are just right for rapid convergence; in fact, after the second iteration, the movement of x is miniscule since it has very nearly reached the minimum, and the weighting function has a very small value, 1.003, as opposed to its value of about 17 when x = 4. Thus the magnitude of the update term was, considering only the value of the weighting function, about 17 times greater at Iteration 1 than it was for Iteration 2, when x had jumped to a value near to zero, the minimum of the cost function. The net value of the update term is the product not only of the weighting function and μ, but also the value of the partial derivative of the cost function with respect to the (single) variable x. It should be intuitively obvious that for an equal change in x (Δx), the cost function changes much more at, say, x = 4 than it does when x equals (say) 1. In fact, the partial derivative of our particular cost function is really just the slope of the cost function curve, which of course changes with x. The slope of our *CostFcn* (which equals $x^2 + 1$) is readily obtained from calculus as simply its derivative:

$$\frac{d(CF)}{dx} = \frac{d(x^2 + 1)}{dx} = 2x \tag{11.9}$$

Thus we can compute the value of the coefficient update term at different values of x to see how it varies with x. The basic coefficient update formula is

$$\Delta C = \mu \left(\partial(CF)/\partial(x[n])\right) CF[n] \tag{11.10}$$

with

$$\frac{\partial(CF)}{\partial(x[n])} = \frac{d(CF)}{dx} = 2x$$

For $x = 4$, $\mu = 0.03$, and $CF[4] = 17$ ($4^2 + 1 = 17$), we get

$\Delta C = (0.03)(2)(4)(17) = 4.08$

$x[2] = x[1] - \Delta C$

$x[2] = 4 - 4.08 = -0.08$

This places $x(2)$ near the minimum of the cost function. If you run the script with these conditions, the actual numbers will vary according to the amount of noise you specify in the function call. It should be clear that if the initial value of x is too large, the initial update term will take the next estimate of x to a value having a magnitude greater than the initial value of x, but with opposite sign. The third value of x will have an even larger magnitude, with opposite sign from the second value of x, and so forth.

- Since we have an explicit formula for the cost function, it is possible to compute values for the variables, such as the initial value of x, μ, etc., which will result in borderline stability, or an immediate jump to the minimum value of the cost function. In most real problems, of course, no such explicit formula for the cost function is available, but this does not present a problem since a number of methods, including coefficient perturbation, are available for estimating the gradient and thus enabling the algorithm to navigate to the bottom of the performance surface.

- The optimization and control of μ (*StepSize*) is an important consideration in adaptive filtering. If μ is too small, convergence is very slow, and if it is just right, a single iteration may place the independent variable very near the optimum value, i.e., where the cost function is minimized. A weighting function can be used to effectively boost the value of μ when the distance from the cost function minimum is great, and reduce it as the cost function is reduced. If μ is larger than optimum, overshoot occurs: x overshoots the target. In fact, if μ is too large, the algorithm will diverge.

- Another reason to reduce μ near convergence is that the steady-state error or misadjustment is larger if μ is larger.

- μ is usually controlled by a weighting function, which must at least maintain stability, and hopefully allow for good convergence speed.

- The weighting function chosen for the script *LVxGradientViaCP* is unregulated, meaning that account has not been taken for all input conditions. For example, the call

LVxGradientViaCP(400,0.03,0.01,1,0.0001,75)

causes divergence because the weighting function is used with a very large starting value of x = 400, which causes the initial update term (considering the value of the weighting function, which is 16,001) to be too large. It should be obvious that the size of the update term must be carefully controlled to maintain adequate convergence speed and stability. We will discuss methods for accomplishing this in practical algorithms, such as the LMS algorithm, and a regulated version called the NLMS algorithm, later in the chapter.

11.7 TWO VARIABLE PERFORMANCE SURFACE

So far, we have examined adaptive problems with just one independent variable. A problem which has two variables is that of fitting a straight line to a set of points that may be distributed randomly in a plane, but which may be modeled by the point-slope form of the equation for a line in the x-y plane:

$$y = Mx + b$$

where M is the slope of the line, and b is its y-axis intercept.

The coefficient perturbation method of gradient estimation is employed in the script

LVxModelPtswLine(TestM, TestYInt,xTest,...

Mu,yMu2MuRatio,MStart,yIntStart,NoIts)

a typical call for which would be:

LVxModelPtswLine(2,0,[-10:1:10],0.005,1,-10,8,40) (11.11)

the results of which, plotted on the performance surface, are shown in Fig. 11.7.

This script allows you to specify the slope and y-intercept values (*TestM* and *TestYInt*) which are used to compute the test points to be modeled. *Mu* is the parameter μ, which scales the update term. To enable speedier convergence, the value of μ to be used for the y-intercept variable is equal to *Mu* multiplied by *yMu2MuRatio*. The fact that different variables have, in general, different optimum step sizes, is an important point that we'll discuss and experiment with below.

MStart and *yIntStart* specify the starting values of the variables. You can experiment with different initial values of the variables to see what happens. In some cases, when an algorithm is being used in specific environments, which may change suddenly but in a historically known way, the values of the variables may be preset to values near what they are likely to eventually converge to, based on prior experience. This technique can greatly reduce convergence time.

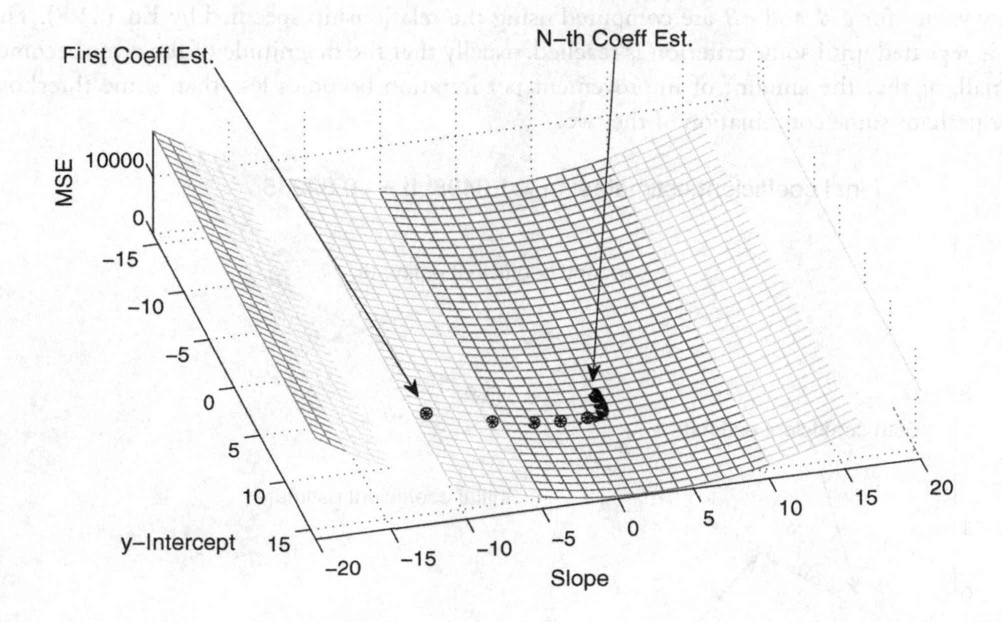

Figure 11.7: Performance Surface with plot of successive coefficient estimates using equal step size weights for both variables in coefficient update, leading to very slow movement down the Performance Surface once the bottom of the gently-sloped trough portion has been reached. This illustrates the Method of Steepest Descent.

Figure 11.8 shows a line to be modeled and the first, second, and fiftieth iterations of the modeling process. The process is very straightforward. The line to be modeled is created using given slope (M) and y-Intercept (B) values. Since discrete mathematics are used, a set of test values are determined, starting with a range of values of x, such as

$$x = [0 : 0.5 : 3]$$

for example. A set of corresponding values of y is generated using the vector x and the given values of M and B. The first estimates for M and B, which we'll call eM and eB, are used to create, using the x vector, a corresponding set of y values, which we can call eY. Then the mean squared error is given as

$$MSE = \frac{1}{N}\sum_{n=0}^{N-1}(y[n] - eY[n])^2$$

From this, estimates of the partial derivative of MSE with respect to eM and eB (the current estimated values of M and B) are made using coefficient perturbation using central differences, and then new values for eM and eB are computed using the relationship specified by Eq. (11.8). The process is repeated until some criterion is reached, usually that the magnitude of the error becomes quite small, or that the amount of improvement per iteration becomes less than some threshold value, or perhaps some combination of the two.

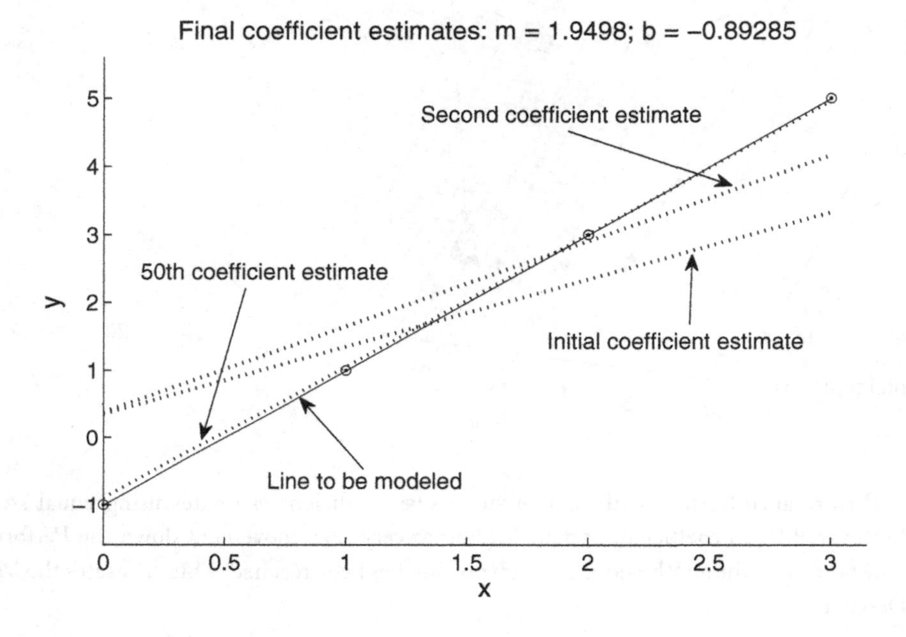

Figure 11.8: Determining the slope and y-Intercept of a line using an iterative method. The line to be modeled is shown as a solid line which passes through the sample or data points which are plotted as circles. Also shown are (estimated) dashed lines based on the first, second, and 50th coefficient estimates.

Figure 11.8 was created using the script

LV_ModelLineLMS(xVec, yVec, Mu, Its)

and in particular, the call

LV_ModelLineLMS([0,1,2,3], [-1,1,3,5], 0.05, 50)

On the other hand, the script *LVxModelPtswLine* provides a separate value of μ for the y-intercept variable by weighting μ for the y-intercept variable relative to μ for the slope variable according to the user-input variable *yMu2MuRatio*. It is easy to see why the coefficient for slope converges more quickly than the coefficient for y-intercept by inspecting Fig. 11.8 and imagining how much a small amount of misadjustment of slope affects the value of MSE compared to an equal amount of misadjustment of y-intercept. Clearly, a slight misadjustment of slope leads to a much larger change in MSE than does the same misadjustment in y-intercept. By artificially boosting the coefficient update term weight for y-intercept, we can make the algorithm converge more quickly.

Let's return to Fig. 11.7, which shows the coefficient estimate track resulting from making the call to the script given at (11.11), plotted with respect to the performance surface, which was itself generated by systematically varying the two coefficients, computing the resulting MSE, and then plotting the resultant data as a surface. Notice in Fig. 11.7 that the performance surface has different slopes in the two dimensions representing slope and y-intercept. Since there are two independent variables, each must converge to its ideal value to minimize the cost function. Since the partial derivative of one of them is very much smaller than the other (at least starting from the initial coefficient estimate, neither coefficient of which is equal to its ideal or test value), it will actually converge much more slowly. In this case, *yIntNow* (the variable representing the estimate for y-intercept in the script) converges very slowly compared to *SlopeNow* (the variable representing the estimate for slope in the script).

The method of steepest descent, which uses the negative of the gradient to estimate the next point to move to on the performance surface, moves along the direction which reduces the MSE by the largest amount for a given very small amount of movement. In Fig. 11.7, the performance surface is much like a river that has relatively steep banks which slope not only down to the river, but which slope slightly downward along the direction of river flow. If a ball were released from the top of such a river bank, its direction of travel would be determined by gravity. It would appear to mostly move directly toward the river, i.e., straight down the river bank to the water. There would, however, under the given assumptions, be a slight bias toward the direction the river is flowing since in fact the banks also have a slight downward slope in the same direction as the river's flow. In fact, the ball would be following gravity along a direction which reduces gravitational potential the greatest amount for a given amount of travel. The path a ball would follow down such a river bank is similar to the path followed in Fig. 11.7 by the coefficient estimate pair *yIntNow* and *SlopeNow* as they move toward the correct values of *TestYInt* and *TestM*. Carrying the analogy further, if the point of lowest gravitational potential (analogous to the minimum value of the cost function) is at the river's mouth, the fastest way to get there (along the shortest path) is one which heads directly for the mouth, not one which first rolls down the banks to the river and then suddenly turns downstream to head for the mouth at a very leisurely pace (this particular situation is analogous to the one shown in Fig. 11.7).

- **The method of steepest descent is not necessarily one which takes the shortest and/or fastest path to the minimum of the cost function. This is a function of the shape of the performance surface.**

Method of Steepest Descent Modified
In the script *LVxModelPtswLine*, the rate at which the slower coefficient converges is modified by the multiplier *yMu2MuRatio*, leading to a more optimal path from starting point to minimum value of cost function.

Example 11.7. Experimentally determine a value of *yMu2MuRatio* which will facilitate rapid convergence of the algorithm in the script *LVxModelPtswLine.*

A good way to start is with a basic call that results in stable convergence, and to gradually increase the value of *yMu2MuRatio.* A little such experimentation results in a useful set of parameters for convergence with *Mu* = 0.009 and *yMu2MuRatio* = 37. The result of the call

LVxModelPtswLine(2,0,[-10:1:10],0.014,37,-10,8,12)

is shown in Fig. 11.9. You should repeat the above call a number of times, varying the parameters *TestM*, *TestYInt*, *MStart*, and *yIntStart*, to observe the result. Does the algorithm still converge immediately?

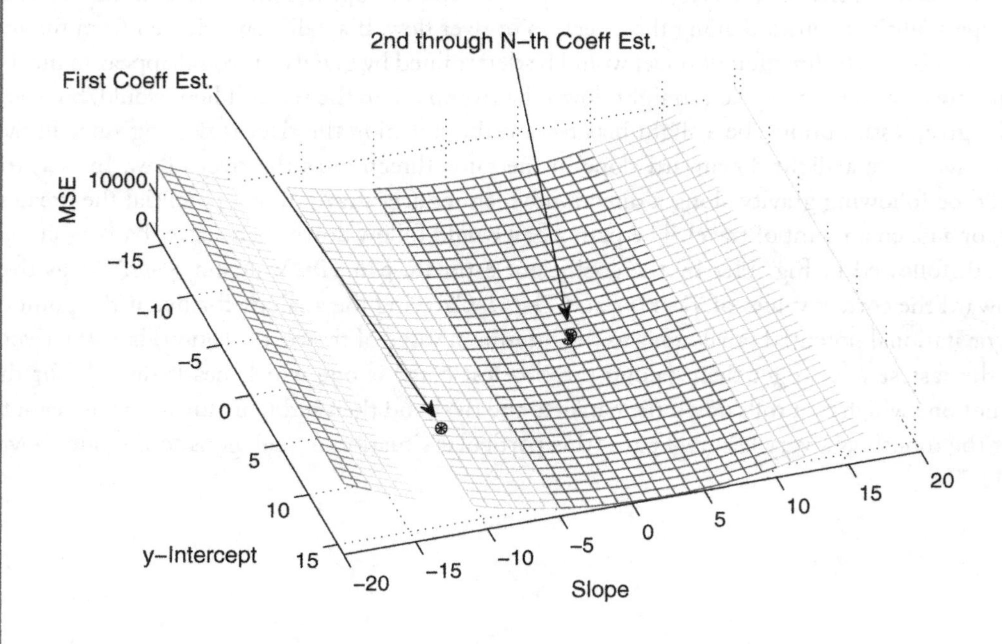

Figure 11.9: Performance Surface with plot of successive coefficient estimates utilizing a weighting function to accelerate the slower-converging of the two coefficients. Note that convergence is extremely rapid since the value of μ for the y-intercept variable has been adjusted to greatly speed the convergence of that variable toward its value of best adjustment.

Using the river analogy, we can see that instead of moving the way gravity would dictate, which would be along a nearly direct path toward the river (with only a slight bias downstream), and then suddenly turning to head directly downstream, the algorithm immediately headed directly for the point of minimized cost function, (the river mouth in our analogy).

In Fig. 11.9, near-convergence occurs in a few iterations, with additional iterations making only very minute improvements.

11.8 AN IMPROVED GRADIENT SEARCH METHOD

Thus far in the chapter we have used coefficient perturbation to estimate the gradient in several adaptive processes; we will now develop an analytic expression to estimate the gradient in the two-variable line-modeling process just discussed. This will be accomplished by writing an analytic expression for the mean squared error, and then obtaining the partial derivative with respect to m and b. This will permit us to write coefficient update equations consisting of analytic expressions.

If we represent the points to be modeled as two vectors $xVec$ and $yVec$ of length M, we can write the mean squared error as

$$MSE[n] = \frac{1}{M}\sum_{i=0}^{M-1}(yVec[i] - (m[n])(xVec[i]) - b[n])^2 \tag{11.12}$$

which is the sum of the mean squared error at each point be modeled. Since $yVec$ and $xVec$ are constants, and b is considered a constant when computing the partial derivative of MSE with respect to m, we get the partial derivative of MSE with respect to m as

$$\frac{\partial(MSE[n])}{\partial(m)} = \frac{2}{M}\sum_{i=0}^{M-1}(yVec[i] - (m[n])(xVec[i]) - b[n])(-xVec[i]) \tag{11.13}$$

and similarly the partial derivative of MSE with respect to b as

$$\frac{\partial(MSE[n])}{\partial(b)} = \frac{2}{M}\sum_{i=0}^{M-1}(yVec[i] - (m[n])(xVec[i]) - b[n])(-1) \tag{11.14}$$

Note that the scalar error $E[i,n]$ at data point (x_i, y_i) at sample index n is

$$E[i,n] = yVec[i] - (m[n])(xVec[i]) - b[n]$$

so we can rewrite Eqs. (11.13) and (11.14) as

$$\frac{\partial(MSE[n])}{\partial(m)} = -\frac{2}{M}\sum_{i=0}^{M-1} E[i,n](xVec[i]) \tag{11.15}$$

$$\frac{\partial(MSE[n])}{\partial(b)} = -\frac{2}{M}\sum_{i=0}^{M-1} E[i,n] \tag{11.16}$$

The corresponding update equations are

$$m[n+1] = m[n] - \mu\frac{\partial(MSE[n])}{\partial(m)} \tag{11.17}$$

and

$$b[n+1] = b[n] - \mu\frac{\partial(MSE[n])}{\partial(b)} \tag{11.18}$$

The above formulas are embodied in m-code in the script (see exercises below)

$$LVxModelLineLMS\_MBX(M, B, xVec, Mu, bMu2mMuRat, NoIts)$$

where M and B are values of slope and y-intercept, respectively, to be used to construct the y values to be modeled from $xVec$; Mu (μ) and $NoIts$ have the usual meaning. The value of Mu is controlled for stability (see exercises below). The rate of convergence of b is much slower than that of m, and as a result the script receives the input variable *bMu2mMuRat*, which is used to generate a boosted-value of Mu for use in the b-variable update equation (11.18). The following call leads to slow, stable convergence:

LVxModelLineLMS_MBX(3,-1,[-10:1:10],0.5,1,150) (11.19)

On the other hand, after determining a suitable value by which to boost μ for the b-coefficient update term, the same thing can be accomplished in ten iterations rather than 150:

LVxModelLineLMS_MBX(3,-1,[-10:1:10],0.5,29,10)

11.9 LMS USED IN AN FIR

11.9.1 TYPICAL ARRANGEMENTS

In order to visualize various aspects of the LMS algorithm as applied to an FIR, we'll use a 2-tap FIR, which allows the graphing of both coefficient values versus MSE.

Figure 11.10 shows a basic 2-Tap adaptive FIR filter which is to model an LTI system represented by the box labeled "Plant." For continuous domain systems, the interfacing to and from the LMS adaptive filter system, is, of course, managed with ADCs and DACs, operated at a suitable sample rate with the required anti-aliasing filter. In most books, such ADCs and DACs are generally not shown, but are understood to be a necessary part of an actual working embodiment.

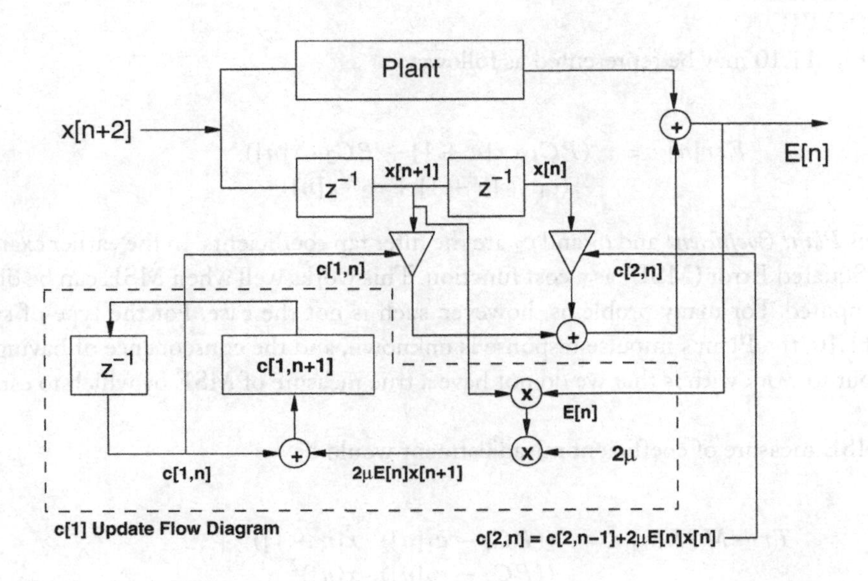

Figure 11.10: A flow diagram of a 2-Tap FIR configured to model a Physical Plant, which presumably has an impulse response of one or two samples length, i.e., a total delay equal or less than that of the adaptive FIR, which must be able to match each delay in the Plant's impulse response to model it. The Plant's impulse response is thus being modeled as a 2-Tap FIR having coefficients PC_1 and PC_2.

In the example shown in Fig. 11.10, the Plant is itself a 2-tap FIR, with unknown tap weights. The LMS adaptive filter's task will be to model the Plant, i.e., determine its impulse response, which consists of the two tap weights.

In general, modeling a particular Plant requires that the FIR have a length at least equal to the longest time delay experienced by the signal travelling through the Physical Plant. In a real-world system, of course, the number of taps involved might be dozens, hundreds, or even thousands, depending on the particular application. A 2-Tap example, however, allows graphing of the coefficients versus MSE or Scalar Error Squared (a cost function which we will investigate shortly), which allows the reader to attain a visual impression of the manner in which the adaptive filter works.

Referring to Fig. 11.10, the test signal enters at the left and splits, one part heading into the Plant, and the other part entering the filter's two series-connected delay stages (labeled z^{-1}). The signal at the output of each delay stage goes to a coefficient multiplier.

11.9.2 DERIVATION

The system of Fig. 11.10 may be represented as follows:

$$\begin{aligned} Err[n] &= (PC_1 \cdot x[n+1] + PC_2 \cdot x[n])... \\ &\quad -(c_1 \cdot x[n+1] + c_2 \cdot x[n]) \end{aligned}$$

where PC means *Plant Coefficient* and c_1 and c_2 are the filter tap coefficients. In the earlier examples, we used Mean Squared Error (MSE) as a cost function. This works well when MSE can be directly obtained or computed. For many problems, however, such is not the case. For the type of system shown in Fig. 11.10, the Plant's impulse response is unknown, and the consequence of having only the Plant's output to work with is that we do not have a true measure of MSE by which to estimate the gradient.

A true MSE measure of coefficient misadjustment would be

$$\begin{aligned} TrueMSE[n] &= ((PC_1 - c_1[n]) \cdot x[n+1])^2 + ... \\ &\quad ((PC_2 - c_2[n]) \cdot x[n])^2 \end{aligned}$$

This cost function has a true global or unimodal minimum since it can only be made equal to zero when both coefficients are perfectly converged to the plant values.

In accordance with our previously gained knowledge, we can write a general formula for the partial derivative of *TrueMSE* with respect to the ith tap coefficient as

$$\frac{\partial(TrueMSE[n])}{\partial(c_i[n])} = -2(PC_i - c_i[n]) \cdot (x[n+N-i]) \tag{11.20}$$

where PC_i means the ith Plant Coefficient, c_i is the ith tap coefficient, N is the total number of taps, and n is the current sample being computed (and remember that the partial derivative of MSE for any Plant coefficient is zero), and the partial derivative of MSE for any given coefficient $c_j[n]$ with respect to coefficient $c_i[n]$ is always zero.

The gradient might be notated as

$$\nabla(TrueMSE[n]) = -2\sum_{i=1}^{N}(PC_i - c_i[n]) \cdot x[n+N-i]) \cdot \hat{c}_i \tag{11.21}$$

Equation (11.21) requires explicit knowledge of the scalar error contributed by each pair of plant and filter coefficients. Returning to Eq. (11.20), what can readily be seen is that the partial derivative for each coefficient is dependent on its individual misadjustment relative to the equivalent plant coefficient, as well as the value of the signal x at the tap, which changes with n. The performance surface for a true-MSE cost function is quadratic; for a two-tap filter, it graphs as a paraboloid of revolution, as shown in Fig. 11.11, in which MSE in decibels is graphed vertically against the two coefficients.

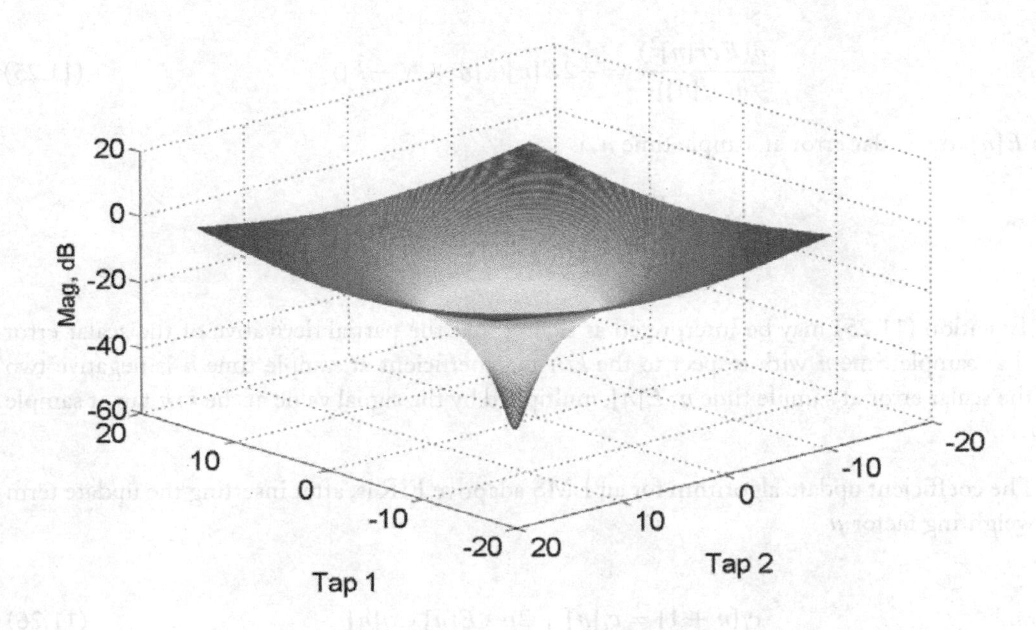

Figure 11.11: Performance surface for a 2-Tap Adaptive filter using a True-MSE algorithm.

The scalar error, the difference between the plant output and the filter output, is all we have access to in most problems, and, for the two-tap problem at hand, we must approximate the true MSE by squaring the scalar error:

$$ApproxMSE[n] = ScalarError^2 = \\ ((PC_1 - c_1[n])x[n+1] + (PC_2 - c_2[n])x[n])^2 \tag{11.22}$$

We can derive an estimate for the partial derivative of *ApproxMSE* with respect to, say, c_1, (as in previous examples) as

$$\frac{\partial(ApproxMSE[n])}{\partial(c_1[n])} = 2((PC_1 - c_1[n])x[n+1]...$$

$$+ (PC_2 - c_2[n])x[n])(-x[n+1]) \tag{11.23}$$

This formula may be generalized for an N-Tap FIR as

$$\frac{\partial(Err[n]^2)}{\partial(c_k[n])} = 2\left[\sum_{i=1}^{N}(PC_i - c_i[n])x[n+N-i]\right](-x[n+N-k]) \tag{11.24}$$

which may be rewritten as

$$\frac{\partial(Err[n]^2)}{\partial(c_k[n])} = -2E[n](x[n+N-k]) \tag{11.25}$$

where $E[n]$, the scalar error at sample time n, is

$$E[n] = \sum_{i=1}^{N}(PC_i - c_i[n])x[n+N-i]$$

Equation (11.25) may be interpreted as saying that the partial derivative of the scalar error squared at sample time n with respect to the kth tap coefficient at sample time n is negative two times the scalar error at sample time n, $E[n]$, multiplied by the signal value at the kth tap at sample time n.

- The coefficient update algorithm for an LMS adaptive FIR is, after inserting the update term weighting factor μ

$$c_i[n+1] = c_i[n] + 2\mu \cdot E[n] \cdot x_i[n] \tag{11.26}$$

where $c_i[n]$ represents the tap coefficient of tap index i and iteration n, μ is a scalar constant which scales the overall magnitude of the update term, $E[n]$ represents the scalar error at iteration n, and $x_i[n]$ is the signal value at tap i at iteration n.

- While the LMS algorithm is elegant, and moderately robust, it is not as robust as a true-MSE algorithm in which the gradient is estimated from knowledge of the actual MSE. Equating squared scalar error to MSE allows the simplicity and usefulness of the LMS algorithm, but there are limitations on the frequency content of the input signal to ensure that the LMS algorithm properly converges. This will be explored below.

11.9.3 LIMITATION ON Mu

One idea of the maximum value that the stepsize Δ (2μ in Eq. (11.26), for example) may assume and remain stable is

$$0 < \Delta < \frac{1}{10NP} \tag{11.27}$$

where N is the filter length and P is the average signal power over M samples of the signal:

$$P = \frac{1}{M}\sum_{i=0}^{M-1}(x[i])^2$$

Thus the quantity NP is a measure of signal power in the filter. The script (see exercises below)

$$LVxLMSAdaptFiltMuCheck(k, NoTaps, C, SigType, Freq, Delay)$$

filters either random noise of standard deviation k, or a cosine of amplitude k and frequency zero (DC), Halfband, Nyquist, or $Freq$, with an LMS adaptive FIR having a length equal to $NoTaps$, and plots the test signal in one subplot, and the filter output in a second subplot. The stepsize for coefficient update is computed as the high limit given by Eq. (11.27), but is also multiplied by the constant C for testing purposes. Variation of k and N has little effect on the convergence rate and stability of the algorithm when stepsize is computed using Eq. (11.27). In practice, a value of stepsize considerably larger may often be used to achieve convergence in a number of iterations approximately equal to ten times the filter length. The parameter C can be experimentally used to observe this. Figure 11.12 shows the error signal for various values of C from 1 to 10 with k = 1 and $NoTaps$ = 100, white noise as the test signal, and $Delay$ = 10 (see the exercises below for a complete description of the input arguments). Figure 11.13 shows the same thing as Fig. 11.12 except that k = 3 and $NoTaps$ = 40. For the examples shown, the best value of C appears to lie between 5.0 and 7.0-this value range allows for both speedy convergence and good stability.

11.9.4 NLMS ALGORITHM

A standard method to assist in controlling the stepsize in Eq. (11.26) is to divide by a factor related to the signal power in the filter, to which is added a small number ϵ in case the signal power in the filter should be zero. We thus get

$$c_i[n+1] = c_i[n] + \frac{\Delta}{\sum x_i[n]^2 + \epsilon} \cdot Err[n] \cdot x_i[n]$$

(where $\Delta = 2\mu$) and refer to this as the Normalized Least Mean Square, or NLMS, algorithm. Note that in Eq. (11.27), the power is averaged over the entire input signal. In the NLMS method, only signal values in the filter are taken into account. The NLMS is thus more adaptive to current signal conditions. The various scripts we will develop using the LMS algorithm in an FIR will generally be NLMS algorithms.

The script (see exercises below)

$$LVxLMSvNLMS(kVec, NoTaps, Delta, Mu, SigType, Freq, Delay)$$

passes a test signal determined by the value of $SigType$ through two FIRs in parallel, one FIR using the LMS algorithm with a stepsize fixed as Mu, and the other FIR using the NLMS algorithm having a stepsize equal to $Delta$ divided by the signal power in the filter (plus a very small constant to avoid potential division by zero). The test signal is given an amplitude profile defined by $kVec$, which specifies a variable number of different equally spaced amplitudes imposed on the test signal.

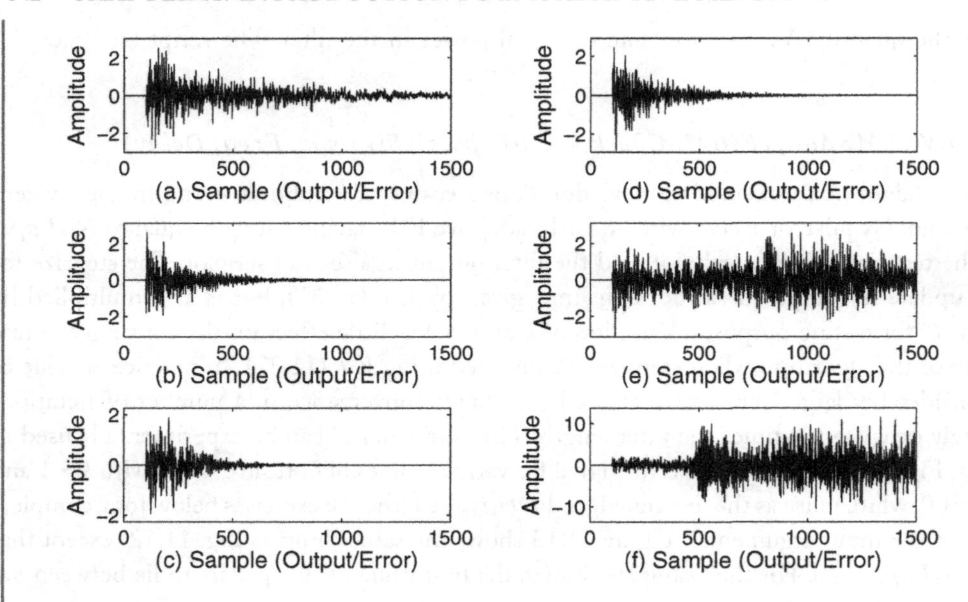

Figure 11.12: (a) Error signal for Stepsize = $C/(10NP)$, with C = 1, k = 1 (see text); (b) Same with C = 3; (c) Same with C = 5; (d) Same with C = 7; (e) Same with C = 9.5; (f) Same with C = 10.

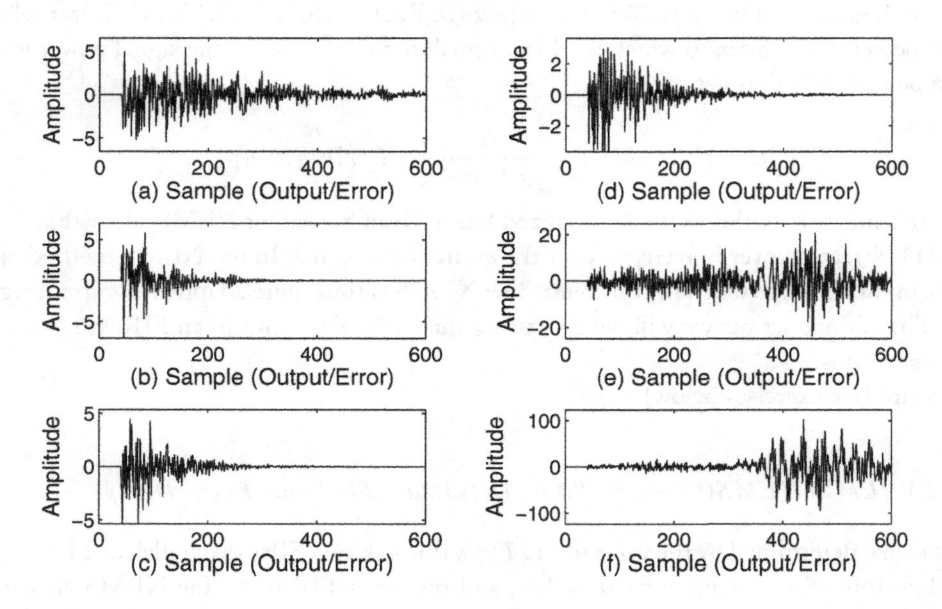

Figure 11.13: (a) Error signal for Stepsize = $C/(10NP)$, with C = 1, k = 3 (see text); (b) Same with C = 3; (c) Same with C = 5; (d) Same with C = 7; (e) Same with C = 9.5; (f) Same with C = 10.

In this manner, the test signal, be it noise or a cosine, for example, can undergo step changes in amplitude that test an algorithm's ability to remain stable.

The call, in which Mu has been manually determined to allow the LMS algorithm to converge,

LVxLMSvNLMS([1,5,15,1],100,0.5,0.00005,1,[],50)

results in Fig. 11.14, while the call

LVxLMSvNLMS([1,5,25,1],100,0.5,0.00005,1,[],50)

in which the test signal amplitude has been boosted in its middle section, results in Fig. 11.15, in which it can be seen that the LMS algorithm has become unstable and divergent, while the NLMS algorithm has remained stable and convergent.

Limitations of the LMS/NLMS Algorithms

We now proceed to study the LMS algorithm in detail for the 2-tap FIR example and observe its behavior with different types of input signals. Referring to Eq. (11.22), the cost function represented by this equation does not have a single minimum; rather, the performance surface (which resembles the lower half of a pipe split along its longitudinal axis) is linear in one direction, and quadratic in a direction perpendicular to the first direction; it has a perfectly flat bottom and defines (along that bottom) an infinite number of pairs of coefficients (c_1, c_2) which will make *ScalarErrorSquared* = 0.

We can derive an expression for this type of performance surface by determining what values of c_1 and c_2 will drive the cost function (*ApproxMSE*) to any arbitrary value ESq. Then, from Eq. (11.22) we get

$$(PC_1 - c_1[n]) \cdot x[n+1] + (PC_2 - c_2[n]) \cdot x[n] = \sqrt{ESq}$$

We proceed by solving for $c_2[n]$ in terms of $c_1[n]$:

$$c_2[n] = -(\frac{x[n+1]}{x[n]}) \cdot c_1[n] + ...$$

$$(PC_2 + PC_1 \cdot \frac{x[n+1]}{x[n]} - \frac{\sqrt{ESq}}{x[n]}) \tag{11.28}$$

Equation (11.28) tells, for any $c_1[n]$, what value of $c_2[n]$ is necessary to force the scalar error to be $\sqrt{ESq}$ (or scalar error squared to be ESq). Equation (11.28) is a linear equation of the familiar form $y = Mx + b$ (or, in the iterative or adaptive sense, $y[n] = M[n]x[n] + b[n]$ in which $y[n]$ equates to $c_2[n]$,

$$M[n] = -\frac{x[n+1]}{x[n]} \tag{11.29}$$

and

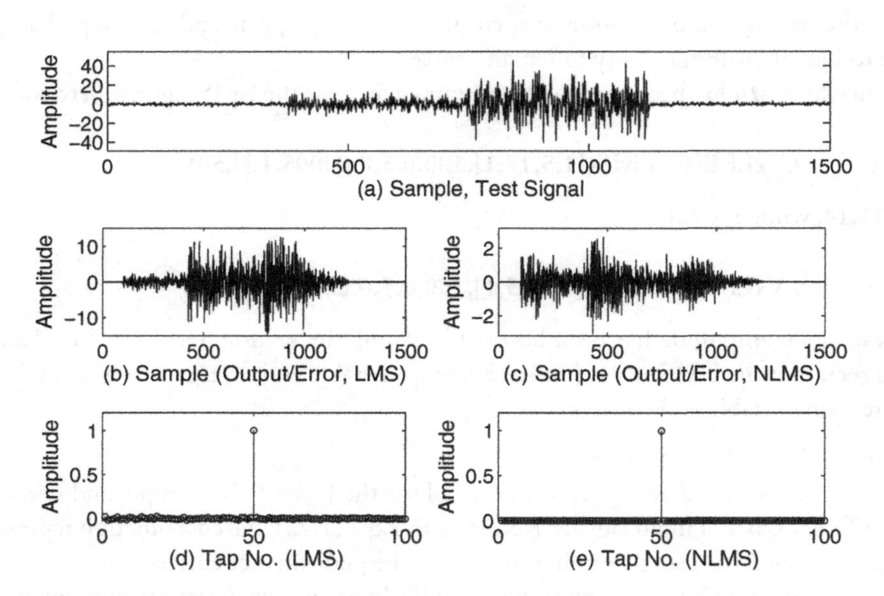

Figure 11.14: (a) Test signal, with amplitude segments scaled by $kVec$; (b) Error signal from the LMS algorithm; (c) Error signal from the NLMS algorithm; (d) Final Tap Weights for the LMS algorithm; (e) Final Tap Weights for the NLMS algorithm.

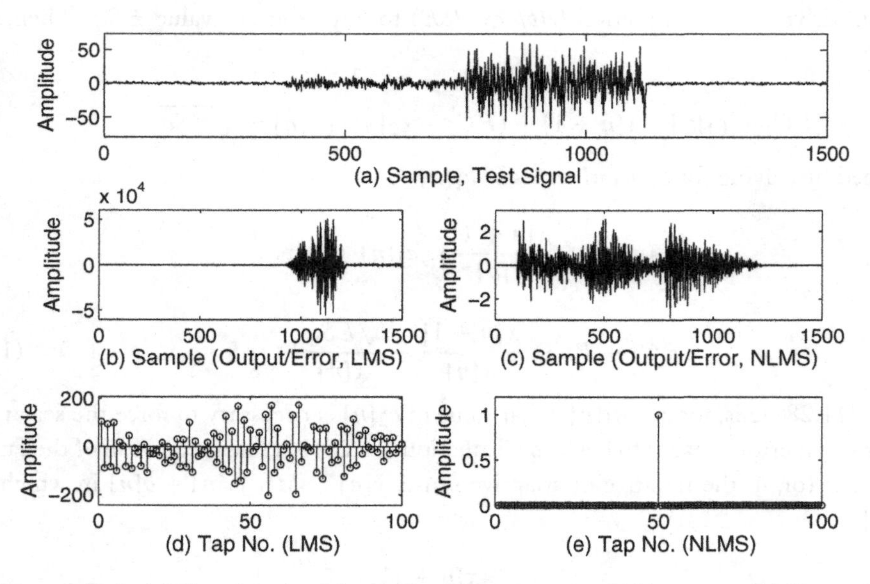

Figure 11.15: (a) Test signal, with amplitude segments scaled by $kVec$; (b) Error signal from the LMS algorithm; (c) Error signal from the NLMS algorithm; (d) Final Tap Weights for the LMS algorithm; (e) Final Tap Weights for the NLMS algorithm.

$$b[n] = (PC_1 + PC_0 \cdot \frac{x[n+1]}{x[n]} - \frac{\sqrt{ESq}}{x[n]}) \tag{11.30}$$

It can be seen that the slope $M[n]$ in Eq. (11.29) is just the ratio of the two samples in the filter. If the input signal is changing, then the slope will also change at each sample time, as will the value of $b[n]$ in Eq. (11.30).

However, no matter what the slope M and intercept value b are, the line defined by the equation always passes through the point of perfect coefficient adjustment, that is, $c_1 = PC_1$ and $c_2 = PC_2$.when $ESq = 0$. This can be readily verified by setting $c_1 = PC_1$ and $ESq = 0$ in Eq. (11.28).

The performance surface generated by Eq. (11.28) is linear in a first direction (lines of equal ESq are parallel to this first direction), and quadratic in the direction perpendicular to the first direction. We'll refer to this performance surface as a *Linear-Quadratic Half Pipe* (the surface is more pipe-like in appearance when magnitude is graphed linearly), and to lines of equal squared error as *IsoSquErrs.* Figure 11.16 shows an example of an LQHP performance surface, Scalar Error squared in dB is plotted against the coefficients.

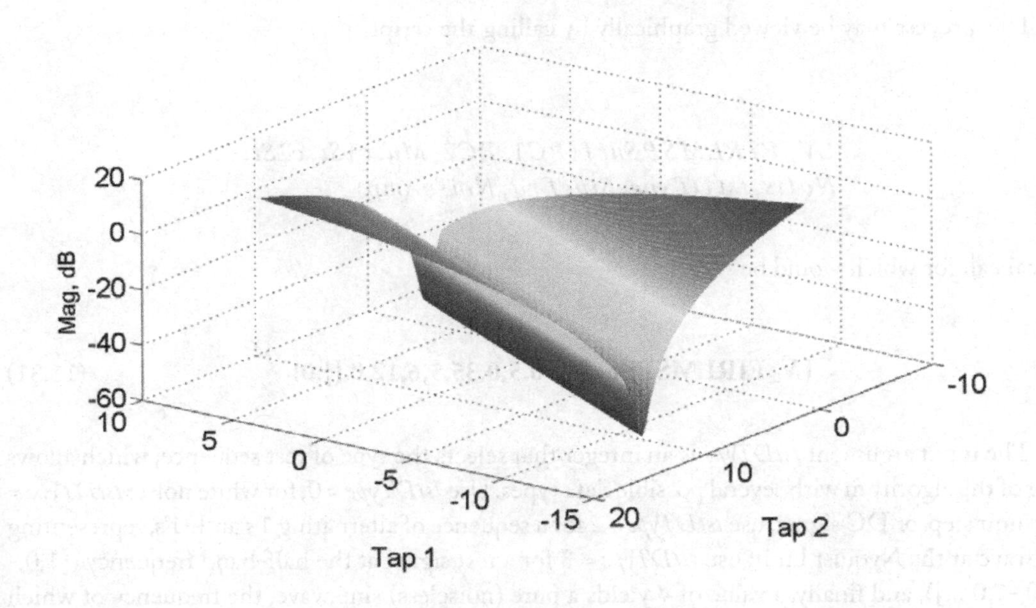

Figure 11.16: Performance surface for a 2-Tap LMS Adaptive filter for a particular instantaneous pair of signal values in the filter; orientation of the "trough" in the performance surface changes as the ratio of the two signal values changes.

There are several important things to note about this:

- The (negative of the) gradient will not, in general, be headed toward the point of best coefficient adjustment, but rather it will be headed along the shortest path (a normal line, possibly corrupted with noise) toward the bottom of a LQHP which varies orientation in the (c_1, c_2) plane (slope- or angle-wise) according to the current values of the input signal in the filter.

- The very bottom of each of the these LQHPs actually passes through the point of ideal coefficient adjustment when $ESQ = 0$, no matter what the slope or angle is of the LQHP in the (c_1, c_2) plane.

- Accordingly, the task of getting to the ideal adjustment point (i.e., when the coefficient estimates properly model the Plant) depends on the orientation of LQHPs changing from one iteration to the next.

- This process may be viewed graphically by calling the script

$$LV\_FIRLMSPSurf(PC1, PC2, Mu, c1St, c2St, ...$$
$$NoIts, tstDType, SineFrq, NoiseAmp)$$

a typical call for which would be

$$\textbf{LV\_FIRLMSPSurf(1,-0.5,0.35,5,6,12,0,[],0)} \quad (11.31)$$

The input argument *tstDType* is an integer that selects the type of test sequence, which allows testing of the algorithm with several possible data types. Use *tstDType* = 0, for white noise; *tstDType* = 1, for a unit step, or DC signal; use *tstDType* = *2* for a sequence of alternating 1's and -1's, representing a sine wave at the Nyquist limit; use *tstDType* = 3 for a test signal at the half-band frequency ([1,0,-1,0,1,0,-1,0 ...]), and finally, a value of *4* yields a pure (noiseless) sine wave, the frequency of which may optionally be specified by entering another number after *tstDType*–if in this case $SineFrq$ is passed as the empty matrix, the sine's frequency defaults to 25 cycles/1024 samples. If you aren't calling for a sine wave as the test signal, you may pass *SineFreq* as the empty matrix, [ ], as shown in the sample call above.

A coefficient track which resulted from the call at (11.31), which specifies the test signal as random noise, is shown in Fig. 11.17.

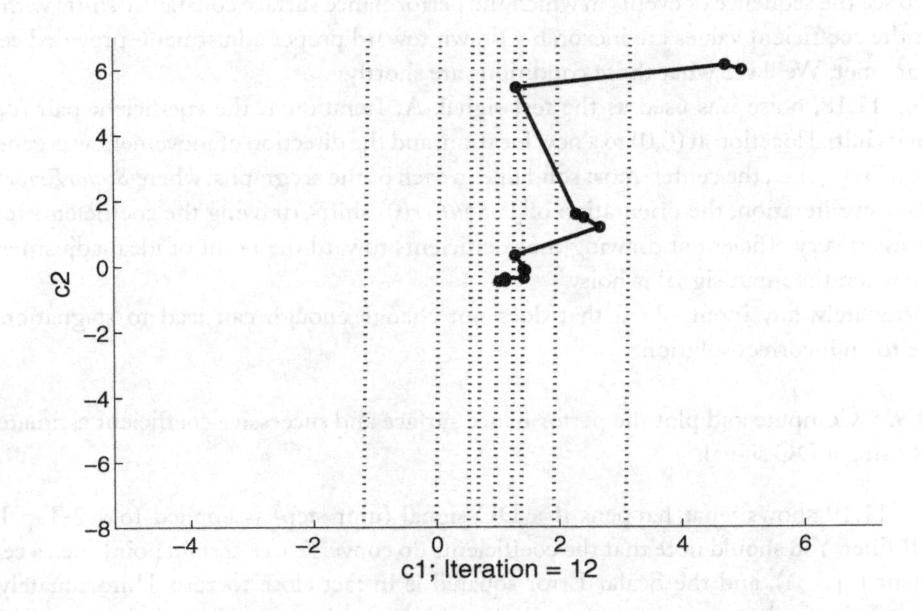

Figure 11.17: Coefficient track for a 2-Tap LMS Adaptive filter using random noise as the test signal. Minimum of cost function is a line (the center-most dotted line in the plot) rather than a point. A random noise test signal causes the orientation of the performance surface to change randomly, resulting in convergence to the correct coefficient values, which are 1, -0.5.

The script

$$LV\_FIRLMS6Panel(PC1, PC2, Mu, c1St, c2St, ...$$
$$NoIts, tstDType, SineFrq, NoiseAmp)$$

is similar to the script *LV_FIRLMSPSurf*, except that it plots the most recent six iterations in one figure.

Example 11.8. Compute and plot the performance surface (several IsoSqErrs will do) and coefficient track for a number of iterations of an adaptive LMS 2-tap FIR, with white noise as the signal.

Figure 11.18, which was produced by the script call

$$\textbf{LV\_FIRLMS6Panel(-2,3,0.25,0,0,6,0,[ ],0)} \quad (11.32)$$

allows one to see the sequence of events in which the performance surface constantly shifts with each sample, but the coefficient values are inexorably drawn toward proper adjustment–provided certain conditions are met. We'll see what those conditions are shortly.

In Fig. 11.18, noise was used as the test signal. At Iteration 1, the coefficient pair (c_1, c_2) moved from its initial location at (0,0) to a new location, and the direction of movement was generally toward *IsoSquErr*(0), i.e., the center-most solid line in each of the six graphs, where $ScalarError^2 = 0$. At each successive iteration, the orientation of *IsoSquErr*(0) shifts, drawing the coefficients toward it. This process is very efficient at drawing the coefficients toward the point of ideal adjustment or convergence when the input signal is noisy.

Unfortunately, any input signal that does not change enough can lead to stagnation and convergence to an incorrect solution.

Example 11.9. Compute and plot the performance surface and successive coefficient estimates for a 2-tap FIR using a DC signal.

Figure 11.19 shows what happens if a DC signal (unit step) is applied to a 2-Tap LMS adaptive FIR filter. You should note that the coefficients do converge to a certain point (i.e., a certain coefficient pair (c_1, c_2)), and the Scalar Error squared is in fact close to zero. Unfortunately, the converged coefficient values are incorrect. A similar thing may happen with other types of highly regular signals, such as a pure (noise-free) sine wave.

Example 11.10. Compute and plot the performance surface and successive coefficient estimates for a 2-tap FIR filtering a sinusoidal signal of frequency 25.

The script call

LV_FIRLMS6Panel(-1,3,0.25,5,3,6,4,25,0)

in which the seventh and eighth input arguments (4, 25) specify a test signal consisting of a pure sine wave of frequency 25 cycles (over 1024 samples), results in the plots shown in Fig. 11.20. The result shown in Fig. 11.20 again shows convergence to an incorrect solution, with, nonetheless, the square of Scalar Error minimized. Logically, a sine wave does change over time, so it is necessary to perform enough iterations so that several cycles of the sine wave are traversed. In general, though, convergence is protracted and uncertain.

Example 11.11. Compute and plot the error signal and coefficient estimates for a 2-tap FIR that models a Plant defined by Plant Coefficients PC1 and PC2, as shown in Fig. 11.10, using a test signal consisting of a sine wave containing, respectively, small, medium, and large amounts of white noise.

The computations may be performed using the script (see exercises below)

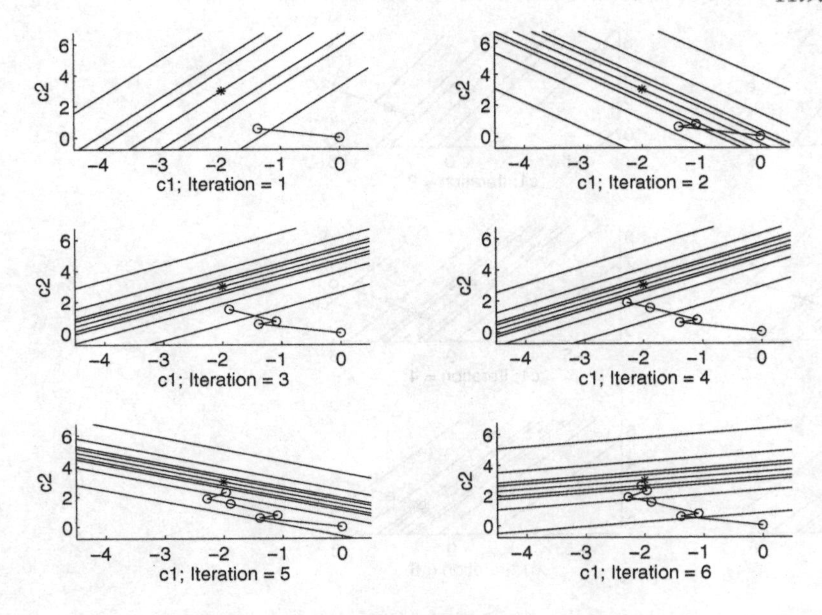

Figure 11.18: Coefficient track in six successive panels/iterations for the two coefficients of an LMS Adaptive filter using random noise as a test signal.

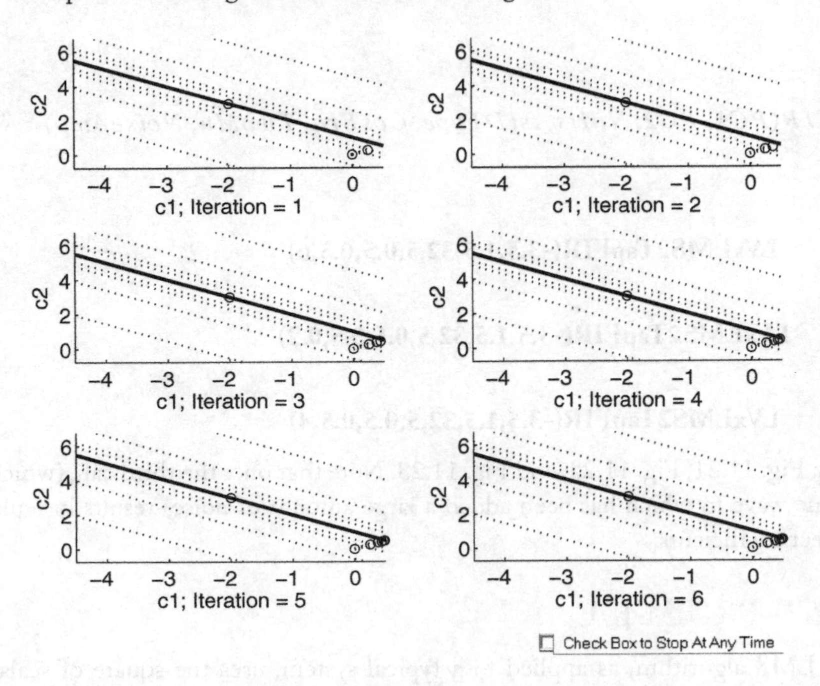

Figure 11.19: Coefficient track in six successive panels/iterations for the two coefficients of an LMS Adaptive filter using a Unit Step (DC) as a test signal.

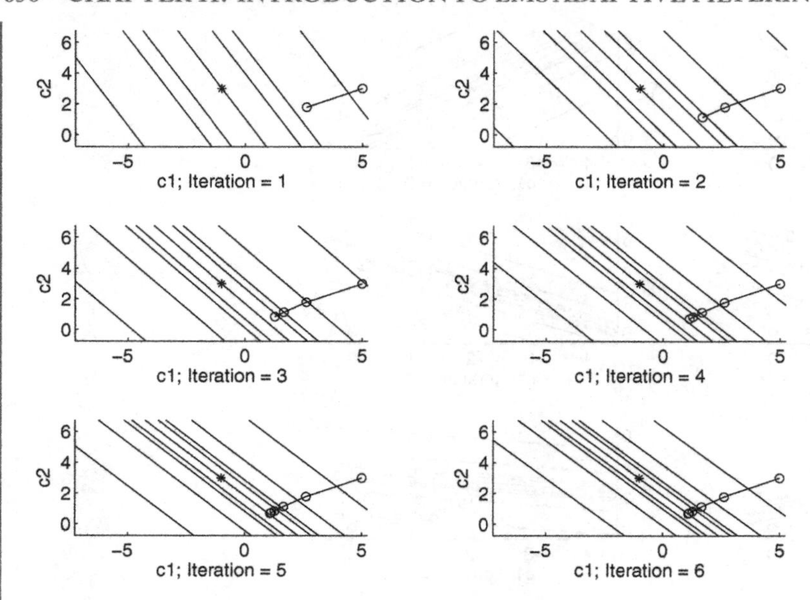

Figure 11.20: Coefficient track in six successive panels/iterations for the two coefficients of an LMS Adaptive filter using a 25 Hz sine wave as a test signal.

$$LVxLMS2TapFIR(PC1, PC2, NoIts, tstDType, CosFrq, TwoMu, NoiseAmp)$$

with the specific calls

LVxLMS2TapFIR(-3.5,1.5,32,5,0.5,0.5,0)

LVxLMS2TapFIR(-3.5,1.5,32,5,0.5,0.5,0.2)

LVxLMS2TapFIR(-3.5,1.5,32,5,0.5,0.5,4)

which yield, respectively, Fig. 11.21, Fig. 11.22, and Fig. 11.23. Note that only the third call, (which uses as a test signal a sine wave to which has been added a large amount of noise) results in rapid convergence to the correct coefficients.

11.10 CONTRAST-TRUE MSE

We have seen that the LMS algorithm, as applied to a typical system, uses the square of scalar error as an estimate of the true MSE in order to estimate the gradient. As a result, the actual

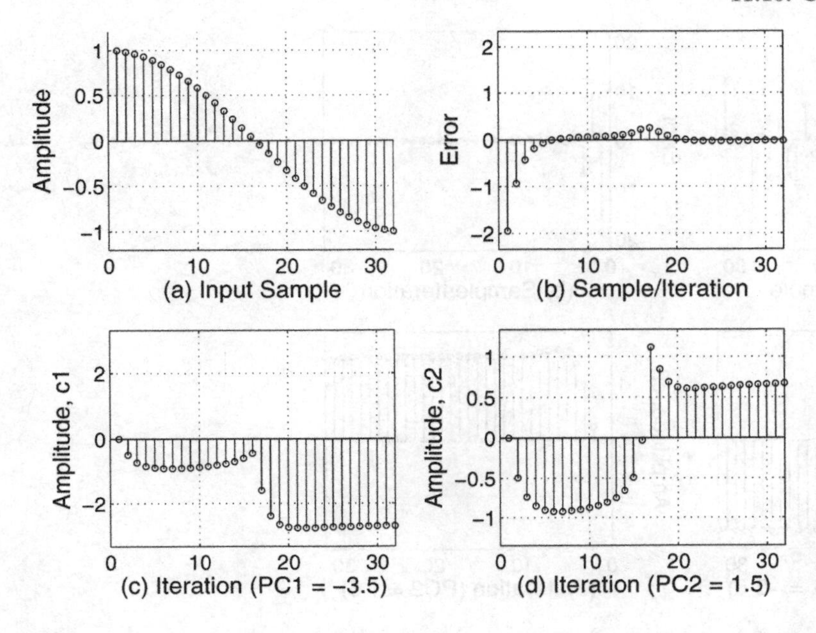

Figure 11.21: (a) Input signal, a sine wave containing a small amount of random noise; (b) Output or error signal; (c) Estimate of 1st coefficient versus iteration; (d) Estimate of 2nd coefficient versus iteration.

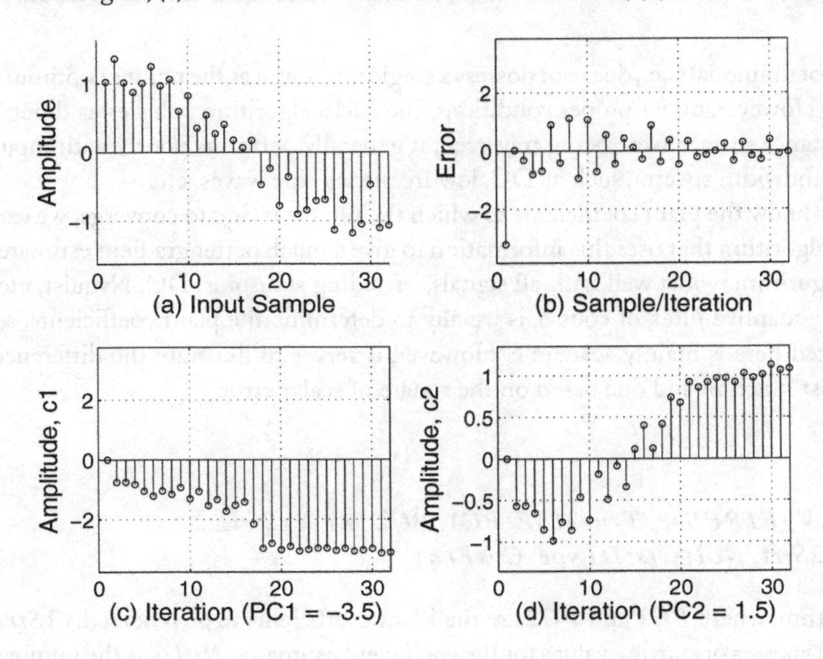

Figure 11.22: (a) Input signal, a sine wave containing a medium amount of random noise; (b) Output or error signal; (c) Estimate of 1st coefficient versus iteration; (d) Estimate of 2nd coefficient versus iteration.

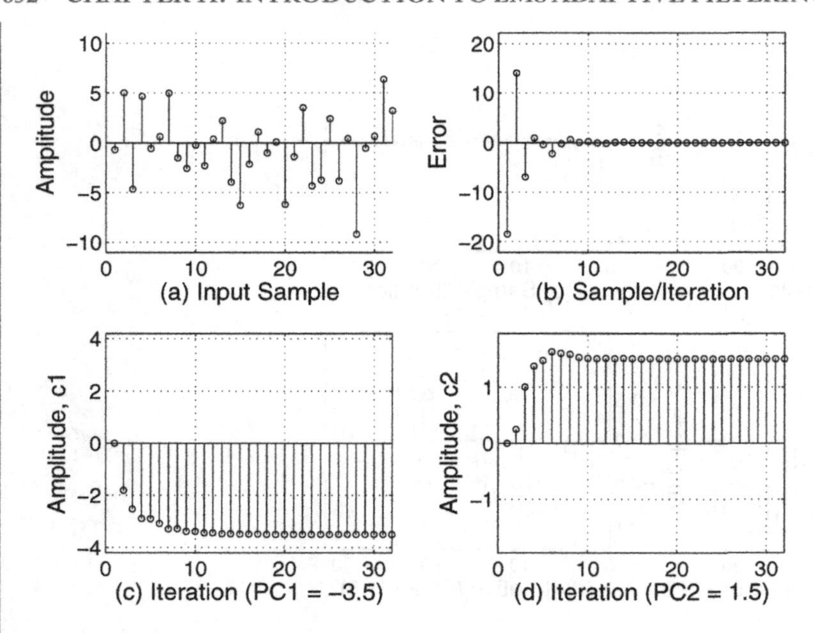

Figure 11.23: (a) Input signal, a sine wave containing a large amount of random noise; (b) Output or error signal; (c) Estimate of 1st coefficient versus iteration; (d) Estimate of 2nd coefficient versus iteration.

performance surface is not unimodal (i.e., does not possess a single minimum at the point of optimum coefficient adjustment). However, under proper conditions, the LMS algorithm behaves as though a true unimodal performance surface were being traversed; it generally performs poorly with input signals having narrow bandwidth spectra, such as DC, low frequency sine waves, etc.

When we actually know the plant coefficients to which the filter is trying to converge, we can construct an LMS-like algorithm that uses this information to give a much better gradient estimate. The result is that the algorithm works well with all signals, including sinusoids, DC, Nyquist, etc. The purpose of using an adaptive filter, of course, is usually to determine the plant coefficients, so the situation hypothesized here is mainly academic. However, it serves to illustrate the difference between a true-MSE cost function and one based on the square of scalar error.

The script

$$LV\_FIRPSurfTrueMSE(PC1, PC2, Mu, c1Strt, ...$$
$$c2Strt, NoIts, tstDType, CosFrq)$$

embodies such an algorithm, where $PC1$ and $PC2$ are the Plant coefficients to be modeled, $c1Strt$ and $c2Strt$ are the initial guesses or starting values for the coefficient estimates, $NoIts$ is the number of iterations to perform, Mu has the usual meaning, and *tstDType* selects the type of test signal. Pass *tstDType* as 0 for white noise, 1 for a unit step (DC), 2 for the Nyquist Limit frequency, 3 for the half-band frequency, and 4 for a cosine of frequency *CosFrq* (may be passed as [] if a cosine is not being used as a test signal).

Example 11.12. Compute and plot the performance surface for a 2-tap FIR using a true-MSE-based coefficient update algorithm; use a sinusoidal input.

Figure 11.24 shows the results after making the call

LV_FIRPSurfTrueMSE(-1,3,1,5,6,7,4,125)

Even though the test signal employed to generate the coefficient track in Fig. 11.24 was a sine wave, convergence is proper.

Example 11.13. Compute and plot the performance surface for a 2-tap FIR using a true-MSE-based coefficient update algorithm; use a DC input.

The script call

LV_FIRPSurfTrueMSE(-1,3,1,5,6,5,1,[])

uses a DC (unit step) signal as the test signal, and results in a coefficient track which forms a straight line from the entry point to the point of ideal adjustment, as shown in Fig. 11.25.

11.11 LMS ADAPTIVE FIR SUMMARY

In general, a pure frequency (a sinusoid) poses difficulties for an LMS algorithm, which uses the square of the scalar error as the cost function rather than the true MSE. However, depending on exact conditions, it is possible to achieve proper convergence. This is more likely with a higher frequency than with a lower frequency–the key word is change. The more rapid the change in test signal values, especially change in sign, the better. Mixtures of sinusoids, to the extent more rapid sign and amplitude change result, improve performance. Recall from some of the examples, however, that an LMS adaptive filter can converge with a narrow bandwidth input (and in fact minimize the scalar error), just as with other, wider-bandwidth signals. The difference is that with narrow bandwidth signals, the filter coefficients do not necessarily model the Physical Plant properly, since there is no unique set of coefficients that will minimize the error (rather, there are many possible sets of coefficients that can minimize the scalar error).

Hence it is, in general, good to ensure that LMS adaptive filters receive signals for processing which are at least somewhat noisy in nature. In many applications, such as noise cancellation, the input signal actually is more or less pure noise. In some applications, such as, for example, feedback cancellation in a hearing aid, it may be necessary to inject noise into the signal to cause the LMS adaptive algorithm to converge. We'll see an application in the next chapter, using a special filter

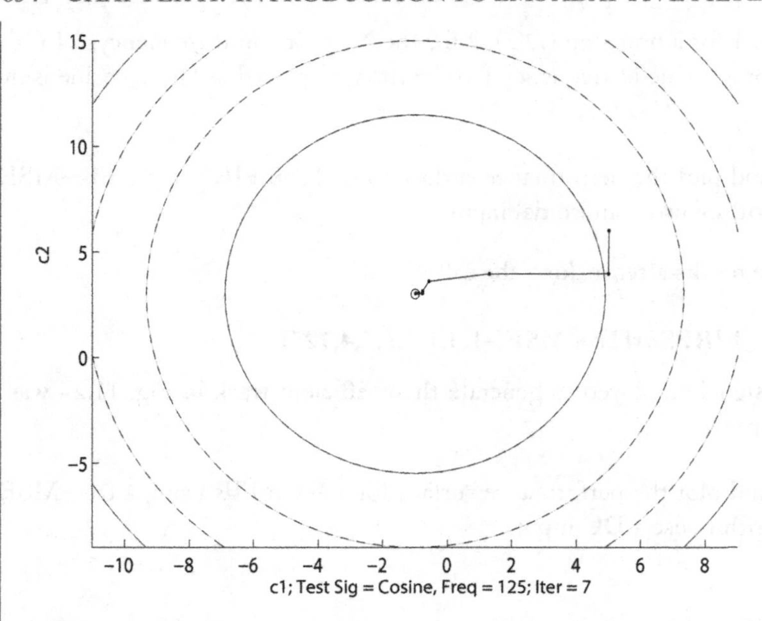

Figure 11.24: Plot of $c1$ and $c2$ for seven iterations of a True-MSE-based gradient algorithm, using as a test signal a cosine of 125 cycles over 1024 samples, i.e., a normalized frequency of a $125/512 = 0.24\pi$. Plant coefficients (-1,3) lie at the center of the plot.

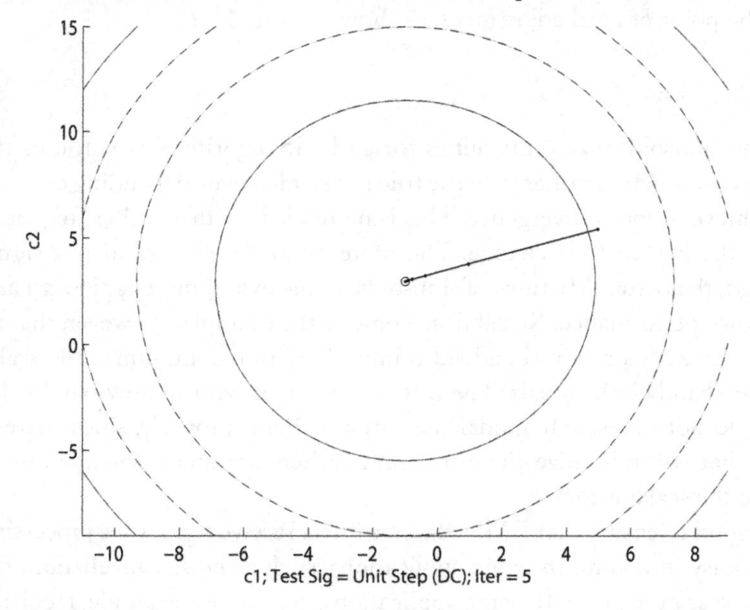

Figure 11.25: Plot of c1 and c2 for 5 iterations of a True-MSE-based gradient algorithm, using as a test signal a 1024 sample unit step. Plant coefficients (-1,3) lie at the center of the plot.

topology, however, in which sinusoidal signals, but not random signals, can be cancelled using an LMS adaptive filter.

We have looked extensively at the inner workings and hidden mechanisms of the LMS algorithm as applied to a simple 2-point FIR. Limiting the number of coefficients to two enabled graphing of the performance surfaces in a three-dimensional space. Adaptive FIRs are employed using large numbers of taps, sometimes in the thousands. It is not possible to visualize the performance surfaces for numbers of taps beyond two, but nonetheless, the LMS algorithm behaves very similarly–that is to say, as a stochastic gradient search algorithm, in which the gradient estimate is noisy. With optimum conditions, typical filters converge in a number of iterations equal to about 10 times the filter length.

The weaknesses uncovered in the 2-Tap FIR are also weaknesses of the general N-Tap LMS adaptive FIR; that is, unreliability with input signals that lack randomness or are of narrow bandwidth. Pure sinusoids, and small combinations of the same can be problematic, and can lead to convergence to incorrect coefficient values.

Reference [1] is a very accessible basic adaptive filtering text; while Reference [2] gives an exhaustive rundown of the LMS adaptive filter and numerous species thereof, along with extensive theoretical discussions, illustrations, and computer exercises.

11.12 REFERENCES

[1] Bernard Widrow and Samuel D. Stearns, *Adaptive Signal Processing*, Prentice-Hall, Englewood Cliffs, New Jersey, 1985.

[2] Ali H. Sayed, *Fundamentals of Adaptive Filtering*, John Wiley & Sons, Hoboken, New Jersey, 2003.

[3] Simon Haykin, *Adaptive Filter Theory, Third Edition*, Prentice-Hall, Upper Saddle River, New Jersey, 1996.

11.13 EXERCISES

1. Write the m-code for the script

LVxGradientViaCP(FirstValX, StepSize,...
NoiseAmp, UseWgtingFcn, deltaX, NoIts)

which is extensively described and illustrated in the text, and test it with the calls below. For each instance, characterize the performance of the algorithm with respect to speed of convergence and stability. Conditions of stability might be, for example: stable, convergent; stable, but nonconvergent, unstable or divergent. For convergent situations, note where there is overshoot (an "underdamped") condition, very slow convergence (an "overdamped" condition) or immediate convergence (a "critically damped" condition).

(a) **LVxGradientViaCP(4,0.05,0.01,0,0.0001,50)**
(b) **LVxGradientViaCP(4,0.03,0.01,0,0.0001,75)**
(c) **LVxGradientViaCP(4,0.9,0.01,0,0.0001,50)**
(d) **LVxGradientViaCP(4,1,0.01,0,0.0001,50)**
(e) **LVxGradientViaCP(4,0.07,0.01,1,0.0001,50)**
(f) **LVxGradientViaCP(4,0.03,0.01,1,0.0001,20)**
(g) **LVxGradientViaCP(400,0.03,0.01,1,0.0001,75)**

2. Using the relationships

$$CF[n] = x^2 + 1$$

$$x[n+1] = x[n] - \Delta C$$

$$\Delta C = \mu\,(\partial(CF)/\partial(x[n]))CF[n]$$

with μ = 0.03, write an expression that is a polynomial in x that can be used to solve for an initial value of x (denoted as $x[1]$) that will result in $x[2]$ being equal or very nearly so to the value of x corresponding to the minimum of the performance curve. There should be three values of x, one of which is 0.0. Test your nonzero values of x with the call

LVxGradientViaCP(x,0.03,0,1,delX,6)

where $delX$ is given the following values:
(a) 0.001;
(b) 0.00001;
(c) 0.00000001
(d) 0.0000000001

3. Write the m-code for the script

LVxModelPtswLine(TestM,TestYInt,xTest,...

Mu,yMu2MuRatio,MStart,yIntStart,NoIts)

as described and illustrated in the text, and test it with the given test calls:

```
function LVxModelPtswLine(TestM,TestYInt,xTest,Mu,...
yMu2MuRatio,MStart,yIntStart,NoIts)
% The line to be modeled is represented in the point-slope form,
% and thus the test line creation parameters are TestM (the slope),
% TestYInt (the test Y-intercept) and the vector xTest.
% Mu is the usual weight for the gradient estimate update
```

```
% term; yMu2MuRatio is a ratio by which to relatively
% weight the partial derivative of y-intercept relative to
% slope; MStart is the initial estimate of slope to use;
% yIntStart is the initial estimate of y-intercept to use,
% and NoIts is the number of iterations to perform.
% The script creates separate 3-D subplots on one figure
% of the performance surface and the coefficient track,
% both plots having the same axis limits enabling
% easy comparison.
% Test calls:
% (a) LVxModelPtswLine(2,0,[-10:1:10],0.014,37,-10,8,12)
% (b) LVxModelPtswLine(2,0,[-10:1:10],0.005,1,-10,8,40)
% (c) LVxModelPtswLine(2,0,[-10:1:10],0.025,1,-10,8,60)
% (d) LVxModelPtswLine(2,0,[-10:1:10],0.025,40,-10,8,30)
% (e) LVxModelPtswLine(20,-90,[-10:1:10],0.005,10,-10,8,40)
% (f) LVxModelPtswLine(2,0,[0:1:10],0.01,50,-10,8,80);
% (g) LVxModelPtswLine(1,-2,[0:1:10],0.01,50,-10,8,80);
```

The results from using LabVIEW for call (c) above, with separate plots for the performance surface and the coefficient track, are shown in Fig. 11.26.

The result from using MATLAB, with the same arguments as in call (c) above, but with the script written to use a single plot having the coefficient track plotted on top of the performance surface, is shown in Fig. 11.27.

4. Write the m-code for a script

$$LVxModelLineLMS\_MBX(M, B, xVec, Mu, bMu2mMuRat, NoIts)$$

in which the algorithm update equations are Eqs. (11.17)–(11.18) in the text. The input arguments M and B are to be used to generate a set of y-values with corresponding to the x-values $xVec$. The input argument *bMu2mMuRat* is a number to boost the value of Mu for updating the estimate of b.

Include code within the iterative loop to control Mu to maintain stability, including code to provide separate values of Mu for updating the estimates of m and b. The following code will return the algorithm to a stable condition if Mu is chosen too high, rolling back the coefficients to the previous values and halving Mu if MSE increases.

```
if Ctr>1
 if MSE(Ctr)>1.02*MSE(Ctr-1)
  m(Ctr+1) = m(Ctr); b(Ctr+1) = b(Ctr);
  Mu = Mu/2
end; end
```

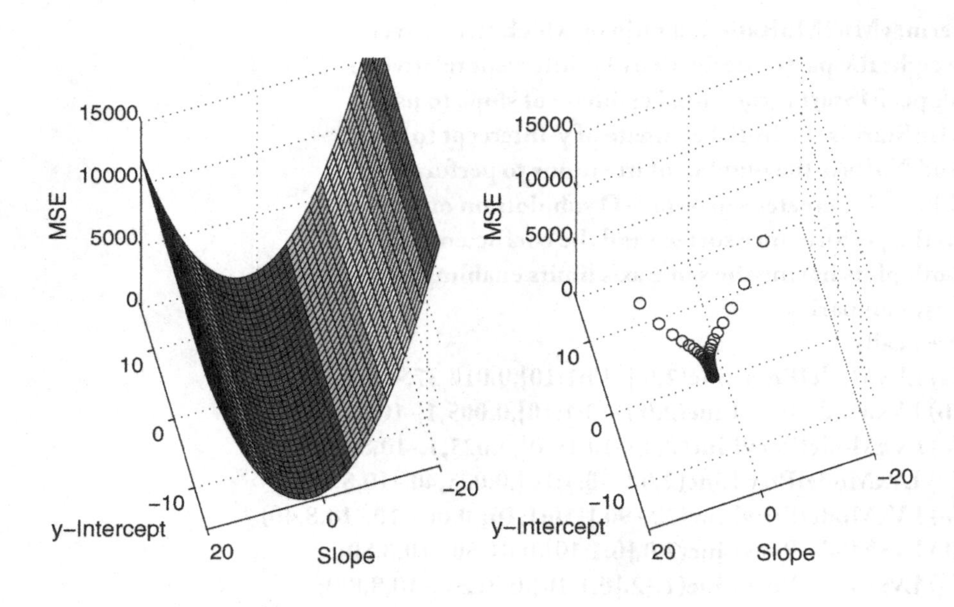

Figure 11.26: (a) Performance Surface; (b) Separate plot of coefficient track along the Performance Surface in (a).

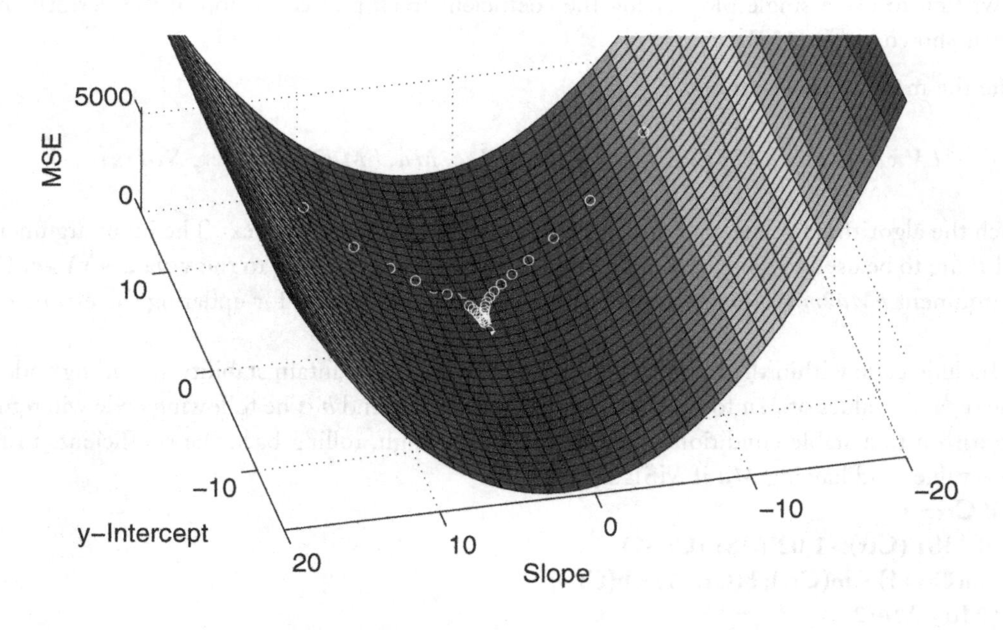

Figure 11.27: Same as the previous figure, except that the coefficient track is plotted onto a plot of the performance surface.

The script should, at the end of each call, create a figure with two subplots, the first of which plots $xVec$ and the corresponding y-values, and a line created using the final estimates of m and b. The second subplot should show MSE v. Iteration.

Test your code with the following calls:

(a) LVxModelLineLMS_MBX(3,-1,[0 1 2 3],1.2, 3,30)
(b) LVxModelLineLMS_MBX(3,-1,[-10:1:10],0.5, 30,5)
(c) LVxModelLineLMS_MBX(3,-1,[0:0.1:2],1.15,3,45)
(d) LVxModelLineLMS_MBX(3,-1,[0:1:2],1.1,2,30)

5. Write the m-code to implement the following script, which creates a two-sample Plant characterized by Plant Coefficients $PC1$ and $PC2$ (as shown in Fig. 11.10) and attempts to model it using a 2-tap LMS adaptive FIR, in accordance with the function specification below. It should create a figure having four subplots which show the test signal, the error signal, and the successive estimates of the plant coefficients, which are given the variable names $c1$ and $c2$. A sample plot, which was created using the call

LVxLMS2TapFIR(4,-3,64,3,[],0.5,0)

is shown in Fig. 11.28.

```
function LVxLMS2TapFIR(PC1,PC2, NoIts,tstDType,CosFrq,...
TwoMu,NoiseAmp)
% PC1 and PC2 are Plant Coefficients, to be modeled by the 2-tap
% FIR; NoIts is the test sequence length in samples
% tstDType selects test signals to pass through the Plant and
% the filter, as follows: 1 = Random Noise; 2 = Unit Step;
% 3 = Nyquist Limit (Fs/2); 4 = Half-Band (Fs/4)
% 5 = cosine of freq CosFrq; 6 = Cos(1Hz)+Cos(Fs/4)+...
% Cos(Fs/2); TwoMu is the update term weight
% Random noise having amplitude NoiseAmp is added to
% the test signal selected by tstDType.
% Test calls:
% LVxLMS2TapFIR(4,-3,64,1,[],0.5,0)
% LVxLMS2TapFIR(4,-3,64,2,[],0.5,0)
% LVxLMS2TapFIR(4,-3,64,3,[],0.5,0)
% LVxLMS2TapFIR(4,-3,64,4,[],0.5,0)
% LVxLMS2TapFIR(4,-3,64,5,[25],0.5,0)
% LVxLMS2TapFIR(4,-3,64,5,[1],0.5,0)
% LVxLMS2TapFIR(4,-3,64,6,[],0.5,0)
% LVxLMS2TapFIR(4,-3,64,0,[],1.15,0)
% LVxLMS2TapFIR(4,-3,7,4,[],1.15,0)
```

6. Write the m-code for the script *LVxLMSAdaptFiltMuCheck* as specified below:

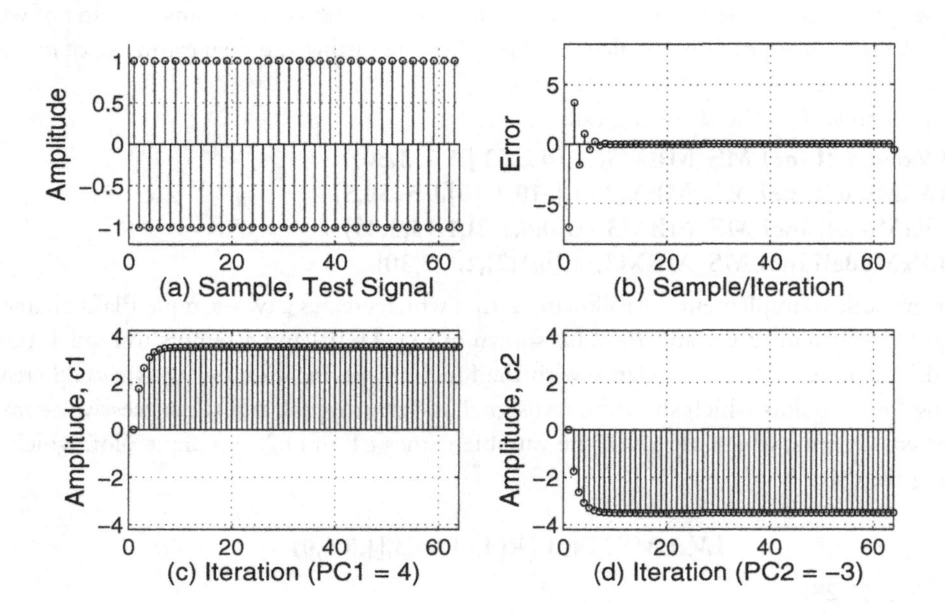

Figure 11.28: (a) Test signal supplied to a two-sample Plant and a 2-Tap LMS Adaptive FIR; (b) Error Signal; (c) Successive estimate plot of first tap coefficient, $c1$; (d) Successive estimate plot of second tap coefficient, $c2$. Note that the test signal is narrow band, and in fact, though the error signal converges to zero, the converged values of $c1$ and $c2$ are incorrect, as expected.

```
function LVxLMSAdaptFiltMuCheck(k,NoTaps,C,SigType,...
Freq,Delay)
% k is the standard deviation of white noise or amplitude of
% a cosine to use as the test signal
% NoTaps is the number of taps to use in the LMS adaptive filter;
% The value for Stepsize is computed as C/(10*P*NoTaps) where
% P = sum(testSig.^2)/length(testSig);
% SigType 0 gives a DC test signal (k*cos(0))
% SigType 1 = random noise, 2 = cosine of frequency Freq
% SigType 3 = Nyquist rate, and 4 = Half-Band Frequency
% The current filter output sample is subtracted from the current
% test signal sample, delayed by Delay samples. Delay must be
% between 1 and NoTaps for the error signal to be able to go
% to zero. Two plots are created, one of the test signal, and
% one of the LMS adaptive FIR error/output signal
% A typical call:
```

```
% LVxLMSAdaptFiltMuCheck(1,100,1,1,[15],50)
% LVxLMSAdaptFiltMuCheck(1,100,4,1,[15],50)
% LVxLMSAdaptFiltMuCheck(1,100,6,1,[15],50)
% LVxLMSAdaptFiltMuCheck(1,100,8,1,[15],50)
% LVxLMSAdaptFiltMuCheck(1,100,9.5,1,[15],50)
% LVxLMSAdaptFiltMuCheck(1,100,10,1,[15],50)
```

7. Write the m-code for the script

$$LVxLMSvNLMS(kVec, NoTaps, Delta, Mu, SigType, Freq, Delay)$$

according to the following function specification, and test it with the given calls.

```
function LVxLMSvNLMS(kVec,NoTaps,Delta,Mu,...
SigType,Freq,Delay)
% kVec is a vector of amplitudes to impose on the test signal as
% a succession of equally spaced amplitude segments over a
% test signal of 15 times the filter length.
% NoTaps is the number of taps to use in the LMS adaptive filter;
% The value for Stepsize is computed as Delta/(P + sm) where
% P = sum(testSig.^2) for the samples of testSig in the filter,
% and this is used in the NLMS algorithm (sm is a small
% number such as 10^(-6). Simultaneously, the test signal is
% processed by an LMS filter of the same length using a
% constant stepsize equal to Mu.
% SigType 0 gives a DC test signal,
% SigType 1 = random noise, 2 = cosine of frequency Freq
% SigType 3 = Nyquist rate, and 4 = Half-Band Frequency
% The current filter output sample is subtracted from the current
% test signal sample, delayed by Delay samples. Delay must
% be between 1 and NoTaps for the error signal to be able to
% go to zero. Five subplots are created, one of the test signal,
% one of the LMS adaptive FIR error/output signal, one of the
% LMS final tap weights, one of the NLMS error signal, and
% one of the final NLMS tap weights.
% Test calls:
% LVxLMSvNLMS([1,5,15,1],100,0.5,0.00005,1,[15],50)
% LVxLMSvNLMS([1,5,25,1],100,0.5,0.00005,1,[15],50)
```

8. Consider the problem of modeling a set of test points $Pt_i = (x_i, y_i)$ with the polynomial

$$y = a_0x^0 + a_1x^1 + a_2x^2 + ... + a_mx^m \tag{11.33}$$

Since Eq. (11.33) will undergo an iterative process, we designate y and the various coefficients as functions of sample or iteration index n.

$$y[n] = a_0[n]x^0 + a_1[n]x^1 + a_2[n]x^2 + ... + a_m[n]x^m \tag{11.34}$$

Equation (11.34) states that the value of y at sample time n, for a certain input x, is equal to the sum of various powers of x, each weighted with a coefficient which is in general a function of iteration n (we seek, of course, to cause all the coefficients, over a number of iterations, to converge to their ideal values, ones which yield the curve which best fits the data points (x_i, y_i)).

We write an expression for total MSE at iteration n, $MSE[n]$, which is the sum of the MSE at each data point:

$$MSE[n] = \frac{1}{N}\sum_{i=1}^{N}(y_i - (\sum_{j=0}^{m} a_j[n]x_i^j))^2 \tag{11.35}$$

Note that $a_j[n]$ and $MSE[n]$ are continuous functions (evaluated at different discrete times or iterations n), and thus may be differentiated. Equation (11.35) may be interpreted as saying that MSE at sample time n is computed by summing, for each test point (x_i, y_i), the square of the difference between the actual test point value y_i and the predicted value of y_i, which is obtained by evaluating the polynomial expression

$$\sum_{j=0}^{m} a_j[n]x_i^j$$

in which x_i is raised in turn to various powers j (0, 1, 2...m), each of which is weighted with the current estimate of the coefficient for that particular power at sample time n, $a_j[n]$.

Using calculus, derive an analytic expression for the partial derivative of MSE with respect to any coefficient a_k and then write an analytic expression for the coefficient update equation to iteratively generate the best estimates of a_k using a true-MSE-based gradient search algorithm, and then test your equations by embodying them in m-code as specified by the function below. Note the additional guidelines given following the function description.

```
function LVxFitCurveViaLMS(cTest,xTest,NoiseAmp,Mu0,...
UseWts,NoIts)
% cTest is a row vector of coefficients for a polynomial in
% xTest, with the first element in cTest being the coefficient
% for xTest to the zeroth power, the second element of cTest
% is the coefficient for xTest to the first power, and so forth;
% NoiseAmp is the amplitude of noise to add to the
% points((xTest[n],yTest[n]) which are generated by the
% polynomial expression using the coefficients of cTest;
% Mu0 is the initial value of Mu to use in the LMS algorithm;
```

```
% UseWts if passed as 1 applies weights to the coefficient
% update equations to speed convergence of the lower power
% coefficients; pass as 0 to not apply any such weights
% NoIts is the maximum number of iterations to perform.
% The script computes the MSE at each iteration, and if it
% increases, the coefficient estimates are rolled back and the
% value of Mu is halved. The brings Mu to an acceptable
% value if it is chosen too large. In order to standardize the
% values of MuO, the initial value of Mu0 is divided by the
% effective power of xTest "in the filter." Additionally, the
% update equation for each coefficient is effectively weighted
% by a factor to cause all coefficients to converge at roughly
% the same rate. The factor is computed as the xTest power
% for the highest power being modeled divided by the
% xTest power for the power whose coefficient is being updated.
%
% Typical calls might be:
%
% LVxFitCurveViaLMS([1, 1, 2],[-5:1:5],0,0.5,1,33)
% LVxFitCurveViaLMS([1, 1, 2],[-5:1:5],0,0.5,0,950)
% LVxFitCurveViaLMS([0, 1, 0, 2],[-5:1:5],0,0.5,1,85)
% LVxFitCurveViaLMS([1, 1, 2],[-1:0.025:1],0,2.75,1,18)
% LVxFitCurveViaLMS([1, 1, -2],[-1:0.025:1],0,4,1,26)
% LVxFitCurveViaLMS([1, 1, -2],[-1:0.025:1],0,1,0,50)
```

The following code can be used to normalize the stepsize to the signal power of $xTest$ (as described in the function specification above) and to generate the relative $xTest$ signal powers to weight the coefficient update equation to accelerate the convergence of the lower powers, which generally converge much more slowly than the highest power (why?). The computations performed by the code are analogous to computing the signal power in the filter as used in the NLMS algorithm. Here, of course, the signal is the constant vector $xTest$.

```
PwrxVec = 0;
for PwrCtr = 0:1:HighestPower
 PartDerPwr(1,PwrCtr+1) = sum(xTest.^(2*PwrCtr));
 PwrxVec = PwrxVec + PartDerPwr(1,PwrCtr+1);
end
PwrxVec = sum(PwrxVec);
Mu(1) = Mu0/PwrxVec; % normalize Mu[1]
% Compute relative signal power for each polynomial coeff
n = 1:1:HighestPower + 1;
```

```
if UseWts==1
MuPwrWts(1,n) = ...
PartDerPwr(1,length(PartDerPwr))./PartDerPwr;
else;
MuPwrWts(1,n) = 1;
end
```

A useful thing in the algorithm is to update Mu each iteration, thus making it a function of discrete time, $Mu[n]$. This can quickly return the algorithm to stability if too large a value of $Mu0$ is chosen. To update $Mu[n]$, compute MSE at each iteration, and if MSE has increased at the current sample, reset (rollback) the coefficient estimates to the previous values, and divide $Mu[n]$ by 2. If MSE is the same or less than for the previous iteration, maintain $Mu[n]$ at the existing value, i.e., let $Mu[N+1] = Mu[n]$. Note that the stepsize for the coefficient update equation, not including the term $MuPwrWts$, is **2*Mu[n]** to work with the values of $Mu0$ in the test calls given above in the function specification.

CHAPTER 12

Applied Adaptive Filtering

12.1 OVERVIEW

In the previous chapter, we explored a number of basic principles of adaptive processes and developed the LMS algorithm for an FIR arranged in a simple system-modeling problem. In this, we'll investigate several applications of the LMS-based adaptive FIR, including:

1. Active Noise Cancellation
2. System Modeling
3. Echo Cancellation
4. Periodic Signal Removal/Prediction/Adaptive Line Enhancement
5. Interference Cancellation
6. Equalization/Inverse Filtering/Deconvolution
7. Dereverberation

We have briefly seen, in the previous chapter, the filter topology (or arrangement) for system modeling, which is the same as that for Active Noise Cancellation (ANC). The converged coefficients of the adaptive filter that model the Plant in a system-modeling problem necessarily produce a filter output the same as that of the Plant. By inverting this signal and feeding it to a loudspeaker in a duct, noise in a duct can be cancelled. Another common use of active noise cancellation is in headphones being used in a noisy environment. Such headphones can be configured to cancel environmental noise entering the ear canal while permitting desired signals from the headphone to be heard.

With Echo Cancellation, useful in duplex communications systems (such as telephony), for example, we introduce a new topology, which involves two intermixed signals, one of which must be removed from the other. With this topology, we introduce the very useful **Dual-H** method, which greatly aids convergence in such systems.

Periodic Signal Removal and Adaptive Line Enhancement (both accomplished with the same filter topology) find applications in communications systems–a sinusoidal interference signal that persists for a significant period of time can be eliminated from a signal that consists of shorter duration components. This is accomplished by inserting a bulk delay (called a decorrelating delay) between the input signal and the adaptive filter so that the filter output and the signal can only correlate for persistent periodic signals. The LMS filter output itself enhances the periodic signal, thus serving as an Adaptive Line Enhancer (ALE). By subtracting the enhanced filter output from the signal (thus creating the error signal), the periodic component of the signal is eliminated.

With Interference Cancellation, we introduce an important concept that can be used in many different environments. An Interference Cancellation arrangement uses two inputs, a signal reference input and a noise reference input. This arrangement, while discussed in this chapter in the context of audio, also applies to any other kind of signal, such as a received RF signal suffering interference. In that case, a signal antenna and a noise antenna serve as the inputs. As the speed and economic availability of higher speed digital sampling and processing hardware increase, applications of digital signal processing at ever-higher RF frequencies will develop. There are nonetheless many applications in the audio spectrum, such as cancelling vehicular noise entering a communications microphone along with the operator's speech, helping to increase the effectiveness and reliability of audio communication.

Another topic finding applicability in communications systems is inverse filtering, which can be used to perform several equivalent functions, such as equalization or deconvolution. Equalization in communications systems, for example, becomes more necessary as systems become more digital, and maintenance of magnitude and phase responses in communications channels becomes critical to accurate data transmission. A common problem in television reception or FM radio reception is ghosting (in the TV environment) or multipath interference (in the FM environment). These are manifestations of the same thing, multiple copies of the signal arriving at the receiving antenna and causing an echo-like response. Since the echos are essentially attenuated duplicates of the signal, it is possible to determine the echo delay times and magnitudes and remove them. To illustrate the general idea, we conclude the chapter with a problem in which we generate a reverberative audio signal (an audio signal mixed with delayed, attenuated copies of itself) using an IIR, and estimate the parameters of the reverberative signal using adaptive filtering and autocorrelation. From these parameters we are able to deconvolve the audio signal, returning it to its original state prior to entering becoming reverberative.

12.2 ACTIVE NOISE CANCELLATION

The script

$$LVxLMSANCNorm(PlantCoeffVec, k, Mu, freq, DVMult)$$

(see exercises below) demonstrates the principle of cancelling noise in a duct or similar Plant, such as that depicted in Fig. 12.1. This kind of simple arrangement can work when the sound travelling down the duct propagates in the simplest mode, with pressure constant across the cross-section of the duct, only varying along the length of the duct. This is accomplished by making sure the duct's cross-sectional dimensions are small enough compared to the longest sound wavelength present to prevent higher propagation modes.

For our simulation of noise cancellation in a duct, we'll use an adaptive LMS FIR filter having 10 taps, with the Plant (duct) Coefficients specified in a vector of 10 values as *PlantCoeffVector*. The input argument *k* specifies an amount of noise to mix with two sinusoids having respective frequencies determined as harmonics of *freq*; *Mu* is the usual scaling factor, and *DVMult* is a vector of amplitudes by which to scale the test signal, the purpose of which is discussed below.

Example 12.1. Simulate active noise cancellation in a duct using the script *LVxLMSANCNorm*.

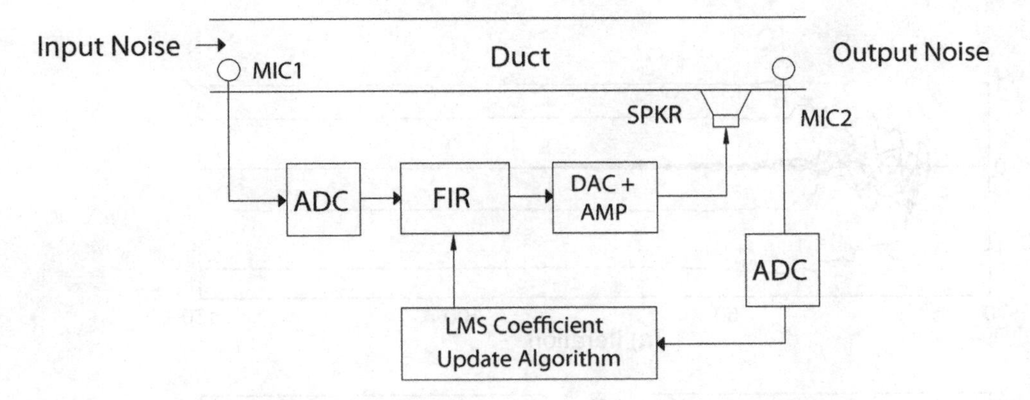

Figure 12.1: In this schematic diagram of a duct noise canceller, MIC1 obtains the noise reference or input for the LMS Adaptive filter (FIR), which is digitized by an analog-to-digital converter (ADC); SPKR (a loudspeaker) emits the negative of the LMS adaptive filter output as a counter-wave to cancel the noise, and MIC2 picks up the residual or error signal, which is also the net duct acoustic output, which is then fed into the LMS coefficient update algorithm block to update the coefficients of FIR.

A typical call might be

LVxLMSANCNorm([0, 0, 1, 0, -0.5, 0.6, 0, 0, -1.2, 0],2,2,3,[1,2,4,8])

which results in Fig. 12.2. It can been seen at plot (a) that all tap coefficients start at a value of zero, and quickly head toward their final values. Tap numbers 3, 5, 6, and 9 can be seen to converge to the same weights found in the Plant's impulse response, *PlantCoeffVec*. Note the random meandering of the coefficient estimates as they slowly, but unerringly, head toward their proper converged values.

The algorithm is an NLMS algorithm (described in the previous chapter), that is to say, it computes the signal power in the filter and uses its reciprocal to scale the coefficient update term; this is what makes the filter stable and immune to the sudden changes in signal amplitude defined by the input argument *DVMult*.

By varying the value of k downward to 0, the deleterious effects on convergence of excessively-low levels of noise in an otherwise-sinusoidal signal can be seen. A typical call that illustrates this, and which results in Fig. 12.3 would be

LVxLMSANCNorm([0,0,1,0,-0.5,0.6,0,0,-1.2,0],0,2,3,[1,2,4,8])

12.3 SYSTEM MODELING

The filter and Plant arrangement in System Modeling is the same as in noise cancellation. Note that in noise cancellation, the noise signal travelled through a system, the Plant, and also through the LMS adaptive filter, which (when properly converged) modeled the Plant's impulse response.

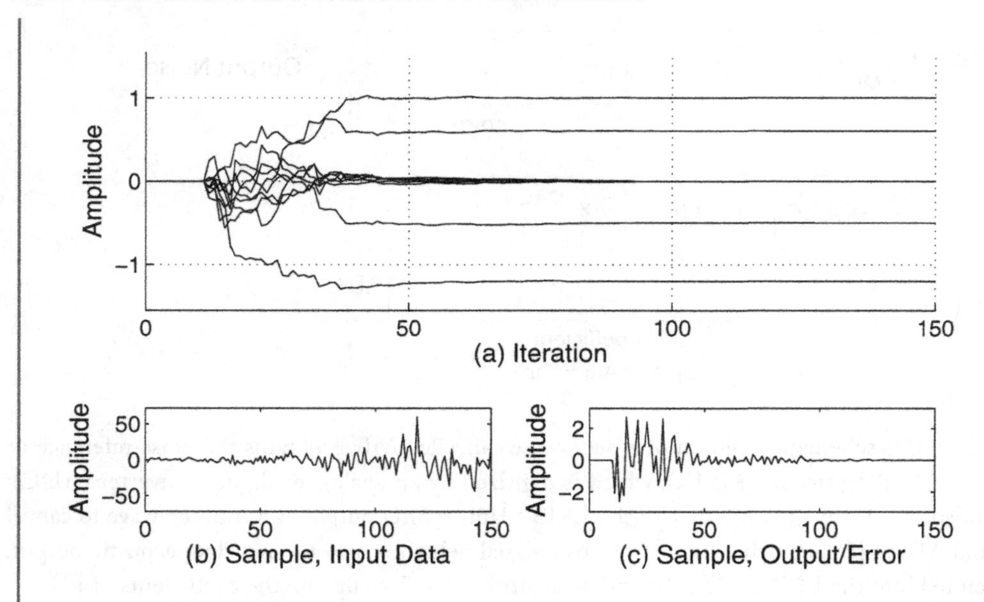

Figure 12.2: (a) All filter Tap Wts versus iteration; (b) Test signal, consisting of noise which has an amplitude profile that increases in steps; (c) The system output (error).

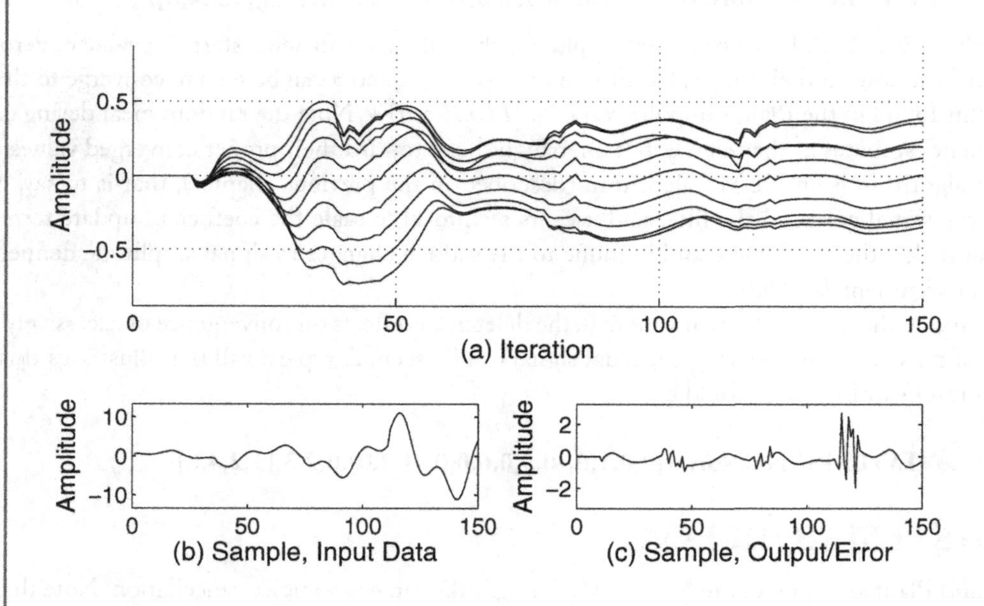

Figure 12.3: (a) Tap Weights versus sample or iteration; (b) Test signal, consisting of sinusoids without noise; (c) The system output, which is the same as the error signal. Note the poor convergence when the input signal is narrowband.

An easy way to obtain any system's impulse response is to send an impulse or other test signal, such as white or pink noise, into the system and record the output. This method is sometimes used in active noise cancellation systems and echo cancellation systems. Often, however, the system cannot be taken offline to determine its impulse response; rather, the system must be modeled (i.e., the impulse response determined) while the system is online, i.e., operating with standard signals rather than test signals.

Example 12.2. A certain system can be accessed only through its input and output terminals. That is to say, we can inject a test signal into its input terminal, and record the resultant output signal, but we cannot otherwise determine what components the system contains. Assume that the system can be modeled as an IIR filter having a specified number of poles and zeros. Devise a method to obtain its poles and zeros.

We first specify a transfer function consisting of one or more poles and zeros to simulate a system (Plant) to be modeled. We then process a test signal, such as noise or speech, with this Plant to produce a corresponding output sequence. Having these two sequences, we can then model a finite (i.e., truncated) version of the impulse response that produced the output sequence from the input sequence. To obtain the poles and zeros, we use Prony's Method. The script (see exercises below)

$$LVxModelPlant(A, B, LenLMS,$$
$$NoPrZs, NoPrPs, Mu, tSig, NAmp, NoIts)$$

allows you to enter a set of filter coefficients *A* and *B* as the numerator and denominator coefficients of the IIR filter that simulates the Plant, the number of LMS FIR filter coefficients *LenLMS* to use to model the Plant, the number of Prony zeros *NoPrZs* and poles *NoPrPs* to use to model the derived LMS coefficients with a pole/zero transfer function, the value of *Mu*, the type of test signal to pass through the Plant and the LMS filter (0 = White Noise, 1 = DC, 2 = Nyquist, 3 = Half-band, 4 = mixed DC, Half-band, and Nyquist), *NAmp*, the amplitude of white noise to mix with the test signal $tSig$, and the number of iterations to do, *NoIts*. The Prony pole and zero estimates are returned in LabVIEW's Output window.

A typical call might be

LVxModelPlant([1,0,0.81],1,100,3,3,0.5,0,0,1000)

which results in Fig. 12.4 and Prony estimates of b = [0.9986, -0.0008, -0.0013] and a = [1, -0.0003, 0.8099] which round off to b = [1] and a = [1, 0, 0.81].

The same call, modified to call for any test signal other than white noise, generally causes the LMS modeling to fail. Injecting white noise into such an input signal will assist convergence.

Using an LMS algorithm to directly model an IIR-based system is possible, although convergence can be protracted and uncertain; performance surfaces can have localized minima leading

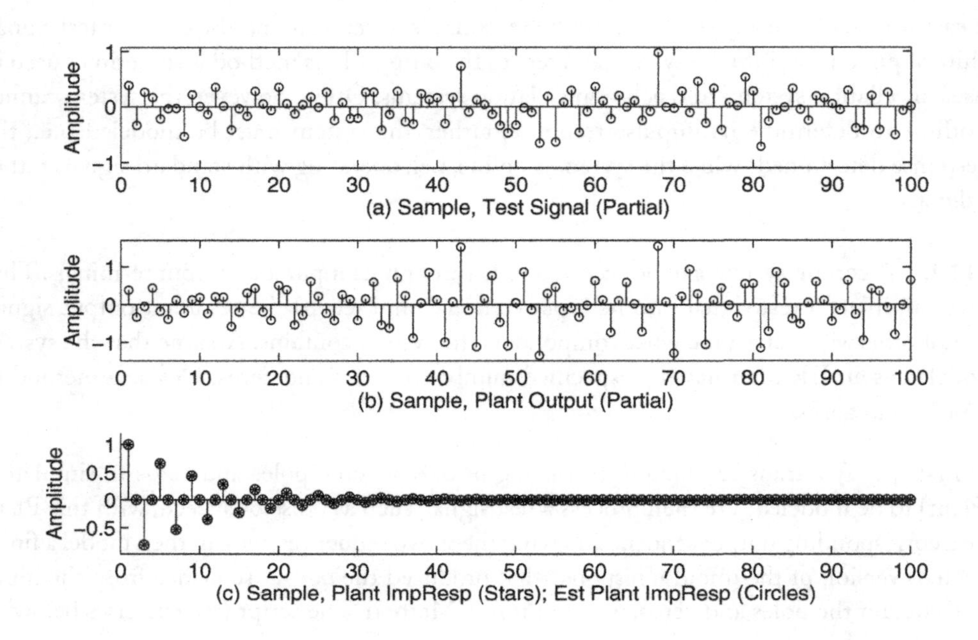

Figure 12.4: (a) Test Signal; (b) Response of Plant to signal at (a); (c) Truncated (actual) impulse response of Plant (stars) and LMS-estimated Plant impulse response (circles).

to incorrect solutions. The method used in this example, although unsophisticated, has the stability and convergence speed of an FIR-based LMS algorithm. Reference [2] supplies several LMS algorithms, including one using an IIR.

12.4 ECHO CANCELLATION

The most common venue for unwanted echo generation is telephone systems, especially loudspeaker telephones. Feedback or howlaround (like the feedback in an auditorium PA system) occurs due to the complete acoustic circuit formed by the loudspeaker and microphone at each end of the conversation. Figure 12.5 shows the basic arrangement.

Referring to Fig. 12.5, we see that a person at the Far End (left side of the Fig.) speaks into microphone M1, the sound is emitted at the Near End by loudspeaker S2, is picked up by microphone M2, and is sent (in the absence of effective echo cancellation at the Near End) down the wire toward the Far End, where it is emitted by loudspeaker S1, and then picked up again by microphone M1, thus forming a complete feedback or echo loop. The adaptive filter ADF2 at the Near End models the acoustic feedback path at the Near End, and trains according to the error signal (the output of the summing junction at the Near End), which is the same as the signal being sent down the wire to the Far End. Since the input to adaptive filter ADF2 is the Far End signal, and

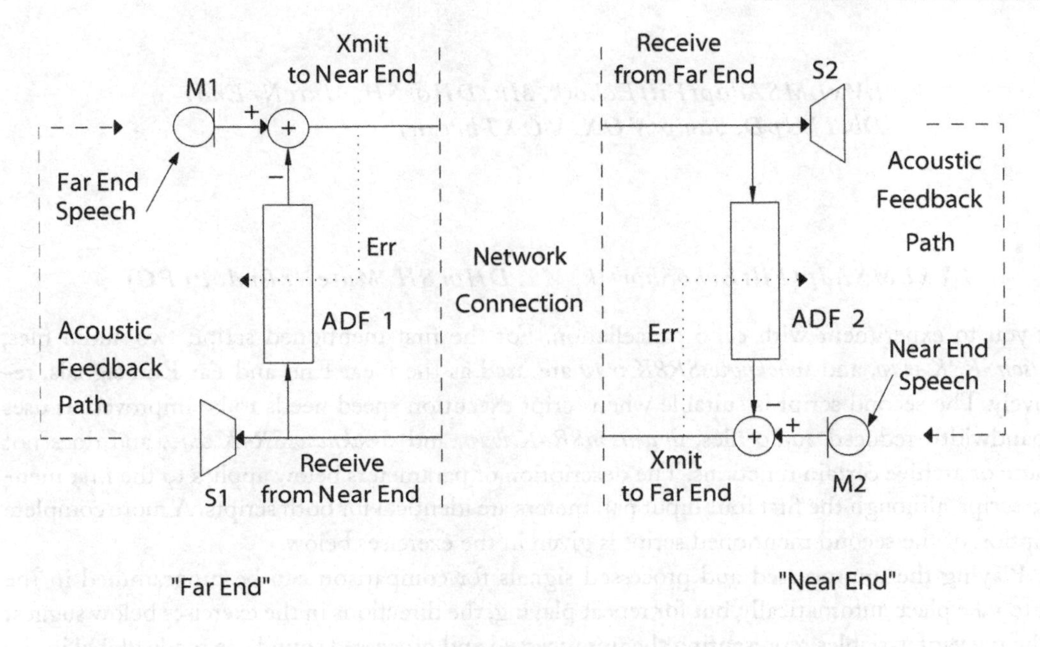

Figure 12.5: A basic echo cancelling system employed in a telephone system in which both subscriber terminal units are loudspeaker phones, thus leading to the likelihood of acoustic echo, necessitating an echo canceller.

only a delayed version of it (we are assuming here that the person at the Near End is not speaking) is entering the error junction, the filter converges well.

Problems arise when the person at the Near End speaks–then we have a signal entering the error junction that is not also entering the adaptive filter. The signal from the error junction output thus does not convey an accurate representation of how well the filter is doing in cancelling the Far End signal. As a result, a perfectly converged set of coefficients, which cancel the Far End signal very well, can be caused to severely diverge due to Near End speech.

The patent literature is replete with patents which attempt to solve or regulate the problem caused by Near End speech. The sheer number of such patents suggest that the problem has not been completely solved. Prior to the advent of echo cancellers, a method was used known as **Echo Suppression**. In this method, the levels of Near and Far end signals are compared, and the dominant (usually louder) direction is designated to be the active circuit, and the gain of the nondominant direction is greatly reduced to the point that howlaround no longer exists.

Returning to the problem of simulating an echo cancellation system, the scripts

$$LVxLMSAdaptFiltEcho(k, Mu, DHorSH, MuteNrEnd, ...$$
$$DblTkUpD, SampsVOX, VOXThresh)$$

and

$$LVxLMSAdptFiltEchoShort(k, Mu, DHorSH, MuteNrEnd, PrPC)$$

allow you to experiment with echo cancellation. For the first mentioned script, two audio files, *drwatsonSR8K.wav*, and *whoknowsSR8K.wav* are used as the Near End and Far End signals, respectively. The second script is suitable when script execution speed needs to be improved; it uses two bandwidth-reduced audio files, *drwatsonSR4K.wav*, and *whoknowsSR4K.wav,* and does not compute or archive certain functions. The description of parameters below applies to the first mentioned script, although the first four input parameters are identical for both scripts. A more complete description of the second mentioned script is given in the exercises below.

Playing the unprocessed and processed signals for comparison can be programmed in the script to take place automatically, but for repeat playing, the directions in the exercises below suggest that the relevant variables representing the unprocessed and processed sounds be made global in the script. Then, by declaring them global on the Command line, it is a simple matter to play them using the function *sound.* To convey to the user the appropriate variable names and the sample rate, these are programmed in the script to be printed in the Command Window in MATLAB or Output Window in LabVIEW. For example, after making the call

LVxLMSAdptFiltEchoShort(0.01,0.2,1,1,50)

in LabVIEW, the following is printed in the Output Window:

global sound output variable is EchoErr & sample rate is 8000

To listen to the processed audio at will, enter the following lines, pressing Return after each line:

global EchoErr
sound(EchoErr,8000)

The method for playing audio signals just described also holds true for the other scripts in this chapter that produce audio outputs. When using MATLAB, it is possible to create pushbuttons on the GUI with callback routines to play the audio files on demand (this method is also implemented using global variables).

Both scripts simulate the acoustic path at the Near End as a single delay with unity gain. In the figures below, the plots of coefficients show, under ideal conditions, one coefficient converging to a value of 1, and the nine other coefficients converging to the value 0. As we'll see, however, conditions are not always ideal and often the coefficients remain misadjusted. Prior to conducting a systematic exploration of why this is and what can be done about it, we'll define the input arguments of the script *LVxLMSAdaptFiltEcho*:

- The argument k allows you to choose an amount of white noise to mix with the Far End signal to assist the LMS adaptive filter in converging to a set of coefficients which models the acoustic path between S2 and M2 as shown in Fig. 12.5.

- The argument Mu is the usual update term weighting constant for the LMS algorithm.

- The argument *DHorSH* tells the script whether or not to use a single LMS adaptive filter (*Single-H*), or to use the *Dual-H* mode, in which an on-line filter and an off-line filter are used, and a figure-of-merit called ERLE is computed to determine how effectively the echo signal is being removed from the error signal, which is the signal being returned to the Far End from the Near End, along with Near End speech. A detailed discussion of the difference between the Single-H and Dual-H modes, along with a mathematical definition of ERLE, is found below.

- *MuteNrEnd* can take on the values 0, 1, or 2; if specified as 0, the Near End signal is emitted continuously into microphone M2. If *MuteNrEnd* is specified as *1*, the Near End signal is muted for a brief period at the beginning of echo cancellation, thus giving the Far End signal a brief period without Near End speech in which to try to converge. If *MuteNrEnd* is specified as 2, the Near End signal is muted once about one-third of the way through. This muting is meant to simulate near silence from the Near End speaker (person); such silence assists the LMS adaptive filter in converging to a set of coefficients which model the acoustic path between the loudspeaker S2 and microphone M2 at the Near End.

- The argument *DblTkUpD*, if passed as 1, prevents coefficient update in Single-H mode whenever Near End speech is detected. If passed as 0, coefficient update is permitted even when Near End speech is present.

- The remaining two arguments, *SampsVOX*, and *VOXThresh*, are two parameters you can manipulate to specify how Near End Speech is detected. The first argument, *SampsVOX*, tells how many samples of the Near End speech signal to use in computing the RMS value, which is then compared to the second argument, *VOXThresh*, which is a threshold below which it is judged that Near End speech is not present.

12.4.1 SINGLE-H

The *Single-H* mode employs a single LMS adaptive filter to cancel a Far End acoustic echo that is "contaminated" with Near End speech. Such an arrangement performs poorly unless coefficient update is stopped whenever Near End speech is detected. The script *LVxLMSAdaptFiltEcho* allows you to specify *Single-H* and whether or not to allow coefficient update during Near End speech (sometimes called "doubletalk," since both Near End and Far End are making sound).

Example 12.3. Demonstrate adaptive echo-cancelling using the Single-H mode.

Suitable calls to use the *Single-H* mode with coefficient update during doubletalk would be

LVxLMSAdaptFiltEcho(0.02,0.2,0,0,0,50,0.02)

LVxLMSAdptFiltEchoShort(0.02,0.2,0,0,50)

Figure 12.6 shows the result of making the first call (the second script call plots only the output/error signal) for *Single-H* echo cancellation with update during doubletalk. Note that in plot (a) the coefficient values in the adaptive filter, which should converge to essentially constant values, are overladen with noise. Note that the Single-H mode does not use the parameter ERLE, but ERLE is computed and displayed for comparison with the Dual-H mode, in which ERLE increases monotonically as better coefficient estimates occur.

Example 12.4. Demonstrate adaptive echo-cancellation using the Single-H mode, but freeze coefficient update during doubletalk.

Freezing coefficient update during doubletalk results in considerable improvement, as can be seen in Fig. 12.7, which resulted from the script call

LVxLMSAdaptFiltEcho(0.02,0.2,0,0,1,50,0.02)

Here we see that the filter coefficients, shown in plot (a), do assume reasonably constant values, but with a much decreased level of noise, since coefficient update during Near End speech was inhibited, as shown in plot (d), where a value of zero at a given sample or iteration means that coefficient update was inhibited for that sample/iteration.

12.4.2 DUAL-H

The Dual-H mode is a simple way to overcome the difficulties caused by Near End speech. In the Dual-H mode, a main filter or on-line filter, which actually generates the counter-echo signal which is subtracted from the Near End's microphone output (M2 in Fig. 12.5), has its coefficients updated only when a second, off-line filter, which adapts continuously (i.e., computes a new set of coefficients for each signal sample), generates a set of filter coefficients which result in a better measure of merit, such as ERLE (Echo Return Loss Enhancement), which is defined here and in the script *LVxLMSAdaptFiltEcho* (and others that follow in this chapter) as the ratio of Far End power (or energy) to Near End (output of LMS error junction) power (or energy). These values can, for better effect, be averaged over a number of samples (in the scripts mentioned above, the

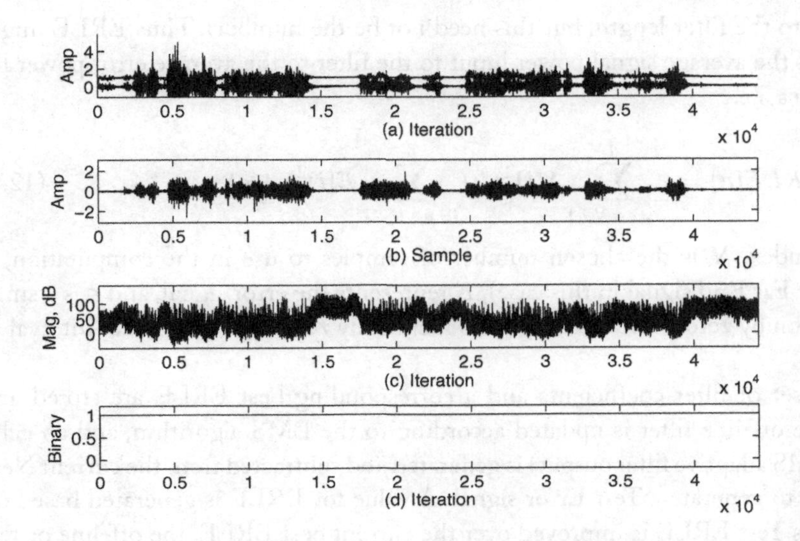

Figure 12.6: (a) All of the adaptive filter coefficients versus sample or iteration; (b) Error signal; (c) ERLE, a figure of merit representing how much of the Far End signal has been removed from the signal; (d) Function "Coefficient Update Permitted?" A value of 0 means that coefficient update was inhibited due to the presence of Near End speech, while a value of 1 means coefficient update was allowed to proceed at the particular iteration or sample number.

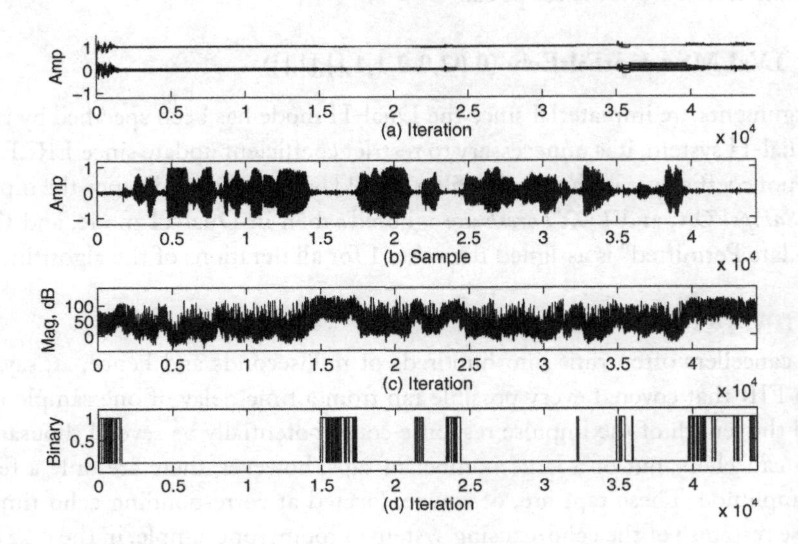

Figure 12.7: (a) All of the adaptive filter coefficients versus sample or iteration; (b) Error signal; (c) ERLE. Note that, due to doubletalk, there was relatively little coefficient update during the course of the conversation, but the performance was improved considerably over the constant coefficient update scenario; (d) Function "Coefficient Update Permitted?"

number chosen is equal to the filter length, but this need not be the number). Thus ERLE might be defined as the ratio of the average signal power input to the filter to the average error power for the corresponding samples, i.e.,

$$ERLE[n] = \left(\sum_{i=n-N+1}^{n} S[i]^2 \right) / \left(\left(\sum_{i=n-N+1}^{n} E[i]^2 \right) + \epsilon \right) \tag{12.1}$$

where n is the sample index, N is the chosen number of samples to use in the computation, S represents the signal (the Far End signal in this case), E represents the error signal, and ϵ is a small number to prevent division by zero if the error signal is identically zero over the subject interval of samples.

The current best set of filter coefficients and a corresponding best ERLE are stored and for each new sample, the off-line filter is updated according to the LMS algorithm, and an echo prediction signal (i.e., LMS adaptive filter output) is generated and subtracted from the current Near End microphone output to generate a Test Error signal. A value for ERLE is generated based on the Test Error, and if this Test ERLE is improved over the current best ERLE, the off-line or test coefficients are stored into the on-line filter and also as the current best coefficients. The best ERLE is replaced by the newly computed Test ERLE.

Example 12.5. Demonstrate adaptive echo-cancelling using the Dual-H mode.

Figure 12.8 shows the results from the script call

LVxLMSAdaptFiltEcho(0.02,0.2,1,1,[],[],[])

in which the last three arguments are immaterial since the Dual-H mode has been specified by the third argument. In the Dual-H system, it is unnecessary to restrict coefficient update since ERLE is used to judge whether or not coefficients in the on-line filter should be updated, and hence the input parameters *DblTkUpD*, *SampsVOX*, and *VOXThresh* are ignored when in Dual-H mode, and the function "Coefficient Update Permitted" is assigned the value 1 for all iterations of the algorithm.

12.4.3 SPARSE COMPUTATION

The plant delay in echo cancellers often runs into hundreds of milliseconds and hence, at, say, a sample rate of 8 kHz, an FIR that covered every possible tap from a time delay of one sample up to and somewhat beyond the length of the impulse response could potentially be several thousand taps long. Often in echo cancellers, out of a large number of taps, however, there are only a few taps having significant amplitude. These taps are, of course, located at corresponding echo times represented by the impulse response of the echo-causing system (a room, for example, in the case of a speakerphone). A common technique to greatly reduce the needed computation is to estimate the locations of the taps having significant amplitudes, and to restrict tap computation to relatively short runs of contiguous taps around each significant tap location. This technique also helps to ensure

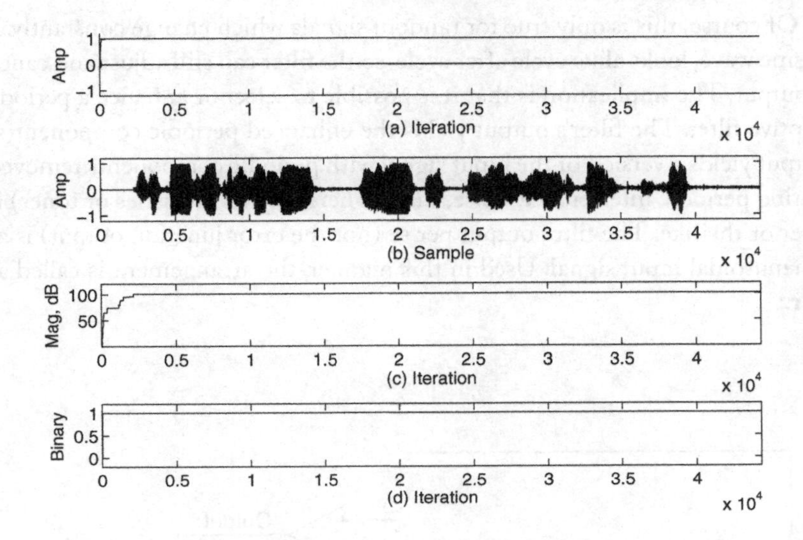

Figure 12.8: (a) All of the adaptive filter coefficients versus iteration; (b) Error signal; (c) Current Best Value of ERLE. Note that the brief introductory portion of the signal was muted, allowing the system to quickly converge to an excellent set of coefficients, which were not improved upon as ERLE remained constant after the initial convergence; (d) Function "Coefficient Update Permitted?"

proper convergence since all noncomputed taps are inherently valued at zero. The selection of taps to be computed, and thus also those not to be computed, is itself a filtering process. One method to estimate the relevant taps is to cross correlate the Far End signal with the signal coming from the Near End microphone, which contains an echoed version of the Far End signal plus the Near End signal.

As sparse computation largely arises from practical and empirical knowledge and practice, an excellent source for information is the website www.uspto.gov, which allows you to do word searches for relevant U.S. Patents, which contain much information not found in other sources. A number of patents relating to sparse computation, which may be viewed or downloaded from www.uspto.gov, are cited below. You may also conduct your own searches. The best (most focused and pertinent) results are usually obtained when a very specific search term is used, such as "pnlms," "nlms," "lms," etc. Advanced searches can also be performed using multiple terms, such as "sparse" and "echo," for example.

12.5 PERIODIC COMPONENT ELIMINATION OR ENHANCEMENT

Figure 12.9 shows the basic arrangement for predictive filtering. Since the input to the filter is delayed, it is not possible for it to adapt to the input signal in time to cancel the input signal at the error summing junction. Of course, this is only true for random signals which change constantly. A periodic signal, such as a sine wave, looks alike cycle after cycle, so the filter can still adjust and cancel it in the error junction's output. The implication is that it is possible to select or enhance a periodic component with the adaptive filter. The filter's output yields the enhanced periodic component(s), and the error junction output yields a version of the input signal with periodic components removed. This is useful for eliminating periodic interference/noise, such as heterodynes (whistles or tones) in a communications receiver or the like. The filter output per se (not the error junction output) is an enhanced version of any sinusoidal input signal. Used in this manner, the arrangement is called an **Adaptive Line Enhancer**.

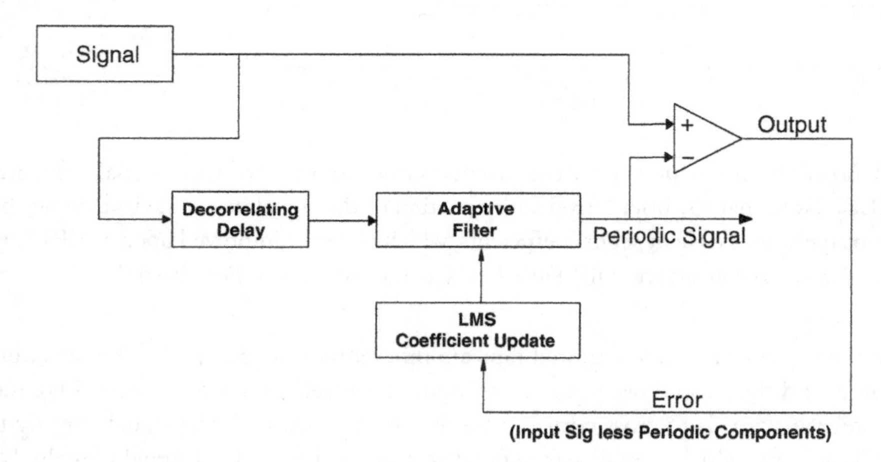

Figure 12.9: A basic arrangement which uses an LMS Adaptive filter to eliminate only periodic components from a signal.

The scripts

$$LVxLMSAdaptFiltDeCorr(k, Mu, freq, NoTaps, DeCorrDelay)$$

and

$$MLx\_LMSAdaptFiltDeCorr(k, Mu, freq, NoTaps, DeCorrDelay)$$

(see exercises below) demonstrate elimination of periodic signals mixed with white noise, leaving mainly the white noise. The first input argument specifies the amplitude of white noise, the second argument is *Mu*, and the third argument, $freq$, is used to generate two sine waves having, respectively, frequencies of *freq* and 3**freq*. The sampling rate is 1024 for this script, so the maximum number you can use for the third argument is about 170 (without aliasing). The fourth argument allows you to specify the number of taps used in the filter, up to 100, and the fifth argument allows you to specify the number of samples of decorrelating delay. In the MATLAB-suitable script *MLx_LMSAdaptFiltDeCorr*, the various signals may be heard by pressing buttons on the GUI created by the script. In the LabVIEW-suitable script, the sounds play automatically, and the sound files are global variables that can be played again by declaring the variables global on the Command line and then giving suitable Commands. An example of this is given immediately below.

Example 12.6. Using the script *LVxLMSAdaptFiltDeCorr*, demonstrate periodic component elimination. Use a decorrelating delay of 2 samples.

Suitable calls would be

LVxLMSAdaptFiltDeCorr(0.2,1,100,10,2)

or

MLxLMSAdaptFiltDeCorr(0.2,1,100,10,2)

which result in Figs. 12.10 and 12.11. Much of the amplitude in the input signal is from the two periodic components. After the filter converges (very quickly in this case), only the residual white noise is left; this is also evident from the respective DFTs, plotted to the right.

After making the script call above, the following information will be printed in the Command Window:

Comment =
global sound variable names are LMSDeCorDataVec, Err, and FiltOut
Comment =
global sound variable sample rate is 8000

This information can be used to listen to the audio files on demand, as described earlier in the chapter. Note that the sample rate in the script is 1024 Hz, but LabVIEW only accepts standard sample rates such as 8000, 11025, 22050, and 44100. Thus the sounds produced by LabVIEW will be higher in pitch by the factor 8000/1024 and shorter in duration by the factor 1024/8000. When writing and running the MATLAB-suited script, you can use the actual sample rate of 1024 and use pushbuttons on the GUI or use the global variable method described above.

Example 12.7. Show that when the decorrelating delay is zero samples, the script *LVxLMSAdaptFiltDeCorr* will no longer decorrelate the filter and algorithm update inputs, and hence the entire input signal will be cancelled.

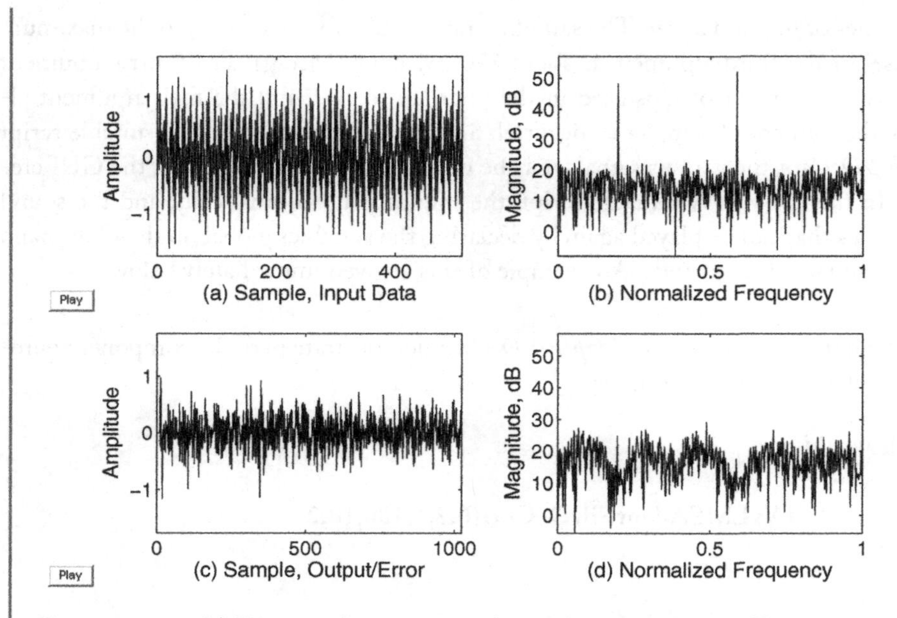

Figure 12.10: (a) Test signal, consisting of periodic components of high magnitude and noise of much lower magnitude; (b) Spectrum of the test signal; (c) Output/Error; (d) Spectrum of Output/Error, showing only noise.

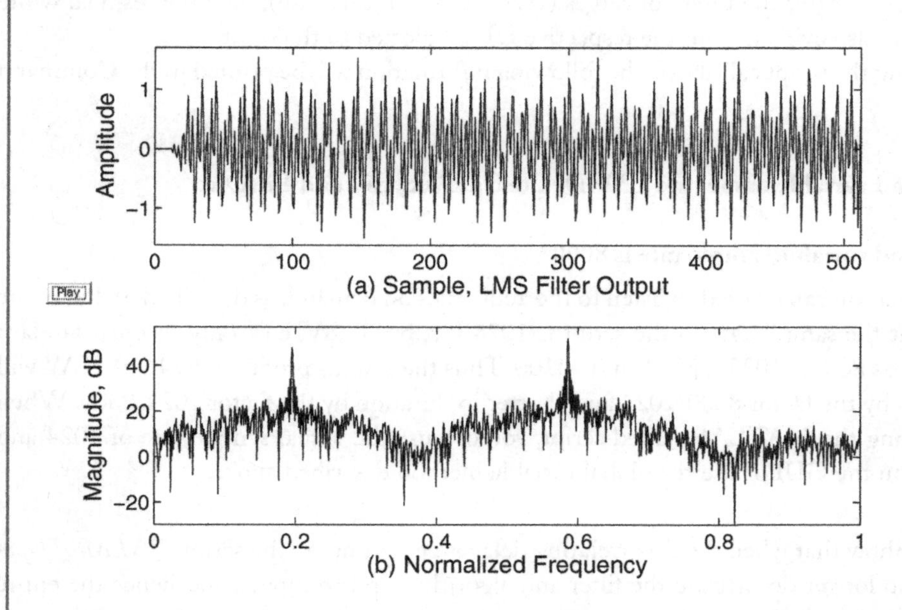

Figure 12.11: (a) The output of the LMS adaptive FIR, showing periodic signals enhanced by the attenuation of noise; (b) Spectrum of signal in (a). Note the attenuation of noise, in contrast to the unattenuated noise-and-signal spectrum shown in Figure 12.10, plot (b).

The call

LVxLMSAdaptFiltDeCorr(0.2,1,100,10,0)

which specifies the decorrelating delay as 0, results in the filter being able to cancel both periodic and nonperiodic components, as shown in Fig. 12.12 (this call essentially configures the filter as an Active Noise Canceller).

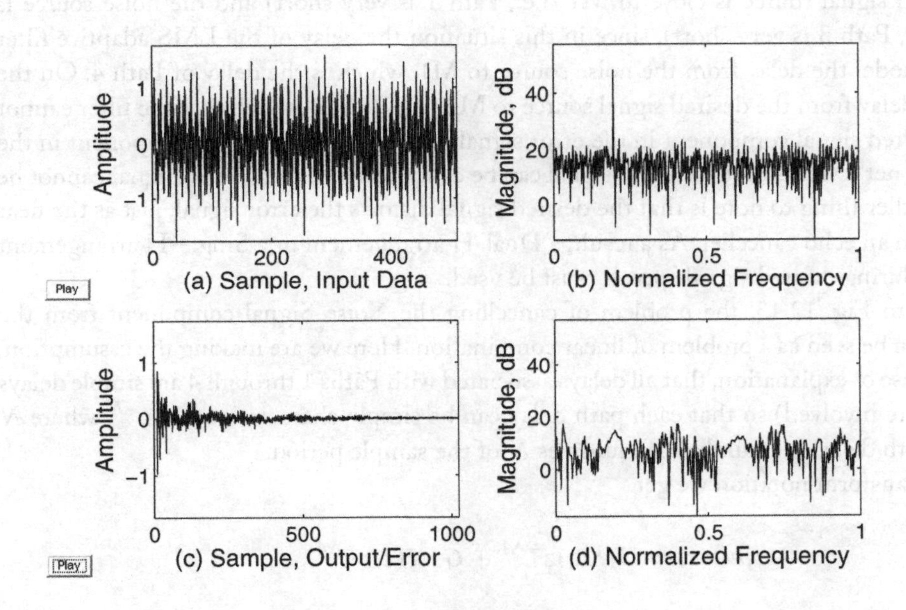

Figure 12.12: (a) Test signal, consisting of periodic components of high magnitude and noise of much lower magnitude; (b) Spectrum of the test signal; (c) Output/Error; (d) Spectrum of Output/Error, showing attenuation of both periodic and aperiodic components.

Two scripts, *LVxLMSDeCorrHeterodyne* and *LVxLMSDeCorrHetDualH*, are developed in the exercises below to explore periodic component elimination in more detail.

12.6 INTERFERENCE CANCELLATION

Another use for adaptive filtering is to cancel interference from a channel signal which is a mixture of some desired signal and an interference signal. A reference signal is obtained for the interference, and it is passed through the adaptive filter and the result subtracted from the channel signal. In the general case, the reference signal also contains a component of the desired signal.

Signal Plus Noise and Noise Plus Signal Channels

Figure 12.13 shows a basic arrangement for interference cancellation. Interference cancellation is implemented by using a separate input which serves as a reference signal for the noise (interference) component. The reference signal, ideally, consists only of the noise signal and is unpolluted with the desired signal component. In many, if not most practical cases, this will not be true, but effective cancellation is usually possible. The general case is shown here, where some of the desired signal mixes with the noise reference, and of course some of the noise mixes with the desired signal (if it didn't, there would be no interference needing to be cancelled). This arrangement works, delay-wise, when the desired signal source is close to M1 (i.e., Path 1 is very short) and the noise source is close to M2 (i.e., Path 3 is very short), since in this situation the delay of the LMS adaptive filter can effectively model the delay from the noise source to M1, which is the delay of Path 4. On the other hand, the delay from the desired signal source to M2 (via Path 2) is such that the filter cannot correlate the desired signal component in the error signal with the desired signal component in the filter. Hence, the net result is that the noise signal can be cancelled, and the desired signal cannot be cancelled. The other thing to note is that the desired signal distorts the error signal, just as the near end signal does in an echo canceller. As a result, a Dual-H arrangement or a Single-H arrangement with no update during desired signal speech must be used.

Referring to Fig. 12.13, the problem of cancelling the Noise Signal component from the Output signal can be seen as a problem of linear combination. Here we are making the assumption, for purposes of ease of explanation, that all delays associated with Paths 1 through 4 are simple delays (i.e., no echoes are involved) so that each path delay can be simply characterized as z^{-N} where N represents the path delay as a number of multiples N of the sample period.

Using z-transform notation we get

$$Output(z) = G_1 S(z) z^{-\Delta 1} + G_4 N(z) z^{-\Delta 4} \ldots$$

$$-(G_3 N(z) z^{-\Delta 3} + G_2 S(z) z^{-\Delta 2}) LMS(z)$$

where G_1, for example, means the gain over Path 1 ($G_i = 1$ means no attenuation, $G_i = 0$ means complete attenuation), $Output(z)$ is the z-transform of the output signal, $S(z)$ means the z-transform of the Desired Signal, $N(z)$ means the z-transform of the noise, $z^{-\Delta 1}$ represents the delay of Path 1, etc., and $LMS(z)$ means the z-transform of the LMS adaptive filter.

When the filter has converged, the Noise component that is mixed with the Desired Signal is cancelled in the output, and hence it is possible to say that

$$G_4 N(z) z^{-\Delta 4} = G_3 N(z) z^{-\Delta 3} LMS(z)$$

which simplifies to

$$LMS(z) = G_4 N(z) z^{-\Delta 4} / (G_3 N(z) z^{-\Delta 3})$$

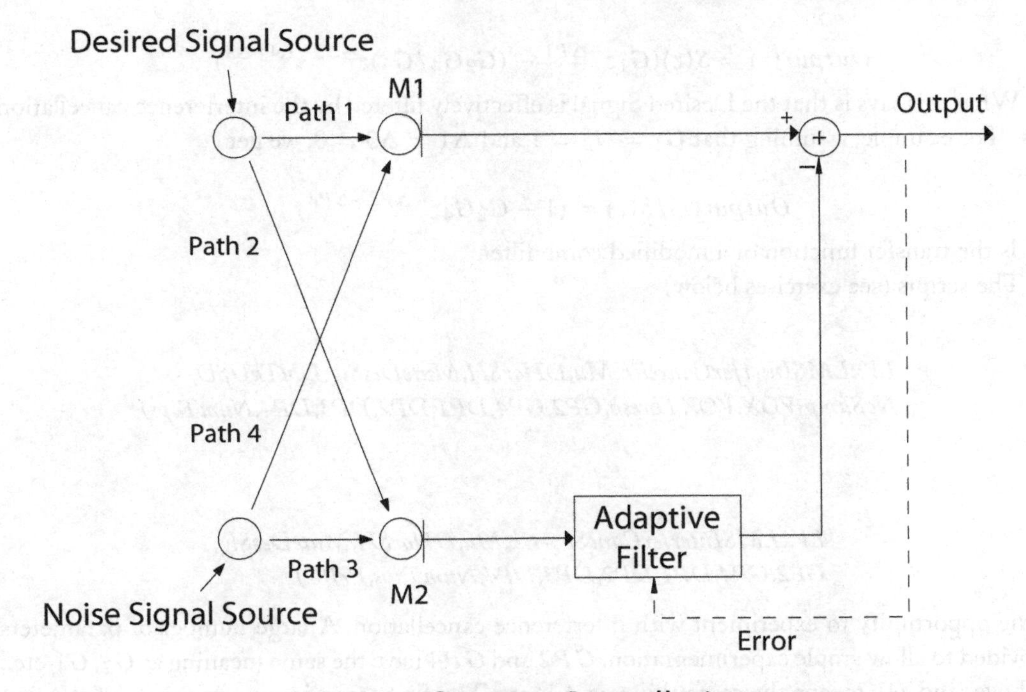

Figure 12.13: A basic interference-cancelling arrangement.

and finally

$$LMS(z) = \frac{G_4}{G_3} z^{-\Delta 4 + \Delta 3}$$

The important thing to note is that the LMS adaptive filter imparts time *delay* only, not time advancement, which means that the quantity $(-\Delta 4 + \Delta 3)$ must remain negative and have a value (in terms of number of samples) that lies within the capability of the filter.

For example, if $\Delta 4$ = 8 samples, and $\Delta 3$ = 1 sample, then the quantity $-\Delta 4 + \Delta 3$ = -7 which means that the LMS adaptive filter must be able to impart a delay of 7 samples. If $\Delta 3$ is greater than $\Delta 4$, of course, the filter would be expected to advance signals in time, which it cannot do (if $\Delta 3 = \Delta 4$, the filter must have a zero-delay tap).

At proper convergence, the Noise Signal is completely cancelled and the output becomes

$$Output(z) = G_1 S(z) z^{-\Delta 1} - G_2 S(z) z^{-\Delta 2} \frac{G_4}{G_3} z^{-\Delta 4 + \Delta 3}$$

which simplifies to

$$Output(z) = S(z)(G_1 z^{-\Delta P1} - (G_2 G_4 / G_3) z^{-\Delta 2 - \Delta 4 + \Delta 3})$$

What this says is that the Desired Signal is effectively filtered by the interference cancellation process. For example, assuming that $G_1 = G_3 = 1$ and $\Delta 1 = \Delta 3 = 0$, we get

$$Output(z)/S(z) = (1 - G_2 G_4 z^{-\Delta P2 - \Delta P4})$$

which is the transfer function of a modified comb filter.

The scripts (see exercises below)

LVxLMSInterferCancel(k,Mu,DHorSH,MuteDesSig,DblTkUpD, NoSampsVOX,VOXThresh,GP2,GP4,DP1,DP2,DP3,DP4,NumTaps)

and

LVxLMSInterferCancShort(k,Mu,DHorSH,MuteDesSig, GP2,GP4,DP1,DP2,DP3,DP4,NumTaps,PrPC)

offer the opportunity to experiment with interference cancellation. A large number of parameters are provided to allow ample experimentation. $GP2$ and $GP4$ have the same meaning as G_2, G_4, etc., used above, and $D1$ means the same thing as $\Delta 1$, etc. The first seven input arguments of the first mentioned script above are the same as the first seven arguments of the script *LVxLMSAdaptFiltEcho*.

For purposes of improving computation speed, the second script, *LVxLMSInterferCancShort,* uses bandwidth reduced audio files ('drwatsonSR4k.wav' and 'whoknowsSR4K.wav') and does not deal with doubletalk or archive parameters such as the figure of merit (IRLE or Interference Return Loss Enhancement). The input argument $PrPC$ is the percentage of the audio files to process when running the script (30% to 50% will usually suffice to demonstrate the required results). More detail on the latter script, *LVxLMSInterferCancShort* will be found in the exercises below, where the student is guided to create the m-code for the script.

The argument, *NumTaps*, found in both scripts, allows you to specify the number of taps used by the adaptive filter. For proper cancellation, it should be equal to at least *DP4* - *DP3*. For example, if *DP4* = 42, and *DP3* = 7, *NoTaps* should be 35 or greater.

For purposes of reducing the number of required input arguments, the gains of paths P1 and P3 are assigned the value 1, with realistic values of G_2 and G_4 being fractions of 1; this corresponds approximately to the situation in which each microphone is very close to its respective sound source, which results in the highest signal-to-noise ratio for M1, and the highest noise-to-signal ratio for M2, which would result in the smallest amount of comb filtering of the desired signal in the output.

Example 12.8. Simulate interference cancellation using the script *LVxLMSInterferCancel.* Make a call that shows interference cancellation with little comb filtering, and make a second call that shows pronounced comb filtering.

For the first task, we make the following call, which has relatively low gain (1/16) for paths 2 and 4, and which specifies use of the Dual-H mode with immediate muting to speed convergence, and which results in Fig. 12.14.

LVxLMSInterferCancel(0.02,0.3,1,1,0,50,0.03,1/16,1/16,6,24,6,24,18)

A call requiring less computation, and which plots only the filtered output signal (see exercises below for further details) is

LVxLMSInterferCancShort(0.02,0.3,1,1,1/16,1/16,6,24,6,24,18,30)

The spectrum of the output signal is shown in Fig. 12.15 in plot (a); this may be contrasted with the spectrum shown in plot (b), which arises from the call

LVxLMSInterferCancel(0.02,0.3,1,1,0,50,0.03,1,1,1,6,1,6,18)

in which the paths P2 and P4 are given gains of 1.0 (i.e., no attenuation). A comb-filtering effect is clearly visible in plot (b), where the cross-paths P2 and P4, like paths P1 and P3, have gains of 1.0. In realistic situations, where the gains of paths P2 and P4 are small compared to those of P1 and P3, the comb-filtering effect is greatly attenuated, as is evident in plot (a) of Fig. 12.15.

Reference [1] gives a basic description of the various LMS filtering topologies discussed in this chapter.

12.7 EQUALIZATION/DECONVOLUTION

It is possible to use an adaptive filter to equalize, or undo the result of convolution; that is to say, a signal that has passed through a Plant (being, perhaps, an amplification channel having a filtering characteristic such as lowpass or bandpass) will have its characteristics changed since it will have been convolved with the Plant's impulse response. A filter that can undo the effects of convolution might be called a deconvolution filter, or an equalization filter, or simply an inverse filter. For simplicity, we'll only use one of these terms in the discussion below, but all of them apply.

The basic arrangement or topology for channel equalization is as shown in Fig. 12.16.

The script

$$LVxChannEqualiz(ChImp, Mu, tSig, lenLMS)$$

allows experimentation with channel equalization using an LMS algorithm in the topology shown in Fig. 12.16. The parameter $ChImp$ is the impulse response of the Channel, Mu has the usual meaning in LMS adaptive filtering, $tSig$ is an integer that selects the type of test signal to pass through the Channel/LMS Filter serial combination to attempt to train the filter to the inverse of the Channel Impulse Response, with 0 yielding random noise, 1 yielding a repeated unit impulse sequence of length equal to the LMS filter, with its phase alternated every repetition, 2 yielding a repeated unit impulse sequence equal in length to the adaptive filter, and 3 yielding a sinusoidal sequence of half-band frequency. The length of the adaptive filter is specified by $lenLMS$.

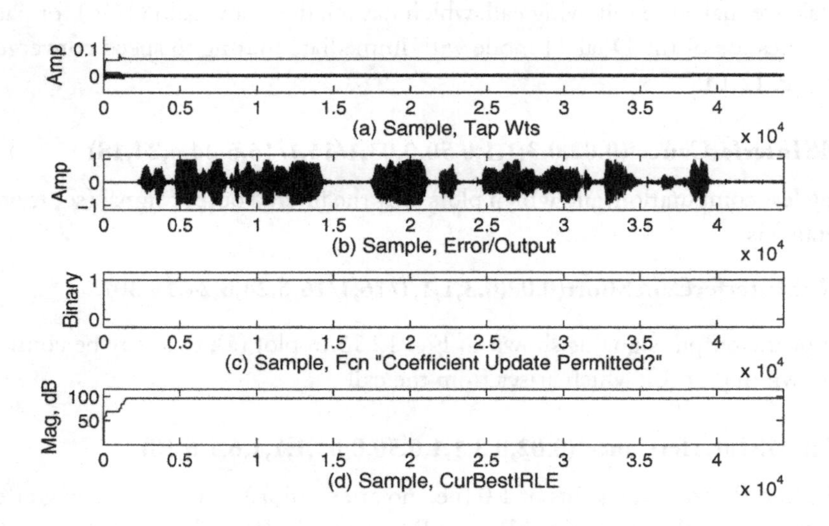

Figure 12.14: (a) All of the adaptive filter coefficients versus sample or iteration, showing good convergence and stability; (b) Output/Error signal; (c) Function "Coefficient Update Permitted?"; (d) Current Best value of Interference Return Loss Enhancement (IRLE) during the course of algorithm execution.

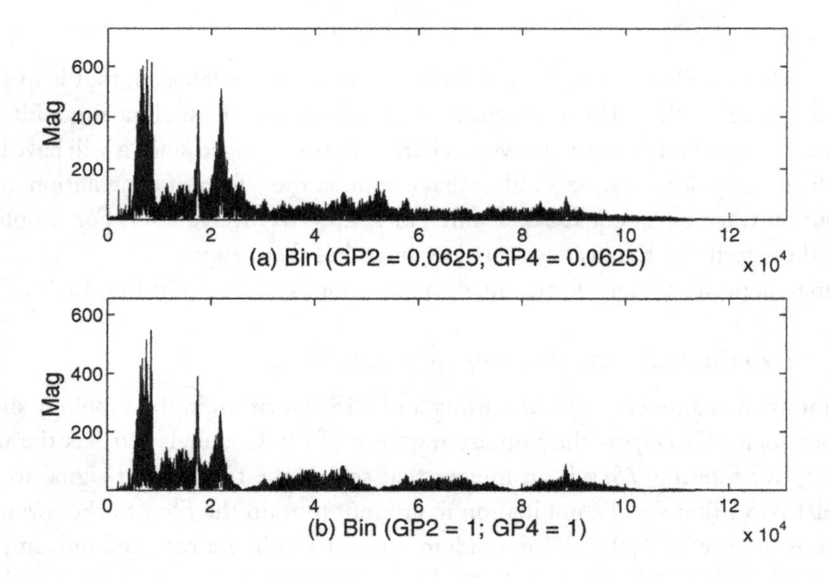

Figure 12.15: (a) Typical output spectrum of signal after noise signal is subtracted, with the condition that significant attenuation exists in the crosspaths P2 and P4; (b) Same as (a), except that there was no attenuation in the crosspaths, leading to significant comb filtering effects in the output signal.

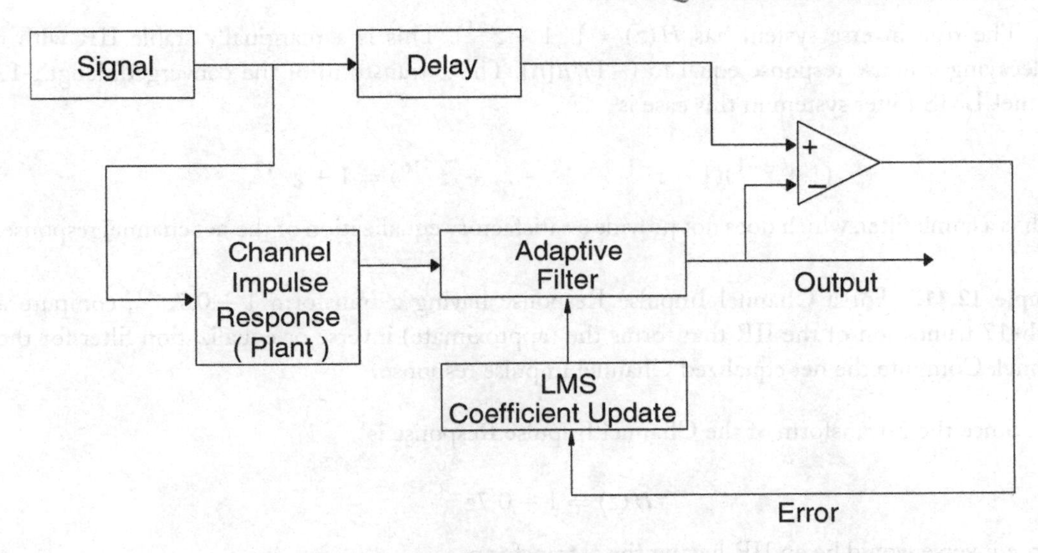

Figure 12.16: A basic arrangement for Channel Equalization.

An important thing to note in the script is that the Channel Impulse Response is always modeled as an FIR, and the adaptive filter is also an FIR. For mathematically perfect inverse filtering, the transfer functions of the two should be the reciprocals of one another. Since both are FIRs, this cannot be. However, an LMS FIR filter of sufficient length can converge to a truncated version of an infinite impulse response.

Example 12.9. A Channel has z-transform 1 - $0.7z^{-1}$. Determine what length of LMS FIR filter should be used if we arbitrarily require that the truncated inverse include all samples having magnitude of 0.01 or greater.

The reciprocal of the Channel's z-transform is $1/(1 - 0.7z^{-1})$ (ROC: $|z| > 0.7$), and the corresponding impulse response is $[1,0.7,0.49,...0.7^n] = (0.7)^n u[n]$. This is an exponentially decaying sequence, and we therefore need the smallest value of n such that

$$0.7^n \leq 0.01$$

We compute $\log(0.01) = n\log(0.7)$ from which n = 12.91 which we round to 13. Thus an LMS FIR length of 13 or more should be adequate under this criterion.

Example 12.10. Determine if the impulse response of an IIR, truncated to a length-17 FIR, can serve as an adequate inverse for a Channel having its impulse response equal to [1, 1].

The true inverse system has $H(z) = 1/(1 + z^{-1})$. This is a marginally stable IIR with a nondecaying impulse response equal to $(-1)^n u[n]$. The z-transform of the converged, length-17 Channel/LMS Filter system in this case is

$$(1 + z^{-1})(1 - z^{-1} + z^{-2} - ... + z^{-16}) = 1 + z^{-17}$$

which is a comb filter, which does not provide a satisfactory equalization of the net channel response.

Example 12.11. For a Channel Impulse Response having z-transform $1 + 0.7z^{-1}$, compute a length-17 truncation of the IIR that forms the (approximate) inverse or equalization filter for the Channel. Compute the net equalized Channel impulse response.

Since the z-transform of the Channel Impulse Response is

$$H(z) = 1 + 0.7z^{-1}$$

the true inverse would be an IIR having the z-transform

$$H(z) = \frac{1}{1 + 0.7z^{-1}}$$

which equates to a decaying Nyquist rate impulse response with values [1,-0.7,0.49,...(-0.7)^n] where n starts with 0.

The z-transform of the net Channel/LMS Filter is

$$(1 + 0.7z^{-1})(1 - 0.7z^{-1} + 0.49z^{-2} - ... + (-0.7)^{16}z^{-16}) = 1 + (0.7)^{17}z^{-17}$$

which simplifies to 1 + $0.0023z^{-17}$; this is a severely modified comb filter with a miniscule delay term having essentially no effect, i.e., the inverse approximation is a very good one as the net combination approaches having a unity transfer function.

The result from making the call

LVxChannEqualiz([1,0.7],2.15,2,17)

(in which Mu is 2.15 and the test signal comprises multiple repetitions of a unit impulse sequence having the same length as the filter, 17) is shown in Fig. 12.17.

Example 12.12. Demonstrate Channel Equalization using the script *LVxChannEqualiz*, with a channel impulse response of $[1, 0, 1]$.

The Channel Impulse Response [1,0,1] is a comb filter with precisely one null at the half-band frequency, making it a simple notch filter at that frequency. The IIR which is the perfect inverse

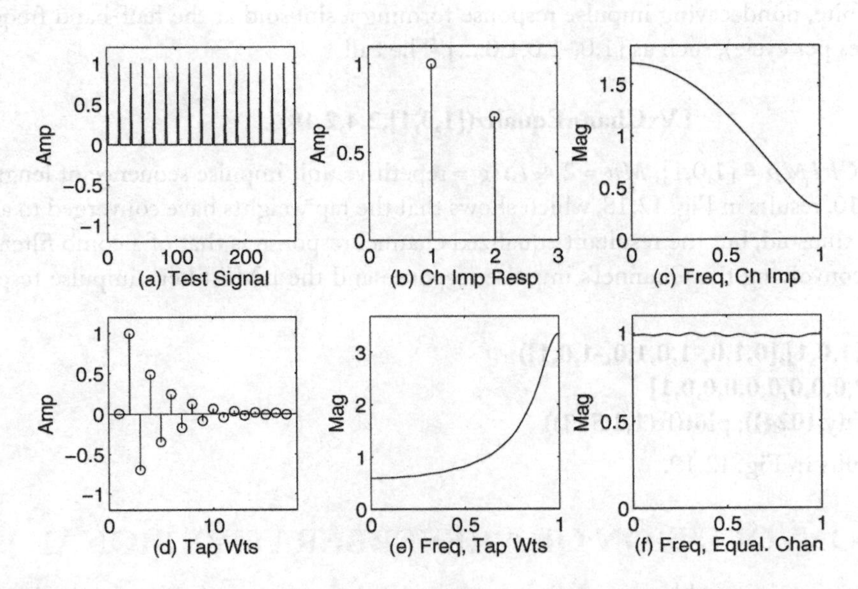

Figure 12.17: (a) The test signal; (b) Channel impulse response; (c) Frequency response of the Channel; (d) Converged tap weights of inverse filter; (e) Frequency response of tap weights shown in (d); (f) Frequency response of net equalized channel.

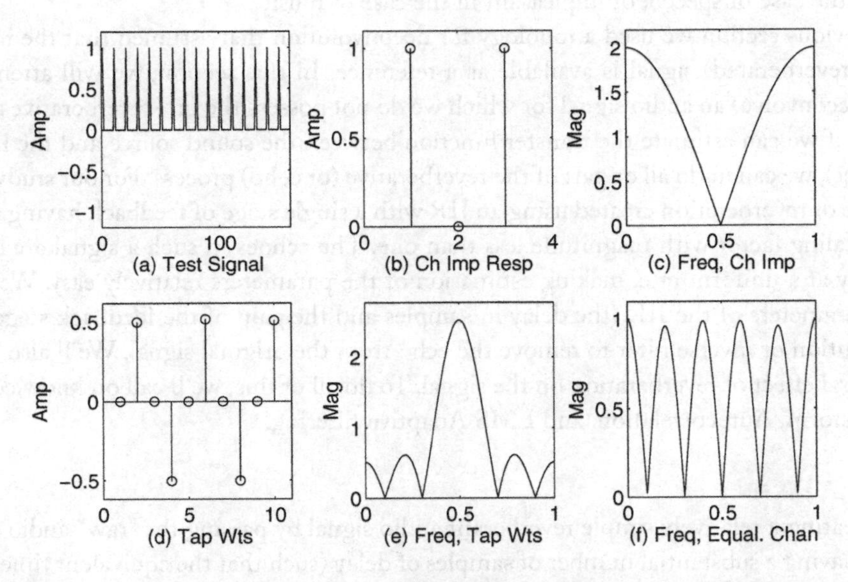

Figure 12.18: (a) The test signal; (b) Channel impulse response; (c) Frequency response of the Channel; (d) Converged tap weights of inverse filter; (e) Frequency response of tap weights shown in (d); (f) Frequency response of equalized channel.

filter has an infinite, nondecaying impulse response forming a sinusoid at the half-band frequency (i.e., four samples per cycle), such as [1,0,-1,0,1,0,...]. The call

```
LVxChannEqualiz([1,0,1],2.4,2,10)
```

which specifies $ChIMp$ = [1,0,1], Mu = 2.4, $tSig$ = repetitive unit impulse sequence of length 10, and $lenLMS$ = 10, results in Fig. 12.18, which shows that the tap weights have converged to a half-band-frequency sinusoid, but the resultant equalized channel response is that of a comb filter. This can be seen by convolving the Channel's impulse response and the LMS filter's impulse response, i.e.,

```
y = conv([1,0,1],[0,1,0,-1,0,1,0,-1,0,1])
% y = [1,0,0,0,0,0,0,0,0,0,1]
fr = abs(fft(y,1024)); plot(fr(1,1:513))
```

which results in Fig. 12.19.

12.8 DECONVOLUTION OF A REVERBERATIVE SIGNAL

An occasionally-occurring problem in audio processing is the presence of echo or reverberation, caused by a highly reflective acoustic environment. While small to moderate amounts of reverberation are often imparted to signals deliberately to make them resemble signals in a reverberative room or hall, large amounts of unwanted or unintended reverberation often make an audio signal difficult to comprehend, in the case of speech, or unpleasant in the case of music.

In the previous section we used a topology for deconvolution that assumed that the unconvolved (i.e., un-reverberated) signal is available as a reference. In this section, we will attempt to dereverberate (deconvolve) an audio signal for which we do not possess the pre-reverberative signal.

In theory, if we can estimate the transfer function between the sound source and the microphone (or listener), we can undo all or part of the reverberative (or echo) process. For our study, we'll use a simple type of reverberation created using an IIR with a single stage of feedback having a long delay and real scaling factor with magnitude less than one. The echoes in such a signal are evenly spaced and decay at a uniform rate, making estimation of the parameters relatively easy. We'll use the estimated parameters of the IIR (the delay in samples and the gain of the feedback stage) and form a deconvolution or inverse filter to remove the echo from the original signal. We'll also take a look at the spectral effect of reverberation on the signal. To do all of this, we'll call on knowledge of IIRs, the z-Transform, Autocorrelation, and LMS Adaptive filtering.

12.8.1 SIMULATION

We'll start by creating a relatively simple reverberant audio signal by passing the "raw" audio signal through an IIR having a substantial number of samples of delay (such that the equivalent time delay is about 50 msec or more), and a single feedback coefficient having a magnitude less than 1.0. In the z-domain such an IIR would have the transfer function

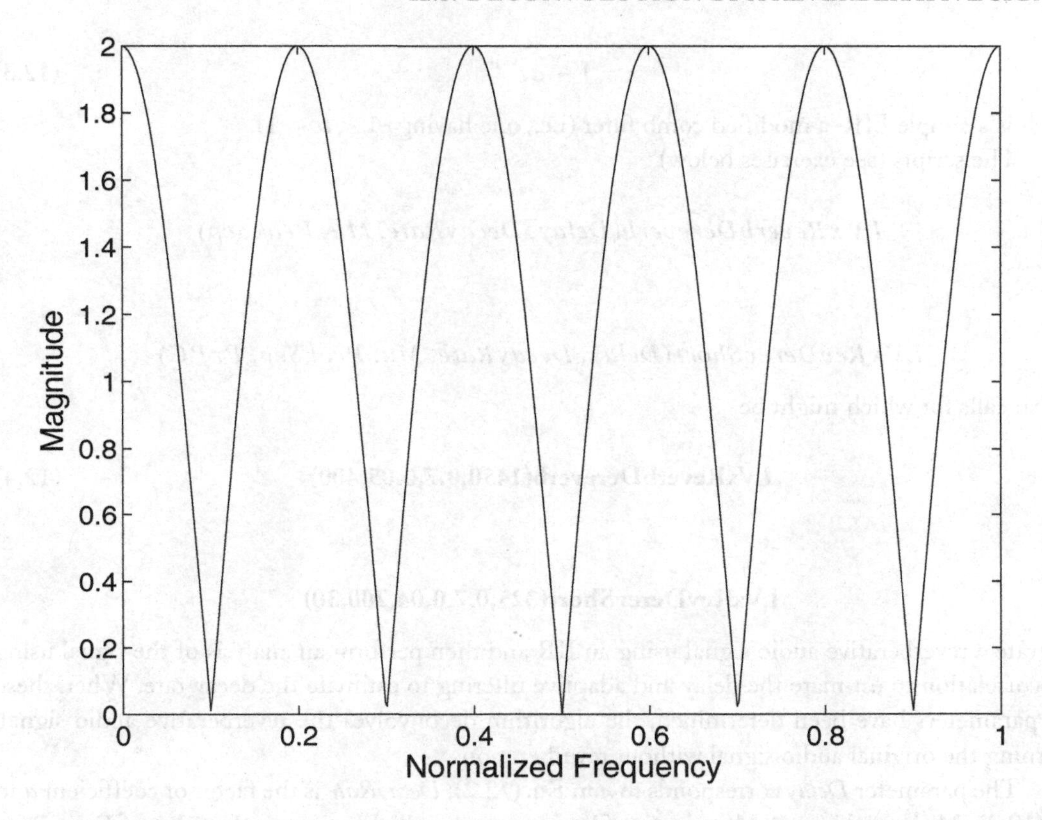

Figure 12.19: The frequency response of the net channel impulse response obtained as the convolution of the Channel Impulse Response [1,0,1] and a set of converged LMS tap weights which form a truncated inverse, [0,1,0,-1,0,1,0,-1,0,1].

$$H(z) = \frac{1}{1 - az^{-n}} \tag{12.2}$$

where a is a real number having magnitude less than 1 and n is chosen such that

$$DesiredEchoPeriod = nT_s$$

where T_s is the sampling period.

Next, we'll estimate the values of a and n in Eq. (12.2) using only the reverberant signal itself, and then we'll design a filter which will undo (deconvolve) the original filter's work and return the audio signal to one without reverberation. The z-transform of such a filter is just the reciprocal of the transfer function that created the reverberant signal, namely

$$1 - az^{-n} \tag{12.3}$$

which is a simple FIR–a modified comb filter (i.e., one having -1 $< a <$ 1).

The scripts (see exercises below)

$$LVxReverbDereverb(Delay, DecayRate, Mu, PeakSep)$$

and

$$LVxRevDerevShort(Delay, DecayRate, Mu, PeakSep, PrPC)$$

typical calls for which might be

$$\textbf{LVxReverbDereverb(1450,0.7,0.05,400)} \tag{12.4}$$

and

LVxRevDerevShort(325,0.7,0.04,200,30)

generate a reverberative audio signal using an IIR and then perform an analysis of the signal using autocorrelation to estimate the delay and adaptive filtering to estimate the decay rate. When these two parameters have been determined, the algorithm deconvolves the reverberative audio signal, returning the original audio signal without reverberation.

The parameter *Delay* corresponds to *n* in Eq. (12.2); *DecayRate* is the factor or coefficient *a* in Eq. (12.2), *Mu* is used in an LMS adaptive filtering process which estimates the value of *DecayRate*; and finally, *PeakSep* specifies the number of samples that must lie between any two adjacent peak values which are picked from the autocorrelation sequence to estimate the value of *Delay*. All of these variables will be discussed below in further detail.

Figure 12.20 shows the reverberative audio signal at (a), which was formed by convolving an audio signal (*drwatsonSR4K.wav* in this case) with an IIR having a single nonzero delay coefficient having *DecayRate* = 0.7 and *Delay* =1450 samples. At (b), the autocorrelation sequence of the reverberated audio signal is shown, and at (c), the dereverberated (or deconvolved) signal is shown. The peaks in the autocorrelation sequence are 1450 samples apart.

12.8.2 SPECTRAL EFFECT OF REVERBERATION

Now that we have both the original audio signal and the reverberated version, let's take a look at the spectral effect of imparting reverberation to the audio signal. One or more delayed versions of an audio signal, when mixed with the original signal at a given listening point (by acoustic superposition, for example) result in a distortion or filtering of the frequency spectrum of the composite audio signal. The same is true here, where the original audio signal is repeatedly mixed with itself at a basic delay of 1450 samples, for example.

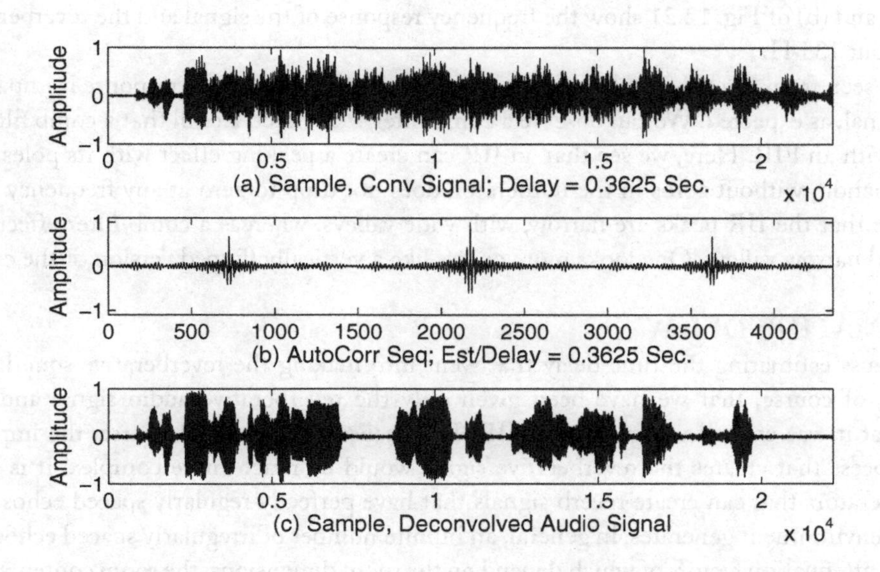

Figure 12.20: (a) Reverberative audio signal, formed by passing original audio signal through a simple IIR; (b) Autocorrelation sequence of reverberative audio signal; (c) Audio signal after being deconvolved using estimates of the IIR's delay and decay parameters using autocorrelation and LMS adaptive filtering.

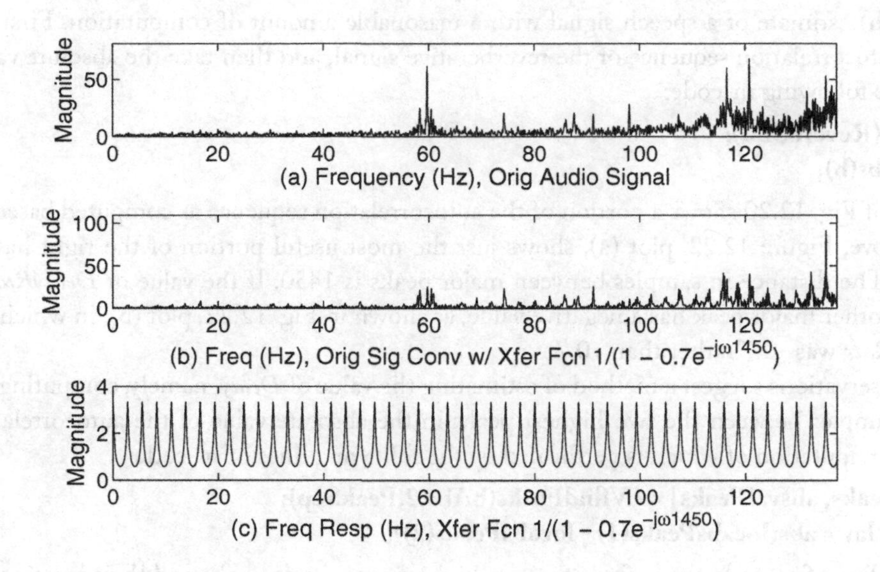

Figure 12.21: (a) Frequency response of original (nonreverberated) audio signal; (b) Frequency response of reverberated audio signal; (c) Frequency response (evaluated using the z-transform) of IIR filter used to generate the reverberative signal.

Plots (a) and (b) of Fig. 12.21 show the frequency response of the signal and the reverberated signal up to about 135 Hz.

It can be seen from Fig. 12.21, plots (b) and (c), that the IIR's frequency response is imparted to the audio signal, as expected. We can observe a comb filter-like effect. Recall that a comb filter is implemented with an FIR. Here, we see that an IIR can create a peaking effect with its poles, but the transfer function, without zeros in the numerator, does not drop to zero at any frequency. You may also notice that the IIR peaks are narrow, with wide valleys, whereas a comb filter effect has broad peaks and narrow valleys. One looks more or less like a vertically flipped version of the other.

12.8.3 ESTIMATING DELAY

Next, we'll discuss estimating the time delay that went into making the reverberative sound. We are pretending, of course, that we have been given only the reverberative audio signal and the information that it was created using a simple IIR. In a realistic situation, of course, the impulse response or process that creates the reverberative signal would be much more complex; it is only artificial reverberators that can create reverb signals that have perfectly regularly spaced echos. An actual acoustic environment generates, in general, an infinite number of irregularly spaced echos the delay times and attenuation factors of which depend on the room dimensions, the room contents, the locations of the source and listening point, and the reflective properties of the room and everything in it. It is very instructive, however, to see how one can estimate a basic delay and attenuation applying simple techniques that we have discussed previously.

The autocorrelation sequence is used extensively in speech analysis for generating an accurate frequency (pitch) estimate of a speech signal with a reasonable amount of computation. First, we compute the autocorrelation sequence of the reverberative signal, and then take the absolute value, as shown by the following m-code:

```
b = xcorr(ReverbSnd);
bAbs = abs(b);
```

Plot (b) of Fig. 12.20 shows a portion of the autocorrelation sequence as computed based on the m-code above. Figure 12.22, plot (a), shows just the most useful portion of the right half of that sequence. The distance in samples between major peaks is 1450. If the value of *DecayRate* is negative, every other major peak has a negative value, as shown in Fig. 12.22, plot (b), in which the value of *DecayRate* was -0.7 rather than +0.7.

These observations suggest a method of estimating the value of *Delay*, namely, computing the separation in samples between the two highest peaks in the absolute value of the autocorrelation sequence, the computation of which is performed by the following lines of m-code:

```
[locabsPeaks, absvalPeaks] = LVfindPeaks(bAbs,2,PeakSep);
absestDelay = abs(locabsPeaks(1) - locabsPeaks(2));
```

The first line of m-code above finds the two highest magnitude peaks and their locations in the signal *bAbs* which are at least *PeakSep* samples apart; in this case, *PeakSep* = 400. *PeakSep* is a parameter which can be estimated by eye if one doesn't have any idea of the actual value of *Delay*,

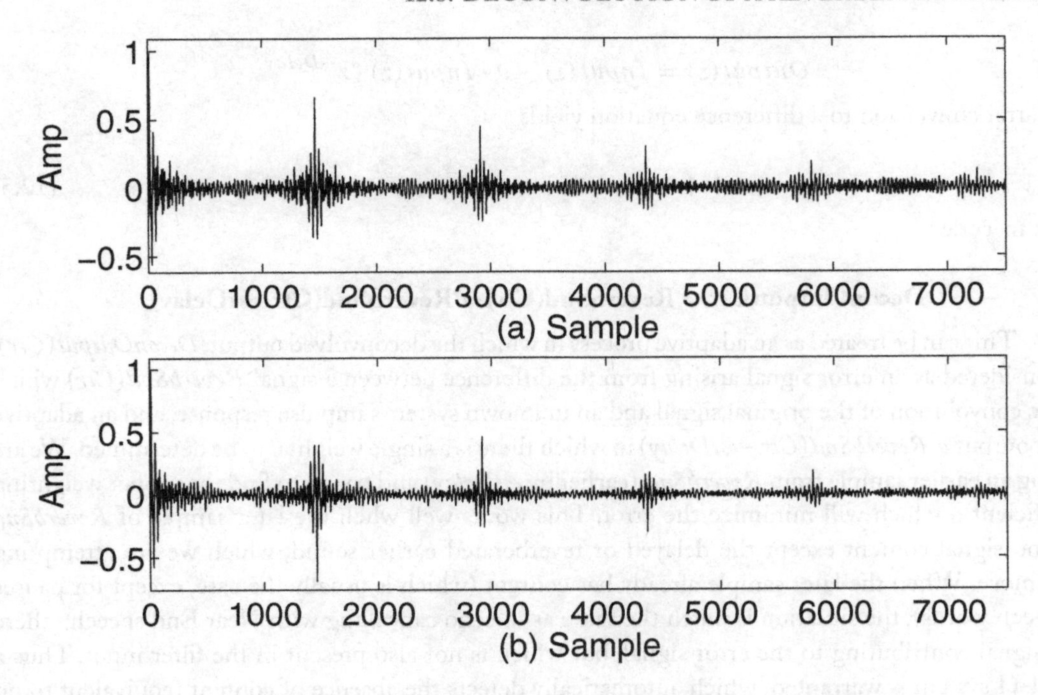

Figure 12.22: (a) Autocorrelation sequence of reverberated audio signal when DecayRate = +0.7; (b) Autocorrelation sequence of reverberated audio signal when DecayRate = -0.7.

which is what we are generally supposing is the case. For purposes of keeping things simple, though, the script allows you to specify the value of *PeakSep*, which you may do by knowing what value of *Delay* you are using. *PeakSep* must be less than the value of *Delay*. To estimate *PeakSep* by eye, you would plot the autocorrelation sequence and examine it for peaks, noting the distance between what appear to be suitable candidate peaks, and then perform the estimation of *Delay* by calling *LVfindPeaks* with *PeakSep* specified as about half the observed candidate peak separation (details of the function *LVfindPeaks* are in the exercises below).

12.8.4 ESTIMATING DECAYRATE

The process that created *ReverbSnd* was an IIR filter; therefore, the inverse or deconvolution process is an FIR filter the z-transform of which is the reciprocal the z-transform of the IIR, as shown above. We have obtained the value for *Delay*, so the z-transform would be

$$\frac{Output(z)}{Input(z)} = 1 - az^{-Delay}$$

which leads to

$$Output(z) = Input(z) - a \cdot Input(z) \cdot z^{-Delay}$$

and after conversion to a difference equation yields

$$Output[n] = Input[n] - a \cdot Input[n - Delay] \tag{12.5}$$

or in m-code

DeconOutput(Ctr) = ReverbSnd(Ctr)-a*ReverbSnd(Ctr-estDelay)

This can be treated as an adaptive process in which the deconvolved output, *DeconOutput(Ctr)*, is considered as an error signal arising from the difference between a signal *ReverbSnd(Ctr)* which is the convolution of the original signal and an unknown system's impulse response, and an adaptive filter output *a·ReverbSnd(Ctr - estDelay)* in which there is a single weight *a* to be determined. We are taking an earlier sample from *ReverbSnd* (earlier by *estDelay*) and trying to find the proper weighting coefficient *a* which will minimize the error. This works well when the later sample of *ReverbSnd* has no signal content except the delayed or reverberated earlier sound, which we are attempting to remove. When the later sample already has content (which is usually the case, except for pauses between words), the situation is much the same as in echo cancelling with Near End speech: there is a signal contributing to the error signal, but which is not also present in the filter input. Thus a Dual-H system is warranted, which automatically detects the absence of content (equivalent to no Near End speech in echo cancelling) by yielding a high value of the ratio of power into the adaptive filter (in this case samples of the reverberated sound) to power output from the error junction. This is similar to the ERLE and IRLE parameters discussed earlier in the chapter; here, we'll call it *Reverb Return Loss Enhancement*, or *RRLE*.

The method we are employing to remove the echo or reverberation from the original signal is very analogous to sparse echo cancellation, which was discussed earlier in the chapter. Note that we used correlation to estimate the relevant tap delay, which was (based on a certain amount of knowledge of the echo or reverberative process), a single delay. We then use only the one tap in the adaptive filter. If we had estimated the delay at, say, 900 samples, we could have used an adaptive filter having, say, 1000 active taps, covering any echo components occurring in the time delay range of 0 to 1000 samples. This, however, not only would result in far more computational overhead, but would likely produce inferior results.

Figure 12.23, plot (a), shows the sequence of estimates of the single coefficient being estimated, the value of *DecayRate* during a trial run of script call (12.4), while plot (b) shows the corresponding sequence of values of *RRLE*.

12.8.5 DECONVOLUTION

After both *Delay* and *DecayRate* have been estimated, the values are used in a deconvolution filter which has as its z-transform the reciprocal of the z-transform of the IIR which was used to generate the reverberated audio signal. The relevant lines of m-code are:

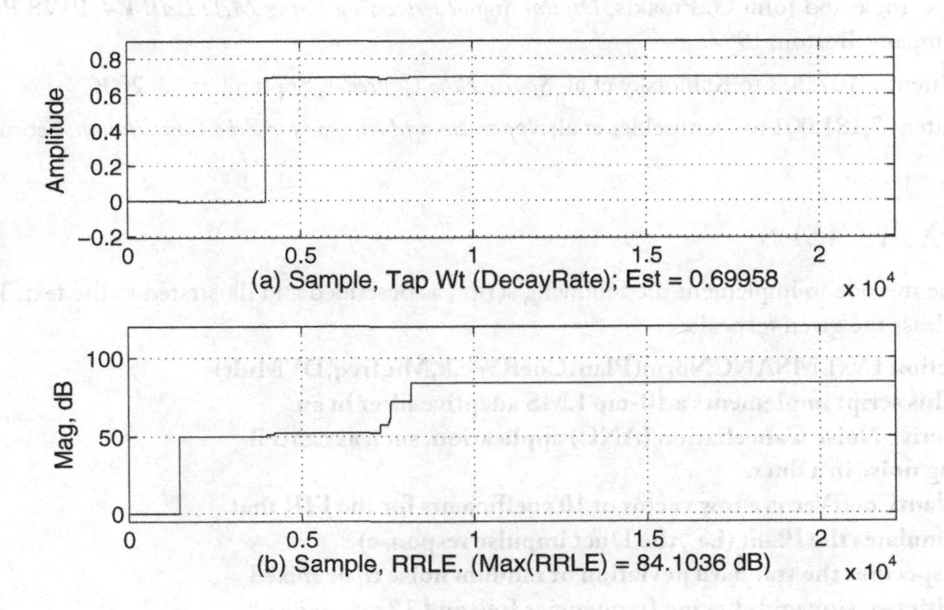

Figure 12.23: (a) Estimated value of DecayRate (the amplitude of the pole of the IIR used to form the reverberative audio signal); (b) Reverb Return Loss Enhancement (RRLE), used to tell the LMS Adaptive filter when to update the coefficient.

```
deconImp = [1,zeros(1,estDelay-1),-estDecayRate];
DereverbSnd = filter(deconImp,1,ReverbSnd)
```

If all has gone well, the dereverberated audio signal will be identical or nearly so to the original audio signal.

12.9 REFERENCES

[1] Bernard Widrow and Samuel D. Stearns, *Adaptive Signal Processing*, Prentice-Hall, Englewood Cliffs, New Jersey, 1985.

[2] Paul M. Embree and Damon Danieli, *C++ Algorithms for Digital Signal Processing*, Prentice-Hall PTR, Upper Saddle River, New Jersey, 1999.

[3] Ali H. Sayed, *Fundamentals of Adaptive Filtering*, John Wiley & Sons, Hoboken, New Jersey, 2003.

[4] John R. Buck, Michael M. Daniel, and Andrew C. Singer, *Computer Explorations in Signals and Systems Using MATLAB, Second Edition*, Prentice-Hall, Upper Saddle River, New Jersey 07548, 2002.

[5] Vinay K. Ingle and John G. Proakis, *Digital Signal Processing Using MATLAB V.4*, PWS Publishing Company, Boston, 1997.

[6] U. S. Patent 7,107,303 to Kablotsky et al, *Sparse Echo Canceller*, September 12, 2006.

[7] U. S. Patent 7,181,001 to Demirekler et al, *Apparatus and Method for Echo Cancellation*, February 20, 2007.

12.10 EXERCISES

1. Write the m-code to implement the following script, as described and illustrated in the text. Test it using at least the given test calls.

```
function LVxLMSANCNorm(PlantCoeffVec,k,Mu,freq,DVMult)
% This script implements a 10-tap LMS adaptive filter in an
% Active Noise Cancellation (ANC) application, such as cancell-
% ing noise in a duct.
% PlantCoeffVec is a row vector of 10 coefficients for the FIR that
% simulates the Plant (i.e., the Duct impulse response).
% k specifies the standard deviation of random noise to be mixed
% with two sinusoids having frequencies freq and 3*freq and
% respective amplitudes of 1.0 and 0.5;
% Mu specifies the tap weight update damping coefficient;
% An NLMS algorithm is used; its effectiveness can be tested
% by giving the test signal various stepped-amplitude profiles
% with the input argument DVMult, which is a vector of
% amplitudes to impose on the test signal as a succession of
% equally spaced amplitude segments over a test signal of 15
% times the filter length.
% Test calls:
% LVxLMSANCNorm([0,0,1,0,-0.5,0.6,0,0,-1.2,0],2,2,27,[1,2,5,8])
% LVxLMSANCNorm([0,0,1,0,-0.5,0.6,0,0,-1.2,0],2,0.2,27,[1,2,5,8])
% LVxLMSANCNorm([0,0,1,0,-0.5,0.6,0,0,-1.2,0],0,2,3,[1,2,5,8])
```

2. Write the m-code to implement the following script, as described and illustrated in the text:

```
function LVxModelPlant(A, LenLMS, NoPrZs, NoPrPs, Mu,...
tSig, NAmp, NoIts)
% A and B are a set of IIR and FIR coefficients to generate the
% Plant response LenLMS is the number of LMS adaptive FIR
% coefficients to use to model the IIR; it should generally be long
% enough to model the impulse response until it has mostly
% decayed away
% NoPrZs is the number of Prony zeros to use to model the
```

```
% LMS-derived impulse response (the converged LMS adaptive
% filter coefficients, which form a truncated version of the Plant
% impulse response); NoPrPs is the number of Prony poles to use
% to model the LMS-derived impulse response; Mu is the standard
% LMS update term weight; tSig if passed as 0 = white noise;
% 1 = DC; 2 = Nyquist; 3 = Half-band; 4 = mixed cosines (DC,
% Half-band, and Nyquist)
% NAmp = amplitude of noise to add to selected tSig
% NoIts is the number of iterations to do in the adaptive process.
% NoIts should be at least 10 times LenLMS.
```

Test the script with at least the following calls:

```
% LVxModelPlant([1,-0.9],[1],100, 2,2, 0.5,0,0,1000)
% LVxModelPlant([1, 0,0.81],[1],100, 3,3,0.5,1, 0,1000)
% LVxModelPlant([1,-1.3,0.81],[1],100, 3,3,0.5,1, 0,1000)
% LVxModelPlant([1,-1,0.64],[1],100, 3,3, 0.5, 1, 0,1000)
```

3. Write the m-code for the following script, which provides a basic illustration of echo cancellation using Single- and Dual-H methods, using a user-selectable percentage of the reduced-bandwidth audio files 'drwatsonSR4K.wav' and 'whoknowsSR4K.wav'. Detailed instructions are given after the function definition.

```
function LVxLMSAdptFiltEchoShort(k,Mu,DHorSH,...
MuteNrEnd,PrPC)
% k is the amount of noise to add to the Far End signal (see below)
% Mu is the usual LMS update term weight;
% DHorSH yields a Dual-H system if passed as 1, or a Single-H
% system if passed as 0; MuteNrEnd mutes the desired signal
% ('drwatsonSR4K') at its very beginning if passed as 1, about
% one-third of the way through if passed as 2, or no muting at
% all if passed as 0.
% The Near End signal consists of the audio file 'drwatsonSR4K.wav'
% The Far End signal consists of the audio file 'whoknowsSR4K.wav'
% plus random noise weighted by k, limited to the length of the
% Near End signal;
% The total amount of computation can be limited by processing
% only a percentage of the audio signals, and that percentage
% is passed as the input variable PrPC.
% Using whoknowsSR4K.wav and drwatsonSR4K.wav as test signals,
% 100%=22184 samps. The adaptive FIR is ten samples long
% and the echo is simulated as a single delay of 6 samples,
% for example. The final filtered output signal is interpolated
```

```
% by a factor of 2 using the function interp to raise its sample
% rate to 8000 Hz so it may be played using the call sound
% (EchoErr,8000) after making the call global EchoErr in
% the Command window. A single figure is created showing
% the filtered output signal
% Test calls:
% LVxLMSAdptFiltEchoShort(0.1,0,0,0,50) % Mu = 0, FarEnd hrd
% LVxLMSAdptFiltEchoShort(0.1,0.2,0,0,50) % S-H, no Mute
% LVxLMSAdptFiltEchoShort(0.1,0.2,0,1,50) % S-H, Mute Imm
% LVxLMSAdptFiltEchoShort(0.1,0.2,1,0,50) % D-H, no Mute
% LVxLMSAdptFiltEchoShort(0.1,0.2,1,1,50) % D-H, Mute Imm
% LVxLMSAdptFiltEchoShort(0.1,0.2,1,2,50) % D-H, Mute del
% LVxLMSAdptFiltEchoShort(0.01,0.2,0,0,50) % S-H, no Mute
% LVxLMSAdptFiltEchoShort(0.01,0.2,0,1,50) % S-H, Mute Imm
% LVxLMSAdptFiltEchoShort(0.01,0.2,1,0,50) % D-H, no Mute
% LVxLMSAdptFiltEchoShort(0.01,0.2,1,1,50) % D-H, Mute Imm
% LVxLMSAdptFiltEchoShort(0.01,0.2,1,2,50) % D-H, Mute del
% LVxLMSAdptFiltEchoShort(0.01,0.005,1,0,60) % Low Mu, can
%  hear Far End gradually diminish
```

Since this script is intended to execute as quickly as possible, the detection of Near End speech has been eliminated, and archiving and plotting of ERLE as well. The code should be written to automatically play the resultant filtered output at least once. Insert in the m-code the statement

global EchoErr

where $EchoErr$ is the filtered output signal. Then, in the Command Window, type **global EchoErr** after the program has run, and you can then play the filtered output at will by making the call

sound(EchoErr, 8000)

In order to avoid retaining older (unwanted) computed values in a global variable, it is good practice is to insert the global statement early in the script prior to computing and assigning new values to the variable, and to follow the global statement immediately with a statement to empty the variable, i.e.,

```
global EchoErr
EchoErr = [];
```

Since the net filtered output is the delayed Far End Signal plus the Near End signal, minus the FIR output, the delayed Far End signal can be added to the Near End Signal prior to the loop filtering operation, resulting in only one array access rather than two to compute the filtered output. The coefficient update normalizing term should be two times Mu, divided by the sum of the squares of the signal in the FIR, and this, too, can be computed ahead of the loop and looked up while in the filtering loop.

Since the audio files were sampled at 4 kHz, the final filtered output signal must be interpolated by a factor of two so that its net final sample rate is 8000 Hz (LabVIEW only allows certain sample rates to be used in the *sound* function, namely, 8000, 11025, 22050, and 44100 Hz).

4. Write the m-code to implement the script *LVxLMSAdaptFiltEcho*, as described and illustrated in the text. The preceding exercise created a much simpler script, which does not use the variables *DblTkUpD*, *SampsVOX*, *VOXThresh*, and which requires a much smaller computation load. This project adds these variables, and archives ERLE and several other variables for purposes of plotting. The preceding project can serve as a good starting basis for this project.

```
function LVxLMSAdaptFiltEcho(k,Mu,DHorSH,MuteNrEnd,...
DblTkUpD,SampsVOX,VOXThresh)
% k is the amount of noise to add to the interference signal
% ('drwatsonSR8K.wav'); Mu is the usual LMS update term
% weight; DHorSH yields a Dual-H system if passed as 1, or
% a Single-H system if passed as 0; MuteNrEnd mutes the
% desired signal ('drwatsonSR8K.wav') at its very beginning
% if passed as 1, about one-third of the way through if passed
% as 2, or no muting at all if passed as 0. In Single-H mode,
% DblTkUpD, if passed as 0, allows coefficient updating any
% time, but if passed as 1 prevents coefficient update when
% the most recent SampsVOX samples of the desired (Near
% End) signal level are above VOXThresh. In Dual-H mode,
% DblTkUpD,SampsVOX,VOXThresh are ignored.
% The Far End signal is the audio file 'whoknowsSR8K.wav',
% limited to the length of 'drwatsonSR8K/wav'
% The adaptive FIR is ten samples long and the echo is
% simulated as a single delay of 6 samples, for example.
% A figure having four subplots is created. The first subplot
% shows the values of all ten FIR coefficients over the entire
% course of computation; the second subplot shows the filtered
% output signal; the third subplot shows as a binary plot
% whether or not coefficient update was allowed at any given
% stage of the computation based on the detection of Near
% End speech; the fourth subplot shows the figure of merit
% (ERLE, as defined in the text) over the course of the
% computation. The filtered output audio signal may be played
% using the call sound(EchoErr,8000) after making the call
% global EchoErr in the Command window. The Near End plus
```

% echoed Far End signal (the signal entering the error junction)
% can be played with the call sound(NrEndPlusFarEnd,8000)
% after making the call global NrEndPlusFarEnd.

Test the script with the following calls, then repeat the calls, changing k to 0.2; compare the audibility of the Far End signal between the same calls using different amounts of noise as specified by k.

% LVxLMSAdaptFiltEcho(0.02,0.2,1,0,[],[],[]) % D-H, no Mute
% LVxLMSAdaptFiltEcho(0.02,0.2,1,1,[],[],[]) % D-H, Mute Im
% LVxLMSAdaptFiltEcho(0.02,0.2,1,2,[],[],[]) % D-H, del Mute
% LVxLMSAdaptFiltEcho(0.02,0.2,0,0,0,50,0.05) % S-H, dbltalk
% update, no Mute
% LVxLMSAdaptFiltEcho(0.02,0.2,0,1,1,50,0.05) % S-H, no
% dbltalk update, Mute Im

5. Write the m-code to implement the following script, as described and illustrated in the text:

function LVxLMSAdaptFiltDeCorr(k,Mu,freq,NoTaps,...
DeCorrDelay)
% k is an amount of white noise to add to a test
% signal which consists of a sinusoid of frequency freq;
% Mu is the LMS update term weight;
% NoTaps is the number of taps to use in the LMS adaptive filter;
% DeCorrDelay is the number of samples to delay the filter input
% relative to the Plant or channel delay.

Test the script with at least the following calls:

% LVxLMSAdaptFiltDeCorr(0.1,0.75,200,20,5)
% LVxLMSAdaptFiltDeCorr(0.1,0.75,200,20,0)

6. Write the m-code to implement the script *LVxLMSInterferCancShort* as specified below.

function LVxLMSInterferCancShort(k,Mu,DHorSH,MuteDesSig,...
GP2,GP4,DP1,DP2,DP3,DP4,NumTaps,PrPC)
% k is the amount of noise to add to the interference signal
% ('whoknowsSR4K.wav'); Mu is the usual LMS
% update term weight;
% DHorSH yields a Dual-H system if passed as 1, or a Single-H
% system if passed as 0. MuteDesSig mutes the desired signal
% ('drwatsonSR4K.wav') at its very beginning if passed as 1,
% about one-third of the way through if passed as 2, or no muting
% at all if passed as 0; GP2 is the gain in the path from the desired
% sound source to the noise reference microphone, and GP4 is the
% gain in the path from the noise source to the desired signal

```
% microphone;
% DP1 is the Delay in samples from the desired sound source to
% the desired sound source microphone; DP2 is the Delay from
% the desired sound source to the noise microphone; DP3 is the
% Delay from the noise source to the noise source microphone;
% DP4 is the Delay from the noise source to the desired sound
% microphone. NumTaps is the number of Delays or taps to use
% in the adaptive filter. The total amount of computation can be
% limited by processing only a percentage of the audio signals,
% and that percentage is passed as the input variable PrPC.
% Using whoknowsSR4K.wav and drwatsonSR4K.wav as test
% signals, 100%=22184 samps. The final filtered output signal
% is interpolated by a factor of 2 using the function interp to
% raise its sample rate to 8000 Hz so it can be played using the
% call sound(ActualErr,8000) after running the script and making
% the call global ActualErr in the Command window
% Typical calls might be:
% LVxLMSInterferCancShort(0.02,0.3,1,0,1,1,1,6,1,6,5,30)
% LVxLMSInterferCancShort(0.02,0.3,1,0,0.16,0.16,...
7,42,7,42,35,30)
% LVxLMSInterferCancShort(0.02,0.3,1,0,0.06,0.06,...
1,6,1,6,5,30)
% LVxLMSInterferCancShort(0.02,0.3,1,0,1,1,6,1,1,6,5,30)
% LVxLMSInterferCancShort(0.02,0.3,1,1,1,1,1,6,1,6,5,30)
% LVxLMSInterferCancShort(0.02,0.2,0,2,1,1,1,6,1,6,5,30)
% LVxLMSInterferCancShort(0.02,0.3,0,1,1,1,1,6,1,6,5,30)
```

7. Write the m-code to implement the following script, as described and illustrated in the text:

```
function LVxLMSInterferCancel(k,Mu,DHorSH,MuteDesSig,...
DblTkUpD,NoSampsVOX,...
% VOXThresh,GP2,GP4,DP1,DP2,DP3,DP4,NumTaps)
% k is the amount of noise to add to the interference signal
% ('whoknowsSR8K.wav'). Mu is the usual LMS update term
% weight; DHorSH yields a Dual-H system if passed as 1, or
% a Single-H system if passed as 0. MuteDesSig mutes the
% desired signal ('drwatsonSR8K.wav') at its very beginning
% if passed as 1, about one-third of the way through if passed
% as 2, or no muting at all if passed as 0; DblTkUpD, if passed
% as 0, allows coefficient updating any time, but if passed as 1
% prevents coefficient update when the most recent NoSamps
```

```
% VOX samples of the desired signal level are above VOXThresh.
% GP2 is the gain in the path from the desired sound source to
% the noise reference microphone, and GP4 is the gain in the
% path from the noise source to the desired signal microphone;
% DP1 is the Delay in samples from the desired sound source
% to the desired sound source microphone;
% DP2 is the Delay from the desired sound source to the noise
% microphone; DP3 is the Delay from the noise source to the
% noise source microphone; DP4 is the Delay from the noise
% source to the desired sound microphone. NumTaps is the
% number of Delays or taps to use in the adaptive filter.
% The filtered audio output can be played using the call
% sound(ActualErr,8000) after running the script and making
% the call global ActualErr in the Command window.
```

Test the script with at least the following calls:

```
% LVxLMSInterferCancel(0.02,0.3,1,0,0,50,0.03,1,1,1,6,1,6,5)
% LVxLMSInterferCancel(0.02,0.3,1,0,0,50,0.03,0.16,...
0.16,7,42,7,42,35)
% LVxLMSInterferCancel(0.02,0.3,1,0,1,50,0.03,0.06,...
0.06,1,6,1,6,5)
% LVxLMSInterferCancel(0.02,0.3,1,0,0,50,0.03,1,1,6,1,1,6,5)
% LVxLMSInterferCancel(0.02,0.3,1,1,0,50,0.03,1,1,1,6,1,6,5)
% LVxLMSInterferCancel(0.02,0.2,0,2,1,80,0.03,1,1,1,6,1,6,5)
% LVxLMSInterferCancel(0.02,0.3,0,1,1,50,0.03,1,1,1,6,1,6,5)
```

8. A certain signal transmission channel has the transfer function

$$H(z) = 1 + 0.9z^{-1} + 0.8z^{-2}$$

Under ideal convergence conditions, what would be the shortest adaptive FIR that could, when well-converged, result in an equalized channel transfer function having a magnitude response within 0.1 dB of unity?

9. Write the m-code to implement the following script, as described and illustrated in the text, creating a figure with subplots as shown, for example, in Fig. 12.17:

```
function LVxChannEqualiz(ChImp,Mu,tSig,lenLMS)
% ChImp is the impulse response of the channel which is to
% be equalized by the LMS FIR;
% Mu is the usual update weight
% tSig designates the type of test signal to be used:
% 0 = random noise
% 1 = Unit impulse sequence of length lenLMS, repeated at
% least ten times, with alternating sign, +,-,+, etc
% 2 = Unit impulse sequence of length lenLMS, repeated at
% least ten times
% 3 = Sinusoid at the half-band frequency
% lenLMS is the length of the adaptive LMS FIR
```

Try two different values for the delay (as shown in Fig. 12.16), namely, (1) one-half the sum of the lengths of the Channel Impulse Response and the LMS FIR, and (2) one-half the length of the Channel Impulse Response. In both cases, round the value obtained in (1) or (2) to the nearest integral value. Test the script with at least the following calls, trying each of the two delay values to obtain the better response.

```
% LVxChannEqualiz([-0.05,0,-0.5,0,-1,0,1,0,0.5,0,0.05],1.5,2,131)
% LVxChannEqualiz([1,0,1],2.1,2,91) % notch
% LVxChannEqualiz([1,0,1],2.4,2,191) % notch
% LVxChannEqualiz([1 1],2.2,2,65) % lowpass
% LVxChannEqualiz([1,0,-1],1.8,2,85) % bandpass
% LVxChannEqualiz([1,-1],2.5,2,179) % highpass
% LVxChannEqualiz([1,-0.8],2.3,2,35) % highpass
% LVxChannEqualiz([1,0.8],2.4,2,35) % lowpass
% LVxChannEqualiz([0.3,1,0.3],2.3,2,9) % lowpass
% LVxChannEqualiz([1,0.7],2.3,2,25) % lowpass
% LVxChannEqualiz([1,0,0,0,0,0,1],1.6,2,179) % comb filter
% LVxChannEqualiz([1,0,0,-0.3,0,0,0.05],2.5,2,23)
% LVxChannEqualiz([1,0,-1],1.5,2,55)
% LVxChannEqualiz( ([1,0,1,0,1,0,1].*hamming(7)'),2.1,2,91)
```

10. Write the m-code for the script *LVxRevDerevShort*, which is a shorter version of the script *LVxReverbDereverb* (see exercise immediately following).

```
function LVxRevDerevShort(Delay,DecayRate,Mu,PeakSep,PrPC)
% Delay is the number of samples of delay used in the single
% feedback stage to generate the reverberated sound file.
% DecayRate is a real number which is feedback gain.
% Mu is the usual LMS update weight term.
% PeakSep the number of samples by which to separate
```

```
% detected peaks in the autocorrelation sequence of the
% reverberated sound. Uses a Dual-H architecture to estimate
% DecayRate, and uses the best estimate of DecayRate to
% block filter the reverberated sound to produce the
% dereverberated sound. Does not archive or plot the values
% of RRLE and current best estimate or DecayRate. Plays
% the reverberated sound once and the dereverberated
% sound once. These sounds may be played again after
% running the script by making the calls global ReverbSnd
% and global DereverbSnd in the Command window and then
% making the calls sound(ReverbSnd,8000) and sound
% (DereverbSnd,8000) at will. Processes only a percentage
% of the audio file equal to PrPC in the adaptive filtering
% loop to estimate DecayRate. Uses the audio file
% 'drwatsonSR4K.wav' as the test signal and at the end
% both the reverberative sound and the dereverberated
% sound are interpolated to a sample rate of 8000 Hz to
% be played by the function sound(x,Fs),which only allows
% Fs = 8000, 11025, 22050, or 44100.
% Sample call:
% LVxRevDerevShort(325,0.7,0.05,200,30)
% LVxRevDerevShort(225,0.7,0.12,200,30)
% LVxRevDerevShort(625,0.7,0.06,200,47)
% LVxRevDerevShort(1625,0.7,0.07,200,47)
```

You will need to extract peak locations for the autocorrelation; suitable code is found below, and the script is provided with the basic software for this book. Figure 12.24 shows the result of the call

[Locs,Pks] = LVfindPeaks((cos(2*pi*2*[0:1:63]/64)),3,16)

with the highest three peaks marked with circles. Note that a zone of ineligible values is set up around each peak that has already been chosen, otherwise peaks might be chosen right next to each other when that is not the desire.

One thing that makes the adaptive LMS algorithm run much more quickly is to recognize the fact that the times when RRLE should increase are when the Near End signal is close to zero in magnitude; such times form a contiguous set of samples. Thus, in stepping through the algorithm, it can be done by using every fifth sample, for example, rather than every sample, and whenever RRLE increases, the algorithm goes back to computing the update coefficients for every sample rather than every fifth sample. In this way, only a little more than 20% of the samples are used to compute the next estimate of $DecayRate$ and the corresponding value of RRLE.

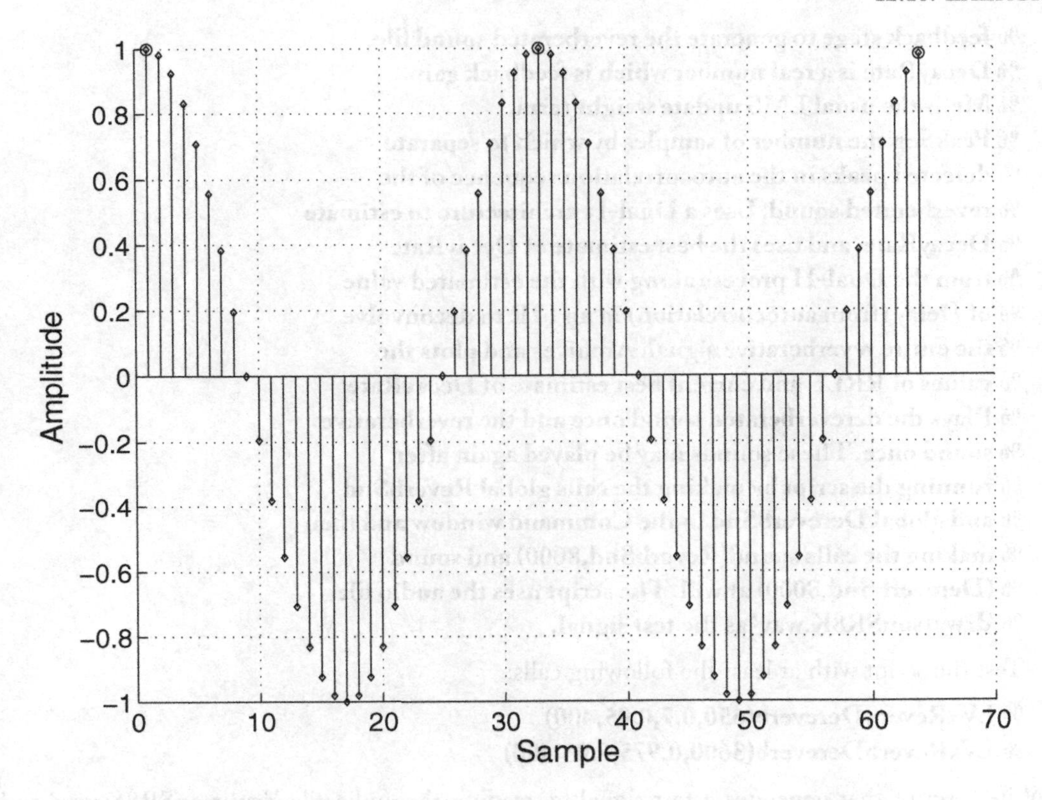

Figure 12.24: Two cycles of a cosine wave over 64 samples with the highest 3 peaks marked with circles (see text).

```
function [Locs, Pks] = LVfindPeaks(Mat,QuanPks,PeakSep)
% Mat is a vector or matrix for which QuanPks peak values
% are sought, each of which is at least PeakSep samples
% distant from the next closest peak value. Locs tells the
% locations of the peak values in the matrix. To make adjacent
% samples of Mat eligible to be peaks, use PeakSep = 0.
% Sample call:
% [Locs, Pks] = LVfindPeaks([0:1:7],8,0)
% [Locs, Pks] = LVfindPeaks([0:1:7],8,1)
% [Locs, Pks] = LVfindPeaks((cos(2*pi*3*[0:1:63]/64)),4,8)
```

11. Write the m-code to implement the following script, as described and illustrated in the text:

```
function LVxReverbDereverb(Delay,DecayRate,Mu,PeakSep)
% Delay is the number of samples of delay used in the single
```

```
% feedback stage to generate the reverberated sound file.
% DecayRate is a real number which is feedback gain.
% Mu is the usual LMS update weight term.
% PeakSep the number of samples by which to separate
% detected peaks in the autocorrelation sequence of the
% reverberated sound. Uses a Dual-H architecture to estimate
% DecayRate, and uses the best estimate of DecayRate
% from the Dual-H process along with the estimated value
% of Delay (from autocorrelation) in an FIR to deconvolve
% the entire reverberative signal. Archives and plots the
% values of RRLE and current best estimate of DecayRate.
% Plays the dereverberated sound once and the reverberative
% sound once. These sounds may be played again after
% running the script by making the calls global ReverbSnd
% and global DereverbSnd in the Command window and then
% making the calls sound(ReverbSnd,8000) and sound
% (DereverbSnd,8000) at will. The script uses the audio file
% 'drwatsonSR8K.wav' as the test signal.
```

Test the script with at least the following calls:

```
% LVxReverbDereverb(650,0.7,0.05,400)
% LVxReverbDereverb(3600,0.975,0.05,400)
```

12. Write a script that generates a test signal by reading the audio file 'drwatsonSR8K.wav' and adding an interfering sinusoid having amplitude A and frequency $Freq$ to it. The script will then use LMS adaptive filtering with a decorrelating delay to remove the sinusoidal tone from the audio signal. Code suitable to create the test signal is

```
[y,Fs,bits] = wavread('drwatsonSR8K.wav');
y = y'; lenFile = length(y);
y = (1/max(abs(y)))*y;
t = [0:1:lenFile-1]/Fs;
LMSDeCorDataVec = y + A*sin(2*pi*t*Freq);
```

The display figure for the script should be like Fig. 12.25, including four subplots that show the input signal (audio file corrupted by a large amplitude sinusoid), the filtered output signal, and the magnitude of the spectrum of each, computed by FFT. Figure 12.25 was generated by making the call

```
LVxLMSDeCorrHeterodyne('drwatsonSR8K.wav',1.5,0.08,500,45,1)
```

Your script should conform to the syntax below; test the script with the given test calls.

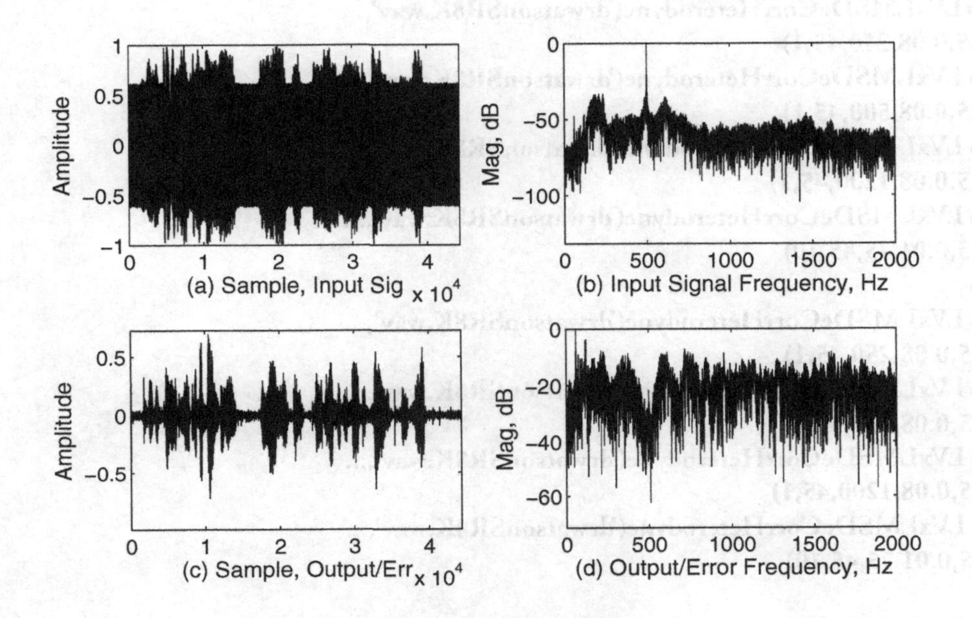

Figure 12.25: (a) Input signal, audio file 'drwatsonSR8K' corrupted with a large amplitude sinusoid; (b) Spectrum of signal in (a); (c) Filtered (output or error) signal, showing original audio signal with no visible interfering sinusoid; (d) Spectrum of signal in (c). Note the deep spectral notch at the frequency of the interfering sinusoid as shown in plot (b).

```
function LVxLMSDeCorrHeterodyne(strWavFile,A,Mu,Freq,...
NoTaps,DeCorrDelay)
% strWavFile is the audio file to open and use as the test signal
% A is the amplitude of an interfering tone (sinusoid) to be added
% to the audio file whose file name or path name (including file
% name and extension) is specified as strWavFile. Mu is the LMS
% update term weight; NoTaps is the number of taps to use in
% the LMS adaptive filter; DeCorrDelay is the number of samples
% to delay the filter input relative to the Plant or channel delay.
% The following plots are created: Frequency response (DTFT)
% of the adaptive filter when converged (magnitude v. frequency
% in Hz), the test signal (amplitude v. sample), the filtered test
% signal (amplitude v. sample), the DTFT of the test signal
% (magnitude v. frequency in Hz, and the DTFT (magnitude v.
% frequency in Hz) of the last 50 % of the filtered test signal..
% Test calls:
```

```
% LVxLMSDeCorrHeterodyne('drwatsonSR8K.wav',...
0.5,0.08,250,45,1)
% LVxLMSDeCorrHeterodyne('drwatsonSR8K.wav',...
0.5,0.08,500,45,1)
% LVxLMSDeCorrHeterodyne('drwatsonSR8K.wav',...
0.5,0.08,1200,45,1)
% LVxLMSDeCorrHeterodyne('drwatsonSR8K.wav',...
0.5,0.01,75,45,10)
%
% LVxLMSDeCorrHeterodyne('drwatsonSR8K.wav',...
1.5,0.08,250,45,1)
% LVxLMSDeCorrHeterodyne('drwatsonSR8K.wav',...
1.5,0.08,500,45,1)
% LVxLMSDeCorrHeterodyne('drwatsonSR8K.wav',...
1.5,0.08,1200,45,1)
% LVxLMSDeCorrHeterodyne('drwatsonSR8K.wav',...
1.5,0.01,75,45,10)
%
% LVxLMSDeCorrHeterodyne('drwatsonSR8K.wav',...
4.5,0.08,250,45,1)
% LVxLMSDeCorrHeterodyne('drwatsonSR8K.wav',...
4.5,0.08,500,45,1)
% LVxLMSDeCorrHeterodyne('drwatsonSR8K.wav',...
4.5,0.08,1200,45,1)
% LVxLMSDeCorrHeterodyne('drwatsonSR8K.wav',...
4.5,0.01,75,45,10)
%
% LVxLMSDeCorrHeterodyne('drwatsonSR8K.wav',...
0.02,0.08,250,45,1)
% LVxLMSDeCorrHeterodyne('drwatsonSR8K.wav',...
0.02,0.08,500,45,1)
% LVxLMSDeCorrHeterodyne('drwatsonSR8K.wav',...
0.02,0.08,1200,45,1)
% LVxLMSDeCorrHeterodyne('drwatsonSR8K.wav',...
0.02,0.01,75,45,10)
```

Use the script *LVx_AnalyzeModWavFile* (developed in the exercises for the chapter devoted to FIR filter design methods in Part III of the book) to make the test file

$$drwat8Kplus400HzAnd740Hz.wav$$

for use in the following test call. The amplitudes of the 400 Hz and 740 Hz waves should be specified as 0.02 when creating the file *'drwat8Kplus400HzAnd740Hz.wav'* using the script *LVx_AnalyzeModWavFile*.

```
% LVxLMSDeCorrHeterodyne('drwat8Kplus400HzAnd740Hz.wav',...
0,0.01,0,50,10)
% LVxLMSDeCorrHeterodyne('drwat8Kplus400HzAnd740Hz.wav',...
0.02,0.01,1100,50,10)
```

13. Write a script that is essentially the same as *LVxLMSDeCorrHeterodyne* with the exception that the new script will use a Dual-H architecture. You may notice that the script *LVxLMSDeCorrHeterodyne* does not work well unless the persistent tone is of relatively high amplitude. When the amplitude of the persistent tone is relatively low, performance of the adaptive filter is poor, leading to distortion and inadequate removal of the persistent tone. You should find that the Dual-H architecture will function much better with low level persistent tones than does the simple Single-H architecture of the script *LVxLMSDeCorrHeterodyne.* The test calls presented with the function specification below are presented in pairs, with the same parameters being used in both scripts, *LVxLMSDeCorrHeterodyne* and *LVxLMSDeCorrHetDualH,* for comparison.

```
function LVxLMSDeCorrHetDualH(strWavFile,A,Mu,Freq,...
NoTaps,DeCorrDelay)
% Has the same input arguments and function as
% LVxLMSDeCorrHeterodyne, but uses a Dual-H
% architectures for better performance.
% strWavFile is the audio file to open and use as the test signal
% A is the amplitude of an interfering tone (sinusoid) to be
% added to the audio file whose file name or path name
% (including file name and extension) is specified as
% strWavFile.
% Mu is the LMS update term weight;
% NoTaps is the number of taps to use in the LMS adaptive filter;
% DeCorrDelay is the number of samples to delay the filter input
% relative to the Plant or channel delay.
% The following plots are created: Frequency response (DTFT)
% of the adaptive filter when converged (magnitude v.
% frequency in Hz), the test signal (amplitude v. sample),
% the filtered test signal (amplitude v. sample), the DTFT
% of the test signal (magnitude v. frequency in Hz, and
% the DTFT (magnitude v. frequency in Hz) of the last
% 50 % of the filtered test signal.
%
% Test calls (pairs to be compared with each other):
```

```
%
% LVxLMSDeCorrHeterodyne('drwatsonSR8K.wav',...
0.012,0.03,850,65,7)
% LVxLMSDeCorrHetDualH('drwatsonSR8K.wav',...
0.012,0.03,850,65,7)
%
% LVxLMSDeCorrHeterodyne('drwat8Kplus400HzAnd740Hz.wav',...
0,0.025,0,73,3)
% LVxLMSDeCorrHetDualH('drwat8Kplus400HzAnd740Hz.wav',...
0,0.025,0,73,3)
%
% LVxLMSDeCorrHeterodyne('drwat8Kplus400HzAnd740Hz.wav',...
0.02,0.025,1150,73,3)
% LVxLMSDeCorrHetDualH('drwat8Kplus400HzAnd740Hz.wav',...
0.02,0.025,1150,73,3)
```

The calls

**LVxLMSDeCorrHeterodyne('drwat8Kplus400HzAnd740Hz.wav',...
0.04,0.025,1150,73,3)**
**LVxLMSDeCorrHetDualH('drwat8Kplus400HzAnd740Hz.wav',...
0.04,0.025,1150,73,3)**

result in Figs. 12.26 and 12.27, respectively, which show the spectrum of the test signals and filtered test signals for the Single-H and Dual-H versions of the persistent-tone removal script.

The call

**LVxLMSDeCorrHetDualH('drwat8Kplus400HzAnd740Hz.wav',...
0.04,0.025,1150,400,7)**

results in Fig. 12.28, in addition to the other figures specified in the function description. Note the peaks in the filter response at the persistent frequencies, and recall that the filtered signal is derived by subtracting the LMS filter output from the test signal.

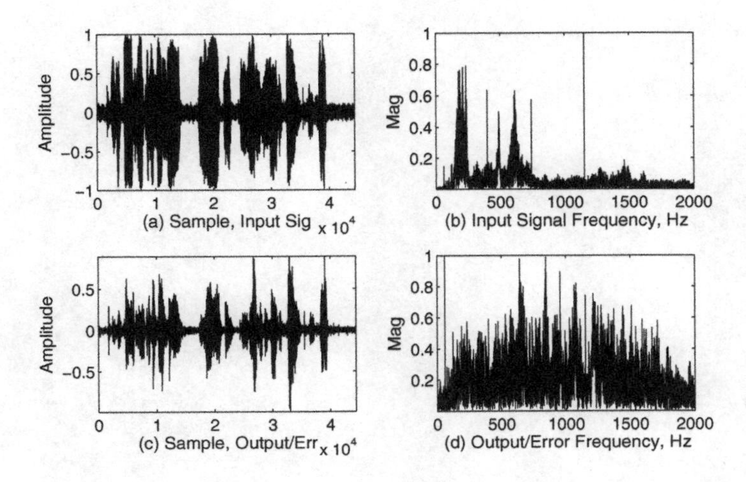

Figure 12.26: (a) Test speech signal having persistent tones added at frequencies of 440 Hz, 740 Hz, and 1150 Hz; (b) Spectrum of test signal in (a); (c) Output/Error signal, i.e., the test signal after being filtered by a Single-H LMS system having a decorrelating delay designed to assist in removing persistent tones; (d) Spectrum of signal in (c).

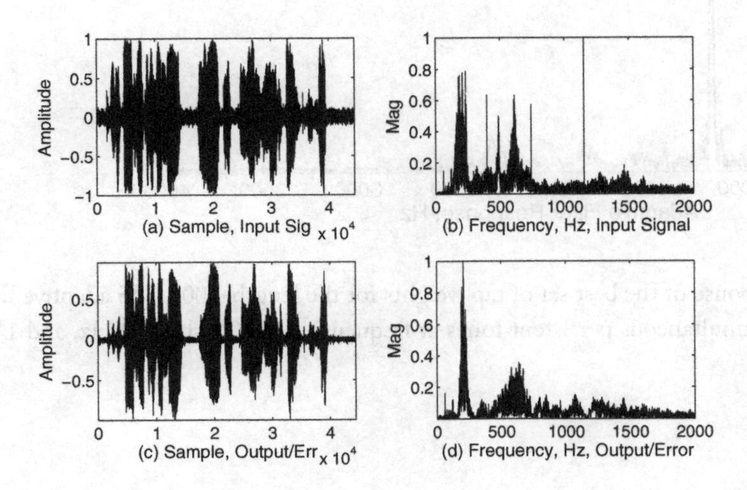

Figure 12.27: (a) Test speech signal having persistent tones added at frequencies of 440 Hz, 740 Hz, and 1150 Hz; (b) Spectrum of test signal at (a); (c) Output/Error signal, i.e., the test signal after being filtered by a Dual-H LMS system using a decorrelating delay designed to assist in removing persistent tones; (d) Spectrum of Output/Error signal. Note the much improved attenuation of the persistent tones compared to that of the previous figure.

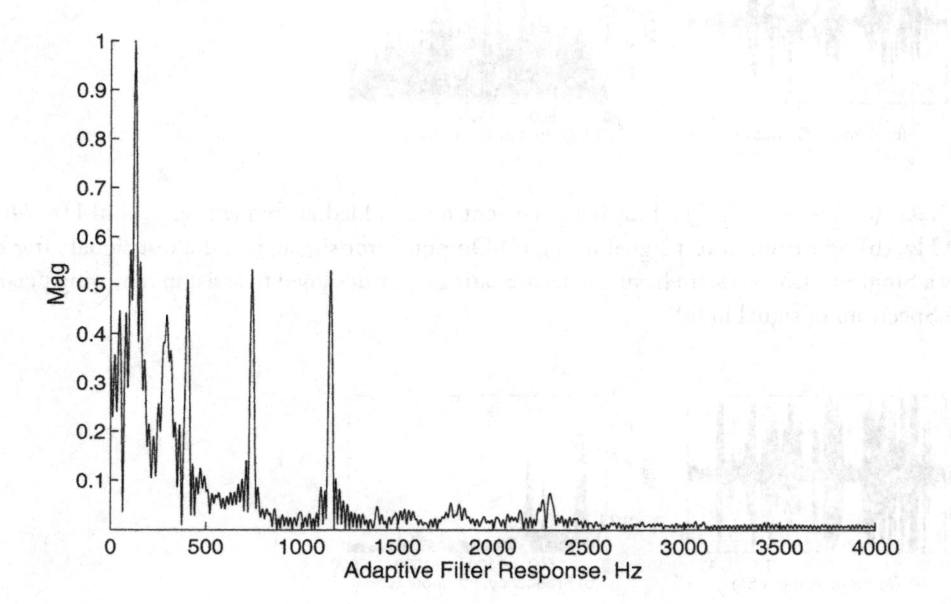

Figure 12.28: Frequency response of the best set of tap weights for the length-400 LMS adaptive filter that is computed to remove simultaneous persistent tones at frequencies of 400 Hz, 740 Hz, and 1150 Hz.

APPENDIX A

Software for Use with this Book

A.1 FILE TYPES AND NAMING CONVENTIONS

The text of this book describes many computer programs or scripts to perform computations that illustrate the various signal processing concepts discussed. The computer language used is usually referred to as **m-code** (or as an **m-file** when in file form, using the file extension **.m**) in MATLAB -related literature or discussions, and as **MathScript** in LabVIEW-related discussions (the terms are used interchangeably in this book).

The MATLAB and LabVIEW implementations of m-code (or MathScript) differ slightly (Lab-VIEW's version, for example, at the time of this writing, does not implement Handle Graphics, as does MATLAB).

The book contains mostly scripts that have been tested to run on both MATLAB and Lab-VIEW; these scripts all begin with the letters **LV** and end with the file extension **.m**. Additionally, scripts starting with the letters **LVx** are intended as exercises, in which the student is guided to write the code (the author's solutions, however, are included in the software package and will run when properly called on the Command Line).

Examples are:

LVPlotUnitImpSeq.m

LVxComplexPowerSeries.m

There are also a small number m-files that will run only in MATLAB, as of this writing. They all begin with the letters *ML*. An example is:

ML_SinglePole.m

Additionally, there are a number of LabVIEW Virtual Instruments (VIs) that demonstrate various concepts or properties of signal processing. These have file names that all begin with the letters *Demo* and end with the file extension *.vi*. An example is:

DemoComplexPowerSeriesVI.vi

Finally, there are several sound files that are used with some of the exercises; these are all in the .wav format. An example is:

drwatsonSR4K.wav

A.2 DOWNLOADING THE SOFTWARE

All of the software files needed for use with the book are available for download from the following website:

http://www.morganclaypool.com/page/isen

The entire software package should be stored in a single folder on the user's computer, and the full file name of the folder must be placed on the MATLAB or LabVIEW search path in accordance with the instructions provided by the respective software vendor.

A.3 USING THE SOFTWARE

In MATLAB, once the folder containing the software has been placed on the search path, any script may be run by typing the name (without the file extension, but with any necessary input arguments in parentheses) on the Command Line in the Command Window and pressing *Return*.

In LabVIEW, from the Getting Started window, select MathScript Window from the Tools menu, and the Command Window will be found in the lower left area of the MathScript window. Enter the script name (without the file extension, but with any necessary input arguments in parentheses) in the Command Window and press *Return*. This procedure is essentially the same as that for MATLAB.

Example calls that can be entered on the Command Line and run are

LVAliasing(100,1002)

LV_FFT(8,0)

In the text, many "live" calls (like those just given) are found. All such calls are in boldface as shown in the examples above. When using an electronic version of the book, these can usually be copied and pasted into the Command Line of MATLAB or LabVIEW and run by pressing *Return*. When using a printed copy of the book, it is possible to manually type function calls into the Command Line, but there is also one stored m-file (in the software download package) per chapter that contains clean copies of all the m-code examples from the text of the respective chapter, suitable for copying (these files are described more completely below in the section entitled "Multi-line m-code examples"). There are two general types of m-code examples, single-line function calls and multi-line code examples. Both are discussed immediately below.

A.4 SINGLE-LINE FUNCTION CALLS

The first type of script mentioned above, a named- or defined-function script, is one in which a function is defined; it starts with the word "function" and includes the following, from left to right:

any output arguments, the equal sign, the function name, and, in parentheses immediately following the function name, any input arguments. The function name must always be identical to the file name. An example of a named-function script, is as follows:

```
function nY = LVMakePeriodicSeq(y,N)
% LVMakePeriodicSeq([1 2 3 4],2)
y = y(:); nY = y*([ones(1,N)]); nY = nY(:)';
```

For the above function, the output argument is *nY*, the function name is *LVMakePeriodicSeq*, and there are two input arguments, *y* and *N*, that must be supplied with a call to run the function. Functions, in order to be used, must be stored in file form, i.e., as an m-file. The function *LVMakePeriodicSeq* can have only one corresponding file name, which is

LVMakePeriodicSeq.m

In the code above, note that the function definition is on the first line, and an example call that you can paste into the Command Line (after removing or simply not copying the percent sign at the beginning of the line, which marks the line as a comment line) and run by pressing *Return*. Thus you would enter on the Command Line the following, and then press *Return*:

```
nY = LVMakePeriodicSeq([1,2,3,4],2)
```

In the above call, note that the output argument has been included; if you do not want the value (or array of values) for the output variable to be displayed in the Command window, place a semicolon after the call:

```
nY = LVMakePeriodicSeq([1,2,3,4],2);
```

If you want to see, for example, just the first five values of the output, use the above code to suppress the entire output,and then call for just the number of values that you want to see in the Command window:

```
nY = LVMakePeriodicSeq([1,2,3,4],2);nY1to5 = nY(1:5)
```

The result from making the above call is

nY1to5 = [1,2,3,4,1]

A.5 MULTI-LINE M-CODE EXAMPLES

There are also entire multi-line scripts in the text that appear in boldface type; they may or may not include named-functions, but there is always m-code with them in excess of that needed to make a simple function-call. An example might be

```
N=54; k = 9; x = cos( 2*pi*k*( 0:1:N-1 )/N);
LVFreqResp(x, 500)
```

Note in the above that there is a named-function (*LVFreqResp*) call, preceded by m-code to define an input argument for the call. Code segments like that above must either be (completely) copied and pasted into the Command Line or manually typed into the Command Line. Copy-and-Paste can often be successfully done directly from a pdf version of the book. This often results in problems (described below), and accordingly, an m-file containing clean copies of most m-code programs from each chapter is supplied with the software package. Most of the calls or multi-line m-code examples from the text that the reader might wish to make are present in m-files such as

McodeVolume1Chapter4.m

McodeVolume2Chapter3.m

and so forth. There is one such file for each chapter of each Part of the book, except Chapter 1 of Part I, which has no m-code examples.

A.6 HOW TO SUCCESSFULLY COPY-AND-PASTE M-CODE

M-code can usually be copied directly from a pdf copy of the book, although a number of minor, easily correctible problems can occur. Two characters, the symbol for raising a number to a power, the circumflex ˆ, and the symbol for vector or matrix transposition, the apostrophe or single quote mark ’, are coded for pdf using characters that are non-native to m-code. While these two symbols may look proper in the pdf file, when pasted into the Command line of MATLAB, they will appear in red.

A first way to avoid this copying problem, of course, is simply to use the m-code files described above to copy m-code from. This is probably the most time-efficient method of handling the problem—avoiding it altogether.

A second method to correct the circumflex-and-single-quote problem, if you do want to copy directly from a pdf document, is to simply replace each offending character (circumflex or single quote) by the equivalent one typed from your keyboard. When proper, all such characters will appear in black rather than red in MATLAB. In LabVIEW, the pre-compiler will throw an error at the first such character and cite the line and column number of its location. Simply manually retype/replace each offending character. Since there are usually no more than a few such characters, manually replacing/retyping is quite fast.

Yet a third way (which is usually more time consuming than the two methods described above) to correct the circumflex and apostrophe is to use the function *Reformat*, which is supplied with the software package. To use it, all the copied code from the pdf file is reformatted by hand into one horizontal line, with delimiters (commas or semicolons) inserted (if not already present) where lines have been concatenated. For example, suppose you had copied

```
n = 0:1:4;
y = 2.ˆn
stem(n,y);
```

where the circumflex is the improper version for use in m-code. We reformat the code into one horizontal line, adding a comma after the second line (a semicolon suppresses computed output on the Command line, while a comma does not), and enclose this string with apostrophes (or single quotes), as shown, where *Reformat* corrects the improper circumflex and *eval* evaluates the string, i.e., runs the code.

```
eval(Reformat('n=0:1:4;y=2.ˆn;stem(n,y)'))
```

Occasionally, when copying from the pdf file, essential blank spaces are dropped in the copied result and it is necessary to identify where this has happened and restore the missing space. A common place that this occurs is after a "for" statement. The usual error returned when trying to run the code is that there is an unmatched "end" statement or that there has been an improper use of the reserved word "end". This is caused by the elision of the "for" statement with the ensuing code and is easily corrected by restoring the missing blank space after the "for" statement. Note that the function *Reformat* does not correct for this problem.

A.7 LEARNING TO USE M-CODE

While the intent of this book is to teach the principles of digital signal processing rather than the use of m-code per se, the reader will find that the scripts provided in the text and with the software package will provide many examples of m-code programming starting with simple scripts and functions early in the book to much more involved scripts later in the book, including scripts for use with MATLAB that make extensive use of MATLAB objects such as push buttons, edit boxes, drop-down menus, etc.

Thus the complexity of the m-code examples and exercises progresses throughout the book apace with the complexity of signal processing concepts presented. It is unlikely that the reader or student will find it necessary to separately or explicitly study m-code programming, although it will occasionally be necessary and useful to use the online MATLAB or LabVIEW help files for explanation of the use of, or call syntax of, various built-in functions.

A.8 WHAT YOU NEED WITH MATLAB AND LABVIEW

If you are using a professional edition of MATLAB, you'll need the Signal Processing Toolbox in addition to MATLAB itself. The student version of MATLAB includes the Signal Processing Toolbox.

If you are using either the student or professional edition of LabVIEW, it must be at least Version 8.5 to run the m-files that accompany this book, and to properly run the VIs you'll need the Control Design Toolkit or the newer Control Design and Simulation Module (which is included in the student version of LabVIEW).

APPENDIX B

Vector/Matrix Operations in M-Code

B.1 ROW AND COLUMN VECTORS

Vectors may be either row vectors or column vectors. A typical row vector in m-code might be [3 -1 2 4] or [3,-1,2, 4] (elements in a row can be separated by either commas or spaces), and would appear conventionally as a row:

$$\begin{bmatrix} 3 & -1 & 2 & 4 \end{bmatrix}$$

The same, notated as a column vector, would be [3,-1,2,4]' or [3; -1; 2; 4], where the semicolon sets off different matrix rows:

$$\begin{bmatrix} 3 \\ -1 \\ 2 \\ 4 \end{bmatrix}$$

Notated on paper, a row vector has one row and plural columns, whereas a column vector appears as one column with plural rows.

B.2 VECTOR PRODUCTS

B.2.1 INNER PRODUCT

A row vector and a column vector of the same length as the row vector can be multiplied two different ways, to yield two different results. With the row vector on the left and the column vector on the right,

$$\begin{bmatrix} 1 & 2 & 3 & 4 \end{bmatrix} \begin{bmatrix} 4 \\ 3 \\ 2 \\ 1 \end{bmatrix} = 20$$

corresponding elements of each vector are multiplied, and all products are summed. This is called the **Inner Product**. A typical computation would be

$$[1, 2, 3, 4] * [4; 3; 2; 1] = (1)(4) + (2)(3) + (3)(2) + (4)(1) = 20$$

B.2.2 OUTER PRODUCT

An **Outer Product** results from placing the column vector on the left, and the row vector on the right:

$$\begin{bmatrix} 4 \\ 3 \\ 2 \\ 1 \end{bmatrix} \begin{bmatrix} 1 & 2 & 3 & 4 \end{bmatrix} = \begin{bmatrix} 4 & 8 & 12 & 16 \\ 3 & 6 & 9 & 12 \\ 2 & 4 & 6 & 8 \\ 1 & 2 & 3 & 4 \end{bmatrix}$$

The computation is as follows:

$$[4; 3; 2; 1] * [1, 2, 3, 4] = [4, 3, 2, 1; 8, 6, 4, 2; 12, 9, 6, 3; 16, 12, 8, 4]$$

Note that each column in the output matrix is the column of the input column vector, scaled by a column (which is a single value) in the row vector.

B.2.3 PRODUCT OF CORRESPONDING VALUES

Two vectors (or matrices) of exactly the same dimensions may be multiplied on a value-by-value basis by using the notation " .* " (a period followed by an asterisk). Thus two row vectors or two column vectors can be multiplied in this way, and result in a row vector or column vector having the same length as the original two vectors. For example, for two column vectors, we get

$$[1; 2; 3]. * [4; 5; 6] = [4; 10; 18]$$

and for row vectors, we get

$$[1, 2, 3]. * [4, 5, 6] = [4, 10, 18]$$

B.3 MATRIX MULTIPLIED BY A VECTOR OR MATRIX

An m by n matrix, meaning a matrix having m rows and n columns, can be multiplied from the right by an n by 1 column vector, which results in an m by 1 column vector. For example,

$$[1, 2, 1; 2, 1, 2] * [4; 5; 6] = [20; 25]$$

Or, written in standard matrix form:

$$\begin{bmatrix} 1 & 2 & 1 \\ 2 & 1 & 2 \end{bmatrix} \begin{bmatrix} 4 \\ 5 \\ 6 \end{bmatrix} = \begin{bmatrix} 4 \\ 8 \end{bmatrix} + \begin{bmatrix} 10 \\ 5 \end{bmatrix} + \begin{bmatrix} 6 \\ 12 \end{bmatrix} = \begin{bmatrix} 20 \\ 25 \end{bmatrix} \quad \text{(B.1)}$$

An m by n matrix can be multiplied from the right by an n by p matrix, resulting in an m by p matrix. Each column of the n by p matrix operates on the m by n matrix as shown in (B.1), and creates another column in the n by p output matrix.

B.4 MATRIX INVERSE AND PSEUDO-INVERSE

Consider the matrix equation

$$\begin{bmatrix} 1 & 4 \\ 3 & -2 \end{bmatrix} \begin{bmatrix} a \\ b \end{bmatrix} = \begin{bmatrix} -2 \\ 3 \end{bmatrix} \tag{B.2}$$

which can be symbolically represented as

$$[M][V] = [C]$$

or simply

$$MV = C$$

and which represents the system of two equations

$$a + 4b = -2$$

$$3a - 2b = 3$$

that can be solved, for example, by scaling the upper equation by -3 and adding to the lower equation

$$-3a - 12b = 6$$

$$3a - 2b = 3$$

which yields

$$-14b = 9$$

or

$$b = -9/14$$

and

$$a = 4/7$$

The inverse of a matrix M is defined as M^{-1} such that

$$MM^{-1} = I$$

where I is called the Identity matrix and consists of all zeros except for the left-to-right downsloping diagonal which is all ones. The Identity matrix is so-called since, for example,

$$\begin{bmatrix} 1 & 0 \\ 0 & 1 \end{bmatrix} \begin{bmatrix} a \\ b \end{bmatrix} = \begin{bmatrix} a \\ b \end{bmatrix}$$

The pseudo-inverse M^{-1} of a matrix M is defined such that

$$M^{-1}M = I$$

System B.2 can also be solved by use of the pseudo-inverse

$$\left[M^{-1}\right][M][V] = \left[M^{-1}\right][C]$$

which yields

$$[I][V] = V = \left[M^{-1}\right][C]$$

In concrete terms, we get

$$\left[M^{-1}\right] \begin{bmatrix} 1 & 4 \\ 3 & -2 \end{bmatrix} \begin{bmatrix} a \\ b \end{bmatrix} = \left[M^{-1}\right] \begin{bmatrix} -2 \\ 3 \end{bmatrix} \tag{B.3}$$

which reduces to

$$\begin{bmatrix} a \\ b \end{bmatrix} = \left[M^{-1}\right] \begin{bmatrix} -2 \\ 3 \end{bmatrix}$$

We can compute the pseudo-inverse M^{-1} and the final solution using the built-in MathScript function *pinv*:

```
M = [1,4;3,-2];
P = pinv(M)
ans = P*[-2;3]
```

which yields

$$P = \begin{bmatrix} 0.1429 & 0.2857 \\ 0.2143 & -0.0714 \end{bmatrix}$$

and therefore

$$\begin{bmatrix} a \\ b \end{bmatrix} = \begin{bmatrix} 0.1429 & 0.2857 \\ 0.2143 & -0.0714 \end{bmatrix} \begin{bmatrix} -2 \\ 3 \end{bmatrix}$$

which yields a = 0.5714 and b = -0.6429 which are the same as 4/7 and -9/14, respectively. A unique solution is possible only when M is square and all rows linearly independent.(a linearly independent row cannot be formed or does not consist solely of a linear combination of other rows in the matrix).

APPENDIX C

Complex Numbers

C.1 DEFINITION

A complex number has a real part and an imaginary part, and thus two quantities are contained in a single complex number.

Real numbers are those used in everyday, non-scientific activities. Examples of real numbers are 0, 1.12, -3.37, etc. To graph real numbers, only a single axis extending from negative infinity to positive infinity is needed, and any real number can be located and graphed on that axis. In this sense, real numbers are one-dimensional.

Imaginary numbers are numbers that consist of a real number multiplied by the square root of negative one ($\sqrt{-1}$) which is usually called i or j, and has the property that $i \cdot i = -1$. In this book, j will typically be used to represent the square root of negative one, although either i or j may be used in m-code. Electrical engineers use j as the imaginary operator since the letter i is used to represent current. Typical imaginary numbers might be $5j$, $-2.37i$, etc.

Since complex numbers have two components, they naturally graph in a two-dimensional space, or a plane, and thus two axes at right angles are used to locate and plot a complex number. In the case of the complex plane, the x-axis is called the real axis, and it represents the real amplitude, and the y-axis is called the imaginary axis and numbers are considered to be an amplitude multiplied by the square root of negative one.

Typical complex numbers might be $1 + i$, $2.2 - 0.3j$, and so forth.

C.2 RECTANGULAR V. POLAR

A complex number can be located in the complex plane using either 1) rectangular coordinates (values for horizontal and vertical axes, such as x and y) or 2) polar coordinates, in which a distance from the origin (center of the plot where x and y are both zero) and an angle (measured counterclockwise starting from the positive half of the real or x-axis) are specified.

Figure C.1 shows the complex number $0.5 + 0.6j$ plotted in the complex plane.

You can convert from rectangular coordinates to polar using these formulas:

$$Magnitude = \sqrt{(\mathrm{Re}(W))^2 + (\mathrm{Im}(W))^2}$$

$$Angle = \arctan\left(\frac{\mathrm{Im}(W)}{\mathrm{Re}(W)}\right) \tag{C.1}$$

Using $0.5 + 0.6j$ as the complex number, and plugging into the formula for magnitude, we get

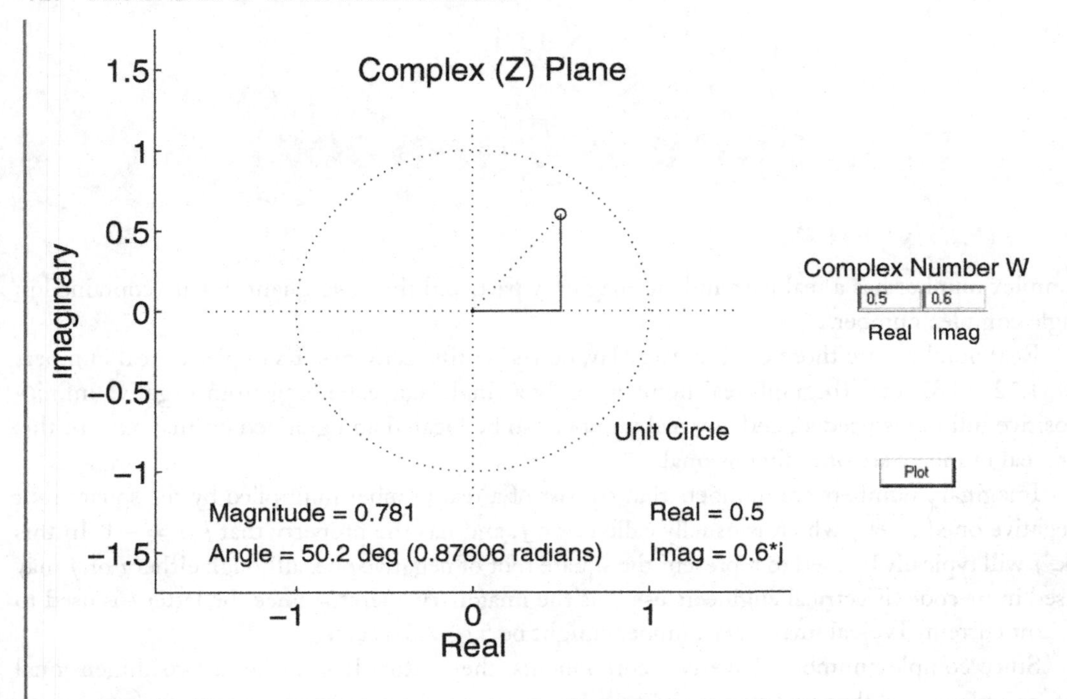

Figure C.1: The complex number -0.5 + 0.6j, plotted in the complex plane.

$$Magnitude = \sqrt{(0.5)^2 + (0.6)^2} = 0.781$$

and the angle would be

$$Angle = \arctan(\frac{0.6}{0.5}) = 0.876\,06\ radian = 50.2^\circ$$

In m-code, use the function $abs(x)$ to obtain the magnitude or absolute value of x, and use $angle(x)$ to obtain the angle in radians.

Example C.1. Compute the magnitude and angle of the complex number 0.5 + 0.6*j.

The call

x = 0.5 + j*0.6; M = abs(x), AngDeg = angle(x)*360/(2*pi)

returns M = 0.781 and AngDeg = 50.194.

Example C.2. Compute the magnitude and angle of the complex number 0.9.

This can be done by inspection since the imaginary part is zero, giving a magnitude of 0.9 and an angle of zero. Graphically, of course, 0.9 is on the positive real axis and thus we also see that its angle must be zero.

Example C.3. Compute the magnitude and angle of the complex number 0 + 0.9j.

By inspection, the magnitude is 0.9 and the angle is 90 degrees or $\pi/2$ radians.

- A common way to express a complex number in polar notation is

$$M\angle\theta$$

where M is the magnitude and θ is the angle, which may be expressed in either degrees or radians.

C.3 ADDITION AND SUBTRACTION

The rule for adding complex numbers is as follows:

$$(a + bj) + (c + dj) = (a + c) + j(b + d)$$

In other words, for addition or subtraction, just add or subtract the real parts and then the imaginary parts, keeping them separate.

C.4 MULTIPLICATION

C.4.1 RECTANGULAR COORDINATES

A real number times a real number is a real number, i.e.,

$$(6)(-2) = -12$$

A real number times an imaginary number is an imaginary number, i.e.,

$$(6)(-2j) = -12j$$

An imaginary number times an imaginary number is negative one times the product of the two remaining real numbers. For example

$$(6j)(-2j) = (6)(j)(-2)(j) = (6)(-2)(j)(j) = (-12)(-1) = 12$$

The general rule is

$$(a + bj)(c + dj) = (ac - bd) + j(ad + bc)$$

C.4.2 POLAR COORDINATES

The product of two complex numbers expressed in polar coordinates is a complex number having a magnitude equal to the product of the two magnitudes, and an angle equal to the sum of the two angles:

$$(M_1\angle\theta_1)(M_2\angle\theta_2) = M_1 M_2\angle(\theta_1 + \theta_2)$$

Example C.4. Multiply the complex number $0.9\angle 90$ by itself.

Multiply the magnitudes and add the angles to get $0.81\angle 180$ which is -0.81.

Example C.5. Multiply $(0.707 + 0.707j)$ and $(0.707 - 0.707j)$.

(0.707 + 0.707j) = $1\angle 45$ and (0.707 - 0.707j) = $1\angle - 45$ (angles in degrees); hence the product is the product of the magnitudes and the sum of the angles = $1\angle 0$ = 1. Doing the multiplication in rectangular coordinates gives a product of $(\sqrt{2}/2)(1 + j)(\sqrt{2}/2)(1 - j) = (2/4)(2) = 1$.

C.5 DIVISION AND COMPLEX CONJUGATE

C.5.1 USING RECTANGULAR COORDINATES

The quotient of two complex numbers can be computed by using complex conjugates. The complex conjugate of any complex number $a + bj$ is simply $a - bj$.

Suppose we wished to simplify an expression of the form

$$\frac{a + bj}{c + dj}$$

which is one complex number divided by another. This expression can be simplified by multiplying both numerator and denominator by the complex conjugate of the denominator, which yields for the denominator a single real number which does not affect the ratio of the real and imaginary parts (now isolated in the numerator) to each other.

$$\frac{a + bj}{c + dj} \cdot \frac{c - dj}{c - dj} = \frac{ac - adj + bcj - bdj^2}{c^2 - cdj + cdj - d^2 j^2} = \frac{(ac + bd) + j(bc - ad)}{c^2 + d^2}$$

The ratio of the real part to the imaginary part is

$$\frac{ac + bd}{bc - ad}$$

If a complex number W is multiplied by its conjugate W^*, the product is $|W|^2$, or the magnitude squared of the number W.

$$WW^* = |W|^2$$

As a concrete example, let

$$W = 1 + j$$

Then

$$W^* = 1 - j$$

and the product is

$$1 - j^2 = 2$$

which is the same as the square of the magnitude of W:

$$(\sqrt{1^2 + 1^2})^2 = 2$$

Example C.6. Simplify the ratio 1/j

We multiply numerator and denominator by the complex conjugate of the denominator, which is $-j$:

$$(\frac{1}{j})(\frac{-j}{-j}) = \frac{-j}{-(-1)} = -j$$

C.5.2 USING POLAR COORDINATES

The quotient of two complex numbers expressed in polar coordinates is a complex number having a magnitude equal to the quotient of the two magnitudes, and an angle equal to the difference of the two angles:

$$(M_1\angle\theta_1)/(M_2\angle\theta_2) = (M_1/M_2)\angle(\theta_1 - \theta_2)$$

Example C.7. Formulate and compute the ratio 1/j in polar coordinates.

The polar coordinate version of this problem is very straightforward since the magnitude and angle of each part of the ratio can be stated by inspection:

$$(1\angle 0)/(1\angle 90) = 1\angle(-90) = -j$$

C.6 POLAR NOTATION USING COSINE AND SINE

Another way of describing a complex number W having a magnitude M and an angle θ is

$$M(\cos(\theta) + j\sin(\theta))$$

This is true since $\text{Re}(W)$ is just $M\cos(\theta)$, and $\cos(\theta)$, by definition in this case, is $\text{Re}(W)/M$; $\text{Im}(W)$ is just $M\sin(\theta)$, and $\sin(\theta) = \text{Im}(W)/M$, so

$$M(\text{Re}(W)/M + j\text{Im}(W)/M) = \text{Re}(W) + j\text{Im}(W)$$

C.7 THE COMPLEX EXPONENTIAL

The following identities are called the Euler identities, and can be demonstrated as true using the Taylor (infinite series) expansions for $e^{j\theta}$, $cos(\theta)$, and $sin(\theta)$:

$$e^{j\theta} = \cos(\theta) + j\sin(\theta)$$

and

$$e^{-j\theta} = \cos(\theta) - j\sin(\theta)$$

where e is the base of the natural logarithm system, 2.718... Such an expression is referred to as a complex exponential since the exponent of e is complex. This form is very popular and has many interesting traits and uses.

For example, by adding the two expressions above, it follows that

$$\cos(\theta) = (e^{j\theta} + e^{-j\theta})/2$$

and by subtracting it follows that

$$\sin(\theta) = (e^{j\theta} - e^{-j\theta})/2j$$

A correlation of an input signal $x[n]$ with both cosine and sine of the same frequency k over the sequence length N can be computed as

$$C = \sum_{n=0}^{N-1} x[n](\cos[2\pi nk/N] + j\sin[2\pi nk/N])$$

or, using the complex exponential notation

$$C = \sum_{n=0}^{N-1} x[n]e^{j2\pi nk/N}$$

The real part of C contains the correlation of $x[n]$ with $\cos[2\pi nk/N]$, and the imaginary part contains the correlation with $\sin[2\pi nk/N]$.

In m-code, the expression

```
exp (x)
```

means e raised to the x power and hence if x is imaginary, i.e., an amplitude A multiplied by $\sqrt{-1}$, we get

$$\exp(jA) = \cos(A) + j\sin(A)$$

Example C.8. Compute the correlation (as defined above) of cos(2π(0:1:3)/4) with the complex exponential exp(j2π(0:1:3)/4).

The call

```
sum( cos (2*pi*(0:1:3)/4).*exp(j*2*pi*(0:1:3)/4))
```

produces the answer 2 + j0.

C.8 USES FOR SIGNAL PROCESSING

- Complex numbers, and in particular, the complex exponential, can be used to both generate and represent sinusoids, both real and complex.
- Complex numbers (and the complex exponential) make it possible to understand and work with sinusoids in ways which are by no means obvious using only real arithmetic. Complex numbers are indispensable for the study of certain topics, such as the complex DFT, the z-Transform, and the Laplace Transform, all of which are discussed in this book.

APPENDIX D

FIR Frequency Sampling Design Formulas

In the formulas below, L represents FIR length in samples, and the values A_k are frequency sample amplitudes as described in Section 9.6 of this book.

D.1 WHOLE-CYCLE MODE FILTER FORMULAS

D.1.1 ODD LENGTH, SYMMETRIC (TYPE I)

The design formula for an odd length linear phase FIR (whole-cycle mode) is

$$h[n] = \frac{1}{L}\left[A_0 + \sum_{k=1}^{M} 2A_k \cos(2\pi(n-M)k/L)\right] \tag{D.1}$$

where the index n for the impulse response $h[n]$ runs from 0 to L - 1, and $M = (L-1)/2$.

D.1.2 EVEN LENGTH, SYMMETRIC (TYPE II)

In the case of even length, $M = (L-1)/2$, is a non-integer (an odd multiple of 1/2) and the design formula is

$$h[n] = \frac{1}{L}\left[A_0 + \sum_{k=1}^{L/2-1} 2A_k \cos(2\pi(n-M)k/L)\right]$$

where the index n for the impulse response $h[n]$ runs from 0 to $L-1$. Note that even length filters need the 1/2 sample offset in the sample index n in order to make the cosine components symmetrical about their midpoint.

D.1.3 ODD LENGTH, ANTI-SYMMETRIC (TYPE III)

$$h[n] = \frac{1}{L}\sum_{k=1}^{M} 2A_k \sin(2\pi(M-n)k/L)$$

where L is the filter length, $M = (L-1)/2$, and n runs from 0 to $L-1$. Note that k may not equal zero here, or rather, the sine of zero is identically zero, so no DC (frequency 0) correlator can be generated using sine waves as the basis, which explains why the Type III and IV filters do not work as lowpass filters.

D.1.4 EVEN LENGTH, ANTI-SYMMETRIC (TYPE IV)

$$h[n] = \frac{1}{L}\left[\sum_{k=1}^{L/2-1} 2A_k \sin(2\pi(M-n)k/L) + A_{L/2}\sin(\pi(M-n))\right]$$

where L is the filter length, $M = (L-1)/2$, and n runs from 0 to $L-1$.

D.2 HALF-CYCLE MODE FILTERS

These filters are built from odd-multiples of half-cycles of cosines or sines.

D.2.1 ODD LENGTH, SYMMETRIC (TYPE I)

$$h[n] = \frac{1}{L}\left[\sum_{k=0}^{M-1} 2A_k \cos(2\pi(n-M)(k+\frac{1}{2})/L) + A_M\cos(\pi(n-M))\right]$$

where $M = (L-1)/2$, and n runs from 0 to $L-1$.

D.2.2 EVEN LENGTH, SYMMETRIC (TYPE II)

$$h[n] = \frac{1}{L}\left[\sum_{k=0}^{N/2-1} 2A_k \cos(2\pi(n-M)(k+\frac{1}{2})/L)\right]$$

where $M = (L-1)/2$, and n runs from 0 to $L-1$.

D.2.3 ODD LENGTH, ANTI-SYMMETRIC (TYPE III)

$$h[n] = \frac{1}{L}\left[\sum_{k=0}^{M-1} 2A_k \sin(2\pi(M-n)(k+\frac{1}{2})/L)\right]$$

where $M = (L-1)/2$, and n runs from 0 to $L-1$.

D.2.4 EVEN LENGTH, ANTI-SYMMETRIC (TYPE IV)

$$h[n] = \frac{1}{L}\left[\sum_{k=0}^{L/2-1} 2A_k \sin(2\pi(M-n)(k+\frac{1}{2})/L)\right]$$

where $M = (L-1)/2$, and n runs from 0 to $L-1$.

D.3 REFERENCES

[1] T. W. Parks and C. S. Burrus, *Digital Filter Design*, John Wiley & Sons, New York, 1987.

Biography

Forester W. Isen received the B.S. degree from the U. S. Naval Academy in 1971 (majoring in mathematics with additional studies in physics and engineering), and the M. Eng. (EE) degree from the University of Louisville in 1978, and spent a career dealing with intellectual property matters at a government agency working in, and then supervising, the examination and consideration of both technical and legal matters pertaining to the granting of patent rights in the areas of electronic music, horology, and audio and telephony systems (AM and FM stereo, hearing aids, transducer structures, Active Noise Cancellation, PA Systems, Equalizers, Echo Cancellers, etc.). Since retiring from government service at the end of 2004, he worked during 2005 as a consultant in database development, and then subsequently spent several years writing the book *DSP for MATLAB and LabVIEW*, calling on his many years of practical experience to create a book on DSP fundamentals that includes not only traditional mathematics and exercises, but "first principle" views and explanations that promote the reader's understanding of the material from an intuitive and practical point of view, as well as a large number of accompanying scripts (to be run on MATLAB or LabVIEW) designed to bring to life the many signal processing concepts discussed in the book.